Sustainable Water Management

Sustainable Water Management

Novel Treatment Technologies and Practices

Edited by

Shiao-Shing Chen
National Taipei University of Technology, Institute of Environmental Engineering and Management, Taipei, Taiwan

Saikat Sinha Ray
Department of Environmental Science and Engineering, School of Engineering and Sciences, SRM University-AP, Amaravati, Andhra Pradesh, India

Nguyen Cong Nguyen
Faculty of Chemistry and Environment, Dalat University, Dalat, Vietnam

Amit Kumar
Nanjing University of Information Science and Technology, School of Hydrology and Water Resources, Nanjing, China

ELSEVIER

Elsevier
Radarweg 29, PO Box 211, 1000 AE Amsterdam, Netherlands
125 London Wall, London EC2Y 5AS, United Kingdom
50 Hampshire Street, 5th Floor, Cambridge, MA 02139, United States

ISBN: 978-0-443-44853-9

For information on all Elsevier publications visit our website at
https://www.elsevier.com/books-and-journals

Publisher: Peter Llewellyn
Acquisitions Editor: Maria Elekidou
Editorial Project Manager: Pratiti Ghosh
Production Project Manager: Kiruthigadevi N.
Cover Designer: Bakyalakshmi S.

Typeset by TNQ Tech

This book is dedicated to young researchers, students at undergraduate and graduate levels, environmental engineers, scientists, policymakers, and water resources planners and managers who are committed to advancing sustainable wastewater treatment and management. It is also dedicated to communities around the world striving for access to safe sanitation and clean water, and to the innovators working tirelessly to transform wastewater into a resource for energy, nutrients, and reclaimed water. Besides, this book is dedicated to achieving the Sustainable Development Goals (SDG-3, 6, 14), whose well-being firmly depends on the collective responsibility of governments, stakeholders, NGOs, and policymakers to protect water resources and build resilient, sustainable systems for a healthier planet.

Contents

PART III Breakthroughs in membrane science for water treatment

PART V Digital innovations and smart monitoring

PART VI Contaminants of emerging concern and health risk management

PART IX Future horizons and transformational pathways in water treatment innovation

Contributors

Young-Ho Ahn
Department of Civil Engineering, Yeungnam University, Gyeongsan, South Korea

Mahboob Alam
Water & Sanitation Service Peshawar, Peshawar, Pakistan

Yazidhi Bamutaze
Makerere University, Kampala, Uganda

Jean-Luc Betrand-Krajewski
INSA Lyon, Villeurbanne, France

Naveen Chand
Environment and Biofuel Research Laboratory, Department of Hydro and Renewable Energy (HRED), Indian Institute of Technology Roorkee, Roorkee, Uttarakhand, India

Surendra Kumar Chandniha
Department of Soil and Water Conservation Engineering, BRSM College of Agricultural Engineering and Technology & Research Station, IGKV, Mungeli, Chhattisgarh, India

Bakim Chandra
National Institute of Technology Manipur, Imphal, Manipur, India

Anandu Chandran
Department of Environmental Science and Engineering, School of Engineering and Sciences, SRM University-AP, Amaravati, Andhra Pradesh, India

Shiao-Shing Chen
National Taipei University of Technology, Institute of Environmental Engineering and Management, Taipei, Taiwan

Aravind D.
Department of Environmental Science and Engineering, School of Engineering and Sciences, SRM University-AP, Amaravati, Andhra Pradesh, India

Kamal Das
National Institute of Technology Agartala, Agartala, Tripura, India

Siddhant Dash
Department of Civil Engineering, School of Engineering and Sciences, SRM University-AP, Guntur, Andhra Pradesh, India

Chinmoy Kanti Deb
Mass Transfer Group, School of Chemical Engineering, Vellore Institute of Technology, Vellore, Tamil Nadu, India

D.U.S.L. Deepthi
Department of Chemical Engineering, MVGR College of Engineering, Vizianagaram, Andhra Pradesh, India

N. Nandini Devi
Bennett University, Greater Noida, Uttar Pradesh, India

Anh Tien Do
Vietnam Meteorological and Hydrological Administration, Hanoi, Vietnam

Quoc-Hoang Do
Faculty of Engineering and Technology, Saigon University, Ho Chi Minh City, Vietnam

Mohamed ElKashouty
Department of Hydrogeology, Faculty of Earth Sciences, King Abdulaziz University, Jeddah, Saudi Arabia

Mahesh Ganesapillai
Mass Transfer Group, School of Chemical Engineering, Vellore Institute of Technology, Vellore, Tamil Nadu, India

Anindita Ganguly
Department of Civil, Geodetic and Environmental Engineering, Division of Environmental Health Sciences, and Sustainability Institute, The Ohio State University, Columbus, OH, United States

Rupanshu Gupta
Department of Civil Engineering, National Institute of Technology, Raipur, Chhattisgarh, India

Nguyen Thi Hong
Master Program in Water Technology, Reuse and Management, Vietnamese-German University, Ho Chi Minh City, Vietnam

Mohd Yawar Ali Khan
Department of Hydrogeology, Faculty of Earth Sciences, King Abdulaziz University, Jeddah, Saudi Arabia

Amit Kumar
Nanjing University of Information Science and Technology, School of Hydrology and Water Resources, Nanjing, China

Sananda Kundu
Department of Geography, Manipur University, Imphal, Manipur, India

Vu-Anh Le
School of Chemical and Biomolecular Engineering, University of Sydney, Darlington, NSW, Australia

Hyung Kae Lee
School of Sustainable Engineering and the Built Environment, Arizona State University, Tempe, AZ, United States

Min Liu
Nanjing University of Information Science and Technology, Nanjing, China

Tran Le Luu
Master Program in Water Technology, Reuse and Management, Vietnamese-German University, Ho Chi Minh City, Vietnam

Nisha Malik
Membrane Technology Group, Faculty of Bioscience Engineering, KU Leuven, Leuven, Belgium

Arun Mondal
Department of Geography, University of Allahabad, Prayagraj, Uttar Pradesh, India

Arkadeep Mukherjee
Department of Biosciences, Graphic Era (Deemed to Be) University, Dehradun, Uttarakhand, India

Diwakar Naidu
Faculty of Agricultural Engineering, SV College of Agricultural Engineering and Technology & Research Station, IGKV, Raipur, Chhattisgarh, India

Hau Thi Nguyen
Faculty of Chemistry and Environment, DaLat University, DaLat, Vietnam

Nguyen Cong Nguyen
Faculty of Chemistry and Environment, Dalat University, Dalat, Vietnam

Van-Truc Nguyen
Faculty of Engineering and Technology, Saigon University, Ho Chi Minh City, Vietnam

Phuong Nghi Nguyen Pham
Faculty of Chemistry and Environment, DaLat University, DaLat, Vietnam

Anupam Panday
Department of Geography, University of Allahabad, Prayagraj, Uttar Pradesh, India

Rajat Pant
Environment and Biofuel Research Laboratory, Department of Hydro and Renewable Energy (HRED), Indian Institute of Technology Roorkee, Roorkee, Uttarakhand, India

Sanjeev Kumar Prajapati
Environment and Biofuel Research Laboratory, Department of Hydro and Renewable Energy (HRED), Indian Institute of Technology Roorkee, Roorkee, Uttarakhand, India

Ch. Ramya
Department of Chemistry, Aditya Institute of Technology and Management (A), Srikakulam, Andhra Pradesh, India

Nirav P. Raval
Department of Environmental Science and Engineering, School of Engineering and Sciences, SRM University-AP, Amaravati, Andhra Pradesh, India

Srinath Ravi
Department of Civil, Urban, Earth and Environmental Engineering, Ulsan National Institute of Science and Technology (UNIST), Ulsan, South Korea

Saikat Sinha Ray
Department of Environmental Science and Engineering, School of Engineering and Sciences, SRM University-AP, Amaravati, Andhra Pradesh, India

Anwar Faizaan Reza
Department of Computer Science and Engineering, School of Engineering and Sciences, SRM University-AP, Amaravati, Andhra Pradesh, India

Deblina Roy
Department of Environmental Science and Engineering, School of Engineering and Sciences, SRM University—AP, Amaravati, Andhra Pradesh, India

Ishita Sarkar
Departament d'Enginyeria Química i Química Analítica, Universitat de Barcelona, Barcelona, Spain

K.V.V. Satyannarayana
Department of Civil Engineering, Yeungnam University, Gyeongsan, South Korea

Rangabhashiyam Selvasembian
Department of Environmental Science and Engineering, School of Engineering and Sciences, SRM University-AP, Amaravati, Andhra Pradesh, India; Center for Interdisciplinary Research, SRM University-AP, Amaravati, Andhra Pradesh, India

Khan Shahzada
University of Engineering & Technology Peshawar, Peshawar, Pakistan

Chenran Shao
Nanjing University of Information Science and Technology, Nanjing, China

Randeep Singh
Department of Basic Sciences, School of Sciences and Humanities, SR University, Warangal, Telangana, India

Vaheed Syed
Department of Environmental Science and Engineering, School of Engineering and Sciences, SRM University-AP, Amaravati, Andhra Pradesh, India

Nguyen Tan Tan
Faculty of Chemical Engineering, Can Tho University, Can Tho City, Vietnam

Ramchandra Taram
Faculty of Agricultural Engineering, SV College of Agricultural Engineering and Technology & Research Station, IGKV, Raipur, Chhattisgarh, India

Van-Anh Thai
HUTECH Institute of Applied Sciences, HUTECH University, Ho Chi Minh City, Vietnam

Thu Uyen Thi Nguyen
Faculty of Chemistry and Environment, DaLat University, DaLat, Vietnam

Aditya Tiwari
School of Engineering, Newcastle University, Newcastle Upon Tyne, United Kingdom

Gulomjon Umirzakov
Hydrometeorological Research Institute, Central Asian University for Environmental and Climate Change Studies, Tashkent, Uzbekistan

Doan Van Hong Thien
Faculty of Chemical Engineering, Can Tho University, Can Tho City, Vietnam

Anil Kumar Vinayak
Department of Chemical and Materials Engineering, University of Alberta, Edmonton, AB, Canada

Thi-Dieu-Hien Vo
Faculty of Applied Science and Technology, Nguyen Tat Thanh University, Ho Chi Minh City, Vietnam

Gete Zeleke
Water and Land Research Center, Addis Ababa, Ethiopia

Preface

In recent decades, rapid urbanization, industrialization, intensive agriculture, and climate change have created huge pressure on water resource quality, quantity, and sustainable water resource management. Among the most critical challenges, wastewater, if inadequately treated, poses aquatic, ecological, and human health risks, and affects overall environmental sustainability. Therefore, sustainable water management demands a comprehensive and scientifically grounded approach to wastewater treatment and resource recovery.

This book mainly focuses on the principles and practices of sustainable wastewater treatment and management under the framework of integrated water resource management. It examines the physico-chemical and biological processes governing wastewater characteristics and treatment performance. Besides, conventional wastewater systems are discussed alongside advanced and emerging technologies, such as membrane technology, adsorption, anaerobic digestion, membrane bioreactors, nutrient recovery systems, constructed wetlands, decentralized treatment solutions, and artificial intelligence (AI) and machine learning (ML)-driven monitoring, modeling, and optimization tools. Further, emphasis is placed on energy efficiency, greenhouse gas (GHG) reduction, sludge management, and the transformation of wastewater from a liability into a valuable resource.

In addition, the book explores circular economy models, water reuse, policy frameworks, and climate-resilient infrastructure to ensure long-term environmental sustainability. This book combines scientific theory with practical applications and aims to equip students, researchers, engineers, and policymakers with the knowledge required to design and implement sustainable wastewater management systems that will help to reduce both ecological and human health risks.

Acknowledgments

The completion of this book entitled *Sustainable Wastewater Treatment"* has been made possible through the joint support of Acquisition Editor, EPM, Editors, and Reviewers throughout the successful completion of the book.

We extend our sincere gratitude and thanks to our academic university, research collaborators, and industry professionals whose insights into wastewater engineering, environmental biotechnology, and sustainable resource management have enriched this work. Their fruitful discussions, innovations, suggestions, technical expertise, and commitment have significantly shaped the scientific depth and practical relevance of this book. We also acknowledge the contributions of worldwide researchers whose pioneering work in wastewater processes, treatment, mitigation, and management, circular water economy, and climate-resilient wastewater systems has laid the foundation for sustainable advancements in this core field.

PART

Vision, challenges, and strategic frameworks

I

CHAPTER

A global overview of sustainable water management challenges, frameworks, and policy-driven technological shifts

1

Saikat Sinha Ray[1], Shiao-Shing Chen[2], Nguyen Cong Nguyen[3] and Amit Kumar[4]

[1]*Department of Environmental Science and Engineering, School of Engineering and Sciences, SRM University-AP, Amaravati, Andhra Pradesh, India;* [2]*National Taipei University of Technology, Institute of Environmental Engineering and Management, Taipei, Taiwan;* [3]*Faculty of Chemistry and Environment, Dalat University, Dalat, Vietnam;* [4]*Nanjing University of Information Science and Technology, School of Hydrology and Water Resources, Nanjing, China*

1. Introduction

Universal access to clean water resources is essential for global sustainability as it underpins human health, supports socio-economic development, and maintains ecosystem integrity. Water scarcity, a key focus of Sustainable Development Goal 6 (SDG 6), under the UN 2030 Agenda for Sustainable Development (UN, 2015), is widely recognized as a growing challenge affecting residential, agricultural, and industrial sectors worldwide (Salehi, 2022). Currently, more than 40% of the global population experiences water scarcity, and over 700 million people still lack access to safe drinking water. At the same time, nearly half of human-generated wastewater is discharged untreated into rivers and oceans, creating serious environmental and public health concerns (Shemer et al., 2023). The UN estimates that by 2025, around 230 million Africans are expected to face water scarcity, and nearly 460 million may live in water-stressed regions (Ingrao et al., 2023; Ssekyanzi et al., 2024). Increasing demand from agriculture and industry further intensifies the problem, accounting for about 70% and 20% of global water withdrawals, respectively (Ssekyanzi et al., 2024). Moreover, by 2050 nearly half of the global urban population may live in water-scarce regions, emphasizing the urgent need to improve water-use efficiency and implement sustainable water management strategies (He et al., 2021).

The industrial revolution significantly advanced science, technology, and living standards; however, it also introduced serious environmental challenges. While the pace of economic development and service coverage expansion through rapid industrialization and urbanization is impressive, it has come at the cost of more intense natural resource exploitation with higher degrees of pollution (Teker & Koc, 2019). Rapid demographic and urban expansion would be the key factors increasing water demand, with estimates indicating that 0.99 (0.829–1.135) billion people could more likely live in cities facing shortages of water supply, or accounting for approximatively 87% of the global increase

Sustainable Water Management. https://doi.org/10.1016/B978-0-443-44853-9.00018-6

in water-stress-population at city level (He et al., 2021). Pollution occurs when the environment is contaminated with such pollutants, such as inorganic substances, organic matter, organometallics, gases, nanoparticles, pathogens and radiation which have destabilizing impacts on the vitality of ecosystems and directly affect human health (Sarker et al., 2021). In 83% of developing regions, including South Asia, water pollution primarily stems from industrial effluents, agricultural runoffs, sewage discharge, oil spills and poor waste management (Dey & Majumdar, 2024). These contaminants lower dissolved oxygen (DO), destroy aquatic habitat and disrupt biodiversity; negatively affecting aquatic organisms (Alao et al., 2023). Contaminated water also spreads diseases like cholera and dysentery and causes economic losses. Hence, suitable pollution regulation and wastewater treatment measures are critical for sustainable water resource management (Saxena, 2025).

Water management is rapidly evolving worldwide as the reliability of conventional water sources declines due to climate change, population growth, and overexploitation of freshwater resources. Although extensive literature exists on water source management, empirical understanding of hybrid water systems management the integration of conventional and non-conventional water sources within a single system, remains limited (Dashkevych & Housh, 2025). Historically, societies relied on decentralized and community-based systems such as stepwells, kinds, tankas, and lakes (Bimrah & Saizen, 2026). As water scarcity intensifies, modern management strategies emphasize optimization, recycling, and integrated water resource management that combine multiple water sources with advanced allocation methods. Technological tools such as remote sensing (RS), geographic information systems (GIS), satellite monitoring, and smart water systems enable real-time assessment of water resources and improve decision-making (Zaigham Abbas Naqvi et al., 2025). The "Blue Revolution" (Onyena & Sam, 2025) further promotes efficient water use, ecological integration, and participatory governance for sustainable water management (Dashkevych & Housh, 2025). Figure 1.1 demonstrates the bibliometric distribution of publications on the *"global water crisis"* AND *"sustainable water management"* obtained from a Scopus advanced search (March 2026). Figure 1.1 (A) indicates the analysis shows that the United States, China, and the United Kingdom are the leading contributors to research output. Figure 1.1 (B) shows the subject-area distribution indicates that environmental science and social sciences dominate the field, followed by agricultural, biological sciences, and engineering, highlighting the multidisciplinary nature of water crisis and sustainable water management research. This chapter summarizes the major global challenges associated with sustainable water management and the evolving frameworks and policies developed to address them. It examines the key drivers of water scarcity, water quality degradation, and climate-induced hydrological extremes that threaten global water security.

The objective of this chapter is to provide a global overview of the major challenges associated with sustainable water use and to synthesize emerging frameworks, policies, and technological innovations developed to address these challenges. It examines the three primary dimensions of the global water crisis, water scarcity, water quality degradation, and climate-related hydrological extremes and discusses how these pressures threaten long-term global water security. The chapter further outlines key governance and resource management frameworks, including Integrated Water Resource Management (IWRM), the water-energy-food (WEF) nexus, the circular water economy, and emerging digital water governance approaches. In addition, it reviews major global policy initiatives, particularly those aligned with SDG 6, and highlights important technological advancements such as advanced water treatment systems, desalination technologies, adsorption-based processes, membrane

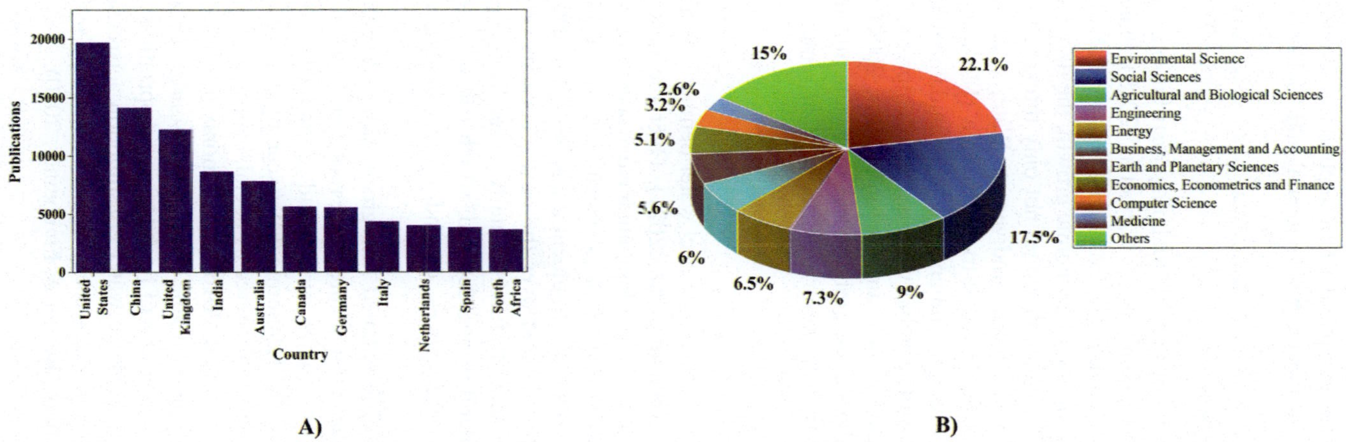

FIGURE 1.1

Bibliometric distribution of publications related to "*global water crisis*" and "*sustainable water management*" till March 2026. (A) The country-wise distribution of publications, (B) the subject-area distribution of the publications.

bioreactors, and smart water management solutions. Overall, the chapter aims to present an integrated conceptual, policy, and technological perspective that supports resilient, efficient, and sustainable water resource management under increasing environmental and socio-economic pressures.

2. Global water management

Global water management refers to the sustainable planning, distribution, and protection of freshwater to meet human, environmental, and economic needs (Ley et al., 2019). In general, water resources are crucial for maintaining a productive environment and ensuring a reliable food supply for all living organisms. With a rising population and rapid economic growth, the demand for freshwater continues to escalate (Vörösmarty et al., 2013). Water shortages not only threaten agricultural productivity but also significantly reduce terrestrial and aquatic biodiversity. Water resources and their management have so far mostly been addressed at the local-to-basin scale (Mishra et al., 2021). It has been acknowledged that water resources are both influenced by and fundamentally linked to global change and globalization. Recently, the idea of a global water system comprising interconnected social and ecological elements has been developed and came in-force (Hoff, 2009). Figure 1.2 represents the interconnections among major challenges in global water management, highlighting four key components: water scarcity, water quality degradation, hydrological extremes, and climate change.

2.1 Water scarcity

Water scarcity, the inability of water resources to meet water demands, is regarded as one of the most significant global risks for society (Huang et al., 2021). Over the past few decades, changes in hydroclimatic patterns and socio-economic development have exacerbated water scarcity at global and regional scales. Currently, many urban populations face water scarcity (Wada et al., 2016). Population growth, urbanization, and socioeconomic development are projected to drive a 50%–80% increase in urban industrial and domestic water demand over the next 3 decades (Flörke et al., 2018; Sperna Weiland et al., 2012). Global water scarcity fundamentally arises from the geographic and temporal mismatch between freshwater availability and demand, which can be assessed either physically or through its social and economic consequences, depending on adaptation capacity (Garrick et al., 2019).

Physical water scarcity can be differentiated into green water scarcity and blue water scarcity. Green water scarcity occurs when the soil moisture in the root zone is too low to support crop growth. In contrast, blue water scarcity happens when surface and groundwater sources cannot meet human water demands (Liu et al., 2017a; 2017b). Economic water scarcity occurs when renewable freshwater resources are physically accessible but are hindered by social, economic, and institutional barriers that prevent their effective use (Gosling & Arnell, 2016). A commonly used water scarcity index measures the ratio of freshwater withdrawal to the available freshwater resources (Rosa et al., 2020). The water footprint (WF) serves as a volumetric measure of freshwater appropriation (Hoekstra, 2016; Kumar et al., 2025). The concept is applied to evaluate water consumption across supply chains, the sustainability of water use within river basins, water-use efficiency, equity of water allocation, and dependence on water within the supply chain (Berger & Finkbeiner, 2010).

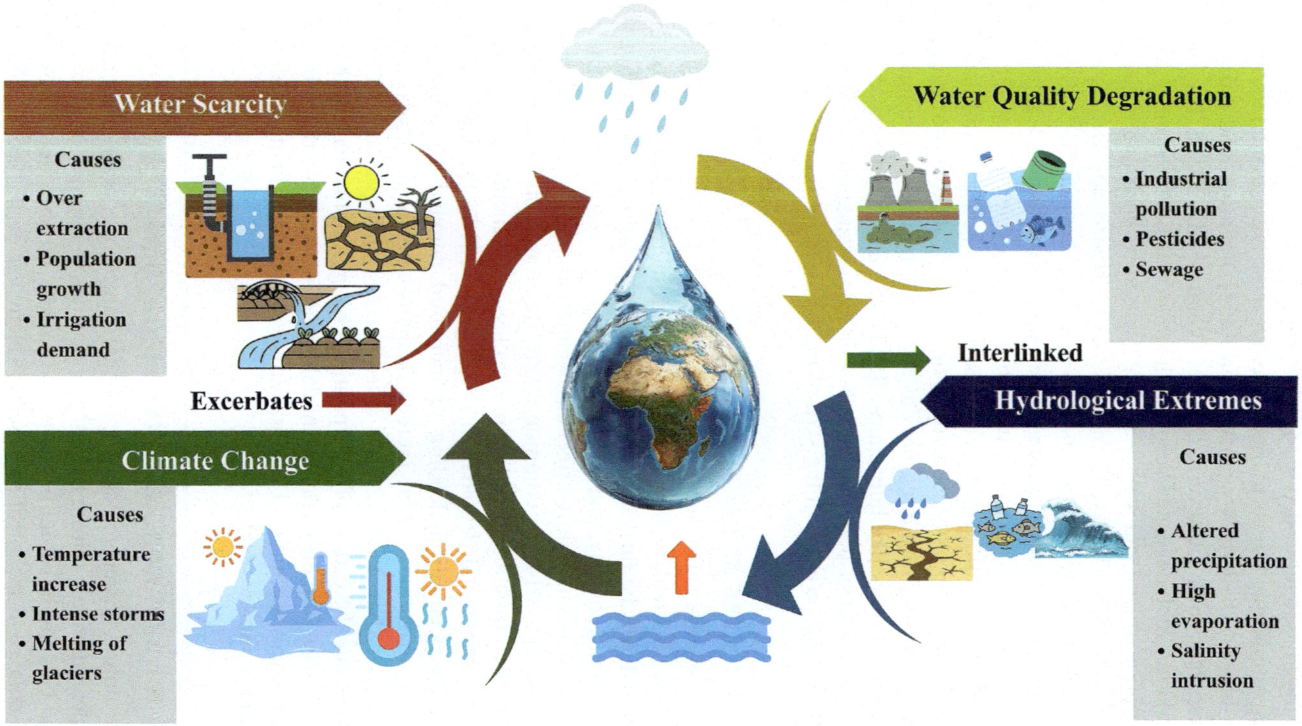

FIGURE 1.2

Schematic representation of the interconnected global water crisis.

2.2 Water quality degradation

Water quality degradation has deteriorated worldwide since the 1990s and is anticipated to accelerate in the coming decades, amplifying threats to human health and environmental integrity (Plessis, 2022). It remains one of the biggest challenges worldwide in the 21st century, and failure to address it will likely have serious repercussions for several SDGs (Akhtar et al., 2021). The major water quality issues causing degradation include water-borne pathogens and noxious and toxic substances. The ongoing rapid deterioration of land and water resources caused by declining water quality may lead to hydrocide for future generations (Issaka & Ashraf, 2017). Hydrogeological and biophysical environments are directly influenced by land use changes and socioeconomic factors, which are primarily driven by human activities and resource management. Apart from human activities, the natural variability of rock and soil interacts with water, impacting water cycles and quality across all domain (Khan et al., 2023). Such modifications can significantly impact human health and the functioning of living organisms. Furthermore, the physicochemical and biological properties, along with the quality, quantity, and availability of water resources, vary as a result of natural and human activities (Peters & Meybeck, 2000). Natural disasters like hurricanes, floods, tsunamis, earthquakes, volcanic eruptions, and landslides significantly impact both water quality and water quantity (P. Gupta et al., 2021). The time scale of the perturbation varies with the size of the disaster.

2.3 Hydrological extremes and climate change

Global warming, caused by the enhanced greenhouse effect, is likely to substantially impact the hydrological cycle. The hydrological cycle will intensify, with increased evaporation and precipitation, but the additional rainfall will be unevenly distributed across the globe (Arnell et al., 2011). Many aspects of the environment, economy, and society rely on water resources, and alterations in the hydrological resource base can significantly affect environmental quality, economic progress, and social well-being (Haddeland et al., 2014). In common terms, there are both supply-side and demand-side pressures. The supply-side pressures include climate change (reducing or increasing the amount of water available) and environmental degradation (Sivakumar, 2011). Demand-side pressures include population growth and concentration, leading to increased demand for domestic, industrial, and agricultural (particularly irrigation) water, increased environmental demands, and changes in how water demand is managed (Nie et al., 2024). Two factors that have complicated the assessment of water scarcity impacts are the pattern of precipitation changes and indicators that are used to define water scarcity (Misra, 2014). Complex climate models generally indicate that climate change leads to changes in precipitation across regions. These patterns vary across climate models, with only a few areas where models show similar results.

3. Frameworks for sustainable water management

Sustainable water resources management assessment frameworks with associated indicators and benchmarks have appeared widely during the last few decades to improve or maintain water resources (Brown et al., 2015). This implies a unified approach, while urban water debates reflect a wide diversity of political, social, and technical viewpoints (Namavar et al., 2023). Five distinct but overlapping frameworks for urban water sustainability are evident in research, policy and practice, reflecting wider environmental theory, politics, and discourse (Alsaeed et al., 2022). Sustainable development focuses on addressing the water and sanitation needs of the urban poor. Ecological modernization focuses on policies to improve water efficiency and treatment through technological innovation (Marcal et al., 2021). Understanding various perspectives on water sustainability establishes the foundation for more productive dialog about water and its role in sustainable cities.

3.1 Integrated water resource management

Integrated Water Resource Management (IWRM) became a widely adopted concept in the water sector during the 20th century. IWRM is recognised as a process that "promotes co-ordinated development and management of water, land and related resources, in order to maximize economic and social welfare in an equitable manner without compromising the sustainability of vital systems" (Abdel-Magid & Ahmed, 2002). IWRM includes objectives, institutions, implementation, and adaptation, reflecting the merging of top-down and bottom-up approaches. Table 1.1 presents the core structural components of the IWRM framework, including the Enabling Environment, Institutional Framework, Management Instruments, and Sustainable Financing. It can assume many institutional forms, and is best implemented at the river basin or subbasin scales (Davis, 2007). Groundwater basins may also

Table 1.1 Key components of the integrated water resources management framework.

Framework component	Framework key elements	Description	Tools	References
Enabling environment	Policies, laws, regulations	Establishes legal and policy framework for water governance	National water policies, transboundary water agreements, water laws, budget allocations	Katusiime and Schütt (2020)
Institutional framework	Coordination mechanisms, stakeholder participation	Integration across sectors, involvement of communities, NGOs, private sector	Water user associations, river basin organizations, coordination between local, basin, national, and transboundary authorities	Medema et al. (2008)
Management instruments	Water resource assessment, planning and allocation	Monitoring availability and demand of water, efficient distribution of water resources	IWRM plans, river basin management plans, water allocation systems, water tariffs, pollution charges, public awareness campaigns.	Apostolaki et al. (2019)
Sustainable financing	Economic instruments, economic outcomes	The budgeting and financial resources secured to develop and manage water resources.	Government budgets, water tariffs, external aid, innovative financing mechanisms	Rahaman and Varis (2005)

cross hydrological drainage basin limits, while integrated surface-coastal management may require addressing multiple hydrological drainage basins (Savenije & Van Der Zaag, 2008). IWRM can provide a potential forum to better address: (a) incorporation of the hydrologic cycle and hydrologic basins into the water supply and demand issues, (b) coordination of surface and groundwater/conjunctive use, (c) multiple water use objectives, (d) improvement of economic and water use efficiency, and (e) reduction/mitigation of water user conflicts. However, IWRM is adopted specifically to target one or more of these items in a strategic and prioritized manner to seek better water use outcomes (Rahaman & Varis, 2005).

3.2 Water energy food nexus

The water–energy–food nexus has become central to discussions regarding the development and subsequent monitoring of the SDGs (Albrecht et al., 2018). The constantly growing world population leads to an ever-surging demand for higher living standards. It generates a huge demand for water, energy, and food, thereby increasing the demand for sustainable development. According to the UN estimate, the population of the Earth will surpass 10 billion by 2100 (Liu et al., 2017a). The Food and Agriculture Organization (FAO) projects that 60% more food and 80% more energy will be required by 2050 to meet global demand; an increase in total global water withdrawals by 50% in developing countries and 18% in developed countries by 2025 is projected (Flammini et al., 2014). Nexus solutions need to clearly and explicitly identify reduced costs and increased benefits for both humans and nature, compared to single-sector strategies for water, land, energy and food (Bhaduri et al., 2015). Figure 1.3 represented the comprehensive circular conceptual framework for the WEF Nexus, the interconnected and interdependent relationships among water, energy, and food systems, while integrating broader dimensions of sustainability, resilience, and global challenges. Governing the nexus necessarily involves governing the trade-offs between the different sectors and dealing with the plurality and interactions of policies that are in place (Cudennec et al., 2017). Nexus governance demands ongoing efforts, since much institutional innovation requires a longer timespan for effectiveness implementation. This also requires political support and constant monitoring of targets to bring about sound, responsible and transparent governance principles that mitigate cross-sectoral competition around resource distribution while addressing both human and environmental concerns toward sustainable development (Endo et al., 2017).

3.3 Circular water economy

The circular economy has also attracted considerable attention in relation to water, an indispensable element for the sustenance of life and a critical input resource for the world economy (Morseletto et al., 2022). According to the FAO of the UN, the agricultural sector, which accounts for over 70% of global freshwater withdrawals, both drives water scarcity and is negatively affected by it. Climate change impacts (e.g., droughts and temperatures) further stress the already limited water resources, impacting food production and food security (FAO, 2024). The concept of sustainable production and consumption was originally derived from waste-to-resource supply chains that reduce emissions (Slootweg, 2020), mitigate waste, and enhance the environmental resilience against climate change (Mannina et al., 2022). Chen et al. defined the circular water

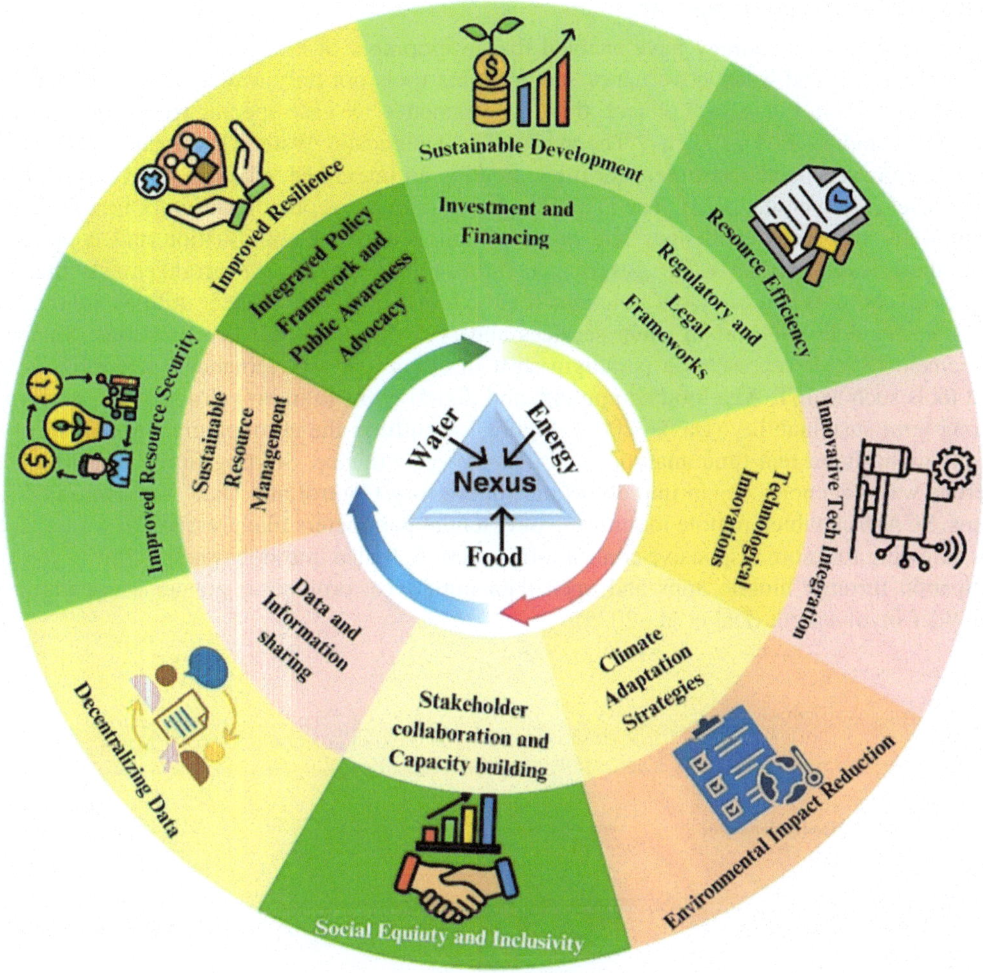

FIGURE 1.3

Integrated circular framework of the water-energy-food nexus.

economy as a business model that enables sustainable production and consumption through water reuse and resource recovery. This transformation is expected to mitigate climate change effects as well as increase economic gains, such as reducing transportation costs and trade of materials (Chen et al., 2021). Although water reuse faces numerous barriers, ranging from public perception to pricing and technological, safety and regulatory challenges, geographical and sector-wide strategies that underpin the circular economy are emerging, and have the potential to transform some of the main barriers to water reuse (Voulvoulis, 2018).

3.4 Digital water governance

Advances in digital technology have enabled the development of numerous digital tools to support decision-making in the water governance sector. Some tools not only assess water governance variables ("Diagnosis") but also connect these assessments to case-specific recommendations for improvement and IWRM as such ("Therapy"). The Diagnostic Water Governance Tool (DWGT) provides a comprehensive instrument for implementing an integrated, adaptive approach to IRWM at the case level (Stein et al., 2023). The DWGT is based on a conceptual framework focused on cross-sectoral water resources challenges and on conditions for better coordination and cooperation to address these issues. Urban water governance refers to the institutional arrangements that engage multiple stakeholders and carry out functions related to water services. Water utilities are the systems responsible for delivering urban water services and performing governance functions. Smart Water Management (SWM) represents a paradigm shift in urban water governance, integrating advanced digital tools such as IoT, AI, cloud computing, and digital twins to improve robustness, performance, and long-term sustainability (Davis, 2007). Figure 1.4 outlines the chronological evolution of SWM with an emphasis on real-time analytics and adaptive governance. Starting in the late 20th century, numerous water agencies began implementing Supervisory Control and Data Acquisition (SCADA) systems. SCADA enabled remote monitoring of essential parameters like pressure, flow rate, and tank levels, allowing for more responsive control infrastructure. It also introduces new pathways for citizen participation through mobile apps and open data initiatives, promoting greater accountability and stakeholder involvement (Dai et al., 2025).

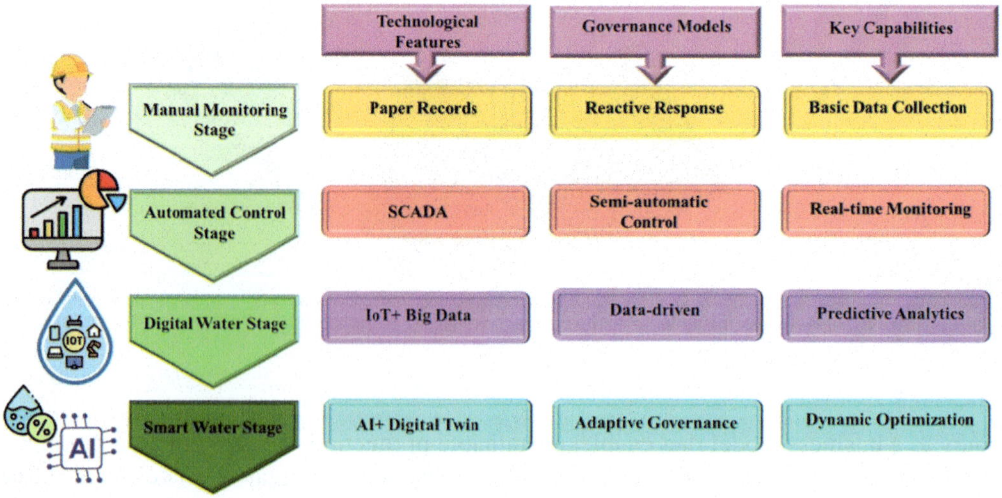

FIGURE 1.4

Evolution of water management systems: From manual monitoring to smart water management.

4. Global policy in water sector

Water is increasingly becoming a limiting factor for sustainable economic growth and development in many countries. It becomes a strategic resource involving conflicts among those who may be affected differently by various policies (Alaerts, 2019). The framework helps assess various linkages among policies and their impacts within individual sectors and at the economy-wide level. Over the past 20 years, total global levels of terrestrial water storage have fallen (Bhaduri, 2025). This is largely due to climate change, both from melting glaciers and groundwater depletion. Global water policy is essential for managing transboundary water resources, mitigating climate-induced scarcity, and ensuring equitable access to drinking water and sanitation for a growing population (Singh & Goyal, 2025). Table 1.2 shows a chronological overview of major international policies, frameworks, and initiatives related to water and sanitation, primarily led by the United Nations (UN-Water), the SDG 6, and the World Bank.

According to The UN World Water Development Report 2026, Water for all people: Equal rights and opportunities, emphasizes available data and actionable solutions to promote gender equality throughout the water sector. The Report highlighted that gender equality, through access to clean, safe, and affordable water and sanitation, along with the equitable and meaningful participation of women in water governance, is essential to poverty reduction and to building healthier, fairer societies with broad social and economic benefits (*United Nations World Water Development Report 2026 | UNESCO*, 2026). According to the World Bank Group, water reuse is an effective strategy for optimizing local water resources, but its implementation is frequently limited by various obstacles and barriers. Inadequate pricing fails to generate enough revenue to cover the full costs of building and maintaining water reuse infrastructure. This ultimately discourages future investment and becomes a major challenge in the long run. To overcome these challenges, a range of targeted strategies is needed. Policy levers, such as tariffs for used water treatment, abstraction and discharge fees, and water resource taxes, can be designed to properly represent the cost of water and reimburse private capital for developing the infrastructure of water reuse. Enhancing quality regulations, adopting technological innovations to achieve cost reduction, and providing consumer education are also important ingredients to make these infrastructures more effective in achieving outcomes (*Continental Drying*, 2025).

Achieving SDG six remains a significant challenge, especially in developing countries where limitations of resources, inadequate infrastructures and environmental pressures are widespread (Arora & Mishra, 2022). This goal goes beyond water supply sanitation and hygiene (WASH) to include all aspects of the water cycle and explicitly recognize that water impacts the entire development agenda, it targets water quantity, quality, water use efficiency and water-related ecosystems (Evaristo et al., 2023). SDG six aims to address the challenge of inadequate access to clean water and sanitation by promoting water conservation, the sustainable management of water resources, and the development of efficient sanitation infrastructure by 2030 (Setty et al., 2020). The combined pressures of climate change, urbanization, and financial limitations will continue to hinder global efforts, underscoring the need for inclusive and sustainable strategies to meet SDG six by 2030 (Aman et al., 2024).

5. Technological shifts in water sector

Technological innovations are increasingly transforming the water sector by improving treatment efficiency and monitoring capabilities. Membrane bioreactor (MBR) technology, which integrates

Table 1.2 Global policies in the water sector: Timeline by UN, UN-water, world bank, and SDG-6.

Year/ Period	Organization/ Framework	Policy/ Initiative	Key features	Influence on countries	References
2003	UN- water	Establishment of UN-Water	Coordination mechanism for global water governance	Helped countries align water policies with UN agencies and improve cooperation.	*Water and Sanitation | Department of Economic and Social Affairs* (2025)
2008	United nations	International year of sanitation	Focus on the sanitation crisis globally	National sanitation campaigns in developing countries.	Lenton et al. (2008)
2015	SDG-6	Clean water and sanitation goal adopted	Universal access to safe water and sanitation by 2030	Countries aligned their national water policies with SDG targets.	Milan (2017)
2015–2030	UN- water	Integrated monitoring initiative	Data-driven monitoring of water indicators	Strengthened evidence-based policymaking globally.	Hjorth and Madani (2023)
2016	World Bank + united nations	High-level panel on water	Global leadership initiative for water crisis	Encouraged national reforms and investments in water governance.	Group (2016)
2020	United nations/ UN-water	SDG-6 global acceleration framework	Fast-tracking progress on water goals	Promoted financing, innovation, and governance reforms	Nations and Water (2021)
2021	UN- water	Global SDG-6 progress update	Assessment of global water indicators	Helped countries identify gaps and policy priorities	Nations and Water (2021)
2023	World bank	Scaling up finance for water: Strategic framework	Mainstreaming water data into policy	Improved national-level planning and implementation	Vetterlein and Schmidtke (2024)

biological wastewater treatment with membrane filtration, has emerged as an advanced system capable of producing high-quality effluent (Jijingi et al., 2024). In addition, nanotechnology-based nano-adsorbents have been developed to enhance contaminant removal and improve the efficiency of water remediation processes (Satyam & Patra, 2024). Growing reliance on desalination technologies, including membrane and thermal processes such as multi-stage flash (MSF) and multi-effect distillation (MED), provides reliable alternative freshwater sources, particularly in water-scarce regions (Alenezi & Alabaiadly, 2025). Furthermore, IoT-based smart water management systems employ sensors, information and communication technology (ICT) (A. D. Gupta et al., 2020), remote sensing, and data analytics to enable real-time monitoring and optimized water use in urban and agricultural systems (Parra-López et al., 2025).

Adsorption is widely used in wastewater treatment because of its ability to operate flexibly with low operational and maintenance costs. It is also extensively employed for the elimination of organic and inorganic pollutants, including anions (fluoride, nitrate, bromate, perchlorate, phosphate). As a process on the surface, ions or molecules bind to the adsorbent surface (Priya et al., 2022). Adsorption efficiency mainly based on the properties of adsorbent material. Different materials (e.g., zeolites, magnetic adsorbents, inexpensive natural materials derived from agricultural wastes, industrial by-products and biopolymers and hydrogels) have been examined for desalination water softening or salt removal because of their affordability and operational flexibility (Alaei Shahmirzadi et al., 2018).

Zhao et al. (2026) developed a composite of chitosan-buckwheat hull biochar (KQB) for removal of Pb^{2+} from wastewater. RSM-BBD optimization (pH 5.38, dosage of 1.06 g L^{-1} and time of 6.02 h) achieved the maximum removal as high as 95.31%. The pseudo-second-order kinetics and Langmuir via surface complexation model described adsorption on the clay. Compared with coagulation KQB exhibited higher efficiency and lower pollution index, as well as good reusability (Zhao et al., 2026).

Desalination has a potential solution to the freshwater scarcity problem, especially in developing and coastal areas. Generally treated as a drought-resilient water source, it works independent of river flows or reservoirs or climatic variability. Desalination is a widely used technology for producing freshwater from seawater, but it can also be employed to treat brackish water, saline groundwater, low-quality surface water and the treated effluents. The latest technological improvements have increased the efficiency levels and lowered the operational prices, making it a much more viable project to pursue as an alternative supply of water. Desalination technologies are generally categorized as membrane-based or thermal processes (Alenezi & Alabaiadly, 2025). Obaid et al. (2026) developed ultra-thin polymeric membranes with subnanometer pores for effective membrane distillation of seawater and brine. The films exhibited 99.8% salt rejection, complete boron removal, high water flux and low energy consumption. They possess high potential for scalable sustainable desalination systems due to chlorine resistance, superhydrophobicity, and polymeric structure (Obaid et al., 2026).

Wastewater treatment using MBR technology combines biological treatment (aerobic or anaerobic processes) with membrane filtration. Unlike the CAS process, where sludge separation occurs through gravity settling, MBR employs microfiltration or ultrafiltration membranes for efficient sludge separation. MBR systems operate with higher Solid Retention Time (SRT) and lower Hydraulic Retention Time (HRT), enabling improved removal of biochemical oxygen demand, suspended solids, and turbidity compared with CAS. These advantages produce high-quality effluent suitable for water reclamation while requiring a smaller treatment footprint (Rahman et al., 2023).

Evaluation of a low-cost Algal Ceramic Membrane Bioreactor (ACMBR) incorporating a microalgae-bacteria consortium and clay–rice bran ceramic membranes was conducted by

Shafiquzzaman et al. for wastewater treatment. At organic loading rates (OLR) $\leq$ 20 g chemical oxygen demand (COD) $m^{-3}{\cdot}d^{-1}$, >90% COD removal was achieved, with moderate nutrient removal, low membrane fouling, and high energy self-sufficiency, indicating its potential for sustainable decentralized wastewater treatment (Shafiquzzaman et al., 2026).

Internet of Things (IoT) sensors monitor water quality, flow, pressure, and volume in reservoirs, dams, and water distribution networks, providing real-time data to cloud-based platforms. These data are analyzed using AI to predict water demand, detect leaks and contaminants, and optimize water distribution under changing climatic conditions (P & Sathiyapriya, 2025). Advanced smart metering technologies, including Advanced Metering Infrastructure (AMI) and Automated Meter Reading (AMR), further enhance leak detection, meter accuracy, and operational efficiency, thereby reducing non-revenue water and supporting sustainable water resource management (Obunga et al., 2025). Table 1.3 presents major technologies including adsorption, desalination, MBR, and smart water management, highlighting their operational principles, advantages, and applications in improving water treatment efficiency and sustainable resource management.

A real-time water quality monitoring system based on IoT was developed by Forhad et al. (2024) for WTPs, to monitor pH, DO, total dissolved solids (TDS), and temperature. The 29-W low-energy system (0.1–0.2 error range) can be used for cloud-based monitoring with several advantages including real-time notifications, autonomous operation, enhanced maintenance and decision-making tools that support sustainable water management (Forhad et al., 2024). Figure 1.5 presents major technological shifts in water. Figure 1.5 **(A)** shows the desalination process where saline water undergoes evaporation and condensation to produce freshwater while leaving concentrated brine. Figure 1.5 (B) presents an MBR system in which biological treatment is coupled with membrane filtration to obtain high-quality effluent. Figure 1.5 **(C)** represents the adsorption process, where pollutants in wastewater are captured on the surface of adsorbent materials, resulting in purified water. Figure 1.5 (D) depicts smart water management using IoT sensors and AI connected to cloud and remote servers for real-time monitoring and management of domestic and industrial water systems. Integrating advanced technologies such as adsorption, membrane bioreactors, desalination, and IoT-based monitoring into water treatment systems, enhance contaminant removal efficiency and resource management, thereby providing more sustainable solutions to global water scarcity and wastewater challenges.

6. Future perspectives and emerging research directions

Recent advances in water systems engineering highlight the need for transdisciplinary collaboration among engineers, policymakers, researchers, industry stakeholders, and communities to co-create sustainable water management solutions. Such integrative approaches facilitate the effective deployment of advanced technologies, including decentralized wastewater treatment systems, AI-driven analytics, smart water grids, and digital monitoring platforms. These innovations significantly enhance water-use efficiency, enable predictive resource management, and support real-time monitoring of water quality and distribution systems. However, despite rapid technological progress, large-scale implementation depends not only on technological maturity but also on the presence of enabling institutional and policy frameworks. Regulatory clarity, policy coherence, and targeted economic incentives are essential to accelerate adoption and ensure long-term operational sustainability. In this context, institutional innovations such as regulatory sandboxes, policy laboratories, and

Table 1.3 Advanced technologies for water and wastewater treatment with their principles, advantages, and applications.

Technology	Principle	Advantages	Applications	References
Adsorption	Adsorption is the phenomenon where molecules of an adsorbate substance attach to the surface of an adsorbent substance through Intermolecular forces.	• Simplicity and cost-effectiveness • Broad pollutant removal • High efficiency and selectivity • Enhanced capacity • Reusability	• Wastewater treatment • Industrial effluents • Broader dye removal	(Alsuhybani et al., 2026; Satyam & Patra, 2024)
Desalination	Distillation is a desalination technique that removes dissolved salts, minerals, and impurities from saline water such as seawater and brackish water. The process produces two outputs, freshwater and concentrated brine as a by-product.	• High performance and efficiency • Scalability • long-term stability and water quality	• Seawater desalination • Industrial-scale and house-hold desalination system	(Isah et al., 2022; S. Liu et al., 2025)
MBR	MBRs combine membrane filtration or separation techniques with established biological treatment methodologies	• High-quality effluent • Compact design • Reduced footprint • Efficient contaminant removal	• Industrial and municipal wastewater treatment • Resource recovery • Specialized wastewater treatment • Nutrient removal • Contaminant removal	(Banerjee et al., 2026; Cicekalan et al., 2025)
Smart water management	Smart water systems, often incorporating IoT, utilize sensors and actuators to collect real-time data on various parameters such as water flow, pressure, and quality	• Enhanced efficiency and waste reduction, • Improved water quality • Monitoring and control • Better resource management • Environmental sustainability	• Water level controlling systems • Smart irrigation systems • Real-time monitoring and control	(Cicekalan et al., 2025; Morchid et al., 2025; Onyena & Sam, 2025)

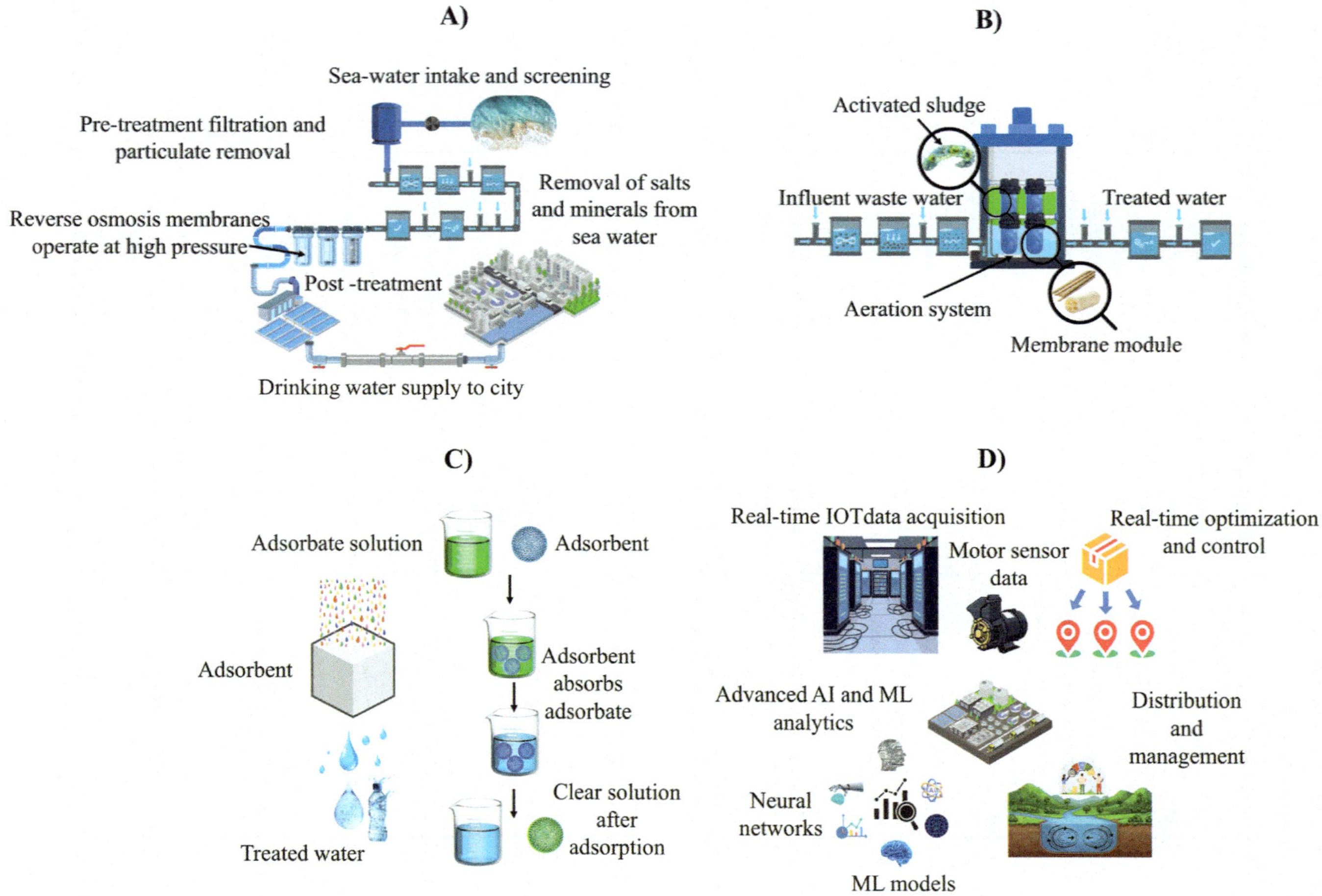

FIGURE 1.5

Schematic illustration of advanced water treatment approaches: (A) Desalination, (B) Membrane bioreactor (MBR), (C) adsorption for contaminant removal and (D) Internet of things (IoTs) and artificial intelligence (AI)-based smart water management.

scenario-based planning tools are gaining prominence as effective mechanisms for testing and scaling emerging technologies under controlled conditions, thereby bridging the gap between technological development and practical implementation.

Future research is expected to increasingly focus on advanced treatment technologies, including membrane distillation, forward osmosis, nanotechnology-enabled filtration materials, and energy-efficient desalination systems. Simultaneously, digital innovations, such as IoT-based sensors, blockchain-enabled governance platforms, and integrated smart water networks, are being developed to improve transparency, data sharing, and operational efficiency across water systems. An important direction in future water management lies in strengthening feedback loops between technological performance and adaptive policy frameworks. Data generated from smart infrastructure and digital monitoring systems can support evidence-based decision-making, enabling continuous system optimization. Furthermore, transforming agricultural systems by prioritizing water-use efficiency will be critical for ensuring resilient and sustainable food production capable of meeting future demands. The integration of advanced technologies, including artificial intelligence, remote sensing, nanotechnology, and next-generation desalination, with regulatory reforms, ecosystem-based approaches, and stakeholder participation will be essential to develop a resilient and adaptive water management framework. Considering escalating challenges such as water scarcity, quality deterioration, climate variability, and inequitable access to safe water resources, a paradigm shift toward integrated, technology-driven, and policy-supported water management is imperative.

7. Conclusion

The global water crisis is increasingly shaped by the convergence of water scarcity, quality deterioration, and climate-driven hydrological changes, placing unprecedented pressure on both natural and engineered systems. Frameworks such as IWRM, the WEF nexus, the circular water economy, and digital water governance represent critical pathways for enhancing resource efficiency, resilience, and adaptive capacity. In parallel, technological advancements, including adsorption processes, membrane-based systems, desalination, and smart water infrastructure, offer significant potential to improve system performance and enable data-driven management by AI and IOT. However, these innovations can only achieve real-world impact when aligned with regulatory frameworks, economic feasibility, and local socio-environmental contexts. Ultimately, the success of these solutions depends on their integration with coherent policy frameworks, strong institutional capacity, and socio-economic viability. Ensuring resilient and comprehensive water security under dynamic global challenges necessitates adaptive, investment-driven policy frameworks capable of replicating context-specific solutions, promote scalable and sustainable outcomes.

References

Abdel-Magid, I. M., & Ahmed, S. E. (2002). Integrated water resources management and global water partnership. https://doi.org/10.13140/RG.2.1.4444.4647.

Akhtar, N., Syakir Ishak, M. I., Bhawani, S. A., & Umar, K. (2021). Various natural and anthropogenic factors responsible for water quality degradation: A review. *Water, 13*(19), 2660. https://doi.org/10.3390/w13192660

Alaei Shahmirzadi, M. A., Hosseini, S. S., Luo, J., & Ortiz, I. (2018). Significance, evolution and recent advances in adsorption technology, materials and processes for desalination, water softening and salt removal. *Journal of Environmental Management, 215*, 324–344. https://doi.org/10.1016/j.jenvman.2018.03.040

Alaerts, G. J. (2019). Financing for water—water for financing: A global review of policy and practice. *Sustainability, 11*(3), 821. https://doi.org/10.3390/su11030821

Alao, J. O., Fahad, A., Abdo, H. G., Ayejoto, D. A., Almohamad, H., Ahmad, M. S., Nur, M. S., Danjuma, T. T., Yusuf, M. A., Francis, O. T., & Joy, A. O. (2023). Effects of dumpsite leachate plumes on surface and groundwater and the possible public health risks. *Science of the Total Environment, 897*, 165469. https://doi.org/10.1016/j.scitotenv.2023.165469

Albrecht, T. R., Crootof, A., & Scott, C. A. (2018). The water-energy-food nexus: A systematic review of methods for nexus assessment. *Environmental Research Letters, 13*(4), 043002. https://doi.org/10.1088/1748-9326/aaa9c6

Alenezi, A., & Alabaiadly, Y. (2025). Emerging technologies in water desalination: A review and future outlook. *Energy Nexus, 17*, 100373. https://doi.org/10.1016/j.nexus.2025.100373

Alsaeed, B. S., Hunt, D. V. L., & Sharifi, S. (2022). Sustainable water resources management assessment frameworks (SWRM-AF) for arid and semi-arid regions: A systematic review. *Sustainability, 14*(22). https://doi.org/10.3390/su142215293

Alsuhybani, M., Alshehri, S., Alquwaizany, A., Alosime, E., Alaeq, R., & Alrehaili, A. (2026). Effective adsorption of methylene blue using natural Saudi zeolite as a low-cost sustainable adsorbent. *Scientific Reports, 16*(1), 211. https://doi.org/10.1038/s41598-025-33491-9

Aman, H., Doost, Z. H., Hejran, A. W., Mehr, A. D., Szczepanek, R., & Gilja, G. (2024). Survey on the challenges for achieving SDG 6: Clean water and sanitation: A global insight. *Knowledge-Based Engineering and Sciences, 5*(3), 21–42. https://doi.org/10.51526/kbes.2024.5.3.21-42

Apostolaki, S., Koundouri, P., & Pittis, N. (2019). Using a systemic approach to address the requirement for integrated water resource management within the water framework directive. *Science of the Total Environment, 679*, 70–79. https://doi.org/10.1016/j.scitotenv.2019.05.077

Arnell, N. W., Van Vuuren, D. P., & Isaac, M. (2011). The implications of climate policy for the impacts of climate change on global water resources. *Global Environmental Change, 21*(2), 592–603. https://doi.org/10.1016/j.gloenvcha.2011.01.015

Arora, N. K., & Mishra, I. (2022). Sustainable development goal 6: Global water security. *Environmental Sustainability, 5*(3), 271–275. https://doi.org/10.1007/s42398-022-00246-5

Banerjee, S., Manoharan, S., Ravikumar, N., & Mahata, C. (2026). Polymeric membrane bioreactors in water treatment: A path to sustainability. *Iranian Polymer Journal (English Edition), 35*(3), 637–660. https://doi.org/10.1007/s13726-025-01528-5

Berger, M., & Finkbeiner, M. (2010). Water footprinting: How to address water use in life cycle assessment? *Sustainability, 2*(4), 919–944. https://doi.org/10.3390/su2040919

Bhaduri, A. (August 28, 2025). India's water policy: Between scarcity, reform, and a sustainable future. *India Water Portal.* https://www.indiawaterportal.org/governance-and-policy/indias-water-policy-between-scarcity-reform-and-a-sustainable-future.

Bhaduri, A., Ringler, C., Dombrowski, I., Mohtar, R., & Scheumann, W. (2015). Sustainability in the water–energy–food nexus. *Water International, 40*(5–6), 723–732. https://doi.org/10.1080/02508060.2015.1096110

Bimrah, K., & Saizen, I. (2026). The evolution and decline of traditional water management systems: A case study of jaipur, India. *Water History.* https://doi.org/10.1007/s12685-026-00386-z

Brown, C. M., Lund, J. R., Cai, X., Reed, P. M., Zagona, E. A., Ostfeld, A., Hall, J., Characklis, G. W., Yu, W., & Brekke, L. (2015). The future of water resources systems analysis: Toward a scientific framework for sustainable water management. *Water Resources Research, 51*(8), 6110–6124. https://doi.org/10.1002/2015WR017114

Chen, C.-Y., Wang, S.-W., Kim, H., Pan, S.-Y., Fan, C., & Lin, Y. J. (2021). Non-conventional water reuse in agriculture: A circular water economy. *Water Research, 199*, 117193. https://doi.org/10.1016/j.watres.2021.117193

Cicekalan, B., Shitreh, S., Cavdar, B., Salimi-Khaligh, S., Yuksekdag, A., Cataltas, A., Musluoglu, A., Guven, H., Koyuncu, I., Ozgun, H., & Ersahin, M. E. (2025). Sustainable treatment of dairy industry wastewater using an integrated anaerobic membrane bioreactor-reverse osmosis system for industrial water reuse. *Chemical Engineering Journal, 521*, 166345. https://doi.org/10.1016/j.cej.2025.166345

Continental drying: A threat to our common future. (2025). Washington, DC: World Bank. https://doi.org/10.1596/978-1-4648-2269-8

Dai, Y., Huang, Z., Khan, N., & Labbo, M. S. (2025). Smart water management: Governance innovation, technological integration, and policy pathways toward economic and ecological sustainability. *Water, 17*(13), 1932. https://doi.org/10.3390/w17131932

Dashkevych, O., & Housh, M. (2025). Management of conventional and non-conventional water sources: A systematic literature review. *Water, 17*(20), 3006. https://doi.org/10.3390/w17203006

Davis, M. D. (2007). Integrated water resource management and water sharing. *Journal of Water Resources Planning and Management, 133*(5), 427–445. https://doi.org/10.1061/(ASCE)0733-9496(2007)133:5(427)

Dey, S., & Majumdar, A. (2024). Current status of pollution in major Rivers and tributaries of India and protection-restoration strategies. In S. Kanhaiya, S. Singh, A. Dixit, & A. K. Singh (Eds.), *Rivers of India: Past, present and future* (pp. 69–93). Springer International Publishing. https://doi.org/10.1007/978-3-031-49163-4_4

Endo, A., Tsurita, I., Burnett, K., & Orencio, P. M. (2017). A review of the current state of research on the water, energy, and food nexus. *Journal of Hydrology: Regional Studies, 11*, 20–30. https://doi.org/10.1016/j.ejrh.2015.11.010

Evaristo, J., Jameel, Y., Tortajada, C., Wang, R. Y., Horne, J., Neukrug, H., David, C. P., Fasnacht, A. M., Ziegler, A. D., & Biswas, A. (2023). Water woes: The institutional challenges in achieving SDG 6. *Sustainable Earth Reviews, 6*(1), 13. https://doi.org/10.1186/s42055-023-00067-2

FAO. (2024). *Water recycling and reuse: Circular water economy and food safety considerations*. FAO. https://openknowledge.fao.org/handle/20.500.14283/cd1686en.

Flörke, M., Schneider, C., & McDonald, R. I. (2018). Water competition between cities and agriculture driven by climate change and urban growth. *Nature Sustainability, 1*(1), 51–58. https://doi.org/10.1038/s41893-017-0006-8

Flammini, A., Puri, M., Pluschke, L., & Dubois, O. (2014). Walking the nexus talk: Assessing the water-energy-food nexus in the context of the sustainable energy for all initiative. *Environment and Natural Resources Management. Working Paper (FAO) Eng No. 58.* https://agris.fao.org/search/en/providers/122621/records/6473966668b4c299a3fb4b4b.

Forhad, H. M., Uddin, M. R., Chakrovorty, R. S., Ruhul, A. M., Faruk, H. M., Kamruzzaman, S., Sharmin, N., Jamal, A. S. I. M., Haque, M. M.-U., & Morshed, A. M. (2024). IoT based real-time water quality monitoring system in water treatment plants (WTPs). *Heliyon, 10*(23), e40746. https://doi.org/10.1016/j.heliyon.2024.e40746

Garrick, D., De Stefano, L., Yu, W., Jorgensen, I., O'Donnell, E., Turley, L., Aguilar-Barajas, I., Dai, X., De Souza Leão, R., Punjabi, B., Schreiner, B., Svensson, J., & Wight, C. (2019). Rural water for thirsty cities: A systematic review of water reallocation from rural to urban regions. *Environmental Research Letters, 14*(4), 043003. https://doi.org/10.1088/1748-9326/ab0db7

Gosling, S. N., & Arnell, N. W. (2016). A global assessment of the impact of climate change on water scarcity. *Climatic Change, 134*(3), 371–385. https://doi.org/10.1007/s10584-013-0853-x

Group, W. B. (2016). *The world bank group A to Z 2016*. World Bank Publications.

Gupta, A. D., Pandey, P., Feijóo, A., Yaseen, Z. M., & Bokde, N. D. (2020). Smart water technology for efficient water resource management: A review. *Energies, 13*(23), 6268. https://doi.org/10.3390/en13236268

Gupta, P., Singh, J., Verma, S., Chandel, A. S., & Bhatla, R. (2021). Impact of climate change and water quality degradation on food security and agriculture. In *Water conservation in the era of global climate change* (pp. 1–22). Elsevier. https://doi.org/10.1016/B978-0-12-820200-5.00006-3

Haddeland, I., Heinke, J., Biemans, H., Eisner, S., Flörke, M., Hanasaki, N., Konzmann, M., Ludwig, F., Masaki, Y., Schewe, J., Stacke, T., Tessler, Z. D., Wada, Y., & Wisser, D. (2014). Global water resources affected by human interventions and climate change. *Proceedings of the National Academy of Sciences, 111*(9), 3251–3256. https://doi.org/10.1073/pnas.1222475110

He, C., Liu, Z., Wu, J., Pan, X., Fang, Z., Li, J., & Bryan, B. A. (2021). Future global urban water scarcity and potential solutions. *Nature Communications, 12*(1), 4667. https://doi.org/10.1038/s41467-021-25026-3

Hjorth, P., & Madani, K. (2023). Adaptive water management: On the need for using the Post-WWII science in water governance. *Water Resources Management, 37*(6), 2247–2270. https://doi.org/10.1007/s11269-022-03373-0

Hoekstra, A. Y. (2016). A critique on the water-scarcity weighted water footprint in LCA. *Ecological Indicators, 66*, 564–573. https://doi.org/10.1016/j.ecolind.2016.02.026

Hoff, H. (2009). Global water resources and their management. *Current Opinion in Environmental Sustainability, 1*(2), 141–147. https://doi.org/10.1016/j.cosust.2009.10.001

Huang, Z., Yuan, X., & Liu, X. (2021). The key drivers for the changes in global water scarcity: Water withdrawal versus water availability. *Journal of Hydrology, 601*, 126658. https://doi.org/10.1016/j.jhydrol.2021.126658

Ingrao, C., Strippoli, R., Lagioia, G., & Huisingh, D. (2023). Water scarcity in agriculture: An overview of causes, impacts and approaches for reducing the risks. *Heliyon, 9*(8), e18507. https://doi.org/10.1016/j.heliyon.2023.e18507

Isah, A. S., Takaijudin, H., & Singh, B. S. M. (2022). Principles and modes of distillation in desalination process. In *Distillation processes—from solar and membrane distillation to reactive distillation modelling, simulation and optimization*. IntechOpen. https://doi.org/10.5772/intechopen.100855

Issaka, S., & Ashraf, M. A. (2017). Impact of soil erosion and degradation on water quality: A review. *Geology, Ecology, and Landscapes, 1*(1), 1–11. https://doi.org/10.1080/24749508.2017.1301053

Jijingi, H. E., Yazdi, S. K., Abakar, Y. A., & Etim, E. (2024). Evaluation of membrane bioreactor (MBR) technology for industrial wastewater treatment and its application in developing countries: A review. *Case Studies in Chemical and Environmental Engineering, 10*, 100886. https://doi.org/10.1016/j.cscee.2024.100886

Katusiime, J., & Schütt, B. (2020). Integrated water resources management approaches to improve water resources governance. *Water, 12*(12), 3424. https://doi.org/10.3390/w12123424

Khan, M. H. R. B., Ahsan, A., Imteaz, M., Shafiquzzaman, M., & Al-Ansari, N. (2023). Evaluation of the surface water quality using global water quality index (WQI) models: Perspective of river water pollution. *Scientific Reports, 13*(1), 20454. https://doi.org/10.1038/s41598-023-47137-1

Kumar, A., Lin, J., Hui, D., & Thakur, T. K. (2025). Structures, challenges, and future perspectives of water footprint in achieving the Sustainable Development Goals. In *Water footprints* (pp. 263–270). Elsevier.

Lenton, R., Lewis, K., & Wright, A. M. (2008). Water, sanitation and the millennium development goals. *Journal of International Affairs, 61*(2), 247–258.

Ley, A. L., Lee, R., Peralta, Y., & Scott, C. (2019). Food and water security in north america's transboundary sonoran desert: A water-exporting dryland. In A. L. Ley, R. Lee, Y. Peralta, & C. Scott (Eds.), *The Oxford handbook of food, water and society* (pp. 343–362). Oxford University Press. https://doi.org/10.1093/oxfordhb/9780190669799.013.22

Liu, S., Li, S., Yang, Q., & Lin, M. (2025). Scale-up of solar interfacial evaporation devices: Advanced optical, thermal, and water management for efficient seawater desalination. *Energy and Environmental Science, 18*(21), 9446–9456. https://doi.org/10.1039/D5EE01958C

Liu, J., Yang, H., Cudennec, C., Gain, A. K., Hoff, H., Lawford, R., Qi, J., Strasser, L. D., Yillia, P. T., & Zheng, C. (2017a). Challenges in operationalizing the water–energy–food nexus. *Hydrological Sciences Journal, 62*(11), 1714–1720. https://doi.org/10.1080/02626667.2017.1353695

Liu, J., Yang, H., Gosling, S. N., Kummu, M., Flörke, M., Pfister, S., Hanasaki, N., Wada, Y., Zhang, X., Zheng, C., Alcamo, J., & Oki, T. (2017b). Water scarcity assessments in the past, present, and future. *Earth's Future, 5*(6), 545–559. https://doi.org/10.1002/2016EF000518

Mannina, G., Gulhan, H., & Ni, B.-J. (2022). Water reuse from wastewater treatment: The transition towards circular economy in the water sector. *Bioresource Technology, 363*, 127951. https://doi.org/10.1016/j.biortech.2022.127951

Marcal, J., Antizar-Ladislao, B., & Hofman, J. (2021). Addressing water security: An overview. *Sustainability, 13* (24), 13702. https://doi.org/10.3390/su132413702

Medema, W., McIntosh, B. S., & Jeffrey, P. J. (2008). From premise to practice: A critical assessment of integrated water resources management and adaptive management approaches in the water sector. *Ecology and Society, 13*(2). https://www.jstor.org/stable/26268004.

Milan, B. F. (2017). Clean water and sanitation for all: Interactions with other sustainable development goals. *Sustainable Water Resources Management, 3*(4), 479–489. https://doi.org/10.1007/s40899-017-0117-4

Mishra, B., Kumar, P., Saraswat, C., Chakraborty, S., & Gautam, A. (2021). Water security in a changing environment: Concept, challenges and solutions. *Water, 13*(4), 490. https://doi.org/10.3390/w13040490

Misra, A. K. (2014). Climate change and challenges of water and food security. *International Journal of Sustainable Built Environment, 3*(1), 153–165. https://doi.org/10.1016/j.ijsbe.2014.04.006

Morchid, A., Et-taibi, B., Oughannou, Z., Alami, R. E., Qjidaa, H., Jamil, M. O., Boufounas, E.-M., & Abid, M. R. (2025). IoT-enabled smart agriculture for improving water management: A smart irrigation control using embedded systems and server-sent events. *Scientific African, 27*, e02527. https://doi.org/10.1016/j.sciaf.2024.e02527

Morseletto, P., Mooren, C. E., & Munaretto, S. (2022). Circular economy of water: Definition, strategies and challenges. *Circular Economy and Sustainability, 2*(4), 1463–1477. https://doi.org/10.1007/s43615-022-00165-x

Namavar, M., Moghaddam, M. R. A., & Shafiei, M. (2023). Developing an indicator-based assessment framework for assessing the sustainability of urban water management. *Sustainable Production and Consumption, 40*, 1–12. https://doi.org/10.1016/j.spc.2023.06.006

Nations, F., & Water, U. N. (2021). *Progress on change in water-use efficiency: Global status and acceleration needs for SDG indicator 6.4.1, 2021*. Food and Agriculture Org.

Nie, T., Jiang, X., Deng, C., Cai, W., Lei, Y., & Gao, S. (2024). Analysis of the evolution of water culture and water security in the weihe river basin over a 100 year-period. *Science of the Total Environment, 920*, 171066. https://doi.org/10.1016/j.scitotenv.2024.171066

Obaid, M., Soukane, S., Qtaishat, M. R., Zhang, Y., Guillerm, V., Mohamed, H. O., Ayach, M., Singaravelu, V., Castaño, P., Maurin, G., Eddaoudi, M., Matsuura, T., & Ghaffour, N. (2026). Subnanoporous hydrophobic thin films for ultrahigh-efficiency seawater and brines desalination using membrane distillation. *Nature Communications, 17*(1), 1416. https://doi.org/10.1038/s41467-025-68158-6

Obunga, P. O., Rwanga, S. S., Dinka, M. O., & Otieno, B. O. (2025). Review of the emerging technologies in the water sector with a focus on the deployment of internet of things solutions. *npj Clean Water, 8*(1), 53. https://doi.org/10.1038/s41545-025-00487-x

Onyena, A. P., & Sam, K. (2025). The blue revolution: Sustainable water management for a thirsty world. *Discover Sustainability, 6*(1), 63. https://doi.org/10.1007/s43621-024-00631-6

Parra-López, C., Ben Abdallah, S., Garcia-Garcia, G., Hassoun, A., Trollman, H., Jagtap, S., Gupta, S., Aït-Kaddour, A., Makmuang, S., & Carmona-Torres, C. (2025). Digital technologies for water use and management in agriculture: Recent applications and future outlook. *Agricultural Water Management, 309*, 109347. https://doi.org/10.1016/j.agwat.2025.109347

Peters, N. E., & Meybeck, M. (2000). Water quality degradation effects on freshwater availability: *Impacts of human activities. Water International, 25*(2), 185–193. https://doi.org/10.1080/02508060008686817

Plessis, A. du (2022). Persistent degradation: Global water quality challenges and required actions. *One Earth, 5* (2), 129–131. https://doi.org/10.1016/j.oneear.2022.01.005

Priya, E., Kumar, S., Verma, C., Sarkar, S., & Maji, P. K. (2022). A comprehensive review on technological advances of adsorption for removing nitrate and phosphate from waste water. *Journal of Water Process Engineering, 49*, 103159. https://doi.org/10.1016/j.jwpe.2022.103159

Rahaman, M. M., & Varis, O. (2005). Integrated water resources management: Evolution, prospects and future challenges. *Sustainability: Science, Practice and Policy, 1*(1), 15–21. https://doi.org/10.1080/15487733.2005.11907961

Rahman, T. U., Roy, H., Islam, M. R., Tahmid, M., Fariha, A., Mazumder, A., Tasnim, N., Pervez, M. N., Cai, Y., Naddeo, V., & Islam, M. S. (2023). The advancement in membrane bioreactor (MBR) technology toward sustainable industrial wastewater management. *Membranes, 13*(2), 181. https://doi.org/10.3390/membranes13020181

Rosa, L., Chiarelli, D. D., Rulli, M. C., Dell'Angelo, J., & D'Odorico, P. (2020). Global agricultural economic water scarcity. *Science Advances, 6*(18), eaaz6031. https://doi.org/10.1126/sciadv.aaz6031

Salehi, M. (2022). Global water shortage and potable water safety; today's concern and tomorrow's crisis. *Environment International, 158*, 106936. https://doi.org/10.1016/j.envint.2021.106936

Sarker, B., Keya, K. N., Mahir, F., Nahiun, K., Shahida, S., & Khan, R. (2021). Scientific review surface and ground water pollution: Causes and effects of urbanization and industrialization in south Asia. *Scientific Review, 7*, 32–41. https://doi.org/10.32861/sr.73.32.41

Satyam, S., & Patra, S. (2024). Innovations and challenges in adsorption-based wastewater remediation: A comprehensive review. *Heliyon, 10*(9), e29573. https://doi.org/10.1016/j.heliyon.2024.e29573

Savenije, H. H. G., & Van Der Zaag, P. (2008). Integrated water resources management: Concepts and issues. *Physics and Chemistry of the Earth, Parts A/B/C, 33*(5), 290–297. https://doi.org/10.1016/j.pce.2008.02.003

Saxena, V. (2025). Water quality, air pollution, and climate change: Investigating the environmental impacts of industrialization and urbanization. *Water, Air, and Soil Pollution, 236*(2), 73. https://doi.org/10.1007/s11270-024-07702-4

Setty, K., Jiménez, A., Willetts, J., Leifels, M., & Bartram, J. (2020). Global water, sanitation and hygiene research priorities and learning challenges under sustainable development goal 6. *Development Policy Review, 38*(1), 64–84. https://doi.org/10.1111/dpr.12475

Shafiquzzaman, M., Rahman, M. S., Hasan, M. M., Azam, M. S., Ahmed, A. T., & Haider, H. (2026). Performance evaluation of low-cost algal ceramic membrane bioreactor for wastewater treatment operating at non-aeration mode and varying organic loading rates. *Bioresource Technology, 445*, 134061. https://doi.org/10.1016/j.biortech.2026.134061

Singh, S., & Goyal, M. K. (2025). A review of India's water policy and implementation toward a sustainable future. *Journal of Water and Climate Change, 16*(2), 493–510. https://doi.org/10.2166/wcc.2025.560

Sivakumar, B. (2011). Global climate change and its impacts on water resources planning and management: Assessment and challenges. *Stochastic Environmental Research and Risk Assessment, 25*(4), 583–600. https://doi.org/10.1007/s00477-010-0423-y

Slootweg, J. C. (2020). Using waste as resource to realize a circular economy: Circular use of C, N and P. *Current Opinion in Green and Sustainable Chemistry, 23*, 61–66. https://doi.org/10.1016/j.cogsc.2020.02.007

Sperna Weiland, F. C., van Beek, L. P. H., Kwadijk, J. C. J., & Bierkens, M. F. P. (2012). Global patterns of change in discharge regimes for 2100. *Hydrology and Earth System Sciences, 16*(4), 1047–1062. https://doi.org/10.5194/hess-16-1047-2012

Ssekyanzi, G., Ahmad, M. J., & Choi, K.-S. (2024). Sustainable solutions for mitigating water scarcity in developing countries: A comprehensive review of innovative rainwater storage systems. *Water, 16*(17), 2394. https://doi.org/10.3390/w16172394

Stein, U., Bueb, B., Knieper, C., Tröltzsch, J., Vidaurre, R., & Favero, F. (2023). The diagnostic water governance tool – supporting cross-sectoral cooperation and coordination in water resources management. *Environmental Science and Policy, 140*, 111–121. https://doi.org/10.1016/j.envsci.2022.11.014

Teker, S., & Koc, T. C. (2019). Industrial revolutions and its effects on quality of life. *Pressacademia, 9*(9), 304–311. https://doi.org/10.17261/Pressacademia.2019.1109

United Nations. (2026). *World water development report 2026*. UNESCO. https://www.unesco.org/en/articles/united-nations-world-water-development-report-2026.

Vörösmarty, C. J., Pahl-Wostl, C., Bunn, S. E., & Lawford, R. (2013). Global water, the anthropocene and the transformation of a science. *Current Opinion in Environmental Sustainability, 5*(6), 539–550. https://doi.org/10.1016/j.cosust.2013.10.005

Vetterlein, A., & Schmidtke, T. (2024). The world bank: A changing organization in a changing world. In A. Vetterlein, & T. Schmidtke (Eds.), *The elgar companion to the world bank* (pp. 2–20). Edward Elgar Publishing. https://doi.org/10.4337/9781802204780.00010

Voulvoulis, N. (2018). Water reuse from a circular economy perspective and potential risks from an unregulated approach. *Current Opinion in Environmental Science and Health, 2*, 32–45. https://doi.org/10.1016/j.coesh.2018.01.005

Wada, Y., Flörke, M., Hanasaki, N., Eisner, S., Fischer, G., Tramberend, S., Satoh, Y., van Vliet, M. T. H., Yillia, P., Ringler, C., Burek, P., & Wiberg, D. (2016). Modeling global water use for the 21st century: The water futures and solutions (WFaS) initiative and its approaches. *Geoscientific Model Development, 9*(1), 175–222. https://doi.org/10.5194/gmd-9-175-2016

Water and Sanitation. (2025). Department of Economic and Social Affairs. https://sdgs.un.org/topics/water-and-sanitation.

Zaigham Abbas Naqvi, S. M., Hussain, S., Awais, M., Tahir, M. N., Saleem, S. R., Al-Yarimi, F. A. M., Ashurov, M., Saidani, O., Khan, M. I., Wu, J., Wei, Z., & Hu, J. (2025). Climate-resilient water management: Leveraging IoT and AI for sustainable agriculture. *Egyptian Informatics Journal, 30*, 100691. https://doi.org/10.1016/j.eij.2025.100691

Zhao, J., Luo, Q., Zhao, W., Li, C., Xiao, J., Li, J., & Zhu, X. (2026). A chitosan-modified buckwheat hull biochar dynamic adsorption column as a sustainable and efficient technology for lead-acid battery wastewater treatment. *Environmental Science: Water Research & Technology*. https://doi.org/10.1039/D5EW01081K

PART

Advanced treatment technologies for sustainable water systems

II

CHAPTER

2

Sustainable desalination approaches and brine management strategies for freshwater production in water-stressed regions

Hau Thi Nguyen, Thu Uyen Thi Nguyen and Phuong Nghi Nguyen Pham
Faculty of Chemistry and Environment, DaLat University, DaLat, Vietnam

1. Introduction

Climate change, rapid population growth, urbanization, and industrial expansion are intensifying global freshwater scarcity. Prolonged droughts and escalating water demand have increased groundwater over-extraction, while seawater intrusion is accelerating the salinization of freshwater aquifers, particularly in arid and coastal regions. Although water covers nearly two-thirds of the Earth's surface, less than 1% is readily accessible as usable freshwater. According to UN estimates, nearly 40% of the world's population currently lives in areas experiencing increasing water stress, and this share is projected to rise to ~60% by 2050 (Gude, 2016).

Desalination has therefore emerged as a strategic solution for augmenting freshwater resources in water-stressed regions. By removing dissolved salts and impurities from seawater or brackish water, desalination provides a climate-resilient water supply independent of rainfall variability. Desalination technologies are broadly classified into thermal processes (e.g., multi-stage flash, multi-effect distillation, vapor compression) and membrane-based processes (e.g., reverse osmosis (RO), electrodialysis (ED), forward osmosis (FO), membrane distillation (MD), and nanofiltration (NF)). Among these, RO currently dominates the global market due to its relatively high efficiency and technological maturity.

Despite significant technological progress, desalination remains energy-intensive and generates concentrated brine streams that pose environmental challenges, especially for inland plants where marine discharge is not feasible. Moreover, reliance on fossil-fuel-based electricity increases greenhouse gas emissions and operational costs. Integrating desalination with renewable energy sources, such as solar or wind, offers a promising pathway to reduce carbon footprint and improve sustainability, although intermittency and storage remain technical constraints.

In addition to energy optimization, sustainable desalination requires effective brine management strategies to minimize environmental impacts and enhance resource recovery. Alternatives to

Sustainable Water Management. https://doi.org/10.1016/B978-0-443-44853-9.00011-3

direct brine discharge, including volume reduction, beneficial reuse, and salt recovery, are gaining attention as pathways toward near-zero liquid discharge (ZLD) and circular water systems. This chapter reviews sustainable desalination approaches and emerging brine management strategies, with particular focus on membrane-based technologies and hybrid systems suitable for water-stressed regions.

2. Potential membrane desalination techniques

To meet the escalating demand for freshwater under climate change, seawater desalination has expanded rapidly worldwide. By 2018, more than 20,000 desalination plants were in operation, with a total installed capacity of approximately 92.5 million m^3/day, and the sector has been dominated by distillation- and membrane-based technologies (Gude, 2016; Nassrullah et al., 2020; Shahrim et al., 2023). Among these, reverse osmosis (RO) is currently the most widely deployed option because it offers high salt rejection and relatively low specific energy consumption. Nevertheless, membrane desalination still faces two persistent barriers, membrane fouling (which increases cleaning frequency and shortens membrane lifetime) and high energy demand (particularly for high-salinity feeds), which ultimately elevate operating costs. Table 2.1 summarizes representative energy requirements and major limitations of key seawater desalination technologies.

Table 2.1 Operating costs and energy consumption of seawater desalination technologies.

Technology	Energy consumption (kWh/m^3)	Disadvantages	References
Reverse osmosis	2.0–4.0	Fouling; energy demand increases with salinity and recovery	Okampo and Nwulu (2020)
Nanofiltration	1–2	Lower seawater desalination efficiency (typically 40%–60% salt rejection)	(Al-Karaghouli & Kazmerski, 2013; Okampo & Nwulu, 2020)
Multi-stage flash Distillation	14–16	High energy consumption	(Al-Karaghouli & Kazmerski, 2013; Okampo & Nwulu, 2020)
Distillation	19–27	High energy consumption	(Al-Karaghouli & Kazmerski, 2013; Okampo & Nwulu, 2020)
Capacitive deionization	0.3–1.0	Limited by feed TDS <20 g/L	Shen et al. (2014)
Electrodialysis	0.7–5.5	Limited by feed TDS <15 g/L	(Al-Karaghouli & Kazmerski, 2013; Okampo & Nwulu, 2020)
Membrane distillation	107	High energy consumption (thermal demand)	Ruiz-Aguirre et al. (2018)

2.1 Reverse osmosis

RO represents a major breakthrough in desalination by eliminating the need for phase change, unlike thermal distillation. It is currently the dominant seawater desalination technology, capable of removing >99% of dissolved salts, heavy metals, and organic contaminants.

RO operates by applying hydraulic pressure (typically 55–85 bar for seawater) to overcome the osmotic pressure (25–30 bar) and force water through a semi-permeable polyamide membrane (Fig. 2.1). Water permeates the dense selective layer, while salts and other solutes are rejected and concentrated in the brine stream. Compared with thermal desalination (10–20 kWh/m^3), modern seawater RO consumes only 3–5 kWh/m^3, making it significantly more energy-efficient and economically competitive (Wang et al., 2025).

In practice, RO performance strongly depends on effective pretreatment (e.g., screening, coagulation–flocculation, filtration or UF/MF) to prevent membrane fouling and scaling; after membrane separation, permeate is often remineralized and pH-adjusted before distribution. Despite its advantages, RO still faces key limitations, including membrane fouling/scaling that increases energy demand, relatively high initial investment for high-pressure equipment, and the production of concentrated brine that must be carefully managed to avoid marine ecological impacts. To address these challenges, recommended solutions include upgrading pretreatment and using optimized antiscalants, adopting energy recovery devices (ERDs) to cut power consumption, developing more fouling-resistant high-permeability membranes, integrating renewable energy for remote/island

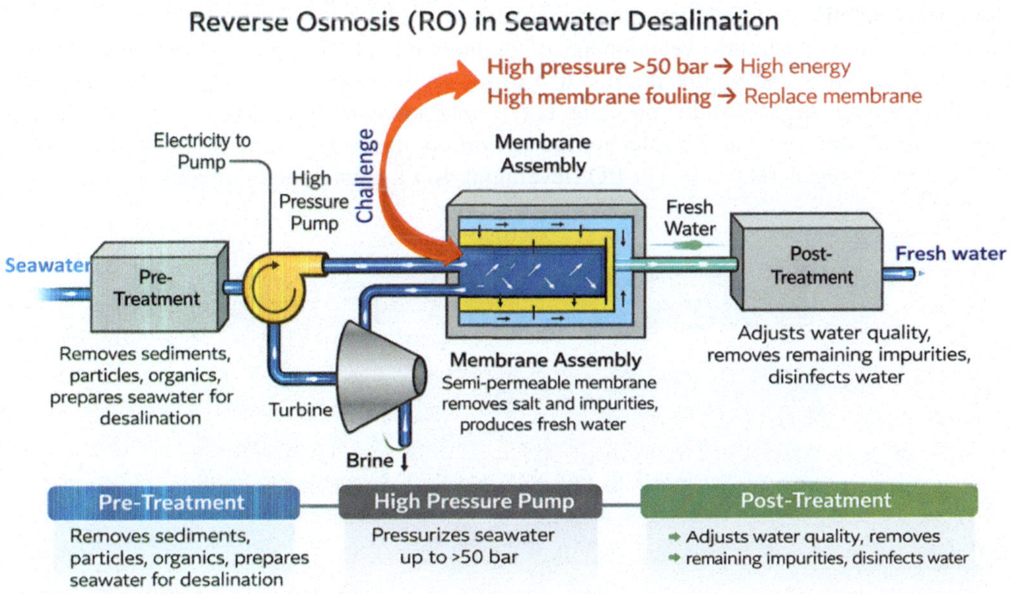

FIGURE 2.1

RO seawater desalination process.

systems, and applying brine minimization or valorization options (e.g., further concentration via RO–MD/RO–ED hybrids, selective salt recovery, or controlled dilution and discharge with environmental monitoring).

2.2 Forward osmosis

FO is an emerging membrane desalination technology with high water flux that exploits the natural osmotic pressure gradient rather than high hydraulic pressure (Fig. 2.2). In a typical FO process, seawater or brackish water (feed) flows along one side of a semi-permeable membrane, while a highly concentrated draw solution (e.g., NaCl, sugars, urea, or engineered solutes) is circulated on the opposite side. The osmotic driving force pulls water from the feed into the draw solution, whereas most salts and larger impurities are rejected by the FO membrane (Ray et al., 2018). Because FO avoids high-pressure pumping, it is often viewed as a potentially more sustainable option, particularly for seawater desalination, brackish water treatment, and water reuse in regions impacted by saltwater intrusion.

Despite these advantages, FO is not yet as commercially established as RO due to several key limitations. The largest barrier is draw-solution regeneration: the diluted draw must be processed to separate freshwater and recycle the draw solute, and this step can dominate the overall energy demand and cost (Nguyen et al., 2015). FO also suffers from internal concentration polarization (ICP) within the porous support layer, which reduces the effective osmotic gradient and lowers water flux. In addition, reverse solute flux (leakage of draw solutes back to the feed) decreases efficiency, increases operating cost, and can create secondary contamination concerns. Membrane fouling may still occur, and long-term stability and scale-up remain challenges.

To address these constraints, recommended solutions include designing FO membranes with ultrathin selective layers and highly porous, low-tortuosity supports to mitigate ICP; selecting draw solutes that provide high osmotic pressure but enable low-energy recovery (e.g., thermolytic or responsive draw solutions); and implementing hybrid configurations such as FO-MD (using waste heat/solar heat for regeneration) or FO-RO (leveraging RO for final polishing and draw recovery). In

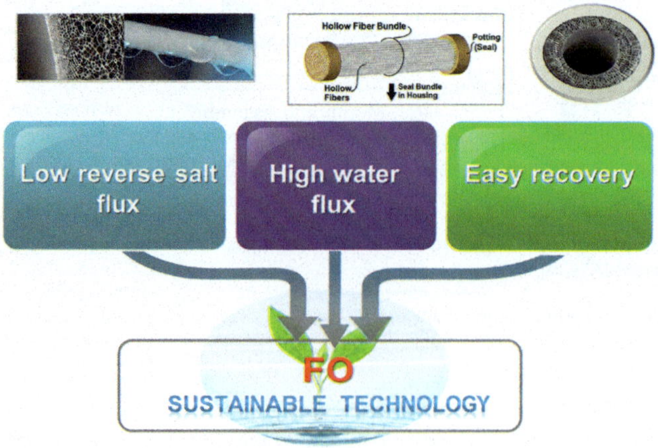

FIGURE 2.2

FO sustainable technology for seawater desalination.

practice, FO is most promising as part of an integrated desalination and brine-management platform, where it can reduce fouling risk in challenging feeds and support energy-efficient water recovery when paired with an optimized draw regeneration strategy.

2.3 Membrane distillation

MD is a thermally driven desalination technology that combines features of distillation and membrane separation. It employs a hydrophobic porous membrane that allows only water vapor to pass while rejecting non-volatile salts and impurities (Nguyen et al., 2025; Thi Nguyen et al., 2025). The driving force is the vapor pressure gradient created by a temperature difference across the membrane, rather than hydraulic pressure. MD typically operates at moderate temperatures (40–80 °C) and can utilize low-grade waste heat or solar thermal energy, making it attractive for coastal, remote, or energy-constrained regions (Fig. 2.3). Salt rejection is typically >99%, even for high-salinity feeds such as seawater or RO brine.

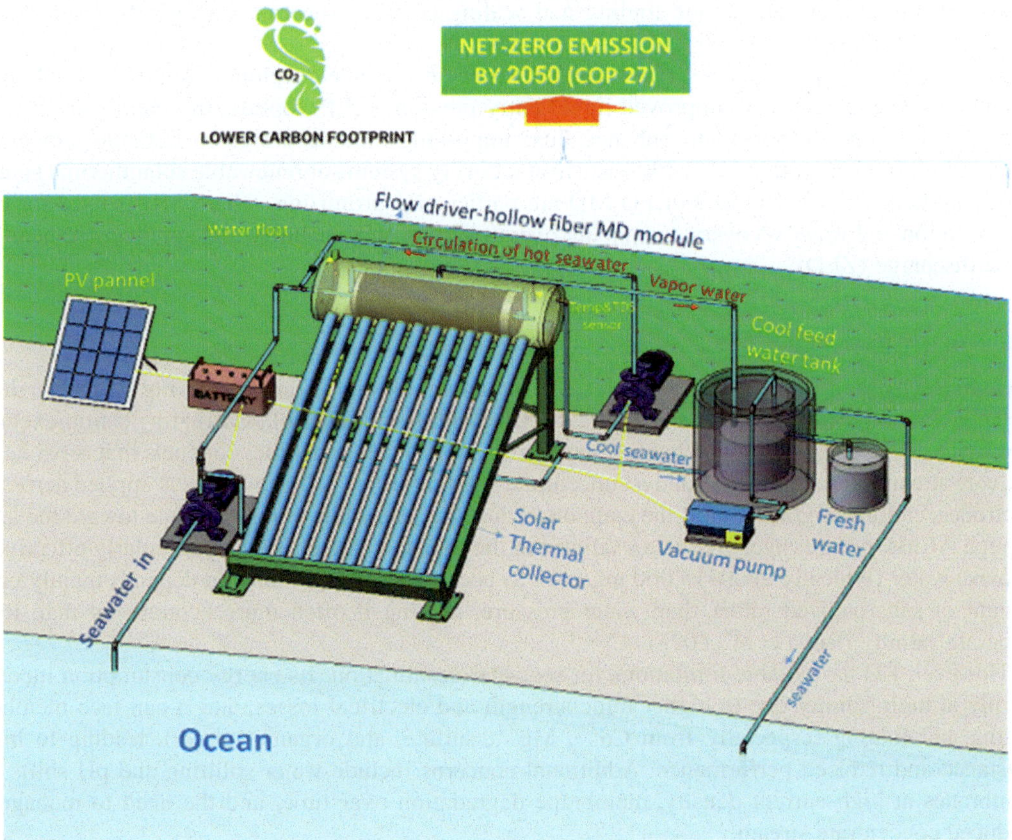

FIGURE 2.3

A 3D illustration of membrane desalination filtration system.

In MD, a heated saline feed flows along one side of a hydrophobic microporous membrane, while a cooler stream (direct liquid contact or an air gap followed by a condenser) is maintained on the opposite side. The resulting temperature difference creates a vapor-pressure gradient at the membrane interfaces: water evaporates at the hot side, water vapor diffuses through the dry membrane pores (typically <0.2 μm), and then condenses on the cold side as purified water. Because the membrane is hydrophobic, the liquid phase is rejected and does not enter the pores (as long as the transmembrane pressure remains below the liquid entry pressure), so mass transfer is governed primarily by the vapor-pressure difference (Dalton's law) and vapor diffusion through the pore network.MD modules are commonly designed as flat-sheet, tubular, spiral-wound, or hollow-fiber configurations (Table 2.2). Among these, flat-sheet modules are currently the most widely used at lab and pilot scale because they are straightforward to fabricate, assemble, inspect, and clean, making them convenient for experimental evaluation and troubleshooting. Figure 2.4

Despite its advantages, MD faces several limitations. The most critical challenge is membrane wetting, which reduces salt rejection and compromises long-term stability. MD also suffers from temperature polarization, lowering the effective driving force and flux. Although it operates at lower temperatures than distillation, its overall thermal efficiency can still be moderate if heat recovery is not optimized. In addition, membrane fouling and scaling remain concerns, particularly when treating concentrated brines.

To overcome these challenges, recommended strategies include developing highly hydrophobic or omniphobic membranes with improved liquid entry pressure (LEP); optimizing membrane structure (high porosity, low tortuosity) to enhance flux; improving module design to mitigate temperature polarization; and integrating MD with waste heat recovery systems or renewable solar thermal sources. Hybrid systems such as RO-MD or FO-MD are particularly promising, where MD is used for brine concentration and water recovery, enhancing overall system sustainability and approaching near-zero liquid discharge (ZLD).

2.4 Electrodialysis

ED is an electrochemical desalination process that removes dissolved salts by driving ions through ion-exchange membranes under a direct-current electric field. In an ED stack, alternating cation-exchange membranes (CEM) and anion-exchange membranes (AEM) create parallel channels that separate into a diluate stream (desalted water) and a concentrate stream (brine). When voltage is applied across end electrodes, cations migrate toward the cathode through CEMs, while anions migrate toward the anode through AEMs, progressively reducing salinity in the diluate channel. ED is particularly effective for brackish water (typically 2000–15,000 mg/L TDS) because its energy demand depends mainly on the amount of salt removed rather than water pressure, making it often more economical than RO at moderate salinity (Ruya et al., 2025).

However, ED has notable limitations for seawater desalination. Its energy consumption increases sharply at high salinity due to higher ionic strength and electrical losses, and it can face membrane scaling and fouling (especially from Ca^{2+}, Mg^{2+}, sulfate, and organic matter), leading to higher resistance and reduced performance. Additional concerns include water splitting and pH shifts near membranes at high current density, membrane degradation over time, and the need to manage the produced concentrate stream.

Table 2.2 Merits and challenges of MD modules.

MD membrane module	Merits	Challenges	References
Flat sheet	- Easy to produce, assemble, use, test, and clean. - Achieve a maximum flow in the MD procedure results	- Low packing density (400 m^2/m^3). - Low mechanical strength leading to more susceptible to damage during operation.	Aliyu et al. (2018)
Tubular	- Better mechanical strength compared to flat sheet modules - Permission for high flow rates. - Permission for high flow rates.	- Low packing density of 300 m^2/m^3 leading to large physical footprint. - High cost of operation, including energy costs and maintenance expenses. - Minimal permeate flux, which can result in lower productivity	Khayet et al. (2011)
Spiral wound	- Effective packing density of 300–1000 m^2/m^3, which can lead to higher productivity. - Reasonable amounts of energy for operation. - Less susceptible to low-temperature polarization.	- Replacing or cleaning the membranes can be challenging as fouling occurs. - Minimal permeability, which can result in lower productivity. - Extremely vulnerable to fouling leading to more frequent maintenance and cleaning.	Winter et al. (2017)
Hollow fiber	- High packing density of 3000 m^2/m^3, which results in high productivity and efficiency. - Capable of working at very high pressures above 100 bars. - Greater effective surface area per volume compared to other modules	- Low flux permeation which can result in lower productivity. - More susceptible to damage and require more frequent maintenance. - Membrane fouling is difficult to control	Thomas et al. (2017)

To address these issues, recommended solutions include using softening or antiscalant pretreatment to control scaling, optimizing operation below the limiting current density to reduce water splitting, applying periodic polarity reversal or cleaning to mitigate fouling, and integrating ED in hybrid systems (e.g., RO for seawater followed by ED/EDR for brackish polishing, or ED as a pre-

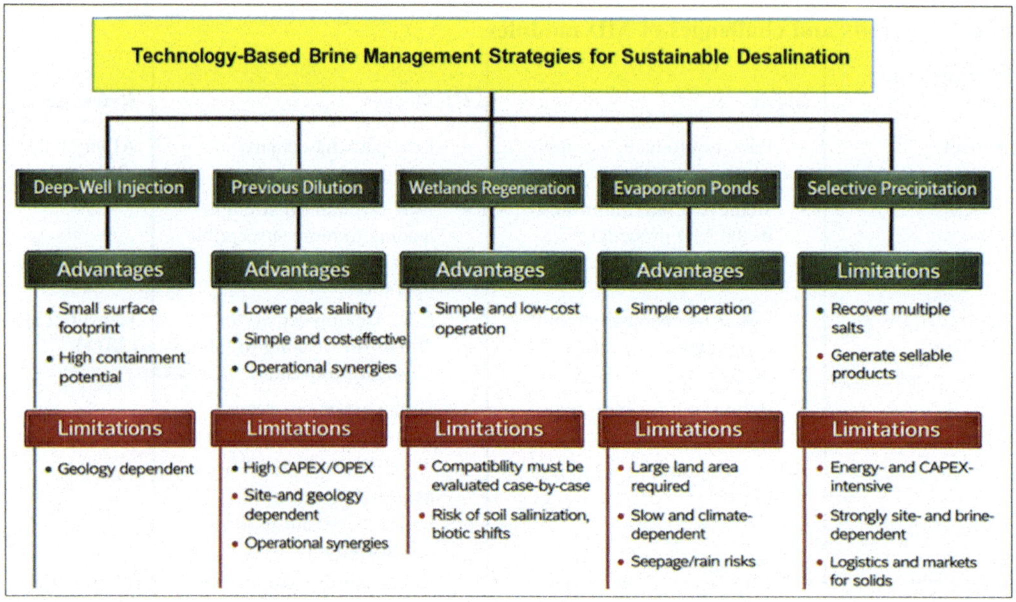

FIGURE 2.4

Main technology-based brine management strategies for sustainable desalination.

concentration step before crystallization/ZLD). In practice, ED is most suitable for brackish water and selective ion removal, while seawater desalination typically requires RO or thermal/MD-based approaches unless ED is used in a targeted hybrid role.

2.5 Hybrid membrane systems for sustainable seawater desalination

Hybrid membrane systems integrate complementary technologies to overcome the inherent limitations of single-process desalination. By combining pressure-driven, osmotic-driven, and thermally driven processes, hybrid configurations aim to improve water recovery, energy efficiency, fouling control, and brine management.

Among these, FO-RO systems are widely investigated. In this configuration, FO acts as a pretreatment step that dilutes seawater before RO, thereby lowering required hydraulic pressure and reducing fouling propensity. Pilot studies report overall energy consumption around 2.8–3.0 kWh/m^3, representing 15%–60% savings compared to standalone RO under optimized conditions (Aboulkacem et al., 2025). FO pretreatment also reduces chemical cleaning frequency and operational costs. However, the main limitation remains draw-solution regeneration, which can offset energy gains if not efficiently designed. Optimizing draw solutes and integrating low-energy regeneration strategies are therefore critical.

Similarly, FO-MD systems combine osmotic extraction with thermally driven separation. FO reduces salinity and fouling load before MD, while MD regenerates the diluted draw solution using low-grade or solar heat. This synergy improves water recovery and supports near-ZLD. The key challenges include membrane wetting in MD, internal concentration polarization in FO, and system complexity. Advanced membrane materials and optimized module design are necessary to ensure long-term stability.

Other innovative integrations include MD-RED systems, where membrane distillation first concentrates brine and reverse electrodialysis (RED) subsequently recovers salinity-gradient energy. Such systems enhance overall recovery (approaching 98% in optimized cases) and improve energy efficiency by utilizing waste heat and extracting energy from concentrated brine. Nevertheless, scaling, membrane durability, and capital cost remain practical constraints.

Overall, hybrid membrane desalination offers clear advantages in energy optimization and brine minimization. However, increased system complexity, higher capital cost, and operational integration challenges must be carefully managed. Future development should focus on low-energy draw regeneration, advanced fouling-resistant membranes, efficient heat recovery, and modular design, enabling economically viable and environmentally sustainable seawater desalination.

3. Brine management strategies

3.1 Sustainable brine management strategies

Most brine-management studies classify available technologies and discuss their pros and cons (e.g., ZLD, reuse, disposal) (Rodríguez-delanuez et al., 2012). However, the diversity of options makes sustainability comparisons difficult and often limited to a few criteria. Because technology-based classifications overlook differences in policy and socio-environmental contexts, they are poorly suited for linking strategies to the SDGs. To address this gap, this book proposes an SDG-based, two-level framework (strategies and objectives) aligned with the SDG Compass approach and adapted to the brine-management industry (Table 2.3).

3.2 Technology-based brine management strategies for sustainable desalination

3.3 Deep-well injection for subsurface brine disposal

Deep-well injection is a subsurface disposal option in which desalination brine is transported and injected into deep geological formations capable of long-term confinement. Originating in the oil and gas sector, this approach has been adapted for industrial and municipal liquid wastes and, in limited cases, for desalination concentrate, most often when brine volumes are small. It has been widely practiced in the United States and has also been reported for RO desalination facilities in Spain (e.g., Alicante, and Campo de Cartagena, Murcia).

Its feasibility depends primarily on the availability of a hydraulically secure geological setting, typically consisting of a permeable injection horizon sealed by a competent, low-permeability caprock that prevents upward migration. The injection program must also be designed to protect other subsurface resources, particularly freshwater aquifers. Although the technology is mature and can be operated under sophisticated monitoring and control systems, it is generally capital- and operation-intensive because it demands rigorous site characterization and strict well-integrity assurance. Key design requirements include corrosion-resistant casing and tubing (due to high salinity), high-quality

Table 2.3 Streamlined framework for sustainable brine management strategies and key indicators.

Strategy pillar	Key purposes	Core indicators	References
1. Reduce marine discharge impacts	Minimize brine volume and salinity stress; reduce pollutant load; avoid harmful additives	Brine volume discharged (m^3/y); discharge salinity (g/L); % discharges with previous dilution; % outfalls with diffusers; % use of green chemicals	Gil-Trujillo and Sadhwani Alonso (2023)
2. Governance, control and monitoring	Ensure clear regulations and effective compliance monitoring	Existence of brine-discharge standard/ permit; % discharges authorized; % plants with environmental monitoring plan + reporting frequency	Gil-Trujillo and Sadhwani Alonso (2023)
3. Energy and carbon reduction	Improve desalination efficiency and decarbonize operations	Specific energy consumption (kWh/m^3); renewable energy share (%)	Leon et al. (2021)
4. Valorization and circular economy	Convert brine from waste to resource through recovery/reuse routes	% Brine treated for reuse/valorization; mass of recovered products (e.g., salts/minerals); by-product management plan	Jiménez-Arias et al. (2022)
5. Innovation and deployment	Scale up brine technologies and cross-sector integration	Pilot/demo installations (e.g., MD, ED, selective precipitation, blue energy); patents/ industry partnerships	Gil-Trujillo and Sadhwani Alonso (2023)

cementation, assessment of formation capacity and expected lifespan, evaluation of potential migration pathways, and routine integrity testing; in many cases, an exploratory well is necessary to reduce geological uncertainty.

In practice, deep-well injection offers a small land footprint and can be a viable alternative where surface discharge is restricted and suitable geology is present. However, it remains highly site-dependent and expensive, with costs driven by drilling, corrosion control, cementation reliability, and long-term monitoring obligations. Critically, it does not eliminate the brine; it relocates it underground, meaning residual environmental risk persists if containment fails through leakage, well degradation, or unforeseen subsurface connectivity (Mace et al., 2005). Consequently, deep-well injection is generally not favored for large seawater desalination facilities and is more defensible for very small brackish-water plants when other disposal or beneficial-use options are not technically or economically feasible.

3.4 Brine mitigation by co-discharge and previous dilution

Previous dilution is a practical alternative to direct brine disposal in which desalination concentrate is blended with another water stream, most commonly treated wastewater (WWTP effluent), seawater, or power-plant cooling water, prior to coastal discharge. The main objective is to reduce the salinity contrast between the discharge and ambient seawater, thereby lowering near-field osmotic stress and minimizing localized ecological impacts at the outfall (Bashitialshaaer et al., 2011). This approach is particularly relevant in sensitive coastal zones with seagrass meadows (e.g., Posidonia oceanica and Cymodocea nodosa), which are vulnerable to salinity fluctuations.

Beyond salinity reduction, previous dilution can provide meaningful synergies depending on the blending stream. Co-discharge with WWTP effluent can improve overall discharge quality because the wastewater stream dilutes salinity while the brine may contribute a partial "sanitizing" effect; blending with cooling water can reduce both salinity and temperature, mitigating thermal and osmotic stress simultaneously. Since it relies on mixing rather than complex treatment, previous dilution is typically straightforward and cost-effective, and it may allow a simpler, lower-cost outfall with improved initial mixing. Reported applications include RO–WWTP co-discharge in Adeje and Arona (Tenerife), seawater-brine blending in Andratx (Mallorca) and Javea, and co-discharge at Sagunto (Valencia) (Malfeito et al., 2005).

However, the applicability of previous dilution is constrained by siting and infrastructure: a compatible dilution stream must be available nearby with sufficient, reliable flow. Importantly, dilution reduces peak concentrations but does not reduce the total salt mass discharged; thus, for large plants or poorly flushed receiving waters, cumulative salinity loading may still drive chronic impacts. In addition, WWTP blending can introduce regulatory and monitoring complexity (e.g., nutrients and trace contaminants), requiring careful outfall design, robust mixing, and sustained environmental surveillance to ensure that gains from salinity reduction are not offset by broader water-quality risks.

3.5 Beneficial use of brine for wetland restoration

Many wetlands and brackish ecosystems are progressively degraded by drought and by pressures from agriculture, livestock activities, and industrial development. In such settings, desalination brine can be considered as a controlled supplementary water source to support wetland regeneration, either (i) for coastal wetlands using concentrate from seawater RO or (ii) for inland brackish wetlands supplied by brine from brackish-water desalination.

When conditions are appropriate, this approach can deliver substantial environmental benefits because it links brine management with ecosystem restoration. However, feasibility and cost–benefit are highly site-specific and depend on the characteristics of both the desalination plant (brine flowrate and chemistry) and the target wetland (natural salinity regime, hydrologic flushing, soil–groundwater interactions, and ecological tolerance). If brine composition and loading are carefully matched to the wetland's baseline brackish conditions, preferably with short conveyance distance and a well-designed distribution system, brine can be transformed from a waste stream into a beneficial input that supports habitat functions and improves social acceptance through visible co-benefits. Conversely, poor matching of salinity/ionic composition or excessive hydraulic loading can cause unintended salinization of soils and groundwater, alter community structure, or mobilize certain constituents, making robust ecological safeguards and long-term monitoring essential.

3.6 Evaporation ponds for brine concentration and salt co-recovery

Evaporation ponds are shallow, open-air basins that concentrate brine through natural evaporation, often producing sea salt as a by-product. They are most suitable in mild, dry climates with low land cost and large available area, because high evaporation requires a large exposed surface. For example, in Oman this option has been widely used for inland desalination plants with capacities of roughly 50–1000 m^3/d (Malfeito et al., 2005).

Ponds can also be integrated into broader valorization schemes. In appropriate settings, they may be coupled with aquaculture (e.g., fish, Artemia, algae) to generate additional revenue, and they can be co-developed with solar ponds for energy use since both require similar climatic and land conditions. Key design variables include the local evaporation rate (climate-driven and influenced by salt content), pond surface area (larger area increases evaporation), and shallow depth; for effective evaporation, basin depth is typically kept below 0.5 m. Because natural evaporation may be too slow for substantial brine volumes, evaporation rates can be enhanced via "forced" measures such as increasing temperature, increasing air flow/wind, aeration, reducing surface tension, or manipulating pressure gradients at the surface.

Evaporation ponds are technologically simple, robust, and easy to implement, and they can reduce brine volume while enabling partial resource recovery (salt) and potential co-benefits through aquaculture or solar-pond integration. However, their application is strongly constrained by climate and land availability: they require very large footprints to handle meaningful brine flows, and even then the overall minimization of disposal can be modest for large desalination plants. In addition, slow natural evaporation, potential seepage risks, and the need for careful environmental management (liners, wildlife protection, and brine handling during rain events) often limit their practicality to small-to-medium inland facilities or niche coastal settings with abundant land and favorable weather.

3.7 Sequential salt recovery via selective precipitation

Selective precipitation is a brine-valorization approach that aims to recover dissolved ions sequentially as marketable salts. In practice, the brine is driven through staged evaporation and/or cooling steps, often supported by the addition of targeted reagents, so that one compound precipitates preferentially before the next. In addition to NaCl, the process can yield products such as gypsum ($CaSO_4.2H_2O$), magnesium hydroxide ($Mg(OH)_2$), calcium chloride ($CaCl_2$), and sodium sulfate (Na_2SO_4), which can achieve commercial quality when crystallization and separation are well controlled.

The viability of selective precipitation is governed primarily by brine composition and concentration, because these determine which salts can be recovered, in what order, and at what yield. Performance and cost are also strongly influenced by local conditions, including climate (temperature, humidity, rainfall affecting natural pre-concentration), energy price and availability (especially for thermal evaporation and crystallization), and land requirements for ponds, storage, and handling of solid products. Since evaporation-crystallization and solid–liquid separation is established industrial operations, the main uncertainty is less about technical feasibility and more about process integration and economics (Rodríguez-delanuez et al., 2012).

Notably, selective precipitation can transform brine from a waste stream into a resource stream, reducing residual brine volume while generating revenue from recovered salts, and it benefits from mature unit operations that are well understood at industrial scale. However, it can become energy- and capital-intensive, particularly when high levels of evaporation are required or when reagents are needed to steer selectivity and purity. The approach is also sensitive to market realities: profitability depends on achieving consistent product purity, securing stable demand, and managing logistics for solids handling and storage. Consequently, selective precipitation is most attractive when brine chemistry is favorable, low-cost heat (e.g., waste heat or solar) is available, and there is a reliable local or regional market for the recovered products.

4. Conclusions and perspectives

Desalination has become a strategic solution for freshwater production in water-stressed regions. Over recent decades, significant advances in membrane materials, system design, and energy recovery have improved efficiency and reduced costs. RO remains the dominant technology due to its relatively low specific energy consumption and technological maturity, while thermal processes remain relevant where waste heat is available. ED offers an efficient option for brackish water treatment. Emerging technologies such as FO and MD provide additional flexibility, particularly in hybrid configurations and high-salinity applications.

Despite these improvements, two critical challenges persist: energy demand and brine management. Brine discharge remains a major environmental concern, especially for inland plants, where dilution and ocean discharge are not feasible. Therefore, sustainable desalination must move beyond water production efficiency toward integrated solutions that minimize environmental impact.

Future development should focus on three interconnected directions. First, advanced membrane engineering, including fouling-resistant, high-permeability, and omniphobic membranes, can reduce energy demand and enhance durability. Second, hybrid system integration (e.g., RO–MD, FO–RO, MD–RED) can improve water recovery, enable energy reuse, and reduce brine volume. Third, brine valorization and zero-liquid discharge (ZLD) strategies, such as selective salt recovery, electrodialysis concentration, and crystallization, support a transition toward a circular water economy.

Ultimately, the optimal desalination strategy must be site-specific, considering feed salinity, energy availability, environmental sensitivity, and economic constraints. By integrating renewable energy, smart process control, hybrid membrane technologies, and resource recovery from brine, future desalination systems can evolve from energy-intensive water producers into sustainable, low-impact water-resource management platforms.

Acknowledgments

This research was funded by Vingroup Innovation Foundation (VINIF) under project code VINIF.2023.DA189 and Vietnam National Foundation for Science and Technology Development (NAFOSTED) under grant number 105.08–2021.54.

References

Aboulkacem, C., Sigue, S., Lachheb, A., Samrane, K., Abushaban, A., & Necibi, M. C. (2025). Seawater desalination and water reuse using FO-RO hybrid system: Short review. *Desalination and Water Treatment, 324*(5), 101487. https://doi.org/10.1039/c8ee00291f

Al-Karaghouli, A., & Kazmerski, L. L. (2013). Energy consumption and water production cost of conventional and renewable-energy-powered desalination processes. *Renewable and Sustainable Energy Reviews, 24*, 343–356. https://www.sciencedirect.com/science/article/abs/pii/S1364032113000208.

Aliyu, U. M., Rathilal, S., & Isa, Y. M. (2018). Membrane desalination technologies in water treatment: A review. *Water Practice and Technology, 13*(4), 738–752. https://doi.org/10.2166/wpt.2018.084

Bashitialshaaer, R., Flyborg, L., & Persson, K. M. (2011). Environmental assessment of brine discharge and wastewater in the Arabian gulf. *Desalination and Water Treatment, 25*(1–3), 276–285. https://doi.org/10.1002/j.1551-8833.1990.tb06962.x

Gil-Trujillo, A., & Sadhwani Alonso, J. J. (2023). Evaluation of brine management strategies from the perspective of the sustainable development goals: Application in the Canary Islands. *Desalination, 554*.

Gude, V. G. (2016). Desalination and sustainability – An appraisal and current perspective. *Water Research, 89*, 87–106. https://www.sciencedirect.com/science/article/abs/pii/S0043135415303390.

Jiménez-Arias, D., Sierra, S. M., García-Machado, F. J., García-García, A. L., Borges, A. A., & Luis, J. C. (2022). Exploring the agricultural reutilisation of desalination reject brine from reverse osmosis technology. *Desalination, 529*, 115644. https://www.sciencedirect.com/science/article/abs/pii/S0011916422000996.

Khayet, M., Souhaimi, M. K., & Matsuura, T. (2011). *Membrane distillation, 41*. Elsevier.

Leon, F., Ramos-Martin, A., & Perez-Baez, S. O. (2021). Study of the ecological footprint and carbon footprint in a reverse osmosis seawater desalination plant. *Membranes, 11*, 377.

Mace, R. E., Nicot, J., Chowdhury, A. H., Dutton, A. R., & Kalaswad, S. (2005). *Please pass the salt: Using oil fields for the disposal of concentrate from desalination plants. Desalination and water purification research and development program. US department of the interior* (pp. 1–105). Bureau of Reclamation.

Malfeito, J. J., Díaz-Caneja, J., Fariñas, M., Fernández-Torrequemada, Y., González-Correa, J. M., Carratalá-Giménez, A., et al. (2005). Brine discharge from the javea desalination plant. *Desalination, 185*(1–3), 87–94.

Nassrullah, H., Anis, S. F., Hashaikeh, R., & Hilal, N. (2020). Energy for desalination: A state-of-the-art review. *Desalination, 491*, 114569. https://www.sciencedirect.com/science/article/abs/pii/S0011916420310304.

Nguyen, H. T., Nguyen, N. C., Chen, S. S., Ngo, H. H., Guo, W., & Li, C. W. (2015). A new class of draw solutions for minimizing reverse salt flux to improve forward osmosis desalination. *The Science of the Total Environment, 538*, 129–136. https://doi.org/10.1016/j.scitotenv.2015.07.156

Nguyen, N. C., Nguyen, T. X. Q., Ngo, H. H., Duong, C. C., Bui, H. M., Le Thuy, T., et al. (2025). A novel anaerobic granular sludge-membrane distillation (AGS-MD) hybrid systems for textile dye wastewater reclamation. *Journal of Water Process Engineering, 80*, 109200. https://linkinghub.elsevier.com/retrieve/pii/S2214714425022731.

Okampo, E. J., & Nwulu, N. I. (2020). Optimal energy mix for a reverse osmosis desalination unit considering demand response. *Journal of Engineering, Design and Technology, 18*(5), 1287–1303. https://www.sciencedirect.com/org/science/article/abs/pii/S1726053120000510.

Ray, S. S., Chen, S. S., Nguyen, N. C., Nguyen, H. T., Dan, N. P., Thanh, B. X., et al. (2018). Exploration of polyelectrolyte incorporated with Triton-X 114 surfactant based osmotic agent for forward osmosis desalination. *Journal of Environmental Management, 209*, 346–353. https://doi.org/10.1016/j.jenvman.2017.12.086

Rodríguez-delanuez, F., Franquiz-Suárez, N., Santiago, D. E., Veza, J. M., & Sadhwani, J. J. (2012). Reuse and minimization of desalination brines: A review of alternatives. *Desalination and Water Treatment, 39*(1–3), 137–148.

Ruiz-Aguirre, A., Andrés-Mañas, J. A., Fernández-Sevilla, J. M., & Zaragoza, G. (2018). Experimental characterization and optimization of multi-channel spiral wound air gap membrane distillation modules for seawater desalination. *Separation and Purification Technology, 205*, 212–222. https://www.sciencedirect.com/science/article/abs/pii/S1383586618309535.

Ruya, P. M., Silva, M. P., Reyniers, G., Ruya, G. A., Lim, S. S., Wenten, I. G., et al. (2025). Intensifying transition metal ion removal and recovery from acidic wastewater via electrodialysis (ED)-Based process. *Water Research, 280*, 123504. https://www.sciencedirect.com/science/article/abs/pii/S0043135425004178.

Shahrim, N. A., Abounahia, N. M., El-Sayed, A. M. A., Saleem, H., & Zaidi, S. J. (2023). An overview on the progress in produced water desalination by membrane-based technology. *Journal of Water Process Engineering, 51*, 103479. https://www.sciencedirect.com/science/article/abs/pii/S2214714422009230.

Shen, J., Xing, Z., Wang, X., & He, Z. (2014). Analysis of a single-effect mechanical vapor compression desalination system using water injected twin screw compressors. *Desalination, 333*(1), 146–153. https://www.sciencedirect.com/science/article/abs/pii/S0011916413004852.

Thi Nguyen, H., Duong, H. C., Chen, S. S., Le Quang, H., Ngo, H. H., Cong, C. D., et al. (2025). Enhancing membrane distillation stability: Isoamyl alcohol coagulation as a novel strategy to mitigate membrane swelling at elevated temperatures. *Environmental Technology & Innovation, 37*, 104029. https://www.sciencedirect.com/science/article/pii/S235218642500015X.

Thomas, N., Mavukkandy, M. O., Loutatidou, S., & Arafat, H. A. (2017). Membrane distillation research and implementation: Lessons from the past five decades. *Separation and Purification Technology, 189*, 108–127. https://www.sciencedirect.com/science/article/abs/pii/S1383586617319469.

Wang, C., Wang, L., Dong, L., Shon, H. K., & Kim, J. (2025). Specific energy consumption of seawater reverse osmosis desalination plants using machine learning. *Desalination, 602*, 118654. https://www.sciencedirect.com/science/article/abs/pii/S0011916425001298.

Winter, D., Koschikowski, J., Gross, F., Maucher, D., Düver, D., Jositz, M., et al. (2017). Comparative analysis of full-scale membrane distillation contactors - Methods and modules. *Journal of Membrane Science, 524*, 758–771. https://www.sciencedirect.com/science/article/abs/pii/S0376738816320580.

CHAPTER

3

Forward osmosis and pressure-retarded osmosis: Emerging membrane platforms for low-energy desalination

Doan Van Hong Thien[1], Nguyen Tan Tan[1] and Hau Thi Nguyen[2]

[1]*Faculty of Chemical Engineering, Can Tho University, Can Tho City, Vietnam;* [2]*Faculty of Chemistry and Environment, Dalat University, Dalat, Vietnam*

1. Introduction

1.1 The global water scarcity situation and the urgent need for energy-efficient desalination technologies

Water shortages have become a critical global issue in the 21st century, significantly influencing economic growth, industrial performance, and overall quality of life. According to recent global assessments, approximately two billion people worldwide still lack access to safe drinking water, while about 3.6 billion individuals live without adequate sanitation facilities (Abney et al., 2021). The intensifying global water crisis is primarily driven by population growth, rapid urbanization, and the accelerating effects of climate change, which collectively disrupt the balance between water demand and natural freshwater availability. This situation underscores the urgent need to identify and implement sustainable technologies capable of providing reliable freshwater supplies for domestic, agricultural, and industrial purposes.

Despite covering nearly 71% of the Earth's surface, about 97% of the global water inventory exists as seawater, unsuitable for direct human consumption or agricultural use due to its high salinity (Musie et al., 2023). As a result, converting seawater and brackish water into potable water has emerged as an essential and viable solution to address the growing deficit of freshwater resources. Converting saline water into freshwater not only addresses immediate water supply issues but also contributes to long-term resource sustainability and climate resilience. In this context, the development of advanced desalination technologies, particularly those based on membrane separation, has gained increasing scientific and industrial significance. These technologies not only ensure water security but also drive innovation in materials science and environmental engineering, promoting cleaner and more energy-efficient water treatment processes.

Currently, a variety of desalination techniques have been implemented worldwide, including thermal (e.g., multi-stage flash evaporation), membrane-based (e.g., membrane distillation), and electrochemical techniques such as electrodialysis and capacitive deionization (Crowe et al., 2016;

Sustainable Water Management. https://doi.org/10.1016/B978-0-443-44853-9.00009-5

Mavukkandy et al., 2019; Missimer et al., 2018). Although these methods have demonstrated effectiveness in ion removal, many of these methods still suffer from relatively high operational costs and energy consumption, particularly when applied to high-salinity feedwaters. As a result, there has been a growing shift toward membrane-based desalination technologies, which offer higher energy efficiency, modularity, and operational simplicity compared to conventional thermal desalination processes and some electrochemical approaches (Cath et al., 2006). Among these emerging technologies, FO and PRO stand out as particularly promising due to their potential for reduced energy consumption at the membrane transport stage, reduced fouling tendencies, and potential integration into hybrid water–energy systems.

1.2 Characteristics of RO, FO, and PRO processes

Reverse osmosis RO, FO, and PRO are membrane-based separation processes that rely on semipermeable membranes to achieve selective water transport. Although these technologies share similar membrane materials and fundamental transport mechanisms, they differ substantially in terms of their driving forces, operational modes, energy demands, and intended applications. Understanding these distinctions is essential for evaluating the role of FO within the broader context of membrane desalination and water treatment technologies.

In the RO process, external hydraulic pressure is applied to the FS to overcome the osmotic pressure difference between the feed and permeate sides, thereby forcing water molecules through a semipermeable membrane toward the low-salinity permeate stream (Fig. 3.1). To achieve a net water flux across the membrane, the applied pressure difference (ΔP) must exceed the osmotic pressure difference ($\Delta\pi$), such that the overall driving force ($\Delta P - \Delta\pi$) is positive (Cath et al., 2006; Touati et al., 2017a). RO has been widely commercialized and remains the dominant technology for seawater and brackish water desalination due to its high salt rejection and technological maturity. Nevertheless, RO is relatively energy-intensive, primarily because of the high operating pressures, typically in the range of approximately 50–80 bar for seawater desalination, required to sustain water flux. As a result, energy consumption constitutes a substantial fraction of the total operating costs of RO desalination

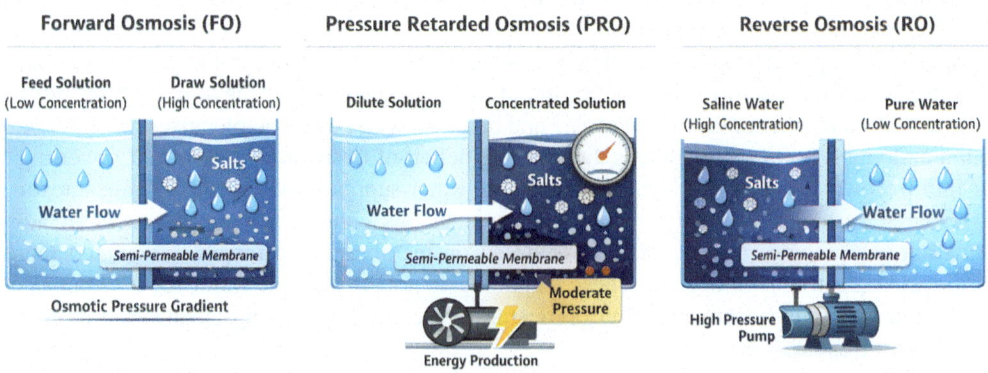

FIGURE 3.1

Schematic illustration of solvent transport mechanisms in FO, PRO, and RO.

plants, in addition to challenges associated with membrane fouling, frequent chemical cleaning, and gradual membrane degradation. These limitations have stimulated extensive research into alternative low-energy desalination technologies.

FO is distinct from RO because it relies exclusively on the intrinsic osmotic pressure difference between two solutions to drive water permeation. In an FO configuration, a dilute FS and a highly concentrated DS are separated by a semipermeable membrane (Fig. 3.1). Owing to the substantial osmotic pressure gradient across the membrane, water migrates spontaneously from the lower-salinity feed side to the higher-salinity draw side in an effort to balance the chemical potential of water on both sides. Importantly, this transport occurs without the application of external hydraulic pressure. Consequently, the magnitude of water flux in FO is governed primarily by the osmotic pressure difference between the draw and feed solutions (Cath et al., 2006).

During FO operation, the membrane selectively allows water molecules to permeate while rejecting the majority of dissolved ions, organic compounds, and suspended contaminants present in the feed solution. Nevertheless, unlike RO, FO is inherently susceptible to reverse solute flux, a phenomenon in which draw solutes diffuse from the draw solution back into the feed solution due to concentration gradients. Reverse solute flux can reduce the effective osmotic driving force, alter feed water chemistry, and contribute to membrane fouling and performance deterioration. Consequently, minimizing reverse solute flux through membrane design and draw solute selection remains a key research focus in FO development. The absence of high hydraulic pressure confers several important advantages to FO systems. First, the energy consumption associated with water transport across the membrane is significantly lower than that of RO, as no high-pressure pumping is required. Second, FO membranes generally exhibit lower fouling propensity, since foulant layers formed on the membrane surface are less compact and more easily removed due to the lack of compressive pressure. These characteristics make FO particularly attractive for treating challenging feed streams, such as highly saline waters, wastewater effluents, and feeds with high fouling potential. Despite these advantages, the practical implementation of FO is constrained by the requirement to regenerate or reconcentrate the draw solution after water extraction. In most applications, the diluted draw solution must undergo an additional separation step, such as reverse osmosis, membrane distillation, or thermal distillation, to recover fresh water and restore the draw solute concentration. This draw solution regeneration step often dominates the overall energy consumption and cost of FO-based systems, limiting the viability of FO as a standalone desalination technology. Consequently, FO has been increasingly explored not as a direct replacement for RO, but as a complementary or hybrid process in integrated water treatment systems. Promising applications include wastewater concentration, nutrient recovery, desalination pretreatment, and the concentration of high-value products in the food and pharmaceutical industries. In these contexts, the unique operating characteristics of FO, low pressure operation, reduced fouling tendency, and high osmotic selectivity, can be effectively leveraged to improve overall process efficiency and sustainability.

The PRO process shares fundamental similarities with FO and can be regarded as a pressure-assisted variant of FO, primarily aimed at energy recovery rather than freshwater production. In PRO, the DS is pressurized to a hydraulic pressure lower than the osmotic pressure difference between the DS and FS, such that a net osmotic-driven water flux from the FS to the DS is maintained (Fig. 3.1) (Abdelkader et al., 2022; Loeb, 2002; Touati et al., 2017b). The resulting water permeation increases both the volumetric flow rate and hydraulic pressure on the draw side, and the pressurized stream is subsequently expanded through a hydraulic turbine or pressure exchanger to recover mechanical energy. The net power output of a PRO system depends on several factors, including the magnitude of the osmotic pressure difference, membrane water permeability and salt rejection, internal concentration polarization, reverse solute flux, and overall system configuration. From a process perspective, PRO can be regarded as an intermediate operation between FO and RO, combining osmotic-driven water transport with partial hydraulic pressurization to enable energy recovery.

Recent studies have demonstrated the potential of PRO as a sustainable energy technology for harvesting salinity gradient energy, often referred to as blue energy, through the controlled mixing of seawater and freshwater across a semipermeable membrane. Nevertheless, similar to FO, the performance of PRO systems is strongly constrained by internal concentration polarization and membrane fouling, which diminish the effective osmotic driving force and power density. Consequently, increasing research efforts have been directed toward the development of advanced thin-film composite (TFC) and thin-film nanocomposite (TFN) membranes with optimized support layer structures and enhanced transport properties, aiming to improve the overall efficiency of both FO and PRO processes (Pandaya et al., 2025; Movahed et al., 2025; Liu et al., 2025; bonsale et al., 2025; Bayrami et al., 2025; Ahmadalipour et al., 2025; Xie et al., 2024; Kwon et al., 2024).

In summary, while RO remains the benchmark technology for large-scale seawater desalination due to its proven performance and technological maturity, FO and PRO are increasingly recognized as promising osmotic membrane processes that offer alternative pathways toward reduced-energy desalination concepts and sustainable salinity-gradient energy recovery. The comparative analysis of these three processes underscores the growing importance of osmotic-based membrane technologies in addressing global challenges related to water security and energy efficiency.

1.3 FO and PRO – emerging platforms for sustainable water treatment

In response to escalating energy and economic constraints, FO and PRO have emerged as promising platforms for sustainable water and energy applications. Both processes utilize natural osmotic pressure gradients as the primary driving force, thereby reducing reliance on externally applied hydraulic pressure.

In conventional desalination configurations, FO does not directly produce purified water. Instead, it generates a diluted DS, which requires an additional separation step for draw solute recovery and clean water production (Chung et al., 2015). Consequently, future FO systems targeting potable water applications must employ draw solutes that are easily separable and regenerable with minimal energy input (McCormick et al., 2008). However, in applications where complete water purification is unnecessary, such as wastewater concentration, resource recovery, and pretreatment, FO offers significant advantages over RO, including lower fouling propensity, reduced mechanical stress, and enhanced operational stability.

A particularly attractive strategy involves the use of fertilizer solutions as the draw solute. In this configuration, the diluted nutrient solution produced after osmosis can be directly applied for fertigation, thereby eliminating the need for DS recovery. Under the osmotic pressure gradient, water permeates into the fertilizer solution, generating a diluted nutrient stream that can be supplied directly to crops through drip irrigation systems. Energy consumption is primarily associated with circulation pumping, as no high hydraulic pressure is required. Importantly, this approach does not inherently increase fertilizer consumption, since nutrients are already essential agricultural inputs. Compared with conventional broadcasting methods, fertigation via irrigation networks improves nutrient use efficiency, reduces operational costs, and facilitates integration with modern agricultural infrastructure (Hanson et al., 2006; Kafkafi et al., 2005; Phuntsho et al., 2011).

Beyond agricultural applications, FO has demonstrated high efficiency in the treatment of oily wastewater, which is a critical challenge in industries such as hydraulic fracturing and petrochemical processing, where oil is often present as stable emulsions that are difficult to separate using conventional techniques. Recent studies have reported that single- or double-layer FO membranes incorporating a sulfonated polymer active layer oriented toward the oily feed stream can achieve oil removal efficiencies exceeding 99%, while maintaining stable water flux and exhibiting reduced membrane fouling (Cath et al., 2006; Phuntsho et al., 2011; Roy, 2010). Moreover, hybrid FO–membrane distillation (FO–MD) systems have been shown to effectively treat complex wastewater streams containing oil, NaCl, and organic acids such as acetic acid, thereby enhancing overall separation efficiency and water recovery (McCutcheon et al., 2006). Owing to its operational simplicity, low energy demand, and the absence of high hydraulic pressure or complex draw-solution recovery processes, FO is increasingly recognized as a promising platform for industrial wastewater treatment and sustainable environmental management.

Alongside FO, the PRO process has attracted considerable attention for its potential to harvest salinity gradient energy, often referred to as "blue energy," which is regarded as a promising renewable energy resource. Thermodynamically, the mixing of 1 m^3 of freshwater with seawater can theoretically release approximately 0.7 kWh of energy, comparable to the electricity generated by a small wind turbine operating for 1 hour (Abdelkader et al., 2022). On a global scale, the exploitable salinity gradient energy has been estimated to reach up to 157 GW (Yip et al., 2012). PRO is particularly attractive due to its stable and continuous energy supply, independence from weather conditions, and high theoretical energy conversion efficiency (54%–56%), which is significantly higher than that of reverse electrodialysis (RED, 18%–38%) (Abdelkader et al., 2022; Yip et al., 2014). In addition, PRO systems can utilize concentrated brine from RO or FO processes as the DS and freshwater from rivers or reclaimed wastewater as the FS, enabling simultaneous power generation and mitigation of environmental impacts associated with brine discharge (Kim et al., 2015; Touati et al., 2017b).

Moreover, increasing the DS concentration can significantly enhance the power density of PRO systems by increasing the osmotic pressure difference across the membrane, particularly when highly concentrated brine from desalination plants is employed as the DS (Kim et al., 2013a). Nevertheless, the commercialization of PRO remains limited due to several technical challenges, including the lack of high-performance semipermeable membranes, internal and external concentration polarization (ICP and ECP), as well as membrane fouling and biofouling, all of which substantially reduce system efficiency and power output (Einarsson et al., 2021; Kim et al., 2012). Overall, PRO represents a clean and stable renewable energy pathway with strong potential to contribute to integrated and sustainable global energy–water management systems.

2. Evaluation of the performance of FO and PRO

2.1 FO process

The operational principle of the FO system is presented schematically in Fig. 3.2. The performance of FO membranes is typically evaluated using several key parameters, including water flux, reverse salt

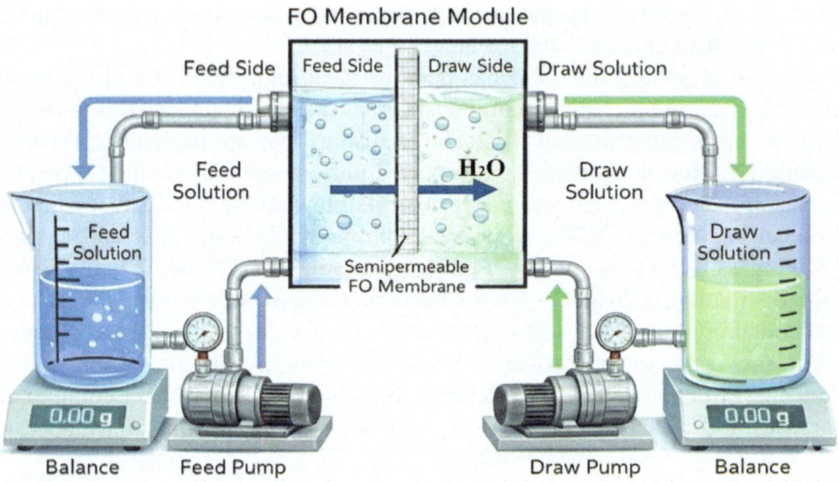

FIGURE 3.2

Schematic diagram of the FO experimental setup.

flux, fouling resistance, and membrane stability during operation. Among these, water flux and reverse salt flux are the most commonly reported indicators in FO studies. The water flux, denoted as J_w ($L\cdot m^{-2}\cdot h^{-1}$, LMH), is determined from the net increase in the volume of the DS over a given time period and is calculated according to Eq. (3.1)

$$J_w = \frac{\Delta V}{A_m.\Delta t} \tag{3.1}$$

In this equation, ΔV (L) represents the change in the volume of the DS, A_m (m^2) is the effective membrane area, and Δt (h) denotes the operating time.

The reverse salt flux, J_s ($g\cdot m^{-2}\cdot h^{-1}$), is calculated based on the increase in salt content in the FS, which is determined from the change in salt concentration over time according to Eq. (3.2).

$$J_s = \frac{C_t.V_t - C_0.V_0}{A_m.\Delta t} \tag{3.2}$$

In this equation, C_0 (mol/L) and V_0 (L) represent the initial concentration and volume of the FS, while C_t (mol/L) and V_t (L) represent the concentration and volume of the FS after operation.

In addition to the main parameters such as water flux and reverse salt flux, several other characteristic parameters also play an important role in evaluating the quality and properties of the membrane, including the water permeability coefficient (A) and the salt permeability coefficient (B). These parameters directly influence the structural parameter (S), which is an important quantity representing the characteristics of the semi-permeable membrane. On this basis, the separation properties of the membrane are determined according to previously published studies (Wu et al.,

2021a; Yu et al., 2020). The pure water flux (J, LMH) and the salt rejection (R, %) are calculated using Equations (3.3) and (3.4).

$$J = \frac{\Delta V}{A_m \times \Delta t} \tag{3.3}$$

$$R = \left(1 - \frac{C_d}{C_f}\right) \times 100\% \tag{3.4}$$

In which ΔV (L) is the permeated water volume, A_m (m^2) is the effective membrane area, Δt (h) is the measurement time, and C_f and C_d represent the concentrations of the FS and the DS, respectively.

After determining the water flux J and the salt rejection R, calculate the water permeability A (L/m^2.h.bar), the salt permeability B (L/m^2.h) and the structural parameter S (μm) using Eqs. (3.5), (3.6) and (3.7).

$$A = \frac{J}{\Delta p} \tag{3.5}$$

$$\frac{1-R}{R} = \frac{B}{A(\Delta p - \Delta \pi)} \tag{3.6}$$

$$S = \frac{D}{J_w} \times \ln\left[\frac{A\pi_d + B}{A\pi_f + J_w + B}\right] \tag{3.7}$$

In which Δp (bar) and Δπ (bar) represent the pressure difference and the osmotic pressure difference across the FO membrane, D is the salt diffusion coefficient, and π_d (bar) and π_f (bar) are the osmotic pressures of the DS and the FS, respectively.

However, during the FO process, the membrane performance is strongly influenced by concentration polarization (CP), which includes external concentration polarization (ECP) and internal concentration polarization (ICP).

ECP commonly occurs in dense symmetric membranes when water moves from the FS to the DS, causing solutes to accumulate on the membrane surface (concentrative ECP) or to be significantly diluted on the draw side (dilutive ECP), thereby reducing the effective osmotic pressure difference (Arjmandi et al., 2020; Sharma et al., 2022). For asymmetric membranes, ECP occurs on the active layer surface when solutes accumulate or become diluted right next to the membrane. Experimental studies have shown that increasing the flow rate can disturb the boundary layer, thus significantly mitigating the impact of ECP (McCutcheon et al., 2006). Meanwhile, ICP mainly occurs in asymmetric membranes due to the reverse diffusion of solutes within the porous support layer, where water permeates but the active layer restricts solute movement, resulting in salt accumulation inside the support layer (Gray et al., 2006; Tan et al., 2013). When the support layer faces the FS, concentrative ICP occurs, whereas when it faces the DS, dilutive ICP takes place. ICP causes a more severe flux decline than ECP and directly depends on the structural parameter $S = t\,\tau/\varepsilon$, where a thinner, more porous, and less tortuous support layer helps reduce ICP (Shi et al., 2019; Tian et al., 2017; Wu et al., 2021a). Overall, both ECP and ICP reduce the effective osmotic pressure and water flux; however, ECP can be controlled through operating conditions, while ICP depends on the intrinsic structure of the membrane and remains the most critical challenge in designing high performance FO membranes.

2.2 PRO process

Figure 3.3 illustrates the operation process of the PRO system. In the PRO process, when the effects of concentration polarization are neglected, the water flux across the semipermeable membrane can be described by Eq. (3.8) (Abdelkader et al., 2022).

$$J_w = A(\Delta\pi - \Delta P) \quad (3.8)$$

The performance is typically evaluated by the power density generated per unit membrane area, which is calculated using Eq. (3.9).

$$W = J_w.\Delta P = A(\Delta\pi - \Delta P).\Delta P \quad (3.9)$$

To determine the optimal ΔP that achieves the maximum power density, Eq. (3.9) is differentiated with respect to ΔP. The maximum power density occurs at $\Delta P = \Delta\pi/2$, and the theoretical maximum power can be expressed as:

$$W_{max} = A.\ \frac{\Delta\pi r^2}{4} \quad (3.10)$$

However, this equation is established under ideal conditions where concentration polarization and salt leakage through the membrane are not considered. In practice, concentration polarization reduces the effective osmotic pressure difference, while salt leakage alters the concentrations of both the feed and DS along the membrane length. The combined effects of these factors cause the actual maximum power output to occur at a lower pressure than the theoretical value, meaning that $\Delta P < \Delta\pi/2$ (Abdelkader et al., 2022).

In the PRO process, the active layer of the membrane faces the DS, while the porous support layer faces the FS. During water permeation through the membrane, concentration polarization is inevitable, leading to a decrease in PRO performance (Kim et al., 2012). Therefore, many studies have focused on evaluating the effects of ICP and ECP on the actual water flux compared to the theoretical value. Notably, Lee et al. (1981) (Lee et al., 1981) developed a solution-diffusion

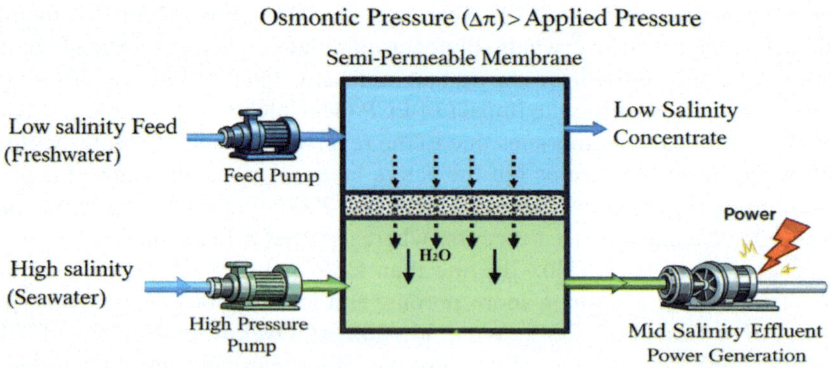

FIGURE 3.3

Schematic PRO system for salinity gradient power generation.

model that accounts for ICP, in which the water flux through the semi-permeable membrane is expressed by Eq. (3.11).

$$J_w = A\left(\pi_{D,b}\frac{1-\frac{C_{F,b}}{C_{D,b}}e^{(J_wK)}}{1+\frac{B}{J_w}\left[e^{(J_wK)}-1\right]}-\Delta P\right) \tag{3.11}$$

In the model proposed by Lee et al. (1981), the effect of ECP was not considered. Therefore, subsequent studies have introduced several modifications to incorporate the influence of ECP into the model.

Achilli's model (Achilli et al., 2009), Eq. (3.12):

$$J_w = A\left(\frac{\pi_{D,b}e^{(J_wK)}-\pi_{F,b}e^{(J_wK)}}{1+\frac{B}{J_w}\left[e^{(J_wK)}-1\right]}-\Delta P\right) \tag{3.12}$$

Yip's model (Yip et al., 2011), Eq. (3.13):

$$J_w = A\left(\frac{\pi_{D,b}e^{(-J_w/k_D)}-\pi_{F,b}e^{(J_wK)}}{1+\frac{B}{J_w}\left[e^{(J_wK)}-e^{(-J_w/k_D)}\right]}-\Delta P\right) \tag{3.13}$$

Touati' model (Touati et al., 2015), Eq. (3.14):

$$J_w = A\left(\left[\pi_{D,b}+\frac{B}{A}\left(1+\frac{A\Delta P}{J_w}\right)\right]e^{(-J_w/k_D)}-\left[\pi_{F,b}+\frac{B}{A}\left(1+\frac{A\Delta P}{J_w}\right)e^{(-J_w/k_F)}\right]-\Delta P\right) \tag{3.14}$$

Previous studies have compared and analyzed the errors among different models to evaluate the influence of ICP and ECP (Abdelkader et al., 2022). The results indicated that when the effect of ECP was considered, the modified models achieved significantly higher accuracy compared to the original one. In contrast, models that neglected both ICP and ECP exhibited an average deviation of approximately 40% from experimental results. The model proposed by Lee et al. after certain modifications, reduced this error to around 17%. Meanwhile, more recent improved models, which incorporated ECP, were able to lower the error to below 1%, meaning the predicted results were almost identical to the experimental data. This finding is highly significant: to accurately simulate and design PRO membranes, it is essential to consider the impact of concentration polarization.

3. Emerging membrane materials for FO and PRO

3.1 The evolution and advancement of TFC and TFN membrane technologies

In FO technology, the semipermeable membrane plays a decisive role in determining filtration performance. Accordingly, continuous research has been dedicated to optimizing membrane architecture and enhancing its physicochemical performance characteristics. Structurally, FO membranes generally consist of two main layers: a thin active layer (AL) responsible for salt rejection, and a support layer

(SL) that provides mechanical stability and facilitates water flow. Among these, the SL is considered the key component because it is where ICP occurs, a phenomenon that significantly reduces water flux. Consequently, most current FO membrane development strategies aim to optimize the support layer by reducing the structural parameter and minimizing ICP.

The semipermeable membranes used in FO systems have undergone several stages of development, with continuous improvements in both structure and performance. FO was first introduced in 1968 using cellulose acetate (CA) membranes (Popper et al., 1968), where fertilizer solutions served as draw solutes that could be directly used for irrigation after filtration. This approach offered benefits in terms of cost, energy efficiency, and agricultural productivity. It was later extended by Moody in 1977 (Phuntsho et al., 2012), opening new opportunities for FO application in irrigation. CA membranes demonstrated high hydrophilicity, excellent antifouling and chlorine resistance, and good mechanical strength. By 2010, Chung et al. applied cellulose membranes in both flat-sheet and hollow-fiber configurations for FO, using phase inversion followed by hot-water treatment to optimize the membrane structure (Zhang et al., 2010). Subsequently, Wang et al. (2012) developed commercial cellulose triacetate (CTA) membranes reinforced with polyester mesh (Wang et al., 2012), which improved mechanical durability and fouling resistance, though salt rejection remained limited due to the asymmetric structure. Since the early 2010s, thin-film composite FO (TFC-FO) membranes have gained increasing attention in both flat-sheet and hollow-fiber forms (Wang et al., 2010; Yip et al., 2010). The first commercial TFC membranes developed by Hydration Technology Innovations (HTI) showed high compatibility with various substrates and promising filtration performance (Blandin et al., 2016; Xie et al., 2013). Compared with CTA membranes, TFC membranes possess an ultrathin selective layer and a porous substrate, enhancing both separation capability and water flux. Ortega-Bravo et al. demonstrated the feasibility of TFC-FO membranes for municipal wastewater concentration (Ortega-Bravo et al., 2016). However, TFC-FO membranes still face challenges such as fouling, cost, and suboptimal selectivity (Wu et al., 2021b). Recent research has focused on improving the support layer to mitigate ICP by enlarging pore size, reducing tortuosity, and optimizing thinner polyamide (PA) active layers. In particular, electrospun nanofiber mats have emerged as a highly promising solution (Shi et al., 2019). Their interconnected porous networks effectively lower the structural parameter and reduce ICP compared with traditional phase-inverted substrates (Pan et al., 2017). Since the introduction of forward osmosis (FO) membranes based on electrospun nanofibers, significant progress and remarkable achievements have been reported. For example, Zeng et al. developed high-performance FO membranes using electrospun chitosan (CS) nanofiber mats as supports (Zeng et al., 2023). Due to their hydrophilicity, low cost, and environmental friendliness, CS nanofiber supports enhanced water flux and reduced reverse salt flux, offering a sustainable pathway for next-generation FO membranes.

In summary, the development of FO membranes has progressed through multiple stages, from CA/CTA to TFC to TFN and finally to nanofiber-based composites. The continuous evolution of porous support layers across these stages has significantly improved the performance of FO membranes. Today, membrane development remains an active research area worldwide. The potential of FO technology can still be further enhanced through various approaches, among which material modification and surface engineering of the support layer remain highly promising directions.

3.2 Surface improvement and membrane support strategies

It can be seen that although the substrate materials have been improved through multiple stages, optimizing the membrane surface still demonstrates significant potential. Electrospun nanofiber substrates, despite their high porosity and good flexibility, remain limited when used independently (Bui et al., 2011). Some studies have incorporated mesoporous silica nanoparticles to enhance mechanical strength (Bui et al., 2016) or modified the surface from hydrophobic to hydrophilic to improve durability (Huang et al., 2016); however, these approaches have not completely overcome the weak interfacial bonding between the active and support layers. In 2019, the development of PAN/CS composite structures marked a significant breakthrough, as the CS coating layer not only enhanced mechanical strength but also filled surface voids, improved interfacial adhesion and hydrophilicity, thereby increasing water flux and reducing reverse salt flux (Shi et al., 2019). Following this direction, Wu et al. (2021) introduced graphene oxide (GO) into the PAN substrate, which limited the penetration of the polyamide (PA) layer, increased crosslinking density, and improved both water flux and salt rejection (Wu et al., 2021a). Meanwhile, Yu et al. (2020) developed a PVDF/GO/O-MWCNTs membrane, in which the intermediate GO/O-MWCNTs layer provided additional water channels and regulated PA layer formation, thus reducing mass transfer resistance and mitigating ICP (Yu et al., 2020). Moreover, incorporating 0.5 wt% amorphous SiO_2 into the PVDF electrospinning process increased water flux nearly threefold while maintaining a salt rejection of 99.7% and significantly reducing the structural parameter S (Obaid et al., 2016). However, excessive SiO_2 content may lead to particle agglomeration and increased reverse salt flux. Another emerging approach to improving FO membrane performance lies in the optimization of fabrication techniques in addition to material modification efforts. A recent study by Tan et al. (2025) investigated the effects of different fabrication methods on water flux and the structural characteristics of porous support layers. The findings highlight that membrane efficiency is influenced not only by material selection but also significantly by the fabrication methodology employed. This work introduced a new perspective, suggesting that achieving optimal filtration performance requires simultaneous consideration of both material development and fabrication optimization.

Membrane performance is primarily evaluated through water permeability and salt rejection, which strongly depend on material selection and the compatibility between the base polymer and additives. Numerous studies have demonstrated the role of organic and inorganic additives in enhancing membrane properties (Ahmad et al., 2013; Qian et al., 2022; Wahab et al., 2019). Among them, polydopamine (PDA) stands out for its ability to increase hydrophilicity, reduce fouling, and enhance water flux. PDA is formed through the oxidation of dopamine in an alkaline environment, producing a strongly adhesive and stable coating on nearly all surfaces, including hydrophobic materials such as PTFE (Lee et al., 2007). The adhesion mechanism of PDA involves the oxidation of catechol to quinone, followed by covalent crosslinking through Michael addition or aryl–aryl coupling reactions. In membrane filtration applications, PDA functions as a "bio-glue" that bonds the active and support layers while enhancing hydrophilicity and antifouling properties due to its abundant –OH and $-NH_2$ functional groups (Arena et al., 2011; Vaselbehagh et al., 2014). This feature is particularly important since most commercial membranes (PVDF, PES, PSf, PEK, PAN, etc.) possess hydrophobic surfaces that are prone to biofouling, reducing both performance and lifespan (Gao et al., 2011). In contrast, the hydrophilic surface of PDA minimizes hydrophobic interactions and prevents microbial attachment through steric hindrance effects (Elimelech et al., 2011; McCloskey et al., 2010). Furthermore, PDA exhibits antibacterial properties by altering surface charge and reducing silver ions (Ag^+) to form silver nanoparticles with strong antimicrobial activity (Sileika et al., 2011). The PDA coating also provides multiple reactive functional groups (catechol, quinone, amine, and carboxyl), facilitating further membrane functionalization with materials such as GO or inorganic nanoparticles to optimize mechanical and surface properties (Feng et al., 2015). However, PDA may also reduce water flux if the coating process is not well controlled, as PDA polymer deposition can clog surface pores (Fang et al., 2017; Kim et al., 2014). Therefore, optimizing coating conditions and selecting appropriate material combinations are critical to achieving the best membrane enhancement performance.

3.3 Advanced materials and nanostructured materials

The Perforene membrane is a unique graphene-based filtration material patented by Lockheed Martin in 2013 (https://news.lockheedmartin.com/2013-03-18-Lockheed-Martin-Achieves-Patent-For-Perforene-TM-Filtration-Solution-Moves-Closer-To-Affordable-Water-Desalination. 2013). Unlike traditional polymer membranes, Perforene utilizes single-layer or multilayer graphene that is nanoperforated with precisely controlled pore sizes (1 nm or smaller), allowing water molecules to pass through easily while blocking salt ions.

According to reports, Perforene can achieve higher filtration efficiency at a much lower cost compared to RO membranes, which has drawn considerable attention from the membrane research community. Previous studies have shown that nanoporous graphene membranes can achieve 2–3 times higher water flux than conventional RO membranes while still maintaining nearly complete salt rejection (Cohen-Tanugi et al., 2014).

In current seawater desalination processes, graphene membranes can be categorized into three main developmental directions. First, single-layer graphene membranes are employed as barrier layers, where sub-nanometer pores must be created on the surface to enable mass transport. Studies have demonstrated exceptional desalination potential, allowing for much higher water permeability compared to conventional RO membranes (Liu et al., 2015). However, several challenges remain, including the large-scale synthesis of single-layer graphene and the precise creation of high-density, uniformly distributed nanopores with controllable sizes. As an alternative, GO nanosheets have been explored to fabricate membranes through two-dimensional stacking. GO possesses several advantages: atomic-level thickness, lateral dimensions of tens of micrometers, and ease of large-scale production, which reduces costs and broadens application potential. The formation of flat nanochannels between stacked GO sheets allows for low-friction water permeation, while functional groups (particularly carboxyl groups) on the GO surface provide ideal sites for further functionalization, enhancing selectivity and pollutant rejection. In addition, GO membranes exhibit good mechanical strength and allow ultrafast vapor transport, attributed to nearly frictionless monolayer water flow within the planar capillaries. Nevertheless, the main limitation of traditional GO membranes lies in their instability in aqueous environments due to their superhydrophilic nature, which causes GO sheet dispersion. This necessitates the development of strategies to form robust interlayer linkages to ensure stability and operational reliability under practical conditions, especially in cross-flow systems (Hu et al., 2013). Consequently, many studies have focused on integrating graphene or its derivatives into polymer matrices, where narrow interfacial gaps between components act as water transport channels, thereby improving filtration efficiency (Shao et al., 2017). These approaches each offer distinct advantages while still facing certain limitations. Nonetheless, graphene membranes remain a highly promising material, attracting considerable research interest for applications in RO, FO, and PRO.

With outstanding characteristics such as antibacterial capability, high stability, and superior filtration performance, graphene has emerged as a highly promising material for FO technology. Owing to its large surface area, excellent mechanical strength, ultrathin nanoscopic channels with high selectivity, and hydrophilic surface enriched with functional groups, graphene membranes exhibit high water flux while simultaneously reducing fouling and biofouling. Furthermore, the versatile functionalization capability of graphene allows it to be combined with TiO_2, polymer matrices, or antibacterial agents to optimize its properties and expand its applications in desalination, wastewater treatment, and biological purification. Thanks to these advantages, graphene not only represents a breakthrough compared to traditional polymer membranes but also offers great potential for future commercialization. Notably, graphene has also been effectively applied in solar-driven desalination due to its outstanding photothermal properties and unique nanostructure (Dai et al., 2022).

4. Direction for the selection and regeneration of DS

In membrane-based separation technologies, the DS plays a central role in determining both the water flux across the membrane and the overall regeneration cost. The selection of a suitable DS is crucial, as an effective one must simultaneously satisfy multiple criteria: generating a sufficiently high osmotic pressure, maintaining low viscosity for easy recirculation, minimizing reverse solute diffusion, having a high diffusion coefficient to mitigate ICP, being readily available and cost-effective, and ensuring environmental and human safety (Johnson et al., 2018).

The earliest DS systems in FO technology involved dissolved gases, notably the NH_3–CO_2 system, described in a patent by Neff in 1964 (Neff, 1964). In this system, NH_3 and CO_2 dissolve to form NH_4HCO_3 on the draw side, which can be easily regenerated by heating. Later studies showed that bicarbonate-based DS produced high osmotic pressures and large water fluxes (McCutcheon et al., 2005, 2006). However, their practical performance was limited by ICP and significant reverse diffusion of NH_3, which can promote biofouling (Johnson et al., 2018). Other compounds such as SO_2, ethanol, and TMA were also proposed as DS candidates, but they suffered from issues like corrosiveness, unpleasant odor, and high reverse diffusion, restricting their commercial viability (Boo et al., 2015; Johnson et al., 2018; McCormick et al., 2008). For inorganic salts, the most commonly used draw solutes are monovalent salts, particularly NaCl, due to their low cost, high availability, strong osmotic pressure, low viscosity, and large diffusion coefficients that help minimize ICP. Consequently, NaCl remains a preferred choice in many FO studies. However, its main limitation is the high reverse solute diffusion caused by its small ionic size, which can exacerbate fouling on the feed side (Johnson et al., 2018). In contrast, multivalent salts such as $MgCl_2$ and $CaCl_2$ exhibit larger hydrated radii, significantly reducing reverse diffusion compared to NaCl (Bowden et al., 2012; Su et al., 2011). Moreover, at equal molar concentrations, they generate higher osmotic pressures because their dissociation produces more ions (Shaffer et al., 2015). Hybrid FO–NF systems employing salts like Na_2SO_4 and $MgSO_4$ have demonstrated superior performance over traditional RO, including lower operating pressure, smaller flux decline, and high flux recovery after cleaning (Zhao et al., 2012a). In addition, seawater has been considered a promising DS for water extraction applications since its diluted effluent can be discharged directly into the environment if cross-contamination is avoided. Inorganic fertilizers also represent a promising class of DSs because they can be reused directly for fertigation, thus reducing both cost and energy consumption. However, the post-FO fertilizer solution is typically too concentrated and requires dilution before use (Johnson et al., 2018). Sugars have also attracted attention as potential DSs for FO due to their abundance and non-toxicity. Monosaccharides such as glucose and fructose have been tested for desalination using cellulose acetate membranes (Kravath et al., 1975), while subsequent studies using disaccharide sucrose achieved similar efficiency for tomato juice concentration (Petrotos et al., 1998). However, like other nutrient-based compounds, sugar leakage into the feed side can promote biofouling. Stimuli-responsive hydrogels represent another class of potential DS materials capable of changing their volume or undergoing gel–solution phase transitions in response to environmental stimuli such as temperature, pH, light, or electric fields. These behaviors allow them to control water absorption and release, showing promise for FO applications. Additionally, magnetic nanoparticles (MNPs) have drawn significant attention due to their magnetic separation and regeneration capability and their reduced reverse diffusion owing to larger particle size. When combined with appropriate ligands, these nanoparticles can form DS with high osmotic pressures. The major drawback, however, is particle aggregation after multiple regeneration cycles, which reduces stability and performance. Although ultrasonication has been proposed to break these aggregates, it may simultaneously deteriorate magnetic properties, limiting long-term recyclability (Johnson et al., 2018).

Recent trends in DS development have shifted from simply maximizing osmotic pressure toward improving regeneration efficiency and optimizing energy cost. Modern DS systems focus on thermally regenerable compounds such as volatile salts or thermo-responsive polymers, which allow the use of low-grade waste heat, or chemically regenerable systems that recover solutes through simple and low-cost precipitation reactions (Johnson et al., 2018; McGinnis et al., 2007).

5. Hybrid FO and PRO systems in low-energy desalination

5.1 Hybrid FO system

Early hybrid FO systems were primarily developed with the goal of recovering the DS, in which FO was often combined with multi-stage heating or distillation processes to separate volatile solutes from pure water. Other approaches, such as the use of superparamagnetic nanoparticles for magnetic recovery of the DS or the application of electric fields for DS regeneration, were also proposed. However, these models faced limitations including high energy consumption, low water flux, significant reverse solute diffusion, and high operational costs, making them impractical at an industrial scale. In more recent stages, research has shifted toward developing more advanced hybrid FO systems, in which FO serves as a pretreatment step for subsequent processes such as RO (Chekli et al., 2016). The first FO–RO hybrid system was proposed by Yaeli in 1992, combining FO with low-pressure RO (LPRO) for glucose recovery, demonstrating higher efficiency than standalone RO due to reduced fouling. Subsequently, Yangali-Quintanilla and colleagues showed that the FO–LPRO system consumed only about 50% of the energy required by conventional RO in seawater desalination (Yangali-Quintanilla et al., 2011). Furthermore, the FO–RO system exhibited reduced fouling and scaling, increased water recovery efficiency, and could particularly remove trace organic compounds (TrOCs) that are difficult to treat with standalone RO, thanks to the "dual-barrier" mechanism of the two membranes (Coday et al., 2014a). More recent studies have focused on improving the initial stage by applying pressure-assisted FO (PAFO), which involves adding a small hydraulic pressure on the FS side. The PAFO–RO configuration increases water flux, reduces reverse solute diffusion, and enhances water recovery rates while saving more energy compared to traditional FO–RO (Blandin et al., 2015). In seawater desalination, this approach also stabilizes operation under fluctuating feed water quality, opening a pathway for more sustainable development of next-generation hybrid FO systems.

In addition to seawater desalination solutions, treating highly saline wastewater is also a major challenge, particularly for RO brine or hydraulic fracturing wastewater in oil and gas extraction, which contain high concentrations of inorganic and organic compounds (Chekli et al., 2016). Oasys Water Inc. implemented a pilot-scale FO–thermal hybrid system to treat shale gas wastewater with salinity above 70,000 ppm TDS, achieving over 60% water recovery and reducing thermal energy consumption by approximately 50% compared to conventional evaporation technologies (Coday et al., 2014b; McGinnis et al., 2013). The produced water met discharge standards for TDS, chloride, barium, and strontium, whereas RO could not effectively treat such hypersaline water, demonstrating the clear advantages of FO for challenging feed waters. Subsequently, Oasys Water developed an FO–membrane brine concentrator (MBC) hybrid system aimed at zero liquid discharge (ZLD), achieving up to 85% water recovery and brine concentration of 25%. In parallel, Altaee and Hilal proposed a two-stage FO/PRO hybrid model to simultaneously dilute the osmotic strength of hydraulic fracturing wastewater and generate power, showing potential to significantly reduce salinity to levels suitable for direct discharge or further treatment via membranes or thermal processes. However, this model has only been tested in simulations, and no experimental data have yet been reported to validate its performance (Chekli et al., 2016).

5.2 Hybrid PRO system

Many studies have shown that the PRO process can be flexibly applied with various FS and DS sources, such as freshwater, seawater, pretreated seawater with concentrated brine (SWRO–PRO hybrid), or wastewater with concentrated brine (SWRO–PRO–wastewater hybrid) (Kim et al., 2013b; Palacin et al., 2013). This combination helps reduce desalination costs and mitigates the environmental impact of brine discharge (Altaee, 2012). Among these configurations, the RO–PRO hybrid system has garnered considerable interest in both the United States and Japan. A notable example is the "Megaton Water System" project (Japan, 2010), where the pilot RO–PRO plant achieved a maximum power density of 13.8 W/m^2 at 30 bar, with 38% of freshwater permeating into the saline solution. The addition of a discharge port helped reduce concentration polarization by enhancing the flushing of leaked salts (Saito et al., 2012). Other studies have also shown that RO–PRO systems offer clear energy-saving advantages compared to RO–RED or other hybrid configurations (Achilli et al., 2014; Feinberg et al., 2013). Additionally, other integrated configurations such as PRO–RO, PRO–FO, PRO–MED, or MDePRO demonstrate the ability to simultaneously produce freshwater and energy, particularly when low-grade waste heat is utilized or membrane design is optimized (Touati et al., 2017a).

The outstanding advantage of PRO hybrid systems lies in energy savings, achieving 30%–40% recovery relative to RO energy consumption, while also reducing environmental impact by diluting the brine stream before discharge. The DS source is typically RO brine, which is readily available, low-cost, and low in fouling potential, thereby extending membrane lifespan. Moreover, PRO can be flexibly combined with other technologies (FO, MED, MD) to maximize resource and waste heat utilization. However, actual performance is still lower than theoretical predictions due to ICP, ECP, and reverse salt flux; fabrication costs and membrane durability remain limiting factors; and operational optimization is needed to balance energy and cost (Altaee et al., 2014). Overall, PRO, especially in RO–PRO configurations, is considered a promising approach for energy-efficient and sustainable desalination systems in the future.

6. Technical challenges and future prospects

6.1 FO process

It can be seen that FO technology is attracting significant attention due to its notable advantages over conventional membrane technologies. The core strength of FO lies in its operation without requiring high hydraulic pressure, resulting in lower energy consumption and minimizing pressure-induced fouling, which is a major limitation of RO membranes (Elimelech et al., 2011; Zhao et al., 2012b). Additionally, fouling in FO is generally lower, more reversible, and easier to control through hydrodynamic optimization (Lee et al., 2010; Mi et al., 2010). The non-porous semipermeable membrane structure also allows FO to achieve high removal efficiency for dissolved organics, nutrients, microorganisms, and trace organic contaminants (TOrCs). Moreover, thanks to the natural osmotic mechanism, FO membranes extend operation cycles between cleaning events, reducing maintenance costs and prolonging membrane lifespan (Zhao et al., 2012b). Notably, FO can preserve the physical properties of FS in the food and pharmaceutical sectors since the process does not require heating or high pressure (Petrotos et al., 2001; Yang et al., 2009), and it is also applied in biomedical fields for controlled drug delivery using osmotic pumps (Shokri et al., 2008; Wang et al., 2005). In the current context, FO holds broad prospects due to the global trend toward sustainable water treatment, energy saving, and resource reuse. With these advantages, FO is expected to become a strategic membrane technology pillar in addressing the global water and energy crisis in the coming decade.

Although the FO process offers numerous advantages, the technology still faces critical challenges that directly impact operational performance and large-scale applicability. Key challenges include concentration polarization, membrane fouling, reverse solute flux (RSF), as well as limitations of membrane materials and DS design. Among these, concentration polarization is one of the most prominent issues in FO, particularly ICP occurring within the membrane support layer. ICP markedly diminishes the effective osmotic pressure gradient, thereby resulting in a decline in water permeation flux. While ECP can be mitigated by increasing flow velocity or temperature, ICP is more difficult to overcome since it is directly related to membrane structural properties. Therefore, developing thin, porous, and highly hydrophilic support layers is a promising approach to reduce ICP and enhance FO performance (Lee et al., 2010; Zhao et al., 2011). The next challenge in FO involves membrane fouling, including organic, inorganic, biological (biofouling), and combined fouling. The formation of fouling layers increases mass transfer resistance, reduces flux, and raises operational costs (Zhao et al., 2012b). Another concern is RSF, which is unavoidable due to the concentration gradient between the DS and the FS. RSF not only decreases salt rejection efficiency but also induces secondary effects such as increased fouling (Lee et al., 2010; Phillip et al., 2010). A further limitation relates to the performance and constraints of commercially available FO membranes. Currently, CTA-FO membranes remain widely used but exhibit several drawbacks, including low flux, high structural parameter causing severe ICP, and poor chemical stability under harsh conditions (Yang et al., 2009). Next-generation TFC-FO membranes with higher A/B ratios and thinner supports have shown improved flux but still face fouling and RSF issues during long-term operation. Therefore, developing advanced FO membranes using nanomaterials, composites, or bio-based materials has become a critical research direction to enhance both selectivity and membrane durability (bonsale et al., 2025; Bayrami et al., 2025; Xie et al., 2024; Zhao et al., 2012b; Zhou et al., 2025; Yousefian-Arani et al., 2025; Panahi et al., 2025; Kazwini et al., 2025; Zhao et al., 2024).

6.2 PRO process

The PRO process is considered one of the next-generation membrane technologies with notable potential due to its ability to simultaneously treat water and recover energy. Its greatest strength lies in exploiting natural or engineered osmotic gradients to produce both clean water and electricity, a benefit that conventional desalination technologies such as RO or thermal processes cannot achieve. In particular, hybridizing PRO with RO or SWRO–PRO has demonstrated the potential to reduce energy consumption by 35%–40% while simultaneously diluting the saline effluent before discharge, thereby minimizing environmental impact. Beyond RO, PRO shows promise when integrated with FO, MED, or MVMD, allowing the use of wastewater or waste heat to improve both water flux and energy efficiency (Touati et al., 2017a). Such strategies are especially critical in light of the growing need for sustainable and energy-efficient water treatment technologies.

Similar to other pressure-driven membrane technologies, PRO is constrained by concentration polarization, reverse solute leakage, and additional factors that adversely affect water flux and membrane longevity (Touati et al., 2017a). Membrane fouling occurs due to convective or diffusive transport of suspended solids, colloids, or biological growth. The fouling layer increases overall mass transfer resistance, significantly reducing performance. Additionally, fouling increases pressure losses along the membrane while decreasing solute rejection (Fritzmann et al., 2007). The first study on PRO fouling was reported by She et al. (2013), showing that the diffusion of Ca^{2+} and Mg^{2+} ions from the DS to the FS exacerbates fouling, as these ions bind with organics and promote deposition. Higher DS concentrations increase salt diffusion, worsening fouling. Scaling occurs when inorganic compounds become supersaturated on the FS side, precipitating on the membrane surface and forming a barrier to mass transfer (Fritzmann et al., 2007). Despite studies by Zhang et al. (2014), scaling in PRO remains underexplored. Concentration polarization is also a key factor affecting PRO flux and power generation. Similar to FO, ICP is typically more severe than ECP, though both can be mitigated through membrane design and operating conditions. Membrane deformation is another critical factor in PRO. Under hydraulic pressure, membranes may deform, reducing water flux and, at high pressures, even collapsing (She et al., 2012). Therefore, the development of mechanically robust membranes is of particular interest for enhancing PRO performance and long-term operational stability.

7. Conclusion

Studies over the past years have demonstrated that FO and PRO are promising membrane technologies for desalination and water resource management with low energy costs. FO stands out for its ability to produce high-quality water, low fouling tendencies, and capability to treat challenging water sources, while PRO offers the opportunity to harness osmotic energy from salinity gradients, adding renewable energy value to water treatment processes. Nevertheless, both technologies still face limitations, primarily related to water flux, reverse solute flux, salt accumulation due to internal concentration polarization, and the energy or cost required for DS regeneration. To address these challenges, advances in membrane technology including new materials, thin and porous support layers, and fouling-resistant designs, along with sustainable DS management that minimizes regeneration energy, and system integration approaches such as FO-RO, FO-MD, and FO/PRO hybrids, have increasingly demonstrated technical feasibility and economic efficiency. These developments not only improve water recovery, save energy, and reduce costs, but also expand applications from seawater desalination, brackish water treatment, and wastewater reuse to green energy recovery. Given current trends, FO and PRO have the potential to become key components in future sustainable water management systems, especially as global demand for water and energy security continues to rise.

Acknowledgments

This research was funded by Ministry of Education and Training of Vietnam under project code B2024-DLA-03 and B2025-TCT-04.

References

Abdelkader, B. A., & Sharqawy, M. H. (2022). Challenges facing pressure retarded osmosis commercialization: A short review. *Energies, 15*(19), 7325.

Abney, S. E., et al. (2021). Toilet Hygiene—Review and research needs. *Journal of Applied Microbiology, 131*(6), 2705–2714.

Achilli, A., Cath, T. Y., & Childress, A. E. (2009). Power generation with pressure retarded osmosis: An experimental and theoretical investigation. *Journal of Membrane Science, 343*(1), 42–52.

Achilli, A., et al. (2014). Experimental results from RO-PRO: A next generation system for low-energy desalination. *Environmental Science and Technology, 48*(11), 6437–6443.

Ahmad, A. L., et al. (2013). Recent development in additives modifications of polyethersulfone membrane for flux enhancement. *Chemical Engineering Journal, 223*, 246–267.

Ahmadalipour, S., & Shakeri, A. (2025). Hydrophilization of ZIF-8 with tannic acid to enhance forward osmosis membranes performance in lithium concentration. *Journal of Molecular Liquids, 430*, Article 127666.

Altaee, A. (2012). Forward osmosis: Potential use in desalination and water reuse. *Journal of Membrane and Separation Technology, 1*(2), 79–93.

Altaee, A., Zaragoza, G., & Sharif, A. (2014). Pressure retarded osmosis for power generation and seawater desalination: Performance analysis. *Desalination, 344*, 108–115.

Arena, J. T., et al. (2011). Surface modification of thin film composite membrane support layers with polydopamine: Enabling use of reverse osmosis membranes in pressure retarded osmosis. *Journal of Membrane Science, 375*(1), 55–62.

Arjmandi, A., et al. (2020). Exploring the use of cheap natural raw materials to reduce the internal concentration polarization in thin-film composite forward osmosis membranes. *Chemical Engineering Journal, 398*, Article 125483.

Bayrami, A., et al. (2025). Amino acid-based nanoparticles-incorporated thin-film nanocomposite forward osmosis membranes for efficient desalination and heavy metal ions rejection. *Desalination, 594*, Article 118312.

Blandin, G., et al. (2015). Opportunities to reach economic sustainability in forward osmosis–reverse osmosis hybrids for seawater desalination. *Desalination, 363*, 26–36.

Blandin, G., et al. (2016). Impact of hydraulic pressure on membrane deformation and trace organic contaminants rejection in pressure assisted osmosis (PAO). *Process Safety and Environmental Protection, 102*, 316–327.

bonsale, R., et al. (2025). Tuning the performance of thin-film nanocomposite forward osmosis membrane using ordered nanoporous silica (LUS-1) with different functional groups. *Applied Surface Science, 698*, Article 163007.

Boo, C., Khalil, Y. F., & Elimelech, M. (2015). Performance evaluation of trimethylamine–carbon dioxide thermolytic draw solution for engineered osmosis. *Journal of Membrane Science, 473*, 302–309.

Bowden, K. S., Achilli, A., & Childress, A. E. (2012). Organic ionic salt draw solutions for osmotic membrane bioreactors. *Bioresource Technology, 122*, 207–216.

Bui, N.-N., et al. (2011). Electrospun nanofiber supported thin film composite membranes for engineered osmosis. *Journal of Membrane Science, 385–386*, 10–19.

Bui, N.-N., & McCutcheon, J. R. (2016). Nanoparticle-embedded nanofibers in highly permselective thin-film nanocomposite membranes for forward osmosis. *Journal of Membrane Science, 518*, 338–346.

Cath, T. Y., Childress, A. E., & Elimelech, M. (2006). Forward osmosis: Principles, applications, and recent developments. *Journal of Membrane Science, 281*(1), 70–87.

Chekli, L., et al. (2016). A comprehensive review of hybrid forward osmosis systems: Performance, applications and future prospects. *Journal of Membrane Science, 497*, 430–449.

Chung, T.-S., et al. (2015). What is next for forward osmosis (FO) and pressure retarded osmosis (PRO). *Separation and Purification Technology, 156*, 856–860.

Coday, B. D., et al. (2014a). Rejection of trace organic compounds by forward osmosis membranes: A literature review. *Environmental Science and Technology, 48*(7), 3612–3624.

Coday, B. D., et al. (2014b). The sweet spot of forward osmosis: Treatment of produced water, drilling wastewater, and other complex and difficult liquid streams. *Desalination, 333*(1), 23–35.

Cohen-Tanugi, D., & Grossman, J. C. (2014). Water permeability of nanoporous graphene at realistic pressures for reverse osmosis desalination. *The Journal of Chemical Physics, 141*(7), Article 074704.

Crowe, A. T., Davidson, M. J., & Nokes, R. I. (2016). Modified reduced buoyancy flux model for desalination discharges. *Desalination, 378*, 53–59.

Dai, Y., et al. (2022). Graphene-based membranes for water desalination: A literature review and content analysis. *Polymers, 14*(19), 4246.

Einarsson, S. J., & Wu, B. (2021). Thermal associated pressure-retarded osmosis processes for energy production: A review. *The Science of the Total Environment, 757*, Article 143731.

Elimelech, M., & Phillip, W. A. (2011). The future of seawater desalination: Energy, technology, and the environment. *Science, 333*(6043), 712–717.

Fang, X., et al. (2017). Internal pore decoration with polydopamine nanoparticle on polymeric ultrafiltration membrane for enhanced heavy metal removal. *Chemical Engineering Journal, 314*, 38–49.

Feinberg, B. J., Ramon, G. Z., & Hoek, E. M. V. (2013). Thermodynamic analysis of osmotic energy recovery at a reverse osmosis desalination plant. *Environmental Science and Technology, 47*(6), 2982–2989.

Feng, K., et al. (2015). A self-protected self-cleaning ultrafiltration membrane by using polydopamine as a free-radical scavenger. *Journal of Membrane Science, 490*, 120–128.

Fritzmann, C., et al. (2007). State-of-the-art of reverse osmosis desalination. *Desalination, 216*(1), 1–76.

Gao, W., et al. (2011). Membrane fouling control in ultrafiltration technology for drinking water production: A review. *Desalination, 272*(1), 1–8.

Gray, G. T., McCutcheon, J. R., & Elimelech, M. (2006). Internal concentration polarization in forward osmosis: Role of membrane orientation. *Desalination, 197*(1), 1–8.

Hanson, B. R., Šimůnek, J., & Hopmans, J. W. (2006). Evaluation of urea–ammonium–nitrate fertigation with drip irrigation using numerical modeling. *Agricultural Water Management, 86*(1), 102–113.

Hu, M., & Mi, B. (2013). Enabling graphene oxide nanosheets as water separation membranes. *Environmental Science and Technology, 47*(8), 3715–3723.

Huang, L., Arena, J. T., & McCutcheon, J. R. (2016). Surface modified PVDF nanofiber supported thin film composite membranes for forward osmosis. *Journal of Membrane Science, 499*, 352–360.

Johnson, D. J., et al. (2018). Osmotic's potential: An overview of draw solutes for forward osmosis. *Desalination, 434*, 100–120.

Kafkafi, U., & Kant, S. (2005). Fertigation. In D. Hillel (Ed.), *Encyclopedia of soils in the environment* (pp. 1–9). Oxford: Elsevier.

Kazwini, T., et al. (2025). Fouling mitigation for cellulose triacetate and thin-film composite forward osmosis membranes for seawater desalination by sodium metasilicate sol-gel. *The Science of the Total Environment, 1001*, Article 180525.

Kim, Y. C., & Elimelech, M. (2013a). Potential of osmotic power generation by pressure retarded osmosis using seawater as feed solution: Analysis and experiments. *Journal of Membrane Science, 429*, 330–337.

Kim, J., et al. (2013b). Reverse osmosis (RO) and pressure retarded osmosis (PRO) hybrid processes: Model-based scenario study. *Desalination, 322*, 121–130.

Kim, K.-Y., et al. (2014). Polydopamine coating effects on ultrafiltration membrane to enhance power density and mitigate biofouling of ultrafiltration microbial fuel cells (UF-MFCs). *Water Research, 54*, 62–68.

Kim, J., et al. (2015). Recent advances in osmotic energy generation via pressure-retarded osmosis (PRO): A review. *Energies, 8*(10), 11821–11845.

Kim, J., Lee, J., & Kim, J. H. (2012). Overview of pressure-retarded osmosis (PRO) process and hybrid application to sea water reverse osmosis process. *Desalination and Water Treatment, 43*(1), 193–200.

Kravath, R. E., & Davis, J. A. (1975). Desalination of sea water by direct osmosis. *Desalination, 16*(2), 151–155.

Kwon, S.-B., et al. (2024). High-performance support-free molecular layer-by-layer assembled forward osmosis membrane incorporated with graphene oxide. *Journal of Membrane Science, 689*, Article 122152.

Lee, K. L., Baker, R. W., & Lonsdale, H. K. (1981). Membranes for power generation by pressure-retarded osmosis. *Journal of Membrane Science, 8*(2), 141–171.

Lee, H., et al. (2007). Mussel-inspired surface chemistry for multifunctional coatings. *Science, 318*(5849), 426–430.

Lee, S., et al. (2010). Comparison of fouling behavior in forward osmosis (FO) and reverse osmosis (RO). *Journal of Membrane Science, 365*(1), 34–39.

Liu, Y., et al. (2025). Tuning interfacial compatibility of LDH nanoparticles and polyamide membranes for improving forward osmosis performance. *Journal of Membrane Science, 735*, Article 124567.

Liu, G., Jin, W., & Xu, N. (2015). Graphene-based membranes. *Chemical Society Reviews, 44*(15), 5016–5030.

Loeb, S. (2002). Large-scale power production by pressure-retarded osmosis, using river water and sea water passing through spiral modules. *Desalination, 143*(2), 115–122.

Mavukkandy, M. O., et al. (2019). Brine management in desalination industry: From waste to resources generation. *Desalination, 472*, Article 114187.

McCloskey, B. D., et al. (2010). Influence of polydopamine deposition conditions on pure water flux and foulant adhesion resistance of reverse osmosis, ultrafiltration, and microfiltration membranes. *Polymer, 51*(15), 3472–3485.

McCormick, P., et al. (2008). Water, salt, and ethanol diffusion through membranes for water recovery by forward (direct) osmosis processes. *Journal of Membrane Science, 325*(1), 467–478.

McCutcheon, J. R., & Elimelech, M. (2006). Influence of concentrative and dilutive internal concentration polarization on flux behavior in forward osmosis. *Journal of Membrane Science, 284*(1), 237–247.

McCutcheon, J. R., McGinnis, R. L., & Elimelech, M. (2005). A novel Ammonia—Carbon dioxide forward (direct) osmosis desalination process. *Desalination, 174*(1), 1–11.

McCutcheon, J. R., McGinnis, R. L., & Elimelech, M. (2006). Desalination by ammonia–carbon dioxide forward osmosis: Influence of draw and feed solution concentrations on process performance. *Journal of Membrane Science, 278*(1), 114–123.

McGinnis, R. L., & Elimelech, M. (2007). Energy requirements of ammonia–carbon dioxide forward osmosis desalination. *Desalination, 207*(1), 370–382.

McGinnis, R. L., et al. (2013). Pilot demonstration of the NH_3/CO_2 forward osmosis desalination process on high salinity brines. *Desalination, 312*, 67–74.

Mi, B., & Elimelech, M. (2010). Organic fouling of forward osmosis membranes: Fouling reversibility and cleaning without chemical reagents. *Journal of Membrane Science, 348*(1), 337–345.

Missimer, T. M., & Maliva, R. G. (2018). Environmental issues in seawater reverse osmosis desalination: Intakes and outfalls. *Desalination, 434*, 198–215.

Movahed, M. F., et al. (2025). Modification of forward osmosis membranes using amine-functionalized hyper-crosslinked polymer nanoparticles: The effect of incorporation strategy on performance and morphology. *Colloids and Surfaces A: Physicochemical and Engineering Aspects, 727*, Article 138324.

Musie, W., & Gonfa, G. (2023). Fresh water resource, scarcity, water salinity challenges and possible remedies: A review. *Heliyon, 9*(8).

Neff, R. A. (1964). *Solvent extractor.* Patent. U.S. Patent No. 3,130,156.

Obaid, M., et al. (2016). Amorphous SiO_2 NP-incorporated poly(vinylidene fluoride) electrospun nanofiber membrane for high flux forward osmosis desalination. *ACS Applied Materials and Interfaces, 8*(7), 4561–4574.

Ortega-Bravo, J. C., et al. (2016). Forward osmosis: Evaluation thin-film-composite membrane for municipal sewage concentration. *Chemical Engineering Journal, 306*, 531–537.

Palacin, L. G., et al. (2013). Evaluation of the recovery of osmotic energy in desalination plants by using pressure retarded osmosis. *Desalination and Water Treatment, 51*(1), 360–365.

Pan, S.-F., et al. (2017). Self-sustained hydrophilic nanofiber thin film composite forward osmosis membranes: Preparation, characterization and application for simulated antibiotic wastewater treatment. *Journal of Membrane Science, 523*, 205–215.

Panahi, M., et al. (2025). Substrate modified thin film composite reverse osmosis membrane with improved desalination performance and fouling-resistant characteristics. *Journal of Environmental Chemical Engineering, 13*(1), Article 115051.

Pandaya, D., et al. (2025). A novel forward osmosis nanocomposite membrane with g-C_3N_4 quantum dots fabricated on polyacrylonitrile electrospun nanofiber membrane for high-permeability and selectivity. *Chemical Engineering Journal, 520*, Article 165909.

Petrotos, K. B., & Lazarides, H. N. (2001). Osmotic concentration of liquid foods. *Journal of Food Engineering, 49*(2), 201–206.

Petrotos, K. B., Quantick, P., & Petropakis, H. (1998). A study of the direct osmotic concentration of tomato juice in tubular membrane – Module configuration. I. The effect of certain basic process parameters on the process performance. *Journal of Membrane Science, 150*(1), 99–110.

Phillip, W. A., Yong, J. S., & Elimelech, M. (2010). Reverse draw solute permeation in forward osmosis: Modeling and experiments. *Environmental Science and Technology, 44*(13), 5170–5176.

Phuntsho, S., et al. (2011). A novel low energy fertilizer driven forward osmosis desalination for direct fertigation: Evaluating the performance of fertilizer draw solutions. *Journal of Membrane Science, 375*(1), 172–181.

Phuntsho, S., et al. (2012). Fertiliser drawn forward osmosis desalination: The concept, performance and limitations for fertigation. *Reviews in Environmental Science and Bio/Technology, 11*, 147–168.

Popper, K., Merson, R. L., & Camirand, W. M. (1968). Desalination by osmosis-reverse osmosis couple. *Science, 159*(3821), 1364–1365.

Qian, X., et al. (2022). A critical review and commentary on recent progress of additive manufacturing and its impact on membrane technology. *Journal of Membrane Science, 645*, Article 120041.

Roy, A. (2010). The role of fertilizers in food production. In *Food security and soil quality* (pp. 109–136). CRC Press.

Saito, K., et al. (2012). Power generation with salinity gradient by pressure retarded osmosis using concentrated brine from SWRO system and treated sewage as pure water. *Desalination and Water Treatment, 41*(1), 114–121.

Shaffer, D. L., et al. (2015). Forward osmosis: Where are we now? *Desalination, 356*, 271–284.

Shao, F., et al. (2017). Layer-by-layer self-assembly TiO_2 and graphene oxide on polyamide reverse osmosis membranes with improved membrane durability. *Desalination, 423*, 21–29.

Sharma, M., et al. (2022). Clean energy from salinity gradients using pressure retarded osmosis and reverse electrodialysis: A review. *Sustainable Energy Technologies and Assessments, 49*, Article 101687.

She, Q., et al. (2013). Organic fouling in pressure retarded osmosis: Experiments, mechanisms and implications. *Journal of Membrane Science, 428*, 181–189.

She, Q., Jin, X., & Tang, C. Y. (2012). Osmotic power production from salinity gradient resource by pressure retarded osmosis: Effects of operating conditions and reverse solute diffusion. *Journal of Membrane Science, 401–402*, 262–273.

Shi, J., et al. (2019). Chitosan sub-layer binding and bridging for nanofiber-based composite forward osmosis membrane. *Applied Surface Science, 478*, 38–48.

Shokri, J., et al. (2008). Swellable elementary osmotic pump (SEOP): An effective device for delivery of poorly water-soluble drugs. *European Journal of Pharmaceutics and Biopharmaceutics, 68*(2), 289–297.

Sileika, T. S., et al. (2011). Antibacterial performance of polydopamine-modified polymer surfaces containing passive and active components. *ACS Applied Materials and Interfaces, 3*(12), 4602–4610.

Su, J., & Chung, T.-S. (2011). Sublayer structure and reflection coefficient and their effects on concentration polarization and membrane performance in FO processes. *Journal of Membrane Science, 376*(1), 214–224.

Tan, N. T., et al. (2025). Effect of fabrication methods on the performance of electrospun PVA/Chitosan FO membranes in seawater desalination. *Journal of Membrane Science, 735*, Article 124513.

Tan, C. H., & Ng, H. Y. (2013). Revised external and internal concentration polarization models to improve flux prediction in forward osmosis process. *Desalination, 309*, 125–140.

Tian, M., et al. (2017). Synthesis and characterization of thin film nanocomposite forward osmosis membranes supported by silica nanoparticle incorporated nanofibrous substrate. *Desalination, 401*, 142–150.

Touati, K., et al. (2015). Effect of the feed and draw solution temperatures on PRO performance: Theoretical and experimental study. *Desalination, 365*, 182–195.

Touati, K., et al. (2017b). *Pressure retarded osmosis: Renewable energy generation and recovery*. Academic Press.

Touati, K., & Tadeo, F. (2017a). Chapter one - Pressure retarded osmosis as renewable energy source. In K. Touati, et al. (Eds.), *Pressure retarded osmosis* (pp. 1–54). Academic Press.

Vaselbehagh, M., et al. (2014). Improved antifouling of anion-exchange membrane by polydopamine coating in electrodialysis process. *Desalination, 332*(1), 126–133.

Wahab, M. Y., et al. (2019). Synergistic effects of organic and inorganic additives in preparation of composite poly (vinylidene fluoride) antifouling ultrafiltration membranes. *Journal of Applied Polymer Science, 136*(27), Article 47737.

Wang, C.-Y., et al. (2005). Asymmetric membrane capsules for delivery of poorly water-soluble drugs by osmotic effects. *International Journal of Pharmaceutics, 297*(1), 89–97.

Wang, R., et al. (2010). Characterization of novel forward osmosis hollow fiber membranes. *Journal of Membrane Science, 355*(1), 158–167.

Wang, Y.-N., et al. (2012). Microscopic characterization of FO/PRO membranes – A comparative study of CLSM, TEM and SEM. *Environmental Science and Technology, 46*(18), 9995–10003.

Wu, W., et al. (2021a). Electrospun nanofiber based forward osmosis membrane using graphene oxide as substrate modifier for enhanced water flux and rejection performance. *Desalination, 518*, Article 115283.

Wu, X., et al. (2021b). Sewage concentration via a graphene oxide modified thin-film nanocomposite forward osmosis membrane: Enhanced performance and mitigated fouling. *Chemical Engineering Journal, 420*, Article 127718.

Xie, M., et al. (2013). Effects of feed and draw solution temperature and transmembrane temperature difference on the rejection of trace organic contaminants by forward osmosis. *Journal of Membrane Science, 438*, 57–64.

Xie, Q., et al. (2024). Tannic acid etched ZIF-8 incorporated thin-film nanocomposite nanofiltration membrane with enhanced performance. *Applied Surface Science, 660*, Article 159975.

Yang, Q., Wang, K. Y., & Chung, T.-S. (2009). A novel dual-layer forward osmosis membrane for protein enrichment and concentration. *Separation and Purification Technology, 69*(3), 269–274.

Yangali-Quintanilla, V., et al. (2011). Indirect desalination of Red Sea water with forward osmosis and low pressure reverse osmosis for water reuse. *Desalination, 280*(1), 160–166.

Yip, N. Y., & Elimelech, M. (2011). Performance limiting effects in power generation from salinity gradients by pressure retarded osmosis. *Environmental Science and Technology, 45*(23), 10273–10282.

Yip, N., & Elimelech, M. (2012). Thermodynamic and energy efficiency analysis of power generation from natural salinity gradients by pressure retarded osmosis. *Environmental Science and Technology, 46*, 5230–5239.

Yip, N. Y., & Elimelech, M. (2014). Comparison of energy efficiency and power density in pressure retarded osmosis and reverse electrodialysis. *Environmental Science and Technology, 48*(18), 11002–11012.

Yip, N. Y., et al. (2010). High performance thin-film composite forward osmosis membrane. *Environmental Science and Technology, 44*(10), 3812–3818.

Yousefian-Arani, M., Sharif, A., & Karimi, M. (2025). Modification of thin film composite membranes substrate by graphene oxide nanosheets and carbon nanotubes for forward osmosis desalination. *Journal of Molecular Liquids, 426*, Article 127485.

Yu, F., et al. (2020). High-performance forward osmosis membrane with ultra-fast water transport channel and ultra-thin polyamide layer. *Journal of Membrane Science, 616*, Article 118611.

Zeng, W., et al. (2023). Electrospun chitosan nanofiber constructing superhigh-water-flux forward osmosis membrane. *International Journal of Biological Macromolecules, 226*, 833–839.

Zhang, S., et al. (2010). Well-constructed cellulose acetate membranes for forward osmosis: Minimized internal concentration polarization with an ultra-thin selective layer. *Journal of Membrane Science, 360*(1), 522–535.

Zhang, S., Sukitpaneenit, P., & Chung, T.-S. (2014). Design of robust hollow fiber membranes with high power density for osmotic energy production. *Chemical Engineering Journal, 241*, 457–465.

Zhao, S., et al. (2012b). Recent developments in forward osmosis: Opportunities and challenges. *Journal of Membrane Science, 396*, 1–21.

Zhao, S., et al. (2024). Structure parameter as an intrinsic property of thin-film composite forward osmosis membrane. *Desalination, 581*, Article 117616.

Zhao, S., & Zou, L. (2011). Relating solution physicochemical properties to internal concentration polarization in forward osmosis. *Journal of Membrane Science, 379*(1), 459–467.

Zhao, S., Zou, L., & Mulcahy, D. (2012). Brackish water desalination by a hybrid forward osmosis–nanofiltration system using divalent draw solute. *Desalination, 284*, 175–181.

Zhou, Y., et al. (2025). Fabrication of high-performance forward osmosis membranes via co-solvent-assisted support-free interfacial polymerization. *Separation and Purification Technology, 379*, Article 134974.

https://news.lockheedmartin.com/2013-03-18-Lockheed-Martin-Achieves-Patent-For-Perforene-TM-Filtration-Solution-Moves-Closer-To-Affordable-Water-Desalination. 2013 [cited 2025 20/09]

PART

III

Breakthroughs in membrane science for water treatment

CHAPTER

4 Advanced antifouling and biodegradable membrane materials: Innovations in material science and fabrication technologies

K.V.V. Satyannarayana[1], Randeep Singh[2], Ch. Ramya[3], D.U.S.L. Deepthi[4], Nisha Malik[5] and Young-Ho Ahn[1]

[1]*Department of Civil Engineering, Yeungnam University, Gyeongsan, South Korea;* [2]*Department of Basic Sciences, School of Sciences and Humanities, SR University, Warangal, Telangana, India;* [3]*Department of Chemistry, Aditya Institute of Technology and Management (A), Srikakulam, Andhra Pradesh, India;* [4]*Department of Chemical Engineering, MVGR College of Engineering, Vizianagaram, Andhra Pradesh, India;* [5]*Membrane Technology Group, Faculty of Bioscience Engineering, KU Leuven, Leuven, Belgium*

1. Introduction

The world's water security policies have become more and more dependent on the exponential growth of membrane technology for desalination, effluent reuse, and heavy metal removal. Many of the large-scale desalination and industrial wastewater treatment facilities and processes are now using pressure-driven membrane processes such as forward osmosis (FO), ultrafiltration (UF), nanofiltration (NF), and reverse osmosis (RO) because of their superior rejection properties and energy efficiency. Nanocomposite and thin-film composite membranes are increasingly being used in industrial applications because of their mechanical strength and scalability, as reported in recent studies on the development of membrane technologies (Laftah & Wan Abdul Rahman, 2025). However, there is an unexplored environmental consequence hidden beneath this success story: the accumulation of nonbiodegradable polymeric membranes derived from materials extracted from fossil fuels, such as polyvinylidene fluoride (PVDF), polyethersulfone (PES), and polysulfone (PSf).

These polymers are heat-resistant, chemically inert, and hydrolytically stable, as thoroughly verified. One of its surprisingly high-quality properties that ensure operational longevity is its resistance to environmental damage. Upon completion of their function as membrane modules, they will inevitably contribute to the rising tide of microplastics and the mountains of solid polymer waste. "Membrane Paradox" is defined by this apparent paradox: while we have succeeded in creating membranes that "last forever," we have also added a new environmental burden to the planet. In our efforts to reduce water pollution, we are inadvertently increasing polymer pollution. A circular

Sustainable Water Management. https://doi.org/10.1016/B978-0-443-44853-9.00016-2

bioeconomy relies on the transition away from fossil fuels and toward renewable energy sources of carbon (Iroegbu & Ray, 2023).

Conventional plastics are to be distinguished by their high molecular weights, resistance to microbial degradation, and the ability to accumulate in terrestrial and marine ecosystems, as well as the ability to impede natural biodegradation pathways (Wu et al., 2024). Therefore, reimagining materials to align with ecological cycles is the next step in membrane research, following closely on the heels of enhancing permeability and selectivity. Moving from "inert and stable" materials to "active and transient" systems is the appropriate transition.

The concept of tunable degradability, which pertains to membranes engineered to maintain hydrolytic stability during operation but intended for enzymatic or compost-driven degradation at the end of their life, is further developed in this chapter. A fundamental change like this calls for new approaches to manufacturing, surface functionalization, morphological control, and also polymer chemistry. The challenge is in achieving a compromise between the three traditionally opposing criteria of biodegradability, antifouling effectiveness, and durability.

Among renewable polymer candidates, polyhydroxyalkanoates (PHAs) and polylactic acid (PLA) have arisen as the principal green backbone. Despite being made from a fermentative lactic acid, PLA has historically been limited due to its brittleness and low temperature tolerance, even though it is compostable and may be scaled up in the industrial sector (Nagarajan et al., 2016). Recent advances in stereo complexation, using PLLA and PDLA chains, have completely transformed PLA's structural performance. Creating a full stereo complex (sc-PLA) in a short amount of time in water can increase the melting point to around 220–230 °C, significantly improve hydrolysis resistance, and keep the strength at or above 78% (Sun et al., 2022). This new research has the potential to prove that fossil fuels may be used to make a biodegradable polymer with properties similar to plastic, including increased durability.

Research on the PLA's impact resistance and crystallization kinetics has progressed in tandem. Adding nucleating chemicals and optimizing mold temperatures drastically reduced crystallization half-time, while increasing heat deflection temperature and impact strength (Nagarajan et al., 2015). Morphological control is required of membrane modules subjected to a cyclic hydraulic stress. The importance of maintaining the renewable carbon content while striking a balance between the stiffness and toughness can be better understood through additional research into long-lasting bioplastics (Nagarajan et al., 2016). Collectively, these studies can provide evidence that biodegradable polymers can be engineered to perform at an engineering-grade level.

Complementing PLA, PHAs represent a bacterially synthesized polyester with an intrinsic biodegradability and tunable mechanical properties (Wu et al., 2024). The biosynthesis pathway governed by the phbCAB operon enables the controlled production of PHB and PHB/PHV copolymers through metabolic engineering regarding (Wu et al., 2024). Biomedical evaluations confirm that excellent biocompatibility and controlled degradation behavior in vivo, reinforcing their suitability for long-term aqueous exposure (Roy & Ragunath, 2018). While high crystallinity in PHB may induce brittleness, blending strategies can also improve toughness, expanding the application potential of filtration substrates.

Yet the material substitute alone is insufficient. Membrane fouling remains a primary limitation in desalination and wastewater treatment systems. Fouling progresses from the conditioning film formation to biofilm maturation, increasing the transmembrane pressure and energy demand (Laftah & Wan Abdul Rahman, 2025). Historically, antifouling strategies have relied on toxic biocides such as tributyltin and copper oxides, which cause ecological harm and are increasingly regulated. Contemporary polymer-based antifouling approaches emphasize nontoxic mechanisms, including hydration-layer formation, and also low-surface-energy interfaces (Song & Feng, 2020).

Zwitterionic functionalization has emerged as the "gold standard" for antifouling surfaces. By grafting zwitterionic motifs onto polymer backbones, a strongly bound hydration shell forms via ion–dipole interactions, effectively repelling proteins and also microbial cells (Shafi et al., 2017). Unlike the PEG-based steric repulsion, zwitterionic systems offer superior long-term stability and also reduce oxidative degradation. Such strategies are particularly important for biobased membranes, which often exhibit intrinsic hydrophobicity.

In addition to the surface chemistry, microstructural integrity plays a decisive role in membrane durability. Studies of marine organic coatings reveal that the microdefects facilitate the permeation of corrosive species and accelerate degradation (Song & Feng, 2020). Thus, biodegradable membranes must be fabricated with strict defect control to ensure that performance parity with the conventional materials.

Sustainable fabrication of processes is equally critical. The production of "green" membranes that use toxic solvents such as NMP or DMF undermines the sustainability of objectives. Recent efforts are focused on replacing these solvents with greener alternatives and also developing solvent-free techniques. Melt-processing compatible with the stereocomplex PLA demonstrates improved hydrolytic stability without a solvent-intensive route (Morales-Jiménez et al., 2023; Sun et al., 2022). Emerging research in sustainable membrane fabrication between the period of 2020 and 2025 emphasizes integrating material innovation with environmentally benign manufacturing pathways (Kim & Nunes, 2021; Laftah & Wan Abdul Rahman, 2025).

Collectively, these advances are forming the foundation of a Circular Membrane Economy: renewable biosourcing, green fabrication, antifouling operation, and also enzymatic degradation. The integration of sc-PLA, PHAs, zwitterionic surface engineering, and solvent-free processing signals a paradigm shift from the membrane stability as the ultimate metric toward membrane sustainability as the defining objective form. To ensure that the water security does not compromise the sustainability of the environment, membrane science has to incorporate high-performance antifouling mechanisms with configurable biodegradability.

This chapter discusses the "Membrane Paradox" arising from the widespread use of nonbiodegradable fossil-based membranes in water purification. Furthermore, the potential of next-generation biopolymers, particularly in stereo complex PLA (sc-PLA) and also PHAs as a sustainable membrane backbone, is to be examined in detail. Consequently, advanced antifouling strategies will include zwitterionic functionalization, and bioinspired surface topographies are highlighted. Finally, this chapter emphasizes green solvents and also solvent-free fabrication techniques to promote the development of a Circular Membrane Economy.

2. The green backbone: Next-generation biodegradable polymers

The quest for green membrane filter materials has advanced from simply replacing cellulose acetate with a petrochemical alternative. Although cellulose acetate provided the first proof of concept, it has significant limitations with regard to thermal stability, chemical resistance, and mechanical properties that have been major drivers toward an expanded selection of biodegradable polymers. This section addresses three of the most important classes of these new materials: polylactic acid (PLA) stereocomplexes, polyhydroxyalkanoate (PHA) bacterial polyesters, and reinforced green composites incorporating nanocellulose and chitin nanowhisker.

2.1 Polylactic acid stereocomplexes

2.1.1 Overcoming the brittleness and low thermal stability of PLA

Polylactic acid has become the first commercially viable bioplastic, made on a large scale by fermenting corn starch and sugarcane. PLA can be derived from a renewable source, is compostable in an industrial setting, and may be processed through all common methods of plastic processing (Zheng et al., 2022). However, PLA does have some limitations that affect its ability to operate in high-performance filtration environments. The glass transition temperature (T_g) of PLA is approximately 55–60°C, and the melting temperature (T_m) is approximately 170°C. In membrane filtration applications, where water sterilization temperatures are frequently raised to 80–121°C during cleaning cycles, and where harsh solvents are used for cleaning, the aforementioned values represent a significant operational limitation (Macnamara et al., 2024). Another major issue with PLA membranes is their brittle nature, with elongations at break typically exceeding 5%–10%, resulting in broken membranes due to mechanical stress imparted by the installation of modules, the cyclic pressure of hydraulic systems, and the backwash process.

2.1.2 Innovations in stereocomplexation for enhanced solvent resistance

Stereocomplex (sc-PLA) is created by equimolar mixtures of the enantiomeric poly(L-lactide) (PLLA) and poly(D-Lactide) (PDLA) under the same thermal history conditions, which form racemic crystallites. The racemic crystallites are packed through van der Waals forces between the methyl group of one PLLA helix and the methyl group of an adjacent PDLA helix. The transition of PLLA/PDLA chains to the stereocomplex structure is a transformation process (Alhaj and Narayan, 2023). In comparison to homo-PLLA crystals, the melting temperature of the stereocomplex crystallites is increased by 50–60°C–230°C. This increase in the melting point is due to the improved crystalline perfection and tighter chain packing of the heterochiral stereocomplex lattice. In addition, the rate of hydrolytic degradation of the stereocomplex under near-neutral pH aqueous conditions is significantly less than that observed for either of the individual enantiomers. This decrease in the hydrolysis rate is an important consideration for membrane materials that require a long period of time to degrade in a controlled fashion after they are no longer required for their intended function. The recent development of novel strategies for stereocomplex formation has led to the realization of additional enhancements in the properties of the resulting material. For example, melt blending of PLLA and PDLA under controlled thermal histories (above the homocrystal melting point but below the stereocomplex melting point) preferentially retains the stereocomplex crystallites. These retained crystallites can then be subjected to secondary processing steps such as electrospinning to produce fibers with finer and more uniform microstructures. A typical example of this strategy was the application of a scalable continuous manufacturing process that utilized corotating twin-screw extrusion to produce sc-PLA with a stereocomplex content of more than 95%, which led to an enhancement of Tm to 240°C, as well as a marked improvement in tensile properties over the two parent homopolymers.

Also, the electrospinning process of sc-PLA fiber membranes has been reported to provide excellent results, with high stereocomplex crystallinity indices of more than 90%, and associated enhancements in hot water dimensional stability, making these materials of interest for the preparation of thermally demanding microfiltration membranes (Kara et al., 2022). From the filtration perspective, the enhanced solvent resistance of sc-PLA presents new opportunities. Conventional PLLA membranes are known to rapidly swell and plasticize in typical organic solvents such as ketones, esters, and aromatics, which are commonly encountered in pharmaceutical and fine chemical process streams. In contrast, sc-PLA membranes show significantly lower solvent absorption, mainly because of the physically compact crystalline structure, which serves as a barrier to solvent diffusion.

2.2 Polyhydroxyalkanoates and PHB

2.2.1 The rise of bacterial polyesters

PHAs are produced within bacterial cells cultured under low-nutrient conditions and exposed to a carbon-rich feedstock, resulting in the formation of intracellular, discrete PHA granules. When bacterial cells are lysed, and the PHA is extracted, it can be converted into a family of thermoplastics of polyester type that vary in physical property from rigid, crystalline solids to elastic rubber-like material based upon the chemical structure of the alkyl side chains attached to the polymer backbone (Kaniuk & Stachewicz, 2021). As a model compound, the first PHA of interest, poly(3-hydroxybutyrate) (PHB), has a high degree of crystallinity, generally greater than 60%, and exhibits mechanical properties that are superficially similar to isotactic polypropylene, with a tensile modulus of approximately 3.5 GPa and a melting temperature near 175°C (Zhang et al., 2024). However, the major limitation of PHB is that it has a very small processing window because it undergoes thermal degradation just above its melting point, is brittle due to secondary crystallization at room temperature, and may exhibit an elongation at break of less than 5%, which does not meet the flexible requirements necessary for use in spiral-wound or hollow-fiber membrane modules.

2.2.2 PHB/PHBV blends for improved mechanical toughness in filtration applications

The incorporation of 3-hydroxyvalerate (HV) comonomers into the PHB polymer chain, forming poly (3-hydroxybutyrate-co-3-hydroxyvalerate) (PHBV), is the primary industrial method used to expand the PHA property envelope (Feijoo et al., 2022). As HV content varies from 0 to approximately 25 mol %, crystalline order in the polymer decreases, the melting temperature drops from about 175° C to approximately 140°C, and perhaps most importantly, the elongation at break rises from >5% to <50%. Mixing PHB with PHBV at weight ratios ranging from 75:25 to 50:50 produces a favorable balance of stiffness and toughness that can be achieved using neither pure polymer separately. These mixtures are partially immiscible in the melt state, followed by phase separation upon cooling. As such, a microstructural arrangement where the PHBV-rich phases act as crack arrest sites, within the stiffer PHB matrix, is developed through a micromechanical toughening process similar to that of rubber-toughened epoxies but completely biodegradable.

Membrane formation of PHB/PHBV solutions through phase inversion has resulted in asymmetric membranes having the desired macrovoid underlayer structure necessary for effective pressure-driven filtration (Feijoo et al., 2022). A detailed analysis of PHBV-based microfiltration membranes generated through both evaporation-induced phase separation (EIPS) and non–solvent-induced phase separation (NIPS) showed the pore size could be varied over the microfiltration and ultrafiltration ranges, while the permeability varied from 50 to 1300 L m^{-2} h^{-1} bar^{-1} based on the solution parameters.

2.3 Green composites

Even the most optimized biopolymer matrices, sc-PLA, PHBV blends, and thermoplastic starch derivatives, struggle to match the specific stiffness and tensile strength of polyethersulfone or polypropylene membranes without additional reinforcement. The introduction of nanostructured bioderived fillers addresses this performance gap while preserving end-of-life biodegradability (Wohlert et al., 2023).

2.3.1 Reinforcing biopolymers with nanocellulose and chitin nanowhiskers to match the strength of fossil-based plastics

The production of cellulose nanocrystals (CNCs) involves the selective acidic hydrolysis of amorphous regions of native cellulose fibers, releasing stiff, needle-shaped crystalline domains having a width of 5–20 nm, an axial elastic modulus of approximately 100–150 GPa, and aspect ratios ranging from 10 to 50, which approaches the theoretical maximum of crystalline cellulose (Yue et al., 2022). When generated through mechanical defibrillation (with or without TEMPO oxidation), cellulose nanofibers (CNFs) also contain both crystalline and amorphous segments, and when formed, they create highly entangled networks possessing exceptionally large specific surface areas, often $>100\ m^2\ g^{-1}$. An overview of the hierarchical arrangement and primary chemical structure of cellulose is shown in Fig. 4.1.

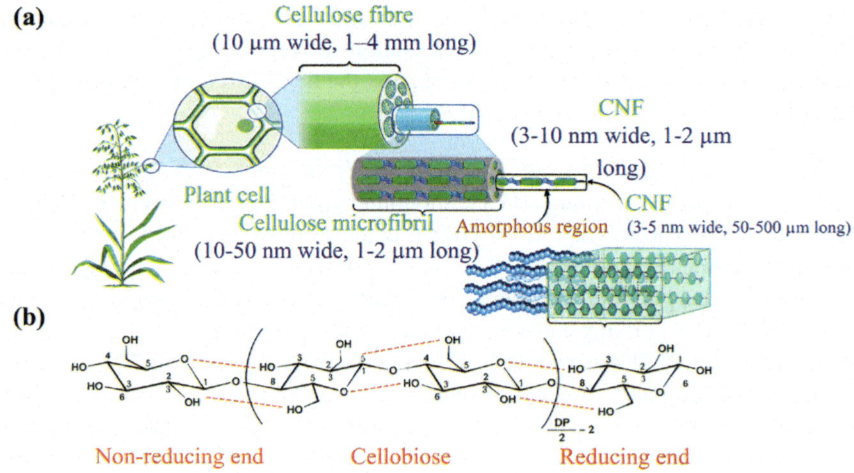

FIGURE 4.1

A diagrammatic overview of the hierarchical arrangement and primary chemical structure of cellulose.

(Adapted from (Casanova et al., 2021), licensed under CC BY 4.0. https://creativecommons.org/licenses/by/4.0/).

Both CNCs and CNFs, at loading levels of 1–10 wt%, have been shown to significantly improve the mechanical properties of the biopolymer matrices. The reinforcement mechanisms of CNC composites occur primarily due to stress transfer across the well-dispersed crystalline rods (Rana et al., 2025). The formation of a percolated network occurs once a critical filler volume has been reached, and this can be related to the aspect ratio of the particles in three dimensions via $V_{th} \sim 0.7/(L/d)$, indicating that CNCs with an aspect ratio of $L/d = 12$ will percolate at a loading level of approximately 9 wt%. At these high loading levels, the CNC network is responsible for carrying most of the applied load, resulting in dramatic increases in modulus of up to 200% in low modulus biopolymer host materials such as PHBV. In addition to the improved bulk mechanical properties for membranes used for filtration applications, the implications extend far beyond that. Due to their inherent high surface hydroxyl content, CNCs and CNFs are inherently hydrophilic; their incorporation into otherwise hydrophobic biopolymer matrices would result in a shift in the membrane surface energy toward more favorable wetting behavior, a second antifouling advantage that would complement the various surface modification methods.

Chitin nanowhiskers (ChNWs) are created through the process of partial deacetylation and acid hydrolysis of chitin, which was recovered primarily from crustacean processing waste. Because of their aspect ratios of 10–20 and surface acetamide groups, axial modulus of approximately 150 GPa, ChNWs could offer the potential to match or surpass the reinforcing efficiency of CNC on a per mass basis due to their surface acetamide groups which allow them to have a natural affinity toward polar biopolymer backbones resulting in interfacial adhesion that can potentially be better than what is obtained when using surface hydroxyl groups on CNCs (Iroegbu & Ray, 2023). The rationale for combining CNC (or CNF) with ChNW's in hybrid composite membranes is mechanistic complementarity. In this regard, the rigid short-range stiffening of CNCs can be used in conjunction with toughening of the membrane by plastic deformation and crack energy absorption provided by ChNWs (Panackal Shibu et al., 2025). In addition, ternary systems consisting of biopolymer matrix/CNC/ChNWs at optimized filler ratios can achieve tensile strengths and elongation at breaks that are very similar to those of commercial polyethersulfones membranes yet still remain fully compostable under ISO 14855 conditions.

3. Defeating the biofilm: Surface engineering for biobased substrates

The engineering challenge confronting any filtration membrane is not only to separate target species from their surrounding fluid matrix but also to sustain that separation function across operational lifetimes measured in months and years without performance-robbing accumulation of biological and organic matter on and within the membrane surface. Biofouling is the colonization of membrane surfaces by microorganisms, followed by the development of EPS matrices forming structured biofilm layers. It is the most common cause of irreversible flux decline in membrane systems. Biofouling adds 30%–50% to the costs associated with membrane system operation through increased energy usage for pumping, cleaning chemicals required to remove foulants, and premature membrane replacement. For membranes based upon biopolymers such as PLA, PHA, and nanocomposites thereof, the fouling problem is complicated by the surface chemistry of these materials. Plasma proteins and bacterial adhesins bind readily to the carbonyl groups, ester linkages, and hydrophobic backbone segments of PLA and PHB. Surface modification techniques are used to develop fouling-resistant membranes that will still be fully biodegradable at the end of their useful life. Three complementary surface modification strategies that have been developed are zwitterionic functionalization, bioinspired topographic patterning, and sacrificial erosion layers.

3.1 Zwitterionic functionalization

The reason zwitterionic polymers are more resistant to fouling than any other type of surface chemistry has been extensively studied for nearly 2 decades. The main reason zwitterionic resist fouling is due to the electrostatically balanced surface charge architecture created by the zwitterionic monomer itself. Each zwitterionic monomer contains one positively charged functional group (usually a quaternary ammonium) and one negatively charged functional group (sulfonate, carboxylate, or phosphate). These two functional groups are tied together by a short tether. When there is no net charge on a polymer chain, it will still exhibit a high local charge density.

In addition to creating this high local charge density, the presence of these two functional groups also creates an extremely dense hydration shell through the electrostatically driven ion–dipole interaction with surrounding water molecules. Because the hydration shell formed around zwitterionic polymers is based upon electrostatic interactions between the water molecules and the zwitterionic moieties, it is far stronger and more ordered than the van der Waals-mediated hydration shell found on nonzwitterionic hydrophilic polymers, like PEG. In fact, the hydration shell around zwitterionic polymers can be so strong that when proteins or bacteria approach the surface of a zwitterionic polymer, they must expend energy to partially desolvate themselves to adsorb onto the surface (He et al., 2025; Lau & Yong, 2021). This makes it very difficult for proteins and bacterial cell wall components to adsorb onto zwitterionic polymer surfaces. Quartz crystal microbalance measurements typically show protein adsorption levels of less than 5 ng cm^{-2} on optimized zwitterionic surfaces. One of the primary distinctions separating zwitterionic antifouling coatings from their predecessors (such as polyethylene glycol (PEG) coatings) is the mechanism used to prevent fouling. While PEG-coated surfaces function primarily through steric repulsion, where the conformational entropy of the grafted polyether chains generates an osmotic pressure barrier preventing approaching macromolecules from adsorbing, the mechanism used by zwitterionic coatings does not rely on the conformational entropy of the chains and therefore is not degraded under oxidative stress (as would occur in ozone- or chlorine-treated water streams). Moreover, PEG-coated surfaces degrade quickly if subjected to oxidative stress, since PEG-coating layers are easily broken down by oxidative chain scission (Wang et al., 2025). Therefore, while PEG-coated surfaces are functional only when the grafted chains are highly mobile, zwitterionic-coated surfaces continue to function even when the grafted chains are immobilized due to high grafting density.

Surface-initiated atom-transfer radical polymerization (SI-ATRP) of zwitterionic polymers such as sulfobetaine methacrylate (SBMA) or carboxybetaine methacrylate (CBMA), initiated from initiator-functionalized PLA or cellulose surfaces, is the most accurate method to create uniform zwitterionic brush layers with high grafting density and well-controlled molecular weights. Initiators can be immobilized on PLA surfaces by reacting aminolytically installed aldehyde functionalities on the surface with bromine-containing ATRP initiators. Cellulose-based membranes that include CNC or CNF have high concentrations of surface hydroxyl groups, which can be used in an esterification reaction with 2-bromoisobutyryl bromide to attach the initiator to the surface (Shalumon et al., 2009). This provides a better initiator density than with PLA and results in higher-density zwitterionic brushes, resulting in $>95\%$ reduction in bovine serum albumin (BSA) adsorption onto these surfaces.

UV-initiated graft polymerization of benzophenone-initiated SBMA or 2-methacryloyloxyethyl phosphorylcholine (MPC) monomers onto cellulose-based membrane surfaces is an attractive option for scaling up membrane fabrication due to its technical simplicity and compatibility with roll-to-roll processes. When optimized with respect to UV fluence and monomer concentration, this process produces sufficient zwitterionic surface coverage to achieve substantial fouling resistance. A single-step initiator-free method to modify hydroxylated cellulose surfaces with zwitterionic polymers involves cerium (IV)-initiated radical grafting of the zwitterionic monomers onto the surface (Shalumon et al., 2009). While this process allows for surface modification without the need for an initiator, it also leads to wider distributions in grafting density compared to other methods.

3.2 Bioinspired surface topographies

Surface topography can impact the dynamics of bacterial adhesion in ways both significant and frequently unrecognized. Sharklet, a unique diamond-shaped microfeature array designed to inhibit fouling, created by mimicking the denticles on the skin of sharks, is the most studied form of physical antifouling topography for biomedical and filtration applications (Arango-Santander, 2022; Choudhury et al., 2018). The Sharklet design consists of elongated diamonds that are typically 2 μm in width and depth, and spaced 2 μm apart, thus creating an energy landscape that discourages the first reversible step of biofilm formation. An experimental study of marine biofouling demonstrated that Sharklet-inspired patterns on polymeric materials can control early colonization patterns, but the amount of change will depend upon the relationship of the dimensions of the Sharklet features to the dimensions of the bacteria (Oopath et al., 2023).

The primary reason why Sharklet's antifouling properties work so well is because of its very skewed surface height distribution, i.e., the peaks of features produce adhesive contact over extremely small projected areas, while the channels formed by the bases of these peaks cannot be accessed by bacteria due to their inability to deform to fit in the submicron radius curvature found at the bottom of these channels (Védie et al., 2022). Up to a 55% decrease in surface area covered by *Escherichia coli* was observed, along with a 76% decrease in average colony size when compared to smooth films at 24 h, for Sharklet AF^{TM} patterned surfaces without any chemical biocides.

Sharklet^{TM} and similar hierarchical topographies can be molded into biodegradable polymer sheets using hot embossing and/or solvent-assisted micromolding against silicon master molds. As both PLA and PHBV exhibit thermoplastic behavior and relatively low glass transitions, they are suitable candidates for thermal nanoimprint lithography, which has resulted in greater than 95% replication fidelity for features at the 1 μm scale. For filtration membranes, micron-scale embossing features are limited to the feed side surface, as the other surface is used as the barrier layer and cannot be embossed.

While this limitation does present some practical constraints to applying topographic antifouling technology to filtration membranes, the feed-side surface remains a viable location for topographic antifouling (Arango-Santander, 2022). This is because the first layer of the conditioning film, a sub-nanometer thick layer of organics that forms immediately after a membrane comes into contact with water, will form less densely on topographically patterned surfaces due to the reduced area of available contact, resulting in a longer time period before cleaning is required and ultimately, a lower need for chemical cleaning agents.

3.3 “Sacrificial” antifouling layers

One of the most distinctively conceptual approaches to biodegradable membrane antifouling utilizes the same characteristics that make the substrate degradable; that is, rather than resisting the degradative characteristics of the substrate, they exploit them. Conventional antifouling paradigms are based on prevention. Therefore, the sacrificial layer antifouling strategy reverses this concept. Instead of preventing the accumulation of fouling substances, the sacrificial layer accumulates fouling substances along with itself and then sheds, taking all of the fouling substances attached to it with it, leaving the fouling free from the surface and exposing the cleaned membrane surface underneath. This dynamic antifouling strategy has been termed surface antifouling and has been articulated in the literature as a contrasting strategy to static repulsive strategies.

Engineering design of sacrificial layers is complex. The sacrificial layer must degrade at a rate that corresponds to the rate of fouling substance accumulation in the application-specific conditions (Wang et al., 2023). That is, the sacrificial layer must degrade quickly enough to inhibit the formation of biofilm and the development of an extracellular polymer matrix (EPS), which makes biofilm removal increasingly difficult, but slowly enough to ensure structural integrity during a period of time that is economically viable for service. In general, for membrane filtration processes in moderately fouling aqueous systems, surface erosion rates of 10–100 nm/day have been found to satisfy these requirements. These rates correspond to a sacrificial layer thickness of a few microns and a service life of weeks to months until the sacrificial layer completely erodes.

One study demonstrated this concept through the use of a coating–decoating–recoating cycle with carboxymethyl chitosan (CMCS) as the biodegradable sacrificial protective layer on nanofiltration membranes (Li et al., 2022). The CMCS layer increased the hydrophilicity of the membrane surface and enhanced fouling resistance against natural organic matter, while the fouled layer was easily and completely removed from the membrane surface by mild acid treatment and was regenerated in situ. Water flux recoveries greater than 88.8% were maintained after each of four sequential coating–decoating cycles, thereby demonstrating sustained long-term performance.

Polyanhydride surface eroders in particular, poly(sebacic acid) and its copolymers with lauric acid or ricinoleic acid, provide alternatives for sacrificial antifouling materials because their degradation occurs only at the surface (i.e., unlike bulk-eroding polyester polymers, such as PLA) and can be controlled by the composition of the copolymer (Wang et al., 2023). Additionally, the degradation products of sebacic acid and fatty acids are both endogenously produced human metabolites and are designated as Generally Recognized as Safe (GRAS) food additives. Thus, there are no toxicological concerns associated with the degradation of these materials and subsequent introduction of the resulting degradation products into the treated water stream.

Designing the relationship between the sacrificial layer’s degradation rate and the rate of degradation of the underlaying biopolymer membrane is another developing aspect of designing for precision. Ideally, a multilayered structure consisting of a sacrificial outer layer/antifouling midlayer/structural-selective membrane/porous support can be engineered such that each layer degrades on timescales of weeks to months to years, thus providing a membrane system capable of renewing its antifouling properties over multiple operational cycles before the last structural layer has reached its end-of-service life and undergoes complete biodegradation. Fig. 4.2 shows the “steric repulsion” (PEG-based) versus “ion-dipole hydration” (Zwitterionic) specially over a biopolymer matrix.

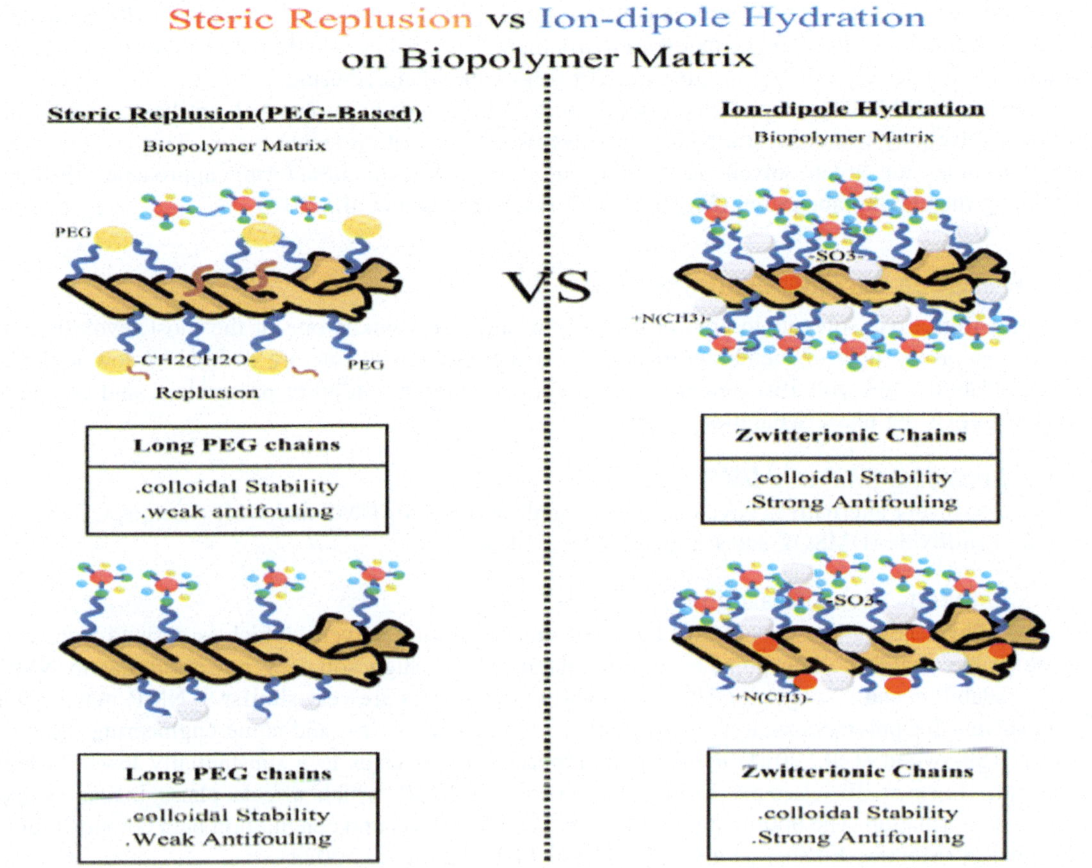

FIGURE 4.2

Mechanism diagram comparing "steric repulsion" (PEG-based) versus "ion-dipole hydration" (Zwitterionic) specifically on a biopolymer matrix.

4. Sustainable fabrication: The end of toxic solvents

The development of advanced antifouling and biodegradable membrane materials is inherently linked to the principles of sustainability. However, a paradox persists in membrane science: many so-called "green" membranes are fabricated using toxic, petrochemical-derived solvents such as N-methyl-2-pyrrolidone (NMP) and N,N-dimethylformamide (DMF) (Figoli et al., 2014; Purkait & Singh, 2018). While the final membrane may be biodegradable or environmentally benign in use, its fabrication often relies on hazardous chemicals that pose risks to human health, ecosystems, and manufacturing workers.

NMP and DMF are powerful aprotic solvents widely used in phase-inversion processes due to their strong solvation capacity for polymers such as PSf, polyethersulfone (PES), PVDF, and, increasingly, biodegradable polymers. However, both solvents are classified as reproductive toxins under European REACH regulations. Their high toxicity, environmental persistence, and energy-intensive recovery systems challenge the claim of sustainability in membrane manufacturing.

Thus, sustainable fabrication demands a paradigm shift, from hazardous solvent-based processing to green solvent systems and, ultimately, to solvent-free fabrication technologies. This section critically discusses the green solvent revolution and solvent-free manufacturing approaches that are redefining membrane fabrication (Figoli et al., 2014; Sherwood et al., 2014).

4.1 The green solvent revolution

The shift from toxic solvents to environmentally friendly solvents is one of the most revolutionary innovations in the field of membrane material science. Green solvents are defined by low toxicity, biodegradability, low volatility, renewability, favorable Hansen solubility parameters, and compatibility with existing phase separation methods.

4.1.1 Replacing NMP and DMF

Several promising alternatives have emerged as replacements for NMP and DMF, such as Cyrene™, dimethyl sulfoxide (DMSO), and Rhodiasolv PolarClean.

4.1.1.1 Cyrene (Dihydrolevoglucosenone)

Cyrene (dihydrolevoglucosenone) is a biobased dipolar aprotic solvent, which is derived from cellulose. It has a high boiling point of approximately 227°C, high solvency power similar to NMP, biodegradability, and low toxicity. Hansen solubility parameters are very similar to NMP, making it a good solvent for polymers such as PLA, PHA, cellulose derivatives, and some engineering thermoplastics (Sherwood et al., 2014). Notably, its biobased origin leads to a substantially lower carbon footprint. However, its viscosity is higher than that of NMP, which affects phase inversion and membrane formation. The optimization of polymer concentration and coagulation bath composition is now critical (Morales-Jiménez et al., 2023; Naziri Mehrabani et al., 2022).

4.1.1.2 Dimethyl sulfoxide

Dimethyl sulfoxide (DMSO) is a polar aprotic solvent with a boiling point of approximately 189°C, low vapor pressure, relatively low toxicity, and high dielectric constant. DMSO has been used in the pharmaceutical and biomedical fields for many years, signifying acceptable biocompatibility. In membrane preparation, DMSO has been shown to be useful for dissolving biodegradable polymers such as PLA and PHA, as well as cellulose acetate (Russo et al., 2020; Venault et al., 2022). Its high affinity for water makes it significantly affect the exchange of solvent and nonsolvent in non–solvent-induced phase separation (NIPS).

4.1.1.3 Rhodiasolv PolarClean

Rhodiasolv PolarClean is also known as methyl-5-(dimethylamino)-2-methyl-5-oxopentanoate, which is a biodegradable solvent with a boiling point of approximately 240°C, low toxicity, a high flash point, and high solubility of polymers. It has shown promising results in the preparation of PVDF and PES membranes with controlled porosity and mechanical strength. The low volatility of the solvent prevents worker exposure and energy requirements for solvent recovery (Marino et al., 2018).

4.1.2 Thermodynamics of phase separation in green solvent systems

The performance of membranes fabricated via phase inversion is governed by the thermodynamics and kinetics of phase separation. Transitioning to green solvents fundamentally alters the phase diagram.

4.1.2.1 Non–solvent-induced phase separation

In NIPS, a thin film of polymer solution is cast and then submerged in a bath of nonsolvents, typically water, triggering fast solvent/nonsolvent exchange. The polymer-rich region solidifies to create the membrane matrix, and the polymer-poor region is removed to create the pore structure (Singh and Purkait, 2016, 2017).

Green solvents have impact on four interrelated parameters that are thermodynamic and transport in nature: (1) the interaction parameter between the polymer and solvent (χ), which controls the scale of the miscibility gap; (2) the affinity of the solvent and nonsolvent, which controls the rate of exchange in the cast film; (3) the viscosity and diffusivity of the solution, which control mass transport during demixing; and (4) the importance of spinodal versus binodal decomposition (as shown in Fig. 4.3), which determines the final membrane morphology, either finger-like macrovoids or sponge-like isotropic structures (Mihir & Purkait, 2018; Nunes & Inoue, 1996).

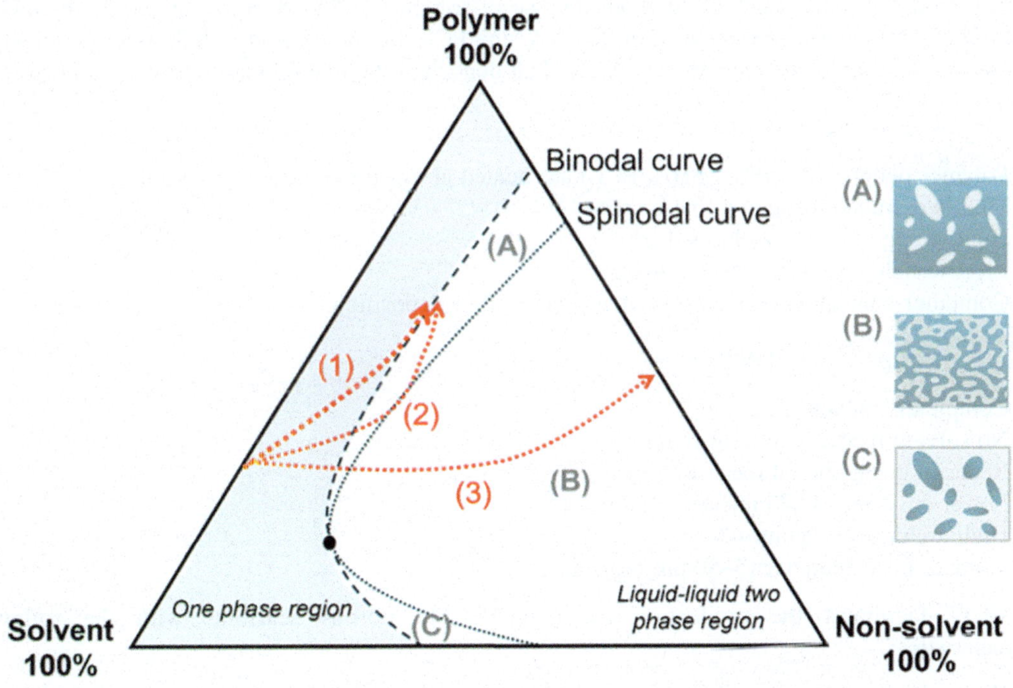

FIGURE 4.3

Concentration gradients and phase separation mechanisms in a ternary system: (1) delayed and (2) instantaneous liquid-liquid (L–L) demixing routes; (3) spinodal decomposition. The diagram identifies specific structural evolutions: (A) nucleation-growth of polymer-deficient domains, (B) spinodal-derived bicontinuous morphologies, and (C) nucleation-growth of polymer-concentrated domains.

(Adapted from (Duarte et al., 2025), licensed under CC BY 4.0. https://creativecommons.org/licenses/by/4.0/).

These conditions correspond to very different structural morphologies based on the solvent choice. DMSO water-based systems have high solvent-solvent interactions, which result in fast demixing and the characteristic finger-like macrovoids typical of instantaneous liquid-liquid phase separation. Polar clean-based systems, on the other hand, tend toward delayed demixing, slowing down the exchange front and resulting in more dense sponge-like structures with smaller pore size distributions (Singh et al., 2019). Cyrene, with its higher viscosity compared to traditional solvents, slows down diffusion across the newly formed membrane film, suppressing the formation of macrovoids and enhancing structural homogeneity, which is highly desirable in applications where robust ultrafiltration membranes are needed (Morales-Jiménez et al., 2023; Naziri Mehrabani et al., 2022).

4.2 Solvent-free techniques

Although green solvents are more environmentally friendly, the most environmentally friendly approach is to avoid solvents altogether. Solventless fabrication is the next generation of biodegradable membrane design.

4.2.1 Melt-electrospinning writing

Melt-electrospinning writing (MEW) is a highly advanced additive manufacturing method that directly processes polymer melts without the use of solvents (Alhaj & Narayan, 2023; Hochleitner et al., 2015). Polymer pellets such as PLA, PHA, and PCL are melted above their melting points and electrospun into microscale fibers, which are then deposited with computer-controlled collectors.

4.2.1.1 Working principle

- Polymer pellets (e.g., PLA, PHA, PCL) are heated above the melting temperature.
- Melt is extruded through a fine nozzle.
- High voltage induces a charged jet.
- Electric field draws microscale fibers.
- Computer-controlled collectors enable precise fiber deposition.

4.2.1.2 Advantages of MEW

- Completely solvent-free
- No solvent recovery system required
- Reduced occupational hazards
- Lower environmental emissions
- High architectural control
- Tunable fiber diameter (5–50 μm typically)

MEW facilitates the production of biodegradable fibrous scaffolds with the following characteristics:

- High porosity
- Controlled pore geometry
- Improved mechanical strength
- Enhanced surface area for antifouling functionalization.

Table 4.1 Comparison of MEW to NIPS/TIPS.

Parameter	NIPS	TIPS	MEW
Solvent use	Yes	Yes	No
Emission risk	High	Moderate	None
Energy demand	Moderate	High	Moderate
Morphology control	Medium	Medium	Very high

These types of scaffolds are of particular interest for membrane distillation substrate supports, filtration scaffolds, and tissue-integrated filtration systems. Comparison of MEW with NIPS/TIPS is given in Table 4.1. Thus, MEW is very much in line with the circular economy and net-zero manufacturing strategies. The comparative matrix for traditional and green solvents is given in Table 4.2. This comparative analysis emphasizes that green solvents have equal solvent capability with reduced hazards.

4.2.1.3 Critical outlook

The future of high-performance antifouling biodegradable membranes will rely not only on material properties but also on sustainable processing routes. The sector is moving from:

- Toxic solvent-based processing
- To green solvent optimization
- Toward solvent-free additive manufacturing

However, challenges remain:

- Higher cost of biobased solvents
- Limited long-term industrial data
- Scale-up challenges in MEW
- Need for reoptimization of phase inversion thermodynamics

Nevertheless, sustainable fabrication is no longer optional; it is a necessity. The integration of green chemistry, polymer science, and advanced processing technologies will shape the next generation of membrane materials.

Table 4.2 Comparative matrix–traditional versus green solvents.

Property	NMP	DMF	Cyrene™	DMSO	PolarClean
Boiling point (°C)	202	153	227	189	~240
Toxicity	Reproductive toxin	Reproductive toxin	Low	Low–Moderate	Low
Renewable origin	No	No	Yes	No	Partially
Biodegradable	No	No	Yes	Yes	Yes
Hansen solubility (PLA compatibility)	Good	Good	Good	Good	Good
Vapor pressure	Moderate	High	Low	Low	Very low

5. Performance and end-of-life analysis

Sustainable membrane materials are to be assessed against two interrelated life cycles: (1) the operational separation performance that determines energy and resource use, and (2) the end-of-life route that determines environmental impact postdisposal. This section provides an honest trade-off analysis and concise end-of-life evidence for the commonly suggested biodegradable membrane polymers (Olonisakin et al., 2025; Werber et al., 2016).

5.1 The trade-off analysis: Permeability versus selectivity versus biodegradability

The membrane design problem is inherently multicriteria. In water and wastewater treatment, the two traditional performance criteria are permeability (flux per unit driving force) and selectivity (solute rejection). Adding biodegradability as a third criterion challenges the engineer to accept some uncomfortable trade-offs (Haider et al., 2019; Werber et al., 2016).

Material structure and permeability: A high permeability is typically found in materials with open, porous microstructures or low-crystallinity amorphous regions that facilitate easy solvent (or water) diffusion. Biodegradable polyesters (such as PLA) tend to have relatively high stiffness and crystallinity relative to amorphous engineering materials; to match their permeability, engineers use thinner selective layers or porous supports. Thinner selective layers enhance hydraulic transport but may compromise mechanical durability and cleaning resistance.

Selectivity and susceptibility to degradation: High selectivity is typically achieved through the use of dense, tortuous diffusion paths (narrow pore-size distributions or highly selective, dense films). These same materials exhibit slower microbial and hydrolytic attack and hence improved service life, but also slower biodegradation rates at end-of-life, thus defeating the selection of a biodegradable material. Alternatively, the design of polymer microstructures to facilitate enzymatic attack (for example, amorphous and hydrophilic regions, or surface functional groups targeted by depolymerases) may improve biodegradability but typically at the expense of salt/solute selectivity.

Operational truth-fouling and cleaning: Biodegradable membranes might be more vulnerable to surface degradation under intense physical or chemical cleaning (e.g., high pH bleach, strong oxidizers). Frequent cleaning will accelerate material loss and potentially release by-products into the process streams. Hence, antifouling engineering (hydrophilic surfaces, zwitterionic brushes, sacrificial surface layers) remains critical even for biodegradable membranes; otherwise, a reduced lifetime may offset the end-of-life benefit.

Practical design principle (honest trade-off): Biodegradability is preferred in applications where replacement and disposal are most problematic (single-use products, hard-to-reach locations, marine environments) and high lifetime performance and energy efficiency are paramount (large-scale potable water reuse MBRs), and then combine with modular end-of-life solutions (disassembly, composting, enzymatic recycling) (Ezugbe & Rathilal, 2020; Werber et al., 2016).

5.1.1 Real-world degradation timelines

Biodegradation is highly dependent on specific conditions (temperature, water, microbial populations, UV light, thickness, etc.). Typical, evidence-based standards can be used to establish reasonable expectations.

Industrial composting (high temperature, ~58°C, controlled humidity): PLA and most aliphatic polyesters can quickly mineralize under industrial composting conditions, with complete degradation times measured in weeks to months. For example, PLA is often cited as undergoing complete degradation under industrial composting conditions (≈28–90 days), depending on crystallinity and thickness.

Natural soil/terrestrial environment—degrades slowly, considerably: The same PLA samples that degrade in industrial composters can take months to years in a temperate soil environment due to lower temperature and lower availability of microbial depolymerases, slowing hydrolysis. Comprehensive reviews highlight the high variability of soil degradation rates and polymer-specificity.

Marine environment: Degradation rate is the slowest. For PHAs biodegradation rates in the marine environment, among the highest marine biodegradation rates for commercial bioplastics, experimental marine biodegradation rates have been measured and used to estimate useful lifetimes. Biodegradation fluxes of ~0.04–0.09 mg/day/cm^2, assuming mean PHA biodegradation rates, suggest that a solid PHA article (such as a bottle wall) could take on the order of 1.5–4 years to degrade in the marine environment; larger articles will take longer, and PLA degrades at orders-of-magnitude slower rates in seawater and can persist effectively. These results indicate that "marine biodegradable" labeling must be polymer- and size/thickness-dependent.

Implication for membranes: Thin films and porous structures will alter the surface area to mass ratio and can thus degrade faster than similar-weight articles of the same polymer, but membranes are also used in aqueous, microbiologically rich environments that can either speed (biofilm-mediated hydrolysis) or slow (antifouling coatings that inhibit enzyme activity) degradation.

5.2 Case study spotlight: Pilot-scale integration of biosourced membrane materials

Bench-scale projects using biodegradable membrane materials (PLA, PHA, cellulose-based films) are relatively common, but pilot-scale projects are less common, especially in terms of longevity. A good example of a type of project that integrates biosourced membrane materials with pilot MBR or recovery processes would be those that involve biosourced membranes.

Recent compact reviews and experiments have demonstrated bench-to-pilot improvements in biosourced and biodegradable membranes for filtration and wastewater treatment, emphasizing positive antifouling properties (hydrophilic surfaces), reasonable permeabilities for low-pressure micro/ultrafiltration, and potential integration scenarios where membrane replacement and disposal can be managed (decentralized sanitation, agricultural runoff treatment) (Dastbaz et al., 2017; Ezugbe & Rathilal, 2020). Although many pilot-scale MBRs continue to use established polymers (PVDF, PES) for their robustness, other pilot projects that combine PHA production with wastewater treatment processes represent a closed-loop process: wastewater → microbial PHA production → processing into membranes or parts → end-of-life biodegradation or composting. Such projects have demonstrated the potential of material loop closure, even if complete commercial substitution is a future goal.

5.2.1 Lessons learned from pilot-scale work

Material economics and scale: PHA and PLA feedstock prices and current production scale make it difficult to compete with commodity fluoropolymers. Blending PHA with wastewater improves economics but adds process steps.

Durability versus disposal control: Pilot systems perform well when end-of-life is planned (collected and industrially composted); uncontrolled disposal (landfill, marine) negates biodegradability benefits.

Regulatory and monitoring needs: Pilots underscore that standard acceptance tests for membranes must be expanded to include controlled biodegradation tests under relevant disposal scenarios, and monitoring for by-products (microplastic fragments, oligomers) is essential.

6. Future outlook: The smart-green nexus

6.1 4D printing and stimuli-responsive biopolymers

The trend that will have the greatest influence on biodegradable membrane technology in the near future is the combination of responsive smart materials and environmentally friendly polymers, which has been termed "four-dimensional" printing, where structures are produced using additive manufacturing that can change their shape or performance in response to changes in their environment, over time [https://doi.org/10.1002/9781119865452.ch19]. Fig. 4.4 represents the various applications of 4D printing and prototypic illustration of the shape-changing architectural skin across thermal extremes.

Polylactic acid is particularly interesting in this regard, since the polylactic acid's glass transition point is approximately 55–60°C, allowing it to be used as a thermally controlled switching element that can be activated by the same hot water pulses used to clean membrane systems. Therefore, the geometry of a 4D printed sc-PLA or PHBV composite feed channel spacer could potentially be programmatically changed during a cleaning process, both mechanically disrupting the residual biofilm and also returning the feed channel to its optimal flow regime, thereby combining two different operations into one.

In addition, looking to the long-term, stimuli responsive biopolymer hydrogels grafted with thermoresponsive grafts, i.e., as poly(N-isopropylacrylamide) PNIPAM on cellulose or PHBV offer the potential for producing membranes with pore openings that can be switched between the tight, selective conformation at temperatures below their lower critical solution temperature and the loose, high flux conformation at temperatures above their lower critical solution temperature. Transferring this ability from model materials to completely compostable biopolymer materials remains an area of active research; however, there is a clear path forward for developing these types of materials based upon the surface modification techniques discussed throughout this chapter.

6.2 From membrane stability to membrane sustainability

The most well-established performance paradigm in membrane science research for five decades has been durability, which is chemically, thermally, and mechanically. It was logical to focus on durability; the membrane needed to endure to function. However, endurance and sustainability represent opposing objectives. Therefore, the field of membrane science is facing the consequences of its past success in maximizing the first objective (endurance). Today, the majority of commercial membranes (overwhelmingly PES, PVDF, and polyamide thin-film composite) are either burned or landfilled when their useful life as a membrane ends, and generate tens of thousands of tons of nonrecycled plastic each year due to the growth rate of over 9% per annum.

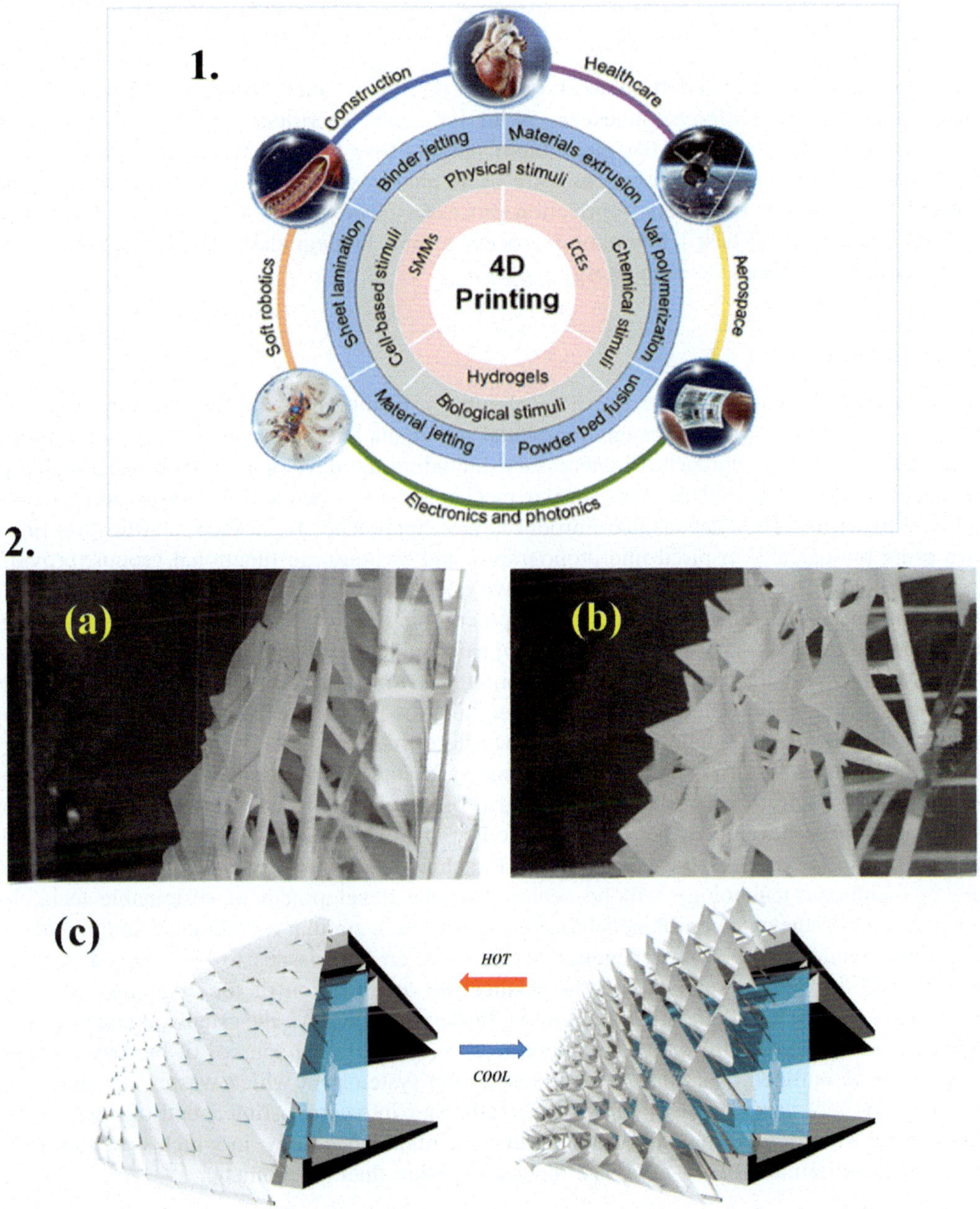

FIGURE 4.4

Schematic of the 4D printing ecosystem, 1. Mapping the integration of stimuli-responsive materials and activation triggers with additive manufacturing techniques and multidisciplinary applications. 2. Prototypical demonstration of the shape-morphing architectural skin across thermal extremes: (a) high-temperature activation, (b) low-temperature recovery, and (c) biomimetic building skin concept inspired by shelf-like structures.

1. (Adapted with permission from (Ding et al., 2025). Copyright 2025 ACS.), 2. (Adapted with permission from [https://doi.org/10.1016/j.jobe.2021.103076]. Copyright 2021 Elsevier.)

The biodegradable materials framework outlined throughout this chapter (sc-PLA, PHBV composites, nanocellulose reinforced matrices, zwitterionic and sacrificial fouling-resistant surfaces) fundamentally changes the central design problem, not how long the membrane lasts before it fails, but how we design the failure to be a benefit. Life cycle analyzes are starting to quantify the benefits, biobased PLA and PHB membrane fabrication processes clearly show lower carbon footprint than traditional petrochemical-based fabrication processes, and biodegradable final products close the material loop with no need for high-energy recycling.

7. Conclusions

This chapter highlights the importance of resolving the membrane paradox, whereby very durable membrane materials are released as environmental pollutants after disposal. To this end, a paradigm shift toward sustainable approaches is necessary, including the development of biodegradable polymers, such as sc-PLA and PHA, with similar mechanical and chemical properties to conventional membrane materials. The chapter also introduces new approaches to surface modification of membranes using biologically inspired microtopography and zwitterionic functional groups, providing a long-term solution to fouling that does not involve toxic chemicals. In addition to surface chemistry modification and biodegradable polymer development, the chapter introduces the shift toward green manufacturing processes. These include the use of safer solvents and/or alternative approaches that do not involve solvents at all, such as melt-electrospinning writing. The aim is to assess the sustainability of membranes from a lifecycle point of view, weighing the performance of the membrane during its service life against the potential of the membrane after it has served its useful life. This will involve assessing alternatives for disposing of the membrane after the end of its service life (e.g., composting or enzymatic recycling). One of the major concerns of the chapter is assessing how smart membrane technologies can be developed that combine biodegradable materials with self-healing capabilities (such as 4D printing and responsive materials) to achieve sustainability. The chapter finds that the future of membrane technology will be centered on the development of sustainable technologies throughout every stage of a membrane's life cycle to achieve a circular membrane economy that offers both environmental and economic benefits. While there are still issues to be resolved to meet the objectives outlined above, such as how to balance the optimization of the crystalline structure of biodegradable materials with the minimization of the degradation time, the establishment of condition-dependent end-of-life schedules, and the identification of accessible sources of bio-based materials, the authors have outlined examples of new closed-loop systems (in which wastewater materials are processed into PHA, then into membranes, and finally back into composting at end-of-life) that clearly demonstrate a way forward to achieve the circular membrane economy, in which water security and environmental sustainability are mutually supportive rather than antagonistic.

References

Alhaj, M., & Narayan, R. (2023). Scalable continuous manufacturing process of stereocomplex PLA by twin-screw extrusion. *Polymers, 15*(4), 922. https://doi.org/10.3390/polym15040922

Arango-Santander, S. (2022). Bioinspired topographic surface modification of biomaterials. *Materials, 15*(7). https://doi.org/10.3390/ma15072383

Casanova, F., et al. (2021). Novel micro-and nanocellulose-based delivery systems for liposoluble compounds. *Nanomaterials, 11*(10), 1–38. https://doi.org/10.3390/nano11102593

Choudhury, R. R., Gohil, J. M., Mohanty, S., & Nayak, S. K. (2018). Antifouling, fouling release and antimicrobial materials for surface modification of reverse osmosis and nanofiltration membranes. *Journal of Materials Chemistry A, 6*(2), 313–333. https://doi.org/10.1039/c7ta08627j

Dastbaz, A., Karimi-Sabet, J., Ahadi, H., & Amini, Y. (2017). Preparation and characterization of novel modified PVDF-HFP/GO/ODS composite hollow fiber membrane for Caspian Sea water desalination. *Desalination, 424*(September), 62–73. https://doi.org/10.1016/j.desal.2017.09.030

Ding, A., Tang, F., & Alsberg, E. (Apr. 2025). 4D printing: A comprehensive review of technologies, materials, stimuli, design, and emerging applications. *Chemical Reviews, 125*(7), 3663–3771. https://doi.org/10.1021/acs.chemrev.4c00070

Duarte, J., Raota, C. S., Baldasso, C., dos Santos, V., & Zeni, M. (2025). Investigation of the ternary system, water/hydrochloric acid/polyamide 66, for the production of polymeric membranes by phase inversion. *Membranes, 15*(1), 7. https://doi.org/10.3390/membranes15010007

Ezugbe, E. O., & Rathilal, S. (2020). Membrane technologies in wastewater treatment: A review. *Membranes, 10* (5). https://doi.org/10.3390/membranes10050089

Feijoo, P., et al. (2022). Development and characterization of fully renewable and biodegradable polyhydroxyalkanoate blends with improved thermoformability. *Polymers, 14*(13), 2527. https://doi.org/10.3390/polym14132527

Figoli, A., et al. (2014). Towards non-toxic solvents for membrane preparation: A review. *Green Chemistry, 16*(9), 4034–4059. https://doi.org/10.1039/c4gc00613e

Haider, T. P., Völker, C., Kramm, J., Landfester, K., & Wurm, F. R. (2019). Plastics of the future? The impact of biodegradable polymers on the environment and on society. *Angewandte Chemie International Edition, 58*(1), 50–62. https://doi.org/10.1002/anie.201805766

He, W., et al. (2025). The advances in zwitterionic materials and their biomedical applications. *International Materials Reviews, 70*(4), 301–349. https://doi.org/10.1177/09506608251323568

Hochleitner, G., et al. (2015). Additive manufacturing of scaffolds with sub-micron filaments via melt electrospinning writing. *Biofabrication, 7*(3). https://doi.org/10.1088/1758-5090/7/3/035002

Iroegbu, A. O. C., & Ray, S. S. (2023). Chitin nanomaterials as multifunctional systems in advanced applications – Progress and challenges toward sustainability. *Macromolecular Materials and Engineering, 308*(9), 1–17. https://doi.org/10.1002/mame.202300053

Kaniuk, Ł., & Stachewicz, U. (2021). Development and advantages of biodegradable PHA polymers based on electrospun PHBV fibers for tissue engineering and other biomedical applications. *ACS Biomaterials Science & Engineering, 7*(12), 5339–5362. https://doi.org/10.1021/acsbiomaterials.1c00757

Kara, Y., Kolos Molnár, Molnár, K., & Hu, M. B. (2022). Decomposition behavior of stereocomplex PLA melt-blown fine fiber mats in water and in compost. *Journal of Polymers and the Environment, 31*(4), 1398–1414. https://doi.org/10.1007/s10924-022-02694-w, 314.

Kim, D., & Nunes, S. P. (2021). Green solvents for membrane manufacture: Recent trends and perspectives. *Current Opinion in Green and Sustainable Chemistry, 28*. https://doi.org/10.1016/j.cogsc.2020.100427

Laftah, W. A., & Wan Abdul Rahman, W. A. (2025). Polymers for anti-fouling applications: A review. *Environmental Science: Advances, 4*(6), 824–841. https://doi.org/10.1039/d5va00034c

Lau, S. K., & Yong, W. F. (2021). Recent progress of zwitterionic materials as antifouling membranes for ultrafiltration, nanofiltration, and reverse osmosis. *ACS Applied Polymer Materials, 3*(9), 4390–4412. https://doi.org/10.1021/acsapm.1c00779

Li, S., Meng, H., Wang, H., Vrouwenvelder, J. S., & Li, Z. (2022). A sacrificial protective layer as fouling control strategy for nanofiltration in water treatment. *Water Research, 219*, Article 118554. https://doi.org/10.1016/j.watres.2022.118554

Macnamara, J. F., Rubino, M., Daum, M., Kathuria, A., & Auras, R. (Feb. 2024). Unlocking the secrets of high-water barrier stereocomplex polylactide blend extrusion films. *Green Chemistry, 26*(4), 2248–2257. https://doi.org/10.1039/d3gc04805e

Marino, T., Blasi, E., Tornaghi, S., Di Nicolò, E., & Figoli, A. (2018). Polyethersulfone membranes prepared with Rhodiasolv®Polarclean as water soluble green solvent. *Journal of Membrane Science, 549*(October 2017), 192–204. https://doi.org/10.1016/j.memsci.2017.12.007

Mihir, R. S., & Purkait, K. (2018). *Membrane technology in separation science*. CRC Press. https://doi.org/10.1201/9781315229263

Morales-Jiménez, M., Palacio, D. A., Palencia, M., Meléndrez, M. F., & Rivas, B. L. (2023). Bio-based polymeric membranes: Development and environmental applications. *Membranes, 13*(7). https://doi.org/10.3390/membranes13070625

Nagarajan, V., Mohanty, A. K., & Misra, M. (2016). Perspective on polylactic acid (PLA) based sustainable materials for durable applications: Focus on toughness and heat resistance. *ACS Sustainable Chemistry & Engineering, 4*(6), 2899–2916. https://doi.org/10.1021/acssuschemeng.6b00321

Nagarajan, V., Zhang, K., Misra, M., & Mohanty, A. K. (2015). Overcoming the fundamental challenges in improving the impact strength and crystallinity of PLA biocomposites: Influence of nucleating agent and mold temperature. *ACS Applied Materials & Interfaces, 7*(21), 11203–11214. https://doi.org/10.1021/acsami.5b01145

Naziri Mehrabani, S. A., Vatanpour, V., & Koyuncu, I. (2022). Green solvents in polymeric membrane fabrication: A review. *Separation and Purification Technology, 298*, Article 121691. https://doi.org/10.1016/j.seppur.2022.121691

Nunes, S. P., & Inoue, T. (Mar. 1996). Evidence for spinodal decomposition and nucleation and growth mechanisms during membrane formation. *Journal of Membrane Science, 111*(1), 93–103. https://doi.org/10.1016/0376-7388(95)00281-2

Olonisakin, K., Mohanty, A. K., Thimmanagari, M., & Misra, M. (2025). Recent advances in biodegradable polymer blends and their biocomposites: A comprehensive review. *Green Chemistry, 27*(38), 11656–11704. https://doi.org/10.1039/d5gc01294e

Oopath, S. V., Baji, A., Abtahi, M., Luu, T. Q., Vasilev, K., & Truong, V. K. (2023). Nature-inspired biomimetic surfaces for controlling bacterial attachment and biofilm development. *Advanced Materials Interfaces, 10*(4), Article 2201425. https://doi.org/10.1002/admi.202201425

Panackal Shibu, R., Jafari, M., Sagala, S. L., & Shamshina, J. L. (Sep. 2025). Chitin nanowhiskers: A review of manufacturing, processing, and the influence of content on composite reinforcement and property enhancement. *RSC Applied Polymers, 3*(5), 1031–1123. https://doi.org/10.1039/d5lp00104h

Purkait, M. K., & Singh, R. (2018). Fouling mechanisms and remedies. *Membrane Technology and Separation Science*, 173–190. https://doi.org/10.1201/9781315229263-7

Rana, R., Ahmad, W., Mishra, A., Kumar, S., & Ahmad, A. (Jul. 2025). Nanocellulose/Nanocellulose-Based membranes in wastewater treatment: A sustainable path forward for environmental protection. *Food Chemistry X, 29*, Article 102654. https://doi.org/10.1016/j.fochx.2025.102654

Roy, S., & Ragunath, S. (2018). Emerging membrane technologies for water and energy sustainability: Future prospects, constraints and challenges. *Energies, 11*(11). https://doi.org/10.3390/en11112997

Russo, F., et al. (2020). Innovative poly (vinylidene fluoride) (PVDF) electrospun nanofiber membrane preparation using DMSO as a low toxicity solvent. *Membranes, 10*(3), 1–17. https://doi.org/10.3390/membranes10030036

Shafi, H. Z., Matin, A., Akhtar, S., Gleason, K. K., Zubair, S. M., & Khan, Z. (2017). Organic fouling in surface modified reverse osmosis membranes: Filtration studies and subsequent morphological and compositional characterization. *Journal of Membrane Science, 527*(November 2016), 152–163. https://doi.org/10.1016/j.memsci.2017.01.017

Shalumon, K. T., et al. (2009). Corrigendum to 'electrospinning of carboxymethyl chitin/poly (vinyl alcohol) nanofibrous scaffolds for tissue engineering applications. *Carbohydrate Polymers, 77*, 863–869. https://doi.org/10.1016/j.carbpol.2020.116897. Carbohydr. Polym., vol. 249, p. 116897, Dec. 2020.

Sherwood, J., et al. (2014). Dihydrolevoglucosenone (cyrene) as a bio-based alternative for dipolar aprotic solvents. *Chemical Communications, 50*(68), 9650–9652. https://doi.org/10.1039/c4cc04133j

Singh, R., & Purkait, M. K. (2016). Evaluation of mPEG effect on the hydrophilicity and antifouling nature of the PVDF-co-HFP flat sheet polymeric membranes for humic acid removal. *Journal of Water Process Engineering, 14*, 9–18. https://doi.org/10.1016/j.jwpe.2016.10.001

Singh, R., & Purkait, M. K. (2017). Role of poly(2-acrylamido-2-methyl-1-propanesulfonic acid) in the modification of polysulfone membranes for ultrafiltration. *Journal of Applied Polymer Science, 134*(37), Article 45290. https://doi.org/10.1002/app.45290

Singh, R., Yadav, V. S. K., & Purkait, M. K. (2019). Cu2O photocatalyst modified antifouling polysulfone mixed matrix membrane for ultrafiltration of protein and visible light driven photocatalytic pharmaceutical removal. *Separation and Purification Technology, 212*, 191–204. https://doi.org/10.1016/j.seppur.2018.11.029

Song, G. L., & Feng, Z. (2020). Modification, degradation and evaluation of a few organic coatings for some marine applications. *Corrosion and Materials Degradation, 1*(3), 408–442. https://doi.org/10.3390/cmd1030019

Sun, M., Lu, S., Zhao, P., Feng, Z., Yu, M., & Han, K. (2022). Scalable preparation of complete stereocomplexation polylactic acid fiber and its hydrolysis resistance. *Molecules, 27*(21). https://doi.org/10.3390/molecules27217654

Védie, E., et al. (2022). Influence of Sharklet-Inspired micropatterned polymers on spatio-temporal variations of marine biofouling. *Macromolecular Bioscience, 22*(11), Article 2200304. https://doi.org/10.1002/mabi.202200304

Venault, A., Aini, H. N., Galeta, T. A., & Chang, Y. (2022). Using the dimethyl sulfoxide green solvent for the making of antifouling PEGylated membranes by the vapor-induced phase separation process. *Journal of Membrane Science Letters, 2*(2), Article 100025. https://doi.org/10.1016/j.memlet.2022.100025

Wang, H., Liu, H., Wang, H., Li, Y., & Yang, J. (2025). Recent advances in high-strength zwitterionic polymer hydrogels: From zwitterionic properties to mechanical reinforcement strategies. *Transactions of Tianjin University, 31*(4), 347–369. https://doi.org/10.1007/s12209-025-00440-6

Wang, M., Ma, C., & Zhang, G. (2023). Dynamic surface antifouling materials. *Chinese Journal of Chemistry, 41*(21), 2881–2888. https://doi.org/10.1002/cjoc.202300321

Werber, J. R., Osuji, C. O., & Elimelech, M. (2016). Materials for next-generation desalination and water purification membranes. *Nature Reviews Materials, 1*. https://doi.org/10.1038/natrevmats.2016.18

Wohlert, J., Chen, P., Berglund, L. A., & Lo Re, G. (Jan. 2023). Acetylation of nanocellulose: Miscibility and reinforcement mechanisms in polymer nanocomposites. *ACS Nano, 18*(3), 1882–1891. https://doi.org/10.1021/acsnano.3c04872

Wu, S., et al. (2024). Research progress of marine anti-fouling coatings. *Coatings, 14*(9). https://doi.org/10.3390/coatings14091227

Yue, S., Wang, H., Xu, H., Liu, H., & Yu, W. (Oct. 2022). Addition of amino acids modulates the in vitro digestibility of corn starch. *Carbohydrate Polymers, 293*, Article 119745. https://doi.org/10.1016/j.carbpol.2022.119745

Zhang, G., et al. (2024). Polyhydroxyalkanoates (PHAs) biological recovery approaches and protein-mediated secretion model hypothesis. *Journal of Cleaner Production, 440*, Article 140851. https://doi.org/10.1016/j.jclepro.2024.140851

Zheng, Y., Xu, S., Yu, C., & Pan, P. (2022). Stereocomplexed materials of chiral polymers tuned by crystallization: A case study on poly(lactic acid). *Accounts of Materials Research, 3*(12), 1309–1322. https://doi.org/10.1021/accountsmr.2c00208

CHAPTER

Understanding membrane fouling: Mechanisms, transport dynamics, and innovative mitigation approaches

5

Vaheed Syed[1], Saikat Sinha Ray[1] and Rangabhashiyam Selvasembian[1, 2]

[1]*Department of Environmental Science and Engineering, School of Engineering and Sciences, SRM University-AP, Amaravati, Andhra Pradesh, India;* [2]*Center for Interdisciplinary Research, SRM University-AP, Amaravati, Andhra Pradesh, India*

Nomenclature and Abbreviations

Abbreviation Full Form

BSA Bovine Serum Albumin
CECP Concentrative External Concentration Polarization
CP Concentration Polarization
DECP Dilutive External Concentration Polarization
DOM Dissolved Organic Matter
EPS Extracellular Polymeric Substances
FO Forward Osmosis
HA Humic Acid
ICP Internal Concentration Polarization
MF Microfiltration
NF Nanofiltration
NOM Natural Organic Matter
PAN Polyacrylonitrile
PE Polyethylene
PES Polyethersulfone
PRO Pressure Retarded Osmosis
PS Polysulfone
PVDF Polyvinylidene Fluoride
PWF Pure Water Flux
RO Reverse Osmosis
TMP Transmembrane Pressure
UF Ultrafiltration

Sustainable Water Management. https://doi.org/10.1016/B978-0-443-44853-9.00002-2

Symbol Description

A Membrane surface area
C Foulant concentration
C_b Foulant concentration in bulk flow
C_m Foulant concentration at membrane surface
Co The quantity of foulant deposited per unit volume of permeate
Cp Foulant concentration in permeate
D Diffusion coefficient
dP Particle diameter
J Permeate flux
J_0 Initial permeate flux
Jc Flux after removing cake layer
Jcp Flux at end of filtration (e.g., with wastewater)
Js Solute flux
Jss Steady-state permeate flux
Jv Solvent flux
k Mass transfer coefficient
K Overall mass transfer coefficient
kb Complete pore blocking coefficient
kc Cake pore blocking coefficient
ki Intermediate pore blocking coefficient
KICP Solute diffusion resistivity due to internal concentration polarization
ks Internal pore blocking coefficient
Lp Pore length or membrane thickness
mc Cake mass per unit area
n Blocking index (fouling mechanism constant)
Pp Pore density (number of pores per unit membrane area)
Ps Solute permeability coefficient
P^- Local solute permeability coefficient
Rad Resistance due to adsorption
Rc Cake resistance
Rcp Resistance due to permanent fouling layer
Rf Fouling resistance (Rcp + Rc + Rad)
Rm Intrinsic membrane resistance
rp Pore radius
rpr Particle radius
Rt Total filtration resistance
Sh Sherwood number
t Filtration time
V Cumulative permeate volume
α Attachment coefficient
δ Boundary layer or cake layer thickness
ε Cake porosity
μ Dynamic viscosity of permeate
π Osmotic pressure
σ Reflection coefficient
τ Tortuosity factor

ΔEb Potential energy barrier from membrane-colloid interaction
ΔEd Hydrodynamic drag energy
ρp Particle density
ΔP Transmembrane pressure (TMP)
ΔΠ Osmotic pressure difference

1. Introduction

Despite the Earth's vast aquatic coverage exceeding 70% of its surface, potable water shortage arose as a primary global threat, driven by the limited accessibility of freshwater resources (Salehi, 2022). However, nearly 97% of this water is saline ocean water, leaving only a small fraction as freshwater resources, 0.007% accessible in rivers and lakes (Le & Nunes, 2016). Rapid population growth and accelerated industrialization have substantially intensified global water demand, over two-thirds of the population experiences water stress, and nearly 1.8 billion people continue to lack reliable access to safe and potable water, with projections indicating further intensification in the future (Tayeh et al., 2025). Availability of clean water is critical in conserving healthy ecosystems, human health and socioeconomic progress. Major factors contributing to the deterioration of the water cycle include population growth, inadequate sanitation, rapid industrialization, and ineffective environmental policies (Bose et al., 2024). Around 80% of wastewater produced globally is discharged into the environment without undergoing sufficient treatment or being reclaimed for reuse (UN World Water Development Report, 2017). Coming to 2050, almost 70% of the world will be urbanized, and many urban areas will have inadequate wastewater systems (Nations, 2022; Kookana et al., 2020). Rapid population growth and degraded ecosystems, it is becoming more difficult to provide people with safe water. Consequently, the advancement of efficient and sustainable wastewater treatment technologies is now recognized as a global imperative (Ray et al., 2021). Particularly in the context of ongoing economic growth and rising standards of living, the problems of energy crises, exhaustion of natural resources, and environmental impairment have been enhanced. The membrane technology is a sustainable method of purifying water, and the world market of membrane material is already estimated to be about $29.7 billion as of 2020, and it is predicted to be almost $39 billion in 2027 (Li et al., 2025a).

Driven by their superior separation efficiency and favorable energetic profiles, membrane-based technologies have seen widespread industrial integration for water purification and wastewater reclamation. These approaches predominantly rely on pressure-driven processes, including microfiltration (MF), ultrafiltration (UF), nanofiltration (NF), and reverse osmosis (RO). In addition, the field has broadened to encompass extensive research on osmotically driven systems, notably forward osmosis (FO) and pressure-retarded osmosis (PRO), as sustainable alternatives (Foorginezhad et al., 2025). While membrane filtration provides higher efficiency, it is not as enduring as the conventional water treatment methods that can last for 3 decades (Li et al., 2025a). Most commercially available organic polymer membranes have operational life less than 10 years in potable water treatment and under 5 years in wastewater applications due to the dependence on operating conditions (Scholes, 2015). About 35,000 tons of polymer membranes were disposed as waste globally in 2022, after their operational life. A traditional disposal process of burning and landfilling deals with a significant resource drain and introduces substantial environmental hazards (Li et al., 2025a; Syed et al., 2025). Table 5.1 provides a comparison of different water treatment membrane technologies: FO, RO, MF, UF, NF and PRO highlighting their major advantages and major limitations.

Table 5.1 Overview of different membrane technologies, advantages and major limitations.

Membrane technology	Advantages	Limitations	References
FO	• FO membrane systems operate by exploiting naturally occurring osmotic gradients to achieve selective separation. • FO's primary benefit is its low Energy consumption.	• Reverse salt flux, together with concentration polarization in FO membrane performance is often compromised by a diminished effective osmotic driving force, which triggers a chronological deterioration of permeate flux throughout prolonged operation	(Farahbakhsh et al., 2024; Ray et al., 2018)
RO	• RO membrane technology produces high-quality water suitable for domestic use, potable water supply, and the treatment of a broad spectrum of wastewater effluents. Membrane scaling result in increased energy consumption as well as reduced membrane life.	• Despite its high efficiency, RO technology is limited by fouling, including colloidal, organic, biological, and scaling issues arising from high salt and contaminant concentrations. • Scaling in the feed water elevates energy demand and reduces membrane service life.	Tayeh (2024)
MF	• MF demonstrates efficacy in eliminating bacteria and viruses. • MF operates under low transmembrane pressure, making it particularly suitable for groundwater treatment and highly cost-effective	• MF is inherently limited in seawater desalination applications owing to its inability to effectively remove dissolved solids and organic contaminants.	Aziz et al. (2024)
UF	• REMOVES water contaminants to produce high-quality effluent. • Provides low capital and operational costs and user-friendly operation design. • Removes macromolecules	• Persistent susceptibility to foulant deposition and subsequent flux attenuation. • Significant yield of secondary sludge • Elevated energy consumption • Insufficient selectivity • Partial removal of contaminants	Shoshaa et al. (2023)
NF	• Versatile design, efficient operation • Compared to RO, NF operates at lower pressure and uses less energy, improving its cost effectiveness.	• High initial capital investment • Membrane fouling issues • require additional post-treatment operations	Samavati et al. (2023)
Pressure retarded osmosis (PROs)	• Electricity production and brine treatment • Reduced ecological impact • Enhanced efficiency and reduced costs in hybrid systems	• Membrane fouling • Process efficiency • Techno-economic viability	Tawalbeh et al. (2021)

Fig. 5.1 presents worldwide trends of membrane fouling research and publication over 1990–2025. (A) presents the publications distribution by country, with China, USA and India noticed as the most active contributors. (B) shows the annual number of peer-reviewed publications, which has been steadily rising since 2000 and spiked after 2010. This trend is a manifestation of the rising worldwide interest in comprehending, modeling and dealing with membrane fouling in water treatment and separation applications. Clogging or fouling of the composite membranes is a significant issue, specific to RO membrane systems where particles (Maeda, 2024) that build up on the membrane. This affects performance, increases operating expenses, reduces productivity, degrades the quality of effluent and elevates pressure demands of system; requires frequent cleaning required, increases downtime and reduces membrane longevity (Das et al., 2022). Membrane fouling is influenced by a combination of factors encompassing feed composition, membrane material and surface properties, and operational parameters (Tayeh et al., 2025). Fig. 5.2 shows a bibliometric co-occurrence network of keywords, according to Scopus (accessed December 2025), visualized using VOSviewer. Red and orange clusters represent important fouling mechanisms "biofouling" and "cake layer," as well as its mitigation strategies such as "antifouling." Green and blue clusters highlight membrane materials, such as "polyethersulfone," while the yellow clusters represent application areas, such as "waste water treatment." Peripheral nodes denote emerging treatment methods such as "photocatalytic membrane" etc. These connections highlight the importance of prioritizing research in membrane fouling for stable and durable membranes over an extended period.

1.1 Fundamentals of membrane fouling

Fouling mechanisms result from complex physicochemical interactions between different feed components and the membrane surface (AlSawaftah et al., 2021; Ray et al., 2021). CP describes the accumulation of solutes near the membrane interface within the mass transfer boundary layer. Fouling occurs when organic particles are transported from the bulk feed through this polarized layer and adhere to the membrane surface or pore interiors, a process governed by advection, diffusion, shear-induced diffusion, and inertial lift forces that vary with particle size (M. Yang et al., 2017). Solute transport through internal concentration polarization (ICP) is therefore retarded by chemical potential gradients and foulant-foulant associations within the CP layer, and electrostatic exerted by membrane toward foulants. Upon reaching the surface, there would be a non-covalent (hydrophobic/electrostatic) interaction and in certain cases covalent binding between membrane-foulants facilitated and the surfaces with binding affinity being affected by steric/spatial constraints (H. Xu et al., 2020).

In desalination membrane, organic fouling occurs if feed water contains natural organic matter (NOM), including humic substances, microbial secretions, plant-based dissolved organic matter (DOM) and animal residues interaction with the membrane surface. In particular, proteins and hydrophobic compounds in NOM are very likely to form gel deposits on the membrane surface causing pore blockage leading to a significant decrease of the filtration performance (Lester et al., 2022). NOM is typically highly reactive, and its propensity to fouling is controlled by the interaction with the membrane surface. Organic fouling can be managed by feedwater pretreatment or selecting membranes with low organic adsorption affinity. The formation of cake deposits on membrane surfaces is well predicted by the cake filtration model (Tayeh et al., 2025).

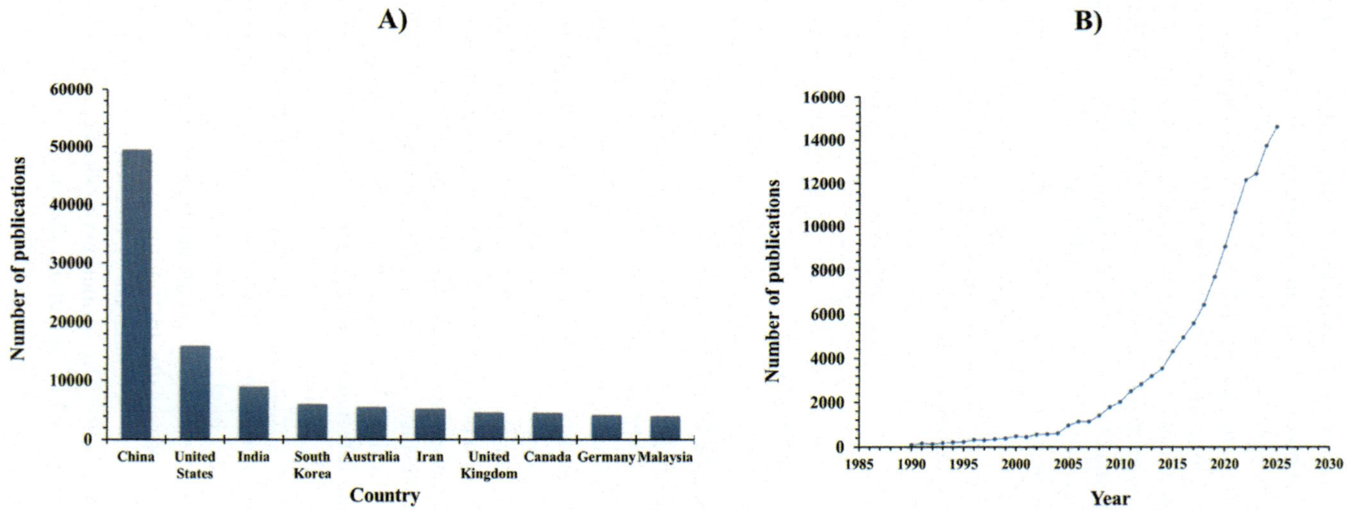

FIGURE 5.1

Membrane fouling research (1990–2025). (A) Distribution of publication counts by country since 1990, derived from scopus advanced scholar search using keywords “membrane” and “fouling,” queried in November 2025. (B) Annual peer-reviewed publication counts since 1990, based on the same scopus search criteria, highlighting yearly research output trends.

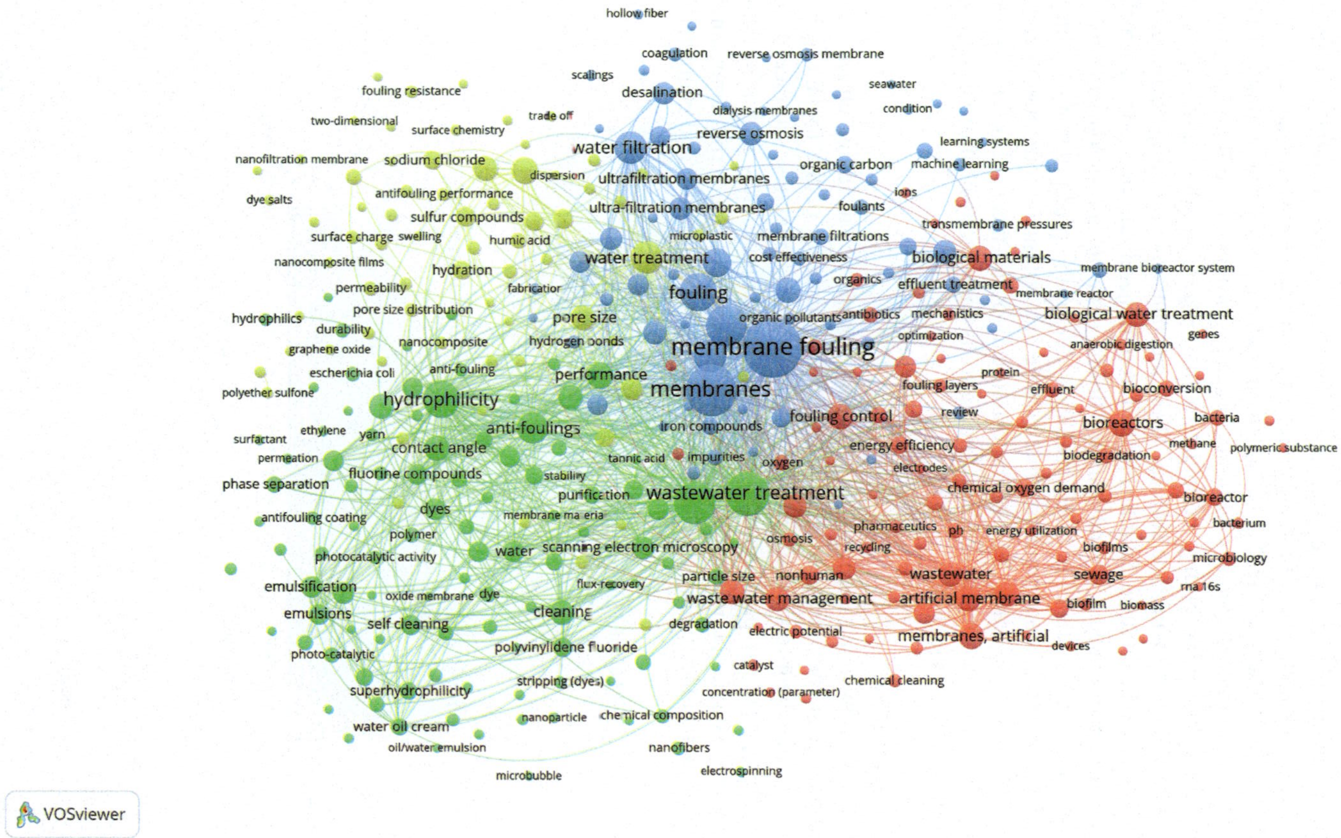

FIGURE 5.2

Co-occurrence network of keywords in membrane fouling publications from the scopus database (search terms: “Membrane” and “fouling”; data retrieved November 2025), visualized with VOSviewer.

1.1.1 Classification of membrane fouling

Membrane fouling primarily occurs in the form of a cake layer, scale or biofilm. Cake-layer is originated by the gradual accumulation of colloidal particles, whether it can be organic, inorganic and biological, on the membrane surface. Fig. 5.3 demonstrates various membrane fouling mechanisms during filtration, depositing microorganisms, colloids, organic substances and inorganic salts on the membrane surface, for instance biofouling, colloidal fouling, organic fouling and scaling resulted when feed stream interacts with membrane surface. These mechanisms of fouling decrease permeate flux and overall membrane performance. Scaling resulted from mineral salts crystalizing which on the membrane as a result of heterogeneous crystallization. Biofilm develops as microorganisms multiply and establishes colonies on membrane. In brief, scaling is due to the precipitation of mineral salts while biofilm formed by bacterial communities which adhere and colonize on the membrane surface (AlSawaftah et al., 2021; Sadr & Saroj, 2015). The classification and mechanism of main types of membrane fouling are summarized in Table 5.2 according to its sources, mechanisms, characteristics and the effects on filtration efficiency.

1.1.2 Influencing factors

The predominant factors influencing fouling are (1) foullants properties including charge, size, concentration and diffusivity of foulants as well as solubility and polarity of the foulant (2) Intrinsic membrane parameters like hydrophobicity, pore structure, surface roughness, surface charge and functional groups. (3) Operational parameters include flux, temperature, flow rate, solution characteristics including composition, pH, ionic strength and the presence of interfering compounds (Koo et al., 2012). Type of membrane fouling depend on the nature, concentration and characteristics of

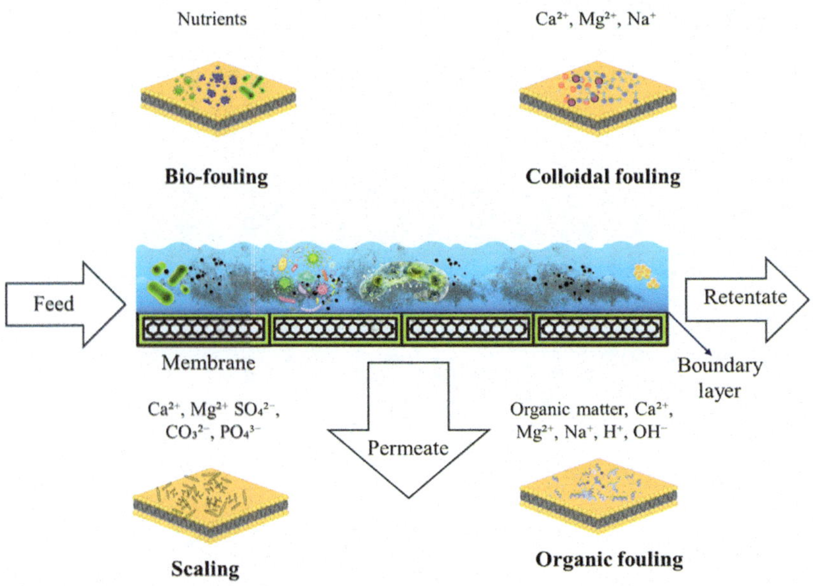

FIGURE 5.3

Overview of primary membrane fouling mechanisms.

Table 5.2 Classification of major of membrane fouling types, sources, characteristics, causes, and impact on membrane performance.

Fouling type	Schematic illustration	Sources	Characteristics	Impacts	References
Organic fouling	Organic matter, Ca^{2+}, Mg^{2+}, Na^{+}, H^{+}, OH^{-} 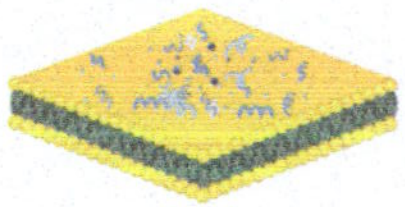	• Natural organic matter (NOM), comprising humic acids (HAs), polysaccharides, and proteins, is ubiquitous in aquatic systems and a primary cause of membrane fouling.	• Synergistic organic-inorganic fouling creates dense interfacial barriers that drastically attenuate permeate flux by elevating hydraulic resistance and reducing effective surface area.	• Membrane fouling gradually reduces permeate flux during filtration. • Fouling degrades operational life of membrane. • The attenuation of flux necessitates rise in operational pressure thereby escalating the energy consumption.	Gao et al. (2023)
Inorganic fouling	P, Ca^{2+}, Mg^{2+}, Na^{+} 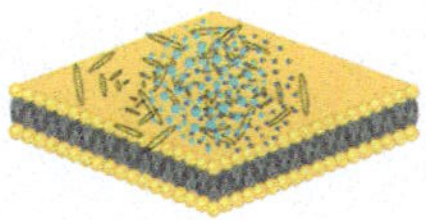	• In RO systems, inorganic fouling arises when sparingly soluble salts surpass their solubility thresholds, resulting in crystal formation and accumulation on the membrane surface.	• Surface mineralization and scaling involving the accumulation of inorganic precipitates on the selective barrier. • Including the deposition of calcium carbonate crystals.	• Pore blockage caused by fouling physically damages membranes, raising operational costs and shortening membrane lifespan. • Scale accumulation in reverse osmosis systems increases operating pressure and lowers flux. • The fouling cake layer from scale creates a significant osmotic pressure buildup, leading to decreased salt rejection.	Shahid et al. (2019)

Continued

Table 5.2 Classification of major of membrane fouling types, sources, characteristics, causes, and impact on membrane performance.—cont'd

Fouling type	Schematic illustration	Sources	Characteristics	Impacts	References
Microbial/Biological fouling	Nutrients 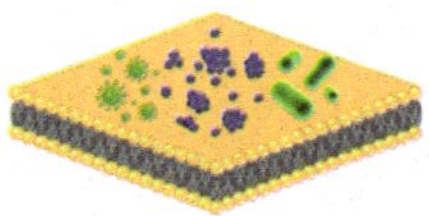	• Bio-growth in water treatment systems is largely dictated by the availability of assimilable organic carbon in influent NOM. • During algae blooms, elevated dissolved AOC levels can trigger biofouling events.	• Biofouling initiates with reversible bacterial attachment and progresses to EPS-mediated colonization, resulting in dense biofilms up to ~200 μm thick at the RO inlet.	• Biofouling significantly inflates the operational expenditure of water treatment facilities; by necessitating higher feed pressures and augmenting energy requirements, it forces more frequent chemical cleaning cycles, which ultimately results in system downtime and premature membrane degradation. • Biofouling can cause biofilm-enhanced concentration polarization (BECP).	Hoek et al. (2022)
Particulate/ Colloidal fouling	Ca^{2+}, Mg^{2+}, Na^{+} 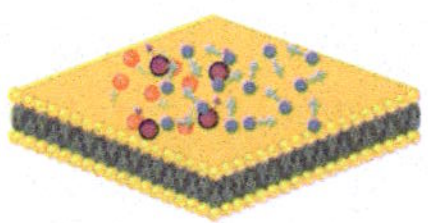	• Colloidal foulants include both inorganic and organic materials. • Inorganic foulants consist of silica, iron oxides/ hydroxides, and aluminum silicate minerals. • Organic foulants include HA, polysaccharides, and proteins.	• Colloidal foulants consist of both inorganic and organic materials. • These foulants are typically fine particles ranging from 1 nm to 1 μm in size.	• Fouling intensifies CP, increasing salt permeation through the membrane and further decreasing salt rejection.	Zhao et al. (2021)

Scaling	Ca^{2+}, Mg^{2+} SO_4^{2-}, CO_3^{2-}, PO_4^{3-} 	• The development of inorganic foulants is governed by a concentration-driven crystallization mechanism. The progressive extraction of water elevates mineral levels beyond their saturation thresholds, triggering the formation of solid scales at the membrane-solution interface and subsequently impairing hydraulic throughput.	• Scale formation occurs via two main crystallization routes. • Bulk crystallization involves salts precipitating in the solution.	• Long-term flux impairment and permanent loss permeability resulting from scaling. • Increased hydraulic resistance from deposited layers contributes to flux reduction. • The cake layer hinders solute back-diffusion, increases interfacial osmotic pressure, and reduces permeate flux.	Matin et al. (2019)

foulants as well as how foulants interact with the membrane. Moreover, inlet temperature, pressure and flow rate are also noted to have direct impact on fouling (Shahzad et al., 2021).

Membrane fouling depends on the pH, at pH 5, organic fouling predominates. This arises from the fact that natural condition pH seven and alkaline condition pH nine are more serious than acid condition combined with inorganic fouling, the coupling of organic-inorganic colloids engendered serious membrane fouling under acid, natural and slight alkalinity conditions (Shi et al., 2025). Ionic strength, particularly increased salinity, induces membrane fouling. This is due to higher salinity causes the diffuse double layer to shrink, which leads to smaller zeta potential and weaker electrostatic repulsive forces. Accordingly, foulants tend to coalesce and accumulate on the membrane when filtering oil-in-water emulsion (Qin et al., 2025). Surface properties of a membrane also play an important role in fouling. Membranes with smoother surfaces have lower fouling tendency. Surface roughness provides increased sites for attachment of foulants. Pore volume in addition to pore size distribution variations impact fouling since they change the water permeability and antifouling performance. Because the hydrophilic membrane can form a hydration layer to repel foulants, it is generally more resistant to be fouled compared with hydrophobic membranes, which are highly susceptible to the adhesion of foulants (Saffarimiandoab et al., 2021). Feed water contained divalent cations significantly impact membrane fouling by reducing critical flux. These cations, particularly calcium ions, Ca^{2+}, have a greater affinity to bind with carboxylic groups of organic matter near the membrane surface than magnesium ions, Mg^{2+}. These interations concentrate fouling close to the membrane compared to that in bulk feed solution (Nguyen et al., 2021). Membrane fouling is also influenced by temperature, elevated temperatures increase permeate fluxes and enhance biodegradation, as well as solute back-diffusion, leading to an attenuation of fouling layer formation. This enhances the life of the membrane and minimizing maintenance (Nguyen et al., 2021).

1.1.2.1 Membrane types and main factors influencing fouling behavior

Fouling susceptibility is governed by membrane-specific factors, including material composition and surface attributes such as roughness, hydrophobicity, and morphological structure (AlSawaftah et al., 2021). Based on membrane materials utilized during fabrication, membranes are primarily categorized into organic, ceramic and metal-based membranes. Organic membranes, including polyethylene (PE), polystyrene (PS), polyethersulfone (PES), polyacrylonitrile (PAN), and polyvinylidene fluoride (PVDF), are widely used owing to their cost-effectiveness and well-established manufacturing techniques; nonetheless, they suffer from limited mechanical strength, reduced service life, and susceptibility to fouling (Sandu et al., 2022). PVDF membranes have better anti-fouling. Ceramic and metal membranes offer better mechanical strength, temperature resistance, and higher flux, but are more difficult and costly to produce. Under similar operating conditions, PAN membranes tend to foul more slowly than PES membranes, whereas metallic membranes show better fouling recovery (Du et al., 2020).

1.1.2.2 Feed characteristics

Feed characteristics are influenced by membrane fouling significantly, including the concentration and properties of suspended solids, pH, ionic strength etc. Increasing the concentration of the feed generally reduces permeate flux as it enables a higher amount of foulants to build up, leading to higher fouling (Cairone et al., 2024). Particles may originate from the feed, or in precipitation of soluble components. They cause fouling by pore blocking, or channel narrowing, or formation of cake layer depending on their size (Shahzad et al., 2021). Large particles produce higher permeate fluxes and larger cakes but are less prone to deposit because of their size. Feed particles are bimodal: fine and coarse particles, both compact the cake together by filling in the void spaces between one another. The size of the particles determines where they can be deposited during high crossflow. Particle shape plays a significant role in determining cake porosity in filtration processes; particles with lower sphericity generally form cakes with higher porosity (Sauer et al., 2024).

Studies have demonstrated that the latex coagulation was inhibited when high pH is maintained in feed, which enhances antifouling performance membrane. On the other hand, at lower pH, NOM size is presumably reduced while the membrane adsorption of these macromolecules is increased, and fouling is still inevitable (Alhweij et al., 2022). Hong and Elimelech found that permeation decreases more at pH 4 than at pH 8, because of the lower membrane charge due to lower pH. Reduced pH is shown to protonate carboxylic group in NOM which decrease organic burden but forming denser fouling layers (Hong & Elimelech, 1997). Organic fouling, on the other hand, increases at lower pH values and in presence of divalent cations, pH also regulates HA/proteins adsorption. Preserving alkaline environment may well suppress the fouling (Ding et al., 2025). Scaling is enhanced with increasing concentration of salts in the feedwater, with increased ionic activity and decreased salt solubility. The primary thermodynamic driver for scale deposition is the supersaturation of salts, in which higher salt contents result in a higher degree of supersaturation and more deposit formation (Hoang, 2022). Further, the presence of more cations adsorbed on the membrane surface will also lead to weakening of the Donnan effect, which might further drive salt precipitation (Asante-Sackey et al., 2021; Li et al., 2025b). Fig. 5.4 illustrates the interaction between feed and draw solution responsible for different fouling mechanisms including biofouling, colloidal fouling, inorganic scaling and organic

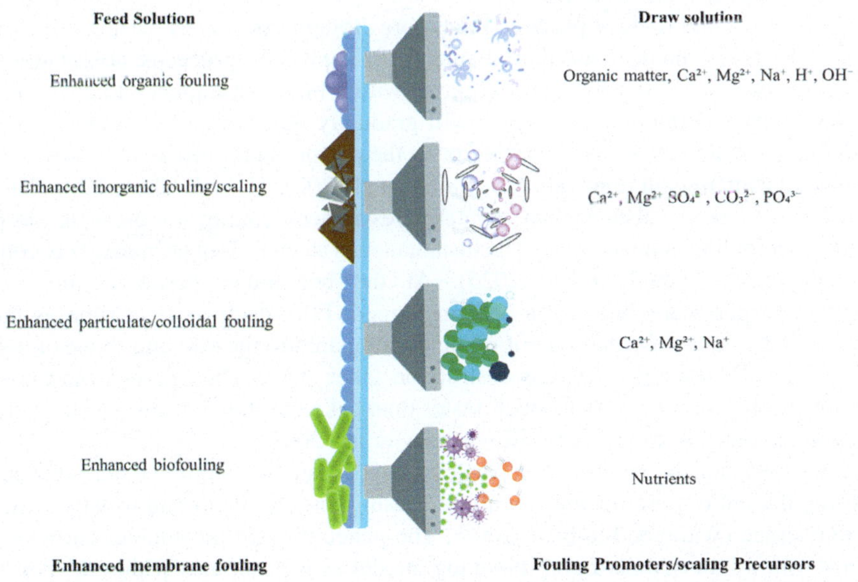

FIGURE 5.4

Illustration of intensified membrane fouling and mineral scaling within osmotic filtration systems.

fouling. Fouling inducing substances, also called as foulants or scalants including nutrients, colloids, organic materials and multivalent ions (Ca^{2+}, Mg^{2+}, SO_4^{2-}, CO_3^{2-}, PO_4^{3-}) migrate across from the feed side to the membrane interface enhancing fouling and scaling process eventually decreasing the efficiency and permeate flux of the membrane.

1.1.2.3 Operating conditions

Researchers reported that elevated temperatures enhance permeate flux and mitigate fouling. Increasing the feed temperature from 20°C to 40°C resulted in flux improvements of up to 60%, primarily attributable to temperature-induced enhancements in permeate diffusion across the membrane (Salahi et al., 2010). Crossflow velocity refers to the superficial flow of the feed that runs parallel to the membrane surface. The effect on permeate flux has been widely investigated for various feed solutions. Greater crossflow velocity can increase mass transfer and result in turbulence close to the membrane, increasing permeate flux. Enhanced mixing suppresses solid materials build-up in the gel layer through infusion into the bulk solution and contributes to relieve CP effects (Abdelrasoul et al., 2013).

TMP, the feed and permeate pressure difference, is an important regulating factor influences permeation flux. Higher TMP results in an enhanced convective force across the membrane resulting higher permeate flux. Although pressure increases flux, it is significant to optimize operating conditions minimize fouling particularly when feed contains high levels of foulants (Miller et al., 2013). Tuning of the operational conditions during filtration is an efficient fouling strategy. This method reduces energy consumption and the amount of chemicals used for membrane cleaning, thereby being sustainable and environmentally friendly (Mittal et al., 2021).

1.1.3 Fouling development mechanism

Particulate fouling is a multi-stage process. First, pore clogging occurs as particles aggregate around the opening of the pores and decrease their diameter. As deposition proceeds, consecutive addition of layers takes place on top of the first deposited layer with a final consequence to cover and fully clog the membrane pores. The membrane surface will gradually be covered by a compact cake layer as operation continues, which intensively compromises the performance, efficiency and selectivity of the membrane toward contaminant removal (AlSawaftah et al., 2021; H. Li & Chen, 2010). The cake layer raises hydraulic resistance also known as cake resistance leading to pressure drop permeate flux reduction and fouling enhancement. The instantaneous steep fall in permeate flux could be due to rapid pore plugging (Ahmed et al., 2023). At the beginning, permeate flux is maximum through membrane pores are fully open and unclogged. Flux declines over time as the pores are partially blocked due to the morphology of particles compared to the size and shape of the membrane pores (Gul et al., 2021). Complete pore obstruction takes place when particle and pore sizes and geometries are similar. This process is much faster than cake buildup, because even a sub-monolayer of particles can completely block pores (Kosvintsev et al., 2002).

The blockage mechanism occurs when particles larger than the membrane pores accumulate on its surface, closing the entire pore openings without passing through them and usually with a complete single-layer coverage (Wang & Tarabara, 2008). The penetration of the smaller particles that fit into pore openings follows the standard pore-blocking model as it penetrates within the pores and builds up inside them, leading to a decrease in pore volumes gradually (Sandoval-García et al., 2025). The intermediate blocking model describes the deposition of particles onto previously accumulated layers,

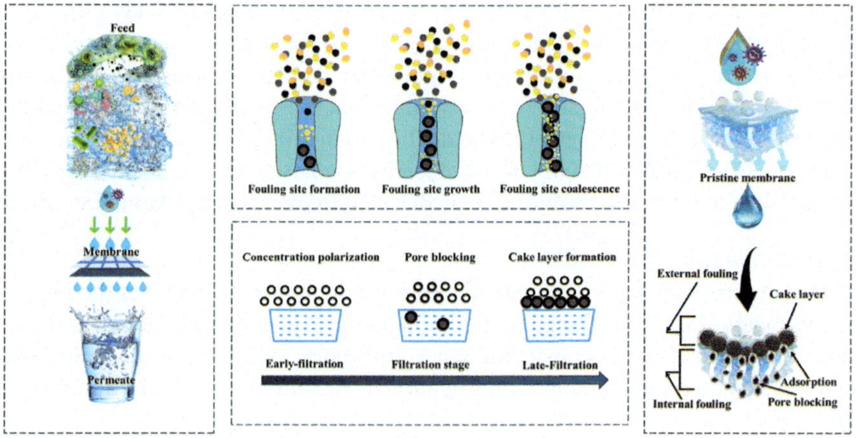

FIGURE 5.5

Accumulation of fouling and development of cake layer during filtration.

which may not directly obstruct pore entrances (P. Kim et al., 2022). The intermediate blocking mechanism, like the complete blocking model, occurs at the membrane surface rather than within the pores. In contrast, the cake filtration model details slowly accumulating particles that are larger than the pores that gradually forms a permeable cake layer on the surface (S. Liu et al., 2024). Fig. 5.5 depicts the chronological evolution of fouling, initiated by pore narrowing through the adsorption of organic matter and fine particulates. This progresses to complete pore occlusion and the eventual maturation of a dense cake layer composed of extracellular polymeric substances (EPS), microbial products, and sludge flocs, which collectively compromise permeate flux. The kinetics of this process are governed by a complex interplay of membrane physicochemical properties (morphology, roughness), feed water chemistry (solute concentration, surface charge), and operational parameters such as hydraulic shear and transmembrane pressure (Alkhatib et al., 2021).

1.2 Transport physics of fouled membranes

Membrane fouling reduces performance by decreasing permeate flux due to temperature polarization and higher thermal resistance. This leads to a smaller effective membrane area and greater mass-transfer tortuosity. These challenges arise from the complex interactions between feed solutes and the membrane material, particularly solute-solute interactions involving molecules adsorbed on the membrane surface (Kujawa et al., 2019). Membrane fouling increases adsorption capacity but hampers mass transport by changing how solutes partition and diffuse across the membrane (P. Xu et al., 2006).

1.2.1 Fundamentals of membrane transport

Fouling impairs membrane performance by decreasing permeate flux, primarily due to heightened temperature polarization and thermal resistance. It also diminishes the effective -solute-solute interactions in the feed, including those involving adsorbed species (Kujawa et al., 2019). All membrane processes encounter boundary-layer mass transfer resistance due to CP. Usually, convective flows parallel to the membrane surface are the dominant factor affecting boundary-layer transfer, surpassing the influence of perpendicular flows (Jaćimovski et al., 2016). CP is a natural result of membrane selectivity, which reduces permeate flux due to gel-layer formation. The elevated concentration of rejected solutes near the membrane surface encourages back-diffusion from this boundary layer into the bulk solution (AlSawaftah et al., 2021). The decline in membrane performance occurs because the transmembrane pressure (TMP) driving force decreases, caused by the osmotic pressure gradient between the permeate and the feed solution near the membrane surface (Ahmed et al., 2023).

1.2.2 Impact on mass transfer dynamics

In numerous mass transfer systems, the concentration gradient primarily drives the process. However, in membrane permeation systems like gas separation and pressure-driven processes (RO, NF, UF, and MF), multiple forces can simultaneously influence transport. When more than one driving force is involved, the typical analysis may require modification to account for these additional factors (Merdaw et al., 2011).

In membrane-based mass transfer, the penetrant must traverse the membrane and boundary layers, which create resistance. This resistance is especially significant when the penetrant is a liquid. Similar to electrical circuits, the total mass transfer resistance is the sum of the resistances of all layers (Dischinger et al., 2024).

$$\frac{1}{K} = \frac{1}{kl1} + \frac{1}{km} + \frac{1}{kl2} \tag{5.1}$$

Here, K represents the overall mass transfer coefficient, with $kl1$, $kl2$, and km denoting the resistances of the boundary layers and the membrane, respectively. For multilayer membranes, the total resistance is the sum of the resistances of each individual layer

$$\frac{1}{Km} = \frac{1}{km1} + \frac{1}{km2} + \frac{1}{km2} \tag{5.2}$$

Here, km represents the overall membrane mass transfer coefficient, while $km1$, $km2$ ... subsequent terms correspond to the resistances of individual composite layers. For high-flux membranes, a two-layer model comprising a dense selective layer and a porous support is commonly applied. The overall mass transfer coefficient is determined by combining the contributions of the respective boundary layer coefficients using the preceding equations (Hwang, 2011).

1.2.2.1 Flux decline

Membrane flux decline results from increased resistance due to pore blockage and cake buildup. Pore blocking raises resistance directly, while the cake forms an extra barrier. These are the main membrane fouling mechanisms. The initial sharp flux drop is from rapid pore blockage when pores are clean. As particles accumulate, pores are blocked, reducing flux. Blocking depends on particle and pore size. Pore blockage occurs quickly even with one particle layer, while cake formation takes longer (Fersi et al., 2009).

Pore blockage is intensified by the maturation of a cake layer, which imposes incremental mass transfer resistance as it densifies. Consequently, the system experiences a steady decline in permeation rates over the course of operation (Abdelrasoul et al., 2013). Fig. 5.6 illustrates membrane fouling during water filtration, emphasizing how a cake layer forms and permeate flow diminishes. It shows

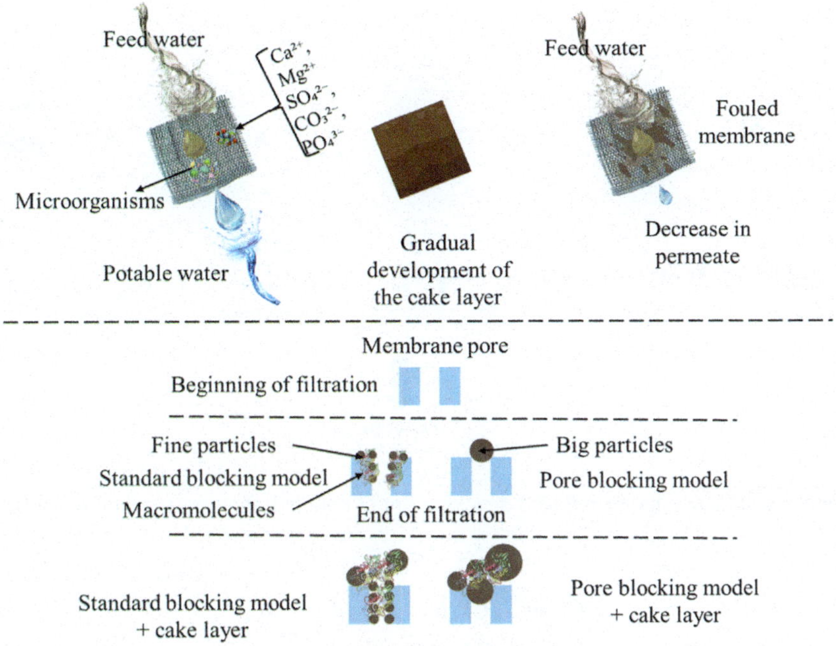

FIGURE 5.6

Depicts the primary fouling mechanisms in membrane filtration, illustrating how different foulants, including inorganic ions, accumulate.

the buildup of foulants, such as organic and inorganic ions, microorganisms, fine particles, and colloids, which progressively accumulate to form a cake layer and reduce permeate flux.

1.2.2.2 Concentration polarization (CP)

CP is an interfacial phenomenon triggered by the membrane's ability to restrict solute transport while permitting solvent flow. This selective behavior results in a localized concentration gradient at the membrane surface, significantly differing from the solute levels in the bulk solution (McCutcheon & Elimelech, 2006). This phenomenon induces an osmotic pressure at the membrane surface that exceeds that of the bulk solution. Consequently, the net driving force is diminished relative to an idealized system, as the elevated interfacial osmotic pressure counteracts the applied pressure, ultimately suppressing the water permeation rate (Bhinder et al., 2017). Unlike conventional pressure-driven processes such as UF and RO, where CP is confined to the feed-side interface, FO is characterized by dual-sided CP. This involves concentrative external CP at the feed-membrane boundary due to solute rejection, and dilutive external CP on the permeate side as the draw solution is diluted by incoming water (McCutcheon & Elimelech, 2006). Fig. 5.7 illustrates the dual impact of concentration polarization and fouling on membrane performance. The accumulation of solutes at the surface provides the initial mass transfer resistance, while subsequent cake formation and pore-narrowing phenomena further impede water transport. Together, these processes drive the observed flux decline and reduce the overall effectiveness of the filtration system (Goi et al., 2025).

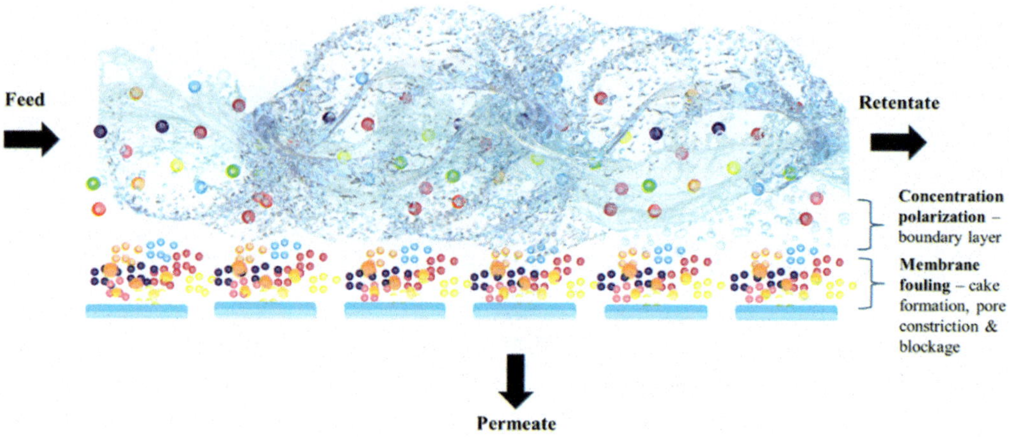

FIGURE 5.7

Schematic overview depicting the synergistic development of concentration polarization and membrane fouling.

$$Cf,m \;/\; Cf,b = exp\;(Jw \;/\; kf) \tag{5.3}$$

$$Cd,b \;/\; Cd,m = exp\;(Jw \;/\; kd) \tag{5.4}$$

Cf,b, Cf,m, Cd, i, Cd,m, and *Cd,b* determines solute concentrations, defined in the bulk feed, the feed-side active layer interface, the boundary between the support layer and the draw solution, and the draw-side membrane surface. To quantify these transport dynamics, the mass transfer coefficient (*k*) is calculated via standard empirical correlations tailored to the system's fluid dynamics

$$k = \frac{ShD}{dh} \tag{5.5}$$

Conversely, the influence of ICP can be described in terms of solute diffusion resistance (KICP) as given below:

$$\mathrm{KICP} = \mathrm{S_S}\,/\mathrm{D} = (1/\ \mathrm{J}_{w,\mathrm{FO}})\ \ln\left((\mathrm{B} + \mathrm{A}_{\pi\mathrm{d,m}})\ /\ (\mathrm{B} + \mathrm{J}_{w,\mathrm{FO}} + \mathrm{A}_{\pi\mathrm{f}})\right) \tag{5.6}$$

Among all CP phenomena, ICP is the main factor responsible for the overall osmotic pressure decrease (Goi et al., 2025).

1.2.2.3 Selectivity loss

Over time, particles deposited on the initial layer build up, sealing and clogging the membrane's pores. This progresses to a second stage, where additional accumulation forms a thick 'cake layer,' reducing the membrane's capacity to effectively and selectively remove contaminants. The buildup of foulants causes a decline in the membrane's selectivity (AlSawaftah, Abuwatfa, Darwish, Husseini, et al., 2021), which is shown as

$$\alpha\ (=\mathrm{A/B}) \tag{5.7}$$

Expressed as the ratio of the permeabilities of two components across the membrane, where A corresponds to the water permeability coefficient and B represents the solute permeability coefficient (Singh, 2015). The performance and durability of membrane systems mainly rely on their selectivity and resistance to fouling. Fouling harms membrane efficiency and structural strength because the buildup of contaminants on the surface or inside the pores reduces permeate flow, damages selectivity and permeability, and ultimately shortens the system's operational life (Mallah et al., 2024). Fouling negatively impacts the stability and selectivity of membrane separation processes by raising hydraulic resistance during filtration (Virga et al., 2021).

1.2.3 Mathematical modeling of membrane fouling

Numerous models analyze fouling, with Hermia's semi-empirical models being the most common in UF (Raspati & Leiknes, 2015). Designed initially for conventional filtration, they have also been adapted for tangential flow. Fouling causes pore blockage and cake formation, necessitating a combined model. Usually, empirical models use Darcy's law to link flux decline to hydraulic resistance (Cifuentes-Cabezas et al., 2023). A few membrane fouling models are discussed below.

1.2.3.1 Cake layer formation

The performance of UF is determined by the membrane's intrinsic hydraulic resistance (*Rm*) and the total resistance (R_t), which includes effects from the development of CP, cake, or gel layers, and fouling, collectively affecting the process. The steady-state permeate flux (*J*), which indicates flux stabilization after forming a uniform cake layer, Darcy's law describes this process through a series resistance model.

$$J = \frac{\Delta P}{\mu(Rm + Rt)} = \frac{1}{A}\frac{dV}{dt} \tag{5.8}$$

Here, ΔP denotes the applied pressure in pascals (Pa), μ represents the dynamic viscosity measured in pascal-seconds (Pa·s), *V* corresponds to the filtrate volume in cubic meters (m^3), t is the filtration time in seconds, and *A* signifies the membrane surface area in m^2. The intrinsic membrane resistance (*Rm*), which depends on the membrane's structural features, remains constant and can be expressed using (8) derived from the Hagen–Poiseu ille law. This principle assumes that the membrane has cylindrical pores of consistent diameter aligned perpendicular to its surface. In contrast, the overall resistance (R_t) is expressed as a function of α

$$R_t = \alpha CoV/A \tag{5.9}$$

Co denotes the mass of deposited material per unit volume of permeate (kg·m^{-3}) and represents the particle or colloidal volume fraction in the feed suspension, marked by xo. The parameter τ represents tortuosity, Lp denotes the pore length or membrane thickness in meters (m), K represents the medium's permeability in square meters (m^2), Pp indicates the pore density or number of pores per unit membrane area (m^{-2}), rp refers to the pore radius in meters (m), and ε corresponds to the cake porosity (Chang & Kim, 2005; Sim et al., 2011; Yiantsios & Karabelas, 2003).

1.2.3.2 Pore blocking

The reduction in permeate flow attributed to pore obstruction can be described as follows:

$$d^2t/dV^2 = k\,(dt/dV)^n \tag{5.10}$$

Here, V denotes the collected permeate volume, and t denotes the filtration period. The terms k and n describe the resistance coefficient and the blocking index, where n is a dimensionless factor that identifies the type of fouling mechanism involved (Chellam & Xu, 2006). Complete pore blocking is described by a condition in which each incoming particle closes a pore upon contact, with the assumption that no particles accumulate on the membrane surface or stack over existing deposits. This mechanism is characterized as complete pore blocking, characterized by $n = 2$.

$$J_{(n\,=\,2)} = J_0\,(k_b A_m)/t \tag{5.11}$$

Here, k_b is the total pore-blocking coefficient, A_m is the membrane area, J_0 is the initial flux, and $J_{n\,=\,2}$ is the flux associated with the $n = 2$ blocking mechanism.

Internal pore blockage occurs when fine particles adhere to the pore walls, progressively constricting the pore channels. This mechanism corresponds to internal pore blocking characterized by $n = 1.5$.

$$J(n=1.5) = \frac{4J_0}{\left[2 + k_s (A_m J_0)^2 t\right]^2} \tag{5.12}$$

Here, k_s denotes the coefficient associated with internal pore blockage, and $J_{n\,=\,1.5}$ denotes the flux predicted by this model. Intermediate pore blocking occurs as particles settle on the membrane surface, contributing both to pore obstruction and to adhesion with earlier deposits. This behavior is characterized by $n = 1$.

$$J(n=1) = \frac{4J_0}{1 + k_i (A_m J_0) t} \tag{5.13}$$

Here, k_i is defined as the intermediate pore-blocking coefficient, while $J_{n\,=\,1}$ denotes the normalized flux predicted by this model.

In the cake filtration regime, particles gather on the membrane surface rather than within the pores, forming a cake layer that increases overall hydraulic resistance. This fouling mechanism is characterized by $n = 0$.

$$J(n=0) = \frac{J_0}{\left[1 + 2k_c (A_m J_0)^2 t\right]^{0.5}} \tag{5.14}$$

Here, k_c represents the cake pore blocking coefficient, and $J_{n\,=\,0}$ signifies the flux (Zhang & Ding, 2015).

1.2.3.3 Combined models

Research reported that fouling mechanisms occurring in sequence: first internal fouling, then pore blockage, and finally cake formation. Initially, during filtration, pore constriction begins as particles cause the inner radius of the pores to narrow. This leads to membrane pore blockage due to particle aggregation, ultimately resulting in a cake layer forming over the blocked areas of the membrane surface (Koonani & Amirinejad, 2019; Nakamura & Matsumoto, 1998).

Standard blocking, intermediate pore obstruction, and cake formation contribute to the total volumetric flow rate, which is the sum of the flow rates through both open and blocked pores.

$$\frac{Q}{Q_0}=\frac{1}{(1+\beta Q_0 C_b t)^2}\left(1-\frac{\alpha'}{\beta A_0(1+\beta Q_0 C_b t)}+\frac{1}{\beta A_0}\right)^{-1} + \int_0^t \frac{\alpha' Q_0 C_b \beta^2 A_0((1+\beta A_0)(1+\beta Q_0 C_b t_c)-\alpha')^{-2} dt_c}{\sqrt{\left[\frac{R_{c0}}{R_m}+(1+\beta Q_0 C_b t_c)^2\right]^2+2\frac{f' R' \Delta P C_b}{\mu R_m{}^2}(t-t_c)}} \tag{5.15}$$

The total flow rate through a fouled membrane is calculated by considering mechanisms such as standard blocking, total pore obstruction, and cake layer development, as described by the Hagen–Poiseuille law. This overall flow reflects the combined permeation through both unblocked and partially or fully obstructed pores.

$$\frac{Q}{Q_0}=\left(1+\frac{2J_0 C_b \alpha_{in}}{\delta_m}t\right)^{-\left(1+\frac{\alpha\delta_m}{2\alpha_{in}}\right)} + \int_0^t \frac{J_0 \alpha C_b\left(\left(1+\frac{2J_0 C_b \alpha_{in}}{\delta_m}\right)t_c\right)^{-\left(1+\frac{\alpha\delta_m}{2\alpha_{in}}\right)}}{\sqrt{\left[\frac{R_{c0}}{R_m}+\left(1+\frac{2J_0 C_b \alpha_{in}}{\delta_m}t_c\right)\right]^2+2\frac{f' R' \Delta P C_b}{\mu R_m{}^2}(t-t_c)}} dt_c \tag{5.16}$$

For standard blocking, which involves intermediate pore blockage and cake formation, the flux can be described using the Hagen–Poiseuille law. The normalized flow rate equation derived from this is:

$$\frac{Q}{Q_0}=\left(1+\frac{2J_0 C_b \alpha_{in}}{\delta_m}t\right)^{-1}\left(1+\frac{\alpha'\delta_m}{2\alpha_{in}}\ln\left(1+\frac{2J_0 C_b \alpha_{in}}{\delta_m}t\right)\right)^{-1} + \int_0^t \frac{\alpha' J_0 C_b\left(\left(1+\frac{2J_0 C_b \alpha_{in}}{\delta_m}\right)t_c\right)^{-(1)}\left(1+\frac{\alpha'\delta_m}{2\alpha_{in}}\ln\left(1+\frac{2J_0 C_b \alpha_{in}}{\delta_m}t_c\right)\right)^{-2}}{\sqrt{\left[\frac{R_{c0}}{R_m}+\left(1+\frac{2J_0 C_b \alpha_{in}}{\delta_m}t_c\right)\right]^2+2\frac{f' R' \Delta P C_b}{\mu R_m{}^2}(t-t_c)}} dt_c \tag{5.17}$$

The total flow rate through the membrane, considering zero-order standard blocking, total pore blockage, and cake layer formation are expressed by:

$$\frac{Q}{Q_0}=(1-K''_{s0}t)^4 \exp\left(\frac{\alpha C_b J_0}{5K''_{s0}}\left((1-K''_{s0}t)^5-1\right)\right) + \int_0^t \frac{\alpha C_b J_0(1-K''_{s0}t)^4 \exp\left(\frac{\alpha C_b J_0}{5K''_{s0}}\left((1-K''_{s0}t_c)^5-1\right)\right)}{\sqrt{\left[\frac{R_{c0}}{R_m}+(1-K''_{s0}t_c)^{-4}\right]^2+2\frac{f' R' \Delta P C_b}{\mu R_m{}^2}(t-t_c)}} dt_c \tag{5.18}$$

The normalized flux equation is derived by including mechanisms such as zero-order standard blocking, partial pore obstruction, and cake layer development (Koonani & Amirinejad, 2019; Nakamura & Matsumoto, 1998):

$$\frac{Q}{Q_0}=\left(1-K''_{s0}t\right)^4\left(1-\frac{\alpha' C_b J_0}{5K''_{s0}}\left(\left(1-K''_{s0}t\right)^5-1\right)\right)^{-1}$$
$$+\int_0^t \frac{\alpha' C_b J_0\left(1-K''_{s0}t_c\right)^4\left(1-\frac{\alpha' C_b J_0}{5K''_{s0}}\left(\left(1-K''_{s0}t_c\right)^5-1\right)\right)^{-2}}{\sqrt{\left[\frac{R_{c0}}{R_m}+\left(1-K''_{s0}t_c\right)^{-4}\right]^2+2\frac{f'R'\Delta P C_b}{\mu {R_m}^2}(t-t_c)}}dt_c \tag{5.19}$$

Here, Q and Q^0 denote the volumetric flow rate through the fouled and clean membrane, respectively, while t is the filtration time and $t_{(c)}$ marks the onset of cake formation. The foulant concentration in the bulk solution is expressed as C_β, and J^0 represents the initial permeate flux of the clean membrane. Resistances associated with membrane fouling are captured through hydraulic resistance of the clean membrane R_m , R_m, C_n combined membrane and pore-constriction resistance, R_c cake-layer resistance, and R_{c0} initial cake resistance, with R' indicating the specific cake resistance. Parameters such as α, α', and αC_n describe complete, intermediate, and internal pore-blocking behaviors, whereas β characterizes the extent of pore constriction. Zero-order internal fouling contributions are incorporated through ${K_s}^{0'}$ and ${K_s}^{0''}$. Additional physical properties include the membrane thickness δ_m, solution viscosity μ, and transmembrane pressure ΔP. The dimensionless term f' represents the fraction of foulant contributing to cake growth (Koonani & Amirinejad, 2019).

1.2.3.4 Concentration polarization model

During membrane filtration, permeate drag moves foulant particles toward the membrane, leading to accumulation and a higher concentration near the surface called CP, which establishes a concentration gradient. Foulant particles tend to diffuse back into the bulk flow because of this gradient within the boundary layer. At steady state, the convective transport flux (JC) is balanced by the back-diffusion term $\left(D\cdot\frac{dC}{dx}\right)$. Porter (1972) introduced the film theory to explain the mechanism behind Cp (Porter, 1972; H. Yang et al., 2022) while $\frac{dC}{dx}$ corresponds to the concentration gradient across the boundary layer. At $x=0$, the foulant concentration is C_m, and at $x=\delta$, it equals C_b (H. Yang et al., 2022).

$$D=\frac{kBT}{3\pi}\mu dP \tag{5.20}$$

Here, kB denotes Boltzmann's constant, T the absolute temperature, μ the solution viscosity, and dP the particle diameter. The ratio D/δ in (20) represents the mass transfer coefficient k (Porter, 1972).

Where δ indicates the boundary layer thickness, and C_m and C_b denote foulant concentrations at the membrane surface and in the bulk stream, respectively. (20) shows that permeate flux J depends on the solute diffusion coefficient D, the concentration ratio $\frac{C_m}{C_b}$, and the boundary layer thickness δ. Operational improvements can be made by increasing D or decreasing δ, with D obtained through the

Stokes–Einstein equation. The ratio *J/k* defines the Péclet number, which quantifies the relative dominance of convective versus diffusive transport within the boundary layer. These models assume that foulant deposits on the membrane are not included in the mass balance. Liu et al. introduced a modified model where foulant convection (*JC*) depends on self-diffusion (*D*), permeate *JC* (*JCp*), and foulant loss through attachment (αJCm). The model is defined by (J. Liu et al., 2021):

$$\frac{J}{k} = \ln\left(C_m - Cp - \frac{\alpha JCm}{C_b} - Cp - \alpha JC_m \right) \tag{5.21}$$

The concentration ratio $\frac{C_m}{C_b}$, also known as the *CP* modulus M, indicates the extent of *CP*. Notably, permeate foulant *Cp* is excluded from the mass balance in (20). Consequently, (21) is valid for modeling *Cp*, when *Cp* is negligible relative to both C_m and C_b a typical scenario in NF/RO systems. When *Cp* is significant, convective solute flux toward the membrane *JC* must balance the sum of permeate solute flux *JCp* and back-diffusive flux $\frac{DdC}{dx}$ away from the surface, yielding (21) in its standard form (S. Kim & Hoek, 2005).(20) and (21) are widely applied *CP* models for predicting scaling in processes including RO, NF, UF, and MF (S. Kim & Hoek, 2005). The attachment coefficient α indicates the probability of foulant deposition during particle-membrane collisions, spanning from 0 (no fouling) to 1 (continuous fouling). Successful attachment occurs when a particle crosses the potential energy barrier ΔEb caused by membrane-colloid interactions, with the hydrodynamic drag energy ΔEd facilitating adhesion. The attachment coefficient α is thus given by (J. Liu et al., 2021):

$$\alpha = \frac{1}{1 + \exp\left(\frac{\Delta Eb}{kBT} - \frac{\Delta Ed}{kBT} \right)} \tag{5.22}$$

1.2.3.5 Critical flux model

The effective driving force decreases as osmotic pressure builds from CP, while the buildup of foulants on the membrane surface or within its pores increases hydraulic resistance beyond that of a clean membrane. These combined effects alter the flux–transmembrane pressure (TMP) relationship. Typically, membrane resistances are modeled as a series of resistances, as explained below.

$$J = J = \Delta P \ / \ (\mu \ (Rm + Rcp + Rc + Rad)) \tag{5.23}$$

Here, *J* indicates the permeate flux ($L/m^2 \cdot h$), μ is the permeate viscosity (Pa·s), and ΔP represents the transmembrane pressure (Pa). R_m corresponds to the permeate flux, *J* ($L/m^2 \cdot h$), which depends on the permeate viscosity μ (Pa·s) and the transmembrane pressure ΔP (Pa). The inherent membrane resistance is *Rm*, and the resistance caused by the accumulated cake layer is *Rc*. Intrinsic membrane resistance, while *Rc* indicates the resistance arising from the cake layer. *Rcp* and *Rad* represent resistances associated with permanent fouling and adsorption, respectively, all expressed in m^{-1}. The total resistance is obtained by summing these individual components.

$$Rt = (Rm + Rf\ ) \tag{5.24}$$

At 25 °C under vacuum, the initial pure water flux (PWF) was recorded over 45 min by tracking the permeate flow. Subsequently, distilled water was replaced with synthetic or real wastewater under the same conditions. During a 300-minute filtration, permeate samples were collected every 30 min, and the final flux was noted as *Jcp*. After filtration, the membrane was rinsed with deionized water for 45 min to remove the CP layer, and the pure water flux was measured again as *Jc*. Next, chemical cleaning with 0.1 M NaOH for 20 min was performed, and the PWF after cleaning was recorded as *Jad*. Since *Rc* and *Rad* are typically eliminated through deionized water rinsing and chemical cleaning, the resistance components were calculated using relevant equations (Jahani et al., 2023; Yuliwati et al., 2022).

$$Rc = (\Delta P \ / \ \mu Jcp) - Rm - Rcp - Rad \tag{5.25}$$

Jw is the initial pure water flux (m/s), *Jc* is the flux after removing the cake layer with deionized water rinse and 0.1 M NaOH cleaning, and *Jcp* is the flux during refinery wastewater filtration at the process end (m/s). Darcy's law describes flux in ultrafiltration and microfiltration (UF/MF) membrane systems (Yuliwati et al., 2022).

$$V \,/\, Adt = TMP \ / \ \mu(RT) \tag{5.26}$$

Where *J* indicates the permeate flux (L/m^2·h), *TMP* is the transmembrane pressure (Pa), μ denotes the permeate viscosity (Pa·s), and R_t represents the total membrane resistance. The critical flux was identified through the flux-step method, which involves gradually increasing the flux while observing changes in *TMP* (W. Li et al., 2018).

A refined critical flux definition differentiates between non-fouling and fouling conditions. In non-fouling conditions, resistance is only due to the membrane's intrinsic resistance *Rm*. In fouling conditions, additional resistance terms, such as *Rc*, are included. This clear separation isolates the effects of each resistance component (Liang et al., 2021). The experiment was performed with a 30-minute filtration period maintained at a low permeate flux between 10 and 20 L/m^2·h, followed by incremental flux increases after each phase. This process was repeated until the transmembrane pressure (TMP) stabilized at a constant flux, thereby allowing determination of the critical flux. At each stage, the initial transmembrane pressure (TMP), representing the onset of filtration resistance, and the final *TMP*, recorded at the end of the stage, were measured to determine the average *TMP* and its rate of change. The critical flux was defined as the point where the rate of *TMP* increase $(\frac{dTMP}{dt}) \geq 0.5$ (Yuliwati et al., 2022).

1.3 Performance metrics

Although membrane technology has made considerable progress, fouling continues to be a major challenge. It hampers system performance by decreasing permeate quality and permeability, while also raising operational costs through increased energy use, higher pressure needs, more frequent cleaning, extra pretreatment steps, and shorter membrane lifespan (Ahmed et al., 2023). For instance, global annual spending on preventive measures against biofouling in desalination is roughly US$15 billion (Flemming, 2011). The predominant fouling mechanism on membranes is influenced by several factors, such as feed water composition, system setup, membrane properties, chemical dosing schedules, and operational procedures (Almalki et al., 2025). The most common fouling types include colloidal buildup, organic fouling, mineral deposits, and biofilm formation, which results from microorganisms interacting with organic material (Ahmed et al., 2023).

1.3.1 Flux decline quantification

The Hermia model is a semi-empirical framework characterized by parameters possessing distinct physical significance and remains the predominant framework for investigating fouling in MF and UF membranes. It effectively elucidates fouling mechanisms and accurately predicts temporal declines in flux (Ghalami Choobar et al., 2019). Hermia proposed four empirical models based on the classical constant-pressure dead-end filtration framework, each representing a different fouling mechanism: The models include complete blocking, intermediate blocking, standard blocking, and cake layer formation. They were later extended to incorporate crossflow filtration systems by adding the steady-state back-transport mass transfer flux. The general governing equation introduced by Hermia is shown below (Vincent Vela et al., 2009).

$$d2\frac{t}{dV}2 = K\left(\frac{dt}{dV}\right)n \tag{5.27}$$

where t denotes the filtration time, V is the total permeate volume, K represents the mass transfer coefficient, and n indicates the fouling index specific to each mechanism.

Modifying the classical Hermia model's general equation leads to the following comprehensive Eq. (5.28) for cross-flow microfiltration (Corbatón-Báguena et al., 2016):

$$\frac{dJ}{dt} = -K\,(J - J_{ss})\,J^{2-n} \tag{5.28}$$

J indicates the permeate flux, while Jss refers to the steady state permeate flux.

The constant K is affected by factors such as transmembrane pressure, permeate viscosity, membrane resistance and the degree of pore area blockage relative to permeate flux (Vincent Vela et al., 2009). The fouling index n depends on the dominant fouling mechanism: complete pore blocking ($n = 2$), intermediate pore blocking ($n = 1$), standard pore blocking ($n = 1.5$), and cake layer formation ($n = 0$) (Ghalami Choobar et al., 2019).

1.3.2 Increased transmembrane pressure

Membrane technology is widely used and proven to be effective while consuming minimal energy. In membrane filtration, various parameters influence performance, with *TMP* being significant (J. Liu et al., 2018). Operating at lower *TMP* reduces fouling and *CP*, but it also lowers permeate flux (Zheng et al., 2018). Conversely, higher *TMP* increases the driving force, which enhances flux and helps prevent fouling blockage. While low *TMP* reduces fouling, it also decreases flux, whereas higher *TMP* boosts flux and mitigates fouling (Jiang et al., 2018; Liang et al., 2012).

Darcy's law states that the total filtration resistance (Rt) of a membrane is given by the following equations:

$$Rt = \frac{\Delta P}{\mu J} \tag{5.29}$$

$$Rt = Rm + Rg \tag{5.30}$$

Jiang et al. investigated how changes in *TMP* affect antifouling performance, using continuous filtration tests over 48 h. They found that membranes running at higher *TMP* had a higher initial flux because of a stronger driving force. However, at a *TMP* of 10 kPa, the flux was more prone to fouling, leading to a sharper decrease in performance (Jiang et al., 2018).

1.3.3 Rejection efficiency

Researchers found that membrane fouling may either enhance or reduce solute rejection, depending on the characteristics of the solute, membrane, and foulant interact. The specific mechanisms by which fouling influences rejection are not completely understood. Organic fouling primarily happens through adsorption, either on the membrane surface or within pores, along with *CP* effects (Park et al., 2019). Using a phenomenological model for rejection analysis offers benefits over evaluating rejection at just one permeate flux. It allows for characterization across a broad flux range where rejection changes, using only two parameters. The Spiegler-Kedem model exemplifies this method (Spiegler & Kedem, 1966) describes the movement of solutes across membranes and is summarized below (Bellona et al., 2010).

$$J_s = -\overline{P}\Delta x\frac{dc}{dx} + (1-\sigma)Jvc \tag{5.31}$$

Here, J_s represents solute flux, P^{-} the local solute permeability coefficient, and the reflection coefficient σ is expressed as a function of *Jv*, solvent flux, and *c* as solute concentration. Through integration, the definition of the solute permeability coefficient P_s (where $P_s = \frac{\overline{P}}{x}$), and algebraic rearrangement, the derived equations facilitate the determination of rejection (*R*).

$$R = \sigma\frac{1-F}{1-\sigma F} \tag{5.32}$$

$$F = exp\,(-1-\sigma\,/\,Ps) \tag{5.33}$$

The model relies on two parameters, *Ps* and, which change depending on the solute and membrane. These can be determined through rejection experiments and a fitting process. The solvent flux is subsequently calculated using the following formula:

$$J_v = Lp(\Delta P - \sigma\Delta\Pi) \tag{5.34}$$

Here, *Lp* stands for the membrane solvent permeability constant, and $\Delta P - \sigma\Pi$ indicates the net driving pressure (Bellona et al., 2010).

1.4 Mitigation strategies for membrane fouling

Research focused significantly on membrane fouling in separation processes to enhance performance. Desired improvements can be achieved by optimizing the operational conditions and tuning membrane characteristics (Alkhatib et al., 2021). Various anti-fouling strategies aim to inhibit biomass accumulation on the surface of membranes. Conventional processes involve operational control strategies, including optimization of loading rates and maintaining the level of flux below critical levels, as well as physical methods such as backwashing, membrane relaxation and use of air scouring to remove foulants (Bagheri & Mirbagheri, 2018). Membrane fouling is inevitable and cannot be eliminated. These control techniques are primarily utilized as mitigation strategies to reduce fouling effects and improve system operation. However, their performance is site-specific depending on the environmental and operational factors. Choosing an inappropriate strategy can result in increased energy consumption, decreased volume of permeate, and poor quality which will compromise the process efficiency (Fortunato et al., 2020). The major strategies for mitigating membrane fouling are depicted in Fig. 5.8. Pre-treatment of the feed comprises coagulation and flocculation for the removal of suspended solids. Surface wettability could also be tuned through membrane modification to achieve antifouling characteristics with the help of surface hydrophilicity or -phobicity. Backwashing and chemical washing apply physical and chemical forces to detach foulants. Finally, the feed solution is chemically conditioned in order to inhibit scaling or fouling on the membrane which can reduce permeation and limit its operational life.

1.4.1 Feed pretreatment

Pre-treatment strategies employed to avoid fouling, the process are based on decreasing levels of foulants in the feed stream by employing different techniques, leading to reduction in foulant deposition on membrane surface and preventing membrane fouling as well (Cairone et al., 2024). Although, pre-treatment is more important for pressure-driven systems, it still serves as a preventive step in membrane processes (Alkhatib et al., 2021). The choice of pre-treatment has a major impact on fouling, permeate quality, membrane productivity and operational life. This is particularly crucial for the real feedwaters, such as waste water or seawater that have various types of foulants (Augusty et al., 2026). Pretreatment is concerned primarily with removal of particulates that would potentially, clog the pores or deposit debris on the membrane surface. Various strategies can be used to reduce the fouling, such as the pretreatment of feed water, improvement of operating parameters and membrane properties (Alkhatib et al., 2021). These strategies are coagulation/flocculation, water softening, antiscalant application, membrane filtration, thermal softening and pH treatment that can be grouped into three main types (Ahmed et al., 2023). Mechanical methods include membrane filtration processes, and chemical methods, such as coagulation/flocculation, pH adjustment, water softening, or the use of antiscalants. Thermal methods involve the heating/softening of water, while hybrid processes may combine mechanisms such as; mechanical-/chemical or mechanical-/thermal for more efficient treatment process (Tayeh et al., 2025).

1.4.2 Cleaning approaches

Cleaning improves performance indicators shift due to10% decrease in permeate flux, 15% increase in differential pressure, or 10% increase in the salinity. Despite pretreatment however, fouling on the RO membrane develops progressively, and requires periodic cleaning to prevent the system failure and reduce membrane replacements (Tayeh et al., 2025). Yet overcleaning can cause the membrane to degrade, resulting in extended down time and potential damage to membrane. Cleaning eliminates particles that cause pores to clog and adhere to surface, addressing fouling in addition to pretreatment. The chemical and physical methods are used primarily to extend operational life of membrane. The performance is evaluated primarily through reduction in resistance and recovery of permeate flux (Zouboulis & Gkotsis, 2020). Chemical cleaning relies on the principle that adding caustic soda, oxidizers, acids, chelating agents and surfactants will disrupt the adherence of foulant to the membrane. These chemicals are often required in large amounts, they have potential to damage the membrane and reduce it operational life, as well as being generated with waste streams that need secondary treatment. Physical cleaning utilizes hydraulic or mechanical effects on solid particles results in removing foulants, cleaning is achieved where deposited materials are successfully dislodged and original performance of the membrane is restored (Gul et al., 2021). The functionality of wet membranes can also recover after cleanings, since wettability influences fouling considerably. If it is to be wetted again, the wetting membranes have to dry which may cause damage. Lowering

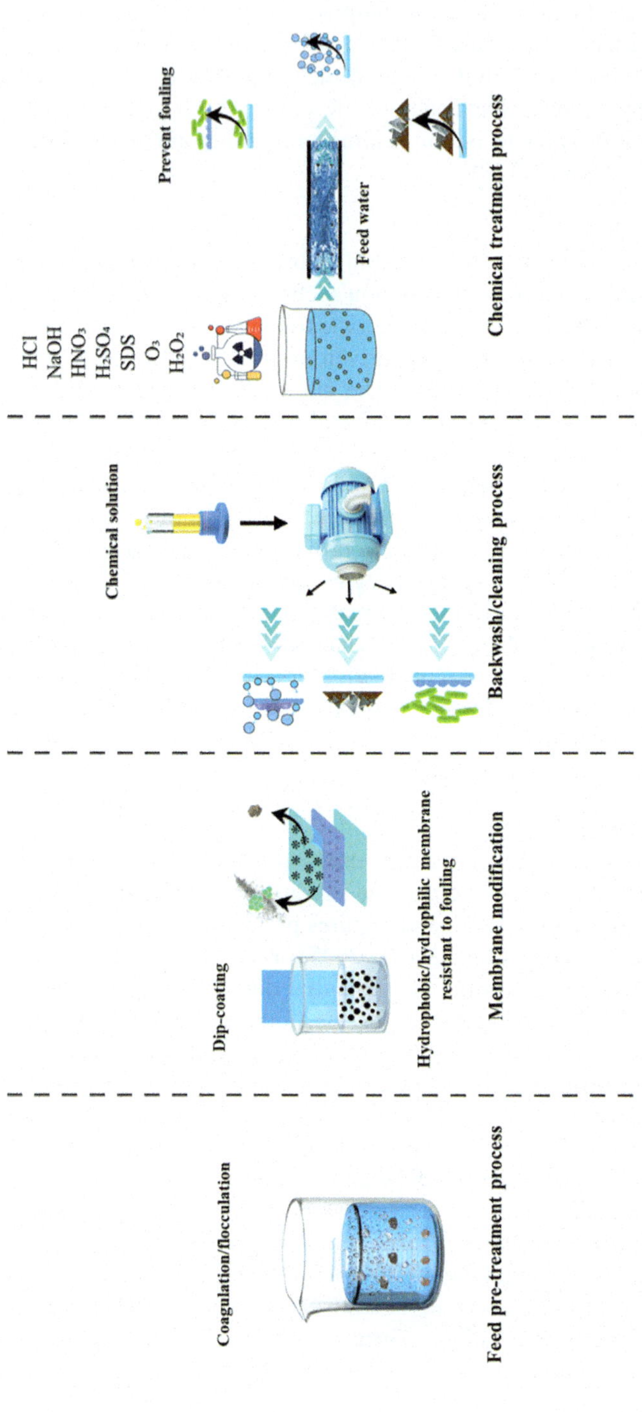

FIGURE 5.8

Schematic representation of various strategies for mitigating membrane fouling.

the hydrostatic pressure alone does not guarantee that the pores return to an un-wet state. In most cases, some solutes are left in the pores after drying and therefore chemical washing along with drying is applied to restore them efficiently (Zulkefli et al., 2021).

1.4.3 Membrane surface modifications

The functionalization or surface modification improves desired characteristics and provides a flexible alternative with blending to maintain core structure. It is involved in an antifouling process by enhancing hydrophilicity (Shen et al., 2018). Standard techniques are for cost effective, surface coating, hydrophilic substance coated and polymer chains grafted onto the surface (Shahkaramipour et al., 2017). These provide hydration layer, that prevent foulants from attaching, enhancing anti-fouling performance (Shahkaramipour et al., 2017). Surface hydrophilization methods such as surface coating, polymerization or grafting and pore-filling grafting processes can effectively improve the resistance to fouling. They produce a hydration layer, in hindrance to the adsorption and accumulation of foulants on the membrane surface and thus reduce fouling (Fang et al., 2020). Membrane surface hydrophilicity regarded as crucial factor and plays significant role in fouling tendency. Hydrophilic membranes usually show superior antifouling performance compared to hydrophobic ones. However, most commercially available membranes are inherently hydrophobic, making them more prone to fouling in practical applications (Ma et al., 2019).

1.4.4 Optimization of operational conditions

In addition to membrane modification and feed pretreatment, the optimization operation conditions are of vital for anti-fouling in oily wastewater treatment. Key hydrodynamic and operational parameters such as type of flow regime, back-pulse duration, temperature, *TMP*, and feed concentration may be adjusted to mitigate fouling significantly (Zulkefli et al., 2021).). In addition, the operating parameters of the membrane by filtering system were optimized by systematic variation and evaluation. The membrane performance was evaluated in permeate flux, fouling resistance and total organic carbon removal efficiency (S. Kim & Park, 2022). Consequently, the filter system was designed to maintain an ideal hydrodynamics in order to minimize fouling. The cross-flow configuration for instance has much less possibility for fouling compared to dead-end configuration (Seyed Shahabadi & Reyhani, 2014). Moreover, employing pulsed feed flows or applying disruptions like continuous or pulsed electrical treatments, surface modification of the membrane can greatly reduce membrane fouling (Zulkefli et al., 2021). Table 5.3 presents an overview of the primary membrane fouling categories: organic, inorganic/scaling, biological, and particulate/colloidal. It details the specific foulants in each category, their impact on membrane structure and performance, and the most effective mitigation methods for each fouling type.

2. Research challenges and future outlook

The mitigation of membrane fouling is a key challenge that must be addressed to enhance the efficiency and reliability of membrane-based processes, especially for the desalination, wastewater treatment and industrial separation processes. To understand fouling in detailed, it is necessary to develop a database of foulants that will be developed in the comprehensive sense of type and nature. Although mathematical and computational models have provided valuable insights in understanding the fouling process, in most instances, they might not be able to capture the nature of the phenomenon

Table 5.3 Types of membrane fouling, reported foulants, impacts on the membrane, effective mitigation strategies.

Type of fouling	Foulants reported	Impact on the membrane	Effective mitigation methods reported for specific fouling	References
Organic fouling	• HA	• Pore wetting • Penetration to the permeate side	• Backwashing, flushing (0.1 M) NaOH	Alkhatib et al. (2021)
	• Protein (bovine serum albumin)	• High-molecular weight compounds cause gel layer formation without pore wetting.	• Flushing	
Inorganic fouling/ scaling	• Alkaline scaling, $CaCO_3$, $Mg(OH)_2$) • Non-alkaline scaling, $CaSO_4$	• Convert the feed into a basic solution	• Thermal softening • Chemical water softening • Acidification • Membrane filtration	Rolf et al. (2022)
Biological fouling	• Biofilm • Bacteria • Fungi	• Pore wetting • Pore blockage	• Chlorination • Coagulation/ Flocculation	J. Mpala et al. (2023)
Particulate/ colloidal fouling	• Silica, silt, clay, corrosion products	• Pore wetting	• Acidification • Antiscalant • Control of aluminum and iron concentrations are commonly employed in conjunction with MF, UF, NF. • Coagulation/ flocculation processes.	Maeda (2024)

of actual fouling process. Some processes cause different types of fouling, and in practice, they tend to intersect and affect each other, causing complex multi-layered interactions between the processes of deposition. Further research should be conducted in order to gain more knowledge about the behavior and interaction of foulant itself and implication on performance of membrane, these insights are very crucial in the formulation of antifouling strategies. Promising prospects using Advanced Predictive models in the sense that the models can be applied to make pre-emptive judgment on membrane material, working conditions, and economics of the system after and before cleaning or alteration. They would assist in designing more economic and sustainable membrane processes. Although current laboratory-scale innovations have been improved, the scaling of antifouling technologies for industrial scale has been challenging due to scalability, manufacturing and the long-term operational stability. Future research should focus on multifunctional membranes that have high fouling resistance and durability and compatible with large scale operations. One of the most radical directions that can significantly change the membrane technology is the addition of smart stimulus-responsive substances, featuring real-time monitoring capabilities for fouling detection and self-cleaning surfaces.

3. Conclusion

MF is of the most significant challenges limiting the scale-up of membrane technologies for wastewater treatment, desalination, and industrial separations remains the persistent problem. This chapter provides a comprehensive discussion of the fundamental principles of membrane fouling and the mechanisms governing its evolution, and an overview of advanced mitigation strategies that can be adopted to limit the fouling of the membranes significantly. It describes different kinds of fouling, such as organic, inorganic, biological, and particulate fouling and evaluates some of their most significant properties, including membrane properties, feed composition, and conditions of operation. It further discusses the complex physicochemical processes which govern the fouling processes. The adverse effects of fouling on mass transfer, reduced flux, CP, and loss of selectivity, prominently highlight the necessity of reliable predictive models, e.g., cake filtration, pore-blocking and critical flux models, to effectively diagnose and mitigate fouling in membrane processes. The chapter identifies the potential of the means to decrease fouling and enhance the performance of membranes, such as practical mitigation strategies such as feed pretreatment, operational controls, and sophisticated modification of membrane surfaces. It supports the physical and chemical cleaning method as significant to the restoration of the functionality of the membranes but mostly stresses the necessity of appropriate balance of the methods to avoid damaging the membranes. However, there are still challenges in the evolution of cost-efficient and antifouling and high-performing membrane. Future research ought to focus on innovative materials of membranes, real-time monitoring, and optimization of the processes so as to enhance the sustainability and efficiency of membrane technologies. These series of challenges should be mitigated to advance membrane technology, enabling sustainable treatment and recovery of water resources, as well as additional vital applications to meet global demands toward securing clean drinkable water.

References

Abdelrasoul, A., Doan, H., Lohi, A., Abdelrasoul, A., Doan, H., & Lohi, A. (2013). Fouling in membrane filtration and remediation methods. In *Mass transfer—advances in sustainable energy and environment oriented numerical modeling*. IntechOpen. https://doi.org/10.5772/52370

Ahmed, M. A., Amin, S., & Mohamed, A. A. (2023). Fouling in reverse osmosis membranes: Monitoring, characterization, mitigation strategies and future directions. *Heliyon, 9*(4), e14908. https://doi.org/10.1016/j.heliyon.2023.e14908

Alhweij, H., Emanuelsson, E. A. C., Shahid, S., & Wenk, J. (2022). Organic matter removal and antifouling performance of sulfonated polyaniline nanofiltration (S-PANI NF) membranes. *Journal of Environmental Chemical Engineering, 10*(3), 107906. https://doi.org/10.1016/j.jece.2022.107906

Alkhatib, A., Ayari, M. A., & Hawari, A. H. (2021). Fouling mitigation strategies for different foulants in membrane distillation. *Chemical Engineering and Processing - Process Intensification, 167*, 108517. https://doi.org/10.1016/j.cep.2021.108517

Almalki, M. A., Mohamed, A. A., Mahmoud, S. A., Ahmed, H. A., Toghan, A., & Ahmed, M. A. (2025). Characterization of fouling layers and their impact on RO membrane performance degradation in diverse water treatment applications. *Water, Air, and Soil Pollution, 236*(13), 852. https://doi.org/10.1007/s11270-025-08503-z

AlSawaftah, N., Abuwatfa, W., Darwish, N., Husseini, G., AlSawaftah, N., Abuwatfa, W., Darwish, N., & Husseini, G. (2021). A comprehensive review on membrane fouling: Mathematical modelling, prediction, diagnosis, and mitigation. *Water, 13*(9). https://doi.org/10.3390/w13091327

Asante-Sackey, D., Rathilal, S., Kweinor Tetteh, E., Ezugbe, E. O., & Pillay, L. V. (2021). Donnan membrane process for the selective recovery and removal of target metal ions—A mini review. *Membranes, 11*(5), 358. https://doi.org/10.3390/membranes11050358

Augusty, A. C., Fan, L., & Kim, S. (2026). Emerging trends in fouling mitigation for membrane distillation and pervaporation: Implications for desalination and wastewater treatment. *Advanced Membranes, 6*, 100177. https://doi.org/10.1016/j.advmem.2025.100177

Aziz, S., Mazhar, A. R., Ubaid, A., Shah, S. M. H., Riaz, Y., Talha, T., & Jung, D.-W. (2024). A comprehensive review of membrane-based water filtration techniques. *Applied Water Science, 14*(8), 169. https://doi.org/10.1007/s13201-024-02226-y

Bagheri, M., & Mirbagheri, S. A. (2018). Critical review of fouling mitigation strategies in membrane bioreactors treating water and wastewater. *Bioresource Technology, 258*, 318–334. https://doi.org/10.1016/j.biortech.2018.03.026

Bellona, C., Marts, M., & Drewes, J. E. (2010). The effect of organic membrane fouling on the properties and rejection characteristics of nanofiltration membranes. *Separation and Purification Technology, 74*(1), 44–54. https://doi.org/10.1016/j.seppur.2010.05.006

Bhinder, A., Shabani, S., Sadrzadeh, M., Bhinder, A., Shabani, S., & Sadrzadeh, M. (2017). Effect of internal and external concentration polarizations on the performance of forward osmosis process. In *Osmotically driven membrane processes—approach, development and current status*. IntechOpen. https://doi.org/10.5772/intechopen.71343

Bose, D., Bhattacharya, R., Kaur, T., Banerjee, R., Bhatia, T., Ray, A., Batra, B., Mondal, A., Ghosh, P., & Mondal, S. (2024). Overcoming water, sanitation, and hygiene challenges in critical regions of the global community. *Water-Energy Nexus, 7*, 277–296. https://doi.org/10.1016/j.wen.2024.11.003

Cairone, S., Hasan, S. W., Choo, K.-H., Li, C.-W., Zorpas, A. A., Ksibi, M., Zarra, T., Belgiorno, V., & Naddeo, V. (2024). Enhancing membrane fouling control in wastewater treatment processes through artificial intelligence modeling: Research progress and future perspectives. *Euro-Mediterranean Journal for Environmental Integration, 9*(4), 1887–1905. https://doi.org/10.1007/s41207-024-00659-0

Chang, I.-S., & Kim, S.-N. (2005). Wastewater treatment using membrane filtration—Effect of biosolids concentration on cake resistance. *Process Biochemistry, 40*(3), 1307–1314. https://doi.org/10.1016/j.procbio.2004.06.019

Chellam, S., & Xu, W. (2006). Blocking laws analysis of dead-end constant flux microfiltration of compressible cakes. *Journal of Colloid and Interface Science, 301*(1), 248–257. https://doi.org/10.1016/j.jcis.2006.04.064

Cifuentes-Cabezas, M., Bohórquez-Zurita, J. L., Gil-Herrero, S., Vincent-Vela, M. C., Mendoza-Roca, J. A., & Álvarez-Blanco, S. (2023). Deep study on fouling modelling of ultrafiltration membranes used for OMW treatment: Comparison between semi-empirical models, response surface, and artificial neural networks. *Food and Bioprocess Technology, 16*(10), 2126–2146. https://doi.org/10.1007/s11947-023-03033-0

Corbatón-Báguena, M.-J., Vincent-Vela, M.-C., Gozálvez-Zafrilla, J.-M., Álvarez-Blanco, S., Lora-García, J., & Catalán-Martínez, D. (2016). Comparison between artificial neural networks and Hermia's models to assess ultrafiltration performance. *Separation and Purification Technology, 170*, 434–444. https://doi.org/10.1016/j.seppur.2016.07.007

Das, S., O'Connell, M. G., Xu, H., Bernstein, R., Kim, J.-H., Sankhala, K., Segal-Peretz, T., Shevate, R., Zhang, W., Zhou, X., Darling, S. B., & Dunn, J. B. (2022). Assessing advances in anti-fouling membranes to improve process economics and sustainability of water treatment. *ACS ES&T Engineering, 2*(11), 2159–2173. https://doi.org/10.1021/acsestengg.2c00184

Ding, M., Zhang, S., Zhou, H., Xu, H., Tao, H., Lin, T., & Gao, L. (2025). Critical review on organic fouling in membrane distillation for desalination and water purification. *Desalination, 606*, 118718. https://doi.org/10.1016/j.desal.2025.118718

Dischinger, S. M., Miller, D. J., Vermaas, D. A., & Kingsbury, R. S. (2024). Unifying the conversation: Membrane separation performance in energy, water, and industrial applications. *ACS ES&T Engineering, 4*(2), 277–289. https://doi.org/10.1021/acsestengg.3c00475

Du, X., Shi, Y., Jegatheesan, V., & Haq, I. U. (2020). A review on the mechanism, impacts and control methods of membrane fouling in MBR system. *Membranes, 10*(2), 24. https://doi.org/10.3390/membranes10020024

Fang, C., Zhang, X., Gong, X., Feng, W., Zhu, L., & Matsuyama, H. (2020). Enhancing membrane surface antifouling by implanting amphiphilic polymer brushes using a swelling induced entrapment technique. *Colloids and Surfaces B: Biointerfaces, 195*, 111212. https://doi.org/10.1016/j.colsurfb.2020.111212

Farahbakhsh, J., Golgoli, M., Khiadani, M., Najafi, M., Suwaileh, W., Razmjou, A., & Zargar, M. (2024). Recent advances in surface tailoring of thin film forward osmosis membranes: A review. *Chemosphere, 346*, 140493. https://doi.org/10.1016/j.chemosphere.2023.140493

Fersi, C., Gzara, L., & Dhahbi, M. (2009). Flux decline study for textile wastewater treatment by membrane processes. *Desalination, 244*(1), 321–332. https://doi.org/10.1016/j.desal.2008.04.046

Flemming, H.-C. (2011). Microbial biofouling: Unsolved problems, insufficient approaches, and possible solutions. In H.-C. Flemming, J. Wingender, & U. Szewzyk (Eds.), *Biofilm highlights* (pp. 81–109). Springer. https://doi.org/10.1007/978-3-642-19940-0_5

Foorginezhad, S., Zerafat, M. M., Ismail, A. F., & Goh, P. S. (2025). Emerging membrane technologies for sustainable water treatment: A review on recent advances. *Environmental Science: Advances, 4*(4), 530–570. https://doi.org/10.1039/D4VA00378K

Fortunato, L., Lamprea, A. F., & Leiknes, T. (2020). Evaluation of membrane fouling mitigation strategies in an algal membrane photobioreactor (AMPBR) treating secondary wastewater effluent. *Science of the Total Environment, 708*, 134548. https://doi.org/10.1016/j.scitotenv.2019.134548

Gao, Q., Duan, L., Jia, Y., Zhang, H., Liu, J., & Yang, W. (2023). A comprehensive analysis of the impact of inorganic matter on membrane organic fouling: A mini review. *Membranes, 13*(10), 837. https://doi.org/10.3390/membranes13100837

Ghalami Choobar, B., Alaei Shahmirzadi, M. A., Kargari, A., & Manouchehri, M. (2019). Fouling mechanism identification and analysis in microfiltration of laundry wastewater. *Journal of Environmental Chemical Engineering, 7*(2), 103030. https://doi.org/10.1016/j.jece.2019.103030

Goi, Y. K., Li, M., & Liang, Y. Y. (2025). A comprehensive review on forward osmosis mass transfer and fouling: Mathematical modeling, mechanism, prediction and optimization. *Journal of Water Process Engineering, 72*, 107677. https://doi.org/10.1016/j.jwpe.2025.107677

Gul, A., Hruza, J., & Yalcinkaya, F. (2021). Fouling and chemical cleaning of microfiltration membranes: A mini-review. *Polymers, 13*(6), 846. https://doi.org/10.3390/polym13060846

Hoang, T. A. (2022). Chapter 2—Mechanisms of scale formation and inhibition. In Z. Amjad, & K. D. Demadis (Eds.), *Water-formed deposits* (pp. 13–47). Elsevier. https://doi.org/10.1016/B978-0-12-822896-8.00031-5

Hoek, E. M. V., Weigand, T. M., & Edalat, A. (2022). Reverse osmosis membrane biofouling: Causes, consequences and countermeasures. *npj Clean Water, 5*(1), 45. https://doi.org/10.1038/s41545-022-00183-0

Hong, S., & Elimelech, M. (1997). Chemical and physical aspects of natural organic matter (NOM) fouling of nanofiltration membranes. *Journal of Membrane Science, 132*(2), 159–181. https://doi.org/10.1016/S0376-7388(97)00060-4

Hwang, S.-T. (2011). Fundamentals of membrane transport. *Korean Journal of Chemical Engineering, 28*(1), 1–15. https://doi.org/10.1007/s11814-010-0493-z

Jaćimovski, D., Garić-Grulović, R., Vučetić, N., Pjanović, R., & Bošković-Vragolović, N. (2016). Mass transfer and concentration boundary layer in a particulate fluidized bed. *Powder Technology, 303*, 68–75. https://doi.org/10.1016/j.powtec.2016.09.025

Jahani, Z., Mosaffa, E., Oroujzadeh, M., & Ghafuri, H. (2023). Performance evaluation of polyethersulfone membranes modified with poly (acrylic acid-co-N-vinyl pyrrolidone) grafted mesoporous carbon nitride for effective removal of cadmium(II) from wastewater. *Polymers for Advanced Technologies, 34*(12), 3803–3817. https://doi.org/10.1002/pat.6183

Jiang, S., Zhang, Y., Zhao, F., Yu, Z., Zhou, X., & Chu, H. (2018). Impact of transmembrane pressure (TMP) on membrane fouling in microalgae harvesting with a uniform shearing vibration membrane system. *Algal Research, 35*, 613–623. https://doi.org/10.1016/j.algal.2018.10.003

Kim, S., & Hoek, E. M. V. (2005). Modeling concentration polarization in reverse osmosis processes. *Desalination, 186*(1), 111–128. https://doi.org/10.1016/j.desal.2005.05.017

Kim, P., Kim, H., Oh, H., Kang, J., Lee, S., & Park, K. (2022). Influence of solute size on membrane fouling during polysaccharide enrichment using dense polymeric UF membrane: Measurements and mechanisms. *Membranes, 12*(2), 142. https://doi.org/10.3390/membranes12020142

Kim, S., & Park, C. (2022). Fouling behavior and cleaning strategies of ceramic ultrafiltration membranes for the treatment and reuse of laundry wastewater. *Journal of Water Process Engineering, 48*, 102840. https://doi.org/10.1016/j.jwpe.2022.102840

Koo, C. H., Mohammad, A. W., Suja', F., & Meor Talib, M. Z. (2012). Review of the effect of selected physicochemical factors on membrane fouling propensity based on fouling indices. *Desalination, 287*, 167–177. https://doi.org/10.1016/j.desal.2011.11.003

Kookana, R. S., Drechsel, P., Jamwal, P., & Vanderzalm, J. (2020). Urbanisation and emerging economies: Issues and potential solutions for water and food security. *Science of the Total Environment, 732*, 139057. https://doi.org/10.1016/j.scitotenv.2020.139057

Koonani, H., & Amirinejad, M. (2019). Combined three mechanisms models for membrane fouling during microfiltration. *Journal of Membrane Science and Research, 5*(4), 274–282. https://doi.org/10.22079/jmsr.2019.95781.1224

Kosvintsev, S., Holdich, R. G., Cumming, I. W., & Starov, V. M. (2002). Modelling of dead-end microfiltration with pore blocking and cake formation. *Journal of Membrane Science, 208*(1), 181–192. https://doi.org/10.1016/S0376-7388(02)00252-1

Kujawa, J., Chrzanowska, E., & Kujawski, W. (2019). Transport properties and fouling issues of membranes utilized for the concentration of dairy products by air-gap membrane distillation and microfiltration. *Chemical Papers, 73*(3), 565–582. https://doi.org/10.1007/s11696-018-0615-3

Le, N. L., & Nunes, S. P. (2016). Materials and membrane technologies for water and energy sustainability. *Sustainable Materials and Technologies, 7*, 1–28. https://doi.org/10.1016/j.susmat.2016.02.001

Lester, Y., Hazut, A., & Spanier, A. (2022). Formation of organic fouling during membrane desalination: The effect of divalent cations and the use of an online visual monitoring method. *Membranes, 12*(12), 1177. https://doi.org/10.3390/membranes12121177

Li, H., & Chen, V. (2010). Chapter 10—Membrane fouling and cleaning in food and bioprocessing. In Z. F. Cui, & H. S. Muralidhara (Eds.), *Membrane technology* (pp. 213–254). Butterworth-Heinemann. https://doi.org/10.1016/B978-1-85617-632-3.00010-0

Li, J., Lu, C., Sun, J., Peng, H., Li, Q., Hosseini, S. S., Zhu, Y., Sun, M., & Ma, B. (2025a). Membrane recycling and resource utilization: Latest progress and prospects. *Journal of Environmental Sciences, 156*, 346–359. https://doi.org/10.1016/j.jes.2024.07.004

Li, X., Xu, M., Liu, X., She, Q., Lau, W. J., & Yang, L. (2025b). Surface-engineered nanofiltration membranes for sustainable lithium recovery from real brine: Addressing fouling and scaling challenges. *Water Research, 278*, 123400. https://doi.org/10.1016/j.watres.2025.123400

Li, W., Wu, W., & Li, Z. (2018). Controlling interlayer spacing of graphene oxide membranes by external pressure regulation. *ACS Nano, 12*(9), 9309–9317. https://doi.org/10.1021/acsnano.8b04187

Liang, S., Zhao, T., Zhang, J., Sun, F., Liu, C., & Song, L. (2012). Determination of fouling-related critical flux in self-forming dynamic membrane bioreactors: Interference of membrane compressibility. *Journal of Membrane Science, 390–391*, 113–120. https://doi.org/10.1016/j.memsci.2011.11.026

Liang, S., Zhu, L., Wang, S., Chen, L., & Fang, H. (2021). Fast reduced graphene-based membranes with high desalination performance. *Membranes, 11*(11), 846. https://doi.org/10.3390/membranes11110846

Liu, S., Shikhov, I., & Arns, C. (2024). Mechanisms of pore-clogging using a high-resolution CFD-DEM colloid transport model. *Transport in Porous Media, 151*(4), 831–851. https://doi.org/10.1007/s11242-024-02072-1

Liu, J., Wang, Z., Tang, C. Y., & Leckie, J. O. (2018). Modeling dynamics of colloidal fouling of RO/NF membranes with A novel collision-attachment approach. *Environmental Science & Technology, 52*(3), 1471–1478. https://doi.org/10.1021/acs.est.7b05598

Liu, J., Zhao, Y., Fan, Y., Yang, H., Wang, Z., Chen, Y., & Tang, C. Y. (2021). Dissect the role of particle size through collision-attachment simulations for colloidal fouling of RO/NF membranes. *Journal of Membrane Science, 638*, 119679. https://doi.org/10.1016/j.memsci.2021.119679

Ma, Z., Zhang, S., Chen, G., Xiao, K., Li, M., Gao, Y., Liang, S., & Huang, X. (2019). Superhydrophilic and oleophobic membrane functionalized with heterogeneously tailored two-dimensional layered double hydroxide nanosheets for antifouling. *Journal of Membrane Science, 577*, 165–175. https://doi.org/10.1016/j.memsci.2019.01.054

Maeda, Y. (2024). Fouling of reverse osmosis (RO) and nanofiltration (NF) membranes by low molecular weight organic compounds (LMWOCs), part 1: Fundamentals and mechanism. *Membranes, 14*(10), 221. https://doi.org/10.3390/membranes14100221

Mallah, N. B., Shah, A. A., Pirzada, A. M., Ali, I., Khan, M. I., Jatoi, A. S., Ullman, J. L., & Mahar, R. B. (2024). Advanced control strategies of membrane fouling in wastewater treatment: A review. *Processes, 12*(12), 2681. https://doi.org/10.3390/pr12122681

Matin, A., Rahman, F., Shafi, H. Z., & Zubair, S. M. (2019). Scaling of reverse osmosis membranes used in water desalination: Phenomena, impact, and control; future directions. *Desalination, 455*, 135–157. https://doi.org/10.1016/j.desal.2018.12.009

McCutcheon, J. R., & Elimelech, M. (2006). Influence of concentrative and dilutive internal concentration polarization on flux behavior in forward osmosis. *Journal of Membrane Science, 284*(1), 237–247. https://doi.org/10.1016/j.memsci.2006.07.049

Merdaw, A. A., Sharif, A. O., & Derwish, G. A. W. (2011). Mass transfer in pressure-driven membrane separation processes, part I. *Chemical Engineering Journal, 168*(1), 215–228. https://doi.org/10.1016/j.cej.2010.12.071

Miller, D. J., Paul, D. R., & Freeman, B. D. (2013). A crossflow filtration system for constant permeate flux membrane fouling characterization. *Review of Scientific Instruments, 84*(3), 035003. https://doi.org/10.1063/1.4794909

Mittal, S., Gupta, A., Srivastava, S., & Jain, M. (2021). Artificial neural network based modeling of the vacuum membrane distillation process: Effects of operating parameters on membrane fouling. *Chemical Engineering and Processing - Process Intensification, 164*, 108403. https://doi.org/10.1016/j.cep.2021.108403

Mpala, T. J., Etale, A., Richards, H., & Nthunya, L. N. (2023). Biofouling phenomena in membrane distillation: Mechanisms and mitigation strategies. *Environmental Science: Advances, 2*(1), 39–54. https://doi.org/10.1039/D2VA00161F

Nakamura, K., & Matsumoto, K. (1998). Adsorption behavior of BSA in microfiltration with porous glass membrane. *Journal of Membrane Science, 145*(1), 119–128. https://doi.org/10.1016/S0376-7388(98)00070-2

Nations, U. (2022). *Human Settlements*. United Nations. https://www.un.org/en/global-issues/human-settlements.

Nguyen, T.-T., Adha, R. S., Lee, C., Kim, D.-H., & Kim, I. S. (2021). Quantifying the influence of divalent cations mass transport on critical flux and organic fouling mechanism of forward osmosis membrane. *Desalination, 512*, 115146. https://doi.org/10.1016/j.desal.2021.115146.

Park, S., Nam, T., You, J., Kim, E.-S., Choi, I., Park, J., & Cho, K. H. (2019). Evaluating membrane fouling potentials of dissolved organic matter in brackish water. *Water Research, 149*, 65–73. https://doi.org/10.1016/j.watres.2018.10.079

Porter, M. C. (1972). Concentration polarization with membrane ultrafiltration. *Product R&D, 11*(3), 234–248. https://doi.org/10.1021/i360043a002

Qin, G., Zhou, H., Tanis, B., Rietveld, L. C., & Heijman, S. G. J. (2025). Impact of ionic strength and surface charge on ceramic membrane fouling by oil-in-water emulsions: A quantitative analysis using DLVO and XDLVO models. *Separation and Purification Technology, 372*, 133424. https://doi.org/10.1016/j.seppur.2025.133424.

Raspati, G. S., & Leiknes, T. (2015). Applicability of blocking laws in Non-Newtonian fluid membrane filtration. *Desalination and Water Treatment, 54*(7), 1771–1781. https://doi.org/10.1080/19443994.2014.891465

Ray, S. S., Chen, S.-S., Sangeetha, D., Chang, H.-M., Thanh, C. N. D., Le, Q. H., & Ku, H.-M. (2018). Developments in forward osmosis and membrane distillation for desalination of waters. *Environmental Chemistry Letters, 16*(4), 1247–1265. https://doi.org/10.1007/s10311-018-0750-7

Ray, S. S., Deb, C. K., Chang, H.-M., Chen, S.-S., & Ganesapillai, M. (2021). Fouling and wetting: A major challenge for membrane distillation. In K. M. Gothandam, S. Ranjan, N. Dasgupta, & E. Lichtfouse (Eds.), *Environmental biotechnology, 3* (pp. 161–192). Springer International Publishing. https://doi.org/10.1007/978-3-030-48973-1_6

Rolf, J., Cao, T., Huang, X., Boo, C., Li, Q., & Elimelech, M. (2022). Inorganic scaling in membrane desalination: Models, mechanisms, and characterization methods. *Environmental Science & Technology, 56*(12), 7484–7511. https://doi.org/10.1021/acs.est.2c01858

Sadr, S. M. K., & Saroj, D. P. (2015). 14—Membrane technologies for municipal wastewater treatment. In A. Basile, A. Cassano, & N. K. Rastogi (Eds.), *Advances in membrane technologies for water treatment* (pp. 443–463). Woodhead Publishing. https://doi.org/10.1016/B978-1-78242-121-4.00014-9

Saffarimiandoab, F., Gul, B. Y., Tasdemir, R. S., Ilter, S. E., Unal, S., Tunaboylu, B., Menceloglu, Y. Z., & Koyuncu, İ. (2021). A review on membrane fouling: Membrane modification. *Desalination and Water Treatment, 216*, 47–70. https://doi.org/10.5004/dwt.2021.26815.

Salahi, A., Abbasi, M., & Mohammadi, T. (2010). Permeate flux decline during UF of oily wastewater: Experimental and modeling. *Desalination, 251*(1), 153–160. https://doi.org/10.1016/j.desal.2009.08.006

Salehi, M. (2022). Global water shortage and potable water safety; today's concern and tomorrow's crisis. *Environment International, 158*, 106936. https://doi.org/10.1016/j.envint.2021.106936

Samavati, Z., Samavati, A., Goh, P. S., Fauzi Ismail, A., & Sohaimi Abdullah, M. (2023). A comprehensive review of recent advances in nanofiltration membranes for heavy metal removal from wastewater. *Chemical Engineering Research and Design, 189*, 530–571. https://doi.org/10.1016/j.cherd.2022.11.042

Sandoval-García, V., Ruano, M. V., Alliet, M., Brepols, C., Comas, J., Harmand, J., Heran, M., Mannina, G., Rodriguez-Roda, I., Smets, I., & Robles, A. (2025). Modeling MBR fouling: A critical review analysis towards establishing a framework for good modeling practices. *Water Research, 268*, 122611. https://doi.org/10.1016/j.watres.2024.122611

Sandu, T., Sârbu, A., Căprărescu, S., Stoica, E.-B., Iordache, T.-V., & Chiriac, A.-L. (2022). Polymer membranes as innovative means of quality restoring for wastewater bearing heavy metals. *Membranes, 12*(12), 1179. https://doi.org/10.3390/membranes12121179

Sauer, F., Löwer, E., Henn, H., Peuker, U., & Hoffner, B. (2024). Displacement washing of filter cakes with a fine particle top layer. https://doi.org/10.1002/ceat.202400394.

Scholes, C. A. (2015). Durability of membrane (lifetime of membrane). In *Encyclopedia of membranes* (pp. 1–2). Berlin, Heidelberg: Springer. https://doi.org/10.1007/978-3-642-40872-4_820-4

Seyed Shahabadi, S. M., & Reyhani, A. (2014). Optimization of operating conditions in ultrafiltration process for produced water treatment via the full factorial design methodology. *Separation and Purification Technology, 132*, 50–61. https://doi.org/10.1016/j.seppur.2014.04.051

Shahid, M. K., Pyo, M., & Choi, Y.-G. (2019). Inorganic fouling control in reverse osmosis wastewater reclamation by purging carbon dioxide. *Environmental Science and Pollution Research, 26*(2), 1094–1102. https://doi.org/10.1007/s11356-017-9008-3

Shahkaramipour, N., Tran, T. N., Ramanan, S., & Lin, H. (2017). Membranes with surface-enhanced antifouling properties for water purification. *Membranes, 7*(1), 13. https://doi.org/10.3390/membranes7010013

Shahzad, A., Oh, J.-M., Azam, M., Iqbal, J., Hussain, S., Miran, W., & Rasool, K. (2021). Advances in the synthesis and application of anti-fouling membranes using two-dimensional nanomaterials. *Membranes, 11* (8), 605. https://doi.org/10.3390/membranes11080605

Shen, S., Hao, Y., Zhang, Y., Zhang, G., Zhou, X., & Bai, R. B. (2018). Enhancing the antifouling properties of Poly(vinylidene fluoride) (PVDF) membrane through a novel blending and surface-grafting modification approach. *ACS Omega, 3*(12), 17403–17415. https://doi.org/10.1021/acsomega.8b02569

Shi, M., Xu, R., Luo, H., Luo, H., Meers, E., He, Q., & Yan, S. (2025). Membrane fouling characterization and cleaning performance in anaerobic digestion effluent treatment by AGMD: Effect of feed pH. *Desalination, 599*, 118438. https://doi.org/10.1016/j.desal.2024.118438.

Shoshaa, R., Ashfaq, M. Y., & Al-Ghouti, M. A. (2023). Recent developments in ultrafiltration membrane technology for the removal of potentially toxic elements, and enhanced antifouling performance: A review. *Environmental Technology and Innovation, 31*, 103162. https://doi.org/10.1016/j.eti.2023.103162

Sim, L. N., Ye, Y., Chen, V., & Fane, A. G. (2011). Investigations of the coupled effect of cake-enhanced osmotic pressure and colloidal fouling in RO using crossflow sampler-modified fouling index ultrafiltration. *Desalination, 273*(1), 184–196. https://doi.org/10.1016/j.desal.2011.01.059

Singh, R. (2015). Chapter 1—Introduction to membrane technology. In R. Singh (Ed.), *Membrane technology and engineering for water purification* (2nd ed., pp. 1–80). Butterworth-Heinemann. https://doi.org/10.1016/B978-0-444-63362-0.00001-X

Spiegler, K. S., & Kedem, O. (1966). Thermodynamics of hyperfiltration (reverse osmosis): Criteria for efficient membranes. *Desalination, 1*(4), 311–326. https://doi.org/10.1016/S0011-9164(00)80018-1

Syed, V., Ray, S. S., & Selvasembian, R. (2025). A review on biopolymer nanocomposite-based membranes for oil/water separation and life cycle assessment. *International Journal of Biological Macromolecules, 332*, 148410. https://doi.org/10.1016/j.ijbiomac.2025.148410

Tawalbeh, M., Al-Othman, A., Abdelwahab, N., Alami, A. H., & Olabi, A. G. (2021). Recent developments in pressure retarded osmosis for desalination and power generation. *Renewable and Sustainable Energy Reviews, 138*, 110492. https://doi.org/10.1016/j.rser.2020.110492

Tayeh, Y. A. (2024). A comprehensive review of reverse osmosis desalination: Technology, water sources, membrane processes, fouling, and cleaning. *Desalination and Water Treatment, 320*, 100882. https://doi.org/10.1016/j.dwt.2024.100882

Tayeh, Y. A., Alazaiza, M. Y. D., Alzghoul, T. M., & Bashir, M. J. (2025). A comprehensive review of RO membrane fouling: Mechanisms, categories, cleaning methods and pretreatment technologies. *Journal of Hazardous Materials Advances, 18*, 100684. https://doi.org/10.1016/j.hazadv.2025.100684

Vincent Vela, M. C., Álvarez Blanco, S., Lora García, J., & Bergantiños Rodríguez, E. (2009). Analysis of membrane pore blocking models adapted to crossflow ultrafiltration in the ultrafiltration of PEG. *Chemical Engineering Journal, 149*(1), 232–241. https://doi.org/10.1016/j.cej.2008.10.027

Virga, E., Žvab, K., & de Vos, W. M. (2021). Fouling of nanofiltration membranes based on polyelectrolyte multilayers: The effect of a zwitterionic final layer. *Journal of Membrane Science, 620*, 118793. https://doi.org/10.1016/j.memsci.2020.118793

Wang, F., & Tarabara, V. V. (2008). Pore blocking mechanisms during early stages of membrane fouling by colloids. *Journal of Colloid and Interface Science, 328*(2), 464–469. https://doi.org/10.1016/j.jcis.2008.09.028

Xu, P., Drewes, J. E., Kim, T.-U., Bellona, C., & Amy, G. (2006). Effect of membrane fouling on transport of organic contaminants in NF/RO membrane applications. *Journal of Membrane Science, 279*(1), 165–175. https://doi.org/10.1016/j.memsci.2005.12.001

Xu, H., Xiao, K., Wang, X., Liang, S., Wei, C., Wen, X., & Huang, X. (2020). Outlining the roles of membrane-foulant and foulant-foulant interactions in organic fouling during microfiltration and ultrafiltration: A mini-review. *Frontiers in Chemistry, 8*. https://doi.org/10.3389/fchem.2020.00417

UN World Water Development Report 2017—*Wastewater, the Untapped*. (2017). https://www.unesco.org/en/wwap/wwdr/2017

Yang, H., Yu, X., Liu, J., Tang, Z., Huang, T., Wang, Z., Zhong, Q., Long, Z., & Wang, L. (2022). A concise review of theoretical models and numerical simulations of membrane fouling. *Water, 14*(21), 3537. https://doi.org/10.3390/w14213537

Yang, M., Yu, D., Liu, M., Zheng, L., Zheng, X., Wei, Y., Wang, F., & Fan, Y. (2017). Optimization of MBR hydrodynamics for cake layer fouling control through CFD simulation and RSM design. *Bioresource Technology, 227*, 102–111. https://doi.org/10.1016/j.biortech.2016.12.027

Yiantsios, S. G., & Karabelas, A. J. (2003). An assessment of the silt density index based on RO membrane colloidal fouling experiments with iron oxide particles. *Desalination, 151*(3), 229–238. https://doi.org/10.1016/S0011-9164(02)01015-9

Yuliwati, E., Ismail, A. F., Othman, M. H. D., Shirazi, M. M. A., Yuliwati, E., Ismail, A. F., Othman, M. H. D., & Shirazi, M. M. A. (2022). Critical flux and fouling analysis of PVDF-mixed matrix membranes for reclamation of refinery-produced wastewater: Effect of mixed liquor suspended solids concentration and aeration. *Membranes, 12*(2). https://doi.org/10.3390/membranes12020161

Zhang, W., & Ding, L. (2015). Investigation of membrane fouling mechanisms using blocking models in the case of shear-enhanced ultrafiltration. *Separation and Purification Technology, 141*, 160–169. https://doi.org/10.1016/j.seppur.2014.11.041

Zhao, S., Liao, Z., Fane, A., Li, J., Tang, C., Zheng, C., Lin, J., & Kong, L. (2021). Engineering antifouling reverse osmosis membranes: A review. *Desalination, 499*, 114857. https://doi.org/10.1016/j.desal.2020.114857

Zheng, Y., Zhang, W., Tang, B., Ding, J., Zheng, Y., & Zhang, Z. (2018). Membrane fouling mechanism of biofilm-membrane bioreactor (BF-MBR): Pore blocking model and membrane cleaning. *Bioresource Technology, 250*, 398–405. https://doi.org/10.1016/j.biortech.2017.11.036

Zouboulis, A. I., & Gkotsis, P. K. (2020). 7—Fouling challenges in ceramic MBR systems. In A. Basile, K. Ghasemzadeh, & E. Jalilnejad (Eds.), *Current trends and future developments on (Bio-) membranes* (pp. 199–217). Elsevier. https://doi.org/10.1016/B978-0-12-816822-6.00007-0

Zulkefli, N. F., Alias, N. H., Jamaluddin, N. S., Abdullah, N., Abdul Manaf, S. F., Othman, N. H., Marpani, F., Mat-Shayuti, M. S., & Kusworo, T. D. (2021). Recent mitigation strategies on membrane fouling for oily wastewater treatment. *Membranes, 12*(1), 26. https://doi.org/10.3390/membranes12010026

CHAPTER

6 Smart and selective membrane technologies for the removal of micropollutants and trace metals from wastewater systems

Srinath Ravi

Department of Civil, Urban, Earth and Environmental Engineering, Ulsan National Institute of Science and Technology (UNIST), Ulsan, South Korea

1. Introduction

The accelerating pace of industrialization, urban expansion, and agricultural intensification has drastically increased the release of persistent chemical contaminants into aquatic environments. Among these, micropollutants such as pharmaceuticals, personal care products, pesticides, surfactants, endocrine disruptors, and per- and polyfluoroalkyl substances (PFAS), and trace metals including chromium (VI), lead (Pb^{2+}), mercury (Hg^{2+}), cadmium (Cd^{2+}), and nickel (Ni^{2+}) represent two of the most challenging classes of pollutants now recognized globally (Deb et al., 2025; Fu et al., 2011; Wołowicz and Munir, 2025). These contaminants are typically present at ng L^{-1} to μg L^{-1} levels, yet even such trace concentrations are sufficient to trigger chronic toxicity, endocrine disruption, mutagenicity, and bioaccumulation throughout the food web (Deb et al., 2025; Gupta et al., 2024; Singh et al., 2025a). Heavy metals, unlike organic pollutants, are non-biodegradable and readily accumulate in sediments and living organisms. Elevated concentrations of Pb, Hg, or Cr(VI) in drinking-water sources have been correlated with neurological, renal, and developmental disorders in humans (Fu et al., 2011). In parallel, organic micropollutants (OMPs) have emerged as a new generation of "invisible contaminants." The global rise in pharmaceutical and personal care product consumption, fueled by expanding healthcare access and population growth, has sharply increased their discharge into municipal and hospital effluents. Worldwide consumption of pharmaceutically active compounds now exceeds 100,000 tons per year, with more than 10,000 formulations approved for human use, and most are excreted unchanged or as active metabolites, eventually entering wastewater systems. These residues persist through sewage treatment processes and are subsequently detected in surface water, groundwater, and even drinking water supplies (Amador et al., 2015; Singh et al., 2025a).

The spatial and temporal persistence of these contaminants is further exacerbated by global water scarcity and the growing demand for wastewater reuse. Studies have shown that emerging organic micropollutants (EOMs) possess high chemical stability, hydrophilicity, and low volatility, which enable them to survive conventional treatment and circulate through aquatic and terrestrial

Sustainable Water Management. https://doi.org/10.1016/B978-0-443-44853-9.00003-1

compartments (Ahamad et al., 2019). Substances such as diclofenac, ibuprofen, triclosan, carbamazepine, and naproxen are frequently reported in effluents across Europe, Asia, and North America. Similarly, antibiotics and hormones in wastewater contribute to the propagation of antibiotic-resistant bacteria and resistance genes, posing a dual chemical-biological hazard (Ahamad et al., 2019; Gupta et al., 2024). In the industrial corridors of developing countries, the coexistence of heavy metals and OMPs magnifies toxicity through synergistic effects. For instance, complexation between Pb^{2+} and organic ligands alters metal mobility, while redox-active species such as Cr (VI) catalyze the oxidative transformation of organics into more persistent intermediates. Because of these interactions, both pollutant classes must be considered simultaneously when designing treatment technologies (Abo-Farha et al., 2009). From a regulatory perspective, agencies such as the U.S. EPA and WHO now classify trace metals and EOMs as priority pollutants requiring monitoring and removal to below ppb levels (Abu-Obaid et al., 2023). Despite significant policy advances, a large fraction of micropollutants remains unregulated or insufficiently monitored, especially in low-income regions. Consequently, the development of high-performance, selective, and sustainable treatment processes has become an urgent research frontier.

Over the past 2 decades, membrane-based separations have gained prominence as sustainable and high-selectivity alternatives for treating complex wastewater matrices. Processes such as ultrafiltration (UF), nanofiltration (NF), reverse osmosis (RO), and membrane distillation (MD) are increasingly adopted because they offer modular operation, small footprints, and tunable selectivity while avoiding the addition of chemical reagents (Al-Jlil et al., 2009; Annamalai et al., 2022; Costa et al., 2023; Covaliu-Mierlă et al., 2023; Gupta et al., 2024). Membranes separate contaminants through a combination of size sieving, electrostatic (Donnan) exclusion, adsorption, and diffusion, allowing efficient removal of multivalent ions and moderately sized organic molecules. Among these, nanofiltration has attracted particular attention for its intermediate molecular-weight-cut-off ($\approx$200–500 Da) and partial charge-based rejection, making it well suited for simultaneous removal of heavy-metal complexes and organic micropollutants. Reverse osmosis achieves near-complete demineralization but at the cost of higher hydraulic pressure and energy input, whereas UF and MD serve as pretreatment or polishing steps within hybrid configurations (Covaliu-Mierlă et al., 2023; Lu et al., 2025; Ravi et al., 2025a, 2025b). A major advantage of membrane processes lies in their inherent selectivity and scalability. They can be combined with oxidation, adsorption, or biological steps to form hybrid systems, for example, powdered-activated-carbon/UF or NF/Advanced Oxidation Process (AOP) couplings, that achieve broader contaminant coverage and mitigate fouling (Gupta et al., 2024; Seo et al., 2025; Wei et al., 2025). Membrane-bioreactor (MBR) technology, which integrates biological degradation with UF or NF, has already demonstrated enhanced removal of antibiotics, hormones, and trace metals compared with conventional activated-sludge units (Asadollahzadeh et al., 2026; Kirthiga et al., 2025; Zhou et al., 2025). Figure 6.1 illustrates an integrated conceptual framework linking the major anthropogenic sources of water contamination with smart membrane-based separation mechanisms and treatment outcomes. Industrial effluents, municipal and hospital wastewater, and agricultural runoff serve as the primary emission sources of organic micropollutants and trace metals, including pharmaceuticals, personal care products, pesticides, endocrine-disrupting compounds, PFAS, and multivalent metal ions such as Cr (VI), Pb^{2+}, Cd^{2+}, and Hg^{2+}. These structurally and chemically diverse contaminants enter wastewater streams and are subsequently treated using advanced membrane-based separation processes. As contaminated feed water passes through selective membrane layers, pollutants are removed through synergistic mechanisms including size exclusion,

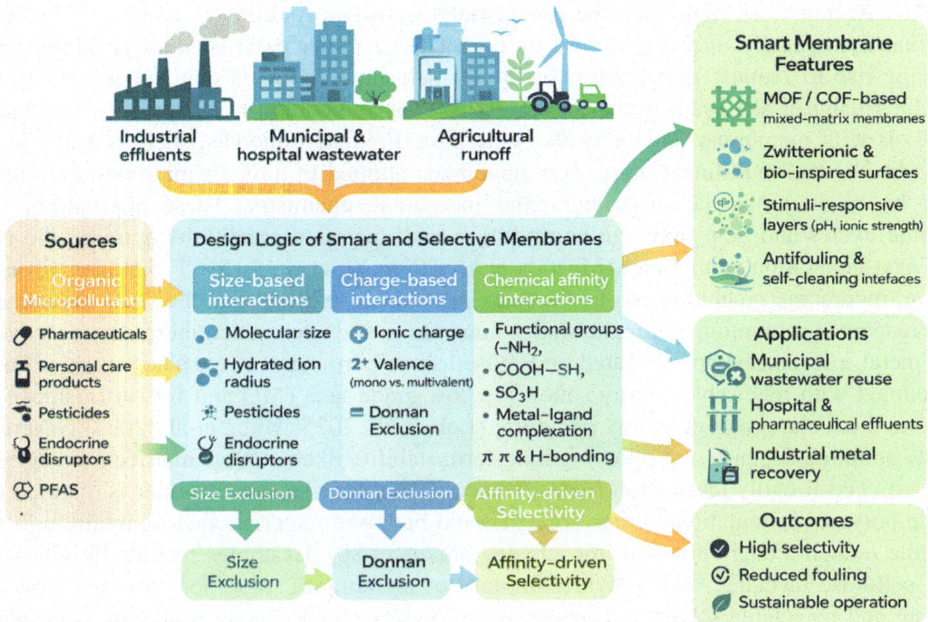

FIGURE 6.1

Conceptual framework linking sources of organic micropollutants and trace metals to smart and selective membrane design, separation mechanisms, and wastewater treatment applications.

electrostatic (Donnan) exclusion, and chemical affinity driven interactions. Membrane processes such as UF, NF, RO, and MD enable the effective separation of organic molecules and multivalent inorganic ions, resulting in high-quality permeate. The figure further highlights the defining features of next-generation smart membranes, such as enhanced selectivity, antifouling behavior, and sustainable operation, achieved through tailored material chemistry and functional design, while also emphasizing reduced chemical consumption and the potential for resource recovery within circular water-treatment systems.

Despite these advantages, two long-standing challenges, fouling and the permeability–selectivity trade-off, continue to limit performance (Bernardes et al., 2025; Xiao et al., 2025a). To address these issues, recent studies have focused on functional and smart membrane design. Incorporation of metal–organic frameworks (MOFs), covalent organic frameworks (COFs), graphene oxide, or other nanomaterials within polymeric matrices improves surface hydrophilicity, charge density, and adsorptive affinity. Such mixed matrix membranes (MMMs) exhibit superior permeability and selective binding toward pollutants such as Cr (VI), Pb^{2+}, and diclofenac (Costa et al., 2023; Islam et al., 2025; Qin et al., 2025; Rahbari-Sisakht et al., 2025). Furthermore, surface modification techniques such as crosslinking, interfacial polymerization, plasma activation, or dopamine-assisted dip-coating introduce functional groups (–COOH, $–NH_2$, $–SO_3H$, or quaternary ammonium) that enable charge-driven rejection of ionic species (Foorginezhad et al., 2025; Huyen et al., 2025; Lu et al., 2025; Yao et al., 2021). Recent work highlights the use of polyelectrolyte multilayers and stimuli-responsive coatings that dynamically alter wettability or pore charge under external cues (pH, temperature, light), giving rise to “smart” membranes capable of self-cleaning or on-demand selectivity (Costa et al., 2023; Huang et al., 2023; Parbat et al., 2023). The integration of data-driven and machine-learning tools with membrane science is also emerging. Predictive models, such as artificial neural networks (ANNs) and multilinear regression, have been applied to forecast micropollutant rejection efficiency based on molecular descriptors and membrane chemistry. These approaches shorten experimental cycles and help optimize composition and operation conditions, guiding the rational design of next-generation membranes (Mustafa et al., 2025; Xu et al., 2024). From an environmental perspective, membrane technologies offer a more sustainable profile than chemical methods. They minimize reagent consumption, eliminate sludge generation, and allow potential resource recovery, for instance, metal ions can be concentrated and reused from permeate or retentate streams (Fu et al., 2011). Coupled with renewable-energy sources or low-grade heat (MD and forward osmosis), they contribute to a circular water economy paradigm (Loh et al., 2025; Voicu et al., 2023).Nevertheless, widespread adoption depends on improving long-term stability in aggressive matrices and developing cost-effective, eco-friendly fabrication routes. Hence, the current research frontier is directed toward sustainable polymer formulations, green solvents, and biobased materials that maintain high performance while reducing environmental impact (dos Santos et al.). Examples include bio-derived polyimides, cellulose acetate, and polybenzimidazole membranes modified through low-energy crosslinking and recyclable solvents. The success of these technologies in removing persistent contaminants such as pharmaceuticals, PFAS, and Cr (VI) illustrates their transformative potential (Balali-Mood et al., 2021; Deb et al., 2025; Singh et al., 2025a). For instance, NF and RO membranes have achieved >95% rejection of diclofenac, ibuprofen, and naproxen (Bilal et al., 2019; Gupta et al., 2024), while functionalized MMM systems incorporating MOFs or GO nanoparticles have shown nearly complete Cr(VI) and Pb^{2+} removal. Such evidence underscores membranes as the cornerstone of future smart treatment systems capable of simultaneously addressing both organic and inorganic pollutants (Jagirani et al., 2024; Ragab et al., 2016; Wei et al., 2025). Collectively, these advances establish smart and selectively engineered membrane systems as a pivotal and unifying platform for overcoming conventional performance limitations and enabling sustainable, high-efficiency removal of organic micropollutants and trace metals from complex wastewater matrices. For clarity and comparison, Table 6.1 presents a concise overview of state-of-the-art smart and selective membranes developed over the last decade, illustrating how functional chemistry and stimuli-responsive design enable efficient removal of micropollutants and trace metals.

2. Micropollutants source and types

Organic micropollutants (OMPs) and trace metals pose a significant global challenge for environmental conservation due to their widespread presence in aquatic environments (Wei et al., 2025). Emerging organic micropollutants (EOMs) are a broad class of chemical compounds increasingly found in water and wastewater sources due to human activities, industrial growth, and other anthropogenic inputs (Wołowicz and Munir, 2025). The discharge of OMPs, such as antibiotics, analgesics, antipyretics, anti-inflammatories, insecticides, antiepileptics, -receptor blockers/agonists, preservatives, and hormones, into the environment has been rising since the 20th century (Pothitou et al.,

Table 6.1 Representative smart and selective membrane systems reported for micropollutant and trace-metal removal.

No.	Smart membrane type	Target pollutants	Smart/selective mechanism	Representative performance	Ref.
1	Temperature-responsive biocatalytic NF membrane	Organic micropollutants (pharmaceuticals, bisphenol)	poly (N-isopropylacrylamide) Temperature gating + enzyme catalysis	~90% OMP removal; stable flux, regenerable activity	Zhang et al. (2022)
2	pH-responsive enzyme-functionalized membrane	Chlorinated organic micropollutants	pH-controlled enzyme activity and adsorption	High degradation efficiency; reusable catalytic layer	Rupam et al. (2017)
3	Stimuli-responsive polymeric smart membranes	Organic micropollutants and metal ions	pH/temperature/light-responsive pore tuning	Tunable rejection and permeability	Mumtaz et al. (2025)
4	MOF-based composite membranes (UF/NF)	Heavy metals, dyes, antibiotics, PFAS	MOF-driven adsorption + size/charge sieving	~95% removal; high adsorption capacity	Mannaf et al. (2025)
5	Magnetic MOF nanocomposite membranes	Pb^{2+}, Hg^{2+}, Cd^{2+}, Cu^{2+}, Cr (VI)	MOF affinity + magnetic recoverability	Pb^{2+} capacity ~ 490 mg g^{-1}; regenerable	Zadehahmadi et al. (2023)
6	Nanocomposite TFC NF membranes	Pharmaceuticals, endocrine disruptors	Tuned MWCO, surface charge, hydrophilicity	~90% rejection with improved permeability	Wu et al. (2022)
7	Functional nanofiller-based MMMs (UF/NF)	Cesium (I) ion (Cs^{+}), strontium (II) ion (Sr^{2+}), (Cu^{2+}), (Cd^{2+})	Nanofiller-induced affinity and charge selectivity	~97%–98% Cu^{2+}/Cd^{2+} removal	Foorginezhad et al. (2025)

2008; Tarigan et al., 2025). They pose potential hazards not only to human health but also to entire ecosystems. Micropollutants are emerging contaminants that include pharmaceuticals, personal care products, steroid hormones, pesticides, industrial chemicals, and microplastics. These persistent pollutants pose risks to ecosystems and human health due to their toxic and bioaccumulative nature (Khanzada et al., 2020; Khoo et al., 2022; Majumder et al., 2019; Sun et al., 2016). They are characterized by persistence, bioaccumulation potential, and toxicity to ecosystems and human health (Majumder et al., 2019; Pothitou et al., 2008). Common pathways include direct release into water bodies, leaching from soils, infiltration into groundwater, and atmospheric deposition via precipitation (Bai et al., 2018; Otálora et al., 2021; Wołowicz and Munir, 2025). Wastewater treatment plants often fail to completely remove these compounds, resulting in persistent environmental contamination (Ajith et al., 2023; Kumar et al., 2024).

2.1 Types of micropollutants

EOMs encompass various classes based on their application or chemical nature (Mannaf et al., 2025). The main groups include pharmaceuticals, personal care products (PCPs), pesticides, industrial chemicals and additives, and by-products of water disinfection and biological toxins. Pharmaceuticals include antibiotics, analgesics, hormones, anticonvulsants, and stimulants, representing one of the most frequently detected classes in wastewater, with physicochemical and biological properties ranging widely. They are persistent, sometimes biologically active, and can bioaccumulate (Chakraborty et al., 2023; Otálora et al., 2021; Ranjan et al., 2022; Wołowicz and Munir, 2025). Personal Care Products (PCPs) are chemicals used for hygiene and beauty, such as fragrances, sunscreens, antimicrobials like triclosan, and detergents. These enter the environment mainly through domestic wastewater and runoff (Dey et al., 2019; Khalid et al., 2021). Pesticides include herbicides, insecticides, fungicides, and rodenticides widely used in agriculture and household applications. They persist in the environment and can contaminate soil and water through runoff and air transport (Alengebawy et al., 2021; Pathak et al., 2022). Industrial chemicals and additives such as bisphenol A, phthalates, and other synthetic compounds are used in manufacturing and consumer products. They reach water bodies through industrial discharges and landfill leachate (Esposito et al., 2020; Loganathan et al., 2023). By-products of water disinfection and biological toxins result from disinfection processes and are produced by microorganisms, respectively, adding to pollutant complexity (Richardson et al., 2011; Wołowicz and Munir, 2025).

The largest source of pharmaceutically active compounds (PhACs) and endocrine-disrupting chemicals (EDCs) are the production processes of the chemical and pharmaceutical industries (Abo-Farha et al., 2009; Abu-Obaid et al., 2023; Taheran et al., 2016). Water serves as a critical resource in chemical and pharmaceutical manufacturing, essential for production, material processing, and cooling operations (Taheran et al., 2016). As a result, the pharmaceutical industry generates large quantities of water containing PhACs and EDCs (Taheran et al., 2016). Notably, an estimated 50% of pharmaceutical wastewater worldwide is discharged without specialized treatment. PhACs and EDCs can also enter aquatic ecosystems through multiple pathways. These contamination routes include septic system failures, sewer line leaks, both permitted and accidental discharges, illegal dumping activities, cross-connections between sanitary and storm sewers, and improper management of pet and livestock waste (Taheran et al., 2016; Wei et al., 2025).

Personal Care Products (PCPs), encompassing a broad spectrum of chemical formulations including disinfectants and cosmetics, are extensively utilized to augment the quality of daily human life. In the context of daily human activities, the utilization of PCPs results in the introduction of these chemical compounds into wastewater treatment facilities in diverse, often complex mixtures. These mixtures are subsequently discharged into the environment via domestic sewage systems. Beyond consumer usage, the disposal of PCPs during the manufacturing processes of daily-use chemicals, medical protection, and food industry additives also contributes to the contamination of freshwater ecosystems, primarily due to the pervasive overuse of preservatives. A substantial proportion of PCPs that are reintroduced into natural ecosystems exhibit non-biodegradable characteristics. This renders them refractory to conventional wastewater treatment methodologies, which typically encompass primary, secondary, and tertiary treatment stages (Krishnan et al., 2021; Moldovan, 2006; Taheran et al., 2016). Table 6.2 categorizes the major types of micropollutants commonly detected in wastewater and natural water bodies. It summarizes their characteristics, primary sources, representative chemical classes, and typical examples, highlighting the diverse and persistent nature of pharmaceuticals, personal care products, pesticides, industrial chemicals, disinfection by-products, biological toxins, and emerging contaminants such as nanomaterials.

2.2 Heavy metals: Classification, toxicity, and environmental significance

Heavy metals are dense metallic substances with high atomic masses and specific densities, generally defined as metals with a density surpassing 5 g per cubic centimeter (Abdullah et al., 2019). However, this definition based on density is not universally accepted, as it lacks biological or chemical relevance (Rahman et al., 2019). Instead, functional classifications based on toxicological effects, environmental persistence, and chemical reactivity are more appropriate in environmental science and public health (Ali et al., 2019; Rahman et al., 2019).Unlike organic contaminants, heavy metals are not biodegradable, remain persistent in aquatic ecosystems, and tend to accumulate in living organisms due to their high water solubility and non-degradability (Briffa et al., 2020; Fu et al., 2011; Oladimeji et al., 2024). They are commonly associated with contamination, toxicity, and ecotoxicity, and pose significant environmental and health risks (Ali et al., 2019). Toxic heavy metals such as cadmium (Cd), chromium (Cr), copper (Cu), nickel (Ni), arsenic (As), lead (Pb), mercury (Hg), and zinc (Zn) are hazardous due to their high solubility in aquatic environments that facilitates absorption by living organisms. These metals accumulate in ecosystems as they do not biodegrade (Fu et al., 2011; Oladimeji et al., 2024).

Heavy metals are classified based on their toxicity into two categories: essential and non-essential metals. Crucial heavy metals, including zinc (Zn), copper (Cu), iron (Fe), and cobalt (Co), are required in trace amounts for biological functions and are non-toxic or minimally toxic at low concentrations (Ali et al., 2019).In contrast, non-essential metals such as cadmium (Cd), mercury (Hg), arsenic (As), and chromium (Cr) have no known biological role and are highly toxic even at low concentrations. Regulatory agencies, including the United States Environmental Protection Agency (USEPA), classify Cr, Cd, Hg, As, Pb, and Cu as priority pollutants due to their persistence and irreversible toxic effects (Kim et al., 2019). Arsenic, lead, cadmium, and mercury are of particular concern for human health, with arsenic ranked highest in toxicity by the US Agency of Toxic Substances and Disease Registry (ATSDR) (Rahman et al., 2019). Sources of heavy metals contamination are both natural and anthropogenic. Anthropogenic sources are predominant and include metal mining, smelting,

Table 6.2 Classification, sources, and examples of Major Micropollutant Groups in Aquatic Environments.

Type of micropollutant	Description	Primary sources	Representative classes	Examples	References
Pharmaceuticals	Synthetic or natural substances used to treat diseases in humans/animals; persistent and bioactive.	Hospital discharges, excretion, accidental spills, veterinary and agricultural waste.	Antibiotics, analgesics, NSAIDs, antiepileptics, antidepressants, β-blockers, hormones.	Amoxicillin, tetracycline, ciprofloxacin, ibuprofen, diclofenac, naproxen, estradiol, atenolol, carbamazepine, metformin.	(Ahamad et al., 2019; Kümmerer, 2010)
Personal care products (PCPs)	Chemicals used for hygiene, beauty, and cleaning purposes; widely available consumer goods.	Household wastewater, bathing, cosmetics industry, detergents, and cleaning agents.	Fragrances, antimicrobials, UV filters, surfactants, preservatives, synthetic musks.	Triclosan, benzophenone-3, parabens, phthalates, siloxanes, DEET, polycyclic musks.	(Halden et al., 2017; Khalid et al., 2021)
Pesticides	Compounds used to control pests in agriculture and public health; many are toxic and persistent.	Agricultural runoff, industrial and household use, improper disposal.	Organochlorines, organophosphates, carbamates, pyrethroids, herbicides, fungicides, insecticides.	DDT, lindane, parathion, malathion, glyphosate, carbaryl, cypermethrin, permethrin.	(Alengebawy et al., 2021; Kaur et al., 2019)
Industrial chemicals	Chemicals used in manufacturing and processing industries; resistant to degradation.	Industrial effluents, urban runoff, manufacturing waste.	Plasticizers, solvents, PFAS, surfactants, dyes, and flame retardants.	Bisphenol A, phthalates, PFOS, PFOA, nonylphenol, toluene, benzene derivatives.	(Qalyoubi et al., 2021a; 2021b)
Disinfection by-products	Compounds formed during chlorination or oxidation of water.	Water treatment and disinfection processes.	Haloacetic acids, trihalomethanes.	Chloroform, bromodichloromethane, dichloroacetic acid.	Bai et al. (2018)
Biological toxins and pathogens	Naturally occurring toxins and microorganisms that cause diseases.	Contaminated water, sewage, agricultural runoff.	Mycotoxins, cyanotoxins, bacteria, viruses, protozoa.	Aflatoxin, microcystin, E. coli, Salmonella, hepatitis a virus.	Ramírez-Castillo et al. (2015)
Nanomaterials	Engineered materials at nanoscale with unique physicochemical properties.	Wastewater, cosmetics, drug delivery, textiles, industrial use.	Carbon-based nanomaterials, metal oxides.	Fullerenes, quantum dots, carbon nanotubes, nanoparticles.	(Hocking, 2006; Malakar et al., 2021)

foundries, electroplating, chemical production, manufacturing, and metal-based industries (Briffa et al., 2020; Oladimeji et al., 2024).

During mining and metallurgical processes, metals are extracted from ores and separated, but a significant amount is lost and discharged into surface water, soil, or the surrounding environment. The steel industry contributes large volumes of wastewater during smelting and rolling processes. Other contamination sources include improper disposal of metallic waste, combustion of fuels and petroleum, and the use of metal-containing pesticides, insecticides, and dyes, all of which contaminate soil, agricultural fields, and water bodies, affecting ecosystems. Heavy metal contamination due to these sources is a major environmental concern as these metals cause toxic effects to humans and aquatic life, including impacts on the respiratory, nervous, reproductive, and digestive systems. Effective removal of these toxic metals from wastewater is vital to safeguard public health and environmental quality (Kaushik et al., 2022; Kim et al., 2019; Oladimeji et al., 2024). Table 6.3 summarizes the key toxic heavy metals commonly detected in contaminated water and industrial environments. It outlines their major exposure pathways, associated human health risks, and the Maximum Contaminant Levels (MCLs) established for drinking water. The listed metals such as arsenic, lead, mercury, and cadmium are among the most hazardous due to their persistence, bioaccumulation, and severe toxicological impacts even at trace concentrations.

Table 6.3 Major toxic heavy metals: exposure sources, health effects, and regulatory limits (Azmi, 2025; Bharti et al., 2022; Cevallos-Mendoza et al., 2022; Oladimeji et al., 2024).

Heavy metal	Common sources of exposure	Primary health effects	Maximum contaminant level (MCL, mg/L)
Arsenic (As)	Contaminated water, air pollution, industrial sites, electroplating, mining, paint, and wood processing industries	Renal dysfunction, neurological complications, dermatitis, keratosis, melanosis, pigmentation, genotoxicity, epigenetic effects	0.010
Lead (Pb)	Contaminated seafood, industrial pollution, PCB manufacturing, electroplating, and mining	Neurodevelopmental effects, central nervous system disruption, cardiovascular disorder, and kidney failure	0.010
Mercury (hg)	Fertilizers, phosphate fertilizers, cigarette smoke, and mining	Disruption of the central nervous system, insomnia, memory loss, tremors, mental retardation, kidney damage, reduced fertility	0.002
Cadmium (Cd)	Contaminated water, fuel burning, industrial sites, the paint industry, and electroplating	Lung and pancreatic cancer, kidney dysfunction, bone damage, pneumonia, autoimmune disease, cardiovascular disorder, lymphoma	0.005

2.3 Health and environmental risks of micropollutants

Micropollutants, including pharmaceuticals, personal care products, pesticides, industrial chemicals, and microplastics, have become a growing global concern due to their persistence, bioaccumulation, and toxicological impacts on ecosystems and human health, as shown in Fig. 6.3 (Tarigan et al., 2025). These substances include pharmaceuticals, personal care products, pesticides, and industrial chemicals that are introduced into water systems through anthropogenic activities such as wastewater discharge, agricultural runoff, and industrial effluents (Bai et al., 2018). Pharmaceuticals and personal care products are among the most commonly detected micropollutants in wastewater, primarily due to their widespread consumption and incomplete removal during wastewater treatment processes (Tarigan et al., 2025).

Organic micropollutants exhibit high resistance to biodegradation, allowing them to persist in aquatic and terrestrial environments for extended periods (Wang et al., 2024). Their persistence and mobility enable long-range environmental transport, posing ecological threats across diverse ecosystems. Pharmaceuticals and personal care products are frequently detected in wastewater and surface waters, particularly in urban and industrial regions, reflecting their widespread use in human activities. Pesticides and industrial chemicals also enter aquatic environments, affecting aquatic flora and fauna through bioaccumulation and biomagnification. Microplastics, another major group of micropollutants, are generated through the breakdown of larger plastic debris and the use of synthetic materials in consumer goods, posing severe risks to aquatic life through ingestion and entanglement. Pesticide residues transported by surface runoff can lead to bioaccumulation and biomagnification across trophic levels, causing toxicity and genetic mutations in aquatic organisms. Emerging organic micropollutants can alter the physiology, reproduction, and survival of aquatic organisms, leading to ecosystem imbalance. For instance, endocrine-disrupting compounds interfere with hormone regulation in fish and amphibians, while antibiotics promote antimicrobial resistance among microbial populations. Once released into the environment, micropollutants can disrupt aquatic ecosystems by altering biological functions such as reproduction, growth, and photosynthesis in aquatic organisms (Tarigan et al., 2025; Wang et al., 2024; Wołowicz and Munir, 2025).

Humans are exposed to micropollutants and heavy metals (Tarigan et al., 2025; Wołowicz and Munir, 2025) primarily through contaminated drinking water, food, and dermal contact. Chronic exposure to emerging organic micropollutants, even at trace concentrations, has been associated with endocrine disruption, neurotoxicity, carcinogenicity, and reproductive toxicity. Non-steroidal anti-inflammatory drugs (NSAIDs), detected in surface and groundwater, can cause long-term ecotoxicological and physiological impacts even at low levels. The widespread presence of antibiotics in the environment contributes to the emergence of antibiotic-resistant bacteria, posing a major global health crisis. Similarly, exposure to antidepressants in contaminated water can lead to neurobehavioral changes in aquatic species and potential neurological effects in humans. Microplastics have been identified in food and air, and their ingestion or inhalation is linked to respiratory, gastrointestinal, and endocrine disorders. Pesticides, including organophosphorus insecticides, are associated with severe health outcomes such as cancer, neurotoxicity, reproductive issues, and metabolic disorders. Epidemiological studies have reported higher incidences of cancers and neurological disorders in populations consuming water contaminated with pharmaceutical and pesticide residues, such as those near the Mississippi River and in urban European regions. Persistent organic pollutants, including bisphenols and synthetic musks, have also been found to bioaccumulate in human tissues, potentially disrupting metabolic and hormonal pathways. Figure 6.2 illustrates the multi-organ health impacts associated with exposure to environmental pollutants and toxicants. The yellow human silhouette highlights the affected organ systems, including the brain, lungs, heart, blood, vascular system, and reproductive organs. Specific physiological and pathological outcomes, such as headaches, cognitive impairment, chronic obstructive pulmonary disease (COPD) exacerbation, altered heart function, thrombosis, atherosclerosis, miscarriage, fetal growth problems, and reduced fertility, are annotated beside each corresponding organ. The figure provides an integrated visualization of how pollutants influence human health through respiratory, cardiovascular, neurological, hematological, and reproductive pathways. The persistence, bioaccumulation, and toxicity of emerging micropollutants and

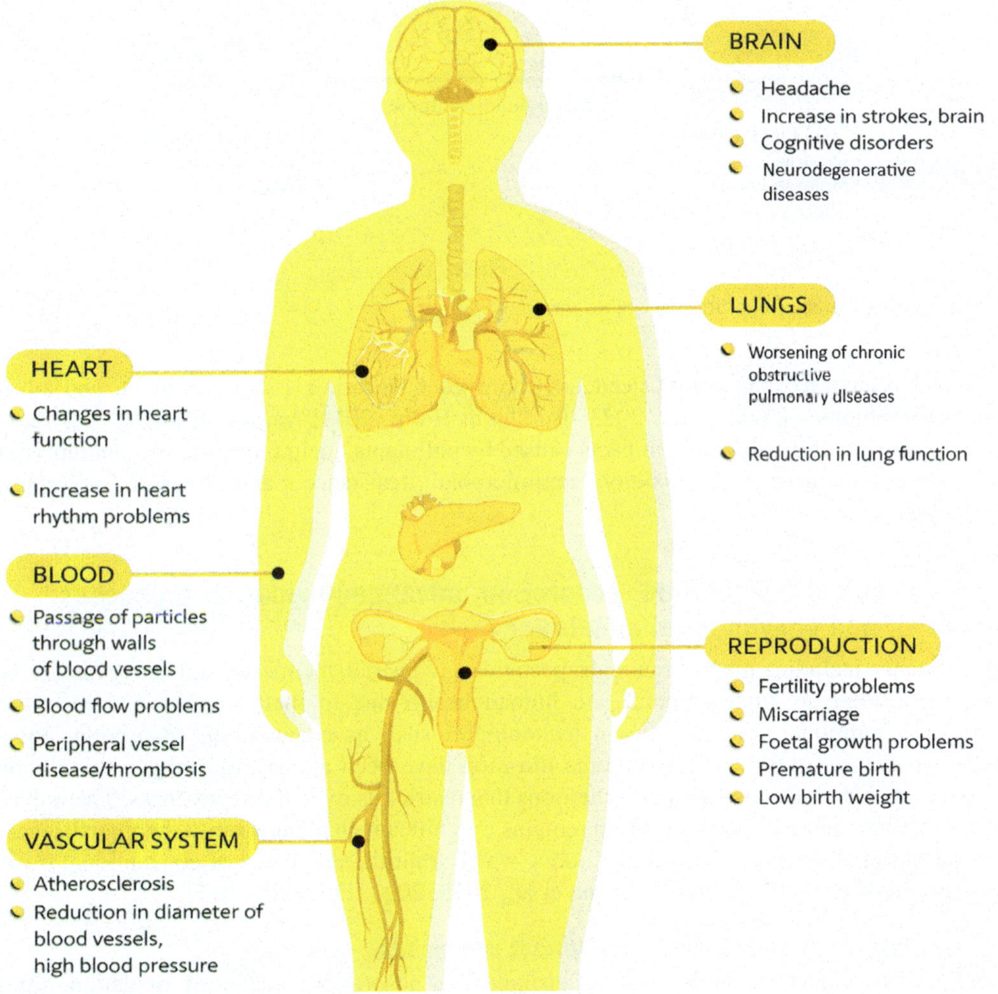

FIGURE 6.2

Systemic health effects of environmental pollutants on major human organs.

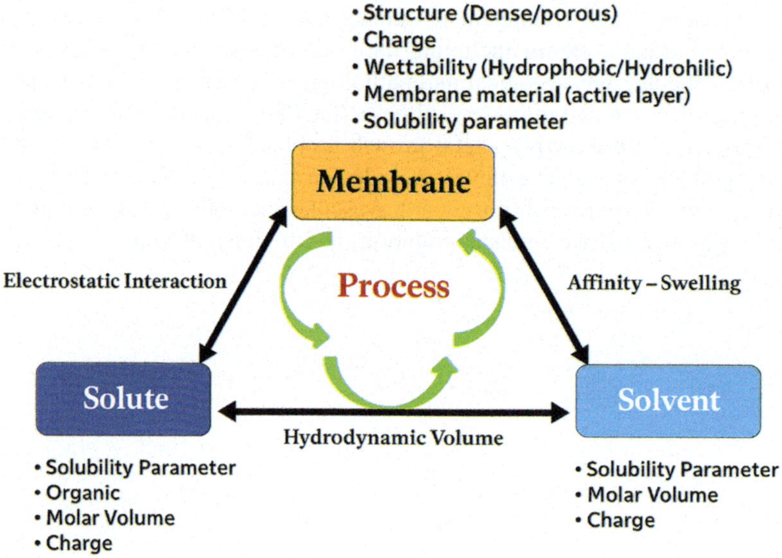

FIGURE 6.3

Interaction triangle between membrane, solute, and solvent.

heavy metals emphasize the urgent need for improved monitoring, risk assessment, and advanced treatment technologies (Aziz et al., 2023; Burkhardt-Holm, 2011; Varjani et al., 2020). Table 6.4 summarizes the major categories of impacts caused by pollutants, including systemic and reproductive health effects, organ-specific toxicity, antimicrobial resistance, and broader environmental consequences.

2.4 Limitations of coagulation, adsorption, advanced oxidation processes (AOPs), and conventional membranes

Micropollutants, including pharmaceuticals, pesticides, industrial chemicals, and heavy metals, pose a persistent challenge to environmental and human health due to their toxicity and resistance to degradation. Traditional water treatment technologies such as coagulation, adsorption, advanced oxidation processes (AOPs), and membrane filtration have been applied to mitigate these contaminants, but each technique has inherent limitations that restrict its overall effectiveness. The removal of emerging organic micropollutants (OMPs) remains a significant challenge due to their persistence, low concentrations, and diverse chemical structures, which limit the efficiency of conventional treatment technologies (Ali et al., 2019; Buonomenna et al., 2015, 2022; Fu et al., 2011).

2.4.1 Limitations of conventional treatment processes

Conventional treatment methods such as coagulation, adsorption, advanced oxidation processes (AOPs), and membrane filtration play vital roles in water purification but each faces inherent

Table 6.4 Classification of pollution-related impact categories on human and environmental health.

Category of effect	Specific adverse effect or consequence	Micropollutant type/Mechanism
Human health (systemic/Chronic)	Long-term (chronic) effects on human health	General micropollutants (e.g., PCs, MPs, EDCs)
	Serious health complications from chronic human exposure	General micropollutants (e.g., PCs, MPs, EDCs)
	Organ damage	Bioaccumulation pathway
	Impaired physiological functions	Bioaccumulation pathway
	Increased risk of cancer	Endocrine disruptors and bioaccumulation pathways
Human health (reproductive/ Hormonal)	Hormone imbalance	Endocrine disruptors
	Fertility problems	Endocrine disruptors
	Sexual dysfunction	Antidepressants
	Unusual sexual development	Endocrine disruptors
Human health (organ/System specific)	Cardiovascular disorders and respiratory disorders	Inhaled microplastics
	Neurological disorders	Bioaccumulation pathway, microplastics
	Liver dysfunction and immune system disruption	Antibiotics
	Bone marrow suppression	Antibiotics
	Gastrointestinal distress and physical irritation	Ingested microplastics
	Contribution to obesity and diabetes	Leached chemicals from plastics (e.g., BPA)
Microbial/Drug resistance	Development and spread of antibiotic resistance	Antibiotics and other pharmaceuticals
	Resistance to drugs, prolonged illness, heightened mortality rates	Drug resistance pathway
Ecosystem/ Environmental	Disruption of aquatic ecosystems	Organic micropollutants
	Environmental degradation	Organic micropollutants
	Persistent damage to ecosystems via bioaccumulation in the food chain	General micropollutants
	Prolonged ecotoxicological effects on ecosystems	Non-steroidal anti-inflammatory drugs (NSAIDs)

limitations in removing emerging micropollutants. Coagulation is effective for suspended solids and colloids but fails to eliminate dissolved organic micropollutants due to their high solubility and low coagulant affinity. The process efficiency depends on factors such as pH, coagulant dosage, and temperature, and generates large volumes of contaminated sludge requiring further disposal, often leading to secondary pollution. Adsorption, though versatile, suffers from adsorbent saturation, competition from natural organic matter, and costly regeneration cycles that reduce long-term efficiency. AOPs such as ozonation, photocatalysis, and Fenton oxidation effectively degrade organic pollutants but are limited by high operational costs, energy consumption, and the potential formation of toxic intermediates during incomplete oxidation. Similarly, conventional membrane processes such as microfiltration, ultrafiltration, and nanofiltration struggle with the removal of small, hydrophilic micropollutants. Membrane fouling, high-pressure requirements, and brine disposal challenges further increase operational costs and sustainability concerns. Collectively, these constraints highlight that single-treatment approaches are insufficient, emphasizing the need for integrated or hybrid systems to achieve efficient and sustainable micropollutant removal (Briffa et al., 2020; Buonomenna et al., 2022; Fu et al., 2011; Khanzada et al., 2020).

2.4.2 Need for selectivity, antifouling, and durability in membranes

For advanced membrane-based purification, selectivity, antifouling, and durability are crucial design goals that directly influence performance and sustainability. Selectivity governs the membrane's ability to discriminate solutes based on size, charge, or chemical affinity, essential for removing specific contaminants like pharmaceuticals and heavy metals without compromising water flux. However, conventional polymeric membranes such as PES and PSf often exhibit poor selectivity due to broad pore size distribution, while efforts to enhance selectivity through nanomaterial modification often reduce permeability, creating a flux–rejection trade-off. Equally important is antifouling capability, as fouling by organic matter, colloids, and microorganisms reduces permeability and shortens membrane lifespan. Hydrophilic and photocatalytic nanomaterial incorporation can mitigate fouling, yet long-term operation still causes physical and chemical deposition. Durability, encompassing mechanical, chemical, and thermal stability, ensures sustained performance under harsh operating conditions; however, polymer degradation, chemical attack, and repeated cleaning cycles accelerate membrane aging. Hybrid membranes incorporating inorganic nanoparticles improve chemical resistance but can introduce interfacial weaknesses that reduce long-term stability. Therefore, developing next-generation membranes that combine high selectivity, strong antifouling resistance, and exceptional durability through hybrid structures and surface functionalization remains a key frontier in achieving sustainable and reliable water treatment (Buonomenna et al., 2022; Kaushik et al., 2022; Oladimeji et al., 2024; Rahman et al., 2019).

3. Fundamentals of smart and selective membrane technologies

The evolution of membrane technology marks a significant shift from conventional filtration systems toward intelligent and adaptive separation platforms. Traditional polymeric membranes, while effective for size-based filtration, often lack the specificity to differentiate between valuable solutes and harmful contaminants. This limitation has prompted the development of smart and selective membranes, which represent a paradigm shift in modern separation science (Ulbricht, 2006). These membranes are engineered with tailored functionalities, allowing them to target and remove specific contaminants while maintaining efficient water permeability. The underlying principle involves the introduction of functional materials or responsive polymers that dynamically adapt to environmental changes such as pH fluctuations, temperature variations, or the presence of specific molecular triggers (Werber et al., 2016). These stimuli-responsive behaviors enable controlled modulation of membrane properties, including pore size, surface charge, and hydrophilicity, to achieve precise and selective separations (Ulbricht, 2006). At the molecular level, smart membranes employ sophisticated design strategies that transcend conventional sieving. Molecularly imprinted polymers (MIPs), for instance, are synthesized with specific cavities that match the size, shape, and functional groups of target molecules, providing a molecular "lock-and-key" mechanism for selective adsorption (Rivas et al., 2003). Similarly, surface functionalization with bio-inspired ligands such as polydopamine or sulfur-containing moieties like thiol groups imparts high affinity toward trace metals, enabling selective complexation and retention of ions such as Hg^{2+} or Cd^{2+} while allowing water molecules to pass freely (Lee et al., 2007; Rivas et al., 2003). These active functional sites operate as selective gates, forming the basis for the intelligent recognition and capture of pollutants from complex aqueous environments. When coupled with stimuli-responsive polymers, these functionalities enable membranes to respond dynamically to external cues, creating self-regulating systems that adjust their selectivity or flux in real time (Baker, 2023; Werber et al., 2016). This transformation from passive filtration to active, intelligent separation marks a defining innovation in water purification, resource recovery, and environmental remediation technologies.

In defining the terminology, "selective" membranes refer to systems designed for molecular-level discrimination. Their function extends beyond physical sieving, relying instead on chemical interactions, ion exchange, and affinity-based recognition processes (Baker, 2023; Ulbricht, 2006). Selectivity arises from engineered binding sites that possess strong affinity toward target contaminants, whether trace metals, dyes, or organic micropollutants, allowing benign solutes and water to permeate (Rivas et al., 2003; Shannon et al., 2008; Werber et al., 2016). The precision of this discrimination is often quantified through separation factors or rejection coefficients that measure a membrane's ability to isolate a given solute amidst coexisting species (Van der Bruggen et al., 2008). On the other hand, a smart membrane exhibits a dynamic and reversible response to environmental stimuli such as temperature, pH, or ionic concentration (Baker, 2023; Ulbricht, 2006). Functional polymers and nanomaterials integrated into the membrane act as molecular switches, altering conformation or charge distribution upon sensing an external trigger (Mendes, 2008; Roy et al., 2013, 2020; Schattling et al., 2014). Such conformational changes can reversibly modulate pore size, surface energy, or hydrophilicity, enabling real-time control of separation processes (Heskins et al., 1968; Ju et al., 2008). For instance, a temperature-sensitive membrane may constrict its pores at elevated temperatures, whereas a pH-responsive system may increase positive charge under acidic conditions to attract and remove negatively charged contaminants (Liu et al., 2010).

The synergy between selectivity and smart responsiveness leads to next-generation membranes capable of adaptive selectivity, their performance can be tuned on demand without structural degradation (He et al., 2017). This dual functionality enables both targeted capture and controlled release of solutes, making the membrane reusable and economically viable (Crini, 2005; Zhao et al., 2015). In essence, smartness introduces a temporal control dimension over selective behavior, converting static membranes into active, intelligent interfaces (Park et al., 2017). This approach aligns with the circular economy framework, wherein membranes not only purify but also recover valuable materials such as metals and nutrients from wastewater streams, promoting sustainable resource management (Tavares et al., 2022).

3.1 Transport and separation mechanisms

The superior performance of smart and selective membranes stems from their ability to exploit multiple transport mechanisms operating simultaneously at micro- and nanoscale interfaces. The primary mechanisms governing solute separation include size-based sieving, electrostatic (charge-based) rejection, and chemical affinity interactions (Mulder, 2012; Strathmann, 2004). The contribution of each mechanism depends on the physicochemical nature of both the membrane and the target solutes. Figure 6.3 illustrates the key interactions governing separation performance in nanofiltration/OSN systems. Membrane properties (structure, charge, wettability, material, and solubility parameter) interact with solute characteristics (size, charge, solubility parameter, and organic nature) and solvent attributes (solubility parameter, molar volume, and charge). The process is driven by three major interactions: electrostatic interactions between membrane and solute, affinity-induced swelling between membrane and solvent, and hydrodynamic volume effects between solute and solvent.

The most fundamental mechanism, steric hindrance or size sieving, separates species based on molecular size and geometry. Molecules or particles larger than the effective pore diameter are rejected, while smaller solutes and solvents traverse the pore network (Fane et al., 1987). This process dominates in microfiltration and ultrafiltration, particularly for the removal of suspended solids, macromolecules, and colloidal matter (Cheryan, 1998). The efficiency of this mechanism depends critically on the membrane's pore size distribution, morphology, and porosity uniformity (Bowen et al., 2002). In nanofiltration, however, sieving operates alongside non-steric mechanisms such as charge and affinity interactions, allowing for the partial rejection of small solutes and ions (Schaep et al., 1998). Here, surface porosity, pore tortuosity, and internal roughness significantly affect the convective and diffusive transport of solutes. Charge-based rejection, often referred to as Donnan exclusion, provides another powerful mode of selectivity, particularly for ionic solutes. Membranes possessing fixed surface charges create an electrostatic potential that repels similarly charged ions (co-ions) while attracting counter-ions, resulting in selective ion exclusion (Szymczyk et al., 2005; Wang et al., 1997). This mechanism underlies the enhanced rejection of multivalent ions like Ca^{2+}, Pb^{2+}, and Cr^{3+} compared to monovalent ions such as Na^{+} or K^{+} (Boussu et al., 2006; Hagmeyer et al., 1998). Donnan exclusion is especially important in the removal of trace metals and charged micropollutants from industrial and municipal wastewater streams (Koyuncu et al., 2008). The extent of this electrostatic repulsion depends on surface charge density, the dielectric constant of the medium, and the ionic strength of the feed solution (Braghetta et al., 1997; Childress et al., 2000). At high ionic strengths, the electric double layer compresses, reducing effective repulsion and thereby diminishing selectivity. Thus, optimizing charge properties is essential for stable performance under variable feed conditions.

A more advanced mechanism is affinity interaction, which relies on specific chemical recognition between membrane functional groups and target molecules. In this process, ligands such as amines, carboxylates, or thiols are immobilized on the membrane surface form reversible complexes with contaminants like heavy metals, pharmaceuticals, or endocrine-disrupting compounds (Charcosset, 1998). Unlike sieving, which physically blocks solutes, affinity separation captures target species through specific chemical bonding or adsorption phenomena. Ion-imprinted or biomimetic ligands further enhance this specificity, enabling selective removal even when molecular sizes are smaller than the pores themselves (Piletsky et al., 1999). The binding of target solutes can induce local structural or charge rearrangements that serve as molecular gates, dynamically controlling permeation and regeneration (Ito et al., 1997; Stayton et al., 1995). In practice, high-performance membranes often combine these mechanisms synergistically. For instance, a negatively charged nanofiltration membrane may employ size exclusion for macromolecules, Donnan exclusion for ionic pollutants, and affinity binding for specific organics (Van der Bruggen et al., 2008). A classic example is a membrane grafted with thiol groups that removes Cd^{2+} and Hg^{2+} ions via strong chemical affinity, while its negatively charged surface simultaneously rejects other divalent cations through electrostatic repulsion (Bellona et al., 2004). By integrating multiple pathways, these membranes achieve precise selectivity, high permeability, and improved antifouling resistance, representing the cutting edge of modern water purification (Nanda et al., 2008). Fig. 6.4., depicts the fundamental working principles of smart and selective membranes used in wastewater treatment. Contaminated feed water containing organic micropollutants and trace metal ions approaches the membrane surface, where multiple separation mechanisms operate simultaneously. Size sieving restricts the passage of solutes larger than the membrane pores, while electrostatic (Donnan) exclusion repels similarly charged ionic species. Functional groups immobilized on the membrane surface enable affinity-based binding of specific contaminants such as heavy metal ions. In addition, stimuli-responsive components allow dynamic modulation of membrane properties (e.g., pore size, surface charge, or hydrophilicity) in response to environmental triggers such as pH or temperature. The combined action of these mechanisms enables selective contaminant removal while allowing clean water to permeate through the membrane.

3.2 Material chemistry and functional modification strategies

The performance, adaptability, and long-term stability of smart and selective membranes are ultimately determined by their material chemistry. The interplay between the hydrophilic/hydrophobic balance and the distribution of functional groups dictates wettability, surface charge, fouling resistance, and selectivity (Ulbricht, 2006; Werber et al., 2016). Tailoring these surface and bulk characteristics through chemical modification or nanocomposite design forms the foundation for next-generation membrane engineering. A proper hydrophilic/hydrophobic balance ensures effective water–membrane interactions and mitigates fouling. Hydrophilic membranes exhibit strong water affinity, leading to enhanced flux and reduced adhesion of hydrophobic contaminants (Miller et al.,

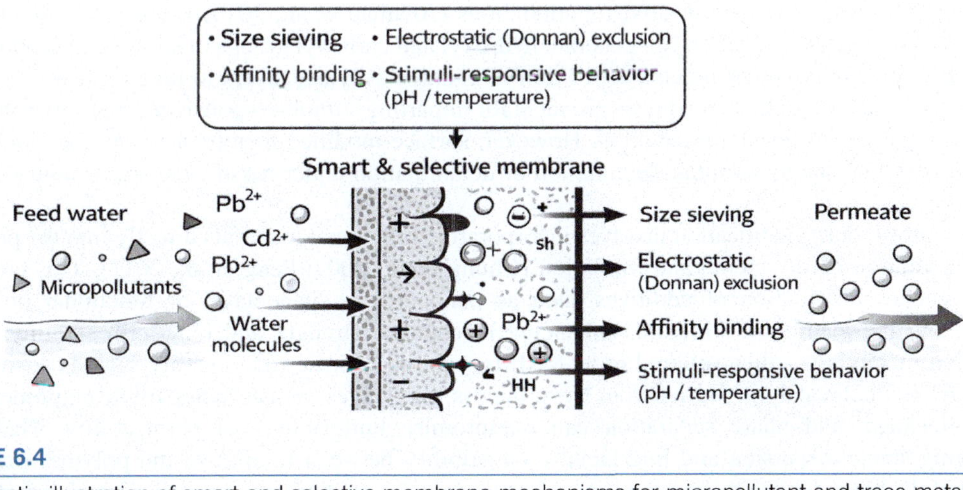

FIGURE 6.4

Schematic illustration of smart and selective membrane mechanisms for micropollutant and trace metal removal.

2011; Ragab et al., 2016; Ravi et al., 2025a). Hydrophilicity is often achieved by incorporating polar monomers or nanofillers such as polyvinyl alcohol (PVA), polyethylene glycol (PEG), or graphene oxide, which improve hydration layer formation and resist organic fouling (Le et al., 2016). Conversely, hydrophobic domains can be intentionally introduced to enhance adsorption of nonpolar organic compounds, thereby increasing selectivity toward specific micropollutants (Elimelech et al., 1997). An amphiphilic structure, containing both hydrophilic and hydrophobic moieties, provides an optimal trade-off between fouling resistance and selective adsorption (Hilal et al., 2004). The surface energy, contact angle, and interfacial tension are key indicators of this balance, directly affecting flux, rejection, and stability under variable operating conditions (Vrijenhoek et al., 2001). Excessive hydrophobicity, however, promotes irreversible fouling, while extreme hydrophilicity may compromise selectivity and mechanical robustness.

Functional groups are the chemical backbone of selective and smart behavior. Through polymerization, grafting, or surface modification, reactive moieties such as carboxylic acid (–COOH) and amine ($–NH_2$) groups can be introduced to impart pH-dependent charge characteristics, enabling charge-based rejection and pH-responsive gating (Gohil et al., 2017). Sulfonic acid ($–SO_3H$) groups confer permanent negative charge, enhancing the removal of multivalent cations via electrostatic repulsion (Oatley-Radcliffe et al., 2017). Hydroxyl (–OH) groups improve hydrophilicity and provide mild complexation sites for metal ions (Zhao et al., 2015), whereas thiol (–SH) groups exhibit high affinity for soft metals such as Hg^{2+} and Cd^{2+} (Crini, 2005). Other functional moieties like amidoxime groups have been successfully used for uranium recovery from seawater due to their exceptional chelating ability (Das et al., 2016). The density and accessibility of these groups determine the overall adsorption capacity and selectivity (He et al., 2017; Park et al., 2017). Thus, molecular-level engineering of functional group distribution is essential for optimizing performance.

To achieve these chemical enhancements, two major modification approaches are employed: surface modification and bulk modification (Ulbricht, 2006; Werber et al., 2016). Surface modification alters only the membrane's top layer or pore walls, maintaining the integrity of the bulk matrix. Techniques such as plasma treatment, surface grafting, and layer-by-layer (LbL) assembly are commonly used to introduce desired functionalities (Ahamad et al., 2019; Kang et al., 2014; Ravi et al., 2025b). Plasma activation can generate hydroxyl and carboxyl groups to increase hydrophilicity, while LbL assembly allows precise deposition of charged polymer layers to tune electrostatic properties. These surface-engineered layers are ideal for imparting stimuli-responsiveness, as their thinness ensures rapid environmental interaction. However, surface-modified membranes may face long-term durability issues due to potential degradation or delamination under harsh cleaning or high-pressure conditions.

In contrast, bulk modification involves incorporating functional materials directly into the polymer casting solution before membrane formation (Chung et al., 2007; Peng et al., 2012). This creates a homogeneous distribution of additives such as nanoparticles, nanoclays, or functional polymers throughout the membrane structure, enhancing mechanical strength and long-term stability. Bulk-modified membranes, often referred to as MMMs, exhibit intrinsic functionality such as adsorption, catalysis, or photocatalytic degradation. Nevertheless, challenges such as nanoparticle agglomeration, pore blockage, and phase separation can compromise uniformity and permeability. Therefore, achieving stable dispersion and interfacial compatibility between fillers and the polymer matrix is essential. Bulk modification is particularly advantageous for membranes exposed to mechanical stress or aggressive chemical environments, as the embedded functionality remains stable during operation (Fane et al., 1987).

The integration of surface and bulk modification represents a promising hybrid strategy. By combining the structural robustness of bulk modification with the precision of surface functionalization, multifunctional membranes with superior performance can be achieved (He et al., 2017; Shannon et al., 2008). For example, a mixed matrix membrane may incorporate TiO_2 nanoparticles within the polymer matrix for durability and photocatalytic self-cleaning, while its surface is grafted with zwitterionic polymers to enhance antifouling properties. Such synergy enables membranes to exhibit high selectivity, strong resistance to fouling, and long-term operational stability under realistic wastewater conditions. Future advancements in membrane science will continue to optimize the interplay between chemistry, structure, and functionality, ultimately leading to self-regulating, selective, and durable membranes that embody the principles of sustainable water and resource management.

4. Membrane materials and modification strategies

4.1 Polymer-based membranes

Polymeric membranes remain the most widely applied class in advanced wastewater purification owing to their adjustable surface chemistry, processability, and chemical stability. Their ability to be tuned at the molecular level allows precise control over permeability, selectivity, and fouling resistance (Mannaf et al., 2025; Shukla et al., 2025). Among the numerous candidates, polyimides (PI), polybenzimidazoles (PBI), polyethersulfone (PES), polyacrylonitrile (PAN), and polyvinylidene fluoride (PVDF) have received the greatest attention. Figure 6.5 categorizes the major classes of membrane materials used in advanced wastewater treatment. The figure highlights three core groups: polymer-based membranes (including PI, PBI, PES, PAN, and PVDF families), inorganic and hybrid membranes (such as ceramic oxides, zeolites, MOF/COF-based MMMs), and modern modification strategies (crosslinking, post-functionalization, fouling control, and sustainable fabrication). Together, these categories summarize the structural diversity, functional enhancements, and engineering approaches that enable high-performance separation of micropollutants and trace metals.

4.1.1 Polyimide and polybenzimidazole families

Aromatic polyimides combine imide linkages with rigid backbones, producing excellent thermal and chemical resistance (Dawam et al., 2025; Foorginezhad et al., 2025). Their dense chain packing ensures size-selective transport, but pristine PI membranes are prone to hydrolysis and compaction in acidic or oxidizing conditions. To address this, covalent crosslinking with multifunctional agents such as diamines, polyethyleneimine (PEI), or cyanuric chloride transforms labile imide carbonyls into stable amide or triazine bridges (Huyen et al., 2025; Ravi et al., 2023; Upadhyaya et al., 2018). This modification suppresses swelling and preserves flux even under strong acid exposure. PBI membranes, formed from benzimidazole units containing imidazole N–H donors, exhibit exceptional tolerance to acids, oxidants, and organic solvents (Ravi et al., 2025a, 2025b; Ray et al., 2023). Exposure to halogen acids (HCl, HBr) or oxidants (NaOCl, KIO_4) induces C–N and N–X crosslinks that restrict chain mobility and create Lewis-basic sites capable of chelating multivalent metal ions (Algieri et al., 2021). Consequently, PBI membranes demonstrate high rejection toward Cr(VI), Pb(II), and Cu(II) ions while

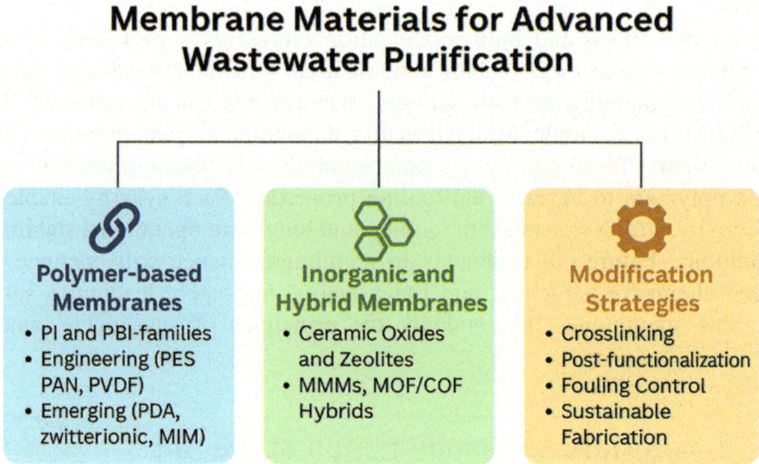

FIGURE 6.5

Classification of membrane materials and modification strategies for advanced wastewater purification.

maintaining mechanical integrity. These findings substantiate PBI's role as a benchmark polymer for chemically aggressive effluents (Zhu et al., 2014).

4.1.2 Engineering polymers (PES, PAN, PVDF)

Polyethersulfone is a commercially dominant matrix because of its mechanical strength and broad pH tolerance. However, its intrinsic hydrophobicity accelerates fouling. Surface sulfonation, carboxylation, or grafting of hydrophilic chains (such as sulfobetaine methacrylate or polyethylene glycol) introduces hydration layers that minimize foulant adhesion (Chen et al., 2024; Horseman et al., 2020). Incorporation of PEI as a crosslinker or PVP as a pore-former further enhances hydrophilicity and mechanical flexibility, producing membranes that remove >95% of dyes and metal ions from synthetic wastewater (Gayatri et al., 2024). PAN membranes, discussed in (Xiao et al., 2025b), exhibit highly polar –CN groups suitable for post-functionalization. Hydrolysis and amination yield –COOH and $-NH_2$ sites that strongly complex soft metal ions (Hg^{2+}, Cd^{2+}). Electrospun PAN nanofibers coated with TiO_2 or polydopamine show fluxes >600 L m^{-2} h^{-1} and >85% Pb^{2+} rejection (Hou et al., 2019; Shukla et al., 2025). PVDF membranes, in contrast, offer high chemical inertness; blending with GO or cellulose nanofibers enhances hydrophilicity and fouling resistance (Foorginezhad et al., 2025).

4.1.3 Mechanistic insights

Crosslinking constrains chain relaxation and prevents compaction, while hydrophilic grafts re-establish water channels through hydrogen-bond networks. Charge modifications induce Donnan exclusion, providing ion-specific selectivity (Gadipelly et al., 2014; Islam et al., 2025). Hence, modern polymer design has shifted from "dense-barrier" philosophy toward chemically responsive architectures that self-stabilize under variable chemical environments (Upadhyaya et al., 2018).

4.2 Inorganic and hybrid membranes

Inorganic membranes offer unparalleled thermal and chemical endurance. Ceramic oxides (Al_2O_3, TiO_2, ZrO_2) and zeolitic structures present uniform pore distributions and hydrophilic surfaces capable of withstanding acidic and alkaline conditions (Ge, 2024; Islam et al., 2025). Zeolite–clay composites achieve >98% Pb^{2+} removal via ion exchange between Si–O⁻ sites and hydrated cations (Lu et al., 2016; Radovanovic et al., 2023). Although their chemical stability is superior, brittleness and costly sintering limit industrial deployment (Islam et al., 2025).

4.2.1 *Mixed-matrix and MOF/COF hybrids*

Hybridization of organic and inorganic phases bridges flexibility and robustness. MOFs and covalent organic frameworks (COFs) have become transformative fillers in mixed-matrix membranes (MMMs). UiO-66-NH_2, MIL-53(Fe), and Cu-BTC MOFs exhibit crystalline porosity and chemically tunable active sites, while melamine–terephthaldehyde and triazine-based COFs provide high surface areas and ordered nanochannels (Islam et al., 2025; Wei et al., 2025; Xiao et al., 2025b). When embedded within PES or PAN matrices, these nanostructures produce dual transport pathways, polymeric free volume for bulk permeation and crystalline nanopores for selective diffusion. As reported in (Shukla et al., 2025), PES/UiO-66-NH_2 MMMs achieved $\approx$ 100% removal of Cu^{2+}, Cd^{2+}, and As(V) with substantial flux improvement. The amine groups within UiO-66 act as coordination centers, and Zr–O clusters enhance oxidative stability (Bai et al., 2023). COF-based MMMs show similar advantages; for example, tannic-acid-modified COFs integrated into polyamide layers increased hydrophilicity and metal-ion adsorption capacity (Mannaf et al., 2025). Such synergies demonstrate that MOF/COF hybrids can surpass the traditional flux-selectivity trade-off common to dense polymers (Islam et al., 2025).

4.2.2 *Graphene-based and other 2D composites*

Graphene oxide (GO) and MXene two-dimensional materials introduce atomic-scale lamellar channels that enable ultrafast water transport while excluding hydrated ions through Donnan and dielectric repulsion (Sana et al., 2020). Their negatively charged oxygen groups enhance antifouling by forming strong hydration shells. GO–TiO_2/PES nanocomposites achieved >95% dye removal and retained flux over extended operation (Leupold et al., 2025). MXene interlayers in thin-film composite membranes have been shown to double permeate flux without reduction in salt rejection (Islam et al., 2025). Additionally, TiO_2 domains within GO hybrids generate reactive oxygen species under UV light, providing in-situ self-cleaning capabilities (Algieri et al., 2021).

4.2.2.1 Interface engineering and limitations

Interfacial incompatibility between organic and inorganic phases remains the primary challenge. Aggregation and micro-void formation reduce selectivity and mechanical cohesion. Surface functionalization using silane coupling agents, PDA coatings, or PEI linkers creates chemical bridges that improve adhesion and dispersion (Foorginezhad et al., 2025; Nibe et al., 2025). Despite these advances, scaling remains difficult because MOF/COF synthesis often involves toxic solvents and precise thermal conditions. Nonetheless, the integration of nanoporous frameworks with functional polymers is emerging as a cornerstone for future reactive filtration platforms combining adsorption, catalysis, and ion-selective transport (Xiao et al., 2025b).

4.3 Emerging modification strategies

Recent developments focus on smart surfaces that self-regulate interactions with foulants and pollutants. Such functional membranes rely on zwitterionic chemistry, molecular imprinting, bio-inspired adhesives, and green synthesis methods (Cao et al., 2016; Gaxela, 2020; Mohamed et al., 2024; Tchekwagep et al., 2022; Wen et al., 2016).

4.3.1 Zwitterionic and charged surfaces

Zwitterionic layers bearing quaternary ammonium and sulfonate groups form hydration shells that repel proteins and microorganisms. Covalent attachment of poly(sulfobetaine methacrylate) or carboxybetaine onto PES or PVDF membranes increased hydrophilicity and reduced fouling by $> 70\%$ (Huang et al., 2015; Upadhyaya et al., 2018). Positively charged PEI coatings promote electrostatic capture of anionic pollutants such as CrO_4^{2-} and PFAS, while negatively charged sulfonated films repel cations (Aryanti et al., 2023; Foorginezhad et al., 2025). Charge asymmetry and hydration synergy thus govern selectivity under varying ionic strengths (Nibe et al., 2025).

4.3.2 Molecular imprinting and ion-selective recognition

Molecularly imprinted membranes (MIMs) contain template-derived cavities that selectively bind target pollutants. Embedding Pb^{2+} or bisphenol A as templates during polymerization creates specific recognition sites which enable selective re-adsorption after template removal. Integration of these MIM sites into PBI and PI matrices provides rigid yet recognition-active structures for trace metal. Such membranes unite filtration and sensing functions in a single platform (Dawam et al., 2025; Shen et al., 2012; Tchekwagep et al., 2022).

4.3.3 Bio-inspired coatings

Bio-inspired chemistry offers universal adhesion and multifunctional reactivity. Polydopamine (PDA) spontaneously forms conformal films under mild alkaline conditions and serves as an anchor for secondary reactions like metal-ion coordination and Schiff-base (Foorginezhad et al., 2025). PDA/UiO-66-NH_2 interlayers in thin-film composite membranes decrease internal concentration polarization and double water flux relative to pristine layers (Xiao et al., 2025b). Tannic acid, another polyphenolic modifier, imparts redox activity and chelating ability for simultaneous antifouling and metal-ion removal (Obotey Ezugbe et al., 2020).

4.3.4 Green and sustainable fabrication

Environmental sustainability has driven a shift toward benign solvents and renewable feedstocks. Water-borne phase-inversion and ionic-liquid-assisted casting eliminate toxic solvents such as N-methyl-2-pyrrolidone (NMP) and N,N-dimethylformamide (DMF) (Prat et al., 2014; Ravi et al., 2025c). Solvent-free initiated chemical vapor deposition (iCVD) allows uniform coating of zwitterionic polymers without volatile organic compounds (VOCs) emission (Upadhyaya et al., 2018). Redox-initiated grafting in aqueous media replaces organohalide catalysts while preserving controlled architecture (Yu et al., 2022). Integrating renewable energy sources into membrane operations—such as solar-assisted forward osmosis—reduces carbon footprint and operating costs (Mannaf et al., 2025).

4.4 Synthesis of insights

The collective evidence from polymeric, inorganic, and hybrid systems confirms that future water-treatment membranes must merge chemical robustness with adaptive selectivity. Crosslinked aromatic polymers (PI, PBI) provide backbone stability; MOF/COF and graphene fillers introduce hierarchical porosity and active sites; and zwitterionic or bio-inspired coatings govern interfacial hydration and fouling dynamics. Sustainable fabrication and AI-assisted design close the loop between materials science and process engineering. Together, these strategies redefine membranes as smart, selective, and sustainable platforms for the removal of micropollutants and trace metals from complex wastewater systems.

5. Case studies in smart and selective membrane applications

5.1 Bismuth-chelate-modified UF/MBR membranes for hospital wastewater

In a comprehensive study, Kose-Mutlu et al. (2025) designed bismuth-chelate-incorporated polymer membranes to enhance the treatment of hospital wastewater containing pharmaceuticals, antibiotics, and resistant microorganisms. The bismuth chelate, synthesized from bismuth nitrate pentahydrate and 2,3-dimercapto-1-propanol (BisBAL), was blended into polyethersulfone (PES) and polyvinylidene fluoride (PVDF) matrices at 150 and 300 μM loadings. Flat-sheet (FS) and hollow-fiber (HF) membranes were fabricated via phase inversion and coupled to a membrane bioreactor (MBR) handling real hospital effluent. The BisBAL incorporation increased surface hydrophilicity and positive zeta potential, promoting stronger electrostatic interaction with anionic micropollutants such as caffeine (CAF), acetaminophen (APAP), and ciprofloxacin (CPFX) (Mackull' et al., 2021). The MM–HF–300 membrane achieved approximately 80% CPFX rejection, outperforming the pristine counterpart. In addition, CAF and APAP removals surpassed 70%, confirming the dual contribution of adsorption and size-based exclusion. Scanning electron micrographs revealed smoother, less-fouled surfaces on BisBAL-doped membranes, attributed to the presence of carboxyl and hydroxyl functional groups forming hydration layers that resisted organic adhesion (Majumder et al., 2021). Flux-decline analyses indicated improved long-term stability despite a slight reduction in permeability. Consequently, the system integrated chemical complexation, electrostatic repulsion, and biofouling resistance within a single low-pressure configuration, ideal for tertiary polishing of medical effluents where trace-level removal is critical. This work exemplifies how metal-organic chelates embedded in polymer networks can transform conventional UF/MBR units into multifunctional barriers for emerging micropollutants (Kose-Mutlu et al., 2025).

5.2 Electrostatically deposited NZVI–PVDF (ENP) membranes for Cr (VI) reduction

Sun et al. (2025) proposed an innovative electrostatic-deposition technique to immobilize nano-zero-valent iron (NZVI) particles on PVDF films, producing an ENP (electrostatically deposited

NZVI–PVDF) membrane with superior reactivity and reuse potential for Cr(VI) detoxification (Balali-Mood et al., 2021). Traditional mixing of NZVI often leads to aggregation and loss of surface activity; by contrast, the deposition method distributes nanoparticles uniformly under a controlled electric field, maximizing their accessibility to pollutants. Batch tests on synthetic 100 mg L^{-1} Cr (VI) solutions revealed that 2 g NZVI-loaded ENP membranes achieved near-complete removal within 60 min, sustaining >98% efficiency over five cycles without significant deterioration of mechanical integrity. SEM and AFM characterizations confirmed that excessive loading (≥2.5 g) caused rough, agglomerated morphologies, whereas optimal dispersion provided abundant reactive sites for electron transfer. The overall removal followed a redox–adsorption pathway: $Fe^0 \rightarrow Fe^{2+} \rightarrow Fe^{3+}$ conversion reduced Cr (VI) to Cr (III), which subsequently precipitated as $FeCr_2O_4$ and $Cr_xFe_{1\text{-}x}(OH)_3$ complexes on the membrane surface. Post-reaction analyses indicated diminished surface roughness (from ~127 to 83 nm Sq) and a thin product layer that passivated but did not block the pores. These results established a balance between reactivity and structural stability. The ENP membranes not only prevented nanoparticle leaching, a key environmental concern, but also allowed simple regeneration by mild acid washing, making them practical for industrial reuse. This case demonstrates how coupling electrochemical deposition with polymer support structures enables high-performance reactive membranes for metal-ion reduction. By eliminating secondary waste and sustaining catalytic efficiency, ENP membranes exemplify circular-use reactive barriers for heavy-metal remediation under real-world conditions.

5.3 NF and RO membrane systems for micropollutant removal under EU directive 2024/3019

A large-scale comparative investigation by Domínguez et al. examined nanofiltration (NF) and reverse osmosis (RO) systems as quaternary treatment stages for municipal wastewater, aimed at compliance with the EU Directive 2024/3019 targeting micropollutant abatement (Yang et al., 2024). The directive mandates at least 80% average removal of 12 indicator contaminants, including pharmaceuticals such as diclofenac, clarithromycin, benzotriazole, and carbamazepine. Four commercial membranes—NF-90 (DuPont) and three RO models (Toray 73AC, 73UAC, 73HA)—were operated at 25–41 bar, pH 2–11, and temperatures up to 45°C (Gurke et al., 2015). The goal was to quantify efficiency across compounds of varying molecular weight and polarity. Among the tested configurations, the RO-AC membrane delivered the highest overall performance, achieving >80% removal for nearly all substances except benzotriazole, whose small size and high hydrophilicity limited rejection to ~67% (Jim et al., 2006). NF membranes demonstrated lower energy demand but weaker exclusion of low-molecular-mass, neutrally charged solutes. Flux and fouling behavior were strongly influenced by total dissolved solids (TDS): lower-salinity feeds enhanced water permeability and reduced cake-layer formation, confirming the role of osmotic pressure and ionic shielding in performance decline (Wu et al., 2010). The study validated that membrane-based quaternary treatment could meet forthcoming European discharge criteria without major infrastructure overhauls, providing a scalable compliance strategy. While NF remains suitable for moderately polar contaminants, RO membranes currently offer the robustness and rejection efficiency required for full alignment with Directive 2024/3019 limits. The findings underscore how policy-driven regulation is accelerating adoption of high-selectivity membrane technologies in municipal water management (Domínguez et al., 2025).

6. Challenges and future perspectives

Despite rapid progress in smart and selective membrane technologies, several scientific, technical, and practical challenges must be addressed before their widespread and sustainable implementation in real-world wastewater treatment systems. One of the most persistent challenges is membrane fouling (She et al., 2016). Organic matter, biofilms, colloids, and inorganic scaling species accumulate on membrane surfaces during operation, leading to flux decline, increased energy demand, and frequent chemical cleaning. Although surface functionalization, zwitterionic coatings, and nanomaterial incorporation have significantly improved antifouling behavior, fouling remains inevitable under long-term operation with complex wastewater matrices (Ibrar et al., 2019; Latif et al., 2024). Future membrane designs must therefore move beyond fouling resistance alone and incorporate self-cleaning, fouling-tolerant, or self-healing functionalities that allow membranes to recover performance without aggressive chemical intervention. Another major challenge lies in balancing permeability and selectivity (Das et al., 2022; Yuan et al., 2025). Enhancing rejection of small organic micropollutants or trace metal ions often leads to reduced water flux due to denser membrane structures or increased interaction with solutes. While mixed-matrix membranes and affinity-based functionalization have partially overcome this trade-off, maintaining high permeability without sacrificing selectivity remains a core design constraint (Das et al., 2022). Future research should focus on hierarchical pore architectures, well-aligned nanochannels, and controlled free-volume engineering that enable fast water transport while preserving molecular-level discrimination. Chemical and mechanical stability under harsh operating conditions is another critical limitation. Wastewater streams often contain strong oxidants, extreme pH conditions, and abrasive particles that accelerate membrane degradation (Rabiee et al., 2023). Although polymers such as polybenzimidazole and crosslinked polyimides offer superior chemical resistance, their fabrication is often energy-intensive and solvent-dependent (Ravi et al., 2025a; Ray et al., 2023). Developing membranes that combine long-term durability with environmentally benign synthesis routes remains an unresolved challenge (Ravi et al., 2025b; Voicu et al., 2023). Sustainability is increasingly central to membrane development. Many high-performance membranes still rely on toxic solvents, petrochemical-based polymers, and energy-intensive fabrication processes. Future perspectives clearly point toward green membrane manufacturing, using bio-based or biodegradable polymers, water-borne solvents, and low-energy crosslinking strategies. The adoption of renewable raw materials and recyclable membrane components will be essential to reduce the environmental footprint of membrane production and disposal (Jiang et al., 2020; Naeem et al., 2024).

From a functional standpoint, future smart membranes are expected to evolve from passive separation tools into multifunctional treatment platforms. Hybrid membranes incorporating metal–organic frameworks (MOFs), covalent organic frameworks (COFs), graphene derivatives, or catalytic nanoparticles can simultaneously enable separation, adsorption, catalytic degradation, and ion recovery (Liang et al., 2019; Singh et al., 2025b; Wang et al., 2023; Yuan et al., 2019). Such multifunctionality supports circular-economy concepts (Loh et al., 2025), where pollutants such as heavy metals are not merely removed but recovered as valuable resources. A rapidly emerging frontier is the integration of digitalization and artificial intelligence into membrane systems. Machine-learning models can predict membrane performance, fouling behavior, and rejection efficiency based on material chemistry and operating conditions. Coupled with real-time sensors, digital twins, and automated control systems, future membrane plants could self-optimize energy consumption, cleaning cycles, and separation efficiency. However, reliable data generation, model transparency, and scalability of AI-driven control remain open challenges (Cao et al., 2024; Ray et al., 2022; Yin et al., 2024). Finally, large-scale deployment requires addressing economic and infrastructural barriers. Laboratory-scale successes often fail during scale-up due to material costs, fabrication complexity, or operational instability. Future research must therefore prioritize pilot-scale validation, standardized testing protocols, and techno-economic assessments to ensure practical feasibility. Interdisciplinary collaboration among materials scientists, engineers, data scientists, and policymakers will be essential to bridge the gap between laboratory innovation and real-world implementation.

7. Conclusion

Smart and selective membrane technologies represent a transformative advancement in modern wastewater treatment, addressing the growing challenge of removing organic micropollutants and trace metals that persist through conventional processes. By integrating tailored chemical functionality, adaptive selectivity, and enhanced structural robustness, these membranes have redefined the role of filtration from simple physical separation to intelligent and responsive treatment. Throughout this chapter, it has been demonstrated that conventional treatment methods such as coagulation, adsorption, and advanced oxidation processes face inherent limitations when dealing with low-concentration, chemically diverse contaminants. In contrast, smart membranes leverage synergistic transport mechanisms including size exclusion, electrostatic interactions, and chemical affinity to achieve high rejection efficiencies while maintaining operational flexibility. Polymeric membranes, particularly those based on polyimides and polybenzimidazoles, have shown exceptional promise due to their chemical stability and tunable surface chemistry. When combined with nanomaterials such as MOFs, COFs, and graphene derivatives, membrane performance is further enhanced through improved selectivity, antifouling behavior, and multifunctionality. The chapter also highlights how surface functionalization strategies, including zwitterionic coatings, bio-inspired modifications, and molecular imprinting, enable precise control over membrane and solute interactions. These advances allow membranes to respond dynamically to environmental stimuli, adjust separation behavior, and extend operational lifetimes. Importantly, case studies from hospital wastewater treatment, heavy-metal remediation, and regulatory-driven municipal applications demonstrate that smart membranes are no longer confined to laboratory research but are increasingly relevant to real-world systems. Looking forward, the future of membrane technology lies in sustainability, intelligence, and integration. Green fabrication routes using bio-based polymers, benign solvents, and low-energy processing will be essential to minimize environmental impact. Multifunctional membranes capable of simultaneous separation, degradation, and resource recovery will support circular water-management systems. At the same time, the integration of machine learning, real-time monitoring, and digital control will transform membranes into adaptive treatment interfaces capable of predictive and optimized operation. In summary, smart membranes are evolving from passive filters into active, intelligent platforms that purify water, sense contaminants, and recover resources simultaneously. Their convergence with green chemistry and artificial intelligence offers a pathway toward low-footprint, high-efficiency treatment technologies capable of meeting increasingly stringent global water-quality regulations. Continued interdisciplinary research and collaboration will be critical to translating these advances into scalable, field-ready solutions, ensuring safer water resources and a more sustainable future.

References

Abdullah, N., et al. (2019). Recent trends of heavy metal removal from water/wastewater by membrane technologies. *Journal of Industrial and Engineering Chemistry, 76*, 17–38.

Abo-Farha, S., et al. (2009). Removal of some heavy metal cations by synthetic resin purolite C100. *Journal of Hazardous Materials, 169*(1–3), 190–194.

Abu-Obaid, S., et al. (2023). Adsorptive membranes for nutrient recovery from wastewater: A novel solution for water purification challenges. *Chemical Engineering Journal, 465*, Article 143057.

Ahamad, A., et al. (2019). Types of water pollutants: Conventional and emerging. In *Sensors in water pollutants monitoring: Role of material* (pp. 21–41). Springer.

Ajith, A., et al. (2023). Application of membrane techniques for water and wastewater treatment. In *Resource recovery in drinking water treatment* (pp. 27–45). Elsevier.

Al-Jlil, S. A., & Alsewailem, F. D. (2009). Saudi Arabian clays for lead removal in wastewater. *Applied Clay Science, 42*(3–4), 671–674.

Alengebawy, A., et al. (2021). Heavy metals and pesticides toxicity in agricultural soil and plants: Ecological risks and human health implications. *Toxics, 9*(3), 42.

Algieri, C., Chakraborty, S., & Candamano, S. (2021). A way to membrane-based environmental remediation for heavy metal removal. *Environments, 8*(6), 52.

Ali, H., Khan, E., & Ilahi, I. (2019). Environmental chemistry and ecotoxicology of hazardous heavy metals: Environmental persistence, toxicity, and bioaccumulation. *Journal of Chemistry, 2019*(1), Article 6730305.

Amador, P. P., et al. (2015). Antibiotic resistance in wastewater: Occurrence and fate of enterobacteriaceae producers of class A and class C β-lactamases. *Journal of Environmental Science and Health, Part A, 50*(1), 26–39.

Annamalai, J., Ummalyma, S. B., & Pandey, A. (2022). Recent trends in the microbial degradation and bioremediation of emerging pollutants in wastewater treatment system. In *Development in wastewater treatment research and processes* (pp. 99–125). Elsevier.

Aryanti, P., et al. (2023). Ultra low-pressure reverse osmosis (ULPRO) membrane for desalination: Current challenges and future directions. *Desalination, 560*, Article 116650.

Asadollahzadeh, M., Torkaman, R., & Torab-Mostaedi, M. (2026). Evaluation of the efficacy of membrane bioreactor in removing hexavalent chromium. *The Canadian Journal of Chemical Engineering, 104*(5), 2652–2663.

Aziz, A., & Aziz, A. (2023). Effects of micropollutants on human health. In *Management of wastewater and sludge* (pp. 151–166). CRC Press.

Azmi, L. S. (2025). Membrane filtration technologies for sustainable industrial wastewater treatment: A review of heavy metal removal. *Desalination and Water Treatment*, Article 101321.

Bai, X., et al. (2018). Occurrence, distribution, and seasonality of emerging contaminants in urban watersheds. *Chemosphere, 200*, 133–142.

Bai, W., et al. (2023). Quantifying and reducing concentration polarization in reverse osmosis systems. *Desalination, 554*, Article 116480.

Baker, R. W. (2023). *Membrane technology and applications*. John Wiley and Sons.

Balali-Mood, M., et al. (2021). Toxic mechanisms of five heavy metals: Mercury, lead, chromium, cadmium, and arsenic. *Frontiers in Pharmacology, 12*, Article 643972.

Bellona, C., et al. (2004). Factors affecting the rejection of organic solutes during NF/RO treatment—a literature review. *Water Research, 38*(12), 2795–2809.

Bernardes, A. F., et al. (2025). Bio-inspired anti-fouling strategies for membrane-based separations. *Chemical Communications, 61*(27), 5064–5071.

Bharti, R., & Sharma, R. (2022). Effect of heavy metals: An overview. *Materials Today: Proceedings, 51*, 880–885.

Bilal, M., et al. (2019). Biocatalytic degradation/redefining "removal" fate of pharmaceutically active compounds and antibiotics in the aquatic environment. *Science of the Total Environment, 691*, 1190–1211.

Boussu, K., et al. (2006). Characterization of commercial nanofiltration membranes and comparison with self-made polyethersulfone membranes. *Desalination, 191*(1–3), 245–253.

Bowen, W. R., & Welfoot, J. S. (2002). Modelling the performance of membrane nanofiltration—critical assessment and model development. *Chemical Engineering and Science, 57*(7), 1121–1137.

Braghetta, A., DiGiano, F. A., & Ball, W. P. (1997). Nanofiltration of natural organic matter: pH and ionic strength effects. *Journal of Environmental Engineering, 123*(7), 628–641.

Briffa, J., Sinagra, E., & Blundell, R. (2020). Heavy metal pollution in the environment and their toxicological effects on humans. *Heliyon, 6*(9).

Buonomenna, M., & Bae, J. (2015). Organic solvent nanofiltration in pharmaceutical industry. *Separation and Purification Reviews, 44*(2), 157–182.

Buonomenna, M. G., et al. (2022). Water cleaning adsorptive membranes for efficient removal of heavy metals and metalloids. *Water, 14*(17), 2718.

Burkhardt-Holm, P. (2011). Linking water quality to human health and environment: The fate of micropollutants. *Institute of Water Policy Working Paper, 3*, 2011.

Cao, X., et al. (2016). Bioinspired multifunctional membrane for aquatic micropollutants removal. *ACS Applied Materials and Interfaces, 8*(44), 30511–30522.

Cao, Z., et al. (2024). Machine learning in membrane design: From property prediction to AI-guided optimization. *Nano Letters, 24*(10), 2953–2960.

Cevallos-Mendoza, J., et al. (2022). Removal of contaminants from water by membrane filtration: A review. *Membranes, 12*(6), 570.

Chakraborty, A., et al. (2023). Pharmaceuticals and personal care products as emerging environmental contaminants: Prevalence, toxicity, and remedial approaches. *ACS Chemical Health & Safety, 30*(6), 362–388.

Charcosset, C. (1998). Purification of proteins by membrane chromatography. *Journal of Chemical Technology & Biotechnology: International Research in Process, Environmental AND Clean Technology, 71*(2), 95–110.

Chen, C.-W., et al. (2024). Exfoliated MXene nanosheet by double layer forces for preparing magnetic adsorbents of toxic metal ions and recycling with hyperthermic effect. *Separation and Purification Technology, 349*, Article 127855.

Cheryan, M. (1998). *Ultrafiltration and microfiltration handbook*. CRC press.

Childress, A. E., & Elimelech, M. (2000). Relating nanofiltration membrane performance to membrane charge (electrokinetic) characteristics. *Environmental Science and Technology, 34*(17), 3710–3716.

Chung, T.-S., et al. (2007). Mixed matrix membranes (MMMs) comprising organic polymers with dispersed inorganic fillers for gas separation. *Progress in Polymer Science, 32*(4), 483–507.

Costa, F. C., dos Santos, C. R., & Amaral, M. C. (2023). Trace organic contaminants removal by membrane distillation: A review on mechanisms, performance, applications, and challenges. *Chemical Engineering Journal, 464*, Article 142461.

Covaliu-Mierlă, C. I., Păunescu, O., & Iovu, H. (2023). Recent advances in membranes used for nanofiltration to remove heavy metals from wastewater: A review. *Membranes, 13*(7), 643.

Crini, G. (2005). Recent developments in polysaccharide-based materials used as adsorbents in wastewater treatment. *Progress in Polymer Science, 30*(1), 38–70.

Das, S., et al. (2016). Extracting uranium from seawater: Promising AF series adsorbents. *Industrial and Engineering Chemistry Research, 55*(15), 4110–4117.

Das, S., et al. (2022). Assessing advances in anti-fouling membranes to improve process economics and sustainability of water treatment. *ACS ES&T Engineering, 2*(11), 2159–2173.

Dawam, M., et al. (2025). Advances in membrane technologies for heavy metal removal from polluted water: A comprehensive review. *Water, Air, and Soil Pollution, 236*(7), 461.

Deb, C. K., et al. (2025). Transition toward the per-and polyfluoroalkyl substance-free environment: Is it possible?. In *PFAS in the environment: Occurrence, characterization, treatment, and management* (pp. 447–474).

Dey, S., Bano, F., & Malik, A. (2019). Pharmaceuticals and personal care product (PPCP) contamination—a global discharge inventory. In *Pharmaceuticals and personal care products: Waste management and treatment technology* (pp. 1–26). Elsevier.

Domínguez, E., Ferre, M., Moya-Llamas, M. J., Ortuño, N., & Prats, D. (2025). Removal of indicator micro-pollutants included in directive (EU) 2024/3019 using nanofiltration and reverse osmosis. *Water, 17*(9), 1269.

dos Santos, C.R., et al., *845–Avaliação de risco de compostos farmacêuticos presentes em ambiente aquático no brasil*.pt.

Elimelech, M., et al. (1997). Role of membrane surface morphology in colloidal fouling of cellulose acetate and composite aromatic polyamide reverse osmosis membranes. *Journal of Membrane Science, 127*(1), 101–109.

Esposito, M., et al. (2020). Organic pollutants (PCBs, PCDD/Fs, PAHs) and toxic metals in farmed mussels from the Gulf of Naples (Italy): Monitoring and human exposure. *Regional Studies in Marine Science, 40*, Article 101497.

Fane, A., & Fell, C. (1987). A review of fouling and fouling control in ultrafiltration. *Desalination, 62*, 117–136.

Foorginezhad, S., et al. (2025). Emerging membrane technologies for sustainable water treatment: A review on recent advances. *Environmental Science: Advances, 4*(4), 530–570.

Fu, F., & Wang, Q. (2011). Removal of heavy metal ions from wastewaters: A review. *Journal of Environmental Management, 92*(3), 407–418.

Gadipelly, C., et al. (2014). Pharmaceutical industry wastewater: Review of the technologies for water treatment and reuse. *Industrial and Engineering Chemistry Research, 53*(29), 11571–11592.

Gaxela, N. N. (2020). *Zwitterion polymer functionalized membranes for the removal of dyes and heavy metals materials in water*. University of Johannesburg (South Africa).

Gayatri, R., et al. (2024). Polymer-based nanocomposite membranes for industrial wastewater treatment: A review. *Journal of Environmental Chemical Engineering, 12*(5), Article 113276.

Ge, S., et al. (2024). A comprehensive review of covalent organic frameworks (COFs) and their derivatives in environmental pollution control. *Chemical Society Reviews, 53*(23), 11259–11302.

Gohil, J. M., & Ray, P. (2017). A review on semi-aromatic polyamide TFC membranes prepared by interfacial polymerization: Potential for water treatment and desalination. *Separation and Purification Technology, 181*, 159–182.

Gupta, A., et al. (2024). Pharmaceutically active micropollutants: Origin, hazards and removal. *Frontiers in Microbiology, 15*, Article 1339469.

Gurke, R., et al. (2015). Occurrence and removal of frequently prescribed pharmaceuticals and corresponding metabolites in wastewater of a sewage treatment plant. *Science of the Total Environment, 532*, 762–770.

Hagmeyer, G., & Gimbel, R. (1998). Modelling the salt rejection of nanofiltration membranes for ternary ion mixtures and for single salts at different pH values. *Desalination, 117*(1–3), 247–256.

Halden, R. U., et al. (2017). The florence statement on triclosan and triclocarban. *Environmental Health Perspectives, 125*(6), Article 064501.

He, Y., et al. (2017). UiO-66 incorporated thin-film nanocomposite membranes for efficient selenium and arsenic removal. *Journal of Membrane Science, 541*, 262–270.

Heskins, M., & Guillet, J. E. (1968). Solution properties of poly (N-isopropylacrylamide). *Journal of Macromolecular Science-Chemistry, 2*(8), 1441–1455.

Hilal, N., et al. (2004). A comprehensive review of nanofiltration membranes: Treatment, pretreatment, modelling, and atomic force microscopy. *Desalination, 170*(3), 281–308.

Hocking, M. B. (2006). *Handbook of chemical technology and pollution control*. Elsevier.

Horseman, T., et al. (2020). Wetting, scaling, and fouling in membrane distillation: State-of-the-art insights on fundamental mechanisms and mitigation strategies. *ACS ES&T Engineering, 1*(1), 117–140.

Hou, X., et al. (2019). Electrospun TiO2 nanofibers coated with polydopamine for enhanced sunlight-driven photocatalytic degradation of cationic dyes. *Surface and Interface Analysis, 51*(2), 169–176.

Huang, X., et al. (2015). Treatment of oily waste water by PVP grafted PVDF ultrafiltration membranes. *Chemical Engineering Journal, 273*, 421–429.

Huang, T., et al. (2023). Advanced stimuli-responsive membranes for smart separation. *Chemical Society Reviews, 52*(13), 4173–4207.

Huyen, D. T. T., Ravi, S., & Kwon, Y.-N. (2025). Modifying polybenzimidazole nanofiber membranes using UiO-66-NH2 metal–organic framework for oil–water separation. *Chemical Engineering Science*, Article 121996.

Ibrar, I., et al. (2019). A review of fouling mechanisms, control strategies and real-time fouling monitoring techniques in forward osmosis. *Water, 11*(4), 695.

Islam, I. U., et al. (2025). MOF and MOF-based membranes: Promising solutions for pharmaceutical wastewater treatment. *Journal of Materials Science, 60*(8), 3634–3662.

Ito, Y., et al. (1997). pH-sensitive gating by conformational change of a polypeptide brush grafted onto a porous polymer membrane. *Journal of the American Chemical Society, 119*(7), 1619–1623.

Jagirani, M. S., & Soylak, M. (2024). Metal-organic framework nanosheets (MONs) a breakthrough in separation and pre-concentration technique: A review. *TrAC, Trends in Analytical Chemistry, 178*, Article 117838.

Jiang, S., & Ladewig, B. P. (2020). Green synthesis of polymeric membranes: Recent advances and future prospects. *Current Opinion in Green and Sustainable Chemistry, 21*, 1–8.

Jim, T. Y., Bouwer, E. J., & Coelhan, M. (2006). Occurrence and biodegradability studies of selected pharmaceuticals and personal care products in sewage effluent. *Agricultural Water Management, 86*(1–2), 72–80.

Ju, X.-J., et al. (2008). A novel thermoresponsive hydrogel with ion-recognition property through supramolecular host– guest complexation. *The Journal of Physical Chemistry B, 112*(4), 1112–1118.

Kümmerer, K. (2010). Pharmaceuticals in the environment. *Annual Review of Environment and Resources, 35*(1), 57–75.

Kang, G.-d., & Cao, Y.-m. (2014). Application and modification of poly (vinylidene fluoride)(PVDF) membranes–a review. *Journal of Membrane Science, 463*, 145–165.

Kaur, R., et al. (2019). Pesticides classification and its impact on environment. *International Journal of Current Microbiology and Applied Sciences, 8*(3), 1889–1897.

Kaushik, P., et al. (2022). Environmental hazards of heavy metal pollution and toxicity: A review. *Flora Fauna, 28*, 315–327.

Khalid, M., & Abdollahi, M. (2021). Environmental distribution of personal care products and their effects on human health. *Iranian Journal of Pharmaceutical Research: IJPR, 20*(1), 216.

Khanzada, N. K., et al. (2020). Removal of organic micropollutants using advanced membrane-based water and wastewater treatment: A review. *Journal of Membrane Science, 598*, Article 117672.

Khoo, Y. S., et al. (2022). Removal of emerging organic micropollutants via modified-reverse osmosis/nanofiltration membranes: A review. *Chemosphere, 305*, Article 135151.

Kim, J.-J., Kim, Y.-S., & Kumar, V. (2019). Heavy metal toxicity: An update of chelating therapeutic strategies. *Journal of Trace Elements in Medicine and Biology, 54*, 226–231.

Kirthiga, S., et al. (2025). Advanced membrane technologies: Membrane bioreactors (MBRs) in resource recovery from wastewater streams. In *Resource resurgence: Mitigating wastewater, smart recycling and novel technologies* (pp. 21–50). Springer.

Kose-Mutlu, B., et al. (2025). Removal of emerging organic micropollutants from real hospital wastewater by modified ultrafiltration membranes. *Separation and Purification Technology, 357*, Article 129960.

Koyuncu, I., et al. (2008). Removal of hormones and antibiotics by nanofiltration membranes. *Journal of Membrane Science, 309*(1–2), 94–101.

Krishnan, R. Y., et al. (2021). Removal of emerging micropollutants originating from pharmaceuticals and personal care products (PPCPs) in water and wastewater by advanced oxidation processes: A review. *Environmental Technology and Innovation, 23*, Article 101757.

Kumar, A., et al. (2024). Organic micropollutants and their effects on the environment and human health. In *Organic micropollutants in aquatic and terrestrial environments* (pp. 87–105). Springer.

Latif, E., & Elmaadawy, K. (2024). Overcoming fouling challenges in ultrafiltration for drinking water treatment. *Engineering Research Express, 6*(4), Article 042101.

Le, N. L., & Nunes, S. P. (2016). Materials and membrane technologies for water and energy sustainability. *Sustainable Materials and Technologies, 7*, 1–28.

Lee, H., et al. (2007). Mussel-inspired surface chemistry for multifunctional coatings. *Science, 318*(5849), 426–430.

Leupold, M. S., et al. (2025). Modular annular photocatalytic membrane reactor for the degradation of micropollutants: Design and application. *MethodsX*, Article 103478.

Liang, B., et al. (2019). Membrane separation in organic liquid: Technologies, achievements, and opportunities. *Advanced Materials, 31*(45), Article 1806090.

Liu, F., & Urban, M. W. (2010). Recent advances and challenges in designing stimuli-responsive polymers. *Progress in Polymer Science, 35*(1–2), 3–23.

Loganathan, P., et al. (2023). Bisphenols in water: Occurrence, effects, and mitigation strategies. *Chemosphere, 328*, Article 138560.

Loh, C. Y., Burrows, A. D., & Xie, M. (2025). Sustainable polymeric membranes: Green chemistry and circular economy approaches. *ACS ES&T Engineering, 5*(8), 1882–1906.

Lu, X., et al. (2016). Adsorption and thermal stabilization of Pb2+ and Cu2+ by zeolite. *Industrial and Engineering Chemistry Research, 55*(32), 8767–8773.

Lu, D., et al. (2025). A smart framework to design membranes for organic micropollutants removal. *Nature Sustainability*, 1–13.

Mackul'ak, T., et al. (2021). Hospital wastewater–source of specific micropollutants, antibiotic-resistant microorganisms, viruses, and their elimination. *Antibiotics, 10*(9), 1–14.

Majumder, A., et al. (2021). A review on hospital wastewater treatment: A special emphasis on occurrence and removal of pharmaceutically active compounds, resistant microorganisms, and SARS-CoV-2. *Journal of Environmental Chemical Engineering, 9*(2), Article 104812.

Majumder, A., Gupta, B., & Gupta, A. K. (2019). Pharmaceutically active compounds in aqueous environment: A status, toxicity and insights of remediation. *Environmental Research, 176*, Article 108542.

Malakar, A., et al. (2021). Nanomaterials in the environment, human exposure pathway, and health effects: A review. *Science of the Total Environment, 759*, Article 143470.

Mannaf, M. M., et al. (2025). Current progress in advanced functional membranes for water-pollutant removal: A critical review. *Membranes, 15*(10), 300.

Mendes, P. M. (2008). Stimuli-responsive surfaces for bio-applications. *Chemical Society Reviews, 37*(11), 2512–2529.

Miller, M., & Bazylak, A. (2011). A review of polymer electrolyte membrane fuel cell stack testing. *Journal of Power Sources, 196*(2), 601–613.

Mohamed, S. I. G. P., & Isloor, A. M. (2024). A review of the effect of ionic/zwitterionic materials on membrane performance for heavy metal removal application. *Membrane Technologies for Heavy Metal Removal from Water*, 320–332.

Moldovan, Z. (2006). Occurrences of pharmaceutical and personal care products as micropollutants in rivers from Romania. *Chemosphere, 64*(11), 1808–1817.

Mulder, M. (2012). *Basic principles of membrane technology*. Springer science and business media.

Mumtaz, F., et al. (2025). Stimuli-responsive membranes—mechanisms, materials and future directions. *Npj Clean Water.*

Mustafa, M., et al. (2025). Innovative approaches to greywater micropollutant removal: AI-driven solutions and future outlook. *RSC Advances, 15*(16), 12125–12151.

Naeem, A., et al. (2024). Sustainable and green membranes for chemical separations: A review. *Separation and Purification Technology, 336*, Article 126271.

Nanda, D., et al. (2008). Effect of solution chemistry on water softening using charged nanofiltration membranes. *Desalination, 234*(1–3), 344–353.

Nibe, R., & Gaikwad, R. (2025). A comprehensive review on application of different natural and chemically modified nanomaterials adsorbents for heavy metals removal. *Journal of Inorganic and Organometallic Polymers and Materials*, 1–21.

Oatley-Radcliffe, D. L., et al. (2017). Nanofiltration membranes and processes: A review of research trends over the past decade. *Journal of Water Process Engineering, 19*, 164–171.

Obotey Ezugbe, E., & Rathilal, S. (2020). Membrane technologies in wastewater treatment: A review. *Membranes, 10*(5), 89.

Oladimeji, T., et al. (2024). Review on the impact of heavy metals from industrial wastewater effluent and removal technologies. *Heliyon, 10*(23). *e40370 CrossRef CAS PubMed.*

Otálora, A., Lerma, T. A., Arrieta-Urango, Y., & Palencia, M. (2021). Emerging organic pollutants in aqueous environments: Detection, monitoring, and removal techniques. *Journal of Science with Technological Applications, 10*, 92–153.

Parbat, D., & Manna, U. (2023). Responsive and reactive layer-by-layer coatings for deriving functional interfaces. *Materials Advances, 4*(1), 35–51.

Park, H. B., et al. (2017). Maximizing the right stuff: The trade-off between membrane permeability and selectivity. *Science, 356*(6343), eaab0530.

Pathak, V. M., et al. (2022). Current status of pesticide effects on environment, human health and it's eco-friendly management as bioremediation: A comprehensive review. *Frontiers in Microbiology, 13*, Article 962619.

Peng, N., et al. (2012). Evolution of polymeric hollow fibers as sustainable technologies: Past, present, and future. *Progress in Polymer Science, 37*(10), 1401–1424.

Piletsky, S., et al. (1999). Receptor and transport properties of imprinted polymer membranes–a review. *Journal of Membrane Science, 157*(2), 263–278.

Pothitou, P., & Voutsa, D. (2008). Endocrine disrupting compounds in municipal and industrial wastewater treatment plants in northern Greece. *Chemosphere, 73*(11), 1716–1723.

Prat, D., Hayler, J., & Wells, A. (2014). A survey of solvent selection guides. *Green Chemistry, 16*(10), 4546–4551.

Qalyoubi, L., Al-Othman, A., & Al-Asheh, S. (2021a). Recent progress and challenges of adsorptive membranes for the removal of pollutants from wastewater. Part II: Environmental applications. *Case Studies in Chemical and Environmental Engineering, 3*, Article 100102.

Qalyoubi, L., Al-Othman, A., & Al-Asheh, S. (2021b). Recent progress and challenges on adsorptive membranes for the removal of pollutants from wastewater. Part I: Fundamentals and classification of membranes. *Case Studies in Chemical and Environmental Engineering, 3*, Article 100086.

Qin, Y., et al. (2025). Metal organic framework for sustainable removal of heavy metals contamination from wastewater: A review. *Green Chemistry Letters and Reviews, 18*(1), Article 2545418.

Rabiee, N., et al. (2023). Green and sustainable membranes: A review. *Environmental Research, 231*, Article 116133.

Radovanovic, D., et al. (2023). Sorption of Pb2+, Zn2+, Cu2+ and Ni2+ ions on na-enriched natural zeolite for wastewater treatment process: A kinetic approach. *Metallurgical and Materials Engineering, 29*(3), 20–35.

Ragab, D., et al. (2016). Micropollutants removal from water using microfiltration membrane modified with ZIF-8 metal organic frameworks (MOFs). *Chemical Engineering Journal, 300*, 273–279.

Rahbari-Sisakht, M., & Ismail, A. F. (2025). Covalent organic framework membranes for water purification: Current status and future research directions. *Separation and Purification Technology, 359*, Article 130759.

Rahman, Z., & Singh, V. P. (2019). The relative impact of toxic heavy metals (THMs)(arsenic (As), cadmium (Cd), chromium (Cr)(VI), mercury (Hg), and lead (Pb)) on the total environment: An overview. *Environmental Monitoring and Assessment, 191*(7), 419.

Ramírez-Castillo, F. Y., Loera-Muro, A., Jacques, M., Garneau, P., Avelar-González, F. J., Harel, J., & Guerrero-Barrera, A. L. (2015). Waterborne pathogens: Detection methods and challenges. *Pathogens, 4*(2), 307–334.

Ranjan, N., Singh, P. K., & Maurya, N. S. (2022). Pharmaceuticals in water as emerging pollutants for river health: A critical review under Indian conditions. *Ecotoxicology and Environmental Safety, 247*, Article 114220.

Ravi, S., et al. (2023). Improvement in acid resistance of polyimide membranes: A sustainable cross-linking approach via green-solvent-based fenton reaction. *Polymers, 15*(2), 264.

Ravi, S., et al. (2025a). Synergistic two-step modification of polybenzimidazole (PBI) nanofiltration membranes for improved molecular separation in acidic and organic environments. *Journal of Industrial and Engineering Chemistry.*

Ravi, S., et al. (2025b). Green solvent-mediated cyanuric chloride crosslinking of polybenzimidazole (PBI) nanofiltration membranes for improved acid stability and organic solvent resistance in molecular separation. *Journal of Applied Polymer Science*, Article e57119.

Ray, S. S., et al. (2022). Exploration of time series model for predictive evaluation of long-term performance of membrane distillation desalination. *Process Safety and Environmental Protection, 160*, 1–12.

Ray, S. S., et al. (2023). Chemical engineering of electrospun nanofibrous-based three-layered nonwoven polymeric protective mask for enhanced performance. *Journal of Applied Polymer Science, 140*(10), Article e53584.

Richardson, S. D., & Postigo, C. (2011). Drinking water disinfection by-products. In *Emerging organic contaminants and human health* (pp. 93–137). Springer.

Rivas, B. L., Pereira, E. D., & Moreno-Villoslada, I. (2003). Water-soluble polymer–metal ion interactions. *Progress in Polymer Science, 28*(2), 173–208.

Roy, D., Brooks, W. L., & Sumerlin, B. S. (2013). New directions in thermoresponsive polymers. *Chemical Society Reviews, 42*(17), 7214–7243.

Roy, A., Mu, L., & Shi, Y. (2020). Tribological properties of polyimide coating filled with carbon nanotube at elevated temperatures. *Polymer Composites, 41*(7), 2652–2661.

Rupam, S., et al. (2017). *Layer-by-Layer-Assembled laccase enzyme on stimuli-responsive membranes for chloro-organics degradation.*

Sana, S. S., et al. (2020). Effects of microplastics and nanoplastics on marine environment and human health. *Environmental Science and Pollution Research, 27*(36), 44743–44756.

Schaep, J., et al. (1998). Influence of ion size and charge in nanofiltration. *Separation and Purification Technology, 14*(1–3), 155–162.

Schattling, P., Jochum, F. D., & Theato, P. (2014). Multi-stimuli responsive polymers–the all-in-one talents. *Polymer Chemistry, 5*(1), 25–36.

Seo, Y., Pooi, C. K., & Ng, H. Y. (2025). Recommendation on the selection of powdered activated carbon as carrier to enhance performance of polymeric UF membrane. *Journal of Membrane Science, 713*, Article 123226.

Shannon, M. A., et al. (2008). Science and technology for water purification in the coming decades. *Nature, 452*(7185), 301–310.

She, Q., et al. (2016). Membrane fouling in osmotically driven membrane processes: A review. *Journal of Membrane Science, 499*, 201–233.

Shen, X., et al. (2012). Molecular imprinting for removing highly toxic organic pollutants. *Chemical Communications, 48*(6), 788–798.

Shukla, B. K., et al. (2025). Polymeric membranes in water treatment: Insights into contaminant removal mechanisms and advanced processes. *Engineering Proceedings, 87*(1), 69.

Singh, R., et al. (2025a). Replacement of legacy per-and polyfluoroalkyl substances: A way forward to mitigate the ill impacts associated with these chemicals. In *PFAS in the environment: Occurrence, characterization, treatment, and management* (pp. 421–445).

Singh, R., et al. (2025b). Emergence of covalent organic frameworks in forever chemicals remediation: A comprehensive and multifaceted review. *Inorganic Chemistry Communications*, Article 114614.

Ravi, S., Shin, S. J., Kim, I. C., Myung, S., & Kwon, Y. N. (2025c). Green Solvent-Mediated Cyanuric Chloride Crosslinking of polybenzimidazole (PBI) nanofiltration membranes for improved acid stability and organic solvent resistance in molecular separation. *Journal of Applied Polymer Science*.

Stayton, P. S., et al. (1995). Control of protein–ligand recognition using a stimuli-responsive polymer. *Nature, 378* (6556), 472–474.

Strathmann, H. (2004). *Ion-exchange membrane separation processes, 9*. Elsevier.

Sun, H., et al. (2016). Tough polymer aerogels incorporating a conformal inorganic coating for low flammability and durable hydrophobicity. *ACS Applied Materials and Interfaces, 8*(20), 13051–13057.

Sun, Y., et al. (2025). Electric field-driven preparation of highly dispersed ENP membrane and its study on Cr (VI) removal from water. *Chemical Engineering Journal, 512*, Article 162581.

Szymczyk, A., & Fievet, P. (2005). Investigating transport properties of nanofiltration membranes by means of a steric, electric and dielectric exclusion model. *Journal of Membrane Science, 252*(1–2), 77–88.

Taheran, M., et al. (2016). Membrane processes for removal of pharmaceutically active compounds (PhACs) from water and wastewaters. *Science of the Total Environment, 547*, 60–77.

Tarigan, M., et al. (2025). The occurrence of micropollutants in the aquatic environment and technologies for their removal. *Processes, 13*(3), 843.

Tavares, T., et al. (2022). Study for recycling water treatment membranes and compnents towards a circular economy—case of macaronesia area. *Membranes, 12*(10), 970.

Tchekwagep, P. M. S., et al. (2022). A critical review on the use of molecular imprinting for trace heavy metal and micropollutant detection. *Chemosensors, 10*(8), 296.

Ulbricht, M. (2006). Advanced functional polymer membranes. *Polymer, 47*(7), 2217–2262.

Upadhyaya, L., Qian, X., & Wickramasinghe, S. R. (2018). Chemical modification of membrane surface—overview. *Current Opinion in Chemical Engineering, 20*, 13–18.

Van der Bruggen, B., Mänttäri, M., & Nyström, M. (2008). Drawbacks of applying nanofiltration and how to avoid them: A review. *Separation and Purification Technology, 63*(2), 251–263.

Varjani, S., & Sudha, M. C. (2020). Occurrence and human health risk of micro-pollutants—A special focus on endocrine disruptor chemicals. In *Current developments in biotechnology and bioengineering* (pp. 23–39). Elsevier.

Voicu, S. I., & Thakur, V. K. (2023). Green polymers-based membranes for water reuse in a circular economy context. *Current Opinion in Green and Sustainable Chemistry, 43*, Article 100852.

Vrijenhoek, E. M., Hong, S., & Elimelech, M. (2001). Influence of membrane surface properties on initial rate of colloidal fouling of reverse osmosis and nanofiltration membranes. *Journal of Membrane Science, 188*(1), 115–128.

Wang, X.-L., et al. (1997). The electrostatic and steric-hindrance model for the transport of charged solutes through nanofiltration membranes. *Journal of Membrane Science, 135*(1), 19–32.

Wang, X., et al. (2023). A mini review of ceramic-based MOF membranes for water treatment. *Membranes, 13*(9), 751.

Wang, F., et al. (2024). Emerging contaminants: A one health perspective. *The Innovation, 5*(4).

Wei, H., et al. (2025). Advanced materials-based nanofiltration membranes for efficient removal of organic micropollutants in water and wastewater treatment. *Membranes, 15*(8), 236.

Wen, T., et al. (2016). Multifunctional flexible free-standing titanate nanobelt membranes as efficient sorbents for the removal of radioactive 90Sr2+ and 137Cs+ ions and oils. *Scientific Reports, 6*(1), Article 20920.

Werber, J. R., Osuji, C. O., & Elimelech, M. (2016). Materials for next-generation desalination and water purification membranes. *Nature Reviews Materials, 1*(5), 1–15.

Wołowicz, A., & Munir, H. M. S. (2025). Emerging organic micropollutants as serious environmental problem: A comprehensive review. *Science of the Total Environment, 958*, Article 177948.

Wu, C., et al. (2010). Uptake of pharmaceutical and personal care products by soybean plants from soils applied with biosolids and irrigated with contaminated water. *Environmental Science and Technology, 44*(16), 6157–6161.

Wu, Y., et al. (2022). Nanocomposite polymeric membranes for organic micropollutant removal: A critical review. *ACS ES&T Engineering, 2*(9), 1574–1598.

Xiao, H., et al. (2025a). Simultaneous flux recovery and trade-off breakthrough: New insights into repeated radical-based membrane cleaning. *Separation and Purification Technology, 363*, Article 132105.

Xiao, M., et al. (2025b). Recent progress in covalent organic framework-based membranes: Design, synthesis, and applications in the fields of energy and the environment. *ACS Macro Letters, 14*(8), 1201–1220.

Xu, R., et al. (2024). Micropollutant rejection by nanofiltration membranes: A mini review dedicated to the critical factors and modelling prediction. *Environmental Research, 244*, Article 117935.

Yang, L., et al. (2024). Removal of antibiotics and estrogens by nanofiltration and reverse osmosis membranes. *Journal of Hazardous Materials, 461*, Article 132628.

Yao, Y., et al. (2021). High performance polyester reverse osmosis desalination membrane with chlorine resistance. *Nature Sustainability, 4*(2), 138–146.

Yin, H., et al. (2024). Machine learning for membrane design and discovery. *Green Energy and Environment, 9*(1), 54–70.

Yu, Y., et al. (2022). Enhanced separation performance of polyamide thin-film nanocomposite membranes with interlayer by constructed two-dimensional nanomaterials: A critical review. *Membranes, 12*(12), 1250.

Yuan, S., et al. (2019). Covalent organic frameworks for membrane separation. *Chemical Society Reviews, 48*(10), 2665–2681.

Yuan, W., et al. (2025). Critical review of membrane fouling in reverse osmosis treatment: Characterizations, models, mechanisms, and controls. *Separation and Purification Technology*, Article 132119.

Zadehahmadi, F., et al. (2023). Removal of metals from water using MOF-based composite adsorbents. *Environmental Science: Water Research and Technology, 9*(5), 1305–1330.

Zhang, H., Luo, J., & Wan, Y. (2022). Regenerable temperature-responsive biocatalytic nanofiltration membrane for organic micropollutants removal. *iScience, 25*(1).

Zhao, F., et al. (2015). EDTA-cross-linked β-cyclodextrin: An environmentally friendly bifunctional adsorbent for simultaneous adsorption of metals and cationic dyes. *Environmental Science and Technology, 49*(17), 10570–10580.

Zhou, C., et al. (2025). Is the pollutant removed by the rejection of nanofiltration (NF) membranes in a NF-based membrane bioreactor (MBRNF)? *Desalination, 596*, Article 118320.

Zhu, W.-P., et al. (2014). Dual-layer polybenzimidazole/polyethersulfone (PBI/PES) nanofiltration (NF) hollow fiber membranes for heavy metals removal from wastewater. *Journal of Membrane Science, 456*, 117–127.

CHAPTER

7

Bio-inspired and multifunctional membranes design strategies for fouling resistance and catalytic water treatment

Hyung Kae Lee[1] **and Randeep Singh**[2]
[1]*School of Sustainable Engineering and the Built Environment, Arizona State University, Tempe, AZ, United States;*
[2]*Department of Basic Sciences, School of Sciences and Humanities, SR University, Warangal, Telangana, India*

1. Introduction

Membrane processes are becoming an increasingly important part of modern water treatment infrastructure (Elimelech & Phillip, 2011; Werber et al., 2016). They are used for large-scale seawater desalination, industrial wastewater upgrading, potable reuse, and decentralized purification. Their appeal is their compact footprint, modularity, and, in many configurations, superior energy efficiency compared with purely thermal or chemical routes (Baker, 2004). However, two systemic limitations persist across feed waters and operating regimes. First, foulants, including natural organic matter, extracellular polymeric substances, proteins, colloids, oil droplets, and sparingly soluble salts, accumulate at or within the membrane, depressing permeate flux, shifting selectivity, increasing transmembrane pressure, and shortening module lifetime. Second, conventional membranes remain fundamentally separation-centric: selectivity is governed by size, charge, or affinity, with little intrinsic interfacial reactivity (Werber et al., 2016). As a result, membranes reject many solutes but do not actively transform trace, recalcitrant, or fast-fouling species that pass through or seed irreversible deposition (Zander & Curry, 2001). To meet tightening regulatory standards and water-reuse/circularity targets without additional energy or chemical inputs, membranes must remain clean for longer periods and provide interfacial reactivity, while preserving baseline permeability and selectivity.

This chapter presents the practical proposal that cleanliness by design and reactivity by design should be pursued concurrently, guided by the physics and chemistry of the membrane-water interface. On the cleanliness side, bio-inspired strategies offer robust, transferable cues from nature's low adhesion surfaces (Lau & Yong, 2021). Hydration layer engineering and low surface energy chemistries reduce the thermodynamic driving forces for adhesion; hierarchical micro/nanotopographies lower the real contact area and make detachment easier under shear; and a responsive surface enables periodic, reversible switching of interfacial wettability/charge, affecting soft, in-situ surface renewal without harsh reagents. On the reactivity side, multifunctional architectures embed catalytic pathways exactly where contaminants first encounter the surface. Photocatalytic layers leverage light to oxidize

Sustainable Water Management. https://doi.org/10.1016/B978-0-443-44853-9.00012-5

organics and disrupt biofilm precursors (Zheng et al., 2017); enzyme functionalized membranes confer reaction selectivity that inorganic catalysts lack (Luo et al., 2020); and electrocatalytic backbones couple convective transport with redox transformations, targeting persistent pollutants at modest overpotentials (Sun et al., 2021). When such functions are integrated with permeable, mechanical sound support, a membrane ceases to be a passive sieve and becomes a reactive interface.

The motivation for this co-design is not merely conceptual; it is operational and economic. Irreversible or slowly reversible fouling forces frequent clean-in-place cycles, elevates specific energy consumption, and increases chemical usage, with knock-on effects on safety, emissions, and brine management (Ahmed et al., 2023; Porcelli & Judd, 2010). Flux decline and selectivity drift complicate plant control, requiring conservative set points and oversizing. From a life cycle perspective, module replacement and downtime dominate costs once a plant reaches steady-state throughput. In parallel, meeting emerging contaminant limits, such as pharmaceuticals, PFAS (Per- and Polyfluoroalkyl Substances) precursors, disinfection byproduct precursors, and antibiotic resistance markers, often demands additional unit operations downstream of membranes (Reid et al., 2023; Wu et al., 2021). Embedding catalytic functionality at the membrane itself promises to curtail those additional ones by attenuating problematic species in situ, reducing the load that would otherwise accumulate on the surface or carry over into post-treatment.

Designing for cleanliness and reactivity at once, however, requires a common language that links adhesion, transport, and chemistry. At the interfacial scale, robust hydration shells, balanced surface charge, and controlled roughness suppress protein bridging and polymer collapse (Okuyama et al., 2024); at the mesoscale, hierarchical textures reduce pinning and lower detachment thresholds under crossflow (Abd Aziz et al., 2022). For catalytic layers, the governing levers depend on modality: photocatalysis hinges on optical access to active sites, rapid boundary layer renewal, and suppressed charge carrier recombination (Wang et al., 2021); enzyme catalysis depends on microenvironment (local pH, cofactor availability, etc.), tether architecture, and diffusional path lengths (Guzik et al., 2014); electrocatalysis demands continuous conduction pathways with minimal ohmic loss and surfaces that resist deactivation by adsorbed intermediates or coprecipitated salts (Eniola et al., 2024). In all cases, polymer catalyst compatibility, adhesion of the functional layer to the support, and the mechanical integrity of thin films under hydraulic and thermal cycling dictate durability.

Measuring success requires metrics that span separation and reaction. On the fouling side, normalized permeability and critical flux, contact angle hysteresis, adsorbed mass, and flux recovery ratios after standardized cleaning protocols provide comparable baselines (Law, 2022; Le-Clech et al., 2006). On the catalytic side, turnover frequency or surface area normalized rates are considered under process-relevant mass transfer conditions, apparent quantum efficiency for light-driven systems, current or energy intensity for electrocatalysis, and, crucially, leaching indices that bind catalyst loss into the permeate or concentrate. Because real feed waters contain radical scavengers, competing adsorbates, and scalants, durability must be probed with mixed matrix challenges and extended runs, not only with model solutes. The most convincing demonstrations report sustained flux and rejection together with stable reaction performance across many cleaning cycles (Argurio et al., 2018).

Material and process choices should anticipate scale. Universal primers (e.g., catechol-based chemistries) help anchor ultrathin functional layers to diverse commercial backbones (Lee et al., 2007). Layer-by-layer assembly, vapor phase functionalization, and roll-to-roll compatible coatings can deliver uniform coverage at the meter scale without pore blockage. For light-driven systems, module geometries must provide optical pathlengths and minimize parasitic absorption by binders (Xu et al., 2025); for electrocatalytic designs, current distribution and mass transport should be co-optimized to avoid hot spots and local pH excursions (Sun et al., 2021). Responsive skins must switch within operational windows and survive cleaning agents (Wu et al., 2025). These constraints favor thin, well-anchored, and permeable architectures whose added resistance is marginal relative to the base membrane.

The scope of the chapter reflects these priorities. To keep comparisons consistent across modalities, a shared operating-point language is used, interfacial metrics ($\Delta\theta$, ζ), transport-reaction coupling (Pe–Sh–Da), and added hydraulic resistance (ΔR_{add}) throughout Sections 2–5. Section 2 establishes this design language that couples adhesion thermodynamics and hydrodynamics with interfacial reaction. Section 3 catalogs bio-inspired fouling control strategies by surface chemistry, topography, and responsiveness, emphasizing mechanisms, durability, and scalable fabrication routes. Section 4 surveys catalytic applications, photocatalysis, biocatalysis, and electrocatalysis with attention to architectures that preserve mass transfer and mechanical integrity. Finally, Section 5 addresses challenges and future directions: catalyst stability and leaching under complex matrices, compatibility with polymer backbones, delivery of light or electric fields within modules, and the techno-economic and life cycle considerations that determine whether a promising film can become commercial technology.

In short, the chapter discusses membranes that are deliberately designed to avoid fouling and to act on contaminants where transport, adhesion, and chemistry converge. When the membrane is treated as a programmable interface rather than a static barrier, module lifetimes are extended, cleaning frequency is reduced, and energy and chemical inputs are lowered, yielding performance compatible with sustainable water-treatment targets.

2. Fundamentals of bioinspired and multifunctional membranes

Designing membranes that both resist fouling and perform chemistry at the interface requires a common language that links adhesion thermodynamics, hydrodynamics, and reaction engineering to materials choices and scalable processing. This section establishes that language by first outlining the interfacial physics that govern the onset and persistence of fouling, then formalizing transport-reaction coupling for "reactive interfaces," and finally closing with materials/process platforms for thin, durable functional layers.

2.1 Interfacial thermodynamics and mechanics of adhesion

Fouling begins at the contact between a foulant and a hydrated surface. Classical Derjaguin, Landau, Verwey, and Overbeek theory (DLVO) interactions, electrostatic double-layer forces, and van der Waals attraction set the baseline; non-DLVO contributions (Lewis acid-base, hydrophobic, hydration, and steric forces) frequently dominate at separations of a few nanometers (Diao et al., 2017). Surfaces that stabilize a dense, mobile hydration layer generate short-range repulsion strong enough to offset attraction and lower the probability of primary-minimum attachment (Valle-Delgado et al., 2005); zwitterionic and polyethylene glycol (PEG)-like chemistries excel by binding water tightly yet dynamically, preserving interfacial free energy without forming glassy and transport-blocking films (Jiang & Cao, 2010).

In realistic waters, pH and ionic strength regulate surface charge (via pKa shifts and Debye length compression); so the same feed can produce different electrostatic landscapes across modules or seasons; charge neutrality (zwitterions) or weak net charge with balanced counterions tends to minimize bridging by polyvalent cations and proteins. Contact-angle hysteresis complements charge metrics: a small advancing-receding gap indicates low pinning and weak solid-liquid heterogeneity, consistent with facile detachment under shear.

Roughness and chemical heterogeneity couple adhesion to mechanics: strongly hydrated surfaces resist nonspecific adsorption (Jiang & Cao, 2010), whereas increasing nanoscale roughness can concentrate stress at asperity tips, lower repulsive barriers, and promote local anchoring and spreading; thus, low roughness helps preserve hydration-layer continuity and minimize pinning, whereas hierarchical topographies (lotus-like asperities or shark-skin-like riblets) decrease the real contact area and guide near-wall flow to promote surface renewal. Time under load further matters: even when initial attachment is weak, conditioning films of low-molecular-weight organics and ions can form, progressively increasing adhesion and altering surface chemistry; viscoelastic foulants (Extracellular polymeric substances (EPS), polysaccharides) relax under shear and spread to occupy low-energy sites, so dynamic metrics such as contact-time distributions, surface-renewal rates, and the critical shear for depinning are as important as static surface energy.

2.2 Transport reaction coupling at a reactive interface

The addition of a catalytic function results in the membrane operation as a small reactor embedded within a separator (Molinari et al., 2017). Three canonical scales emerge: (i) optical or electrical access to active sites; (ii) mass transfer across the hydrodynamic boundary layer and within porous films; and (iii) intrinsic reaction at those sites. Their interplay is conveniently summarized by non-dimensional groups. The Péclet number (Pe) compares advection to diffusion and shapes boundary layer thickness (Rapp and Rapp, 2017); the Sherwood number (Sh) relates convective mass transfer coefficients to diffusion (Luft et al., 2001); the Damköhler number (Da) compares the reaction rate to transport (Otálvaro-Marín and Machuca-Martínez, 2020). In practice, catalytic membranes should be operated away from transport-limited regimes (i.e., at moderate Da) so that additional photons or electrons translate into higher conversion.

Photocatalysis: Light must reach the semiconductor without excessive parasitic absorption by binders, polymer backbones, or overcoats (Hu et al., 2025). Ultrathin, well-anchored films (tens to a few hundred nanometers) typically deliver the best compromise between optical penetration and mechanical integrity (Heredia Deba et al., 2023). Internally, charge carrier recombination competes with interfacial redox; heterojunctions, dopants, or cocatalysts can lengthen carrier lifetimes and lower activation barriers. Externally, mass transfer control via crossflow, spacers, or pulsed light to "shake" the boundary layer keeps reactants and short-lived radicals supplied at the interface. Apparent quantum efficiency, therefore, folds three efficiencies: photon delivery, charge separation, and interfacial mass transfer.

Enzymatic catalysis: Immobilized enzymes follow Michaelis–Menten kinetics modified by confinement (Niyaz Ahmed and Praveen, 2024). The effective Michaelis constant can rise if substrate access is hindered; the Thiele modulus and effectiveness factor capture diffusion reaction coupling within pores or gels. Activity and half-life depend on the microenvironment: tether length and flexibility, local pH and ionic strength, and the presence of cofactors or stabilizers. Co-immobilization (e.g., enzyme plus protective polymer or radical scavenger) can mitigate deactivation by oxidants or proteases, while porous supports with controlled pore size limit diffusional barriers without letting enzymes leach.

Electrocatalysis: Reactive electrochemical membranes (REMs) require continuous conductive pathways with minimal ohmic drop and uniform current distribution (Soestbergen, 2012). Electrode kinetics follow Butler-Volmer behavior; Tafel behavior emerges in the high overpotential limit. Selectivity and energy intensity are often governed by mass transport to the reactive surface and by side reactions (e.g., oxygen evolution) at high overpotential. Porous, conductive backbones can enhance spacetime yield by aligning convective flow with electron pathways, but must resist fouling and scaling that block pores or passivate sites (Trainham & Newman, 1981). Local pH shifts and gas evolution within pores add design constraints on pore size, tortuosity, and mechanical robustness. Across modalities, boundary layer engineering is the unifying theme: keep active sites accessible; refresh the interface faster than fouling can create a barrier; and avoid thick, resistive overcoats. This logic often favors ultrathin functional skins and mixed-matrix designs with percolated pathways, so catalytic function rides on transport rather than resisting it. For placement, Table 7.1 provides a concise cross-reference linking interfacial physics ($\Delta\theta$, ζ, texture), transport-reaction coupling (Pe–Sh–Da), and downstream materials/process levers at a single operating point.

2.3 Materials and process platforms for scalable functional layers

Building on the operating-point levers summarized in Table 7.1, this subsection translates them into materials and processing routes that deliver ultrathin, well-anchored, and permeable functional layers at scale. Real membranes live in modules, not on glass slides; functional layers must therefore be thin, well-anchored, permeable, and manufacturable. Figure 7.1 shows materials and various processes for the synthesis of scalable functional layers in membranes and their applications in various fields.

Base polymers/backbones: Commercially relevant polymers (e.g., PVDF, PES/PSf, PAN, cellulose-based supports, thin-film polyamide composites) differ in surface energy, chemical resistance, and thermal properties (Elimelech & Phillip, 2011). Base polymer selection sets the platform for primers, solvents, and curing (Petersen, 1993). Porosity and skin thickness bound the added hydraulic resistance tolerated by any functional overlayer (Freger, 2003).

Anchors and primers: Universal adhesion layers, catechol-rich coatings (e.g., mussel-inspired polydopamine), silane or diazonium grafts, plasma-activated surfaces create reactive sub-groups for subsequent growth (Lee et al., 2007). The primer must be conformal and ultrathin to avoid pore blockage and maintain base permeability, yet robust to clean-in-place (CIP) conditions (oxidants, acids/alkali) and temperature excursions.

Coating and growth methods: The LbL assembly affords nanometer-scale control over thickness/charge but demands careful rinsing and counter-ion management to avoid salt trapping. Surface-initiated polymerization (e.g., Surface-initiated atom transfer radical polymerization (SI-ATRP), Reversible Addition-Fragmentation chain-Transfer (RAFT)) grows dense brushes for hydration-layer engineering; initiator density/termination must avoid brittle, glassy films (Liu et al., 2009). Sol-gel/colloidal routes enable photocatalyst deposition (e.g., TiO_2) but can crack or flake unless crosslinked and mechanically matched to the substrate (Gartner et al., 2023). Vapor-phase processes (Initiated Chemical Vapor Deposition (iCVD), Atomic Layer Deposition (ALD)) offer conformality at low thickness, though throughput and capital cost can limit deployment (Min et al., 2018). Mixed-matrix

Table 7.1 Fundamentals of bio-inspired and multifunctional membranes, concise cross-reference from interfacial physics to transport-reaction coupling and scalable materials/process levers.

Topic	Key idea	Primary levers	Metrics (minimum)	Key findings	Ref
Interfacial thermodynamics and mechanics	DLVO + non-DLVO; hydration continuity; pinning ↔ depinning	Zwitterion/PEG brushes; charge balance; nano asperities/riblets; ultrathin chemistry	$\Delta\theta$ ($\leq$10–15 degrees); ζ-potential (mV); AFM roughness (nm); FRR (%)	Avoid oil/EPS traps; check oxidant/UV stability	Jiang and Cao (2010)
Transport reaction (Pe–Sh–Da)	Operate near Da $\approx$1 with Sh ↑; modality-specific couplings	Crossflow/spacers; ultrathin skins; heterojunctions (photo); η aware loading (enzyme); flow through + low R contacts (electro)	AQE (spectrum/irradiance/path); η with ϕ, k_{cat}/KM; EE/O, FE; common: J_v, rejection, ΔR_{add}, FRR	Flag low-Sh/long-dwell regions; watch contact-resistance hot spots	Werber et al. (2016)
Materials and processes	Thin, well-anchored, permeable, scalable	Backbones (PVDF, PSf, PAN, cellulose, TFC PA); primers (PDA, silane, diazonium); coatings (layer-by-layer (LbL), SI polymer, solgel, iCVD/ALD, mixed matrix)	$\Delta R_{add} \leq$ 10–20%; permeability/rejection retention; adhesion/stress tests; leaching index; byproduct panel	Beware of solvent attack; salt trapping (LbL); solgel cracking; PEG oxidation	Zhang et al. (2016)
Roles: Bio-inspired versus multifunctional	Bio-inspired lowers adhesion; multifunctional adds reactivity	Preserve access; design for renewal; ensure compatibility (CIP (clean-in-place), conductive paths)	Co report: Interfacial ($\Delta\theta$, ζ) + hydraulics (ΔP, J_v, ΔR_{add}) + modality (AQE/η/EE/O, FE) + FRR, leaching, byproducts	Avoid burying catalysts; limit fatigue; verify at module relevant Sh/Da	Molinari et al. (2002)

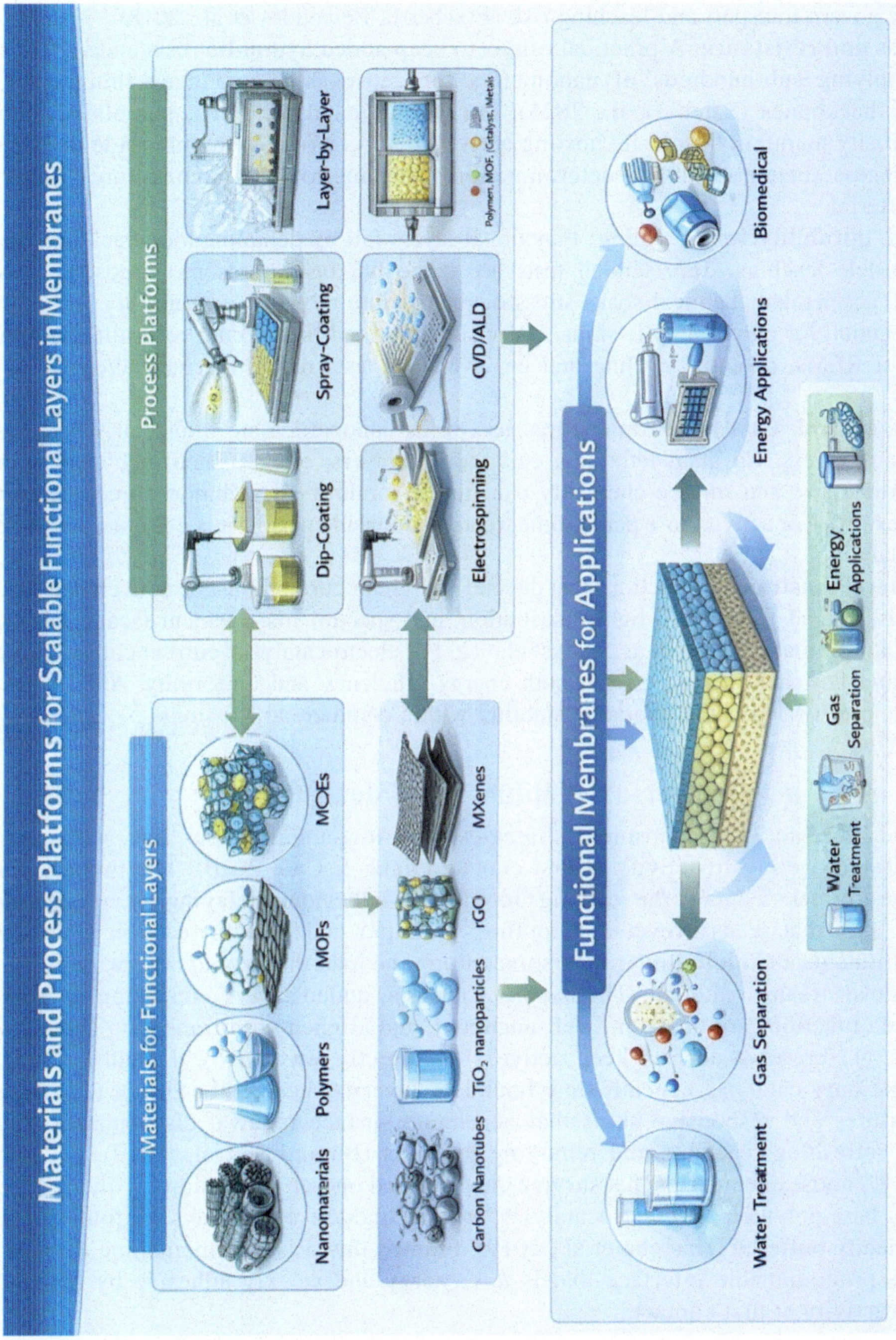

FIGURE 7.1
Materials, processes, and applications of membranes incorporating scalable functional layers.

casting disperses catalytic particles within a polymer matrix; percolation and particle-polymer compatibility govern transport and leaching risk (Etxeberria-Benavides et al., 2018).

Thickness and resistance: A practical rule is to keep added hydraulic resistance $\lesssim$10–20% of baseline, implying sub-hundreds of nanometers for dense skins and even thinner for low-permeability backbones (Saleh et al., 2025). For porous catalytic films, percolated pathways and high porosity maintain flux while hosting active sites. As a rule of thumb, trade thickness for site effectiveness (better catalysts, heterojunctions, enzyme-loading architecture) rather than piling on material.

Adhesion, durability, and cleaning: Functional layers fail by delamination, cracking, chemical aging, or particle leaching. Tape/scratch tests are crude but fast; ultrasonic exposure, crossflow abrasion, and accelerated chemical soaks stress distinct failure modes. Compatibility with CIP conditions is essential; for photocatalytic skins, UV/visible protocols should not embrittle the primer or backbone. For REMs, potential cycling and gas-evolution tests reveal mechanical/electrical weak points.

Leaching control: Catalysts in mixed matrices or as nanoparticulate films can shed under hydraulic/chemical stress. Covalent tethering, ceramic interlayers, or core-shell architectures reduce leaching; particle size and surface chemistry can further minimize detachment forces. Reporting a leaching index (mass or activity loss per volume treated) alongside performance is critical to assessing real-world viability.

Module-level constraints: Functions that depend on light or current must be delivered at scale. For photocatalysis, optical pathlength, light distribution, and spectral management (scattering spacers, reflective backings) matter as much as catalyst choice. For electrocatalysis, current collectors, contact resistances, and flow-field design govern both energy efficiency and uniformity. All designs must accommodate pressure drop and hydraulic stability within commercial housings.

2.4 Roles of bio inspired versus multifunctional design

As organized in Table 7.1, bio-inspired strategies set a low-adhesion baseline, whereas multifunctional design overlays reactivity at first contact (Jiang & Cao, 2010). The two are complementary: the former reduces the driving forces for adhesion, delaying or preventing the conditioning film that seeds irreversible fouling; the latter adds in-situ reaction pathways that neutralize pollutants and biofouling precursors, cutting the load that would otherwise accumulate or demand downstream treatment. The apparent tension, added layers increasing resistance, is resolved when functions are ultrathin, well-anchored, and aligned with transport. Three working rules follow. (1) Preserve access: keep active sites directly washed by the bulk; avoid thick overcoats that bury catalysts beneath slow boundary layers (Chen et al., 2001). (2) Design for renewal: textures and responsive skins that accelerate surface renewal also enhance catalytic turnover by refreshing reactants and removing products (Barambu et al., 2019). (3) Engineer compatibility: choose chemistries that survive cleaning and operating windows without degrading the primer or base polymer; for electrocatalysis, ensure conductive pathways are fouling-resistant and mechanically buffered (Hashiba et al., 2019). Framed this way, the membrane is not a static barrier but a programmable interface that is low-energy and weakly adhesive by default, with controlled reactivity at first contact.

3. Bio-inspired strategies for fouling control

Bio-inspired fouling control translates the performance of naturally low-adhesion surfaces into membrane-compatible chemistries, textures, and dynamic skins. The design goal is to reduce thermodynamic and mechanical drivers for attachment, keep the interface "wet and weakly adhesive" under crossflow, and preserve permeability through ultrathin, durable modifications. This section organizes strategies into surface chemistry (section 3.1), topographical/hierarchical engineering (section 3.2), and responsive/dynamic skins (section 3.3), with emphasis on mechanisms, scalable fabrication, durability under cleaning, and practical metrics. Figure 7.2 schematically depicts the three routes discussed here: (a) hydration-based chemistry, (b) flow-compatible textures, and (c) stimulus-responsive skins, used to reduce attachment and to facilitate shear- or stimulus-assisted detachment.

3.1 Surface chemistry modifications

Mechanistic basis: Antifouling chemistry centers on stabilizing a dense, mobile hydration layer and minimizing specific/non-specific interactions with proteins, polysaccharides, microbes, and oils. Hydrophilic brushes (PEG-like chains) and zwitterionic moieties (sulfobetaine, carboxybetaine, phosphorylcholine) bind water via ion-dipole/hydrogen-bond networks, creating short-range repulsion

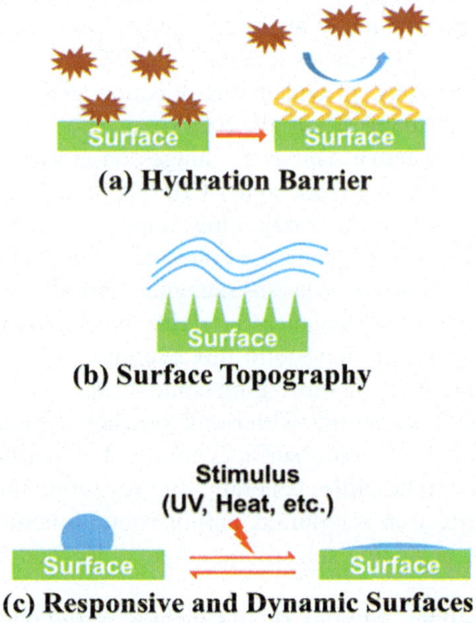

FIGURE 7.2

Schematic overview of bio-inspired fouling control routes. (a) Hydration barrier: Hydrophilic/zwitterionic layers retain bound water at the interface and repel approaching foulants at near contact. (b) Surface topography: Micro/nano textures limit real contact to asperity tips and promote shear-assisted depinning. (c) Responsive and dynamic surfaces: An external stimulus (UV, heat, electric field, etc.) periodically switches the interfacial state to release weakly adhered species.

that offsets van der Waals and acid-base attractions at nanometer separations (Leng et al., 2015). Compared with neutral hydrophiles, zwitterions typically exhibit higher hydration enthalpy and better salt tolerance, mitigating cation bridging and protein adsorption at elevated ionic strength (Liu et al., 2022). Reduced contact-angle hysteresis ($\Delta\theta \leq 10$–15 degrees) and near-neutral ζ-potential across the operating pH window correlate with low pinning and weak adhesion (Schmidt et al., 2004).

Primers and anchors: Translating these chemistries to commercial backbones requires thin, conformal, and robust primers. Mussel-inspired polydopamine (PDA) and catechol/amine primers adhere to polyvinylidene fluoride (PVDF), polysulfone (PSf), polyacrylonitrile (PAN), cellulose, and thin film composite (TFC) polyamide (PA) via covalent/coordination interactions, offering sub-groups (catechol, quinone, and amine) for grafting. Alternatives include plasma-activated hydroxyl/carboxyl surfaces for silanization, diazonium-based aryl grafts, and ultrathin ALD/iCVD interlayers. Primer thickness should remain in the tens-of-nanometers regime to avoid pore blockage while achieving high graft density.

Grafting and growth modes: Grafting/growth routes include surface-initiated polymerization, surface-initiated-ATRP (SI-ATRP), surface-initiated-RAFT (SI-RAFT) to form dense, chain-length-controlled zwitterionic (Sulfobetaine methacrylate (SBMA), carboxybetaine methacrylate (CBMA)) or hydrophilic (2-Hydroxyethyl methacrylate (HEMA), poly(ethylene glycol) methacrylate (PEGMA)) brushes (Kuang & Messersmith, 2012); LbL deposition of polyelectrolytes (e.g., Poly (sodium 4-styrenesulfonate) (PSS)/Poly(allylamine hydrochloride) (PAH), Poly(acrylic acid) (PAA)/Poly(N, N-dimethylaminoethyl methacrylate) (PDMAEMA) with attention to counter-ion management to prevent salt trapping (Gutfreund et al., 2023); and "click" reactions (thiol-ene, azide-alkyne) for efficient post-functionalization (Ghosal et al., 2024). In TFC PA, interfacial polymerization can co-introduce hydrophiles or zwitterions, provided salt rejection and skin integrity are preserved via careful control of diffusion time scales (Li et al., 2021).

Durability and cleaning resilience: Durability hinges on covalent anchoring and oxidative/mechanical robustness. Physisorbed layers erode under CIP (alkali, acids, oxidants) and thermal swings; amide/ester/triazole linkages and mild crosslinking improve retention (Kyriakou et al., 2020); catechol-metal coordination (e.g., Ti, Fe) can raise cohesive strength (Kafkopoulos et al., 2025). PEG-like chains are radical/chlorine-sensitive, whereas many zwitterionic brushes and PDA-based anchors tolerate moderate oxidant exposure. Mechanical integrity under crossflow can be screened by ultrasonic delamination, abrasion tests, and long-term flux cycling.

Oil and biofouling contexts: Oil and biofouling contexts demand underwater superhydrophilicity and low oil adhesion, favoring hydrophilic/zwitterionic brushes without roughness amplification that traps oil. For biofouling, strongly cationic surfaces may reduce early settlement yet promote EPS adsorption; zwitterionic/neutral hydrophiles generally suppress both initial adhesion and conditioning-film stabilization. Where disinfection is required, antimicrobial functions (quaternary ammonium, N-halamine) may be co-grafted, balancing selectivity, cytotoxicity, and by-product control (Lainioti & Druvari, 2024).

Characterization and metrics: Metrics should include static/dynamic contact angles (and hysteresis), ζ-potential, X-ray Photoelectron Spectroscopy (XPS)/Time-of-Flight Secondary Ion Mass Spectrometry (ToF-SIMS) for graft verification, Quartz Crystal Microbalance with Dissipation monitoring (QCM-D)/Surface Plasmon Resonance (SPR) adsorption with model foulants (Bovine serum albumin (BSA), alginate, humic acids, etc.), bacterial adhesion assays, and flux-recovery ratio (FRR) after standardized fouling/cleaning. Report areal adsorbate mass, added hydraulic resistance (ΔR_{add}), and retention of baseline selectivity (e.g., NaCl rejection for TFC).

Scale-up considerations: Scale-up favors roll-to-roll dip/spray priming followed by aqueous grafting or LbL, minimizing solvent use and cure temperature to preserve morphology. Catalyst lifetime (oxygen control in ATRP, ligand/catalyst recovery) and waste minimization become important beyond bench scale.

As illustrated in Fig. 7.2 (a), hydration-dominated interfaces establish a near-contact repulsive barrier; in practice, low $\Delta\theta$ and ζ near neutrality accompany reduced adsorbed mass (Γ) and higher FRR after cleaning, provided the added layer remains ultrathin to limit ΔR_{add}.

3.2 Surface topography engineering

Mechanistic basis: Topographical features lower the real contact area and reduce detachment barriers. Two effective scale bands are commonly reported: nano-textures (~10–200 nm) that leverage hydration/steric repulsion to suppress pinning, and submicron-micron riblets/grooves (~0.5–5 μm) aligned with flow to promote shear-induced depinning (Cassie & Baxter, 1944). For oily feeds, re-entrant geometries can further hinder underwater oil impalement; with typical oil-water interfacial tension and high underwater oil contact angles, keeping the characteristic throat radius small (≈0.5–1 μm or below) raises the capillary entry pressure and reduces oil penetration under module-relevant pressure perturbations (Donaldson et al., 2008). In practice, the design target in membrane modules is underwater superhydrophilicity with low pinning, rather than air-trapping superhydrophobicity.

Design rules: Practical rules: (i) minimize contiguous flats larger than foulant footprints; (ii) avoid deep crevices that become sorption sinks; and (iii) align textures with flow to facilitate shear-induced depinning. Hierarchical roughness (micro-features bearing nano-asperities) prevents multiscale locking of soft foulants and shortens contact-time distributions. For emulsified oils, smooth nano-asperities plus strong hydration hinder coalescence/spreading; for microbes, sub-micrometer riblets and low-hysteresis chemistry reduce initial adhesion and remove sheltered niches.

Fabrication routes: Fabrication options span phase-inversion templating (control of non-solvent influx, humidity, pore-formers), micro-embossing/hot-press molding from Ni/Si masters, breath-figure templating (ordered pores; requires crosslinking/overcoats), additive micro/nanopatterning (inkjet, nanoimprint, direct laser writing), and plasma/ALD/iCVD overcoats to "lock-in" texture while adding chemistry.

Hydrodynamic implications: Hydrodynamically, textures interact with spacer-induced local shear and secondary flows. Properly oriented riblets can dampen wall-shear fluctuations and lower the critical depinning force, but overly tall features raise pressure drop or create dead zones. Co-design with spacer geometry (filament diameter, pitch, orientation) ensures that texture enhances boundary-layer renewal instead of disrupting it.

Durability and cleaning: Durability under crossflow, backwash, and particle impact is paramount. Embossed/cast-in textures generally outperform particulate overlays. Cleaning agents can round nano-asperities or oxidize peaks; thin ceramic or crosslinked overcoats mitigate erosion. Track AFM roughness, feature-height distributions, critical shear for detachment (microfluidics), and FRR after standardized foulants (protein, alginate, oil-in-water) (Ibrahim & Hilal, 2023).

Oil-rich and biofouling matrices: In oily feeds, underwater superhydrophilic texture + zwitter-ionic chemistry prevents oil impalement/coalescence; in biofouling matrices, riblets discourage settlement and groove-alignment, weakening biofilm anchoring. Avoid re-entrant pits that capture EPS and resist cleaning.

Scalability considerations: For scale-up, embossing and phase-inversion are most R2R-friendly; laser/nanoimprint suits high-value modules or localized patterning (e.g., feed entrance) (Ibrahim & Hilal, 2023). Masters must withstand repeated cycles without drift; verify replication fidelity by profilometry/AFM across web width.

Consistent with Fig. 7.2 (b), flow-aligned micro/nanotextures confine contact to asperity tips and increase the probability of shear-induced depinning, while excessive feature height incurs a pressure-drop penalty ($\Delta P/L$); practical designs therefore target aspect ratios that maximize detachment at acceptable hydraulic cost.

3.3 Responsive and dynamic surfaces

Mechanistic basis and switching chemistry: Responsive skins modulate wettability, charge, stiffness, or permeability on demand, interrupting adhesion kinetics and refreshing the interface without aggressive chemicals. The operative idea is a periodic micro-reset that releases weakly adsorbed foulants before they reorganize into persistent films. Chemistries include thermoresponsive polymers that cycle between hydrated coils and collapsed globules, altering hydration, chain-scale roughness, and free volume to disrupt weak adsorption (Lanzalaco & Armelin, 2017; Singh et al., 2020); photoresponsive motifs (azobenzene, spiropyran) that reversibly isomerize or ring-open (Delorme et al., 2005), and shift polarity/charge, with hybrids such as TiO_2-induced UV superhydrophilicity or visible-driven g-C_3N_4 enabling wetting transitions paired with ROS-assisted soft cleaning (Hollingsworth et al., 2019; Wei et al., 2023); pH-responsive brushes (PAA, PDMAEMA, polybetaines) that expand/collapse with ionization when pKa is placed near operating pH; and electroresponsive coatings (polypyrrole, PEDOT: PSS) that toggle oxidation state and charge density, and sometimes surface roughness, under small potentials to enable electrically assisted detachment (del Castillo et al., 2022).

Design tradeoffs: Thin, permeable layers are preferred; excess thickness slows switching and buries the base skin. Reversibility and fatigue resistance over thousands of cycles determine practical value. Heating or illumination should be delivered efficiently within modules (conductive backings, waveguides, reflective housing). pH excursions must fit the backbone's tolerance and align with existing CIP, avoiding added chemical burden.

Integration with antifouling chemistry and texture: Pairing responsiveness with zwitterionic or neutral-hydrophilic baselines maintains low adhesion while preserving the ability to reset. Texture limits real contact area, so switching only needs to overcome weak pinning at asperity tips. In photocatalytic hybrids, light-triggered wettability changes can coincide with mild oxidative cleaning if radical exposure is buffered to protect primers and backbones.

Durability and performance readouts: Key risks include thermal plasticization or embrittlement, photobleaching of photochromic groups, and counter-ion loss or reorganization in polyelectrolytes. Mitigation relies on covalent anchoring, protective interlayers, and oxidant-tolerant chemistries for chlorine or peroxide service. Performance should be benchmarked under standardized switching protocols (amplitude, frequency, duration) and realistic foulant loads, with core readouts including switching speed (t_{90}), on/off contrast in contact angle or ζ-potential, reversibility over N cycles, FRR without chemicals, and energy per reset ($W\ h\ m^{-2}$). QCM-D can clarify mass uptake and release, and microfluidics can quantify detachment probability as a function of trigger amplitude and shear.

Module level considerations: Hardware includes low-voltage bus bars (electro-skins), fiber or LED arrays (photo-skins), and heat-exchange schemes (thermo-skins) that avoid bulk temperature swings (Saberi et al., 2019). The most practical implementations piggyback on routine operations, for example, brief localized heating during backwash or short light pulses during flush, so responsiveness supplements maintenance rather than complicates it.

Table 7.2 Bio-inspired fouling control, single-point summary of design levers, and minimum metrics.

Strategy	Mechanism	Scalable route	Minimum metrics	Notes	Ref
Surface chemistry (zwitterion/PEG)	Dense hydration; low $\Delta\theta$; balanced ζ	PDA/catechol + SI polymer/LbL; click; controlled IP (TFC)	$\Delta\theta$ ($\leq$10–15 degrees); ζ; QCM-D/SPR mass; ΔR_{add}; rejection; FRR	Ultrathin; oxidant/UV stable; no pore blocking	Zhang et al. (2016)
Topography (nano asperities/riblets)	Contact↓, depinning force↓; underwater superhydrophilicity	Phase inversion; embossing; nanoimprint/laser	AFM roughness/height; critical shear; FRR (protein/alginate/oil); ΔR_{add}	Co-design with spacer; avoid reentrant pits; control ΔP	Bixler and Bhushan (2012)
Responsive skins (thermal/photo/Electro/pH)	Periodic micro reset → early release	PNIPAm; TiO_2/g C_3N_4 switch; PPy/PEDOT:PSS; pH brush	t_{90}; on/off $\Delta\theta$ or ζ; FRR (no chemicals); reset energy; ΔR_{add}	Thin and permeable; fatigue check; efficient heat/light/bias	Ulbricht (2006)

As shown in Fig. 7.2 (c), responsive skins use brief stimuli (UV/thermal/electrical) to switch interfacial state and purge weakly adhered foulants; efficacy and cost should be reported as t_{90}, chemical-free FRR, and E_{reset}, alongside cycle-to-cycle durability.

To provide a concise cross-reference, Table 7.2 compiles the core levers and minimum readouts for each route, chemistry, topography, and responsiveness, at a single operating point. The intent is to align the mechanism with what must be measured ($\Delta\theta$, ζ, FRR, ΔR, t_{90}, depinning shear, E_{reset}) so that results remain comparable across feeds, modules, and cleaning protocols.

4. Multifunctional strategies for catalytic applications in water treatment

This section builds reactivity by design on top of cleanliness by design, reframing the membrane as a reactive interface rather than a purely passive sieve. Across modalities, three design axes recur. First, catalytic skins should remain ultrathin and securely anchored, keeping the added hydraulic resistance within $\Delta R_{add} \leq$ 10–20% of the baseline (Nambi Krishnan et al., 2022). Second, boundary-layer renewal (high Sh) and operation at moderate Damköhler number (Da $\lesssim$ O(1)) help convert photons or electrons into reaction rather than saturating transport. Third, primers, catalytic layers, and polymer backbones must be mutually compatible to withstand CIP conditions, pressure/thermal cycling, and scavenger-rich waters. Figure 7.3 provides a schematic overview of the three modalities discussed below: (a) ultrathin photocatalytic skins, (b) hydrated and immobilized enzyme skins, and (c) electrocatalytic/REM configurations operated with a counter-electrode to drive flow-through redox along the convective path.

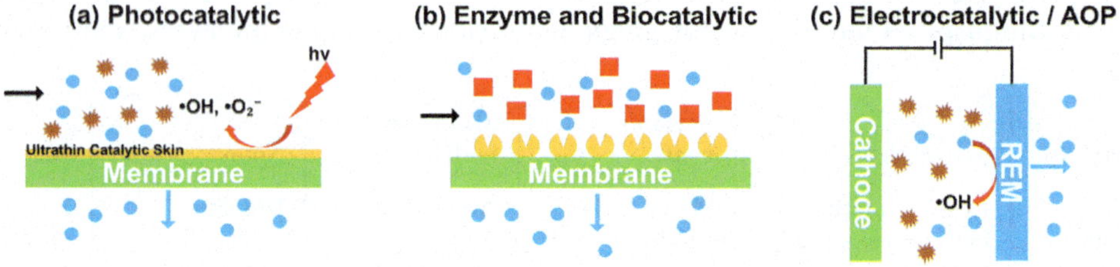

FIGURE 7.3

Modalities for catalytic membranes at the water-membrane interface. (a) Ultrathin photocatalytic skin generates oxidants. (b) Immobilized enzymes convert solutes at the interface with covalent/multipoint anchoring to minimize leaching. (c) Reactive electrochemical membrane (REM) operated with a counter-electrode drives flow-through redox along the convective path.

4.1 Photocatalytic membranes

Design logic and objectives: Photocatalytic skins bring photons, charge carriers (e^-/h^+), and short-lived oxidants ($\bullet OH$, $\bullet O_2^-$, $^1O_2$) to the water-membrane interface so contaminants are addressed at first contact (Purkait et al., 2018; Zhang et al., 2024). Practical aims are to maintain optical access to active sites with minimal parasitic absorption, suppress recombination so carriers survive to the interface, and refresh reactants/radicals faster than they are consumed or quenched. Running at $Da \lesssim 1$ avoids transport plateaus where extra irradiance no longer increases conversion (Visan et al., 2019). Any gain must uphold separation, that is, low ΔR_{add}, stable permeance, and preserved rejection for TFC PA skins. As sketched in Fig. 7.3 (a), an ultrathin semiconductor overlayer atop a permeable membrane absorbs light without incurring large ΔR_{add} and generates oxidants at the interface.

Architecture: thin, adhered, and permeable: Ultrathin semiconductor films (tens to a few hundred nanometers; TiO_2, g-C_3N_4, ZnO, $BiVO_4$) keep diffusion paths short while presenting accessible sites. Mixed-matrix skins with heterojunction particles (e.g., TiO_2@g-C_3N_4, TiO_2@rGO) can pair charge separation with mechanical compliance if water pathways remain percolated and particle-polymer compatibility prevents embrittlement or leaching (Acharya & Parida, 2020). Charge-extraction interlayers (ultrathin rGO, MoS_2, carbon dots, etc.) shorten recombination routes and tune interfacial redox (Gillespie & Martsinovich, 2019). Photothermal-photocatalytic hybrids use modest, surface-confined warming to thin the boundary layer without heating the bulk (Meng et al., 2024).

Transport and optics: In a Pe–Sh–Da view, spacers and crossflow raise Sh and shrink the diffusion layer that throttles interfacial reactions (Li et al., 2005). Spectral matching between LEDs and bandgaps maximizes useful photon flux; reflective housings, scattering spacers, and shallow illuminated channels improve photon utilization at the module scale. Duty-cycled illumination can "shake" the boundary layer and limit heat build-up. Because many primers/binders absorb UV, visible-light materials reduce parasitic losses and pair with high wall-plug-efficiency sources.

Fouling interplay and soft cleaning: Hydration-layer primers (zwitterionic or PEG-like, avoiding glassy films) delay conditioning layers that quench radicals and occlude sites (Schlenoff, 2014). Light-triggered wettability shifts, for example, UV-induced superhydrophilicity on TiO_2 or visible-driven transitions with g-C_3N_4 hybrids, synchronize with brief permeate flushes, restoring flux without harsh chemicals and improving FRR (Madaeni et al., 2011). Lower contact-angle hysteresis reduces the shear needed for depinning; gentle nano-textures that do not impale oil aid both antifouling and light-assisted resets.

Durability and reporting: Extended irradiations can yellow or embrittle primers; TFC PA is vulnerable to reactive oxygen species (ROS) and chlorine, motivating oxidation-tolerant overcoats or an ultrathin ceramic interlayer (Hashiba et al., 2019). Mixed-matrix and particulate skins should report a leaching index (mass or activity loss per m^3 treated). Minimum co-reporting: AQE with spectral irradiance; surface-area-normalized rates or TOF at module-relevant Sh; ΔR_{add} and retention of permeance/rejection; FRR after standardized fouling/cleaning; leaching index; and a byproduct panel where halogen chemistry is plausible (Hoque & Guzman, 2018).

4.2 Enzyme and biocatalytic membranes

Role and advantages: Enzymes offer molecular selectivity under mild conditions (ambient temperature and near-neutral pH), suiting oxidant-sensitive compounds, biofilm precursors, and trace organics that challenge inorganic catalysts (Bachosz et al., 2022). Immobilized near the interface, they act at first contact, reducing the load that would otherwise be built as conditioning films or pass downstream. Figure 7.3 (b) depicts this architecture: enzymes anchored within a hydrated interphase convert target solutes as permeate passes through the membrane. Figure 7.4 illustrates that the morphology of biocatalytic PEI-PSS membranes is dictated by casting solution pH and operational temperature (Li et al., 2024). The dominant finger-like macrovoids indicate a process of instantaneous demixing. This study shows that structures remain consistent at pH 11.4 and 10.9; a shift to pH 10.5 suppresses macrovoid growth and thickens the top layer, reducing porosity from over 80% to approximately 66%. This transition is driven by higher solution viscosity and a slower precipitation rate, which collectively densify both the control and lysozyme-integrated membranes.

Immobilization and anchoring strategies: Covalent coupling to catechol/amine/silane primers (amide, imine, and click) resists leaching and survives CIP when active-site geometry is preserved. Affinity or multipoint strategies (e.g., His-tag/Ni-NTA; boronate-cis-diol) give strong yet reversible attachment for facile refresh (Castro-Hinojosa et al., 2023). Ultrathin gels or porous hosts balance retention with diffusion; mesoporous silica or MOF-like scaffolds can stabilize conformation (Carlsson et al., 2014). Co-immobilizing protective polymers/osmolytes or scavengers suppresses unfolding and oxidative loss (Lee et al., 2015).

Kinetics-transport coupling (ϕ-η perspective): Confinement raises apparent KM when access is hindered; the Thiele modulus (ϕ) and effectiveness factor (η) capture internal diffusion-reaction coupling (Fink et al., 1973). Target $\eta \rightarrow 1$ with ultrathin, highly hydrated layers and external Sh high enough to avoid mass-transfer control, particularly for macromolecular substrates.

Microenvironment and cascades: Activity and half-life depend on tether length/flexibility, local pH/ionic strength, and cofactor availability. Oxidoreductases often benefit from closed cofactor loops ($NADH/NAD^+$, FAD) via co-immobilized partners or electro-enzymatic regeneration (Di Spiridione et al., 2022). In cascades, nanoscale spacing and linkers promote substrate channeling and selectivity. Gentle buffering and hydration stabilize active conformations during shear and temperature excursions.

Antifouling compatibility and reporting: Zwitterionic or neutral-hydrophilic baselines limit nonspecific adsorption. Cleaning should prefer low-oxidant/low-pH protocols; brief photo-/thermal assists can reset surfaces without denaturation when compatible. Minimum co-reporting: η with ϕ context; k_{cat}/KM on-membrane versus in solution; half-life across cleaning cycles; cofactor strategy; ΔR_{add} and retention of permeance/rejection; FRR; leaching (mass or activity); and biofouling assays under realistic matrices.

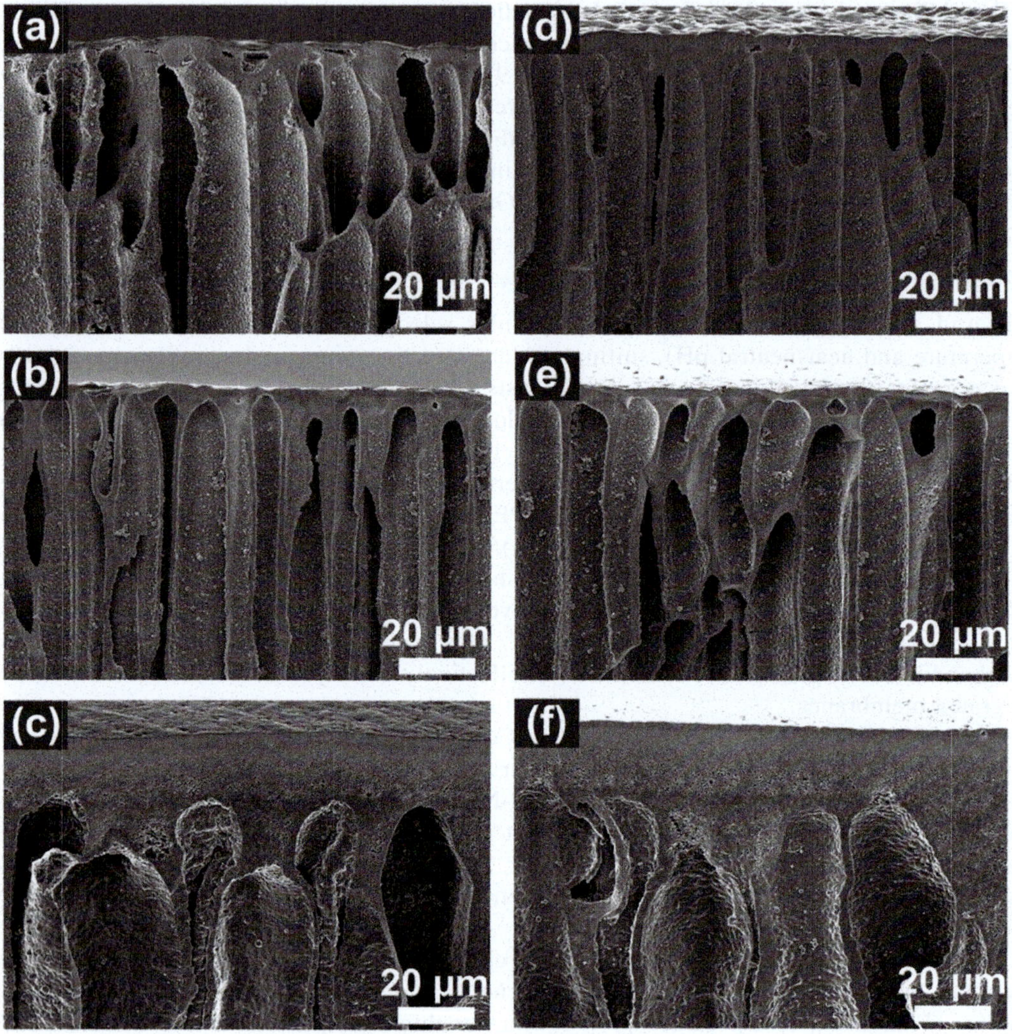

FIGURE 7.4

Morphological impact of pH and temperature on biocatalytic PEI-PSS membranes. Cross-sectional SEM images of membranes formed via aqueous phase separation without lysozyme (a–c) and with lysozyme (d–f) at pH 11.4, 10.9, and 10.5, respectively (Li et al., 2024).

Application sketches: Representative transformations include quorum-signal degradation, phenolic oxidations (laccase, peroxidase in situ H_2O_2), selective ester/amide cleavage, and detoxification of specific algal or bacterial toxins. In all cases, reporting should connect space-time yield under LMH-relevant fluxes with ΔR/FRR to show that catalysis and separation remain co-optimized.

4.3 Electrocatalytic and advanced oxidation membranes

Modes and design rationale: Reactive electrochemical membranes (REMs) couple flow-through convection with electron delivery for high space-time yields via: (i) direct electron-transfer reactions at the surface, (ii) mediated oxidants (active chlorine, electro-Fenton in situ H_2O_2/•OH, persulfate activation), and (iii) photo-assisted variants where small bias steers photocarriers. Performance hinges on uniform current distribution, low ohmic loss, bubble management (O_2/H_2), and controlled local pH inside pores. As shown in Fig. 7.3 (c), a REM operated adjacent to a counter-electrode aligns electron paths with convection in a flow-through channel, making gas management and local pH control central to stable operation. Figure 7.5 showcases the schematics for the function and processes of electrocatalytic and advanced oxidation membranes, along with key processes involved.

Materials and architecture: Conductive carbons/graphite provide continuous pathways with permeability; B/N doping or quinone-type sites bias routes toward direct oxidation or radicals (Awad & de Lannoy, 2024). Porous metals/oxides (Ti, Magnéli phases, TiO_2 nanotube arrays) form robust anodes but must resist passivation in hard waters. Ultrathin conductive polymers (Poly(3,4-ethylenedioxythiophene) (PEDOT): polystyrene sulfonate (PSS), polypyrrole) act as charge-transfer interlayers on TFC PA, improving adhesion while keeping ΔR_{add} low (Ouyang et al., 2017).

Distribution, mass transfer, gas: Uniform current prevents hot spots and large pH swings. Flow-through geometries that align electron paths with convection shorten diffusion distances and raise Sh. Gas evolution can intermittently block pores; size and tortuosity should vent bubbles without sacrificing strength. Contact resistances at collectors and interfaces need to be low and stable.

Energy, selectivity, and byproducts: EE/O (kWh m^{-3}-log) tracks energy intensity; Faradaic efficiency links charge to transformation. In chloride-rich matrices, oxy-chlorine species (chlorate, perchlorate) require control via potential, residence time, and pH. Pulsed or asymmetric waveforms and surface doping steer selectivity while suppressing parasitic oxygen evolution at high overpotential (Liu et al., 2024).

Durability and reporting: Monitor contact resistance and ohmic losses over long runs; potential cycling reveals electrical or mechanical weak points. Carbonate/silica scaling passivates sites, so acid/chelant cleans and reversible potential swings are useful. Compatibility with TFC PA must be preserved to avoid local pH/oxidant damage. Minimum co-reporting: EE/O, Faradaic efficiency, polarization (V-i) under module-relevant flow; ΔR_{add} and retention of permeance/rejection; FRR after fouling/cleaning; byproduct panels (with emphasis on oxy-chlorine where relevant); and time-courses for contact resistance/ohmic loss and stability.

Table 7.3 consolidates the cross-cutting design constraints and minimum reporting set used across modalities: ΔR_{add} ($\leq$10–20% of baseline), boundary-layer renewal (high Sh), operation at moderate Damköhler number (Da $\lesssim O(1)$), and module-level metrics (AQE/TOF, η-ϕ, EE/O, Faradaic efficiency, FRR, leaching index, by-products), to enable like-for-like comparison.

5. Challenges and future perspectives

Reactive antifouling membranes will be credible only if catalytic functions remain thin, durable, and compatible with commercial backbones while delivering measurable reductions in energy, cleaning

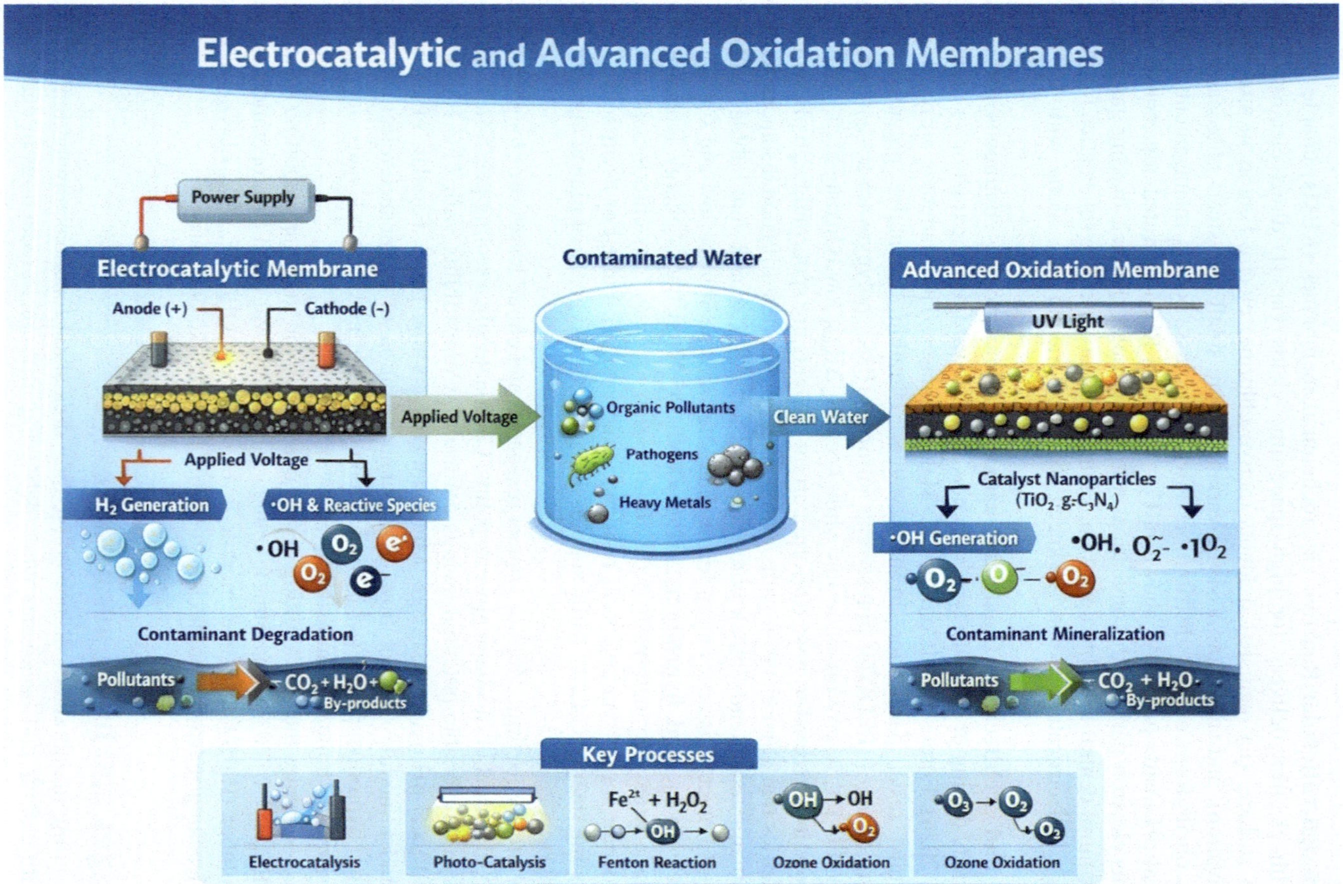

FIGURE 7.5

Schematic illustrations of electrocatalytic and advanced oxidation membranes highlighting the key underlying processes.

Table 7.3 Concise design checklist for catalytic membranes.

Modality	Thin layer (ΔR_{add} target) (%)	Transport/field (Sh, Da)	Key levers	Must-report (minimum)	Durability/ retrofit	Ref
Photo-Catalytic	Ultrathin TiO_2/g-C_3N_4 (tens-few 100 nm); strong primer. $\Delta R_{add} \leq$ 10–20	High Sh (crossflow/ spacers); shallow light path; moderate Da.	Hetero junctions/ cocatalysts; visible-light materials; duty-cycled light.	AQE (with spectrum/ irradiance), rate or TOF, ΔR_{add} and rejection, FRR, leaching, byproducts.	UV/ROS-stable tie-coat; long light-soak; NOM/ bicarbonate stress; reflector/ light-plate retrofit.	Molinari et al. (2002)
Enzymatic	Covalent immobilization; ultrathin hydrogel/porous host. $\Delta R_{add} \leq$ 10–20	Sh↑ externally; check ϕ-η internally; avoid transport plateaus.	Tether length/ flexibility; micro-pH/ionic strength; co-immobilization/ cofactor recycle.	η, k_{cat}/KM (membrane vs. solution), half-life (with cleaning), ΔR_{add} and rejection, FRR, leaching	Mild-CIP protocols; protease/oxidant stress; swapable enzyme skin (cartridge).	Sheldon and van Pelt (2013)
Electro Catalytic/ AOP	Conductive backbone + ultrathin PEDOT: PSS/ PPy/Ti-based layer. $\Delta R_{add} \leq$ 10–20	Flow-through alignment; low iR drop; Sh↑, Da ≈ O(1); gas management.	Pulsed/ asymmetric waveforms; mediator (electro-Fenton/ persulfate) tuning; low-loss collectors.	EE/O, FE, j-V polarization, ΔR_{add} and rejection, FRR, by-products (e.g., chlorate)	Scaling/ohmic tracking; potential cycling; clip-on busbars/ collectors (no vessel change).	Sun et al. (2021)

frequency, and downstream burden. To organize priorities, Table 7.4 links the main risks to failure signatures, design levers, and a minimum reporting set with practical cues for targets. The remaining gaps cluster into six areas: (1) stability and leaching under real waters; (2) chemical-mechanical compatibility with TFC PA and porous supports; (3) uniform delivery of light or current at module scale; (4) byproducts, scaling, and radical scavenging in complex matrices; (5) standardized co-reporting that links reaction and separation; and (6) techno-economic and life-cycle performance that supports retrofit or greenfield adoption.

Table 7.4 Commercial readiness framework for reactive antifouling membranes.

Challenge	Typical failure mode	Primary design lever	Must-report metric	Practical target	Ref
Catalyst stability and leaching	Activity loss; nanoparticle/active-site shedding under shear/CIP	Covalent tethering; ultrathin ceramic interlayers; core-shell particles; surface chemistry tuned for low detachment	Leaching index (mass or activity per m^3), activity retention (%)	Ppb-ppm range over 100–500 h continuous run	Kajau et al. (2021)
Polymer-catalyst compatibility	PA embrittlement; pore blockage; delamination	Solvent-safe primers; low-temperature deposition; adhesion promoters	ΔR_{add}, permeance/rejection retention	$\Delta R_{add} \leq$ 10–20%; no rejection loss	Zhou et al. (2018)
Optical delivery (photo)	Parasitic absorption; recombination; hot spots	Spectral matching; reflective/light-guide hardware; heterojunctions	AQE with spectrum; irradiance map	Stable AQE at module depth; uniformity documented	Molinari et al. (2017)
Current distribution (electro)	Hot spots; local pH swings; high ohmic drop	Collector/contact optimization; flow-through alignment; waveform control	EE/O, faradaic efficiency; polarization	Low EE/O at uniform current; minimal polarization drift	Sun et al. (2021)
Enzyme lifetime	Denaturation; proteolysis; cofactor loss	Co-immobilization; nanoconfinement; gentle microenvironment; cofactor regeneration	η, half-life, FRR	Tens-hundreds of cleaning cycles with η retained	Sheldon and van Pelt (2013)
Real-water byproducts	Oxy-chlorine; aldehydes/organics	Potential/time/pH control; selective catalysts; quenchers	Byproduct panel (+limits)	Below regulatory limits in target matrices	Lin et al. (2020)

5.1 Catalyst stability and leaching

Mixed-matrix and particulate skins can lose mass or activity under hydraulic shear, thermal swings, CIP conditions, or ligand competition (Cunha et al., 2018). Enzymes denature or are proteolyzed; photocatalysts detach from weak binders; electrocatalysts passivate or dissolve under bias. Near-term remedies include covalent tethering, ultrathin ceramic interlayers (ALD/iCVD), core-shell architectures, and particle surface chemistries that reduce detachment forces (Nikkola et al., 2014). A leaching index (mass or activity per m^3 treated) should be reported alongside ΔR_{add}, FRR, and catalytic metrics (AQE, η, and EE/O). As a pragmatic cue, sustained ppb-low-ppm loss over 100–500 h continuous operation is desirable.

5.2 Polymer-catalyst compatibility and mechanical integrity

Functional layers must adhere to PVDF, PES/PSf, PAN, cellulose supports, and TFC PA without embrittling the selective layer or blocking pores (Nikkola et al., 2014). Primers (PDA/catechol, silane, and diazonium) and conductive/ceramic interlayers improve adhesion but can add UV absorption (photo) or contact resistance (electro). Low-temperature processes that preserve morphology are preferred. Keep added resistance within $\Delta R_{add} \leq$ 10–20% and pass mechanical cycling (crossflow abrasion, ultrasonic delamination, and pressure/thermal ramps) and CIP-representative chemical soaks.

5.3 Field delivery inside modules: Light and current

For photocatalysis, LED spectrum/irradiance, reflective housing, scattering spacers, and channel-scale optical pathlength are as consequential as the catalyst. Designs should minimize parasitic absorption by binders and supports, maintain shallow illuminated depths, and document irradiance at the membrane plane. For electrochemical membranes, uniform current with low ohmic loss relies on low-resistance collectors, reliable contacts, and flow-fields that align convection with electron pathways while venting gas. In both cases, operate at high Sh and avoid transport plateaus (large Da). Practical cues include textured skins or shallow channels to lift mass transfer, collector/waveform optimization to level current density, and verification by irradiance or potential-drop mapping. Performance should be reported at plant-relevant hydraulics (flow, ΔP, channel dimensions) with $\Delta R_{add} \leq$ 10–20% and co-reported reaction/separation metrics.

5.4 Complex matrices: Scavengers, scaling, and byproducts

Natural organic matter, bicarbonate, halides, and hardness ions quench radicals, shift local pH, and drive scaling or passivation (Hübner et al., 2024). Photocatalysis should be validated beyond clean dyes; electrocatalysis must quantify oxy-chlorine species in chloride-rich feeds; enzyme layers should be tested with competing substrates and proteases. Stress protocols should combine mixed-matrix challenges, scaling precursors, and extended runs. Report conversion together with byproduct panels and space-time yield under LMH-relevant fluxes.

5.5 Reporting standards and comparability

To make results comparable across studies, the purpose is explicit: all performance should be co-reported at the same operating point as the module conditions stated. A minimum set comprises: (i) reaction metrics, including apparent quantum efficiency (AQE), effectiveness factor (η), electrical energy per order (EE/O) and Faradaic efficiency (FE); (ii) separation metrics, including added hydraulic resistance (ΔR_{add}), permeance/rejection retention, and flux-recovery ratio (FRR) under a stated fouling/cleaning protocol; (iii) a leaching index (mass or activity loss per treated volume); (iv) byproduct panels matched to the matrix; and (v) module descriptors (spacer, crossflow, ΔP, channel geometry, and irradiance or current density). Claims that rely on low Sherwood numbers (Sh) or long dwell times should be flagged, and data that hold performance near Da $\approx$1 with elevated Sh are most credible for deployment. To operationalize this, the one-page checklist below (Table 7.5) specifies units, test conditions, and pass/cue criteria without prescribing narrow thresholds, enabling authors and reviewers to apply the same yardstick across modalities and modules.

Table 7.5 One-page minimum reporting checklist for catalytic antifouling membranes (photocatalytic, enzymatic, electrocatalytic; single operating point).

Category	What to report	Unit	Test condition	Pass/cue	Ref
Hydraulics	ΔP, ΔR_{add}, Jv (LMH)	Bar, %, LMH	Module flow with spacer; steady state	$\Delta R_{add} \leq$ 10–20% versus baseline	Schock and Miquel (1987)
Interfacial state	θ hysteresis (Δθ), ζ-potential	°, mV	Operating pH/ionic strength	$\Delta\theta \leq$ 10–15 degrees; ζ stable	Jiang and Cao (2010)
Fouling resilience	FRR after standardized foul/clean cycles	%	BSA/alginate/oil and protocol noted	Target FRR met with mild cleaning	Vrouwenvelder et al. (2009)
Photocatalysis	AQE with spectrum, area-normalized rate/TOF	%, (mol m^{-2} s^{-1})	LED irradiance and pathlength stated	AQE stable at module depth	Hisatomi and Domen (2019)
Enzymatic	η (effectiveness), k_{cat}/KM, half-life	-, (M^{-1} s^{-1}), h	Mixed-matrix water; cleaning cycles	η retained over tens-hundreds of cycles	Sheldon and van Pelt (2013)
Electrocatalysis	EE/O, faradaic efficiency, polarization	kWh m^{-3}-log, %	Uniform current; flow-through	Low EE/O; minimal drift	Sun et al. (2021)
Leaching	Mass/activity loss per volume	mg m^{-3} (or activity m^{-3})	Long-run under the target matrix	Ppb-ppm range over 100–500 h	Mecha et al. (2023)
Byproducts	Targeted panel (e.g., oxy-chlorine, aldehydes)	μg L^{-1}	Realistic chloride/NOM; pH noted	Below regulatory limits	Radjenovic and Sedlak (2015)

5.6 Techno-economic analysis (TEA)/life cycle assessment (LCA) and retrofit pathways

Adoption depends on net cost and footprint once energy (LED wall-plug, EE/O), ΔP-driven pumping, cleaning frequency, replacement cycles, and embodied impacts are included. Two routes are most practical: (a) roll-to-roll ultrathin skins that convert existing modules with minimal ΔR; and (b) modular add-ons (light plates, reflective backings, serviceable current collectors) without opening the pressure vessel. TEA should benchmark "membrane + catalytic layer" against "membrane + downstream AOP" under identical treatment goals and matrices. In summary, designs that hold ΔR_{add} within 10%–20%, maintain reaction and separation performance in mixed-matrix feeds, document uniform field delivery, and offer field-serviceable retrofit kits present the clearest path to commercial deployment.

6. Conclusion

Reactive antifouling membranes can reconcile cleanliness with interfacial reactivity when functions are kept ultrathin, securely anchored, and aligned with transport. Across photocatalytic, enzymatic, and electrocatalytic skins, the same operating logic holds: maintain $\Delta R_{add} \leq 10$–20%, operate at elevated Sh and Da ≈ 1 so reaction benefits from convective renewal rather than stalling behind diffusion limits, and use primers or interlayers compatible with TFC polyamide and porous supports under realistic cleaning windows. This synthesis integrates the foundations in section 2 (adhesion thermodynamics and Pe–Sh–Da coupling), the bio-inspired cleanliness levers in section 3, the modality-specific reactivity architectures in section 4, and the deployment criteria in section 5.

Credible demonstrations co-report reaction and separation at a single operating point under module-relevant hydraulics and field delivery (light or current). Beyond instantaneous rates, bankability depends on durability in complex waters (low leaching over extended runs and transparent byproduct accounting), so catalytic gains do not come at the expense of stability or safety. Adoption ultimately turns on standardized reporting (AQE, η, EE/O, FE with permeance/rejection, FRR, ΔR_{add}, leaching index, byproduct panels, and module descriptors per section 5.5) and techno-economic and life-cycle performance that meets retrofit or greenfield constraints (section 5.6).

Near-term practice points are concrete: roll-to-roll compatible ultrathin skins (nm-scale sol-gel/iCVD/ALD on universal primers) that add negligible resistance; module-level field equalization (reflective backings, low-loss current collectors, shallow illuminated channels) that sustains flux and uniformity; and standardized stress tests in mixed-matrix waters (NOM, bicarbonate, halides, hardness) that reveal scaling and passivation risks. Advanced together, these elements reduce cleaning frequency and downstream polishing, extend module lifetimes, and lower energy and chemical footprints consistent with water-reuse targets.

References

Abd Aziz, M. H., Othman, M. H. D., Tavares, J. R., Pauzan, M. A. B., Tenjimbayashi, M., Lun, A. W., Alias, N. H., Ismail, A. F., Rahman, M. A., & Jaafar, J. (2022). Self-cleaning and anti-fouling superhydrophobic hierarchical ceramic surface synthesized from hydrothermal and fluorination methods. *Applied Surface Science, 598*, Article 153702. https://doi.org/10.1016/j.apsusc.2022.153702

Acharya, R., & Parida, K. (2020). A review on TiO2/g-C3N4 visible-light- responsive photocatalysts for sustainable energy generation and environmental remediation. *Journal of Environmental Chemical Engineering, 8*, Article 103896. https://doi.org/10.1016/j.jece.2020.103896

Ahmed, M. A., Amin, S., & Mohamed, A. A. (2023). Fouling in reverse osmosis membranes: Monitoring, characterization, mitigation strategies and future directions. *Heliyon, 9*, Article e14908. https://doi.org/10.1016/j.heliyon.2023.e14908

Argurio, P., Fontananova, E., Molinari, R., & Drioli, E. (2018). Photocatalytic membranes in photocatalytic membrane reactors. *Processes, 6*. https://doi.org/10.3390/pr6090162

Awad, A. M., & de Lannoy, C.-F. (2024). Production of stable electrically conductive PVDF membranes based on polydopamine-Polyethyleneimine—Assisted deposition of carbon nanotubes. *Membranes, 14*. https://doi.org/10.3390/membranes14040094

Bachosz, K., Vu, M. T., Nghiem, L. D., Zdarta, J., Nguyen, L. N., & Jesionowski, T. (2022). Enzyme-based control of membrane biofouling for water and wastewater purification: A comprehensive review. *Environmental Technology and Innovation, 25*, Article 102106. https://doi.org/10.1016/j.eti.2021.102106

Baker, R. W. (2004). Overview of membrane science and technology. In *Membrane technology and applications* (pp. 1–14). https://doi.org/10.1002/0470020393.ch1

Barambu, N. U., Bilad, M. R., Wibisono, Y., Jaafar, J., Mahlia, T. M., & Khan, A. L. (2019). Membrane surface patterning as a fouling mitigation strategy in liquid filtration: A review. *Polymers, 11*. https://doi.org/10.3390/polym11101687

Bixler, G. D., & Bhushan, B. (2012). Biofouling: Lessons from nature. *Philosophical Transactions of the Royal Society A: Mathematical, Physical and Engineering Sciences, 370*, 2381–2417. https://doi.org/10.1098/rsta.2011.0502

Carlsson, N., Gustafsson, H., Thörn, C., Olsson, L., Holmberg, K., & Åkerman, B. (2014). Enzymes immobilized in mesoporous silica: A physical–chemical perspective. *Advances in Colloid and Interface Science, 205*, 339–360. https://doi.org/10.1016/j.cis.2013.08.010

Cassie, A. B. D., & Baxter, S. (1944). Wettability of porous surfaces. *Transactions of the Faraday Society, 40*, 546–551. https://doi.org/10.1039/TF9444000546

Castro-Hinojosa, C., Del Sol-Fernández, S., Moreno-Antolín, E., Martín-Gracia, B., Ovejero, J. G., de la Fuente, J. M., Grazú, V., Fratila, R. M., & Moros, M. (2023). A simple and versatile strategy for oriented immobilization of his-tagged proteins on magnetic nanoparticles. *Bioconjugate Chemistry, 34*, 2275–2292. https://doi.org/10.1021/acs.bioconjchem.3c00417

Chen, D., Li, F., & Ray, A. K. (2001). External and internal mass transfer effect on photocatalytic degradation. *Catalysis Today, 66*, 475–485. https://doi.org/10.1016/S0920-5861(01)00256-5

Cunha, D., Kuznetsov, A., Achete, C., Da Hora Machado, A. E., & Marques, M. (2018). Immobilized TiO2 on glass spheres applied to heterogeneous photocatalysis: Photoactivity, leaching and regeneration process. *PeerJ, 6*. https://doi.org/10.7717/peerj.4464

del Castillo, G. F.-D., Kyriakidou, M., Adali, Z., Xiong, K., Hailes, R. L. N., & Dahlin, A. (2022). Electrically switchable polymer brushes for protein capture and release in biological environments. *Angewandte Chemie International Edition, 61*, Article e202115745. https://doi.org/10.1002/anie.202115745

Delorme, N., Bardeau, J.-F., Bulou, A., & Poncin-Epaillard, F. (2005). Azobenzene-containing monolayer with photoswitchable wettability. *Langmuir, 21*, 12278–12282. https://doi.org/10.1021/la051517x

Di Spiridione, C., Aresta, M., & Dibenedetto, A. (2022). Improving the enzymatic Cascade of reactions for the reduction of CO2 to CH3OH in water: From enzymes immobilization strategies to cofactor regeneration and cofactor suppression. *Molecules, 27*. https://doi.org/10.3390/molecules27154913

Diao, Y., Han, M., Lopez-Berganza, J. A., Valentino, L., Marinas, B., & Espinosa-Marzal, R. M. (2017). Reconciling DLVO and non-DLVO forces and their implications for ion rejection by a polyamide membrane. *Langmuir, 33*, 8982–8992. https://doi.org/10.1021/acs.langmuir.7b02306

Donaldson, E. C., & Alam, W. (2008). Chapter 2 - Surface forces. In E. C. Donaldson, & W. Alam (Eds.), *Wettability* (pp. 57–119). Gulf Publishing Company. https://doi.org/10.1016/B978-1-933762-29-6.50008-9

Elimelech, M., & Phillip, W. A. (2011). The future of seawater desalination: Energy, technology, and the environment. *Science, 333*, 712–717. https://doi.org/10.1126/science.1200488

Eniola, J. O., Kujawa, J., Nwokoye, A., Al-Gharabli, S., Avornyo, A. K., Giwa, A., Amusa, H. K., Yusuf, A. O., & Okolie, J. A. (2024). Advances in electrochemical membranes for water treatment: A comprehensive review. *Desalination and Water Treatment, 319*, Article 100450. https://doi.org/10.1016/j.dwt.2024.100450

Etxeberria-Benavides, M., David, O., Johnson, T., Łozińska, M. M., Orsi, A., Wright, P. A., Mastel, S., Hillenbrand, R., Kapteijn, F., & Gascon, J. (2018). High performance mixed matrix membranes (MMMs) composed of ZIF-94 filler and 6FDA-DAM polymer. *Journal of Membrane Science, 550*, 198–207. https://doi.org/10.1016/j.memsci.2017.12.033

Fink, D. J., Na, T., & Schultz, J. S. (1973). Effectiveness factor calculations for immobilized enzyme catalysts. *Biotechnology and Bioengineering, 15*, 879–888. https://doi.org/10.1002/bit.260150505

Freger, V. (2003). Nanoscale heterogeneity of polyamide membranes formed by interfacial polymerization. *Langmuir, 19*, 4791–4797. https://doi.org/10.1021/la020920q

Gartner, M., Szekeres, A., Stroescu, H., Mitrea, D., & Covei, M. (2023). Advanced nanostructured coatings based on doped TiO2 for various applications. *Molecules, 28*. https://doi.org/10.3390/molecules28237828

Ghosal, K., Bhattacharyya, S. K., Mishra, V., & Zuilhof, H. (2024). Click chemistry for biofunctional polymers: From observing to steering cell behavior. *Chemical Reviews, 124*, 13216–13300. https://doi.org/10.1021/acs.chemrev.4c00251

Gillespie, P. N. O., & Martsinovich, N. (2019). Origin of charge trapping in TiO2/Reduced graphene oxide photocatalytic composites: Insights from theory. *ACS Applied Materials and Interfaces, 11*, 31909–31922. https://doi.org/10.1021/acsami.9b09235

Gutfreund, P., Higy, C., Fragneto, G., Tschopp, M., Felix, O., & Decher, G. (2023). Molecular conformation of polyelectrolytes inside layer-by-layer assembled films. *Nature Communications, 14*, 4076. https://doi.org/10.1038/s41467-023-39801-x

Guzik, U., Hupert-Kocurek, K., & Wojcieszyńska, D. (2014). Immobilization as a strategy for improving enzyme properties-application to oxidoreductases. *Molecules, 19*, 8995–9018. https://doi.org/10.3390/molecules19078995

Hübner, U., Spahr, S., Lutze, H., Wieland, A., Rüting, S., Gernjak, W., & Wenk, J. (2024). Advanced oxidation processes for water and wastewater treatment – Guidance for systematic future research. *Heliyon, 10*, Article e30402. https://doi.org/10.1016/j.heliyon.2024.e30402

Hashiba, K., Nakai, S., Ohno, M., Nishijima, W., Gotoh, T., & Iizawa, T. (2019). Deterioration mechanism of a tertiary polyamide reverse osmosis membrane by hypochlorite. *Environmental Science and Technology, 53*, 9109–9117. https://doi.org/10.1021/acs.est.9b00663

Heredia Deba, S. A., Wols, B. A., Yntema, D. R., & Lammertink, R. G. H. (2023). Advanced ceramics in radical filtration: TiO2 layer thickness effect on the photocatalytic membrane performance. *Journal of Membrane Science, 672*, Article 121423. https://doi.org/10.1016/j.memsci.2023.121423

Hisatomi, T., & Domen, K. (2019). Reaction systems for solar hydrogen production via water splitting with particulate semiconductor photocatalysts. *Nature Catalysis, 2*, 387–399. https://doi.org/10.1038/s41929-019-0242-6

Hollingsworth, N. R., Wilkanowicz, S. I., & Larson, R. G. (2019). Salt- and pH-induced swelling of a poly(acrylic acid) brush via quartz crystal microbalance w/dissipation (QCM-D). *Soft Matter, 15*, 7838–7851. https://doi.org/10.1039/C9SM01289C

Hoque, MdA., & Guzman, M. I. (2018). Photocatalytic activity: Experimental features to report in heterogeneous photocatalysis. *Materials, 11*. https://doi.org/10.3390/ma11101990

Hu, Z.-Y., Liu, T., Yang, Y.-R., An, A. K., Liew, K. M., & Li, W.-W. (2025). Binder-free immobilization of photocatalyst on membrane surface for efficient photocatalytic H2O2 production and water decontamination. *Nano-Micro Letters, 17*, 301. https://doi.org/10.1007/s40820-025-01822-0

Ibrahim, Y., & Hilal, N. (2023). A critical assessment of surface-patterned membranes and their role in advancing membrane technologies. *ACS ES&T Water, 3*, 3807–3834. https://doi.org/10.1021/acsestwater.3c00564

Jiang, S., & Cao, Z. (2010). Ultralow-fouling, functionalizable, and hydrolyzable zwitterionic materials and their derivatives for biological applications. *Advanced Materials, 22*, 920–932. https://doi.org/10.1002/adma.200901407

Kafkopoulos, G., Martinho, R. P., Padberg, C. J., Duvigneau, J., Wurm, F. R., & Vancso, G. J. (2025). Metal ions in polydopamine coatings enhance polymer–metal adhesion. *ACS Applied Polymer Materials, 7*, 2408–2418. https://doi.org/10.1021/acsapm.4c03551

Kajau, A., Motsa, M., Mamba, B. B., & Mahlangu, O. (2021). Leaching of CuO nanoparticles from PES ultrafiltration membranes. *ACS Omega, 6*, 31797–31809. https://doi.org/10.1021/acsomega.1c04431

Kuang, J., & Messersmith, P. B. (2012). Universal surface-initiated polymerization of antifouling zwitterionic brushes using a mussel-mimetic peptide initiator. *Langmuir, 28*, 7258–7266. https://doi.org/10.1021/la300738e

Kyriakou, N., Merlet, R. B., Willott, J. D., Nijmeijer, A., Winnubst, L., & Pizzoccaro-Zilamy, M.-A. (2020). New method toward a robust covalently attached cross-linked nanofiltration membrane. *ACS Applied Materials and Interfaces, 12*, 47948–47956. https://doi.org/10.1021/acsami.0c13339

Lainioti, G. C., & Druvari, D. (2024). Designing antibacterial-based Quaternary ammonium coatings (surfaces) or films for biomedical applications: Recent advances. *International Journal of Molecular Sciences, 25*. https://doi.org/10.3390/ijms252212264

Lanzalaco, S., & Armelin, E. (2017). Poly(N-isopropylacrylamide) and copolymers: A review on recent progresses in biomedical applications. *Gels, 3*. https://doi.org/10.3390/gels3040036

Lau, S. K., & Yong, W. F. (2021). Recent progress of zwitterionic materials as antifouling membranes for ultrafiltration, nanofiltration, and reverse osmosis. *ACS Applied Polymer Materials, 3*, 4390–4412. https://doi.org/10.1021/acsapm.1c00779

Law, K.-Y. (2022). Contact angle hysteresis on smooth/flat and rough surfaces. Interpretation, mechanism, and origin. *Accounts of Materials Research, 3*, 1–7. https://doi.org/10.1021/accountsmr.1c00051

Le-Clech, P., Chen, V., & Fane, T. A. G. (2006). Fouling in membrane bioreactors used in wastewater treatment. *Journal of Membrane Science, 284*, 17–53. https://doi.org/10.1016/j.memsci.2006.08.019

Lee, H., Dellatore, S. M., Miller, W. M., & Messersmith, P. B. (2007). Mussel-inspired surface chemistry for multifunctional coatings. *Science, 318*, 426–430. https://doi.org/10.1126/science.1147241

Lee, J., Ko, J. H., Lin, E.-W., Wallace, P., Ruch, F., & Maynard, H. D. (2015). Trehalose hydrogels for stabilization of enzymes to heat. *Polymer Chemistry, 6*, 3443–3448. https://doi.org/10.1039/C5PY00121H

Leng, C., Hung, H.-C., Sun, S., Wang, D., Li, Y., Jiang, S., & Chen, Z. (2015). Probing the surface hydration of nonfouling zwitterionic and PEG materials in contact with proteins. *ACS Applied Materials and Interfaces, 7*, 16881–16888. https://doi.org/10.1021/acsami.5b05627

Li, L., Baig, M. I., de Vos, W. M., & Lindhoud, S. (2024). Biocatalytic PEI–PSS membranes through aqueous phase separation: Influence of casting solution pH and operational temperature. *Soft Matter, 20*, 5425–5434. https://doi.org/10.1039/D4SM00311J

Li, F., Meindersma, W., de Haan, A. B., & Reith, T. (2005). Novel spacers for mass transfer enhancement in membrane separations. *Journal of Membrane Science, 253*, 1–12. https://doi.org/10.1016/j.memsci.2004.12.019

Li, X., Wang, Z., Han, X., Liu, Y., Wang, C., Yan, F., & Wang, J. (2021). Regulating the interfacial polymerization process toward high-performance polyamide thin-film composite reverse osmosis and nanofiltration membranes: A review. *Journal of Membrane Science, 640*, Article 119765. https://doi.org/10.1016/j.memsci.2021.119765

Lin, M.-H., Bulman, D. M., Remucal, C. K., & Chaplin, B. P. (2020). Chlorinated byproduct formation during the electrochemical advanced oxidation process at magnéli phase Ti4O7 electrodes. *Environmental Science and Technology, 54*, 12673–12683. https://doi.org/10.1021/acs.est.0c03916

Liu, H., Li, M., Lu, Z.-Y., Zhang, Z.-G., & Sun, C.-C. (2009). Influence of surface-initiated polymerization rate and initiator density on the properties of polymer brushes. *Macromolecules, 42*, 2863–2872. https://doi.org/10.1021/ma802817r

Liu, S., Tang, J., Ji, F., Lin, W., & Chen, S. (2022). Recent advances in zwitterionic hydrogels: Preparation, property, and biomedical application. *Gels, 8*. https://doi.org/10.3390/gels8010046

Liu, F., Zhou, R., Zhang, C., Wu, Z., Ren, H., & Ng, H. Y. (2024). Critical review on the pulsed electrochemical technologies for wastewater treatment: Fundamentals, current trends, and future studies. *Chemical Engineering Journal, 479*, Article 147588. https://doi.org/10.1016/j.cej.2023.147588

Luft, G., Laurent, A., Recasens, F., Trabelsi, F., & Weickert, G. (2001). Chapter 5 - Industrial reaction units. In A. Bertucco, & G. Vetter (Eds.), *Industrial chemistry library* (pp. 243–350). Elsevier. https://doi.org/10.1016/S0926-9614(01)80023-X

Luo, J., Song, S., Zhang, H., Zhang, H., Zhang, J., & Wan, Y. (2020). Biocatalytic membrane: Go far beyond enzyme immobilization. *Engineering in Life Sciences, 20*, 441–450. https://doi.org/10.1002/elsc.202000018

Madaeni, S. S., Ghaemi, N., Alizadeh, A., & Joshaghani, M. (2011). Influence of photo-induced superhydrophilicity of titanium dioxide nanoparticles on the anti-fouling performance of ultrafiltration membranes. *Applied Surface Science, 257*, 6175–6180. https://doi.org/10.1016/j.apsusc.2011.02.026

Mecha, A. C., Chollom, M. N., Babatunde, B. F., Tetteh, E. K., & Rathilal, S. (2023). Versatile silver-nanoparticle-impregnated membranes for water treatment: A review. *Membranes, 13*. https://doi.org/10.3390/membranes13040432

Meng, X., Wang, X., Yin, K., Jing, Y., Gu, L., Tao, Z., Ren, X., Tang, M., Shao, X., Sun, L., Sun, Y., Dai, Y., & Xiong, Y. (2024). Integration of photothermal water evaporation with photocatalytic microplastics upcycling via nanofluidic thermal management. *Proceedings of the National Academy of Sciences, 121*, Article e2317192121. https://doi.org/10.1073/pnas.2317192121

Min, J.-H., Chen, Y.-A., Chen, I.-T., Sun, T., Lee, D. T., Li, C., Zhu, Y., O'Connor, B. T., Parsons, G. N., & Chang, C.-H. (2018). Conformal physical vapor deposition assisted by atomic layer deposition and its application for stretchable conductors. *Advanced Materials Interfaces, 5*, Article 1801379. https://doi.org/10.1002/admi.201801379

Molinari, R., Borgese, M., Drioli, E., Palmisano, L., & Schiavello, M. (2002). Hybrid processes coupling photocatalysis and membranes for degradation of organic pollutants in water. *Catalysis Today, 75*, 77–85. https://doi.org/10.1016/S0920-5861(02)00047-0

Molinari, R., Lavorato, C., & Argurio, P. (2017). Recent progress of photocatalytic membrane reactors in water treatment and in synthesis of organic compounds. A review. *Catalysis Today, 281*, 144–164. https://doi.org/10.1016/j.cattod.2016.06.047

Nambi Krishnan, J., Venkatachalam, K. R., Ghosh, O., Jhaveri, K., Palakodeti, A., & Nair, N. (2022). Review of thin film nanocomposite membranes and their applications in desalination. *Frontiers in Chemistry, 10*. https://doi.org/10.3389/fchem.2022.781372

Nikkola, J., Sievänen, J., Raulio, M., Wei, J., Vuorinen, J., & Tang, C. Y. (2014). Surface modification of thin film composite polyamide membrane using atomic layer deposition method. *Journal of Membrane Science, 450*, 174–180. https://doi.org/10.1016/j.memsci.2013.09.005

Niyaz Ahmed, L., & Praveen, T. (2024). A study using a semi-analytical approach to the reaction kinetics that describes the behavior of heterogeneous reaction–diffusion process with a glyoxyl-agarose immobilized enzyme (penicillin G acylase) in a batch reactor. *Chemical Engineering Journal, 496*, Article 153858. https://doi.org/10.1016/j.cej.2024.153858

Okuyama, H., Sugawara, Y., & Yamaguchi, T. (2024). Machine-learning-aided understanding of protein adsorption on zwitterionic polymer brushes. *ACS Applied Materials and Interfaces, 16*, 25236–25245. https://doi.org/10.1021/acsami.4c01401

Otálvaro-Marín, H. L., & Machuca-Martínez, F. (2020). Sizing of reactors by charts of Damköhler's number for solutions of dimensionless design equations. *Heliyon, 6*. https://doi.org/10.1016/j.heliyon.2020.e05386

Ouyang, L., Wei, B., Kuo, C., Pathak, S., Farrell, B., & Martin, D. C. (2017). Enhanced PEDOT adhesion on solid substrates with electrografted P(EDOT-NH2). *Science Advances, 3*, Article e1600448. https://doi.org/10.1126/sciadv.1600448

Petersen, R. J. (1993). Composite reverse osmosis and nanofiltration membranes. *Journal of Membrane Science, 83*, 81–150. https://doi.org/10.1016/0376-7388(93)80014-O

Porcelli, N., & Judd, S. (2010). Chemical cleaning of potable water membranes: A review. *Separation and Purification Technology, 71*, 137–143. https://doi.org/10.1016/j.seppur.2009.12.007

Purkait, M. K., Sinha, M. K., Mondal, P., & Singh, R. (2018). Photoresponsive membranes. In *Interface science and technology* (pp. 115–144). Elsevier. https://doi.org/10.1016/B978-0-12-813961-5.00004-8

Radjenovic, J., & Sedlak, D. L. (2015). Challenges and opportunities for electrochemical processes as next-generation technologies for the treatment of contaminated water. *Environmental Science and Technology, 49*, 11292–11302. https://doi.org/10.1021/acs.est.5b02414

Rapp, B. E. (2017). Chapter 9 - Fluids. In B. E. Rapp (Ed.), *Microfluidics: Modelling, mechanics and mathematics* (pp. 243–263). Oxford: Elsevier. https://doi.org/10.1016/B978-1-4557-3141-1.50009-5

Reid, E., Igou, T., Zhao, Y., Crittenden, J., Huang, C.-H., Westerhoff, P., Rittmann, B., Drewes, J. E., & Chen, Y. (2023). The minus approach can redefine the standard of practice of drinking water treatment. *Environmental Science and Technology, 57*, 7150–7161. https://doi.org/10.1021/acs.est.2c09389

Saberi, A., Jabbari, F., Zarrintaj, P., Saeb, M. R., & Mozafari, M. (2019). Electrically conductive materials: Opportunities and challenges in tissue engineering. *Biomolecules, 9*. https://doi.org/10.3390/biom9090448

Saleh, E. A. M., Kumar, A., Alghazali, T., Ganesan, S., Shankhyan, A., Sharma, G. C., Satyam Naidu, K., & Rahbari-Sisakht, M. (2025). Recent advances in thin film composite (TFC) membrane development: Materials and modification methods. *Environmental Science: Water Research & Technology, 11*, 1059–1085. https://doi.org/10.1039/D4EW01011F

Schlenoff, J. B. (2014). Zwitteration: Coating surfaces with zwitterionic functionality to reduce nonspecific adsorption. *Langmuir, 30*, 9625–9636. https://doi.org/10.1021/la500057j

Schmidt, D. L., Brady, R. F., Lam, K., Schmidt, D. C., & Chaudhury, M. K. (2004). Contact angle hysteresis, adhesion, and marine biofouling. *Langmuir, 20*, 2830–2836. https://doi.org/10.1021/la035385o

Schock, G., & Miquel, A. (1987). Mass transfer and pressure loss in spiral wound modules. *Desalination, 64*, 339–352. https://doi.org/10.1016/0011-9164(87)90107-X

Sheldon, R. A., & van Pelt, S. (2013). Enzyme immobilisation in biocatalysis: Why, what and how. *Chemical Society Reviews, 42*, 6223–6235. https://doi.org/10.1039/C3CS60075K

Singh, R., Sinha, M. K., & Purkait, M. K. (2020). Stimuli responsive mixed matrix polysulfone ultrafiltration membrane for humic acid and photocatalytic dye removal applications. *Separation and Purification Technology, 250*, Article 117247. https://doi.org/10.1016/j.seppur.2020.117247

Soestbergen, M. (2012). Frumkin-butler-volmer theory and mass transfer in electrochemical cells. *Russian Journal of Electrochemistry, 48*. https://doi.org/10.1134/S1023193512060110

Sun, M., Wang, X., Winter, L. R., Zhao, Y., Ma, W., Hedtke, T., Kim, J.-H., & Elimelech, M. (2021). Electrified membranes for water treatment applications. *ACS ES&T Engineering, 1*, 725–752. https://doi.org/10.1021/acsestengg.1c00015

Trainham, J. A., & Newman, J. (1981). A comparison between flow-through and flow-by porous electrodes for redox energy storage. *Electrochimica Acta, 26*, 455–469. https://doi.org/10.1016/0013-4686(81)87024-7

Ulbricht, M. (2006). Advanced functional polymer membranes. *Polymer, 47*, 2217–2262. https://doi.org/10.1016/j.polymer.2006.01.084

Valle-Delgado, J. J., Molina-Bolívar, J. A., Galisteo-González, F., Gálvez-Ruiz, M. J., Feiler, A., & Rutland, M. W. (2005). Hydration forces between silica surfaces: Experimental data and predictions from different theories. *The Journal of Chemical Physics, 123*, Article 034708. https://doi.org/10.1063/1.1954747

Visan, A., van Ommen, J. R., Kreutzer, M. T., & Lammertink, R. G. H. (2019). Photocatalytic reactor design: Guidelines for kinetic investigation. *Industrial and Engineering Chemistry Research, 58*, 5349–5357. https://doi.org/10.1021/acs.iecr.9b00381

Vrouwenvelder, J. S., Graf von der Schulenburg, D. A., Kruithof, J. C., Johns, M. L., & van Loosdrecht, M. C. M. (2009). Biofouling of spiral-wound nanofiltration and reverse osmosis membranes: A feed spacer problem. *Water Research, 43*, 583–594. https://doi.org/10.1016/j.watres.2008.11.019

Wang, D., Mueses, M. A., Márquez, J. A. C., Machuca-Martínez, F., Grčić, I., Peralta Muniz Moreira, R., & Li Puma, G. (2021). Engineering and modeling perspectives on photocatalytic reactors for water treatment. *Water Research, 202*, Article 117421. https://doi.org/10.1016/j.watres.2021.117421

Wei, Y., Wu, Q., Meng, H., Zhang, Y., & Cao, C. (2023). Recent advances in photocatalytic self-cleaning performances of TiO2-based building materials. *RSC Advances, 13*, 20584–20597. https://doi.org/10.1039/D2RA07839B

Werber, J. R., Osuji, C. O., & Elimelech, M. (2016). Materials for next-generation desalination and water purification membranes. *Nature Reviews Materials, 1*, Article 16018. https://doi.org/10.1038/natrevmats.2016.18

Wu, J., Cao, M., Tong, D., Finkelstein, Z., & Hoek, E. M. V. (2021). A critical review of point-of-use drinking water treatment in the United States. *npj Clean Water, 4*, 40. https://doi.org/10.1038/s41545-021-00128-z

Wu, Y., Wang, Z., Zhou, J., Gu, Q., & Zhong, Z. (2025). Stimulus-responsive membranes: A mini review on principles, preparation methods, and emerging applications. *Separations, 12*. https://doi.org/10.3390/separations12080219

Xu, R., Qin, S., Lu, T., Wang, S., Chen, J., & He, Z. (2025). Engineering photocatalytic membrane reactors for sustainable energy and environmental applications. *Catalysts, 15*. https://doi.org/10.3390/catal15100947

Zander, A. K., & Curry, N. K. (2001). Membrane and solution effects on solute rejection and productivity. *Water Research, 35*, 4426–4434. https://doi.org/10.1016/S0043-1354(01)00169-5

Zhang, R., Liu, Y., He, M., Su, Y., Zhao, X., Elimelech, M., & Jiang, Z. (2016). Antifouling membranes for sustainable water purification: Strategies and mechanisms. *Chemical Society Reviews, 45*, 5888–5924. https://doi.org/10.1039/C5CS00579E

Zhang, J., Wu, H., Shi, L., Wu, Z., Zhang, S., Wang, S., & Sun, H. (2024). Photocatalysis coupling with membrane technology for sustainable and continuous purification of wastewater. *Separation and Purification Technology, 329*, Article 125225. https://doi.org/10.1016/j.seppur.2023.125225

Zheng, X., Shen, Z.-P., Shi, L., Cheng, R., & Yuan, D.-H. (2017). Photocatalytic membrane reactors (PMRs) in water treatment: Configurations and influencing factors. *Catalysts, 7*. https://doi.org/10.3390/catal7080224

Zhou, X., Zhao, Y.-Y., Kim, S.-R., Elimelech, M., Hu, S., & Kim, J.-H. (2018). Controlled TiO2 growth on reverse osmosis and nanofiltration membranes by atomic layer deposition: Mechanisms and potential applications. *Environmental Science and Technology, 52*, 14311–14320. https://doi.org/10.1021/acs.est.8b03967

PART IV

Complementary treatment technologies for complex contaminants

CHAPTER

8 Advanced adsorption and ion exchange technologies for removing complex and emerging pollutants from water sources

Aravind D., Anandu Chandran and Nirav P. Raval

Department of Environmental Science and Engineering, School of Engineering and Sciences, SRM University-AP, Amaravati, Andhra Pradesh, India

1. Introduction

Global water scarcity is a critical challenge in the 21st century, with over 40% of the world's population facing water stress and four billion people experiencing severe water scarcity at least 1 month per year (Mekonnen & Hoekstra, 2016). UNICEF reports that by 2040, roughly one in four children worldwide will be living in areas of extremely high-water stress (https://www.unicef.org/wash/water-scarcity) due to climate change, population growth, and intensive agriculture. In response, water reuse and desalination technologies have expanded rapidly. Among the techniques, reverse osmosis (RO) membranes, a cornerstone of modern water reuse schemes, are widely used to convert brackish water, ground water, or seawater into potable freshwater (Devaisy et al., 2023b; Wu et al., 2024). RO produces high-quality permeate suitable for drinking and industrial use by removing a wide range of contaminants, including dissolved salts (ions), heavy metals, bacteria, organic molecules, and pathogens (Vingerhoeds et al., 2016). However, RO has significant drawbacks, such as RO brine or RO concentrate (ROC), which contains concentrated pollutants rejected by the membrane during filtration (Valdés et al., 2021).

In addition to salinity, ROC often contains trace levels of complex organic chemicals like contaminants of emerging concern (CECs). The term "emerging" reflects that these pollutants have been increasingly detected recently due to advances in instrumentation. Studies show that many CECs, such as pharmaceuticals, hormones, steroids, pesticides, personal care products, endocrine disruptors, PFAS (per- and polyfluoroalkyl substances), and microplastics, are present in water sources and accumulated in RO membranes (Konradt et al., 2021; Romeyn et al., 2015). These CECs are not removed by conventional treatments and pose a risk to the ecosystem and human health when discharged. Simply disposing of ROC can release persistent pollutants to the environment (Harish & Jegatheesan, 2025; Matesun et al., 2024; Rezaei et al., 2024). This highlights the requirement for the closed-loop approaches, such as zero-liquid discharge (ZLD) and minimal-liquid discharge (MLD), for ROC management with the aim of recovering all the water from the brine. While these approaches can eliminate

Sustainable Water Management. https://doi.org/10.1016/B978-0-443-44853-9.00017-4

effluent in principle, it concentrates or crystallize the pollutants, insufficiently addressing the accumulation of CECs in the concentrated waste streams (Finnerty et al., 2024; Tong & Elimelech, 2016).

In this context, the present chapter focuses on evaluating the occurrence of CECs in widely generated ROC streams by examining their distribution, associated challenges, and implications for environmental and treatment complexity. It then assesses existing remediation approaches, with particular emphasis on adsorption and ion exchange (IEX) techniques, by analyzing the performance of diverse materials in terms of efficiency, selectivity toward trace-level CECs, operational parameters, and resource-recovery potential. Finally, the chapter explores opportunities for integrating adsorption and IEX processes into closed-loop treatment systems, emphasizing their role in enhancing water recovery, enabling resource reuse, and supporting sustainable brine management in alignment with SDG-6 and SDG-12.

2. Unidentified issues and challenges in emerging contaminants in ROC

CECs are synthetic or naturally occurring contaminants comprises of broad, heterogeneous group of chemicals detected in water that are not fully understood, which includes pharmaceuticals, hormones, steroids, pesticides, personal care products, endocrine disruptors, PFAS, and microplastics (Feng et al., 2023; X. Li et al., 2024). They are present at trace levels (ng/L to μg/L). A study conducted in Indonesia monitored 22 CECs in influent and effluent of a municipal wastewater treatment plant (WWTP) with 14 CECs in the influent and seven in the effluent, with a total concentration of 29.8 ± 0.4 μg/L and 0.5 ± 0.0 μg/L (Astuti et al., 2023). Due to the trace-level concentration, identifying CECs in water is challenging, requiring highly sensitive methods such as LC-MS (liquid chromatography–mass spectrometry) or HRMS (high-resolution mass spectrometry) (Meher & Zarouri, 2025; Ng et al., 2020). Hundreds of potential CECs and their degradation products are still without analytical standards, making it a significant challenge in environmental monitoring and risk assessment. The high salinity of ROC further complicates the analysis by interfering with ionization in MS (Mass Spectrometry) and sample preparation due to the competition between excess ions from salts and analyte ions, making routine monitoring of ROC for CECs rare. Studies show high concentrations of salts in ROC can lead to ion suppression effects during ESI-MS (electrospray ionization–mass spectrometry), reducing the sensitivity and accuracy of detecting low-abundance CECs (Wouters et al., 2022). In practice, assessments often focus on a few indicator compounds rather than the full spectrum, which causes an overlook of other potentially harmful CECs present in brines (Zhang et al., 2021).

CECs are very recalcitrant; some of the CECs, such as PFAS, are called as "Forever chemicals" due to their slow degradation. PFAS are extremely hard to degrade due to the high bond dissociation energy of the carbon–fluorine (C–F) bond (R. Schlesinger et al., 2022). Studies indicate that PFOA has a half-life of approximately 92 years in water under neutral pH conditions, demonstrating its long-term persistence in aquatic environments (Skinner et al., 2025). CECs are very persistent, which raises the risk of bioaccumulation and long-term exposure. Microplastics act as carriers for these persistent pollutants by adsorbing various CECs from the surrounding environment, leading to the bioaccumulation (Manjarrés-López et al., 2025). Studies revealed the presence of six CECs, namely diclofenac, atenolol, ibuprofen, 4-acetamidophenol, bisphenol A, and 2-mercaptobenzothiazole, on microplastic and also showed that the rate of adsorption of 4-acetamidophenol onto macro-polystyrene was the fastest among all CECs and microplastics evaluated (Tseng et al., 2022). Even though CECs are at trace levels, CECs like endocrine-disrupting chemicals (EDCs) can cause harm at even trace levels. For example, a study analyzed 14 multiclass EDCs in drinking water sources and found their presence in both source and tap water, highlighting the potential for human exposure to these chemicals, even at low concentrations (Wee et al., 2020). Exposure to EDCs, even at low levels, can increase the risk of various diseases, including reproductive impairments, cognitive deficits, and certain cancers (La Merrill et al., 2020).

Discharge of ROC containing CECs poses multiple risks. ROC from municipal wastewater treatment plants can have concentrations of CECs significantly higher than in the influent, making their discharge potentially harmful to aquatic life (Finnerty et al., 2024). The discharge of ROC has been associated with chronic ecotoxicity risks. A study highlighted that the concentrated waste stream resulting from RO treatment can pose such risks if discharged to surface waters or shallow coastal areas (Scholes et al., 2021). ROC can contain antiscalants and heavy metals, which can cause adverse effects in the environment, such as mobilization of heavy metals (Romeyn et al., 2015). Crucially, most CECs have no binding water quality standards. In practice, this means ROC discharges are unchecked for CECs, posing a hidden contamination issue (Fig. 8.1). SDG-6 and SDG-12 cannot be met by RO membrane, an efficient desalination and purification technology, without addressing this issue. This requires advanced CEC removal techniques, paving the way for closed-loop approaches.

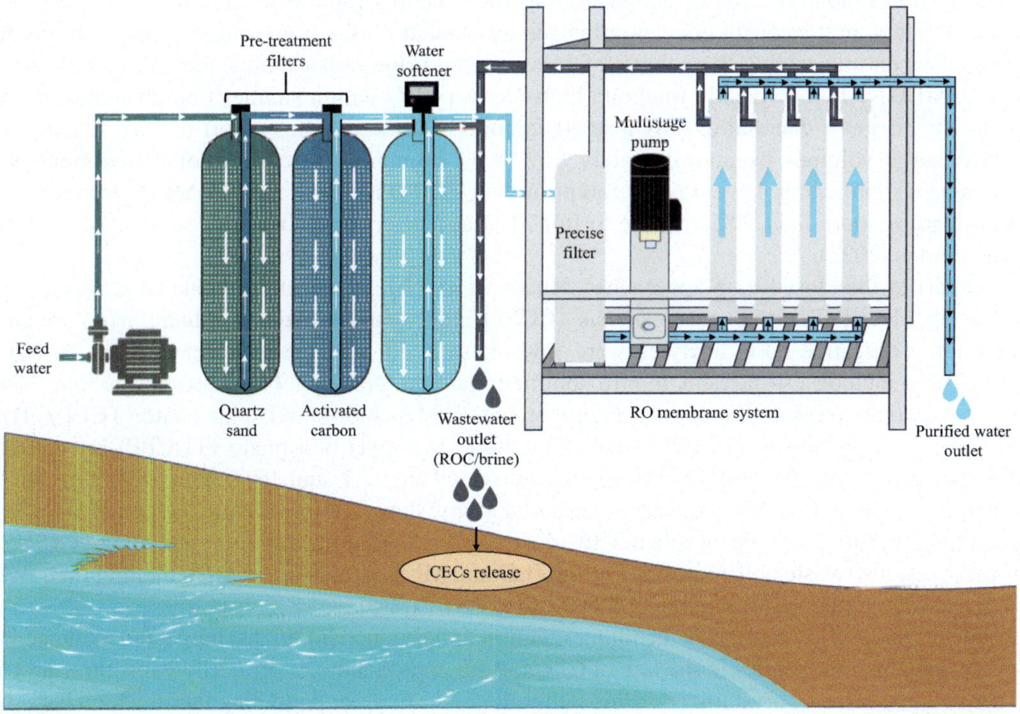

FIGURE 8.1

Schematic representation of CECs release into the environment through ROC from the RO plant, emphasizing the need for optimized regeneration and polishing processes to ensure sustainable ROC management.

3. Closed-loop management approaches with emphasis on CECs

In response to water scarcity and stricter discharge practices, closed-loop strategies like ZLD and MLD are gaining attention. In ZLD, all wastewater is reused or evaporated so that no liquid is discharged off-site. ZLD uses advanced pretreatment followed by membrane concentration and then thermal processes to recover all the water. The end products of ZLD are purified water and crystallized salt or sludge wastes. By concentrating and immobilizing contaminants, ZLD reduces the volume, mobility, and environmental exposure risk of diverse CECs. Many pollutant-generating industries have adopted these techniques, with the semiconductor industry being one of the earliest adopters of ZLD and MLD concepts, as its wastewater contains a wide range of contaminants which including CECs such as PFAS (often >1000 ng L^{-1}), tetramethylammonium hydroxide, etc., if discharged, could pose ecological risks. Previous studies have reported that, as part of ZLD implementation, high-efficiency RO systems operated at pH above 10, paired with extensive pretreatment to prevent silica scaling and maximize water recovery, generate a concentrated stream that is either sent to lined brine ponds or further processed in a brine concentrator (Subramani & Jacangelo, 2014; Tong et al., 2025). Reported performance data show that ZLD can achieve up to 97.04% water recovery with an energy demand of 17.53 kWh/m^3 (Panagopoulos & Giannika, 2024). Another study demonstrated water recovery of 99.14% using a brine crystallizer and 85.75% with WAIV (wind-aided intensified evaporation), with respective energy requirements of 22.15 kWh/m^3 and 15.34 kWh/m^3, leaving less than 1 m^3/day of solid brine (Panagopoulos, 2022), demonstrating the system's high efficiency in maximizing water recovery while simultaneously concentrating the associated CECs from wide exposure. In contrast, MLD focuses on minimizing the volume of concentrated brine rather than achieving complete water recovery. MLD systems typically integrate high-recovery RO with a smaller final evaporation stage, enabling discharge or disposal of only a small fraction of brine. Studies report that MLD can reduce brine discharge volumes by approximately 85%–95% compared to conventional treatment, while maintaining a significantly lower energy footprint than ZLD (Plata et al., 2022). MLD also reduces the thermal load by more than 70% relative to full ZLD and offers operating cost savings of 30%–40% (Panagopoulos, 2022).

As a part of the closed-loop approaches, Huang et al. (2025) recently conducted a 5-month pilot study using closed-circuit reverse osmosis (CCRO) for treating tertiary effluent from municipal wastewater. Thirty-five CEC analytes were investigated; 13 were entirely rejected by the CCRO system. These include 1,4-dioxane, N-nitrosodimethylamine (NDMA), N-nitrosomorpholine (NNM), primidone, carbamazepine, fluoxetine, meprobamate, Tris(2-chloroethyl) phosphate (TCEP), Tris(1-chloro-2-propyl) phosphate (TCPP), Tris(1,3-dichloro-2-propyl) phosphate (TDCPP), atenolol, cotinine, and phenytoin. Notably, NDMA was undetected at 92% and 93% recovery rates but was completely rejected at a 95% recovery rate, and a study showed that the molecular weight and hydrophilicity attribute a significant role in CEC removal. These findings reinforce the role of CCRO as a high-recovery, low-waste polishing step within ZLD and MLD frameworks, strengthening their overall capacity to minimize CEC discharge and environmental exposure.

While all these, ZLD and MLD techniques significantly reduce liquid discharge and enhance water recovery, a major concern lies in the fate of CECs. Both approaches are primarily concentrating CECs into the residual brine or crystallized solids rather than removing or degrading them, which becomes a secondary source of environmental contamination (Table 8.1). This increases the risk of handling the

Table 8.1 Tabulated illustration of closed-loop treatment approaches designed for ROC management and wastewater discharge minimization, along with their respective limitations in controlling CECs.

Sl. No.	Closed-loop approaches	Illustrations	Principle	Limitations	References
1	Crystallizer	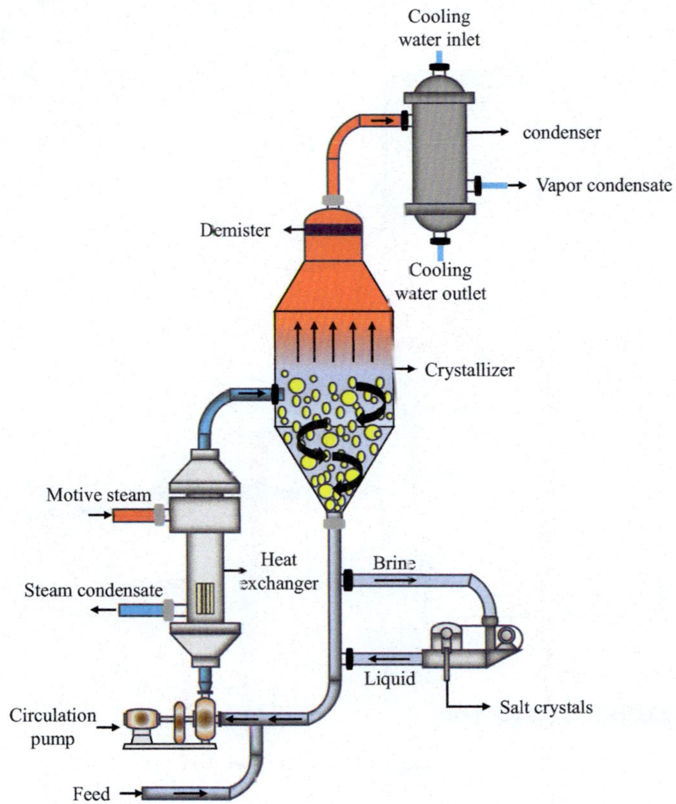	Evaporates water from concentrated brine to recover salts or solids, producing near-zero liquid discharge	• Operates primarily on phase separation, not contaminant degradation. • CECs can volatilize, thermally degrade into toxic byproducts, or remain entrapped in residual solids. • High energy requirement limits scalability for continuous ROC treatment. • Poor removal of nonionic and volatile organic CECs.	(Backer et al., 2022; Q. Chen et al., 2021; Liao et al., 2025; Tong & Elimelech, 2016; F. Wang et al., 2024)

Continued

Table 8.1 Tabulated illustration of closed-loop treatment approaches designed for ROC management and wastewater discharge minimization, along with their respective limitations in controlling CECs.—cont'd

Sl. No.	Closed-loop approaches	Illustrations	Principle	Limitations	References
2	Brine concentrator	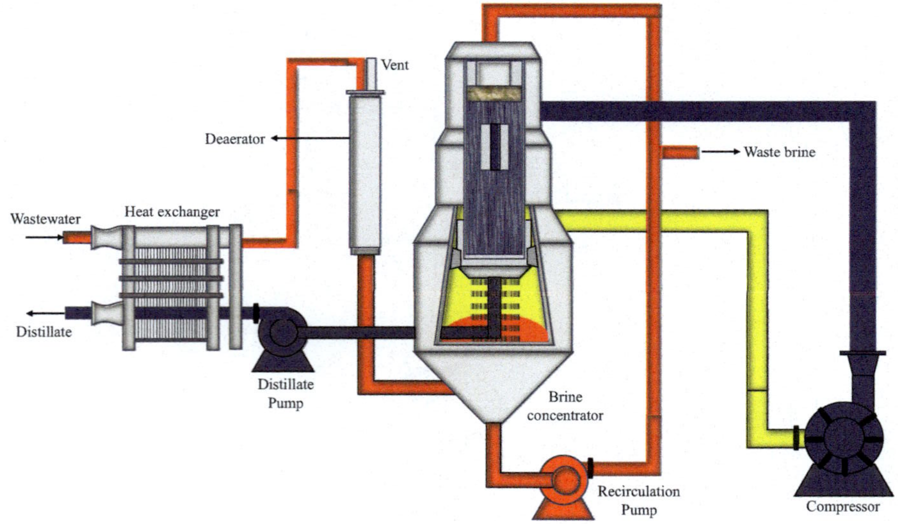	Uses thermal or mechanical vapor recompression to further concentrate RO brine before crystallization.	• Concentrates CECs rather than removing them; most are retained in the liquid phase. • Some volatile or semivolatile CECs may escape with vapor. • No transformation or mineralization occurs. • High operational cost and scaling/fouling issues reduce long-term efficiency.	(Backer et al., 2022; O'Connell et al., 2024; Romeyn et al., 2015; Vane, 2017)
3	Coagulation-flocculation	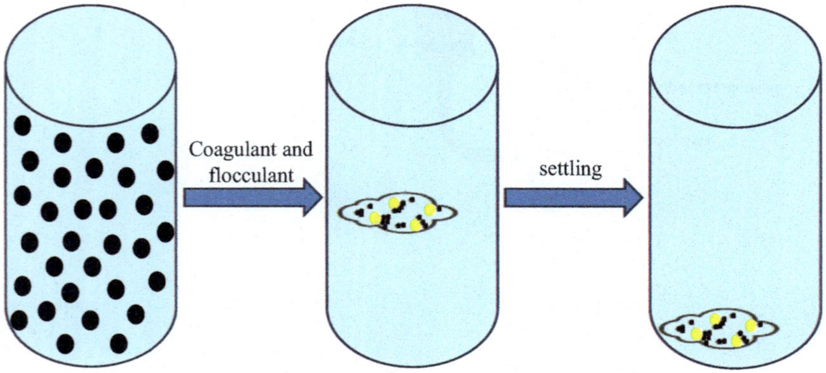	Destabilizes suspended colloids and organic matter via metal salts (e.g., $FeCl_3$, $Al_2(SO_4)_3$).	• Ineffective for dissolved or low-molecular-weight CECs. • Only partial removal of hydrophobic compounds through co-precipitation. • CECs associated with dissolved organic matter can remain in solution. • Generates chemical sludge that may contain concentrated CECs, requiring further treatment.	(Katrivesis et al., 2019; Kooijman et al., 2020; Lucas et al., 2025; Mustereț et al., 2021; Łukasiewicz, 2025)

residual streams and toxicity. Thus, both approaches demand safe additional treatment and safe handling of those concentrates to truly address the contaminant removal. Also, these approaches are energy-intensive and require high capital costs. Given these challenges, achieving both sustainability and CEC management requires a complementary closed-loop system with targeted treatment of pollutants. Promising technologies include IEX and adsorption integrated systems for better contaminant management.

4. Promising technologies for CEC removal

As mentioned earlier, the high toxicity of CECs has intensified research focus on identifying effective remediation techniques, leading to various explorations of both conventional and advanced treatment technologies for their removal from contaminated water. Conventional methods such as coagulation and flocculation have demonstrated poor removal efficiencies (<20%) for CECs, including acetaminophen, diclofenac, estrone, ethinylestradiol, musk ketone, and oxybenzone (Westerhoff et al., 2005). Likewise, chlorination usually reaches ~40% removal, and activated sludge operations often yield low removal rates for substances such as trimethoprim, macrolide antibiotics, and penicillin antibiotics. The removal of other CECs is also limited, with extents ranging from 11% to 35% and 26%–40% for quinolone antibiotics and carbamazepine, respectively (Gros et al., 2007; Lin et al., 2009). The complex molecular structure, nonbiodegradability, and low concentrations of emerging contaminants (ECs) are the significant reasons that limit their removal by conventional treatment methods; also, these compounds often cannot be metabolized by microorganisms and may even inhibit their activity (Radwan et al., 2023).

In contrast, advanced treatment technologies have demonstrated substantially higher remediation efficiencies. Membrane-based processes such as ultrafiltration (UF) achieve up to 75% removal of bisphenol and 85% of 17α-ethinylestradiol, while nanofiltration (NF) removes 51%–97% of bisphenol A and 76.5%–98% of 17α-ethinylestradiol (Radwan et al., 2023). Other advanced oxidation processes (AOPs) such as ozonation, Fenton reactions, and photocatalysis also possess excellent performance in degrading a wide range of CECs, along with emphasizing their mechanical reliability, operation stability, and applicability. Membrane distillation has also been reported with higher removals for hydrophilic trace organics (Costa et al., 2023). Adsorption and IEX techniques are also widely explored due to their broad sorption, affinity, rapid surface interactions, and ability to target both ionic and nonionic CECs even at trace concentrations. Table 8.2 provides a comparative overview of several CECs, the treatment methods evaluated, and their respective removal efficiencies. However, unlike adsorption and IEX methods, several other advanced technologies face challenges associated with secondary waste generation, formation of toxic by-products, high energy inputs, and increased costs with scale-up (Mahlangu et al., 2023; Ramrakhiani et al., 2022; Tokoyi et al., 2025). Various mechanisms involved in the remediation of CECs via adsorption and IEX technique are illustrated in Fig. 8.2.

4.1 Adsorption-based approaches

Adsorption is a surface phenomenon in which molecules or ions from a liquid or gaseous phase accumulate on the solid material (adsorbent) surface. In general, adsorption is a net accumulation of a

Table 8.2 Comparative overview of various CECs, their evaluated treatment methods, and removal efficiencies.

Sl. No.	Solution	Technique	Materials used	CECs type	Removal efficiency (%) or q_{max} (mg/g)	References
1	Synthetic wastewater	Filtration	Nano filtration (NF)	➢ N,N-Diethyl-m-toluamide ➢ Benzophenone ($C_{13}H_{10}O$) ➢ Carbamazepine	➢ 79.34% ➢ 94.14% ➢ 89.32%	Piekutin et al. (2025)
2	Multiple wastewater Types	AOPs	Electrochemical oxidation	➢ PFHxA ➢ PFOS ➢ PFOA	➢ 90% ➢ 87.8%–97.6% ➢ 70.6%–85%	Capodaglio (2025)
			Sonolysis	➢ PFOS ➢ PFOA	➢ 28%–95% ➢ 63%–72%	
3	Synthetic wastewater	Adsorption	Granular activated carbon (GAC)	➢ Carbamazepine ➢ Phenacetin ➢ Pentoxifylline	➢ 97.66 mg/ g ➢ 97.91 mg/ g ➢ 97.75 mg/ g	Lee et al. (2023)
4	WTP ROC	IEX	Purolite® A502PS	➢ Diclofenac ➢ Triclosan ➢ Carbamazepine	➢ 96% ➢ 87% ➢ 98%	Devaisy et al. (2023a)Devaisy, Kandasamy, Aryal, et al. (2023)
5	Synthetic wastewater	AOPs	Ozonation	➢ Ciprofloxacin	➢ 64.3%	Lu et al. (2020)
6	Synthetic wastewater	AOPs (photo-Fenton process)	$Cu_{0.5}Mn_{0.5}Fe_2O_4$	➢ Bisphenol-A	➢ 95%	Yang et al. (2020)
7	Synthetic wastewater	AOPs (Sono-catalysis)	FeCeOx	➢ Diclofenac	➢ 80%	Chong et al. (2017)
8	Synthetic wastewater	AOPs (sono-catalysis)	Fe-Cu layered double hydroxide biochar nanocomposites	➢ Cefazolin	➢ 97.6%	Gholami et al. (2020a)

9	Synthetic wastewater	AOPs (sono-photocatalytic degradation)	Zn-Cu-Mg mixed metal hydroxide/g-C_3N_4 composite	➢ Sulfadiazine	➢ 93%	Gholami et al. (2020b)
10	WWTP effluent	AOPs (photocatalysis)	UV/H_2O_2	➢ 4-Nonylphenols ➢ Bisphenols	➢ 90% ➢ 90%	Rott et al. (2018)
11	Synthetic wastewater	Membrane filtration	MgO/diatomite ceramic membrane	➢ Tetracycline	➢ 99.7%	Meng et al. (2016)
12	Synthetic wastewater	AOPs	Photocatalysis (TiO_2 (Degussa P25)) supported bed photoreactor	➢ Paracetamol	➢ 99%	Borges et al. (2015)

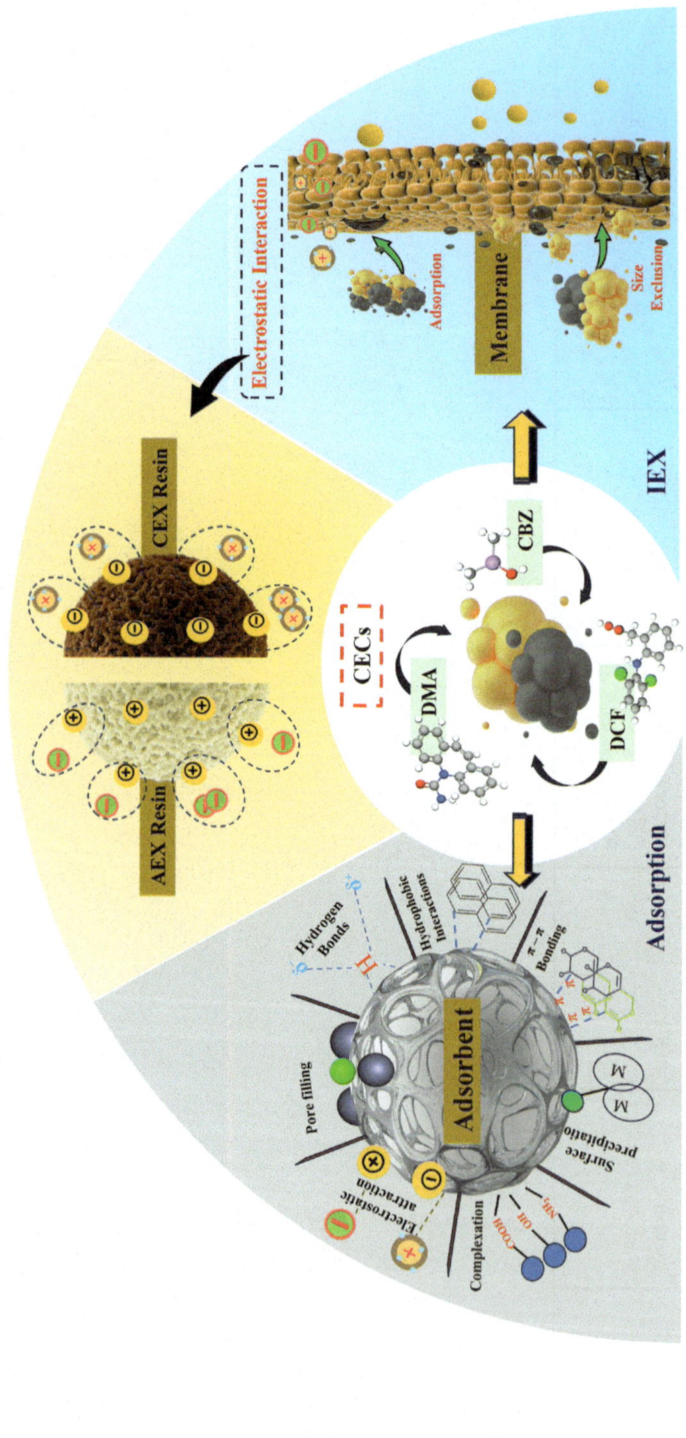

FIGURE 8.2

Schematic illustration of the mechanisms by which adsorbent and IEX materials remove CECs.

substance at the interface between two contiguous phases (Pérez-Botella et al., 2022; Sadegh et al., 2017). Adsorption works by binding the contaminants onto the solid surfaces through physical and chemical interactions. The adsorption mechanism involves van der Waals forces, pore filling, hydrophobic partitioning, IEX, complexation, precipitation, electrostatic attraction, and π–π stacking (Ambaye et al., 2021; Raji et al., 2023). This process can be described by isotherms (e.g., Langmuir and Freundlich) and influenced by adsorbent surface area, pore structure, and chemistry. The higher the surface area and porosity, the higher the adsorption. For instance, the maximum capacity for methylene blue (MB) adsorption of 1000 mg/g was achieved by the activated carbon (AC) with a specific surface area of 1773 m^2/g and mesopore volume of 0.474 cm^3/g, which was 44.1% of the total pore volume and the result showed Redlich-Peterson model provided the overall best fitting isotherm (Lawtae & Tangsathitkulchai, 2021). Apart from AC, there are various materials used for adsorption, especially in CECs.

AC is a processed carbonaceous, highly porous form of carbon with an exceptionally large internal surface area, created through physical or chemical activation of carbon-rich materials (Blachnio et al., 2020). Its high surface area and hydrophobic graphitic surfaces make it effective for nonpolar or moderately polar organics. A study evaluating 83 sets of micropollutants (pharmaceuticals, PFASs, personal care products, artificial sweeteners, parabens, pesticides, stimulants) using various sorbents, including granular activated carbon (GAC) and GAC + Polonite. Results showed GAC and GAC + P provided the higher removal rates, averaging values above 97% (Rostvall et al., 2018). Biochar is a carbonaceous material produced through the pyrolysis of biomass with limited oxygen. It is similar to the AC but cheaper and less processed. Many studies depend on feedstock, and pyrolysis biochar can adsorb CECs. Biochar derived from slow pyrolysis (300, 500, and 800°C) of waste fiberboard showed the highest aromatization degree and surface area with abundant pyridinic N (N-6) at 800°C, with 68.6% removal efficiencies of tetracycline (D. Xu et al., 2020). Nanomaterials such as carbon nanotubes (CNTs), graphene oxide (GO), etc., have also been used in the CECs adsorption process due to their extremely high surface areas and tunable chemistry. A study used graphene oxide/single-walled carbon nanotube (GO–SWCNT) composite membranes for removing three nonsteroidal anti-inflammatory drug (NSAIDs) (diclofenac, ketoprofen, naproxen), where the highest capacities were achieved at 75 wt.% graphene oxide, specifically 118 $\pm$ 2 mg/ g for diclofenac, 116 $\pm$ 2 mg/ g for ketoprofen, and 126 $\pm$ 3 mg/ g for naproxen at pH 4 (Anegbe et al., 2024). On the other hand, CNTs have strong adsorption affinity for CECs due to their π–π interactions, van der Waals forces, hydrophobic interactions, and sometimes charge-assisted hydrogen bonding. Sulfide-doped magnetic carbon nanotubes show maximum adsorption capacity (Langmuir model) of 116.27 mg/ g for tetracycline and 105.26 mg/ g for cefixime (Sereshti et al., 2022).

Adsorbents are commonly evaluated based on their equilibrium adsorption capacity (q_{max}, in mg of contaminant per g of adsorbent) and adsorption kinetics, which together determine their efficiency and practical applicability in contaminant removal. A study reported a maximum adsorption capacity of 3871 μg/g for AC in removing CECs particularly sulfamethoxazole, methamphetamine, trimethoprim, and caffeine were 100% removed, while carbamazepine, benzotriazole, atrazine, methaqualone, tramadol, cetirizine, efavirenz, 1,7 dimethylxanthine, benzoylecgonine, and venlafaxine were greatly reduced 94%–99% and adsorption process typically follows pseudo-second-order kinetics, indicating chemisorption (Dikobe et al., 2024). A study on cerium oxide (CeO_2) nanoparticles (NPs) reported an increase in adsorption capacity from 0.62 to 5.05 mg/g and removal efficiency from 35% to 54% over 240 min (Ramirez Arenas et al., 2023).

Overall, adsorption provides highly effective removal of many organic CECs at low concentrations using relatively simple equipment such as columns or contact tanks. Its adaptability through the selection of appropriate adsorbents and comparatively low energy requirements makes it an attractive treatment option. However, adsorption only transfers pollutants from the aqueous phase to a solid phase and does not achieve complete destruction. Therefore, it must be coupled with safe disposal or regeneration strategies within a closed-loop system.

4.2 IEX technologies

IEX is a reversible stoichiometric exchange with similarly charged ions, which is bound or immobilized on the surface of the IEX materials (solid phase) through electrostatic interactions, without altering the solid structure. This exchange occurs without altering the physical structure of the solid medium, thereby enabling efficient pollutant removal and multiple regeneration cycles (Bustillos et al., 2020; Z. Gao et al., 2025). IEX is a highly effective and widely adopted technology for the removal of charged or ionizable CECs from water due to its high selectivity, operational simplicity, cost-effectiveness, and ease of regeneration (Oladimeji et al., 2024; Prakash et al., 2022; Somya & Varshney, 2025). The major challenging classes of CECs, such as PFAS, pharmaceuticals and personal care products (PPCPs), EDCs, and pesticides etc., are effectively addressed by IEX due to its ability to target persistent, charged, and ionizable pollutants even at trace concentrations (Dixit et al., 2019; 2021; Wang et al., 2016). For instance, a hybrid IEX-microfiltration system using Purolite A502PS resin achieved 68%–99% removal of various CECs, including 99% diclofenac, 96% ibuprofen, and 91% benzotriazole from ROC, confirming its high efficiency for persistent, charged, and ionizable pollutants (Devaisy et al., 2023a; Devaisy, Kandasamy, Aryal, et al., 2023).

IEX uses insoluble cross-linked polymer resins bearing fixed ionic (acidic or basic) functional groups that are covalently attached to the polymer matrix. These fixed sites carry counter-ions which are exchanged, on a stoichiometric basis, with target ions from the solution via electrostatic interactions and ion-selectivity mechanisms (Badia et al., 2023; Gupta et al., 2010). In electrostatic IEX, cation-exchange resins, containing sulfonic groups or carboxylic groups, exchange positively charged contaminants (e.g., metal ions, ammonium) for H^+ or Na^+ ions, while anion-exchange resins, bearing quaternary ammonium functional groups, target negatively charged species such as nitrate, perchlorate, and various organic acids (Faraz Ahmer & Kashif Uddin, 2024).

IEX resins, therefore, serve as the primary solid phase in IEX processes and are classified based on the strength and type of their functional groups. The main categories include (A) strong acid cation (SAC) resins—fully dissociated over a wide pH range, and typically used for water softening and demineralization; (B) weak acid cation (WAC) resins—exhibit exchange capacity dependent on pH and effective for hardness and alkalinity removal; (C) strong base anion (SBA) resins—capable of exchanging all anions including weak acids, used for demineralization and organic acid removal; (D) weak base anion (WBA) resins—effective in removing strong acids like chloride and sulfate. IEX ions are often selective, for example, SBA resin (quaternary amine) is versatile for PFAS, some pesticides, and deprotonated acids, while WBA resins (tertiary amine) target specific anions like phenols. Mixed-bed and chelating resins are also available for multi-ion removal. The study targeting 11 PFASs, including PFOA, PFOS, PFHxA, PFHpA, PFNA, PFDA, PFUnDA, PFDoDA, PFBS, PFHxS, and PFHpS for removal, demonstrated that WBA exchange resins (WB-AER) achieved >99% removal of long-chain PFASs (PFOA, PFOS) and 85%–95% for short-chain PFASs, with adsorption capacities up to 0.72 mmol/ g (Kassar et al., 2022). Another study using 2 SB-AER, named Purolite A-500P (polystyrenic) and A-860 (polyacrylic), both with quaternary ammonium groups (0.8 eq/L), showed >93% PFHpA, PFOA, PFNA removal by A-500P, while A-860 removed <15% in natural water due to sulfate competition. Both resins removed about 75% dissolved organic carbon (DOC), with A-860 being more efficient for humic substances (Rahman et al., 2022). An experiment testing anion-exchange resins (5 polystyrene-divinylbenzene (PS-DVB): A592E, PFA694E, DOWEX1, TAN1, IRA910; two polymethacrylate: MIEX GOLD, DOC; two polyacrylic: IRA67, IRA958) at 5 mL/L dose shows PS-DVB resins achieved >90% removal of 35 out of 40 PFAS (including PFCA, PFSA, PFEA, FTSA, FASA, and zwitterionic species) within 24 h, while polyacrylic resins showed minimal or negative removal (Liu & Sun, 2021). Apart from the selectivity of the IEX resins for targeted contaminants, this also showcases that IEX can achieve very high removal of specific CECs if the resin chemistry matches the contaminant.

A key advantage of IEX technology is the regenerability of resins, which enables long-term use and cost efficiency. During regeneration, the exhausted resin is restored to its original ionic form using a concentrated solution of the regenerant (e.g., NaCl, HCl, or NaOH), which displaces the adsorbed ions. In other words, regeneration is done with brine. The recent studies have shown the 100% of regeneration efficiencies for polystyrene resins and polyacrylic resins with various concentrated regenerants, including the brine solution. Liu & Sun (2021) reported that 10% NaCl/NH_4Cl brine recovered up to 94% short-chain PFAS, but long-chain PFAS and PFAS-specific resins (A592E, PFA694E) were nonregenerable. This cycle allows for the resin's reuse over multiple operational phases. Regeneration efficiency improved with higher salt concentration, especially for short-chain PFAS, which are more weakly bound. However, they noted that very little of long-chain PFAS was removed by simple brines unless methanol was added, pointing to a limitation in the regenerability of some compounds. In general, strong-base resins for anions are regenerated with NaCl (or NaOH), and strong-acid cation resins with HCl. Regeneration thus produces a small volume of concentrated brine laden with the desorbed contaminants, which itself must be handled, and resin performance can degrade over time due to fouling, oxidation, or mechanical breakdown, leading to reduced exchange capacity. Moreover, regeneration produces a very salty regenerant that must be treated. The regeneration efficiency of IEX was summarized in Table 8.3. While the recyclable nature and ion-selectivity of these resins have an advantage, there are certain disadvantages also which needed to be overlooked. IEX is ineffective for neutral organics, as these molecules do not possess the ionic charge needed for electrostatic attraction with the resin's functional groups, which leads to the fouling or pore blockage as the neutral organics pass through the resin matrix without removal (Devaisy et al., 2023b).

In recent years, hybrid IEX systems have emerged as advanced treatment technologies for CECs. These systems combine the high selectivity of IEX with complementary processes such as microfiltration, adsorption, or electrodialysis to enhance removal performance and reduce secondary waste generation. For example, in desalination plants, integrated membrane electrodialysis systems have achieved effective rejection of trace organic contaminants (TrOCs), a key subset of CECs (Nghiem & Fujioka, 2016). Similarly, Devaisy et al. (2023a; 2023b) demonstrated that coupling Purolite A502PS IEX resin with microfiltration significantly improved the removal of emerging micropollutants, including pharmaceuticals, personal care products, and pesticides from ROC (Devaisy et al., 2023a). Such hybrid configurations exemplify the adaptability and scalability of IEX technology in addressing complex contamination scenarios involving mixtures of ions, nutrients, and organic micropollutants. By integrating physical or electrochemical processes with IEX, hybrid systems achieve enhanced

Table 8.3 Comparison of different IEX's regeneration efficiency along with the solution used for regeneration.

Sl. No.	CECs type	Materials used	Solution	Regeneration efficiency (%)	References
1	Bisphenol-A (BPA)	Purolite®	Ozone gas (3 mL/min)	70	Alves et al. (2025)
2	Ciprofloxacin	Polystyrene-based	Fenton-like oxidation	65.3	Song et al. (2023)
3	GenX	IRA67	0.5% NaCl	95	W. Wang et al. (2019)
4	PFBS	A860	20% NaCl	96	Dixit et al. (2020)
5	PFHxS	Sorbix A3F	Brine + organic solvent	100	Woodard et al. (2017)
6	PFHpS	Sorbix A3F	Brine + organic solvent	100	Woodard et al. (2017)
7	8:2 FS	Sorbix A3F	Brine + organic solvent	100	Woodard et al. (2017)
8	PFOS	IRA67	1% NaCl+ 70% MeOH	97	Y. Gao et al. (2017)
9	F–53B	IRA67	1% NaCl+ 90% MeOH	98	Y. Gao et al. (2017)
10	Diclofenac	Dowex 22	5% NaCl	100	Landry et al. (2015)
11	Ibuprofen	Dowex 22	5% NaCl	64	Landry et al. (2015)
12	Ketoprofen	Dowex 22	5% NaCl	74	Landry et al. (2015)
13	Naproxen	Dowex 22	5% NaCl	100	Landry et al. (2015)

contaminant removal efficiencies, operational sustainability, and reduced waste production, positioning IEX as a cornerstone in next-generation water treatment solutions.

In conclusion, adsorption and IEX represent a powerful yet complementary approach for the ROC remediation and CECs removal. While adsorption provides a broad-spectrum treatment, effectively targeting a wide range of organic CECs, IEX offers high selectivity toward charged and ionizable species. Effective integration of both processes within ZLD/MLD frameworks requires careful design considerations and proper management of the secondary waste streams they generate. To better understand their relative strengths and practical applicability, the following section presents a comparative evaluation of adsorption and IEX processes based on performance efficiency, selectivity, regeneration potential, scalability, and environmental compatibility.

5. Comparative evaluation of promising technologies

The management of CECs in ROC requires the implementation of sustainable closed-loop systems that minimize waste discharge, resource recovery, and advanced treatment technologies. A comparative evaluation of the closed-loop approaches is essential to assess their technical feasibility, operational stability, and long-term sustainability.

5.1 Efficiency, selectivity, and operational parameters in removing CECs

Among closed-loop treatment strategies, adsorption-based methods and IEX technologies have emerged as the most practical and widely studied approaches for the removal of CECs from ROC. These two processes form the core of advanced ROC treatment, with performance governed by water chemistry, adsorbent or resin characteristics, and regeneration efficiency. The comparative evaluation of their removal efficiencies, key operating parameters, and application potential is presented in Table 8.4.

5.2 Resource recovery

Resource recovery from ROC has gained attention as a sustainable approach to mitigate secondary pollution and reclaim valuable materials from concentrated waste streams. Both adsorption and IEX technologies have shown potential for the removal and subsequent recovery of CECs, such as pharmaceuticals, personal care products, and heavy metals. Study shows polymeric ligand exchanger resin removed and then recovered >85% of the adsorbed phosphorus by precipitating struvite ($MgNH_4PO_4{\cdot}6H_2O$), which can be used as a slow-release fertilizer (Kumar et al., 2007). A study using self-regenerating anion-exchange resin stage ahead of RO in brackish desalination achieved 80% recovery for a brackish water composition after anion exchange combined with RO, demonstrating recovery rather than disposal (Smith & SenGupta, 2015).

In adsorption systems, recovery largely depends on the reversibility of adsorbate–adsorbent interactions. Regeneration using mild eluents or pH adjustment can release bound compounds, enabling partial recovery and reuse of both adsorbent and contaminants. Materials like biochar, AC, and metal–organic frameworks (MOFs) can sometimes be regenerated multiple times with minimal capacity loss, but the recovery of CECs themselves often remains limited due to their complex, nonionic nature and the heterogeneity of ROC matrices. Experimental studies found very high regeneration efficiency (~99–100%) for four cycles of iron-modified biochar through thermal regeneration for a toxin (microcystin-LR) in lake water, highlighting the reuse potential of the adsorbents (Zeng & Kan, 2021). A review highlights that AC derived from agricultural and household biomass wastes not only effectively removes various contaminants but also possesses strong resource recovery potential due to its regenerability and reusability. The study emphasizes that spent AC can be recovered through thermal, wet air oxidation, chemical, or solvent regeneration methods to desorb adsorbed pollutants and restore porosity with minimal structural damage. Among these, thermal regeneration (200–1000 °C) enables desorption of mixed adsorbates but is energy-intensive, while chemical regeneration can be performed in situ, avoiding carbon loss and allowing the recovery of adsorbates via distillation. Solvent-based regeneration, including the use of ethanol or supercritical CO_2, has

Table 8.4 Comparative evaluation of the removal efficiencies of CECs using different adsorptive and IEX materials.

Sl. No.	Approach	Technology type	Nature of the material	Functional group	Dosage (g/L)	Studied water type	Emerging contaminant types	Removal efficiency (%) Or q_{max} (mg/g)	References
1	IEX	Micro-filtration–IEX resin hybrid system (Purolite® A502PS)	Strong base anion exchange resin	Polystyrenic macroporous quaternary/ammonium with chloride ion	20	WTP ROC	Various emerging contaminants including PPCPs, pesticides, etc.	• Diclofenac 96% • Triclosan 87% • Carbamazepine 98%	Devaisy et al. (2023a)
2		Combined nanofiltration (NF) with anion exchange (AIX) resins.; AIX: Purolite A600 (type I strong-base resin, quaternary ammonium functional group, polystyrene-divinylbenzene gel-type) and purofine PFA694 (complex amino functional group, polystyrene-divinylbenzene).	Anion-exchange resin	Quaternary ammonium ($-N^+(CH_3)_3$) and complex amino groups facilitate electrostatic attraction of PFAS anions.	10	Nanofiltration (NF) concentrate from Bäcklösa drinking water treatment plant, Uppsala, Sweden.	37 PFASs analyzed, including: PFBA, PFPeA, PFHxA, PFHpA, PFOA, PFNA, PFDA (PFCAs), and PFBS, PFPeS, PFHxS, PFHpS, PFOS.	>99% PFOS and PFOA removal observed initially; PFHxA and PFBS broke through faster.	Franke et al. (2021)
3		GAC—Purolite A502PS column study	Strong base anion exchange resin	Polystyrene cross-linked with divinylbenzene quaternary/ammonium with chloride ion	777	Water reclamation plant ROC	Various CECs, including PPCPs, pesticides, etc. • Atrazine • Benzotriazole • Carbamazepine • Diclofenac • Diuron • Gemfibrozil • Ibuprofen • Ketoprofen • Naproxen • Paracetamol • Saccharin • Simazine • Triclocarban • Trimethoprim	>99%	Jamil et al. (2020)

4		Polymeric hydrochar from Purolite	Strong acid cation exchange resin.	Styrene–divinylbenzene based, sulfonated form	3.0	synthetic	• Bisphenol-A • Diclofenac	• BPA—12.87 mg/g • DCF—8.21 mg/g	Alves et al. (2025)
5		MIEX® GOLD resin	Anionic exchange resins	Macroporous methacrylic with quaternary ammonium with chloride ion	115	Synthetic wastewater	• PFBA • PFBS, • PFHxA, • PFHxS • PFOA • PFOS	>99%	Tamanna et al. (2023)
6		Fluorophilic ion-exchange material along with SPE cartridge	Anion-exchange type,	F-polymer with perfluorohexyl-ethyl-tethered polyvinylimidazolium chloride crosslinked by EDMA	1	Effluent wastewater	• PFOA • GenX	• 504 mg/ g • 84.3 mg/ g	Freilinger et al. (2025)
7		Modified catalyst study (CuFeO2@CR composite from the waste D001 resin)	Strong acid cation exchange (SAC) resin	Sulfonic acid group	0.3	Simulated wastewater	Carbamazepine	91.3%	Q. Li et al. (2025)
8		Cationic porphyrin-based organic Nanocages (Hybrid IEX with strong acid cations)	$[oNC]_8PF_6$ $[Co^{2+}\text{-}oNC]_8PF_6$	Hexafluorophosphate anion.		Simulated wastewater	• PFMPA • PFEESA • PFBS • PFPeA • PFMBA • NFDHA • PFPeS • 3:3 FTCA • GenX • 4:2 FTS • PFHxA • PFHxS	• 67 ± 3% • 70 ± 5% • 90 ± 2% • 95 ± 0.3% • 82 ± 3% • 82 ± 4% • 86 ± 5% • 94 ± 4% • 78 ± 2% • 90 ± 4% • 92.5 ± 6% • 94 ± 1% • 98.5 ± 1% • 73 ± 4% • 82 ± 6% • 97.5 ± 1%	Sanchez-Lievanos et al. (2025)

Continued

Table 8.4 Comparative evaluation of the removal efficiencies of CECs using different adsorptive and IEX materials.—cont'd

Sl. No.	Approach	Technology type	Nature of the material	Functional group	Dosage (g/L)	Studied water type	Emerging contaminant types	Removal efficiency (%) Or q_{max} (mg/g)	References
			$[Co^{3+}(N \equiv O)\text{-}oNC]_8PF_6$					• 96 ± 1% • 84 ± 0.5% • 93 ± 3% • 97 ± 1% • 96 ± 1% • 88 ± 2% • 88 ± 7% • 98 ± 1% • 94 ± 1.5% • 99 ± 0.1% • 84 ± 1% • 89 ± 2% • 98 ± 0.5% • 99 ± 1% • 92 ± 2% • 92 ± 3% • 97 ± 1% • 99 ± 0.5% • 83 ± 2% • 98 ± 2% • 99 ± 1% • 95 ± 1% • 99 ± 0.5%	
9		Linker bifunctional anion exchange covalent organic frameworks	Cationic, bifunctional anion-exchange COF	Quaternary ammonium ($-NR_3^+$)		Simulated wastewater	ReO_4^-	1593.21 mg/g	Yuan et al. (2025)
10		MIEX® DOC	Strong-base anion	Quaternary ammonium ($-NR_3^+$),	44	River water	• Methylparaben • Ethylparaben • Propylparaben • Butylparaben • Bisphenol A	• 51%, • 61% • 68% • 80% • 66%	López-Ortiz et al. (2018)

11		Commercial polystyrene-divinylbenzene (PS-DVB)–based anion exchange resins with varying porosity and functional groups	Strong-base type I (quaternary ammonium), strong-base type II (dimethylethanolamine), and weak-base (tertiary amine).	Type I: Quaternary ammonium ($-N^+(CH_3)_3Cl^-$), type II: Quaternary ammonium with hydroxyethyl substituent ($-N^+(CH_3)_2CH_2CH_2OHCl^-$), and weak base: Tertiary amine ($-N(CH_3)_2$).	1.0	Synthetic aqueous PFAS solutions prepared using ultrapure water	Per- and polyfluoroalkyl substances (PFASs): Perfluorooctanoic acid (PFOA) Perfluorooctane sulfonate (PFOS) Perfluorobutane sulfonate (PFBS) Perfluorohexanoic acid (PFHxA) Perfluorononanoic acid (PFNA)	Langmuir qmax: up to 289.86 mg/g for PFOS and 235.42 mg/g for PFOA (strong-base resin).; removal efficiency: >99% for long-chain PFASs (PFOS, PFOA) and 75%–90% for short-chain PFASs (PFBS, PFHxA).	Riegel et al. (2023)
12		IRA910	Strong base anion exchange	Quaternary ammonium ($-NR_3^+$)	0.1	AFFF impacted groundwater	PFHxS	1400.175 mg/g	Maimaiti et al. (2018)
13	Adsorption	Batch adsorption	Biochar from pinewood (PW)	Mainly carboxylic and hydroxyl functional groups	2	Deionized water (miliq)	PFASs: PFHxA, GenX, PFOA, PFOS, PFDA, PFTeDA	Long-chain PFASs (PFOS, PFDA, PFTeDA) showed highest sorption; PFOS > PFTeDA > PFDA > PFOA > PFHxA > GenX; qe (qmax) ≈ 0.2–0.6 mg/g	Petrangeli Papini et al. (2024)
14		Batch adsorption	Biochar from date seeds (DSs)	Mainly carboxylic and hydroxyl functional groups	2	Deionized water (miliq)	PFASs: PFHxA, GenX, PFOA, PFOS, PFDA, PFTeDA	Adsorption higher for long-chain PFASs (PFDA > PFTeDA > PFOS > PFOA); short-chain (PFHxA, GenX) least adsorbed; qe (qmax) ≈ 0.15–0.5 mg/ g	Petrangeli Papini et al. (2024)
15		Coupling granular activated carbon (GAC) adsorption with ozone (GAC/O_3 system) for removal and in-situ regeneration.	Coconut shell granular activated carbon (CGAC)—20–30 mesh; high surface area (822.47 → 852.49 m^2/g after ozone regeneration), mesoporous–microporous structure.	Surface oxygen-containing groups such as phenolic hydroxyl (–OH) and carboxylic (–COOH) groups activate ozone to form •OH radicals for degradation. Also includes C=O, C–O, and aromatic C=C	10	Reverse osmosis concentrate (ROC) from Yangzhou Chemical Industry Park (China)	Complex organic matter in ROC, including humic acids and soluble microbial by-products, represents refractory and aromatic organic pollutants.	Static adsorption (CGAC): 78.7% removal of organics (DOC); Langmuir qmax: 26.207 mg/g; dynamic adsorption: DOC reduced from 44.18 mg/L to 16.09 mg/L	Y. Chen et al. (2025)

Continued

Table 8.4 Comparative evaluation of the removal efficiencies of CECs using different adsorptive and IEX materials.—cont'd

Sl. No.	Approach	Technology type	Nature of the material	Functional group	Dosage (g/L)	Studied water type	Emerging contaminant types	Removal efficiency (%) Or q_{max} (mg/g)	References
				groups contributing to π–π interactions.				(≈63.6% removal) over five cycles; ozone consumption reduced 9.4× (4.19 mg O_3/mg DOC) compared to ozone alone.	
16		Combined nanofiltration (NF) with granular activated carbon (GAC)	Filtrasorb 400 (bituminous coal-based, 1050 $m^{2/}g$, iodine number 1000 mg/g) and Norit 1240 W (coal-based, 1150 m^2/g, iodine number 1020 mg/g).	Surface oxygen groups (phenolic –OH, C=O) enabling hydrophobic and π–π interactions.	500 mL of adsorbent per column; flow rate 100 mL/min; equivalent to 10 g/L	Nanofiltration (NF) concentrate from bäcklösa drinking water treatment plant, Uppsala, Sweden.	37 PFASs analyzed, including: PFBA, PFPeA, PFHxA, PFHpA, PFOA, PFNA, PFDA (PFCAs), and PFBS, PFPeS, PFHxS, PFHpS, PFOS.	Nanofiltration: >98% PFAS removal from raw water; GAC (filtrasorb 400, norit 1240 W): PFAS loading 8.4–18 μg/ g; DOC removal 38–33 mg/g; overall: >99% PFOS and PFOA removal observed initially; PFHxA and PFBS broke through faster.	Franke et al. (2021)
17		Granular activated carbon (GAC) adsorption system	Bituminous coal–based granular activated carbon (Calgon Filtrasorb 400)—widely used commercial GAC; average particle size 0.55 mm, surface area ~1050 m^2/g.	Surface oxygen-containing functional groups (–OH, C=O, –COOH) promote hydrophobic interactions and electrostatic attraction with PFASs	10	Groundwater and surface water spiked with selected perfluoroalkyl acids (PFAAs)	Perfluoroalkyl acids (PFAAs): Perfluorooctanoic acid (PFOA), perfluorooctane sulfonate (PFOS), perfluorobutanoic acid (PFBA), perfluorohexanoic acid (PFHxA), perfluoroheptanoic acid (PFHpA), perfluorononanoic acid (PFNA)	Qmax (Langmuir isotherm): 2.3 mg/ g for PFOS and 1.8 mg/ g for PFOA; removal efficiency: >95% PFOS and PFOA removal observed at initial concentrations of 1 μg /L; short-chain PFAAs (PFBA, PFHxA) showed significantly lower adsorption (<40%).	Kempisty et al. (2022)

18		Activated carbon filtration (ACF) process operated in parallel with low-Pressure reverse osmosis (LPROs)	Granular activated carbon (GAC)—virgin norit row supra 0.8 carbon used as adsorbent.	Oxygen-containing surface groups on activated carbon surface (–OH, –COOH, C=O) responsible for hydrophobic, π–π, and electrostatic interactions	10–20	Rhine river bank filtrate employing a semitechnical plant used as drinking water source; spiked with 32 representative trace organic contaminants (TrOCs).	32 TrOCs divided into classes: • Polyfluorinated aliphatic compounds: PFBA, PFBS, PFHxA, PFHxS, PFOA, PFOS • Pharmaceuticals and pesticides: Metformin, isoproturon, terbutaline, sulfamethoxazole, carbamazepine, caffeine • Volatile organic compounds (VOCs): Dichloromethane (DCM), 1,2-dichloroethane (DCE), trichloromethane (TCM), trichlorethylene (TRI), benzene, methyl tert-butyl ether (MTBE), ethyl tert-butyl ether (ETBE) • Nitrosamines: N-nitrosodimethylamine (NDMA), N-nitrosomorpholine (NMOR) • Aminopolycarboxylic acids: EDTA, diatrizoic acid (ATZ) • others: 1H-benzotriazole (BTA), dimethylsulfamide (DMS), acesulfame.	Removal efficiency: • High removal (>99%) for many hydrophobic TrOCs-benzene, MTBE, ETBE, acesulfame, PFOS, PFHxS, 1H-BTA. • Moderate to high removal (63%–97%) for metformin, NDMA, DCE, DCM, TCM, TRI. • Lower removal for highly polar compounds (DMS, EDTA, PFBA) showing breakthrough beyond 8.1 m^3/kg. • Overall range: 50–>99% removal efficiency, depending on compound polarity (LOG K_ow −1.2 to 5.0).	Konradt et al. (2021)
19		Batch adsorption process using coking coal	Coking coal (1/3 coking coal) obtained from Linhuang Coking Co., Anhui, China. BET surface area: 2.794 m^2/g; average pore size: 11.189 nm; mainly mesoporous and macroporous structure.	Surface functional groups identified by FTIR and XPS include: –OH, –COOH, C=C, C–O–C, and C–N.	80	ROC from a coal chemical wastewater treatment plant (two-stage RO brine).	Complex organics present in ROC, mainly: • Chlorinated hydrocarbons • Heterocyclic organics • Nitrogen- and oxygen-containing species • Humic macromolecular aromatics (refractory organic compounds).	TOC removal = 70.18% UV254 = 70.15% Chromaticity = 59.55% Langmuir model fit with R^2 = 0.9997; qmax (TOC) = 24.527 mg/g; qmax (UV254) = 1.067 Abs/g.	H. Xu et al. (2019)

Continued

Table 8.4 Comparative evaluation of the removal efficiencies of CECs using different adsorptive and IEX materials.—cont'd

Sl. No.	Approach	Technology type	Nature of the material	Functional group	Dosage (g/L)	Studied water type	Emerging contaminant types	Removal efficiency (%) Or q_{max} (mg/g)	References
20		Granular activated carbon (GAC) adsorption—batch and column studies	Coal-based premium grade GAC (MDW4050CB) with particle size 0.42–1.68 mm, iodine number 1000 mg/g, BET surface area 1000 m^2/g, micropore volume 25%–27%, mesopore volume 73%–75%.	Hydrophobic carbon surface with anionic character at neutral pH; interactions via hydrophobic bonding, H-bonding, π–π interactions, and anion exchange	0.1, 0.5, and 3 g/L in batch experiments	ROC from an advanced water reclamation plant (treating stormwater and biologically treated sewage effluent) in Sydney, Australia	18 micro-organic pollutants (MOPs): Pharmaceuticals, personal care products, endocrine disruptors, pesticides, and industrial by-products, including atenolol, benzophenone, benzotriazole, bisphenol A, caffeine, carbamazepine, diclofenac, diuron, gemfibrozil, ibuprofen, naproxen, Paracetamol, saccharin, sulfamethoxazole, triclocarban, triclosan, trimethoprim, Verapamil	Batch adsorption: DOC removal 40%–60%; individual MOPs 73%–100%. Column adsorption: MOP removal efficiencies grouped as: Group A (positive/ neutral, log kow >3.5): 95%–100% Group B (positive/ neutral, log kow <3.5): 90%–99% Groups C and D (negatively charged): 73%–94%. Amounts adsorbed: DOC fractions 0.01–3 mg/g; MOPs 0.01–2.5 mg/g; partition coefficients for MOPs 0.21–21.6 L/g	Jamil et al. (2019)
21		Batch adsorption using granular activated carbon (GAC)	Commercial Norit1240 granular activated carbon (GAC), derived from bituminous coal; BET surface area 930.5 m^2/g, micropore area 676 m^2/g, total pore volume 0.571 cm^3/g, micropore volume 0.311 cm^3/g, particle size 0.4–1.7 mm, pHpzc = 10.0	Carbon surface with predominantly hydrophobic and π–π interaction sites; adsorption involves physical–chemical interactions, including π–π bonding, hydrophobic forces, and weak electrostatic interactions	0.01–2.0 g in 50 mL solution (equivalent to 0.2–40 g/L)	Synthetic aqueous solutions of nine emerging contaminants prepared in deionized water at 100 mg/L initial concentration; pH adjusted to 7.0	Pharmaceuticals and pesticides: Carbamazepine (CBZ), Phenacetin (PHE), Pentoxifylline (PEN), norfloxacin (NOR), iprobenfos (IPB), isoprothiolane (ISP), metolachlor (MET), tebuconazole (TEB), hexaconazole (HEX)	CBZ = 97.66 mg/g, PHE = 97.91 mg/g, PEN = 97.75 mg/g, NOR = 77.32 mg/g, IPB = 80.03 mg/g, ISP = 99.29 mg/g, MET = 69.77 mg/g, TEB = 99.07 mg/g, HEX = 75.37 mg/g	Lee et al. (2023)

22		Activated carbon adsorption and advanced AOPs tested individually and in combination, including: • UVC/TiO_2 (photocatalysis) • O_3 oxidation • ultrasound (US) • H_2O_2 oxidation • coupled systems: UVA/TiO_2/O_3, US/h_2O_2/O_3, UVA/H_2O_2/O_3, etc. • pretreatments: $FeCl_3$ coagulation and activated carbon (AC) adsorption.	Activated carbon (AC): Granular (GAC, Darco® 8–12 mesh) and powdered (PAC, Darco® G-60, −100 mesh, Sigma–Aldrich).; Photocatalyst: Degussa P-25 TiO_2 (anatase:rutile = 75:25; surface area 50–60 m^2/g; particle size 21 nm).	AC surface groups: Oxygenated (–OH, –COOH), responsible for adsorption.	0.1–5.0	Reverse osmosis (ROs) concentrate obtained from a municipal wastewater reclamation plant (Singapore), treating secondary effluent from a municipal wastewater treatment plant.	The RO concentrate contained refractory organics such as: • natural organic matter (NOM) • refractory micropollutants: Pharmaceuticals, personal care products, endocrine disruptors, and pesticides • soluble microbial products from biological treatment (high molecular weight hydrophilic fractions).	GAC: Up to 88% DOC removal at 5 g/L dosage. PAC: Up to 95% DOC removal at 5 g/L dosage. AOPs (without pretreatment): Up to 52% DOC removal (UVA/TiO_2/O_3). AOPs + coagulation pretreatment: Up to 68% DOC removal after 1 h; extended 95% DOC removal after 6 h (UVC/TiO_2 + $FeCl_3$).	Zhou et al. (2011)
23	Combined adsorption and ion exchange	Dynamic fixed-bed column system: ROC passed through a granular activated carbon (GAC) column, followed by anion exchange resin (Purolite A502PS) column.	GAC: Coal-based; particle size 0.42–1.68 mm; BET surface area 1000 m^2/g; pore volume 0.69 cm^3/g; average pore diameter 2.7 nm. Purolite A502PS resin: Polystyrene matrix cross-linked with divinylbenzene; BET surface area 21 m^2/g; particle size 0.43–0.60 mm.	Purolite: Quaternary ammonium ions ($–N^+(CH_3)_3Cl^−$) GAC: Contains carboxylic and phenolic surface groups; hydrophobic, negatively charged surface enabling π–π bonding, hydrogen bonding, and van der Waals interactions with organic compounds.	80 g GAC and 122 g Purolite in 2-cm diameter × 50 cm bed height columns; upward flow at 40 mL/min (6.31 m/h)	ROC from the Sydney Olympic Park water reclamation plant, treating stormwater and biologically treated sewage effluent. ROC characteristics: DOC = 45.8 mg/L; hydrophilic DOC = 72% of total; pH = 7.78; EC = 2.68 mS/cm.	17 micro-organic pollutants (MOPs) including: Atrazine, benzotriazole, caffeine, carbamazepine, diclofenac, diuron, gemfibrozil, ibuprofen, Ketoprofen, naproxen, Paracetamol, saccharin, simazine, Sulfamethoxazole, TCEP, triclocarban, trimethoprim.	MOPs: 95%–100% removal by GAC column after 2880 bed volumes. DOC fractions: Hydrophobic DOC 78%–90% at 606 BVs and 8%–40% at 2880 BVs (GAC alone); combined GAC + Purolite removed 90%–100% of DOC fractions (humics 100%). Partition coefficients (kp): DOC fractions 0.1–25 L/g (indicating strong sorption for LMW neutrals and humics).	Jamil et al. (2020)

demonstrated up to 92%–96% regeneration efficiency, making it a viable approach for maintaining adsorption capacity and reducing waste generation (Ani et al., 2020).

On the other hand, because IEX depends on reversible electrostatic interactions between ionic species and functional groups on resin surfaces, it provides more direct opportunities for selective recovery. By targeting particular ionic medications, ammonium, or heavy metal ions found in ROC, advanced IEX resins enable the regeneration of both resin and captured solutes via salt or acid–base elution. The study showed that Purolite S930 and S920 resins removed Ni, Cr, and Hg nearly 100%, reducing Ni from 89.3 mg/ L to <0.1 mg /L, and allowed regeneration via 10% HCl elution, producing concentrated eluates with Ni = 246 mg /L and Cr = 673 mg /L, confirming selective recovery through reversible IEX (Czupryński et al., 2022). Another study reported that short-chain PFAS are 10% recovered up to 94% using NaCl/NH_4Cl brine (Liu & Sun, 2021). Moreover, coupling IEX with downstream concentration or crystallization processes facilitates the recovery of nutrients or metals in usable form, improving circularity. The study showed that coupling IEX with struvite crystallization enabled >99% P and ~95% N recovery from wastewater at pH ≈ 9 (N:Mg:P = 1:1.1:1.1–1:1.3:1.3), producing a pure, reusable fertilizer that enhances circular resource use and nutrient recycling (Siciliano et al., 2020).

These studies collectively demonstrate that both adsorption and IEX not only enable efficient contaminant removal but also present viable routes for resource recovery from ROC through regeneration and valorization of captured compounds. While such closed-loop approaches advance circular economy objectives and minimize secondary waste, their real-world implementation depends strongly on the economic feasibility, scalability, and operational robustness of these systems under complex ROC conditions.

6. Cost, scalability, and operational considerations

While implementing closed-loop approaches in real-world ROC management, the two main obstacles are cost-effectiveness and scalability. In this context, adsorption and IEX are generally much cheaper and simpler compared with ZLD and AOP technologies, which require complex reactor designs and substantial energy input, even though they offer higher treatment efficiency and recovery rates. Moreover, both adsorption and IEX are eco-friendly and cost-effective processes (Vecino & Reig, 2022). The electro-Fenton process reported by Mousset et al. (2021) is an example, as it requires specialized equipment and specific chemical inputs, resulting in a treatment cost ranging between €108 and €125 per cubic meter, depending on the degree of mineralization achieved. For ZLD, costs reach USD 0.1 to one million per m^3/h which is 2–5 times more than the conventional treatment for ZLD with thermal units consuming 60%–70% of total equipment cost along with up to 60 kWh m^{-3} energy and overall capital expenditure (CAPEX) is 2–5 times higher, which recovered salts show 90%–95% purity. Resource recovery adds extra CAPEX despite limited economic return, making affordability a key barrier, especially in developing countries (Liang et al., 2021).

In contrast, adsorption and IEX technologies are generally more cost-effective at the pilot scale. Among them, those using AC, biochar, or advanced nanocomposites are easier to operate and can handle a wide range of contaminants. Their operating and maintenance costs depend on factors like how often the adsorbent needs to be regenerated, the chemicals required for oxidation, electrode durability, and the level of system automation. Likewise, IEX technologies that use selective resins are highly effective at removing ionic or charged pollutants, including PFAS, pharmaceuticals, and endocrine-disrupting compounds. However, the regeneration process produces concentrated brines, which then require additional treatment (Devaisy et al., 2023a). Adsorption and IEX can serve as effective pretreatment steps in MLD/ZLD systems because they polish the ROC and reduce membrane or evaporator fouling, making the downstream process easier to manage. An integration of the AOP step after adsorption will help to break down any remaining organic residues. While the membrane processes combined with IEX can even replace or at least reduce the need for a full AOP treatment train (Kim et al., 2022; Loganathan et al., 2023; Yan et al., 2023). In general, adsorption and IEX offer economic viability and simple implementation, whereas other advanced technologies provide higher treatment and recovery but at a much greater cost and complexity.

The main bottleneck in application and implementation lies in scalability, particularly for nanostructured material-based technologies that perform well at the laboratory scale but face early challenges related to synthesis cost, environmental benignity, and large-volume production (Abady et al., 2025). Closed-loop treatment systems demand more sophisticated skill-based, monitoring, and design capability to deal with the changing ROC composition. However, they offer long-term benefits, such as the minimization of liquid brine production, improvement of resource utilization, and a positive impact on environmental control. From a cost-benefit perspective, it has been demonstrated that hybrid systems that combine adsorption, IEX technologies, and AOPs can result in the best compromise as far as their use is concerned when placed into decentralized treatment schemes, particularly for industrial and municipal ROC applications (Ganesh Kumar et al., 2023). Moreover, combining IEX with other membrane filtration processes can solve problems such as membrane fouling and the high cost of operation, which will competitively improve ROC treatment systems (Gebreeyessus, 2019).

7. Future research and technological outlook

The rapid pace of urbanization, global population growth, and the widespread adoption of modern lifestyles have collectively intensified the generation and release of various emerging contaminants with toxic nature. This CEC's presence in widely generating ROCs underscores the urgent need for next-generation treatment strategies that emphasize efficiency, sustainability, and resource recovery. Hybrid systems that couple adsorption with IEX techniques represent a promising approach for the simultaneous degradation of toxic pollutants and resource recovery. Future research should focus on advancing such integrated systems by combining physical, chemical, and biological processes to achieve synergistic removal of CECs, supported by real-time monitoring tools and sensor-based feedback mechanisms for enhanced process optimization and control.

Advances in material development have also demonstrated improved performance. Studies reported that the modified MIEX resin magnetically enhanced anionic polyacrylic, which outperformed the conventional polystyrene-based SBA resin (A-72MP) in terms of reduced environmental impact and lower toxic brine generation (Amini et al., 2015). Future science will enable selective, regenerable adsorbents and ion-exchange resins, including nanocomposites and bio-inspired sorbents, capable of multiple reuse cycles. Coupling electrochemical oxidation with membrane-based IEX can reduce energy demand and secondary waste. Closed-loop frameworks should emphasize resource recovery, transforming ROC from waste into valuable water, ions, and nutrients. Integration of AI and machine learning can optimize treatment under variable conditions. Techno-economic assessments, life-cycle analyses, and cross-sector collaborations are crucial for translating laboratory innovations into scalable, sustainable solutions aligned with a circular water economy.

8. Conclusion

The integration of advanced treatment and management strategies for ROC remains critical for sustainable wastewater reclamation, particularly in addressing CECs. Industrial water reclamation strategies such as ZLD and MLD have emerged as responses to water scarcity and strict discharge norms. However, both face significant challenges in terms of cost, energy intensity, and treatment efficiency. ZLD, while achieving >90–95% water recovery, demands 60%–70% of total CAPEX for thermal units and consumes up to 60 kWh/m, making it economically unsustainable. MLD, though less energy-intensive, still produces ROC rich in salinity (up to 200 g/ L) and CECs such as pharmaceuticals, dyes, and surfactants, often exceeding μg/L levels. To address these, adsorption and IEX technologies have shown strong potential as polishing treatments. Adsorbents such as biochar, AC, and functionalized resins achieve removal efficiencies up to 90%–98% for hydrophobic and ionic CECs. IEX resins further enhance selectivity for charged micropollutants (e.g., quaternary ammonium or sulfonated species), and when integrated with membrane units, they can lower chemical oxygen demand (COD) from >100 mg/ L to <20 mg/ L, enabling sustainable ROC reuse. Integration of adsorption–IEX hybrids with membrane or electrochemical units reduces chemical sludge generation and overall energy by 25%–40% relative to conventional ZLD, improving scalability for industrial deployment. Future efforts should focus on low-cost, regenerable sorbents and selective exchangers tailored for mixed CEC–salinity matrices. Such hybrid systems represent a pragmatic pathway toward cost-efficient, low-emission water recovery and CEC control within MLD/ZLD frameworks.

9. Statements and declarations

The authors declare that they have no known competing financial interests or personal relationships that could have appeared to influence the work reported in this paper.

References

Abady, M. M., Mohammed, D. M., Soliman, T. N., Shalaby, R. A., & Sakr, F. A. (2025). Sustainable synthesis of nanomaterials using different renewable sources. *Bulletin of the National Research Centre, 49*(1), 24. https://doi.org/10.1186/s42269-025-01316-4

Alves, A. C. F., Barbosa, J. G., de Oliveira, S. B., & Scalize, P. S. (2025). Polymeric hydrochars produced from ion exchange resin residue applied in the removal of emerging contaminants in water. *International Journal of Environmental Science and Technology, 22*, 79–98. https://doi.org/10.1007/s13762-024-05666-w

Ambaye, T. G., Vaccari, M., van Hullebusch, E. D., Amrane, A., & Rtimi, S. (2021). Mechanisms and adsorption capacities of biochar for the removal of organic and inorganic pollutants from industrial wastewater. *International Journal of Environmental Science and Technology, 18*(10), 3273–3294. https://doi.org/10.1007/s13762-020-03060-w

Amini, A., Kim, Y., Zhang, J., Boyer, T., & Zhang, Q. (2015). Environmental and economic sustainability of ion exchange drinking water treatment for organics removal. *Journal of Cleaner Production, 104*, 413–421. https://doi.org/10.1016/j.jclepro.2015.05.056

Anegbe, B., Ifijen, I. H., Maliki, M., Uwidia, I. E., & Aigbodion, A. I. (2024). Graphene oxide synthesis and applications in emerging contaminant removal: A comprehensive review. *Environmental Sciences Europe, 36*(1), 15. https://doi.org/10.1186/s12302-023-00814-4

Ani, J. U., Akpomie, K. G., Okoro, U. C., Aneke, L. E., Onukwuli, O. D., & Ujam, O. T. (2020). Potentials of activated carbon produced from biomass materials for sequestration of dyes, heavy metals, and crude oil components from aqueous environment. *Applied Water Science, 10*(2), 69. https://doi.org/10.1007/s13201-020-1149-8

Astuti, M. P., Notodarmojo, S., Priadi, C. R., & Padhye, L. P. (2023). Contaminants of emerging concerns (CECs) in a municipal wastewater treatment plant in Indonesia. *Environmental Science and Pollution Research, 30*(8), 21512–21532. https://doi.org/10.1007/s11356-022-23567-8

Backer, S. N., Bouaziz, I., Kallayi, N., Thomas, R. T., Preethikumar, G., Takriff, M. S., Laoui, T., & Atieh, M. A. (2022). Review: Brine solution: Current status, future management and technology development. *Sustainability, 14*(11), 6752. https://doi.org/10.3390/su14116752

Badia, J. H., Soto, R., Ramírez, E., Bringué, R., Fité, C., Iborra, M., & Tejero, J. (2023). Role of ion-exchange resins in hydrogenation reactions. *Catalysts, 13*(3), 624. https://doi.org/10.3390/catal13030624

Blachnio, M., Derylo-Marczewska, A., Charmas, B., Zienkiewicz-Strzalka, M., Bogatyrov, V., & Galaburda, M. (2020). Activated carbon from agricultural wastes for adsorption of organic pollutants. *Molecules, 25*(21), 5105. https://doi.org/10.3390/molecules25215105

Borges, M. E., García, D. M., Hernández, T., Ruiz-Morales, J. C., & Esparza, P. (2015). Supported photocatalyst for removal of emerging contaminants from wastewater in a continuous packed-bed photoreactor configuration. *Catalysts, 5*(1), 77–87. https://doi.org/10.3390/catal5010077

Bustillos, S., Alturki, A., Prentice, D., La Plante, E. C., Rogers, M., Keller, M., Ragipani, R., Wang, B., Sant, G., & Simonetti, D. A. (2020). Implementation of ion exchange processes for carbon dioxide mineralization using industrial waste streams. *Frontiers in Energy Research, 8*. https://doi.org/10.3389/fenrg.2020.610392

Capodaglio, A. G. (2025). Prospects of novel technologies for PFAS destruction in water . And wastewater. *Applied Sciences, 15*(17), 9311. https://doi.org/10.3390/app15179311

Chen, Q., Burhan, M., Shahzad, M. W., Ybyraiymkul, D., Akhtar, F. H., Li, Y., & Ng, K. C. (2021). A zero liquid discharge system integrating multi-effect distillation and evaporative crystallization for desalination brine treatment. *Desalination, 502*, 114928. https://doi.org/10.1016/j.desal.2020.114928

Chen, Y., Gong, X., Yang, P., Zhu, M., Zhou, H., & Pan, S. (2025). Efficient removal of organic pollutants in reverse osmosis concentrate by coupling granular activated carbon adsorption with ozone. *Chemical Engineering Journal Advances, 22*, 100736. https://doi.org/10.1016/j.ceja.2025.100736

Chong, S., Zhang, G., Wei, Z., Zhang, N., Huang, T., & Liu, Y. (2017). Sonocatalytic degradation of diclofenac with FeCeOx particles in water. *Ultrasonics Sonochemistry, 34*, 418–425. https://doi.org/10.1016/j.ultsonch.2016.06.023

Costa, F. C. R., dos Santos, C. R., & Amaral, M. C. S. (2023). Trace organic contaminants removal by membrane distillation: A review on mechanisms, performance, applications, and challenges. *Chemical Engineering Journal, 464*, 142461. https://doi.org/10.1016/j.cej.2023.142461

Czupryński, P., Płotka, M., Glamowski, P., Żukowski, W., & Bajda, T. (2022). An assessment of an ion exchange resin system for the removal and recovery of Ni, Hg, and Cr from wet flue gas desulphurization wastewater—A pilot study. *RSC Advances, 12*(9), 5145–5156. https://doi.org/10.1039/D1RA09426B

Devaisy, S., Kandasamy, J., Aryal, R., Johir, M. A. H., Ratnaweera, H., & Vigneswaran, S. (2023a). Removal of organics with ion-exchange resins (IEX) from reverse osmosis concentrate. *Membranes, 13*(2), 136. https://doi.org/10.3390/membranes13020136

Devaisy, S., Kandasamy, J., Nguyen, T. V., Ratnaweera, H., & Vigneswaran, S. (2023b). Membranes in water reclamation: Treatment, reuse and concentrate management. *Membranes, 13*(6), 605. https://doi.org/10.3390/membranes13060605

Dikobe, P. I., Tekere, M., Masindi, V., & Foteinis, S. (2024). Removal of contaminants of emerging concern from drinking water using bio-based activated carbon. *Journal of Water Process Engineering, 68*, 106313. https://doi.org/10.1016/j.jwpe.2024.106313

Dixit, F., Barbeau, B., Mostafavi, S. G., & Mohseni, M. (2019). PFOA and PFOS removal by ion exchange for water reuse and drinking applications: Role of organic matter characteristics. *Environmental Science: Water Research and Technology, 5*(10), 1782–1795. https://doi.org/10.1039/C9EW00409B

Dixit, F., Barbeau, B., Mostafavi, S. G., & Mohseni, M. (2020). Removal of legacy PFAS and other fluorotelomers: Optimized regeneration strategies in DOM-Rich waters. *Water Research, 183*, 116098. https://doi.org/10.1016/j.watres.2020.116098

Dixit, F., Dutta, R., Barbeau, B., Berube, P., & Mohseni, M. (2021). PFAS removal by ion exchange resins: A review. *Chemosphere, 272*, 129777. https://doi.org/10.1016/j.chemosphere.2021.129777

Faraz Ahmer, M., & Kashif Uddin, M. (2024). Structure properties and industrial applications of anion exchange resins for the removal of electroactive nitrate ions from contaminated water. *RSC Advances, 14*(45), 33629–33648. https://doi.org/10.1039/D4RA03871A

Feng, W., Deng, Y., Yang, F., Miao, Q., & Ngien, S. K. (2023). Systematic review of contaminants of emerging concern (CECs): Distribution, risks, and implications for water quality and health. *Water, 15*(22), 3922. https://doi.org/10.3390/w15223922

Finnerty, C. T. K., Childress, A. E., Hardy, K. M., Hoek, E. M. V., Mauter, M. S., Plumlee, M. H., Rose, J. B., Sobsey, M. D., Westerhoff, P., Alvarez, P. J. J., & Elimelech, M. (2024). The future of municipal wastewater reuse concentrate management: Drivers, challenges, and opportunities. *Environmental Science and Technology, 58*(1), 3–16. https://doi.org/10.1021/acs.est.3c06774

Franke, V., Ullberg, M., McCleaf, P., Wålinder, M., Köhler, S. J., & Ahrens, L. (2021). The price of really clean water: Combining nanofiltration with granular activated carbon and anion exchange resins for the removal of per- and polyfluoralkyl substances (PFASs) in drinking water production. *ACS ES&T Water, 1*(4), 782–795. https://doi.org/10.1021/acsestwater.0c00141

Freilinger, J., Back, J. O., Plangger, R., Schottenberger, H., Huck, C. W., Rupprich, M., & Bakry, R. (2025). Development of a fluorophilic ion-exchange material with dual binding mechanism for solid-phase extraction of PFAS. *Journal of Hazardous Materials Letters, 6*, 100158. https://doi.org/10.1016/j.hazl.2025.100158

Ganesh Kumar, P., Kanmani, S., Senthil Kumar, P., & Vellingiri, K. (2023). Efficacy of simultaneous advanced oxidation and adsorption for treating municipal wastewater for indirect potable reuse. *Chemosphere, 321*, 138115. https://doi.org/10.1016/j.chemosphere.2023.138115

Gao, Z., Chu, J., Xu, W., Liu, X., Zhou, C., Ren, F., Qin, B., Shan, Q., & Liu, G. (2025). Research on the mechanism of ion exchange based on thermodynamic equilibrium. *Minerals, 15*(4), 367. https://doi.org/10.3390/min15040367

Gao, Y., Deng, S., Du, Z., Liu, K., & Yu, G. (2017). Adsorptive removal of emerging polyfluoroalkyl substances F-53B and PFOS by anion-exchange resin: A comparative study. *Journal of Hazardous Materials, 323*, 550–557. https://doi.org/10.1016/j.jhazmat.2016.04.069

Gebreeyessus, G. D. (2019). Status of hybrid membrane–ion-exchange systems for desalination: A comprehensive review. *Applied Water Science, 9*(5), 135. https://doi.org/10.1007/s13201-019-1006-9

Gholami, P., Dinpazhoh, L., Khataee, A., Hassani, A., & Bhatnagar, A. (2020a). Facile hydrothermal synthesis of novel Fe-Cu layered double hydroxide/biochar nanocomposite with enhanced sonocatalytic activity for degradation of cefazolin sodium. *Journal of Hazardous Materials, 381*, 120742. https://doi.org/10.1016/j.jhazmat.2019.120742

Gholami, P., Khataee, A., Vahid, B., Karimi, A., Golizadeh, M., & Ritala, M. (2020b). Sonophotocatalytic degradation of sulfadiazine by integration of microfibrillated carboxymethyl cellulose with Zn-Cu-Mg mixed metal hydroxide/g-C3N4 composite. *Separation and Purification Technology, 245*, 116866. https://doi.org/10.1016/j.seppur.2020.116866

Gros, M., Petrović, M., & Barceló, D. (2007). Wastewater treatment plants as a pathway for aquatic contamination by pharmaceuticals in the Ebro River basin (northeast Spain). *Environmental Toxicology and Chemistry, 26*(8), 1553–1562. https://doi.org/10.1897/06-495R.1

Gupta, S., Benien, P., & Sahoo, P. K. (2010). Ion exchange resins transforming drug delivery systems. *Current Drug Delivery, 7*(3), 252–262. https://doi.org/10.2174/156720110791560955

Harish, H., & Jegatheesan, V. (2025). A review of sources, worldwide legislative measures and the factors influencing the treatment technologies for contaminants of emerging concern (CECs). *Current Pollution Reports, 11*(1), 44. https://doi.org/10.1007/s40726-025-00357-7

Huang, J., Li, S., Delagah, S., Stone, S., Mouawad, J., & Sharbatmaleki, M. (2025). Assessing closed-circuit reverse osmosis (CCRO) efficiency in removing contaminants of emerging concern (CECs) in a high-recovery water reclamation pilot study. *Water Practice and Technology, 20*(2), 314–323. https://doi.org/10.2166/wpt.2025.019

Jamil, S., Loganathan, P., Kandasamy, J., Listowski, A., McDonald, J. A., Khan, S. J., & Vigneswaran, S. (2020). Removal of organic matter from wastewater reverse osmosis concentrate using granular activated carbon and anion exchange resin adsorbent columns in sequence. *Chemosphere, 261*, 127549. https://doi.org/10.1016/j.chemosphere.2020.127549

Jamil, S., Loganathan, P., Listowski, A., Kandasamy, J., Khourshed, C., & Vigneswaran, S. (2019). Simultaneous removal of natural organic matter and micro-organic pollutants from reverse osmosis concentrate using granular activated carbon. *Water Research, 155*, 106–114. https://doi.org/10.1016/j.watres.2019.02.016

Kassar, C., Graham, C., & Boyer, T. H. (2022). Removal of perfluoroalkyl acids and common drinking water contaminants by weak-base anion exchange resins: Impacts of solution pH and resin properties. *Water Research X, 17*, 100159. https://doi.org/10.1016/j.wroa.2022.100159

Katrivesis, F. K., Karela, A. D., Papadakis, V. G., & Paraskeva, C. A. (2019). Revisiting of coagulation-flocculation processes in the production of potable water. *Journal of Water Process Engineering, 27*, 193–204. https://doi.org/10.1016/j.jwpe.2018.12.007

Kempisty, D. M., Arevalo, E., Spinelli, A. M., Edeback, V., Dickenson, E. R. V., Husted, C., Higgins, C. P., Summers, R. S., & Knappe, D. R. U. (2022). Granular activated carbon adsorption of perfluoroalkyl acids from ground and surface water. *AWWA Water Science, 4*(1), e1269. https://doi.org/10.1002/aws2.1269

Kim, S., Nam, S.-N., Jang, A., Jang, M., Park, C. M., Son, A., Her, N., Heo, J., & Yoon, Y. (2022). Review of adsorption–membrane hybrid systems for water and wastewater treatment. *Chemosphere, 286*, 131916. https://doi.org/10.1016/j.chemosphere.2021.131916

Konradt, N., Kuhlen, J. G., Rohns, H.-P., Schmitt, B., Fischer, U., Binder, T., Schumacher, V., Wagner, C., Kamphausen, S., Müller, U., Sacher, F., Janknecht, P., Hobby, R., ElSherbiny, I. M. A., & Panglisch, S. (2021). Removal of trace organic contaminants by parallel operation of reverse osmosis and granular activated carbon for drinking water treatment. *Membranes, 11*(1), 33. https://doi.org/10.3390/membranes11010033

Kooijman, G., de Kreuk, M. K., Houtman, C., & van Lier, J. B. (2020). Perspectives of coagulation/flocculation for the removal of pharmaceuticals from domestic wastewater: A critical view at experimental procedures. *Journal of Water Process Engineering, 34*, 101161. https://doi.org/10.1016/j.jwpe.2020.101161

Kumar, M., Badruzzaman, M., Adham, S., & Oppenheimer, J. (2007). Beneficial phosphate recovery from reverse osmosis (RO) concentrate of an integrated membrane system using polymeric ligand exchanger (PLE). *Water Research, 41*(10), 2211–2219. https://doi.org/10.1016/j.watres.2007.01.042

La Merrill, M. A., Vandenberg, L. N., Smith, M. T., Goodson, W., Browne, P., Patisaul, H. B., Guyton, K. Z., Kortenkamp, A., Cogliano, V. J., Woodruff, T. J., Rieswijk, L., Sone, H., Korach, K. S., Gore, A. C., Zeise, L., & Zoeller, R. T. (2020). Consensus on the key characteristics of endocrine-disrupting chemicals as a basis for hazard identification. *Nature Reviews Endocrinology, 16*(1), 45–57. https://doi.org/10.1038/s41574-019-0273-8

Landry, K. A., Sun, P., Huang, C.-H., & Boyer, T. H. (2015). Ion-exchange selectivity of diclofenac, ibuprofen, ketoprofen, and naproxen in ureolyzed human urine. *Water Research, 68*, 510–521. https://doi.org/10.1016/j.watres.2014.09.056

Lawtae, P., & Tangsathitkulchai, C. (2021). The use of high surface area mesoporous-activated carbon from longan seed biomass for increasing capacity and kinetics of methylene blue adsorption from aqueous solution. *Molecules, 26*(21), 6521. https://doi.org/10.3390/molecules26216521

Lee, S.-H., Kim, N., & Park, D. (2023). Insights into the adsorption performance of emerging contaminants on granular activated carbon. *Separations, 10*(9), 501. https://doi.org/10.3390/separations10090501

Li, X., Shen, X., Jiang, W., Xi, Y., & Li, S. (2024). Comprehensive review of emerging contaminants: Detection technologies, environmental impact, and management strategies. *Ecotoxicology and Environmental Safety, 278*, 116420. https://doi.org/10.1016/j.ecoenv.2024.116420

Li, Q., Zong, X., Li, H., Li, X., Yang, K., Wu, J., Wu, L., Han, Z., Wang, R., Ye, Y., & Pan, F. (2025). The efficient degradation of carbamazepine by Cu-Fe bimetallic composite carbon derived from the waste cation exchange resins: Mechanism, ecotoxicity, and continuous flow catalysis. *Separation and Purification Technology, 356*, 129813. https://doi.org/10.1016/j.seppur.2024.129813

Liang, Y., Lin, X., Kong, X., Duan, Q., Wang, P., Mei, X., & Ma, J. (2021). Making waves: Zero liquid discharge for sustainable industrial effluent management. *Water, 13*(20), 2852. https://doi.org/10.3390/w13202852

Liao, B., Zeng, X., Ling, Z., Zhao, S., Li, B., & Han, X. (2025). Recent advances in zero discharge treatment technologies for desulfurization wastewater in coal-fired power plants: A mini-review. *Processes, 13*(4), 982. https://doi.org/10.3390/pr13040982

Lin, A. Y.-C., Yu, T.-H., & Lateef, S. K. (2009). Removal of pharmaceuticals in secondary wastewater treatment processes in Taiwan. *Journal of Hazardous Materials, 167*(1), 1163–1169. https://doi.org/10.1016/j.jhazmat.2009.01.108

Liu, Y.-L., & Sun, M. (2021). Ion exchange removal and resin regeneration to treat per- and polyfluoroalkyl ether acids and other emerging PFAS in drinking water. *Water Research, 207*, 117781. https://doi.org/10.1016/j.watres.2021.117781

Loganathan, P., Kandasamy, J., Ratnaweera, H., & Vigneswaran, S. (2023). Submerged membrane/adsorption hybrid process in water reclamation and concentrate management—A mini review. *Environmental Science and Pollution Research, 30*(15), 42738–42752. https://doi.org/10.1007/s11356-022-23229-9

López-Ortiz, C. M., Sentana-Gadea, I., Varó-Galvañ, P. J., Maestre-Pérez, S. E., & Prats-Rico, D. (2018). Effect of magnetic ion exchange (MIEX®) on removal of emerging organic contaminants. *Chemosphere, 208*, 433–440. https://doi.org/10.1016/j.chemosphere.2018.05.194

Lu, P., Lin, K., & Gan, J. (2020). Enhanced ozonation of ciprofloxacin in the presence of bromide: Kinetics, products, pathways, and toxicity. *Water Research, 183*, 116105. https://doi.org/10.1016/j.watres.2020.116105

Lucas, M. S., Teixeira, A. R., Jorge, N., & Peres, J. A. (2025). Industrial wastewater treatment by coagulation–flocculation and advanced oxidation processes: A review. *Water, 17*(13), 1934. https://doi.org/10.3390/w17131934

Łukasiewicz, E. (2025). Coagulation–sedimentation in water and wastewater treatment: Removal of pesticides, pharmaceuticals, PFAS, microplastics, and natural organic matter. *Water, 17*(21), 3048. https://doi.org/10.3390/w17213048

Mahlangu, O., Nthunya, L., Motsa, M., Richards, H., & Mamba, B. (2023). *Treatment of trace organics and emerging contaminants using traditional and advanced technologies* (pp. 243–264). https://doi.org/10.1016/B978-0-323-95684-0.00002-6

Maimaiti, A., Deng, S., Meng, P., Wang, W., Wang, B., Huang, J., Wang, Y., & Yu, G. (2018). Competitive adsorption of perfluoroalkyl substances on anion exchange resins in simulated AFFF-Impacted groundwater. *Chemical Engineering Journal, 348*, 494–502. https://doi.org/10.1016/j.cej.2018.05.006

Manjarrés-López, D. P., Martínez-Megías, C., Vitale, D., Picó, Y., Rico, A., & Pérez, S. (2025). Occurrence, bioaccumulation, and ecological risk assessment of contaminants of emerging concern using native and invasive species as biomonitors. *Emerging Contaminants, 11*(3), 100533. https://doi.org/10.1016/j.emcon.2025.100533

Matesun, J., Petrik, L., Musvoto, E., Ayinde, W., & Ikumi, D. (2024). Limitations of wastewater treatment plants in removing trace anthropogenic biomarkers and future directions: A review. *Ecotoxicology and Environmental Safety, 281*, 116610. https://doi.org/10.1016/j.ecoenv.2024.116610

Meher, A. K., & Zarouri, A. (2025). Environmental applications of mass spectrometry for emerging contaminants. *Molecules, 30*(2), 364. https://doi.org/10.3390/molecules30020364

Mekonnen, M. M., & Hoekstra, A. Y. (2016). Four billion people facing severe water scarcity. *Science Advances, 2* (2), e1500323. https://doi.org/10.1126/sciadv.1500323

Meng, X., Liu, Z., Deng, C., Zhu, M., Wang, D., Li, K., Deng, Y., & Jiang, M. (2016). Microporous Nano-MgO/diatomite ceramic membrane with high positive surface charge for tetracycline removal. *Journal of Hazardous Materials, 320*, 495–503. https://doi.org/10.1016/j.jhazmat.2016.08.068

Mousset, E., Loh, W. H., Lim, W. S., Jarry, L., Wang, Z., & Lefebvre, O. (2021). Cost comparison of advanced oxidation processes for wastewater treatment using accumulated oxygen-equivalent criteria. *Water Research, 200*, 117234. https://doi.org/10.1016/j.watres.2021.117234

Mustereț, C. P., Morosanu, I., Ciobanu, R., Plavan, O., Gherghel, A., Al-Refai, M., Roman, I., & Teodosiu, C. (2021). Assessment of coagulation–flocculation process efficiency for the natural organic matter removal in drinking water treatment. *Water, 13*(21), 3073. https://doi.org/10.3390/w13213073

Ng, K. T., Rapp-Wright, H., Egli, M., Hartmann, A., Steele, J. C., Sosa-Hernández, J. E., Melchor-Martínez, E. M., Jacobs, M., White, B., Regan, F., Parra-Saldivar, R., Couchman, L., Halden, R. U., & Barron, L. P. (2020). High-throughput multi-residue quantification of contaminants of emerging concern in wastewaters enabled using direct injection liquid chromatography-tandem mass spectrometry. *Journal of Hazardous Materials, 398*, 122933. https://doi.org/10.1016/j.jhazmat.2020.122933

Nghiem, L., & Fujioka, T. (2016). Removal of emerging contaminants for water reuse by membrane technology. https://doi.org/10.1016/B978-0-444-63312-5.00009-7.

O'Connell, M. G., Rajendran, N., Elimelech, M., Gilron, J., & Dunn, J. B. (2024). Analysis of energy, water, land and cost implications of zero and minimal liquid discharge desalination technologies. *Nature Water, 2*(11), 1116–1127. https://doi.org/10.1038/s44221-024-00327-1

Oladimeji, T. E., Oyedemi, M., Emetere, M. E., Agboola, O., Adeoye, J. B., & Odunlami, O. A. (2024). Review on the impact of heavy metals from industrial wastewater effluent and removal technologies. *Heliyon, 10*(23), e40370. https://doi.org/10.1016/j.heliyon.2024.e40370

Panagopoulos, A. (2022). Techno-economic assessment of zero liquid discharge (ZLD) systems for sustainable treatment, minimization and valorization of seawater brine. *Journal of Environmental Management, 306*, 114488. https://doi.org/10.1016/j.jenvman.2022.114488

Panagopoulos, A., & Giannika, V. (2024). A comprehensive assessment of the economic and technical viability of a zero liquid discharge (ZLD) hybrid desalination system for water and salt recovery. *Journal of Environmental Management, 359*, 121057. https://doi.org/10.1016/j.jenvman.2024.121057

Pérez-Botella, E., Valencia, S., & Rey, F. (2022). Zeolites in adsorption processes: State of the art and future prospects. *Chemical Reviews, 122*(24), 17647–17695. https://doi.org/10.1021/acs.chemrev.2c00140

Petrangeli Papini, M., Senofonte, M., Cuzzola, R. A., Remmani, R., Pettiti, I., Riccardi, C., & Simonetti, G. (2024). Adsorption technology for PFAS removal in water: Comparison between novel carbonaceous materials. *Materials, 17*(17), 4169. https://doi.org/10.3390/ma17174169

Piekutin, J., Bolińska, M. I., Kotowska, U., Koszelnik, P., & Puchlik, M. (2025). Research of the possibility of removing organic pollutants from water by membrane methods and purification of the obtained concentrate by chemical methods. *Economics and Environment, 92*(1), 1105. https://doi.org/10.34659/eis.2025.92.1.1105

Plata, S. L., Devenport, C. L., Miara, A., Sitterley, K. A., Evans, A., Talmadge, M., Van Allsburg, K. M., Kurup, P., Cox, J., Kerber, S., Howell, A., Breckenridge, R., Manygoats, C., Stokes-Draut, J. R., Macknick, J., & Childress, A. E. (2022). Zero liquid discharge and water reuse in recirculating cooling towers at power facilities: Review and case study analysis. *ACS ES&T Engineering, 2*(3), 508–525. https://doi.org/10.1021/acsestengg.1c00377

Prakash, O., Panchal, D., Sharma, A., & Pal, S. (2022). Advanced treatment technologies for removal of contaminants of emerging concern. In *Contaminants of emerging concerns and reigning removal technologies*. CRC Press.

Radwan, E. K., Abdel Ghafar, H. H., Ibrahim, M. B. M., & Moursy, A. S. (2023). Recent trends in treatment technologies of emerging contaminants. *Environmental Quality Management, 32*, 7–25. https://doi.org/10.1002/tqem.21877

Rahman, M. F., Anderson, W. B., Peldszus, S., & Huck, P. M. (2022). Ion-exchange treatment of perfluorinated carboxylic acids in water: Comparison of polystyrenic and polyacrylic resin structures and impact of sulfate on their performance. *ACS ES&T Water, 2*(7), 1195–1205. https://doi.org/10.1021/acsestwater.1c00501

Raji, Z., Karim, A., Karam, A., & Khalloufi, S. (2023). Adsorption of heavy metals: Mechanisms, kinetics, and applications of various adsorbents in wastewater remediation—A review. *Waste, 1*(3), 775–805. https://doi.org/10.3390/waste1030046

Ramirez Arenas, L., Le Coustumer, P., Ramseier Gentile, S., Zimmermann, S., & Stoll, S. (2023). Removal efficiency and adsorption mechanisms of CeO2 nanoparticles onto granular activated carbon used in drinking water treatment plants. *Science of the Total Environment, 856*, 159261. https://doi.org/10.1016/j.scitotenv.2022.159261

Ramrakhiani, L., Ghosh, S., & Majumdar, S. (2022). *Emerging contaminants in water and wastewater: Remediation perspectives and innovations in treatment technologies* (pp. 253–284). https://doi.org/10.1007/978-981-19-1847-6_11

Rezaei, L., Alipour, V., Hassani, A. H., & Dehghani Ghantghestani, M. (2024). Environmental health risk of brine disposal for reverse osmosis water desalination plants: A case study in Bandar Abbas city, Iran. *Regional Studies in Marine Science, 77*, 103606. https://doi.org/10.1016/j.rsma.2024.103606

Riegel, M., Haist-Gulde, B., & Sacher, F. (2023). Sorptive removal of short-chain perfluoroalkyl substances (PFAS) during drinking water treatment using activated carbon and anion exchanger. *Environmental Sciences Europe, 35*(1), 12. https://doi.org/10.1186/s12302-023-00716-5

Romeyn, T. R., Harijanto, W., Sandoval, S., Delagah, S., & Sharbatmaleki, M. (2015). Contaminants of emerging concern in reverse osmosis brine concentrate from indirect/direct water reuse applications. *Water Science and Technology, 73*(2), 236–250. https://doi.org/10.2166/wst.2015.480

Rostvall, A., Zhang, W., Dürig, W., Renman, G., Wiberg, K., Ahrens, L., & Gago-Ferrero, P. (2018). Removal of pharmaceuticals, perfluoroalkyl substances and other micropollutants from wastewater using lignite, xylit, sand, granular activated carbon (GAC) and GAC+Polonite® in column tests – Role of physicochemical properties. *Water Research, 137*, 97–106. https://doi.org/10.1016/j.watres.2018.03.008

Rott, E., Kuch, B., Lange, C., Richter, P., Kugele, A., & Minke, R. (2018). Removal of emerging contaminants and estrogenic activity from wastewater treatment plant effluent with UV/Chlorine and UV/H2O2 advanced oxidation treatment at pilot scale. *International Journal of Environmental Research and Public Health, 15*(5), 935. https://doi.org/10.3390/ijerph15050935

R. Schlesinger, D., McDermott, C., Le, N. Q., Ko, J. S., Johnson, J. K., Demirev, P. A., & Xia, Z. (2022). *Destruction of per/poly-fluorinated alkyl substances by magnetite nanoparticle-catalyzed UV-Fenton reaction*. https://doi.org/10.1039/D2EW00058J

Sadegh, H., Ali, G. A. M., Gupta, V. K., Makhlouf, A. S. H., Shahryari-ghoshekandi, R., Nadagouda, M. N., Sillanpää, M., & Megiel, E. (2017). The role of nanomaterials as effective adsorbents and their applications in wastewater treatment. *Journal of Nanostructure in Chemistry, 7*(1), 1–14. https://doi.org/10.1007/s40097-017-0219-4

Sanchez-Lievanos, K. R., Zhang, D., Simpson, S. M., Wijayahena, M. K., Rizzo, G., Aguilar, J. M. N., Abaya, L. M., Dovi, J. M., Sirotkin, H. I., Crawley, M. R., Cook, T. R., & Aga, D. S. (2025). Synthesis and evaluation of cationic porphyrin-based organic nanocages for the removal of 38 PFAS from water:

Experimental, theoretical, and eco-toxicological insights. *ACS ES&T Engineering, 5*(3), 701–713. https://doi.org/10.1021/acsestengg.4c00639

Scholes, R. C., Stiegler, A. N., Anderson, C. M., & Sedlak, D. L. (2021). Enabling water reuse by treatment of reverse osmosis concentrate: The promise of constructed wetlands. *ACS Environmental Au, 1*(1), 7–17. https://doi.org/10.1021/acsenvironau.1c00013

Sereshti, H., Beyrak-Abadi, E., Esmaeili Bidhendi, M., Ahmad, I., Shahabuddin, S., Rashidi Nodeh, H., Sridewi, N., & Wan Ibrahim, W. N. (2022). Sulfide-doped magnetic carbon nanotubes developed as adsorbent for uptake of tetracycline and cefixime from wastewater. *Nanomaterials, 12*(20), 3576. https://doi.org/10.3390/nano12203576

Siciliano, A., Limonti, C., Curcio, G. M., & Molinari, R. (2020). Advances in struvite precipitation technologies for nutrients removal and recovery from aqueous waste and wastewater. *Sustainability, 12*(18), 7538. https://doi.org/10.3390/su12187538

Skinner, J. P., Raderstorf, A., Rittmann, B. E., & Delgado, A. G. (2025). Biotransforming the "forever chemicals": Trends and insights from microbiological studies on PFAS. *Environmental Science and Technology, 59*(11), 5417–5430. https://doi.org/10.1021/acs.est.4c04557

Smith, R. C., & SenGupta, A. K. (2015). Integrating tunable anion exchange with reverse osmosis for enhanced recovery during inland brackish water desalination. *Environmental Science and Technology, 49*(9), 5637–5644. https://doi.org/10.1021/es505439p

Somya, A., & Varshney, A. P. (2025). Ion exchange and zeolites: Sustainable approaches to water pollution mitigation. In *Advancements in zeolites and micro-meso porous hierarchical materials* (pp. 151–180). IGI Global Scientific Publishing. https://doi.org/10.4018/979-8-3373-0320-8.ch005

Song, Y., Zeng, Y., Jiang, T., Chen, J., & Du, Q. (2023). Efficient removal of ciprofloxacin from contaminated water via polystyrene anion exchange resin with nanoconfined zero-valent iron. *Nanomaterials, 13*(1), 116. https://doi.org/10.3390/nano13010116

Subramani, A., & Jacangelo, J. G. (2014). Treatment technologies for reverse osmosis concentrate volume minimization: A review. *Separation and Purification Technology, 122*, 472–489. https://doi.org/10.1016/j.seppur.2013.12.004

Tamanna, T., Mahon, P. J., Hockings, R. K., Alam, H., Raymond, M., Smith, C., Clarke, C., & Yu, A. (2023). Ion exchange MIEX® GOLD resin as a promising sorbent for the removal of PFAS compounds. *Applied Sciences, 13*(10), 6263. https://doi.org/10.3390/app13106263

Tokoyi, V., Deenadayalu, N., Tokoyi, V., & Deenadayalu, N. (2025). *Ti-Bimetallic MOFs as photo-reformers and degraders of emerging water contaminants*. IntechOpen. https://doi.org/10.5772/intechopen.1009930

Tong, T., & Elimelech, M. (2016). The global rise of zero liquid discharge for wastewater management: Drivers, technologies, and future directions. *Environmental Science and Technology, 50*(13), 6846–6855. https://doi.org/10.1021/acs.est.6b01000

Tong, T., Xu, L., Horseman, T., Westerhoff, P., Xu, P., Yao, Y., Zhang, X., Alghanayem, R., & Lin, S. (2025). Brine management with zero and minimal liquid discharge. *Nature Reviews Clean Technology, 1*(3), 185–200. https://doi.org/10.1038/s44359-025-00036-2

Tseng, L. Y., You, C., Vu, C., Chistolini, M. K., Wang, C. Y., Mast, K., Luo, F., Asvapathanagul, P., Gedalanga, P. B., Eusebi, A. L., Gorbi, S., Pittura, L., & Fatone, F. (2022). Adsorption of contaminants of emerging concern (CECs) with varying hydrophobicity on macro- and microplastic polyvinyl chloride, polyethylene, and polystyrene: Kinetics and potential mechanisms. *Water, 14*(16), 2581. https://doi.org/10.3390/w14162581

Valdés, H., Saavedra, A., Flores, M., Vera-Puerto, I., Aviña, H., & Belmonte, M. (2021). Reverse osmosis concentrate: Physicochemical characteristics, environmental impact, and technologies. *Membranes, 11*(10), 753. https://doi.org/10.3390/membranes11100753

Vane, L. M. (2017). Water recovery from brines and salt-saturated solutions: Operability and thermodynamic efficiency considerations for desalination technologies. *Journal of Chemical Technology and Biotechnology, 92*(10), 2506–2518. https://doi.org/10.1002/jctb.5225

Vecino, X., & Reig, M. (2022). Wastewater treatment by adsorption and/or ion-exchange processes for resource recovery. *Water, 14*(6), 911. https://doi.org/10.3390/w14060911

Vingerhoeds, M. H., Nijenhuis-de Vries, M. A., Ruepert, N., van der Laan, H., Bredie, W. L. P., & Kremer, S. (2016). Sensory quality of drinking water produced by reverse osmosis membrane filtration followed by remineralisation. *Water Research, 94*, 42–51. https://doi.org/10.1016/j.watres.2016.02.043

Wang, W., Li, X., Yuan, S., Sun, J., & Zheng, S. (2016). Effect of resin charged functional group, porosity, and chemical matrix on the long-term pharmaceutical removal mechanism by conventional ion exchange resins. *Chemosphere, 160*, 71–79. https://doi.org/10.1016/j.chemosphere.2016.06.073

Wang, W., Maimaiti, A., Shi, H., Wu, R., Wang, R., Li, Z., Qi, D., Yu, G., & Deng, S. (2019). Adsorption behavior and mechanism of emerging perfluoro-2-propoxypropanoic acid (GenX) on activated carbons and resins. *Chemical Engineering Journal, 364*, 132–138. https://doi.org/10.1016/j.cej.2019.01.153

Wang, F., Xiang, L., Sze-Yin Leung, K., Elsner, M., Zhang, Y., Guo, Y., Pan, B., Sun, H., An, T., Ying, G., Brooks, B. W., Hou, D., Helbling, D. E., Sun, J., Qiu, H., Vogel, T. M., Zhang, W., Gao, Y., Simpson, M. J., … Tiedje, J. M. (2024). Emerging contaminants: A one health perspective. *The Innovation, 5*(4), 100612. https://doi.org/10.1016/j.xinn.2024.100612

Wee, S. Y., Aris, A. Z., Yusoff, F. M., & Praveena, S. M. (2020). Occurrence of multiclass endocrine disrupting compounds in a drinking water supply system and associated risks. *Scientific Reports, 10*(1), 1–12. https://doi.org/10.1038/s41598-020-74061-5

Westerhoff, P., Yoon, Y., Snyder, S., & Wert, E. (2005). Fate of endocrine-disruptor, pharmaceutical, and personal care product chemicals during simulated drinking water treatment processes. *Environmental Science and Technology, 39*(17), 6649–6663. https://doi.org/10.1021/es0484799

Woodard, S., Berry, J., & Newman, B. (2017). Ion exchange resin for PFAS removal and pilot test comparison to GAC. *Remediation Journal, 27*(3), 19–27. https://doi.org/10.1002/rem.21515

Wouters, S., Pijpers, I., Vanden Haute, N., Meston, D., Dillen, L., Cuyckens, F., & Eeltink, S. (2022). Making high salt concentrations for optimal chromatography compatible with electrospray ionization mass spectrometry using an ion exchange membrane suppressor: Analysis of biomarkers for transporter protein inhibition as a case study. *Frontiers in Analytical Science, 2*. https://doi.org/10.3389/frans.2022.1002935

Wu, S., Peng, L. E., Yang, Z., Sarkar, P., Barboiu, M., Tang, C. Y., & Fane, A. G. (2024). Next-generation desalination membranes empowered by novel materials: Where are we now? *Nano-Micro Letters, 17*(1), 91. https://doi.org/10.1007/s40820-024-01606-y

Xu, D., Gao, Y., Lin, Z., Gao, W., Zhang, H., Karnowo, K., Hu, X., Sun, H., Syed-Hassan, S. S. A., & Zhang, S. (2020). Application of biochar derived from pyrolysis of waste fiberboard on tetracycline adsorption in aqueous solution. *Frontiers in Chemistry, 7*. https://doi.org/10.3389/fchem.2019.00943

Xu, H., Qin, Q., Zhang, C., Ning, K., Zhao, R., Wang, P., Deng, J., & Huang, G. (2019). Adsorption of organic constituents from reverse osmosis concentrate in coal chemical industry by coking coal. *Processes, 7*(1), 44. https://doi.org/10.3390/pr7010044

Yan, S., Lai, X., Fan, L., Wang, T., Yao, Y., & Wang, W. (2023). Integrating adsorption and in situ advanced oxidation for the treatment of organic wastewater by 3D carbon aerogel embedded with Fe-doped carbonitrides. *Environmental Science and Pollution Research, 30*(1), 1386–1398. https://doi.org/10.1007/s11356-022-22275-7

Yang, J., Zhang, Y., Zeng, D., Zhang, B., Hassan, M., Li, P., Qi, C., & He, Y. (2020). Enhanced catalytic activation of photo-Fenton process by Cu0·5Mn0·5Fe2O4 for effective removal of organic contaminants. *Chemosphere, 247*, 125780. https://doi.org/10.1016/j.chemosphere.2019.125780

Yuan, L., Zhang, H., Yu, H., Xu, R., Ji, C., & Zhang, W. (2025). Efficient capture of 99TcO4-/ReO4- via node and linker bifunctional anion exchange covalent organic frameworks. *Journal of Hazardous Materials, 487*, 137289. https://doi.org/10.1016/j.jhazmat.2025.137289

Zeng, S., & Kan, E. (2021). Adsorption and regeneration on iron-activated biochar for removal of microcystin-LR. *Chemosphere, 273*, 129649. https://doi.org/10.1016/j.chemosphere.2021.129649

Zhang, S., Gitungo, S., Dyksen, J. E., Raczko, R. F., & Axe, L. (2021). Indicator compounds representative of contaminants of emerging concern (CECs) found in the water cycle in the United States. *International Journal of Environmental Research and Public Health, 18*(3), 1288. https://doi.org/10.3390/ijerph18031288

Zhou, T., Lim, T.-T., Chin, S.-S., & Fane, A. G. (2011). Treatment of organics in reverse osmosis concentrate from a municipal wastewater reclamation plant: Feasibility test of advanced oxidation processes with/without pretreatment. *Chemical Engineering Journal, 166*(3), 932–939. https://doi.org/10.1016/j.cej.2010.11.078

CHAPTER

9

Application of electrochemical processes for microplastics and other recalcitrant pollutant removals in water

Nguyen Thi Hong and Tran Le Luu

Master Program in Water Technology, Reuse and Management, Vietnamese-German University, Ho Chi Minh City, Vietnam

List of Abbreviations

AEMs	Anion-Exchange Membranes
AOPs	Advanced Oxidation Processes
AOX	Adsorbable Organic Halides
BDD	Boron-doped diamond
BMED	Bipolar Membrane Electrodialysis
BOD	Biological Oxygen Demand
CEMs	Cation-Exchange Membranes
COD	Chemical Oxygen Demand
DBPs	Disinfection By-Products
DSA	Dimensionally Stable Anode
EAOPs	Electrochemical Advanced Oxidation Processes (EAOPs)
EC	Electrocoagulation
EDIS	Electrodisinfection
EDL	Electrodialysis
EF	Electro-Fenton
EFLT	Electrofloatation
EO	Electrooxidation
LCA	Life-cycle assessment
MMO	Mixed Metal Oxide
PE	Polyethylene
PET	Polyethylene terephthalate
PP	Polypropylene
PPCP	Pharmaceuticals and Personal Care Products
PS	Polystyrene
PVC	Polyvinyl Chloride
TDSs	Total Dissolved Solids
UV	Ultraviolet
WWT	Wastewater treatment

Sustainable Water Management. https://doi.org/10.1016/B978-0-443-44853-9.00004-6

1. Introduction

The global demand for clean water continues to rise, while aquatic ecosystems are increasingly impacted by pollution from municipal and industrial activities. Industrial effluents often contain complex mixtures of hazardous contaminants, including microplastics and persistent organic compounds (Salehi, 2022). Many of these substances are resistant to conventional water and wastewater treatment technologies (Amin, 2023). Typical recalcitrant pollutants include microplastics, synthetic dyes, pharmaceuticals, pesticides, and complex petrochemical compounds (Zainul et al., 2024). These contaminants are often poorly removed by biological treatment processes and can pass through treatment systems unchanged (Amin, 2023).

Advanced oxidation processes (AOPs) have been widely investigated to address these challenges (Amin, 2023). Among them, electrochemical technologies have attracted increasing attention due to their operational flexibility and strong oxidation capability. Electrocoagulation is effective for removing suspended solids, colloids, and microplastics through in-situ coagulant generation (Körbahti, 2014; Rakhmania et al., 2022).

Electrochemical oxidation, in contrast, focuses on the degradation of dissolved organic contaminants through direct and indirect oxidation pathways. These processes generate powerful oxidizing species, such as hydroxyl radicals, capable of treating compounds that are difficult to degrade biologically (Güven et al., 2012; Zhu et al., 2011). Electrochemical methods can operate under mild conditions, require limited chemical addition, and can be integrated into modular or decentralized treatment systems (Martínez-Huitle et al., 2015).

Collectively, existing studies highlight the vital role of electrochemical processes within sustainable water and wastewater treatment strategies and the broader framework of a circular economy (Huang et al., 2016; Radjenović & Sedlak, 2015).

2. Classification, characteristics and mechanisms of electrochemical processes

Electrochemical water treatment encompasses a variety of processes that remove or transform contaminants through mechanisms driven by electrical energy. These processes can be classified into electrocoagulation/electroflotation (EC/EF), electrooxidation (EO) including direct and mediated (indirect) pathways, electroreduction, electrodisinfection, and integrated hybrid approaches that pair electrochemical steps with other treatment modalities (e.g., UV, membranes, biological pretreatment) (Demir & Gören, 2019; Klidi et al., 2019; Nath, 2024). Each category relies on specific electrode configurations, materials, and operating conditions, which influence pollutant removal mechanisms, energy requirements, by-products formation, and suitability for different water and wastewater types (Demir & Gören, 2019; Feng et al., 2016; Klidi et al., 2019). Table 9.1 summarizes the classification of electrochemical processes with corresponding mechanisms, application fields, advantages and limitations.

Table 9.1 Classification of electrochemical processes.

Process class	Core mechanism	Application fields	Advantages	Limitations/ challenges	References
Direct electrochemical oxidation (DEO)	Direct electron transfer between pollutant and anode surface	Wastewater treatment; organic pollutant degradation	Simple operation; no chemical additives; high controllability	Limited to electroactive compounds; electrode fouling	Panizza and Cerisola (2009)
Indirect electrochemical oxidation (IEO)	Electro-generation of oxidants (e.g., Cl_2, HClO, persulfates) that oxidize pollutants in bulk	Industrial wastewater; textile effluents; saline waste streams	High oxidation capacity; effective for refractory pollutants	Formation of toxic by-products (e.g., chlorinated organics)	Martínez-Huitle and Brillas (2009)
Electrochemical advanced oxidation processes (EAOPs)	In situ generation of •OH radicals (e.g., BDD anodes)	Water and wastewater treatment; micropollutant removal	Non-selective mineralization; strong oxidizing power	High energy demand; expensive electrode materials	Sirés et al. (2014); Miklos et al. (2018)
Electro-fenton (EF)	Electro-generation of H_2O_2 and Fe^{2+} to produce •OH via Fenton reaction	Treatment of pharmaceuticals, dyes, pesticides	High degradation efficiency; operates at mild conditions	Requires acidic pH; iron sludge management	Oturan and Aaron (2014)
Photo-electro-fenton (PEF)	UV/visible light enhances Fe^{3+} reduction and radical regeneration	Advanced wastewater treatment; trace organic removal	Faster kinetics; improved mineralization	Higher operational complexity; light source costs	
Electro-coagulation (EC)	Sacrificial anodes generate metal hydroxides for pollutant aggregation	Industrial wastewater; heavy metals; oily wastewater	Low chemical input; effective solid–liquid separation	Sludge production; electrode consumption	Mollah et al. (2004)
Electro-flotation (EFlt)	Electro-generation of fine H_2/O_2 bubbles for flotation	Oil removal; algae harvesting; sludge thickening	High separation efficiency; compact systems	Energy consumption; limited oxidation capability	Chen (2004)
Electrodialysis (EDL)	Ion migration through selective membranes under electric field	Desalination; ion recovery; wastewater reuse	High selectivity; chemical-free separation	Membrane fouling; capital cost	Strathmann (2010)

Continued

Table 9.1 Classification of electrochemical processes.—cont'd

Process class	Core mechanism	Application fields	Advantages	Limitations/ challenges	References
Microbial electrochemical technologies (METs)	Bio-electrochemical oxidation/ reduction using electroactive microorganisms	Wastewater treatment; energy recovery; nutrient removal	Energy-positive potential; low sludge production	Slow kinetics; scale-up challenges	Logan et al. (2019)
Electrocatalytic conversion	Selective electrochemical transformation of substrates on catalysts	CO_2 reduction; plastic upcycling; hydrogen production	High selectivity; supports circular economy	Catalyst deactivation; system complexity	Cho et al. (2023)
Electrochemical reduction processes	Cathodic electron transfer (e.g., dechlorination, nitrate reduction)	Groundwater remediation; nitrate removal	Effective for halogenated compounds; reductive pathways	Competing hydrogen evolution; mass transfer limits	Rajeshwar et al. (2008)

2.1 Electrocoagulation and electroflotation (EC/EF)

Electrocoagulation generates coagulant species in situ at sacrificial anodes (often Al or Fe), which destabilize colloids and precipitate contaminants, allowing for their removal through flotation or sedimentation. Electoflocculation refers to the concurrent collection of flocs by buoyant forces. Key design parameters include electrode type, electrode material, current density, and treatment duration, which significantly affect the removal efficiency of pollutants such as microplastics, natural organic matter and dyes (Demir & Gören, 2019; Mansouri et al., 2012).

The approach is widely recognized for its effectiveness in treating microplastics, textile, paper, and industrial wastewater due to its simplicity and low chemical addition needs (Ardhianto & Bagastyo, 2019; Mahmoodi & Dalvand, 2013; Nath, 2024). Al–or Fe-based sacrificial electrodes are common, with their dissolution behavior and speciation directly affecting emerging contaminants, pollutant removal efficiency, and by-product management (Ardhianto & Bagastyo, 2019; Mansouri et al., 2012). Hybrid configurations coupling EC with EO or membrane processes have been shown to enhance performance in challenging effluents (Asaithambi et al., 2022; Sala & Gutiérrez-Bouzán, 2012).

EC has been applied to microplastics, textile, paper mill, and produced waters, achieving significant reductions in color and COD compared to conventional coagulation (Asaithambi et al., 2022; Mahmoodi & Dalvand, 2013). Hybrid EC/EO and EC–UV systems have been explored for improved reclamation and disinfection performance (Huang et al., 2016; Sala & Gutiérrez-Bouzán, 2012). Anodic dissolution of sacrificial electrodes (commonly Al or Fe) releases metal ions that hydrolyze to form coagulant species (e.g., $Al(OH)_3$, $Fe(OH)_3$), destabilizing colloids and allowing aggregation and flotation of microplastics or sedimentation of pollutants. Electroflotation enhances removal through buoyant transport of flocs (Ardhianto & Bagastyo, 2019; Demir & Gören, 2019).

General reactions:

- Al or Fe dissolution: Me → Me^{n+} (electrode dissolution under current)
- Hydrolysis to coagulants: $Me^{n+} + H_2O$ → Me(OH)x + protons; Me(OH)z precipitates destabilize colloids like microplastics
- Flocculation and removal: destabilized particles and microplastics aggregate and are removed by sedimentation or flotation

Effective removal of microplastics, natural organic matter, color and turbidity, metals, and certain dyes with low chemical dosing (Ardhianto & Bagastyo, 2019; Cotillas et al., 2014; Mansouri et al., 2012). EC/EF hybrids and coupling with EO or membranes exploit complementary mechanisms, improving overall removal and enabling disinfection or mineralization in challenging effluents (Asaithambi et al., 2022; Cotillas et al., 2014).

2.2 Electrooxidation (EO)

Electrooxidation or EO includes direct oxidation at the anode surface (direct EO) and indirect oxidation via in-situ formed oxidants such as hydroxyl radicals, ozone, persulfate, or other mediators (indirect EO). This modular approach enables the mineralization or transformation of a wide range of organic like microplastics and inorganic contaminants with tunable selectivity and mineralization potential. Common anode materials include boron-doped diamond (BDD) and dimensionally stable anodes (DSA) like $Ti/RuO_2–IrO_2$, each offering distinct reactivity, durability, and energy efficiency (Demir & Gören, 2019; Ogundele et al., 2023).

EO performance depends on electrode material, current density, mass transfer, and wastewater matrix. Energy intensity can be high for diluting streams. For this reason, staged configurations and integration with biological or physical pretreatment are often required (Radjenović & Sedlak, 2015). EO has proven effective in treating microplastics, textile dyes, tannery waste, paper mill effluents, and produced waters, effectively achieving color and COD reductions and enabling decentralized treatment strategies when combined with hydrogen generation or other resource recovery methods (Zhu et al., 2008).

Pollutants are oxidized either directly at the anode surface (direct EO) or indirectly via in situ generated oxidants (indirect EO), such as hydroxyl radicals, ozone, or persulfates. This leads pollutants to mineralization to CO_2, nitrate, sulfate, and other inorganic end products, or transformation to more biodegradable intermediates (García-Montoya et al., 2015; Ogundele et al., 2023). Highly reactive materials such as boron-doped diamond (BDD) and dimensionally stable anodes (e.g., $Ti/IrO_2–RuO_2$) promote robust hydroxyl radical generation and high mineralization potential. However, their energy intensity and costs vary depending on the wastewater matrix and pollutant load (Ogundele et al., 2023; Zhu et al., 2008).

Reaction pathways:

- Direct EO: Pollutant such as microplastics + electrode surface leads to electron transfer, generating oxidized intermediates.
- Indirect EO: Water oxidation or mediators produce hydroxyl radicals (•OH) that non-selectively attack pollutants.

EO is capable of breaking down chemically robust organics (microplastics, dyes, phenolics, pesticides, pharmaceuticals) but can be energy-intensive in diluting wastewater; integration with pretreatment or hybridization can mitigate energy demands while preserving mineralization potential (Bazrafshan et al., 2013; Ogundele et al., 2023; Zhu et al., 2008).

2.3 Electrodisinfection (EDIS)

Electrodisinfection involves in situ oxidation or reduction processes that inactivate pathogens, leveraging EO pathways, EC/EF by-products, or direct electrode reactions. This modality is particularly applicable to municipal and industrial effluents where pathogen load poses a concern, with performance contingent on electrode materials and configurations that favor microbial oxidation or reduction while minimizing harmful by-products (Akarsu et al., 2021; Gonzalez-Rivas et al., 2019). Recent reviews highlight the effectiveness of electrocoagulation, direct and indirect electrooxidation in pathogen removal from various water sources, emphasizing the importance of electrode selection and process integration in ensuring effective disinfection (Akarsu et al., 2021; Gonzalez-Rivas et al., 2019).

One significant advantage of electrodisinfection is its capacity to achieve high pathogen inactivation efficiencies with minimal contact times, making it particularly suitable for dynamic treatment scenarios, such as those found in wastewater treatment and drinking water purification (Hasanbiki et al., 2016; Raut et al., 2019). Furthermore, electrodisinfection does not rely solely on the addition of chemical disinfectants, which can lead to harmful by-products. Instead, it generates disinfectants in situ through electrochemical reactions (Radjenović & Sedlak, 2015). For instance, studies have demonstrated the effectiveness of electrochemical processes in achieving significant reductions in various microbial species, including *Escherichia coli*, within several minutes of treatment (Hasanbiki et al., 2016; Raut et al., 2019).

Different configurations of electrochemical systems can enhance the performance of electrodisinfection. For example, the use of boron-doped diamond (BDD) electrodes has been shown to improve the efficiency of bacterial inactivation by providing a stable and highly oxidative environment conducive to disinfection (Göltz et al., 2019). The coupling of electrocoagulation with electrochemical oxidation has also been identified as a robust method for water purification, resulting in both the removal of suspended particles and the inactivation of pathogens, thus optimizing the overall disinfection process (Chen et al., 2020, 2021).

Moreover, electrochemical technologies are evolving to address the challenges posed by antibiotic-resistant bacteria, which pose a significant public health threat. Innovative electrochemical disinfection systems have shown promise in effectively removing intracellular antibiotic resistance genes along with the pathogen populations themselves, thereby reducing the prevalence of antibiotic resistance (Chen et al., 2020; Mandal, 2024). This dual action not only contributes to pathogen control but also mitigates the spread of resistance within microbial communities (Chen et al., 2020).

Electrodisinfection stands out as a promising alternative to conventional disinfection methods such as chlorination and UV treatment. Its ability to generate disinfectants without residual chemicals makes it an environmentally friendly and potentially more effective choice for ensuring safe drinking water and treating wastewater (García-Espinoza et al., 2021). Continued research and optimization of electrochemical systems will enhance their application in various contexts, ranging from municipal water supplies to industrial effluent treatment. Pathogens are inactivated via oxidative and/or reductive electrochemical processes, leveraging EO pathways or EC/EF by-products. Pathogen destruction is enhanced by strong oxidants, localized pH changes, and reactive chlorine species when applicable, with performance dependent on electrode materials and design that favor microbial oxidation while minimizing toxic by-products (Akarsu et al., 2021; Gonzalez-Rivas et al., 2019).

Selection of materials (e.g., well-suited coatings and electrodes) and reactor design to maximize microbial inactivation while controlling disinfection by-products stays a focus of recent reviews and engineering studies (Akarsu et al., 2021; Gonzalez-Rivas et al., 2019).

2.4 Electro-Fenton (EF)

The electro-Fenton (EF) technique represents a significant advancement in the field of advanced oxidation processes (AOPs), effectively utilized for the degradation of microplastics and organic pollutants in various wastewater sources. This method enhances the conventional Fenton reaction by integrating electrochemical principles, facilitating the in-situ generation of hydrogen peroxide (H_2O_2). The electrochemical reduction of oxygen occurs at the cathode, producing H_2O_2, which subsequently reacts with ferrous ions (Fe^{2+}) to yield hydroxyl radicals (·OH) - powerful oxidants capable of mineralizing a wide array of organic contaminants without producing harmful by-products (Davarnejad et al., 2019; Midassi et al., 2020). Moreover, the electro-Fenton system offers distinct advantages over traditional Fenton methods, including improved operational control and the potential to produce less sludge due to lower amounts of iron precipitates compared to homogeneous processes (Davarnejad et al., 2019). For instance, the treatment efficiency for chloroquine degradation showcases the electro-Fenton method's ability to effectively combat emerging contaminants within pharmaceutical waste streams (Midassi et al., 2020). Performance optimization remains critical in electro-Fenton applications, as several factors can significantly influence degradation efficiency. These factors include the concentration of Fe^{2+}, the current density, the initial pH of the solution, and the duration of electrolysis (Davarnejad et al., 2019).

Recent studies have employed methodologies such as response surface methodology to optimize these operational parameters, enhancing overall treatment effectiveness and economic viability (Davarnejad et al., 2019; Ghernaout et al., 2020). EF relies on the situ generation of Fe^{2+} and H_2O_2 to produce hydroxyl radicals (•OH) via Fenton chemistry, often enhanced by electrochemical regeneration of the Fe^{2+}/Fe^{3+} redox couple at the cathode. Heterogeneous EF employs Fe-functionalized catalysts to work at near-neutral pH, improving catalyst stability and reusability (Alavi et al., 2019; Bazrafshan et al., 2013; Duesterberg et al., 2008). Representative reactions:

- $Fe^{2+} + H_2O_2 \rightarrow Fe^{3+} + {}^{\bullet}OH + OH^-$ (classic Fenton step)
- Regeneration: $Fe^{3+} + e- \rightarrow Fe^{2+}$ (electrochemical reduction at the cathode)
- H_2O_2 generation can be supplemented by electrochemical means or in situ from O2 reduction.

EF enables mineralization of refractory organics (e.g., dyes, phenolics, pesticides) under milder pH conditions than classical Fenton, with opportunities for heterogeneous catalysts to improve process integration with EC and EO (Bazrafshan et al., 2022; Duesterberg et al., 2008; Zhang et al., 2015). EF is frequently discussed in the context of integrated or hybrid AOPs for water and wastewater treatment.

2.5 Electrodialysis (EDL)

Electrodialysis (EDL) is a significant electrochemical membrane process used for water desalination and the treatment of distinct types of wastewater. Utilizing the principle of selective ion transport through ion-exchange membranes under the influence of an electric field, EDL facilitates the separation of ions from water, making it an essential technology for resource recovery and water purification (Gurreri et al., 2020; Xu et al., 2021). One of the primary advantages of electrodialysis is its ability to selectively concentrate specific ions or contaminants while allowing for adjustable desalination levels based on operational parameters. This tunability enables the precise control of ion composition in the treated water by adjusting factors such as voltage, flow rates, and the configuration of the electrodialysis stack (Li et al., 2025).

For instance, research has shown that electrodialysis can reclaim municipal wastewater for irrigation and effectively reduce total dissolved solids (TDSs) by adjusting operational metrics (Xu et al., 2021). This makes it particularly suitable for treating highly saline industrial wastewater and recovering valuable resources, thereby enhancing the sustainability of water usage (Gurreri et al., 2020; Xu et al., 2023).

Electrodialysis has also been applied to the recovery of specific nutrients and pollutants from wastewater. Studies indicate that this technology can effectively concentrate nitrogenous compounds from farm effluents, demonstrating promising results in fertilizer production (Tahmid et al., 2024). Moreover, ongoing research emphasizes the potential to utilize electrodialysis in the recovery of valuable metals from industrial wastewater, contributing to waste management and resource efficiency strategies (Cerrillo-Gonzalez et al., 2024).

The development of advanced ion-exchange membranes, such as composite and modified membranes designed to reduce fouling and enhance selectivity, has further improved the efficiency and effectiveness of electrodialysis (Müller et al., 2017). Acknowledging the challenge of membrane fouling, recent studies focus on enhancing membrane lifetime and operational effectiveness by investigating materials that minimize scaling and other deposit formations (Choi et al., 2025). Electrodialysis (EDL) is an electromembrane separation process that relies on the migration of ions through selective ion-exchange membranes under the influence of an electric field.

The process primarily consists of cation-exchange membranes (CEMs) and anion-exchange membranes (AEMs) arranged alternately within a modular configuration to form distinct compartments where ion concentration changes occur (Yan et al., 2016). This arrangement facilitates the selective migration of cations toward the cathode and anions toward the anode, leading to the formation of concentrated and dilute solutions as a result of ion transport driven by the applied electric potential difference (Rodrigues et al., 2008; Yan et al., 2016).

The efficiency of electrodialysis hinges upon several factors, including the properties of the membranes, the nature of the ionized solutions, and operational parameters such as flow rates and current density (Mourad & Béchir, 2016). Cation and anion exchange membranes are engineered to exhibit high permeability for their respective ions while maintaining high selectivity against ions of opposite charge, which underpins the effectiveness of the separation process (Falina, I. et al., 2021).

Furthermore, the conductivity of the solution influences the current utilization and overall energy efficiency, thus impacting the process's viability in practical applications (Mourad & Béchir, 2016). Electrodialysis is particularly advantageous due to its continuous nature, contrasting with batch methods, which allows for enhanced process sustainability when paired with other membrane technologies (Haerens, K. et al., 2012).

Recent advancements have introduced bipolar membrane electrodialysis (BMED), which allows for the generation of H^+ and OH^- ions from water at the membrane interface. This significantly enhances the versatility of educts without the additional need for external chemical reagents (Baek et al., 2018; Balster, J. et al., 2010). BMED can also facilitate the recovery of valuable organic acids and other compounds, thus expanding its application range beyond traditional desalination to areas such as wastewater treatment and resource recovery (Handojo et al., 2019; Xu & Huang, 2008). The ability to efficiently recover and concentrate specific materials from a solution makes electrodialysis a compelling option in industrial processes, including those in chemical manufacturing, food science, and biofuel production (Handojo et al., 2019; Pagáč et al., 2022).

Figure 9.1 gives comparison of main electrochemical processes including EC/EF, EO, EF and EDL for water treatment. In this figure, illustrations of the four primary electrochemical treatment mechanisms are (A) Electrocoagulation/Electroflotation (EC/EF) with sacrificial electrode dissolution and floc formation; (B) Electrooxidation (EO) showing direct and indirect oxidation pathways; (C) Electro-Fenton (EF) with in-situ H_2O_2 and hydroxyl radical generation; (D) Electrodialysis (EDL) with selective ion transport through ion-exchange membranes.

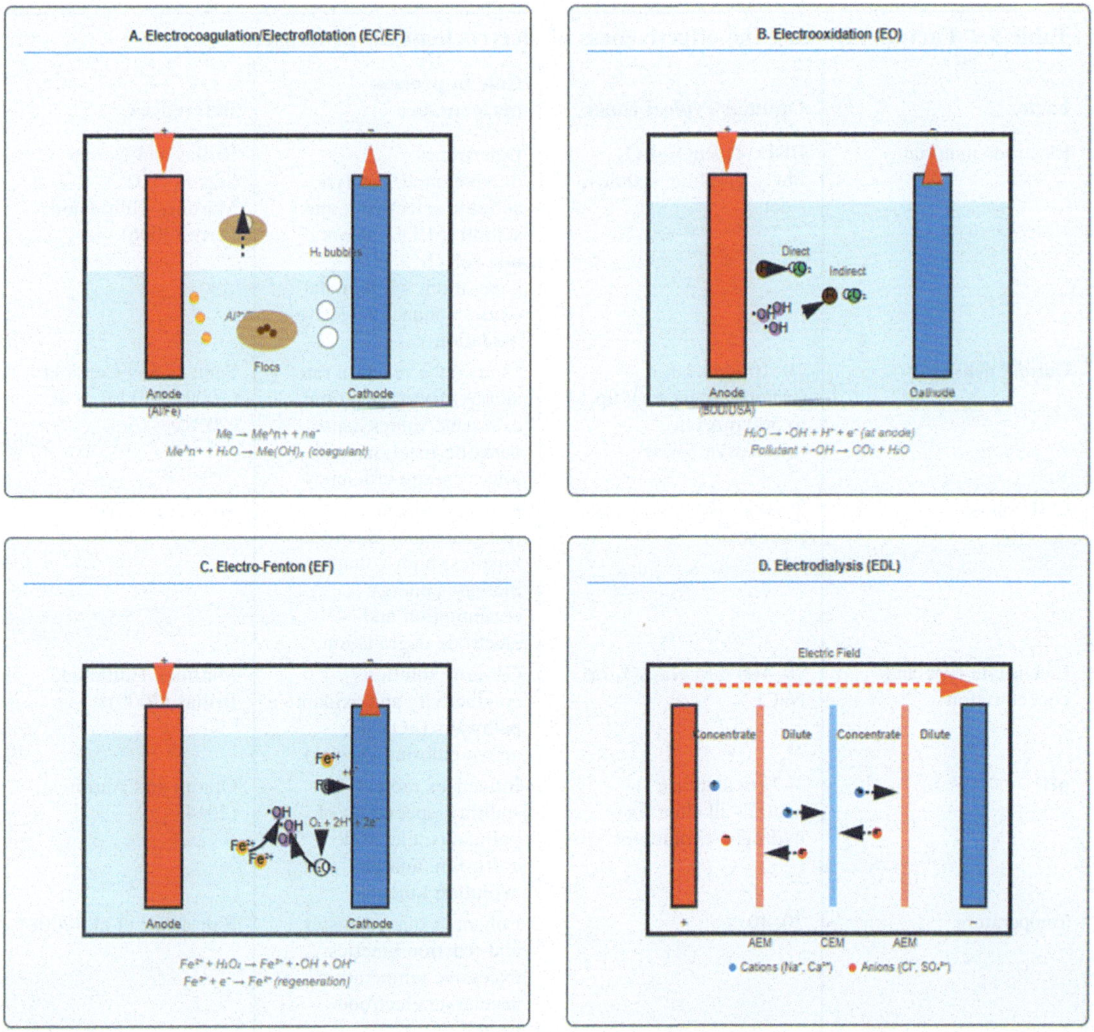

FIGURE 9.1

Schematic comparison of main electrochemical water treatment processes.

3. Factors affecting the effectiveness of electrochemical water treatment

Electrochemical water treatment leverages EC/EF, EO, electrodisinfection, and hybrid configurations, with performance controlled by a set of interrelated factors. Robust design and operation require understanding how electrode materials, cell geometry, operating conditions, and wastewater characteristics couple to influence contaminant removal as microplastics, energy use, and by-product formation (Demir & Gören, 2019; Klidi et al., 2019; Nath, 2024). Table 9.2 summaries all factors that affect the effectiveness of electrochemical processes.

Table 9.2 Factors affecting the effectiveness of electrochemical processes.

Factor	Optimal/Typical range	Role in process performance	References
Electrode material	BDD, Ti/RuO_2–IrO_2, PbO_2, graphite, stainless steel	Determines overpotential, catalytic activity, selectivity, and stability; BDD favors non-selective •OH generation, while metal oxides enhance selective oxidation	Brillas and Garcia-Segura (2025); Martínez-Huitle and Ferro (2006)
Current density	10–100 mA cm^{-2} (water/wastewater); up to 500 mA cm^{-2} (electrosynthesis)	Control the reaction rate and oxidant generation; excessive values cause parasitic reactions and lower energy efficiency	Panizza and Cerisola (2009); Miklos et al. (2018)
Cell voltage	2–10 V (typical lab and pilot scale)	Reflects system resistance and electrode kinetics; high voltage increases energy consumption and electrode degradation	Sirés et al. (2014)
Electrolyte type and concentration	10–100 mM Na_2SO_4 or NaCl	Governs solution conductivity and oxidant pathways (•OH vs. active chlorine species)	Martínez-Huitle and Brillas (2009)
pH	3–7 (oxidation); neutral–alkaline for hydrogen evolution	Influences radical stability, speciation of pollutants, electrode corrosion, and gas evolution kinetics	Oturan and Aaron (2014)
Temperature	20–40 °C	Enhances mass transfer and reaction kinetics; excessive temperatures accelerate electrode wear	Rajeshwar et al. (2008)

Table 9.2 Factors affecting the effectiveness of electrochemical processes.—cont'd

Factor	Optimal/Typical range	Role in process performance	References
Interelectrode distance	0.5–2.0 cm	Affects ohmic resistance and energy efficiency; shorter distances reduce voltage losses	Muddemann et al. (2019)
Hydraulic retention time (HRT)	10–120 min (flow systems)	Determines contact time between pollutants and reactive species	Miklos et al. (2018)
Mass transfer conditions	Turbulent flow or stirring (Re > 2000)	Enhances transport of pollutants to electrode surface, reducing diffusion limitations	Panizza and Cerisola (2009)
Pollutant concentration	Process-specific (mg L^{-1} to g L^{-1})	Influences apparent kinetics; high concentrations may reduce current efficiency	Sirés et al. (2014)
Presence of radical scavengers	Low bicarbonate/natural organic matter	Competes with target pollutants for •OH, reducing degradation rates	Miklos et al. (2018)
Electrode surface area/ Geometry	High specific surface (porous, mesh, 3D electrodes)	Improves current distribution and reaction rates	Walsh and Ponce de León (2018)
Operating mode	Galvanostatic preferred scale-up	Provides stable oxidant generation and reproducibility	Brillas and Garcia-Segura (2025)

3.1 Electrode materials and configurations

The effectiveness of electrochemical water treatment processes is influenced by multiple interdependent operational and design parameters that collectively determine treatment efficiency, energy consumption, and pollutant removal rates. Material choice for anodes and cathodes significantly shapes the efficiency, selectivity, and energy intensity of electrochemical wastewater treatment. Boron-doped diamond (BDD) electrodes are notable for generating highly reactive hydroxyl radicals and achieving strong mineralization of organics, albeit with higher capital costs and energy demands that must be balanced by process intensity and pollutant load (Jiang et al., 2008; Kim, 2021; Zhu et al., 2008).

Alternative anodes such as dimensionally stable anodes (DSA) and metal oxide coatings (e.g., Ti/IrO_2–RuO_2) have demonstrated effective oxidation with favorable durability and cost profiles for specific contaminants (Gu & Zheng, 2015; Martínez-Huitle & Ferro, 2006).

On the cathodic side, inert materials and configurations that support hydrogen evolution or coupling with hydrogen production workflows can be helpful in integrated systems (Dermentzis et al., 2022; Jiang et al., 2008). EC configurations rely on sacrificial metal electrodes to generate coagulation-active species in situ, with current density, electrode type, and treatment time identified as key design parameters influencing removal efficiency for targeted pollutants (e.g., microplastics, heavy metals) (Ardhianto & Bagastyo, 2019; Demir & Gören, 2019).

Electrode choice governs dissolution behavior (EC), coagulant generation, and oxidant production (EO), as well as durability and side-reactions. Configuration (monopolar, bipolar, galvanic pairs) affects current distribution, mass transfer, and coupled mechanisms (EC/EF with EO, etc.) (Bote, 2021; Demir & Gören, 2019; Ogundele et al., 2023).

Sacrificial anodes (Al, Fe) in EC provide in situ coagulants ($Al(OH)_3$, $Fe(OH)_3$); dissolution rate and speciation control removal efficiency (Mansouri et al., 2012; Muthukumar et al., 2004). In EO, inert/oxide-coated anodes (e.g., $Ti/IrO_2–RuO_2$) or BDD offer different oxidant generation and mineralization capabilities, influencing energy intensity and pollutant scope (Jiang et al., 2008; Ogundele et al., 2023). Hybrid and galvanic arrangements can improve overall efficiency and enable concurrent pollutant removal as microplastics and energy/resource recovery (Cotillas et al., 2014).

3.2 Current density

Primary driver of electrode reactions; determines rate of coagulant release in EC, oxide formation, and oxidant generation in EO, and overall electron transfer rates (Bazrafshan et al., 2013; Bote, 2021). Higher current density accelerates pollutant removal and mass transfer but raises energy consumption and may increase excessive metal dissolution or formation of undesirable by-products; there is often an optimum region beyond which gains plateau increases substantially (Bote, 2021). Use design-of-experiments to locate operating windows where removal is maximized with acceptable energy use; monitor electrode dissolution as a proxy for coagulant dose and energy expenditure (Bote, 2021).

3.3 Treatment time (contact time)

Controls total exposure of pollutants to coagulants, oxidants, and radical species; dictates cumulative removal for EC/EF and EO processes. Longer times improve microplastics, color, COD removal and metal/organic capture but increase energy use; diminishing returns occur beyond process-specific optimum (Bote, 2021). Optimize time to achieve target removals with minimal energy; employ response-surface or design of experiment (DOE) -based optimization for each wastewater matrix (Bote, 2021; Sudibyo et al., 2016).

3.4 Inter-electrode distance and reactor design

Determine solution resistance, mass transfer, bubble management (gas evolution in EC/EF/EO), and distribution of current density (Bazrafshan et al., 2013). Shorter distances reduce ohmic losses and energy demand but may require more careful gas management and mixing, geometry influences flocculation efficiency and oxidant delivery to pollutants. Engineer for low energy consumption per unit COD removed (kWh/m^3) and confirm with pilot tests to balance mass transfer and reactor hydraulics (Bazrafshan et al., 2013).

3.5 Conductivity and electrolyte composition

Conductivity governs current efficiency and the distribution of electrochemical reactions; electrolyte ions can participate in or influence coagulation chemistry and EO pathways (Bote, 2021). Increased conductivity lowers cell voltage for a given current but can alter coagulant speciation; electrolyte additives can buffer pH and modulate by-products. Increase conductivity to enable desired current densities while maintaining effluent quality; monitor potential formation of secondary pollutants from electrolyte ions (Bote, 2021).

3.6 pH and wastewater matrix

pH affects metal hydroxide speciation, coagulation efficiency, and EO/EF reaction pathways; matrix composition (microplastics, organic load, salts, ammonia, chlorides) interacts with mechanism efficiency (Bazrafshan et al., 2013; Bote, 2021). EC with Al/Fe coagulants often performs best at pH ranges where hydroxide coagulants predominate; EO oxidant generation efficiency and selectivity can be pH-dependent; complex matrices may require buffering or pretreatment to maintain optimal conditions (Bazrafshan et al., 2013). Implement pH control or buffering when necessary; tailor operating pH to the chosen electrode/coagulant system and target contaminants including microplastics; anticipate matrix effects in real wastewater (Bazrafshan et al., 2013; Bote, 2021).

3.7 Hydraulic and mass-transfer conditions

Flow regime, mixing intensity, and residence time influence pollutant such as microplastics transport to reactive zones and flocculation kinetics; poor mass transfer limits EO and EC performance, especially for dilute streams (Bazrafshan et al., 2013). Enhanced mixing improves pollutant delivery to electrodes and flocculation but increases energy use; in EO, mass transfer limitations can bottleneck radical generation and pollutant such as microplastics encounter rates (Bazrafshan et al., 2013). Design reactors to ensure adequate mixing without excessive energy penalties; consider pulsed or staged current strategies to optimize mass transfer while reducing energy demand (Bote, 2021).

3.8 Wastewater composition and contaminants

Matrix composition (as microplastics, dyes, organics, heavy metals, hardness, turbidity, salinity) determines which mechanisms are most effective and how they should be operated (Bazrafshan et al., 2013; Demir & Gören, 2019; Nath, 2024). Microplastics or textile effluents with dyes may respond well to EC/EF and EO depending on coagulant performance and dye chemistry; produced waters with high salinity or metals may require tailored electrode choices and pretreatment (Ardhianto & Bagastyo, 2019; Demir & Gören, 2019). Characterize target pollutants and conduct batch/mini-pilot studies to name dominant removal pathways and energy footprints; use hybrid configurations when single mechanisms are insufficient (Asaithambi et al., 2022; Bazrafshan et al., 2013).

3.9 Hybridization and energy coupling

Combining EC/EF with EO, membranes, UV, or biological pretreatment can shift parameter sensitivities and improve overall performance; energy-coupled or solar-driven setups offer opportunities for energy recovery (e.g., hydrogen) (Asaithambi et al., 2022; Cotillas et al., 2014; Dermentzis et al., 2022). Hybrids can reduce the required intensity of any single process, lower energy per unit pollutant removed, and enable treatment of microplastics or complex wastewater; they may introduce modern design variables (flow split between modules, inter-stage pH, inter-stage mixing) (Asaithambi et al., 2022; Bazrafshan et al., 2013). When designing hybrids, treat current density, time, and inter-stage transfer separately as tunable levers within an integrated control strategy; assess overall energy balance and potential co-product streams (e.g., H_2) (Asaithambi et al., 2022; Dermentzis et al., 2022).

The performance of any electrochemical technique is not inherent to the method alone but is critically dependent on a set of interrelated operating parameters. Optimizing these conditions is essential for achieving high efficiency, minimizing energy consumption, and reducing operational costs as in Table 9.1.

4. Application of electrochemical technologies for water treatment

4.1 Purposes of application of electrochemical processes

The application of electrochemical processes for water and wastewater treatment spans various fields, showcasing their versatility and efficiency in removing pollutants and microplastics to enhance water quality. The following outlines key areas in which these innovative techniques are increasingly utilized, supported by recent studies and findings.

- **Electrocoagulation**: This technique combines electrochemical processes with coagulation to remove microplastics, suspended solids, heavy metals, and organic dyes from wastewater. For instance, Ma et al. demonstrated that electrocoagulation can effectively purify water by inducing a coagulation process through current application, highlighting its potential for treating industrial effluents laden with contaminants (Ma et al., 2018). The integration of advanced oxidation processes with electrocoagulation has shown remarkable results in removing arsenic from groundwater (Oreščanin et al., 2013).
- **Advanced Oxidation Processes (AOPs)**: Electrochemical AOPs, which generate strong oxidants like hydroxyl radicals through the electrochemical activation of oxidants, have emerged as powerful tools for degrading persistent organic pollutants in wastewater as microplastics. Recent reviews indicate that this technique is effective against a wide range of pollutants, including pharmaceuticals and industrial chemicals (Alkhadra et al., 2022; Sirés et al., 2014). Studies also confirm the utility of AOPs in producing reactive oxidants on-site, thus improving overall process efficiency (Chen et al., 2018).
- **Desalination and Ion Separation**: Electrochemical methods play a critical role in water desalination and ion separation processes, particularly through electrodialysis. This technique employs ion-exchange membranes to selectively transport ions, facilitating the removal of contaminants from water. For example, electrodialysis has been effectively applied in treating brackish water and for resource recovery purposes (Gurreri et al., 2020; Yang et al., 2020). The technology not only recycles water but also enhances water purification without introducing hazardous chemicals (Simon et al., 2018).

- **Disinfection of water**: The electrochemical disinfection of drinking water and wastewater has gained prominence, with studies highlighting the effectiveness of electrochemical methods in inactivating microorganisms. Methods such as the generation of reactive oxygen species have been shown to provide effective disinfection results in various water matrices (Jeong et al., 2006; Moreno-Andrés et al., 2018). These processes are especially important for treating wastewater from industrial processes, where chemical disinfectants may be inadequate or harmful.
- **Heavy metal removal**: Focused electrochemical techniques have shown significant promise in the removal of heavy metals from effluents, including galvanic wastewater. Helesh et al. reported that electrochemical methods could effectively purify wastewater from nickel ions, demonstrating the technology's applicability for treating heavy metal-laden industrial discharges (Helesh et al., 2023). The ease of integrating these techniques with existing wastewater treatment infrastructures presents a distinct advantage.
- **Use of nanomaterials in electrochemical processes**: Recent advances in materials science have led to the development of nanomaterials that enhance the performance of electrochemical techniques. For instance, carbon-based materials and titanium oxide photocatalysts have been explored for their role in improving the efficiency of EOPS in wastewater treatment (Gayen et al., 2018; Zaky & Chaplin, 2013). These innovative materials contribute to better electrocatalytic performance and higher energy efficiency, facilitating the degradation of complex organic pollutants (Zhang et al., 2015).
- **Bioelectrochemical systems**: Integrating biological processes with electrochemical methods offers a novel approach for wastewater treatment. Bioelectrochemical systems combine microbial metabolism with electrochemical reactions to treat contaminated waters while potentially generating energy and valuable bioproducts (Agudelo-Escobar & Erazo Cabrera, 2024). This dual functionality enhances sustainability and resource recovery, marking a significant trend in wastewater management.
- **Microplastic Removal**: When aluminum or iron electrodes are operated under low current densities, metal hydroxides are generated directly in the water. These species destabilize suspended microplastics by adsorption and enmeshment, leading to the formation of larger aggregates that can be removed by settling or flotation (Holt et al., 2005; Mollah et al., 2004). Electroflotation is used simultaneously, as the fine gas bubbles produced at the electrodes attach to the aggregates and enhance their separation at the water surface (Perren et al., 2018). Reported removal efficiencies for common polymers such as polyethylene and polypropylene are typically above 80%, even at small particle sizes (Xiahong et al., 2022).

Electrochemical oxidation is not applied as a primary removal pathway. Instead, limited oxidation is considered only where surface modification improves aggregation behavior. Complete electrochemical breakdown of polymers is avoided due to high energy demand and uncertain by-products (Martínez-Huitle & Brillas, 2009; Sirés et al., 2014). The main advantages of the electrochemical approach are operational control and reduced chemical dosing. However, electrode consumption and electricity demand, must be addressed at the design stage (Miklos et al., 2018). For this reason, the electrochemical unit is treated as a modular, supporting process, integrated with downstream solid removal and containment rather than as a standalone solution.

4.2 Application fields

Electrochemical approaches have proved applicability to a broad spectrum of wastewater types, including microplastics, textile effluents, papermaking and plastic industry effluents, refinery and produced water, and highly concentrated streams with refractory organics. For textile and dye-containing wastewater, EAOPs and EC have shown strong removal performance with the potential for color removal and COD reduction, while avoiding excessive chemical consumption and secondary pollution (Drušković et al., 2021; Mahmoodi & Dalvand, 2013). In industrial and electronic contexts, electrooxidation with proper electrode materials has achieved effective treatment of high-load organics and pollutants, sometimes enabled simultaneous production of hydrogen and oxidized by-products, and provided a path toward decentralized or modular treatment platforms (Huang et al., 2016).

Integrated approaches combining biological pretreatment with electrochemical oxidation have also been explored to address highly toxic microplastics or recalcitrant pesticides and other organics, illustrating the potential to couple biological and electrochemical steps for enhanced treatment efficiency (Liu et al., 2010). One significant contribution to the field is the work of Güven et al., who reported that electrochemical oxidation processes, while energy-intensive, are effective for treating wastewater from the food industry. Their findings emphasized the importance of considering economic viability alongside treatment efficiency when implementing electrochemical methods (Güven et al., 2012).

Gopal et al. focused on the removal of chromium (III) from tannery wastewater using electrochemical peroxidation, achieving notable reductions in pollutant concentrations, showcasing the versatility of electrochemical techniques in addressing various organic pollutants (Gopal et al., 2022). The adaptability of electrochemical methods is further highlighted by Karichappan et al., who optimized a continuous electrocoagulation technique for treating rice mill wastewater. Their study provides a framework for modeling and improving the treatment process through response surface methodologies (Karichappan et al., 2013). The urgency of addressing pharmaceutical contaminants in wastewater has also been emphasized by various researchers. Moreover, Lan et al. presented data on the efficacy of using boron-doped diamond electrodes for the electrochemical oxidation of wastewater laden with pharmaceuticals, effectively reducing micropollutants (Lan et al., 2017).

In textile wastewater treatment, electrochemical techniques have shown substantial benefits. Studies by Miyata et al. and Kariyajjanavar et al. noted that electrochemical oxidation achieved high removal rates for persistent organic contaminants, such as microplastics, tetracycline and reactive azo dyes, respectively, underscoring the technique's applicability to challenging pollutants (Karichappan et al., 2013; Miyata et al., 2011). These findings demonstrate that electrochemical treatments can provide solutions for waste streams that are typically difficult to manage through conventional biological or physical methods.

The potential of electrochemical methods extends beyond organic pollutants to include the removal of heavy metals. Hewehy et al. explored the application of carbon and aluminum electrodes for treating electroplating wastewater, illustrating that electrochemical approaches can achieve significant metal reduction and function as a sustainable alternative to traditional treatment methods (Hewehy et al., 2016). Similarly, Zhang et al. introduced a multistage electrochemical flow-through mode, emphasizing innovative frameworks that can enhance operational efficiency and pollutant removal in wastewater systems (Zhang et al., 2021). Electrochemical approaches (EO, EC/EF, EF, electro-disinfection, and hybrids) address a wide spectrum of pollutants, including dyes, refractory organics, microplastics, pharmaceuticals, pesticides, metals, ammonia/nitrogen species, and organic micropollutants.

This broad applicability is documented across plastic industry, textile, tannery, paper, dairy, and municipal wastewater, with electrode material and reactor design being primary levers to tailor performance (Ganiyu et al., 2018; Martínez-Huitle et al., 2015). The breadth is reinforced by demonstrations of simultaneous pollutant removal and mineralization in diverse matrices and configurations (Ganiyu et al., 2018; Medel et al., 2019). The emergence of hybrid configurations that couple electrochemical steps with membranes, biological pretreatment, or photocatalysis is a recurring finding, reported to improve removal efficiency, broaden the pollutant spectrum, and enable energy/resource recovery in appropriate setups (Ganiyu et al., 2018; Medel et al., 2019).

These hybrids are particularly highlighted for industrial effluents where single-step electrochemical approaches struggle with complex matrices (Ensano et al., 2016; Medel et al., 2019). Direct EO and indirect EO (mediated by in situ oxidants such as hydroxyl radicals) provide complementary oxidation pathways, enabling both selective pollutant transformation and broad mineralization in a single platform. The choice of anode material (e.g., BDD, Ti-based DSA, PbO2) strongly modulates oxidant generation, mineralization potential, and energy intensity, guiding electrode selection for specific contaminant classes (Ganiyu et al., 2018; Simon et al., 2018).

This mechanistic flexibility is echoed in reviews that categorize electrochemical processes into EC/EF, EO, EF, and their hybrids, highlighting the ability to tune oxidation and coagulation pathways to match wastewater composition (Martínez-Huitle et al., 2015; Ganiyu et al., 2018). EC/EF operates via in situ generation of coagulants ($Al(OH)_3$, $Fe(OH)_3$) from sacrificial anodes, destabilizing colloids and enabling removal via flotation or sedimentation. This mechanism has been shown to be effective for color, COD, turbidity, and metal removal in microplastics, textile, dairy, refinery, and stormwater streams composition (Ensano et al., 2016; Martínez-Huitle et al., 2015). Hybrid EC/EF with EO or membranes often yields improved performance for recalcitrant streams (Ensano et al., 2016). EF relies on in situ generation of hydroxyl radicals through electro-Fenton chemistry, enabling mineralization of organics under a wide pH range, particularly with heterogeneous catalysts that enable near-neutral operation. This mechanism is highlighted for microplastics, dyes, pesticides, pharmaceuticals, and other complex organics, with ongoing development of heterogeneous EF catalysts to improve stability and integration with EC and EO steps (Huang et al., 2016; Martínez-Huitle et al., 2015).

4.3 Microplastic removal applications

For microplastic-contaminated water, electrochemical technologies are applied primarily for physical removal rather than chemical degradation. Electrocoagulation coupled with electroflotation promotes aggregation and separation of microplastic particles, enabling removal through sedimentation or flotation (Holt et al., 2005; Mollah et al., 2004; Perren et al., 2018). Reported removal efficiencies for common polymers such as polyethylene and polypropylene frequently exceed 70%–90% under optimized laboratory and pilot-scale conditions, depending on particle size and operating parameters (Akarsu et al., 2021; Xiahong et al., 2022).

Electrochemical oxidation is not applied as a primary microplastic degradation pathway due to high energy demand and uncertainty regarding transformation products. Instead, oxidation is limited to surface conditioning where improved aggregation or downstream separation is required (Martínez-Huitle & Brillas, 2009; Sirés et al., 2014). Consequently, electrochemical systems for microplastic control are best implemented as modular supporting units integrated with solid separation and containment processes rather than as standalone treatment solutions (Miklos et al., 2018).

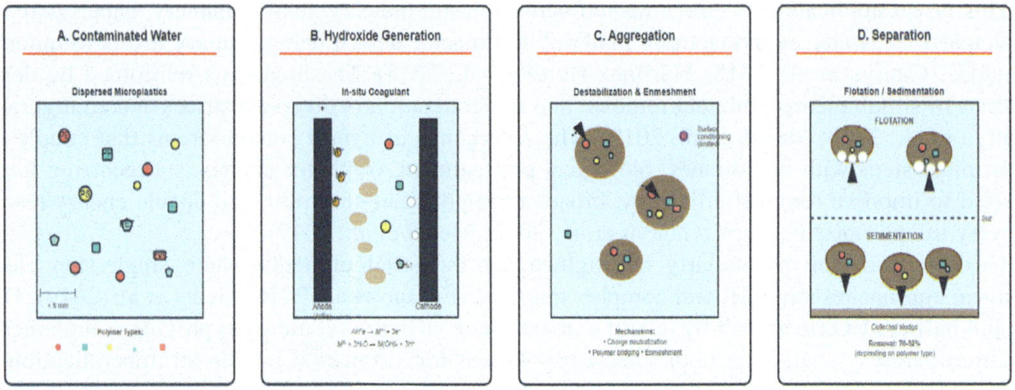

FIGURE 9.2

Microplastics removal mechanisms in electrochemical systems.

Figure 9.2 illustrates sequential removal pathway for microplastics via electrocoagulation/electroflotation: (A) Dispersed microplastics of various polymer types in contaminated water; (B) In-situ generation of metal hydroxides from sacrificial electrodes; (C) Destabilization, aggregation, and enmeshment of microplastics particles; (D) Separation via flotation (bubble-assisted) or sedimentation. Surface conditioning via limited electrooxidation may enhance aggregation efficiency but is not the primary removal mechanism. Reported removal efficiencies: PE (70%–92%), PP (75%–90%), PS (65%–88%), PET (60%–85%).

4.4 Practical performance trends and constraints

Despite demonstrated effectiveness, several constraints limit large-scale deployment of electrochemical technologies. Energy consumption, particularly for oxidation-based processes treating dilute streams, remains a primary limitation (Ganiyu et al., 2018; Muddemann et al., 2019). Electrode consumption and replacement costs, especially in systems employing sacrificial electrodes, significantly influence operating expenses and life-cycle impacts (Martínez-Huitle et al., 2015).

Wastewater matrix effects, including salinity, organic load, and suspended solids, strongly influence performance and energy efficiency. As a result, electrochemical processes are most effective when integrated with pretreatment or hybridized with biological, membrane, or physical separation technologies (Ensano et al., 2016; Medel et al., 2019). Life-cycle and techno-economic assessments are therefore essential to identify viable application scenarios and guide scale-up decisions (Ganiyu et al., 2018).

The trade-off between capital expenditure and operating expenditure is frequently highlighted in comparative studies and reviews (Ganiyu et al., 2018; Muddemann et al., 2019). By-product formation and safety concerns - particularly for indirect EO in chlorinated or saline matrices and for EF pathways - are recurring themes. Proper control of oxidant generation, reactor design, and process sequencing is necessary to minimize disinfection by-products and other toxic by-products, with reviews calling for careful assessment of matrix effects and by-product formation potential (Ensano et al., 2016; Martínez-Huitle et al., 2015).

Table 9.4 provides a comparative analysis of how different electrochemical techniques are applied to treat wastewater from various industries, highlighting the specific pollutants targeted and the effectiveness of the methods.

4.5 Process design considerations and optimization

Electrode material selection and reactor configuration: Choice of sacrificial (Al, Fe) versus inert or coated anodes; stack configurations (bipolar, tubular, disc-stack) influence coagulant generation, oxidant production, and mass transfer; energy efficiency is critical design considerations across microplastics, textile, tannery, and municipal wastewater (Körbahti, 2014; Oturan et al., 2015; Szpyrkowicz et al., 2005).

- Pretreatment and sequencing: Pre-treatment enhances electrochemical efficiency; higher pretreatment levels correlate with improved EO/EC performance for microplastics, tannery and textile effluents, while sequencing EO after EC or integrating EF aids mineralization of organics (Meiramkulova et al., 2020; Szpyrkowicz et al., 2005).
- Energy and sustainability metrics: Reports stress the importance of life-cycle and techno-economic evaluations, energy consumption per unit pollutant removed, and the potential for co-produced hydrogen or other value streams to improve overall viability (Szpyrkowicz et al., 2005).
- Scale-up and decentralization: Studies emphasize modular, compact systems suitable for decentralized deployment, with emphasis on robust electrode materials, energy efficiency, and minimal secondary pollution; challenges include energy intensity, mass transfer limitations, and by-product formation in real-water matrices (Oturan et al., 2015).

4.6 Integrated and hybrid electrochemical processes

In this section, selected examples of integrated electrochemical systems are discussed to illustrate broader principles of process hybridization and electrochemical integration; examples not related to water or wastewater treatment are included solely for conceptual comparison and do not represent primary treatment applications. Integrated and hybrid electrochemical processes have gained significant attention in recent years due to their ability to synergistically combine electrochemical techniques with other chemical or mechanical processes, enhancing efficiency, selectivity, and effectiveness in various applications, as can be seen in Fig. 9.1. Many practical systems utilize hybrid approaches that couple electrochemical steps with biological treatment, advanced oxidation, or membrane filtration to address complex wastewater containing high concentrations of recalcitrant organics as microplastics or toxicity.

Examples include combining biological pretreatment with EO/EC steps or marrying UV irradiation with electrocoagulation to enhance reclamation and disinfection outcomes (Asaithambi et al., 2022; Cotillas et al., 2014). Further innovations in hybrid electrochemical processing include advancements in the integration of microfluidics with electrochemical sensors, facilitating the miniaturization and field deployment of analytical platforms. These systems combine sample pre-treatment and detection in a compact format, significantly enhancing the speed and efficiency of electrochemical measurements (Huang et al., 2011; Kim, 2021).

The utilization of nanotechnology in electrochemical biosensing devices also exemplifies how integrating various scientific realms can lead to new innovations in health diagnostics and environmental monitoring (Kim, 2021). Hybrids combine EC/EF with EO, membranes, or biological pretreatment to leverage complementary mechanisms, enabling efficient pollutant removal and potentially recovering energy or valuable by-products (e.g., hydrogen production) (Asaithambi et al., 2022; Cotillas et al., 2014; Sala & Gutiérrez-Bouzán, 2012).

Hybridization often modulates key parameters (current density, residence time, pH control, and electrode configuration) to balance rapid removal with energy efficiency, enabling treatment of complex industrial effluents (Asaithambi et al., 2022; Bazrafshan et al., 2013). Wastewater composition strongly modulates mechanism efficiency, including pH sensitivity for EF and the formation of reactive oxygen species in EO. Different wastewater types (microplastics, textile, paper, refinery, and municipal) require varied electrode materials and strategies for oxidant generation to achieve targeted removals under manageable energy costs (Demir & Gören, 2019; Duesterberg et al., 2008; Feng et al., 2016).

Material choices (e.g., BDD, DSA, Fe/Al for EC, Ti-based coatings for EO) and electrode arrangements (monopolar or bipolar configurations) directly determine mechanistic pathways, efficiency, and by-product formation, emphasizing the need for a material-process match in design (Demir & Gören, 2019; Ogundele et al., 2023). Reaction rates for coagulation, radical generation, and mineralization are governed by current density, inter-electrode distance, conductivity, and hydraulic conditions; these parameters set the balance between removal efficiency and energy consumption (Bazrafshan et al., 2022; Bote, 2021).

4.7 Advantages and disadvantages of electrochemical techniques for water treatment

Electrochemical approaches for water and wastewater treatment include electrocoagulation/electroflotation (EC/EF), electrooxidation (EO), electrodisinfection, electro-Fenton (EF), and a range of hybrid configurations. These technologies have been investigated across diverse application contexts, including microplastics-containing waters, textile and tannery effluents, dairy and paper industry wastewaters, refinery streams, municipal wastewater, and stormwater. A balanced assessment of their advantages and limitations is necessary to support appropriate technology selection, reactor design, and integration within multi-stage treatment trains. Reported benefits commonly include reduced reliance on chemical reagents, modular implementation, opportunities for energy or resource recovery, and improved removal of persistent contaminants. However, limitations frequently noted in the literature relate to energy demand, mass-transfer constraints during scale-up, electrode durability, and the need to control by-product formation.

4.7.1 Advantages

Electrochemical systems can address a broad spectrum of contaminants, including microplastics, dyes, refractory organics, metals, ammonia, and emerging micropollutants, with removal outcomes ranging from separation to partial transformation or mineralization depending on the process configuration and electrode materials (Martínez-Huitle et al., 2015; García-Montoya et al., 2015). Several reviews highlight their applicability to complex industrial effluents—such as those from textile, tannery, paper, and refinery operations—through the use of tailored electrode materials (e.g., BDD, DSA, Ti-based coatings) and adaptable reactor configurations (Gu & Zheng, 2015; Martínez-Huitle & Ferro, 2006).

A key operational advantage is the reduced requirement for external chemical addition. In EC/EF, coagulant species are generated in situ, while EO and electrochemical advanced oxidation processes can produce oxidants electrochemically, thereby decreasing dependence on externally supplied reagents and potentially reducing secondary waste generation compared with conventional coagulation or chemical oxidation routes (Martínez-Huitle et al., 2015; García-Montoya et al., 2015).

Electrochemical treatment units can be compatible with decentralized or modular deployment. Their compact footprint and scalability through electrode stacking support "modular deployment" implementation in industrial settings and remote locations, particularly where conventional centralized infrastructure is limited (Gu & Zheng, 2015; García-Montoya et al., 2015). In addition, integrated designs that couple pollutant removal with energy or resource recovery—such as hydrogen co-production or renewable-energy-assisted operation—have been proposed as potential pathways to improve overall process sustainability in suitable applications (Gu & Zheng, 2015; Martínez-Huitle & Ferro, 2006).

Mechanistically, electrochemical platforms offer tunability through complementary pathways. Direct EO, indirect EO mediated by electro-generated oxidants, EC/EF-based separation via in situ coagulant formation, and EF-based hydroxyl radical generation can be configured to prioritize selective removal or deeper oxidation depending on electrode choice, current density, and reactor design (Martínez-Huitle et al., 2015; García-Montoya et al., 2015). These systems are also well suited to automation and real-time control, enabling operational optimization to balance removal efficiency with energy consumption (Martínez-Huitle et al., 2015; Martínez-Huitle & Ferro, 2006). Finally, hybridization—such as coupling EC/EF with EO, membranes, UV, or biological pretreatment—has been widely reported to enhance overall performance, broaden the treatable contaminant range, and reduce energy demand per unit of pollutant removed in several wastewater contexts (García-Montoya et al., 2015; Gu & Zheng, 2015).

4.7.2 Disadvantages

A commonly reported limitation of electrochemical treatment is energy intensity, particularly for diluting wastewaters or streams containing strongly colored and refractory organics. During scale-up, mass-transfer limitations and ohmic losses can become more pronounced, requiring careful reactor design, optimized hydrodynamics, and advanced electrode arrangements to maintain efficiency (Crini & Lichtfouse, 2018; Ling et al., 2015).

Electrode durability and material costs also remain important constraints. In EC/EF, sacrificial electrodes (typically Al or Fe) are consumed during operation, increasing material demand and generating metal-rich sludge that requires downstream management. In contrast, inert or coated anodes (e.g., BDD, Ti-based systems) can reduce dissolution-related issues but often involve higher capital costs and may face long-term stability concerns under aggressive operating conditions (Crini & Lichtfouse, 2018; Gu & Zheng, 2015).

By-product formation represents an additional challenge, particularly for indirect EO and EF-based systems where secondary oxidation products may form. In chloride-containing matrices, the potential formation of chlorinated species and disinfection by-products necessitates careful assessment of water chemistry, reactor operation, and post-treatment needs (Martínez-Huitle & Ferro, 2006; García-Montoya et al., 2015).

Finally, performance in real wastewater is often more variable than in synthetic or model solutions due to complex matrix effects associated with salinity, organic load, suspended solids, and co-contaminants. These factors can reduce treatment efficiency, increase energy consumption, and introduce risks of secondary pollution, highlighting the importance of pretreatment, appropriate electrode selection, and robust process control strategies for field implementation (Oturan et al., 2015; García-Montoya et al., 2015).

Overall, while electrochemical wastewater treatment offers practical advantages in terms of flexibility and contaminant coverage, practical deployment is constrained by energy requirements, scale-up limitations, and by-product management. Current reviews emphasize the need to reduce energy consumption, improve mass transfer, and strengthen control of secondary products, particularly for decentralized applications and complex real-water matrices (Radjenović & Sedlak, 2015). Future progress will likely depend on improved electrode and catalyst materials, optimized integrated configurations (including systems coupled with hydrogen generation), and compact automated reactor designs that support reliable "rapid on-site implementation" operation in industrial settings (Demir & Gören, 2019; Drušković et al., 2021).

5. Evaluating the performance and further research

The accumulating evidence supports electrochemical technologies as viable components of a sustainable wastewater management strategy, with the potential to reduce chemical usage, enable on-site treatment, and contribute to circular economy objectives through energy and resource recovery. This position is reinforced by reviews and empirical studies that articulate the versatility, robustness, and environmental compatibility of direct and indirect electrochemical processes, as well as the broader implications for microplastics, industrial wastewater remediation and decentralized treatment networks (Drušković et al., 2021; Radjenović & Sedlak, 2015). The field continues to mature through advances in electrode materials, reactor design, and integration with complementary treatment steps, aiming to deliver scalable, cost-effective, and environmentally benign wastewater solutions (Drušković et al., 2021; Martínez-Huitle & Ferro, 2006).

Standardized metrics center on removal efficiencies (COD, TOC, color), complete surface oxidation or conditioning of microplastics, energy consumption (kWh/m^3 or per kg COD removed), electrode wear, and co-product yields (e.g., hydrogen in energy-coupled systems) (Ganiyu et al., 2018; Muddemann et al., 2019). Design of experiments (DOE) and life-cycle assessment (LCA) are recommended to capture interactions among current density, electrode material, reactor geometry, pretreatment, and wastewater matrix effects in a robust, generalizable manner (Ensano et al., 2016; Ganiyu et al., 2018). Comparability across studies is enhanced by reporting consistent operating conditions including current density, electrode area, inter-electrode distance, solution conductivity, pH, temperature and by providing energy-intensity benchmarks for specific wastewater types, enabling cross-study synthesis and scale-up planning (Muddemann et al., 2019).

Table 9.3 provides a comparison between electrochemical techniques and other conventional and advanced methods, evaluating them on various performance and operational criteria. Table 9.5

Table 9.3 Key factors and operating conditions in electrochemical water treatment.

Factors/Operating conditions	Impact on effectiveness	Typical values/Ranges	Sources of citations
pH	Influences the speciation of pollutants, the form of metal hydroxides (in EC), the generation of oxidizing species (like •OH), and electrode surface chemistry. The optimal pH varies significantly between different electrochemical processes.	EC: Typically, 4–9. The final pH often shifts toward neutral. EO/EF: Often most effective in acidic conditions (pH 2–4) to favor the production of hydroxyl radicals.	(Chen et al., 2018; Mollah et al., 2004)
Current density/Applied voltage	Directly controls the rate of coagulant dosing (EC), the production rate of oxidants, and the speed of pollutant removal such as microplastics. It is the primary driver of energy consumption.	Highly variable, from 10 to >500 A/m^2. Higher densities lead to faster treatment but can decrease current efficiency and increase energy costs.	(Koparal & Öğütveren, 2002; Martínez-Huitle et al., 2015)
Electrode materials	The choice of anode and cathode material determines the primary reactions, the type of oxidants generated, the electrode's stability (lifespan), and overall process cost.	Anodes: Sacrificial (Fe, Al for EC); Non-active (Graphite, Pt, Boron-Doped Diamond - BDD); Active (Mixed Metal Oxides - MMOs like RuO_2/TiO_2, IrO_2/TiO_2). Cathodes: Often Stainless Steel, Carbon, or the same as the anode.	(Martínez-Huitle et al., 2015; Panizza & Cerisola, 2009)
Electrode spacing	Affects the internal resistance (Ohmic drop) of the electrochemical cell. Smaller spacing reduces energy consumption but can increase the risk of clogging or short-circuiting.	Typically, 0.5–5 cm. The optimal distance is a trade-off between energy savings and operational practicality.	(Chen, 2004; Holt et al., 2005)
Reaction time (electrolysis duration)	The duration of the process directly impacts the extent of pollutant removal and mineralization. Longer times lead to better removal but also increase energy consumption.	Varies widely from minutes to several hours, depending on the initial pollutant concentration as microplastics and the desired final quality.	(Koparal & Öğütveren, 2002; Mollah et al., 2004)

Continued

Table 9.3 Key factors and operating conditions in electrochemical water treatment.—cont'd

Factors/Operating conditions	Impact on effectiveness	Typical values/Ranges	Sources of citations
Supporting electrolyte (conductivity)	The presence of ions (like Cl^-, SO_4^{2-}) in the wastewater increases its conductivity, which reduces the cell voltage and energy consumption. Some ions (e.g., Cl^-) can also participate in indirect oxidation.	Conductivity >1 mS/cm is often desirable. NaCl or Na_2SO_4 are commonly added if the microplastics's natural conductivity is too low.	(Brillas & Garcia-Segura, 2025; Martínez-Huitle et al., 2015)
Temperature	Affects reaction kinetics, mass transport of pollutants such as microplastics to the electrode surface, and the conductivity of the solution.	Typically operated at ambient temperature (20–40 °C). Higher temperatures can improve kinetics but may not be economically viable to control and can decrease the solubility of gases.	(Chen, 2004; Vik, E. A. et al., 1984)
Flow rate/Mixing	In continuous systems, the flow rate determines the hydraulic retention time. Mixing (stirring) is crucial for ensuring effective mass transport of pollutants including microplastics to the electrode surface, preventing concentration gradients.	Stirring speeds of 100–500 rpm are common in lab-scale batch reactors. In flow systems, turbulence is desired.	(Holt et al., 2005; Mollah et al., 2004)

5.1 Evaluating the environmental impacts of electrochemical processes

Electrochemical processes for wastewater treatment possess significant potential to promote environmental sustainability through improved contaminant removal. Nonetheless, their environmental impacts, particularly associated with energy consumption and chemical by-products, must be addressed through optimized practices and technological innovations. The environmental impacts of these processes for wastewater treatment involve a range of factors, including energy consumption, chemical by-products formation, water quality, and overall sustainability. A thorough understanding of these impacts is crucial to advance electrochemical treatment technologies and to develop strategies for mitigating adverse effects.

Table 9.4 Comparative analysis of electrochemical water treatment across industries.

Industry/ Application	Main pollutants	Applicable electrochemical techniques	Performance and engineering assessment	References
Microplastics pollution	Microplastics (PE, PP, PS, PVC; <5 mm), synthetic fibers	EC/EF; hybrid EC–membrane; EO as conditioning step	EC/EF physically removes microplastics through in-situ hydroxide-induced aggregation and flotation/ sedimentation. EO is limited to surface conditioning due to high energy demand and potential by-products. EC-based systems are preferred for cost and energy efficiency, with hybrids improving capture and sludge handling.	(Akarsu et al., 2021; Holt et al., 2005; Martínez-Huitle & Brillas, 2009; Mollah et al., 2004; Perren et al., 2018)
Textile industry	Azo/reactive dyes, COD, BOD, suspended solids, metals (Cr, Cu), surfactants	EC (primary); EO, EF (polishing/ AOPs)	EC is robust and cost-effective for color and turbidity removal, often used as a primary step. EO and EF enable oxidative degradation and partial mineralization of dissolved dyes but with higher energy demand. Hybrid EC–EO configurations reduce overall energy intensity.	(Kobya et al., 2003; Mahmoodi & Dalvand, 2013; Panizza & Cerisola, 2009)

Continued

Table 9.4 Comparative analysis of electrochemical water treatment across industries.—cont'd

Industry/ Application	Main pollutants	Applicable electrochemical techniques	Performance and engineering assessment	References
Paper and pulp industry	Lignin, AOX, COD, BOD, color, suspended solids	EO (BDD, MMO); EC (pretreatment)	EO is effective for detoxification of chlorinated organics and lignin, achieving high COD and AOX reductions. EC is typically applied as pretreatment to reduce solids and color prior to biological or EO polishing	(El-Ashtoukhy et al., 2009; Martínez-Huitle et al., 2015)
Tannery industry	Cr(III)/Cr(VI), sulfides, COD, BOD, phenols, ammonia	EC, EO, EF	EC efficiently removes chromium and suspended solids via hydroxide precipitation. EO and EF oxidize residual organics and sulfides, reducing toxicity. Hybrid configurations improve overall performance and effluent compliance prior to discharge or reuse.	(Körbahti, 2014; Szpyrkowicz et al., 2005)
Refinery and produced water	Oil and grease, phenols, refractory organics, salinity	EO, EF, hybrid systems	EO and EF reduce refractory organics and phenols, but energy demand is strongly matrix-dependent (salinity, conductivity). Hybridization and staged treatment are critical for technical and economic feasibility at scale.	(Ganiyu et al., 2018; Oturan et al., 2015)

Table 9.4 Comparative analysis of electrochemical water treatment across industries.—cont'd

Industry/ Application	Main pollutants	Applicable electrochemical techniques	Performance and engineering assessment	References
Municipal and stormwater	COD, nutrients, pathogens, suspended solids, microplastics	EC/EF; EO (selective)	EC/EF is effective for suspended solids, pathogen reduction, and microplastics capture in decentralized applications. EO is applied selectively due to energy considerations and potential disinfection by-products. Integration with conventional treatment is recommended.	Akarsu et al. (2021)

Table 9.5 Comparative analysis of wastewater treatment techniques.

Parameters	Techniques (EC, EO, EF)	Biological treatment	Chemical coagulation	Advanced oxidation processes
Principle	Uses electrical current to drive oxidation, coagulation, or reduction reactions.	Uses microorganisms to biodegrade organic pollutants.	Uses chemical coagulants (e.g., alum, ferric chloride) to destabilize and aggregate colloids.	Uses highly reactive species (e.g., hydroxyl radicals) to oxidize pollutants.
Effectiveness on recalcitrant organics	High. Physical removal of microplastics via aggregation and separation	Low. Ineffective against non-biodegradable or toxic organic compounds.	Low to Moderate. Primarily removes suspended solids and some dissolved organics via adsorption.	High. Specifically designed to physical removal of microplastics; mineralization of dissolved organics
Color removal	High. Highly effective at breaking down microplastics or dye molecules.	Low. Poor, especially for microplastics or synthetic dyes.	Moderate to High. Effective for colloidal color but less so for microplastics, soluble dyes.	High. remarkably effective at destroying chromophores.

Continued

Table 9.5 Comparative analysis of wastewater treatment techniques.—cont'd

Parameters	Techniques (EC, EO, EF)	Biological treatment	Chemical coagulation	Advanced oxidation processes
Heavy metal removal	High. EC is particularly effective at removing heavy metals via precipitation and adsorption.	Low. Some biosorption can occur, but it is not a primary removal mechanism.	Moderate. Can remove metals but may require significant pH adjustment.	None. AOPs do not remove metals.
Disinfection	High. Direct oxidation and generation of disinfectants (like chlorine) are effective against pathogens.	Low. Can reduce pathogen load but is not a dedicated disinfection process.	Low. Some removal through floc entrapment, but not reliable for disinfection.	High. Ozone and UV are powerful disinfectants.
Operational footprint	Small. Reactors are compact and modular.	Large. Requires large basins and aeration tanks.	Moderate. Requires mixing tanks, flocculators, and clarifiers.	Small to Moderate. Reactor sizes vary but are compact.
Robustness	High. Less sensitive to toxicity and variations in flow or pollutant concentration.	Low. Extremely sensitive to toxic shocks, temperature, and organic loading variations.	Moderate. Performance can be affected by pH and wastewater chemistry.	Moderate. Can be affected by turbidity (UV) or scavengers of radicals.
Operating cost	Moderate to High. Primarily driven by electricity costs.	Low to Moderate. Primarily driven by aeration energy.	Moderate. Driven by chemical costs.	High. Driven by energy (UV lamps, ozone generation) and chemical (H_2O_2) costs.
Sources of citation	Martínez-Huitle et al. (2015)	(Von Sperling, M., 2007; Tchobanoglous et al., 2014)	Bratby, J. (2016).	(Glaze et al., 1987; Poyatos et al., 2010)

5.1.1 Environmental impacts of electrochemical procsses

5.1.1.1 Energy consumption and greenhouse gas emissions

- Electrochemical processes such as electrocoagulation, electrooxidation, and advanced electrochemical oxidation processes are energy-intensive. Liu et al., emphasized that the energy demand for these technologies is significant, contributing up to 2.7% of global energy consumption, subsequently increasing greenhouse gas emissions (Liu et al., 2016). This energy usage fuels discussions about the sustainability of these techniques, especially if powered by fossil fuels.

- Electrochemical processes, particularly those related to water desalination have demonstrated lower energy requirements through strategic innovations. For example, research has suggested that water splitting can yield hydrogen with a minimal carbon footprint when coupled with renewable energy sources (Huang et al., 2020; Li & Zheng, 2017).

5.1.1.2 Chemical by-products

- The generation of hazardous chemical by-products is a critical concern. For instance, electrochemical treatments can lead to the formation of disinfection by-products (DBPs), such as trihalomethanes and other chlorinated compounds when chlorine is employed. However, the specific generation of DBPs in the context mentioned was not the focus of Tarpeh et al.'s study (Tarpeh et al., 2018). Therefore, further studies are needed to understand the implications of DBPs in electrochemical solutions as current references lack direct evidence.
- One notable application of electrochemical processes is in the upcycling of plastic waste. Research has shown that electrocatalytic methods can convert various forms of plastic, such as polyethylene terephthalate (PET), into functional commodity chemicals and hydrogen fuel. This transformation not only reduces plastic waste but generates useful products that can further contribute to the circular economy (Cho et al., 2023). The decomposition of solid plastics through these processes is garnering interest as a means to facilitate cleaner waste management practices (Cho et al., 2023).
- Additionally, electrochemical processes can effectively treat organic waste, yielding valuable by-products such as carboxylic acids. Studies indicate that electrolytic extraction techniques can efficiently produce chemicals from biorefinery side streams, which significantly contributes to the reduction of organic waste volumes as microplastics (Andersen et al., 2014). This approach highlights a dual benefit: alleviating waste management challenges and generating high-value chemicals, thus addressing both environmental and economic concerns (Andersen et al., 2014).
- Another example of electrochemical waste conversion is found in e-waste recycling. Electrochemical methods enable the selective recovery of metals from electronic waste, thereby producing purified metal by-products such as copper, gold, and silver. These recovered materials reduce the need for virgin resource extraction, making the process both economically viable and environmentally friendly (Rai et al., 2021). Moreover, such techniques can reduce solvent use and energy consumption compared to traditional hydrometallurgical methods (Rai et al., 2021).
- Electrochemical approaches have also been employed for the production and recovery of biochar from organic waste. Biochar can serve multiple purposes, including soil enhancement and carbon sequestration. This process involves transforming organic feedstocks through electrochemical reforming, leading to reduced emissions while producing a stable carbon-rich product that can substitute conventional fertilizers (Yang & He, 2025). The efficiency of these processes showcases their potential contributions to sustainable agricultural practices and carbon management (Li et al., 2025).

5.1.1.3 Toxic residuals

- The treatment of certain contaminants may not result in complete mineralization, leaving residual toxic compounds that are even more concentrated post-treatment. Research has shown that while electrochemical techniques can effectively reduce chemical oxygen demand (COD), they can sometimes concentrate other harmful substances, raising additional environmental risks (Yang et al., 2017).

- One of the main concerns of electrochemical processes is gas evolution, which can lead to the release of toxic gases such as chlorine and ozone, especially during the electrolysis of aqueous solutions. For instance, during the electrochemical oxidation of organic pollutants such as microplastics, side reactions can produce hazardous gases, including chloroform or other halogenated compounds if halides are present in the system (Zhou et al., 2005). Such gaseous by-products can contribute to air pollution and pose inhalation risks to individuals in proximity to the facility.
- Moreover, certain electrochemical methods can lead to the generation of toxic chemical intermediates. For example, during reactions involving microplastics, phenolic compounds, intermediates like benzoquinone can form, exhibiting toxicity significantly higher than that of the original substrate. This illustrates that the selection of starting materials and operating conditions in electrochemical processes can influence the formation of harmful residues.
- Heavy metals, frequently linked to electrochemical processes, also raise concerns. The use of electrode materials, such as lead in lead-acid batteries, can lead to the leaching of toxic metals into the environment if not managed properly. Lead and other heavy metals can accumulate in ecosystems, leading to bioaccumulation and posing substantial health threats to wildlife and humans (Funakawa et al., 2005; Wang, H. D. et al., 2014). Moreover, the recovery processes for valuable metals from electronic waste often involve electrochemical methods, which may produce leachates containing heavy metals, further exacerbating environmental contamination issues.
- Electrochemical processes can also give rise to toxic organic by-products. The decomposition of organic materials through electrochemical oxidation can produce hazardous substances, including dioxins and other persistent organic pollutants such as microplastics. These compounds are particularly concerning due to their long-term environmental persistence and potential carcinogenic properties (Yuan & Lei, 2020). The careful management of reaction conditions and materials utilized is essential to minimize the generation of such toxic residues.

5.1.2 Solutions for mitigating environmental impacts

- **Energy source optimization**: Integrating renewable energy sources (e.g., solar or wind power) into electrochemical processes can reduce greenhouse gas emissions associated with energy consumption. Research indicates that utilizing renewable energy can enhance the environmental sustainability of treatment facilities (Lam et al., 2022; Nath, 2024).
- **Optimizing electrode materials**: Advancements in electrode design and material selection can limit energy usage and enhance reaction efficiency. For example, using boron-doped diamond (BDD) electrodes has been shown to improve oxidation performance in electrochemical processes while reducing harmful by-product generation (García-Montoya et al., 2015; Liu et al., 2016). Research into new electrode materials continues to minimize negative environmental impacts and enhance treatment efficiencies.
- **Real-time monitoring and control**: Implementing advanced monitoring systems can help track and manage chemical by-product formation during electrochemical processes, allowing for real-time adjustments to reduce unwanted by-product concentrations. Sensors can be employed to gauge toxic compound levels in effluents, leading to timely interventions when necessary (Mathew et al., 2024).

- **Life-cycle assessment (LCA)**: The application of LCA provides a holistic approach to assess the environmental footprints of electrochemical wastewater treatments (Dapkienė, 2022; Nath, 2024). Regularly conducted LCAs can guide facilities in making operational decisions that reduce overall environmental impacts (Buscio et al., 2019; Vidal et al., 2017).
- **Integrated treatment approaches**: Combining electrochemical processes with biological treatment or other advanced technologies can enhance overall treatment effectiveness and minimize environmental impacts. The sequential application of electrocoagulation followed by EAOPs has been shown to optimize pollutant removal while minimizing harmful side effects (Hijrah et al., 2023; Sudarno et al., 2024). Integrating different methodologies can effectively target a broader range of contaminants while decreasing the risk of residual toxicity.

5.2 Implications for future work and deployment

Research emphasis is shifting toward hybrid architectures that combine EO/EF/EC with membranes, UV, or biological pretreatment to tackle complex effluents while enabling energy recovery (e.g., hydrogen generation) and reducing overall energy intensity (Ensano et al., 2016; Muddemann et al., 2019). There is growing interest in decentralized and modular electrochemical systems for on-site treatment, with emphasis on robust electrode materials, low-footprint reactors, and scalable stack configurations that maintain performance under fluctuating wastewater compositions as microplastics (Ensano et al., 2016; Simon et al., 2018).

Mechanistic understanding and process control are essential for minimizing by-products and ensuring safe, compliant effluents, particularly for systems operating near neutral pH or in saline matrices; continued work on reactor design, electrode development, and real-water testing is needed to translate laboratory successes into reliable field implementations (Martínez-Huitle et al., 2015; Muddemann et al., 2019).

- Scope and versatility: EO, EC/EF, EF, and hybrids cover a broad contaminant spectrum; applicability spans microplastics, textile, tannery, dairy, paper, refinery, stormwater, municipal wastewater; hybrids enhance performance and enable energy/resource recovery (Ensano et al., 2016; Ganiyu et al., 2018; Martínez-Huitle et al., 2015).
- Mechanisms and materials: Direct/indirect EO, EC/EF coagulants, EF radicals; electrode materials (BDD, Ti-based DSA) critically affect efficiency and energy use; hybrids leverage complementary mechanisms (Muddemann et al., 2019; Simon et al., 2018).
- Energy and scale-up: Energy intensity is a central constraint; scale-up requires advanced reactor designs and mass-transfer optimization; life-cycle analyses are essential for evaluating feasibility and sustainability (Ganiyu et al., 2018; Medel et al., 2019).
- By-products and safety: Need for careful management of disinfection by-products and oxidation by-products; matrix effects mandate tailored process control and pretreatment in real wastewater scenarios (Ensano et al., 2016; Martínez-Huitle et al., 2015).

Pathways to deployment: Decentralized/modular systems, energy-coupled configurations, and integration with membranes offer practical routes to implementation; hydrogen and other co-products can improve economics in suitable setups (Ensano et al., 2016; Ganiyu et al., 2018; Muddemann et al., 2019).

6. Conclusions

This chapter reviewed electrochemical technologies for the treatment of microplastics and recalcitrant water pollutants, with emphasis on process mechanisms, performance drivers, and deployment constraints. Electrochemical technologies have been increasingly investigated for the treatment of microplastics and other recalcitrant pollutants in water and wastewater. Reported outcomes in the literature indicate that high removal or degradation efficiencies can be achieved under optimized conditions; however, performance remains strongly dependent on pollutant type, water matrix composition, and reactor configuration.

Across studies, electrode material selection emerges as a critical determinant of efficiency and energy demand. Boron-doped diamond (BDD) anodes are frequently associated with high degrees of oxidation and mineralization, but they may require relatively high energy input and capital cost, which can limit their feasibility for large-scale deployment. In contrast, electro-Fenton systems can improve cost-effectiveness by generating H_2O_2 in situ and promoting hydroxyl radical formation under milder conditions, although their performance is sensitive to pH, iron speciation, and mass-transfer limitations. Photo-assisted systems (PEC/PEF) have been proposed to enhance oxidation pathways and reduce treatment time or energy consumption in suitable configurations, but their practical advantages depend on light delivery efficiency and reactor design at full-scale implementation.For microplastics, electrocoagulation/electroflotation (EC/EF) is most commonly applied as a separation step that promotes aggregation and subsequent flotation or sedimentation, whereas oxidation-based processes are generally more relevant for dissolved recalcitrant organic contaminants.

Despite the demonstrated potential of these approaches, several constraints continue to limit broader implementation. Key challenges include electricity demand, electrode durability and material availability, scale-up limitations related to hydrodynamics and ohmic losses, and overall economic competitiveness relative to established treatment technologies. In chloride-rich waters, the possible formation of chlorinated oxidants and disinfection by-products requires careful control of operating conditions and appropriate process sequencing to ensure effluent safety and regulatory compliance. From a deployment perspective, the current evidence supports electrochemical processes as useful components of integrated treatment trains rather than universal standalone solutions. Future work should prioritize durable and lower-cost electrode materials, reactor designs that improve mass transfer and reduce energy intensity, and systematic evaluation using life-cycle assessment and techno-economic analysis. Pilot-scale demonstrations under realistic wastewater conditions will be essential to validate long-term performance, quantify by-product risks, and define application niches where electrochemical systems can be implemented reliably and sustainably.

Acknowledgments

The present study was sponsored by The Vietnam Ministry of Education and Training (MOET) under grant number B2024-VGU-05.

References

Agudelo-Escobar, M.,L., & Erazo Cabrera, S. (2024). *Microorganisms and microbial communities in bio-electrochemical systems for wastewater bioremediation and energy generation*. IntechOpen. https://doi.org/10.5772/intechopen.112470

Akarsu, C., Kumbur, H., & Kideys, A. E. (2021). Removal of microplastics from wastewater through electrocoagulation-electroflotation and membrane filtration processes. *Water Science and Technology, 84*(7), 1648–1662. https://doi.org/10.2166/wst.2021.356

Alavi, N., Dehvari, M., Alekhamis, G., Goudarzi, G., Neisi, A., & Babaei, A. A. (2019). Application of electro-Fenton process for treatment of composting plant leachate: Kinetics, operational parameters, and modeling. *Journal of Environmental Health Science and Engineering, 17*(1), 417–431. https://doi.org/10.1007/s40201-019-00361-2

Alkhadra, M. A., Su, X., Suss, M. E., Tian, H., Guyes, E. N., Shocron, A. N., Conforti, K. M., De Souza, J. P., Kim, N., Tedesco, M., Khoiruddin, K., Wenten, I. G., Santiago, J. G., Hatton, T. A., & Bazant, M. Z. (2022). Electrochemical methods for water purification, ion separations, and energy conversion. In *Chemical reviews* (pp. 13547–13635). American Chemical Society. https://doi.org/10.1021/acs.chemrev.1c00396. Vol. 122, Issue 16.

Amin, A. R. (2023). Treatment of industrial wastewater through the process of electrocoagulation: A review. *Evolutions in Mechanical Engineering, 5*(1). https://doi.org/10.31031/eme.2023.05.000603

Andersen, S. J., Hennebel, T., Gildemyn, S., Coma, M., Desloover, J., Berton, J., Tsukamoto, J., Stevens, C., & Rabaey, K. (2014). Electrolytic membrane extraction enables production of fine chemicals from biorefinery sidestreams. *Environmental Science and Technology, 48*(12), 7135–7142. https://doi.org/10.1021/es500483w

Ardhianto, R., & Bagastyo, A. (2019). Personal care wastewater treatment with electrocoagulation and electro-oxidation. *E3S Web of Conferences, 125*, Article 03008. https://doi.org/10.1051/e3sconf/201912503008

Asaithambi, P., Yesuf, M. B., Govindarajan, R., Hariharan, N. M., Thangavelu, P., & Alemayehu, E. (2022). A review of hybrid process development based on electrochemical and advanced oxidation processes for the treatment of industrial wastewater. *International Journal of Chemical Engineering, 2022*. https://doi.org/10.1155/2022/1105376. Hindawi Limited.

Baek, J., Kim, C., Song, Y. E., Im, H. S., Sakuntala, M., & Kim, J. R. (2018). Separation of acetate produced from C1 gas fermentation using an electrodialysis-based bioelectrochemical system. *Energies, 11*(10). https://doi.org/10.3390/en11102770

Balster, J., Srinkantharajah, S., Sumbharaju, R., Pünt, I., Lammertink, R. G. H., Stamatialis, D. F., & Wessling, M. (2010). Tailoring the interface layer of the bipolar membrane. *Journal of Membrane Science, 365*(1–2), 389–398. https://doi.org/10.1016/j.memsci.2010.09.034

Bazrafshan, E., Moein, H., Kord Mostafapour, F., & Nakhaie, S. (2013). Application of electrocoagulation process for dairy wastewater treatment. *Journal of Chemistry.* https://doi.org/10.1155/2013/640139

Bazrafshan, E., Zakeri, H. R., Vieira, M. G. A., Derakhshan, Z., Mohammadi, L., Mohammadpour, A., & Mousavi Khaneghah, A. (2022). Slaughterhouse wastewater treatment by integrated chemical coagulation and electro-fenton processes. *Sustainability, 14*(18), Article 11407. https://doi.org/10.3390/su141811407

Bote, M.E.. Studies on electrode combination for COD removal from domestic wastewater using electro-coagulation. (2021). *Heliyon*, Volume 7, Issue 12, e08614. https://doi.org/10.1016/j.heliyon.2021.e08614

Bratby, J. (2016). *Coagulation and flocculation in water and wastewater treatment*. IWA publishing. https://doi.org/10.2166/9781780407500

Brillas, E., & Garcia-Segura, S. (2025). Electrochemical advanced oxidation processes for real wastewater remediation: Systems, performance, and generated oxidants. *ACS Electrochemistry.* https://doi.org/10.1021/acselectrochem.5c00311

Buscio, V., López-Grimau, V., Álvarez, M. D., & Gutiérrez-Bouzán, C. (2019). Reducing the environmental impact of textile industry by reusing residual salts and water: ECUVal system. *Chemical Engineering Journal, 373*, 161–170. https://doi.org/10.1016/j.cej.2019.04.146

Cerrillo-Gonzalez, M.d. M., Villen-Guzman, M., Rodriguez-Maroto, J. M., & Paz-Garcia, J. M. (2024). Metal recovery from wastewater using electrodialysis separation. In *Metals*. Multidisciplinary Digital publishing Institute (MDPI). https://doi.org/10.3390/met14010038. Vol. 14, Issue 1.

Chen, G. (2004). Electrochemical technologies in wastewater treatment. *Separation and Purification Technology, 38*(1), 11–41. https://doi.org/10.1016/j.seppur.2003.10.006

Chen, S., Chen, Z., Siahrostami, S., Roy Kim, T., Nordlund, D., Sokaras, D., Nowak, S., To, J. W. F., Higgins, D., Sinclair, R., Nørskov, J. K., Jaramillo, T. F., & Bao, Z. (2018). *ACS Sustainable Chemistry and Engineering, 6* (1), 311–317. https://doi.org/10.1021/acssuschemeng.7b02517

Chen, L., Gao, S., Lou, L., & Zhou, Z. (2021). Removal of intracellular and extracellular antibiotic resistance genes from swine wastewater by sequential electrocoagulation and electro-fenton processes. *Environmental Engineering Science, 38*(2), 74–80. https://doi.org/10.1089/ees.2020.0203

Chen, L., Zhou, Z., Shen, C., & Xu, Y. (2020). Inactivation of antibiotic-resistant bacteria and antibiotic resistance genes by electrochemical oxidation/electro-fenton process. *Water Science and Technology, 81*(10), 2221–2231. https://doi.org/10.2166/wst.2020.282

Cho, J., Kim, B., Kwon, T., Lee, K., & Choi, S. I. (2023). Electrocatalytic upcycling of plastic waste. *Green Chemistry, 25*(21), 8444–8458. https://doi.org/10.1039/d3gc03337f

Choi, H. J., Tahmid, M., Mondal, S., & Hatzell, M. C. (2025). Concentrating ammonia from wastewater with electrodialysis. *ACS ES&T Water, 5*(9), 5720–5727. https://doi.org/10.1021/acsestwater.5c00721

Cotillas, S., Canizares, P., Martin De Vidales, M., Saez, C., Rodrigo, M., & Llanos, J. (2014). Electrocoagulation-UV irradiation process for urban wastewater reuse. *Chemical Engineering Transactions, 41*, 133–138. https://doi.org/10.3303/CET1441023

Crini, G., & Lichtfouse, E. (2018). Wastewater treatment: An overview. In G. Crini, & E. Lichtfouse (Eds.), *Green adsorbents for pollutant removal* (pp. 1–25). Springer. https://doi.org/10.1007/978-3-319-92111-2_1

Dapkienė, M. (2022). Operational efficiency and environmental impact of wastewater treatment plants in Lithuanian rural areas. *Rural Development*. https://doi.org/10.15544/RD.2021.013, 2019.

Davarnejad, R., Sabzehei, M., Parvizi, F., Heidari, S., & Rashidi, A. (2019). Study on soybean oil plant wastewater treatment using the electro-fenton technique. *Chemical Engineering and Technology, 42*(12), 2717–2725. https://doi.org/10.1002/ceat.201800765

Demir, A., & Gören, N. (2019). Investigation of electrocoagulation and electrooxidation methods of real textile wastewater treatment. *Anadolu University Journal of Science and Technology-A Applied Sciences and Engineering*, 80–91. https://doi.org/10.18038/aubtda.445716

Dermentzis, K., Karakosta, K., Kokkinos, N., Mitkidou, S., Stylianou, M., & Agapiou, A. (2022). Photovoltaic-driven electrochemical remediation of drilling fluid wastewater with simultaneous hydrogen production. *Waste Management & Research the Journal for a Sustainable Circular Economy, 41*(1), 155–163. https://doi.org/10.1177/0734242x221105414

Drušković, M., Vouk, D., Šiljeg, M., & Maldini, K. (2021). Treatment of wastewater from separators for rainfall runoff using electrochemical oxidation processes. *Environmental Engineering, 8*(1–2), 32–38. https://doi.org/10.37023/ee.8.1-2.5

Duesterberg, C. K., Mylon, S. E., & Waite, T. D. (2008). pH effects iron-catalyzed oxidation using Fenton's reagent. *Environmental Science and Technology, 42*(22), 8522–8527. https://doi.org/10.1021/es801720d

El-Ashtoukhy, E. S. Z., Amin, N. K., & Abdelwahab, O. (2009). Treatment of paper mill effluents in a batch-stirred electrochemical tank reactor. *Chemical Engineering Journal, 146*(2), 205–210. https://doi.org/10.1016/j.cej.2008.05.037

Ensano, B. M. B., Borea, L., Naddeo, V., Belgiorno, V., de Luna, M. D. G., & Ballesteros, F. C. (2016). Combination of electrochemical processes with membrane bioreactors for wastewater treatment and fouling control: A review. In *Frontiers in environmental science*. Frontiers Media S.A. https://doi.org/10.3389/fenvs.2016.00057. Vol. 4, Issue AUG).

Falina, I., Loza, N., Loza, S., Titskaya, E., & Romanyuk, N. (2021). Permselectivity of cation exchange membranes modified by polyaniline. *Membranes, 11*(3). https://doi.org/10.3390/membranes11030227

Feng, Y., Yang, L., Liu, J., & Logan, B. E. (2016). Electrochemical technologies for wastewater treatment and resource reclamation. In *Environmental science: Water research and technology*. Royal Society of Chemistry. https://doi.org/10.1039/c5ew00289c. Vol. 2, Issue 5, pp. 800–831.

Funakawa, A., Yamanaka, I., & Otsuka, K. (2005). Active control of methanol carbonylation selectivity over Au/carbon anode by electrochemical potential. *Journal of Physical Chemistry B, 109*(18), 9140–9147. https://doi.org/10.1021/jp0448752

García-Montoya, M. F., Gutiérrez-Granados, S., Alatorre-Ordaz, A., Galindo, R., Ornelas, R., & Peralta-Hernández, J. M. (2015). Application of electrochemical/BDD process for the treatment wastewater effluents containing pharmaceutical compounds. *Journal of Industrial and Engineering Chemistry, 31*, 238–243. https://doi.org/10.1016/j.jiec.2015.06.030. ISSN 1226-086X.

Göltz, M., Koch, M., Detsch, R., Karl, M., Burkovski, A., & Rosiwal, S. (2019). Influence of in-situ electrochemical oxidation on implant surface and colonizing microorganisms evaluated by scanning electron microscopy. *Materials, 12*(23), 3977. https://doi.org/10.3390/ma12233977

Güven, G., Perendeci, A., Özdemir, K., & Tanyolaç, A. (2012). Specific energy consumption in electrochemical treatment of food industry wastewater. *Journal of Chemical Technology and Biotechnology, 87*(4), 513–522. https://doi.org/10.1002/jctb.2739

Ganiyu, S. O., Vieira dos Santos, E., Tossi de Araújo Costa, E. C., & Martínez-Huitle, C. A. (2018). Electrochemical advanced oxidation processes (EAOPs) as alternative treatment techniques for carwash wastewater reclamation. *Chemosphere, 211*, 998–1006. https://doi.org/10.1016/j.chemosphere.2018.08.044

García-Espinoza, J., Robles, I., Duran-Moreno, A., & Godinez, L. (2021). Photo-assisted electrochemical advanced oxidation processes for the disinfection of aqueous solutions: a review. *Chemosphere, 274*, Article 129957. https://doi.org/10.1016/j.chemosphere.2021.129957

Gayen, P., Chen, C., Abiade, J. T., & Chaplin, B. P. (2018). Electrochemical oxidation of atrazine and clothianidin on Bi-doped SnO 2 -Ti n O 2n-1 electrocatalytic reactive electrochemical membranes. *Environmental Science and Technology, 52*(21), 12675–12684. https://doi.org/10.1021/acs.est.8b04103

Ghernaout, D., Elboughdiri, N., & Ghareba, S. (2020). Fenton technology for wastewater treatment: Dares and trends. *OALib. 07*, 1–26. https://doi.org/10.4236/oalib.1106045

Glaze, W. H., Kang, J. W., & Chapin, D. H. (1987). The chemistry of water treatment processes involving ozone, hydrogen peroxide, and ultraviolet radiation. *Ozone: Science and Engineering, 9*(4), 335–352. https://doi.org/10.1080/01919518708552148

Gonzalez-Rivas, N., Reyes-Pérez, H., & Barrera-Díaz, C. E. (2019). Recent advances in water and wastewater electrodisinfection. In *ChemElectroChem* (pp. 1978–1983). Wiley-VCH Verlag. https://doi.org/10.1002/celc.201801746. Vol. 6, Issue 7.

Gopal, S., Somanathan, A., Jeyakumar, R. B., & Fernandez, C. (2022). A bench-scale electrochemical peroxidation reactor performance on removal of organic pollutants from tannery industrial wastewater. *Desalination and Water Treatment, 277*, 120–135. https://doi.org/10.5004/dwt.2022.29039

Gu, Q., & Zheng, J. (2015). *Study on the electrochemical oxidation treatment of coking wastewater by DSA anode*. https://doi.org/10.2991/ap3er-15.2015.71

Gurreri, L., Tamburini, A., Cipollina, A., & Micale, G. (2020). Electrodialysis applications in wastewater treatment for environmental protection and resources recovery: A systematic review on progress and perspectives. In *Membranes* (pp. 1–93). MDPI AG. https://doi.org/10.3390/membranes10070146. Vol. 10, Issue 7.

Haerens, K., De Vreese, P., Matthijs, E., Pinoy, L., Binnemans, K., & Van Der Bruggen, B. (2012). Production of ionic liquids by electrodialysis. *Separation and Purification Technology, 97*, 90–95. https://doi.org/10.1016/j.seppur.2012.02.017

Handojo, L., Wardani, A. K., Regina, D., Bella, C., Kresnowati, M. T. A. P., & Wenten, I. G. (2019). Electro-membrane processes for organic acid recovery. *RSC Advances, 9*(14), 7854–7869. https://doi.org/10.1039/C8RA09227C

Hasanbiki, A., Rezaee, A., Masoumbigi, H., & Hossini, H. (2016). Removal of Escherichia coli from contaminated water using bipolar electrochemical systems. *Journal of North Khorasan University of Medical Sciences, 7*, 551–560. https://doi.org/10.29252/jnkums.7.3.551

Helesh, A., Mudrynets, P., Kalymon, Y., Kindzera, D., & Hnativ, V. (2023). Establishment of technologically feasible modes of electrocoagulation purification of wastewater from nickel ions. *Technology Audit and Production Reserves, 2*(3/70), 25–30. https://doi.org/10.15587/2706-5448.2023.278006

Hewehy, M., Razek, M. A., Taha, Hamid, A., & Monir, R. (2016). Electrochemical treatment of electroplating wastewater using carbon and aluminum electrodes. *Journal of Environmental Science, 33*, 25–41. https://doi.org/10.21608/jes.2016.25102

Hijrah, N., Syakir, M. I., & Syah, B. I. (2023). Quantification of environmental impact potentials within the processes of industrial wastewater treatment system. *IOP Conference Series: Earth and Environmental Science, 1167*(1). https://doi.org/10.1088/1755-1315/1167/1/012019

Holt, P. K., Barton, G. W., & Mitchell, C. A. (2005). The future for electrocoagulation as a localised water treatment technology. *Chemosphere, 59*(3), 355–367. https://doi.org/10.1016/j.chemosphere.2004.10.023

Huang, C. J., Chien, H. C., Chou, T. C., & Lee, G. B. (2011). Integrated microfluidic system for electrochemical sensing of glycosylated hemoglobin. *Microfluidics and Nanofluidics, 10*(1), 37–45. https://doi.org/10.1007/s10404-010-0644-x

Huang, H., Cho, A., Kim, S., Jun, H., Lee, A., Han, J. W., & Lee, J. (2020). Structural design of amorphous CoMoPx with abundant active sites and synergistic catalysis effect for effective water splitting. *Advanced Functional Materials, 30*(43). https://doi.org/10.1002/adfm.202003889

Huang, X., Qu, Y., Cid, C. A., Finke, C., Hoffmann, M. R., Lim, K., & Jiang, S. C. (2016). Electrochemical disinfection of toilet wastewater using wastewater electrolysis cell. *Water Research, 92*, 164–172. https://doi.org/10.1016/j.watres.2016.01.040

Jeong, J., Kim, J. Y., & Yoon, J. (2006). The role of reactive oxygen species in the electrochemical inactivation of microorganisms. *Environmental Science and Technology, 40*(19), 6117–6122. https://doi.org/10.1021/es0604313

Jiang, J., Chang, M., & Pan, P. (2008). Simultaneous hydrogen production and electrochemical oxidation of organics using boron-doped diamond electrodes. *Environmental Science and Technology, 42*(8), 3059–3063. https://doi.org/10.1021/es702466k

Körbahti, B. K. (2014). Finite element modeling of continuous flow tubular electrochemical reactor for industrial and domestic wastewater treatment. *Journal of the Electrochemical Society, 161*(8), E3225–E3234. https://doi.org/10.1149/2.025408jes

Karichappan, T., Venkatachalam, S., Jeganathan, P. M., & Sengodan, K. (2013). Treatment of rice mill wastewater using continuous electrocoagulation technique: Optimization and modelling. *Journal of the Korean Chemical Society, 57*(6), 761–768. https://doi.org/10.5012/jkcs.2013.57.6.761

Kim, T. H. (2021). Toward emerging innovations in electrochemical biosensing technology. In *Applied sciences (Switzerland)*. MDPI AG. https://doi.org/10.3390/app11062461. Vol. 11, Issue 6.

Klidi, N., Proietto, F., Vicari, F., Galia, A., Ammar, S., Gadri, A., & Scialdone, O. (2019). Electrochemical treatment of paper mill wastewater by electro-Fenton process. *Journal of Electroanalytical Chemistry, 841*, 166–171. https://doi.org/10.1016/j.jelechem.2019.04.022

Kobya, M., Can, O. T., & Bayramoglu, M. (2003). Treatment of textile wastewater by electrocoagulation using iron and aluminum electrodes. *Journal of Hazardous Materials, 100*(1–3), 163–178. https://doi.org/10.1016/S0304-3894(03)00102-X

Koparal, A. S., & Öğütveren, Ü. B. (2002). Removal of nitrate from water by electroreduction and electrocoagulation. *Journal of Hazardous Materials, 89*(1), 83–94. https://doi.org/10.1016/S0304-3894(01)00301-6

Lam, K. L., Solon, K., Jia, M., Volcke, E. I. P., & Van Der Hoek, J. P. (2022). Life-cycle environmental impacts of wastewater-derived phosphorus products: An agricultural end-user perspective. *Environmental Science and Technology, 56*(14), 10289–10298. https://doi.org/10.1021/acs.est.2c00353

Lan, Y., Coetsier, C., Causserand, C., & Groenen Serrano, K. (2017). On the role of salts for the treatment of wastewater containing pharmaceuticals by electrochemical oxidation using a boron doped diamond anode. *Electrochimica Acta, 231*, 309–318. https://doi.org/10.1016/j.electacta.2017.01.160

Li, Q., Gunawan, D., Jiang, L., Gunawan, R., Gunasekara, G., Sarmin, S., … Scott, J. (2025). Recent advances in electrochemical organic waste reforming: Highlights on anodic chemistry, materials design, and system integration. *Acs Applied Engineering Materials, 3*(1), 21–43. https://doi.org/10.1021/acsaenm.4c00705

Li, J., & Zheng, G. (2017). One-dimensional earth-abundant nanomaterials for water-splitting electrocatalysts. *Advanced Science, 4*(3). https://doi.org/10.1002/advs.201600380

Ling, Y., Xu, H., & Chen, X. (2015). Continuous multi-cell electrochemical reactor for pollutant oxidation. *Chemical Engineering Science, 122*, 630–636. https://doi.org/10.1016/j.ces.2014.10.031

Liu, Y., Hejazi, M., Kyle, P., Kim, S. H., Davies, E., Miralles, D. G., Teuling, A. J., He, Y., & Niyogi, D. (2016). Global and regional evaluation of energy for water. *Environmental Science and Technology, 50*(17), 9736–9745. https://doi.org/10.1021/acs.est.6b01065

Liu, L., Zhao, G., Pang, Y., Lei, Y., Gao, J., & Liu, M. (2010). Integrated biological and electrochemical oxidation treatment for high toxicity pesticide pollutant. *Industrial and Engineering Chemistry Research, 49*(12), 5496–5503. https://doi.org/10.1021/ie100333v

Logan, B. E., Rossi, R., Ragab, A., et al. (2019). Electroactive microorganisms in bioelectrochemical systems. *Nature Reviews Microbiology, 17*, 307–319. https://doi.org/10.1038/s41579-019-0173-x

Müller, F., Andretta, R., de Oliveira Meneguzzi, L., & Arthur Ferreira, C. (2017). Development of quaternarized poly(vinyl alcohol) anion-exchange membranes for applications in electrodialysis. *Journal of Applied Polymer Science, 134*(31). https://doi.org/10.1002/app.44946

Ma, B., Chi, J., & Liu, H. (2018). Fabric-based ion concentration polarization for pump-free water desalination. *ACS Sustainable Chemistry & Engineering, 6*(1), 99–103. https://doi.org/10.1021/acssuschemeng.7b03679

Mahmoodi, N. M., & Dalvand, A. (2013). Treatment of colored textile wastewater containing acid dye using electrocoagulation process. *Desalination and Water Treatment, 51*(31–33), 5959–5964. https://doi.org/10.1080/19443994.2013.791769

Mandal, T. K. (2024). Nanomaterial-enhanced hybrid disinfection: A solution to combat multidrug-resistant bacteria and antibiotic resistance genes in wastewater. *Nanomaterials, 14*(22), 1847. https://doi.org/10.3390/nano14221847

Mansouri, K., Hannachi, A., Abdel-Wahab, A., & Bensalah, N. (2012). Electrochemically dissolved aluminum coagulants for the removal of natural organic matter from synthetic and industrial wastewater. *Industrial and Engineering Chemistry Research, 51*(5), 2428–2437. https://doi.org/10.1021/ie202188m

Martínez-Huitle, C. A., & Brillas, E. (2009). Decontamination of wastewaters containing synthetic organic dyes by electrochemical methods: A general review. *Applied Catalysis B: Environmental, 87*(3–4), 105–145. https://doi.org/10.1016/j.apcatb.2008.09.017

Martínez-Huitle, C. A., & Ferro, S. (2006). Electrochemical oxidation of organic pollutants for the wastewater treatment: Direct and indirect processes. *Chemical Society Reviews, 35*(12), 1324–1340. https://doi.org/10.1039/b517632h

Martínez-Huitle, C. A., Rodrigo, M. A., Sirés, I., & Scialdone, O. (2015). Single and coupled electrochemical processes and reactors for the abatement of organic water pollutants: A critical review. In *Chemical reviews* (pp. 13362–13407). American Chemical Society. https://doi.org/10.1021/acs.chemrev.5b00361. Vol. 115, Issue 24.

Mathew, J. T., Inobeme, A., Musah, M., Azeh, Y., Abdullahi, A., Shaba, E. Y., Salihu, A. M., Muhammad, E. B., Josiah, J. G., Jibrin, N. A., Ismail, H., Muhammad, A. I., Maurice, J., Mamman, A., & Ndamitso, M. M. (2024). A critical review of green approach on wastewater treatment strategies. *Journal of Applied Sciences & Environmental Management, 28*(2), 363–391. https://doi.org/10.4314/jasem.v28i2.9

Medel, A., Ramírez, J. A., Cárdenas, J., Sirés, I., & Meas, Y. (2019). Evaluating the electrochemical and photoelectrochemical production of hydroxyl radical during electrocoagulation process. *Separation and Purification Technology, 208*, 59–67. https://doi.org/10.1016/j.seppur.2018.05.021

Meiramkulova, K., Jakupova, Z., Orynbekov, D., Tashenov, E., Kydyrbekova, A., Mkilima, T., & Inglezakis, V. J. (2020). Evaluation of electrochemical methods for poultry slaughterhouse wastewater treatment. *Sustainability, 12*(12), 5110. https://doi.org/10.3390/su12125110

Midassi, S., Bedoui, A., & Bensalah, N. (2020). Efficient degradation of chloroquine drug by electro-Fenton oxidation: Effects of operating conditions and degradation mechanism. *Chemosphere, 260*. https://doi.org/10.1016/j.chemosphere.2020.127558

Miklos, D. B., Remy, C., Jekel, M., Linden, K. G., Drewes, J. E., & Hübner, U. (2018). Evaluation of advanced oxidation processes for water treatment. *Water Research, 139*, 118–131. https://doi.org/10.1016/j.watres.2018.03.042

Miyata, M., Ihara, I., Yoshid, G., Toyod, K., & Umetsu, K. (2011). Electrochemical oxidation of tetracycline antibiotics using a Ti/IrO 2 anode for wastewater treatment of animal husbandry. *Water Science and Technology, 63*(3), 456–461. https://doi.org/10.2166/wst.2011.243

Mollah, M. Y. A., Morkovsky, P., Gomes, J. A. G., Kesmez, M., Parga, J., & Cocke, D. L. (2004). Fundamentals, present and future perspectives of electrocoagulation. *Journal of Hazardous Materials, 114*(1–3), 199–210. https://doi.org/10.1016/j.jhazmat.2004.08.009

Moreno-Andrés, J., Ambauen, N., Vadstein, O., Hallé, C., Acevedo-Merino, A., Nebot, E., & Meyn, T. (2018). Inactivation of marine heterotrophic bacteria in ballast water by an electrochemical advanced oxidation process. *Water Research, 140*, 377–386. https://doi.org/10.1016/j.watres.2018.04.061

Mourad, B. S. A., & Béchir, H. (2016). Development of a predictive model of the limiting current density of an electrodialysis process using response surface methodology. *Membrane Water Treatment, 7*(2), 127–141. https://doi.org/10.12989/mwt.2016.7.2.127

Muddemann, T., Haupt, D., Sievers, M., & Kunz, U. (2019). Electrochemical reactors for wastewater treatment. *ChemBioEng Reviews, 6*(5), 142–156. https://doi.org/10.1002/cben.201900021

Muthukumar, K., Sundaram, P. S., Anantharaman, N., & Basha, C. A. (2004). Treatment of textile dye wastewater by using an electrochemical bipolar disc stack reactor. *Journal of Chemical Technology and Biotechnology, 79* (10), 1135–1141. https://doi.org/10.1002/jctb.1104

Nath, S. (2024). Electrochemical wastewater treatment technologies through life-cycle assessment: A review. In *ChemBioEng reviews*. John Wiley and Sons Inc. https://doi.org/10.1002/cben.202400016. Vol. 11, Issue 4.

Ogundele, O. D., Oyegoke, D. A., & Anaun, T. E. (2023). Exploring the potential and challenges of electrochemical processes for sustainable wastewater remediation and treatment. *Acadlore Transactions on Geosciences, 2*(2), 80–93. https://doi.org/10.56578/atg020203

Oreščanin, V., Kollar, R., Vratarić, D., Nadj, K., Medunić, G., Lončar, G., Halkijević, I., & Kuspilić, M. (2013). The electrocoagulation/advanced oxidation treatment of the well water from ŽUPANJA. *GLASNIK HEMIČARA, TEHNOLOGA I EKOLOGA REPUBLIKE SRPSKE, 4*(9). https://doi.org/10.7251/ghte1309037o

Oturan, M. A., & Aaron, J. J. (2014). Advanced oxidation processes in water/ wastewater treatment: Principles and applications: A review. *Critical Reviews in Environmental Science and Technology, 44*, 2577–2641. https://doi.org/10.1080/10643389.2013.829765

Oturan, N., Hullebusch, E., Zhang, H., Mazéas, L., Budzinski, H., Ménach, K., … Oturan, M. (2015). Occurrence and removal of organic micropollutants in landfill leachates treated by advanced electrochemical oxidation processes. *Environmental Science and Technology, 49*(20), 12187–12196. https://doi.org/10.1021/acs.est.5b02809

Pagáč, J., Kovář, P., & Slouka, Z. (2022). Electric potential profiles in a model single-path electrodialysis unit. *Membranes, 12*(11). https://doi.org/10.3390/membranes12111136

Panizza, M., & Cerisola, G. (2009). Direct and mediated anodic oxidation of organic pollutants. *Chemical Reviews, 109*(12), 6541–6569. https://doi.org/10.1021/cr9001319

Perren, W., Wojtasik, A., & Cai, Q. (2018). Removal of microbeads from wastewater using electrocoagulation. *Water Research, 139*, 326–334. https://doi.org/10.1021/acsomega.7b02037

Poyatos, J. M., Muñio, M. M., Almecija, M. C., Torres, J. C., Hontoria, E., & Osorio, F. (2010). Advanced oxidation processes for wastewater treatment: State of the art. *Water, Air, and Soil Pollution, 205*(1–4), 187–204. https://doi.org/10.1007/s11270-009-0065-1

Radjenović, J., & Sedlak, D. L. (2015). Challenges and opportunities for electrochemical processes as next-generation technologies for the treatment of contaminated water. In *Environmental science and technology* (pp. 11292–11302). American Chemical Society. https://doi.org/10.1021/acs.est.5b02414. Vol. 49, Issue 19.

Rai, V., Liu, D., Xia, D., Jayaraman, Y., & Gabriel, J. C. P. (2021). Electrochemical approaches for the recovery of metals from electronic waste: A critical review. In *Recycling*. MDPI AG. https://doi.org/10.3390/recycling6030053. Vol. 6, Issue 3.

Rajeshwar, K., Mcconnell, R., & Licht, S. (2008). Solar hydrogen generation: Toward a renewable energy future. *Environmental Science, Engineering*.

Rakhmania, Kamyab, H., Yuzir, M. A., Abdullah, N., Quan, L. M., Riyadi, F. A., & Marzouki, R. (2022). Recent applications of the electrocoagulation process on agro-based industrial wastewater: A review. In *Sustainability (Switzerland)*. MDPI. https://doi.org/10.3390/su14041985. Vol. 14, Issue 4.

Raut, A. S., Parker, C. B., Klem, E. J. D., et al. (2019). Reduction in energy for electrochemical disinfection of *E. coli* in urine simulant. *Journal of Applied Electrochemistry, 49*, 443–453. https://doi.org/10.1007/s10800-019-01292-4

Rodrigues, M. A. S., Amado, F. D. R., Xavier, J. L. N., Streit, K. F., Bernardes, A. M., & Ferreira, J. Z. (2008). Application of photoelectrochemical-electrodialysis treatment for the recovery and reuse of water from tannery effluents. *Journal of Cleaner Production, 16*(5), 605–611. https://doi.org/10.1016/j.jclepro.2007.02.002

Sala, M., & Gutiérrez-Bouzán, M. C. (2012). Electrochemical techniques in textile processes and wastewater treatment. In *International journal of photoenergy*. Hindawi Limited. https://doi.org/10.1155/2012/629103. Vol. 2012.

Salehi, M. (2022). Global water shortage and potable water safety; Today's concern and tomorrow's crisis. *Environment International, 158*. https://doi.org/10.1016/j.envint.2021.106936, 106936, ISSN 0160-4120.

Simon, R. G., Stöckl, M., Becker, D., Steinkamp, A. D., Abt, C., Jungfer, C., Weidlich, C., Track, T., & Mangold, K. M. (2018). Current to clean water – Electrochemical solutions for groundwater, water, and wastewater treatment. In *Chemie-ingenieur-technik* (pp. 1832–1854). Wiley-VCH Verlag. https://doi.org/10.1002/cite.201800081. Vol. 90, Issue 11,.

Sirés, I., Brillas, E., Oturan, M. A., et al. (2014). Electrochemical advanced oxidation processes: Today and tomorrow. A review. *Environmental Science & Pollution Research, 21*, 8336–8367. https://doi.org/10.1007/s11356-014-2783-1

Strathmann, H. (2010). Electrodialysis, a mature technology with a multitude of new applications. *Desalination, 264*(3), 268–288. https://doi.org/10.1016/j.desal.2010.04.069

Sudarno, Wibowo, M. A., Andarini, P., & Al Qadar, S. (2024). Assessing the environmental implications of water and wastewater production using life-cycle assessment. *Ecological Engineering and Environmental Technology, 25*(4), 70–80. https://doi.org/10.12912/27197050/183183

Sudibyo, Burhani, D., Kusumaningrum, A., Apriyana, W., Aji, R. P. H., Fathnan, A. A., Rajani, A., Syafarina, I., Abka, A. F., Majid, N., & Nurjaman, F. (2016). *Applied Mechanics and Materials, 851*, 226–231. https://doi.org/10.4028/www.scientific.net/amm.851.226

Szpyrkowicz, L., Kaul, S. N., & Neti, R. N. (2005). Tannery wastewater treatment by electro-oxidation coupled with a biological process. *Journal of Applied Electrochemistry, 35*, 381–390. https://doi.org/10.1007/s10800-005-0796-7

Tahmid, M., Choi, H. J., Ganapavarapu, S. T., Scott, J., & Hatzell, M. C. (2024). Concentrating nitrogen waste with electrodialysis for fertilizer production. *Environmental Science and Technology Letters, 11*(12), 1413–1419. https://doi.org/10.1021/acs.estlett.4c00595

Tarpeh, W. A., Barazesh, J. M., Cath, T. Y., & Nelson, K. L. (2018). Electrochemical stripping to recover nitrogen from source-separated urine. *Environmental Science and Technology, 52*(3), 1453–1460. https://doi.org/10.1021/acs.est.7b05488

Tchobanoglous, G., Stensel, H. D., Tsuchihashi, R., & Burton, F. L. (2014). *Wastewater engineering: Treatment and resource recovery* (5th ed.). McGraw-Hill Education.

Vidal, R., Alberola-Borràs, J. A., Gómez-Cordón, J., Moliner, E., Ortega, A., & Verdejo, E. (2017). LCA to evaluate the environmental impact for chemical pre-treatment in plastics metallization. *Journal of Polymers and the Environment, 25*(4), 961–972. https://doi.org/10.1007/s10924-016-0872-6

Vik, E. A., Carlson, D. A., Eikum I, A. S., Gjessing, E. T., & Scandinavian Exchange Program, –Nalle. (1984). Electrocoagulation of potable water. *Water ICes, 18*(Issue 11).

Von Sperling, M. (2007). *Basic principles of wastewater treatment*. IWA publishing.

Walsh, F. C., & Ponce de León, C. (2018). Progress in electrochemical reactor engineering. *Chemical Engineering Journal, 344*, 29–47. https://doi.org/10.1016/j.electacta.2018.05.027

Wang, H. D., Gao, K., & Lu, B. (2014). Electrochemical catalytic oxidation treatment of coking wastewater RO brine. *Advanced Materials Research, 838–841*, 2751–2758. https://doi.org/10.4028/www.scientific.net/AMR.838-841.2751

Xiahong, S., Zhang, X., Gao, W., Zhang, Y., & He, D. (2022). Removal of microplastics from water by magnetic nano-Fe3O4. *Science of the Total Environment, 802*, Article 149838. https://doi.org/10.1016/j.scitotenv.2021.149838. ISSN 0048-9697.

Xu, X., He, Q., Ma, G., Wang, H., Nirmalakhandan, N., & Xu, P. (2021). Pilot demonstration of reclaiming municipal wastewater for irrigation using electrodialysis reversal: Effect of operational parameters on water quality. *Membranes, 11*(5), 333. https://doi.org/10.3390/membranes11050333

Xu, T., & Huang, C. (2008). Electrodialysis-based separation technologies: A critical review. In *AIChE journal* (pp. 3147–3159). https://doi.org/10.1002/aic.11643. Vol. 54, Issue 12.

Xu, S., Zhao, H., Xie, L., Wang, K., & Zhang, W. (2023). Study on the treatment of refined sugar wastewater by electrodialysis coupled with upflow anaerobic sludge blanket and membrane bioreactor. *Membranes, 13*(5). https://doi.org/10.3390/membranes13050527

Yan, H., Xu, C., Li, W., Wang, Y., & Xu, T. (2016). Electrodialysis to concentrate waste ionic liquids: Optimization of operating parameters. *Industrial and Engineering Chemistry Research, 55*(7), 2144–2152. https://doi.org/10.1021/acs.iecr.5b03809

Yang, K., & He, Z. (2025). Integrating oxidation and reduction processes in electrochemical wastewater treatment for contaminant removal with by-product control. *Water Research, 282*. https://doi.org/10.1016/j.watres.2025.123632

Yang, M., Jing, B., Chen, W., Li, Q., & Yin, X. (2017). Experimental study on COD composition and electrochemical degradation of wastewater in offshore oilfields. *Journal of the Chinese Chemical Society, 64*(1), 73–79. https://doi.org/10.1002/jccs.201600244

Yang, J., Monnot, M., Ercolei, L., & Moulin, P. (2020). Membrane-based processes used in municipal wastewater treatment for water reuse: State-of-the-art and performance analysis. *Membranes, 10*(6), 1–56. https://doi.org/10.3390/MEMBRANES10060131

Yuan, Y., & Lei, A. (2020). Is electrosynthesis always green and advantageous compared to traditional methods?. In *Nature communications*. Nature Research. https://doi.org/10.1038/s41467-020-14322-z. Vol. 11, Issue 1.

Zainul, R., Chaya, R., Wijaya, K., Rahmawati, F., Laghari, I. A., Rahmadiawan, D., & Abdullah, M. (2024). Advances in electrochemical technologies for sustainable wastewater treatment and chemical synthesis: Mechanisms, challenges, and prospects. *Journal of Medicinal and Pharmaceutical Chemistry Research, 6*(2), 140–156. https://doi.org/10.48309/JMPCR.2024.424054.1031

Zaky, A. M., & Chaplin, B. P. (2013). Porous substoichiometric TiO2 anodes as reactive electrochemical membranes for water treatment. *Environmental Science and Technology, 47*(12), 6554–6563. https://doi.org/10.1021/es401287e

Zhang, C., Zhao, X., Wang, C., & Pan, S. (2021). Novel multistage electrochemical flow-through mode (EFTM) with porous electrodes for reclaimed wastewater treatment in pipes. *ACS ES&T Water, 1*(3), 653–660. https://doi.org/10.1021/ACSESTWATER.0C00165

Zhang, C., Zhou, M., Yu, X., Ma, L., & Yu, F. (2015). Modified iron-carbon as heterogeneous electro-Fenton catalyst for organic pollutant degradation in near neutral pH condition: Characterization, degradation activity, and stability. *Electrochimica Acta, 160*, 254–262. https://doi.org/10.1016/j.electacta.2015.01.092

Zhou, M., Dai, Q., Lei, L., Ma, C., & Wang, D. (2005). Long life modified lead dioxide anode for organic wastewater treatment: Electrochemical characteristics and degradation mechanism. *Environmental Science and Technology, 39*(1), 363–370. https://doi.org/10.1021/es049313a

Zhu, X., Ni, J., Wei, J., Xing, X., & Li, H. (2011). Destination of organic pollutants during electrochemical oxidation of biologically-pretreated dye wastewater using boron-doped diamond anode. *Journal of Hazardous Materials, 189*(1–2), 127–133. https://doi.org/10.1016/j.jhazmat.2011.02.008

Zhu, X., Tong, M., Shi, S., Zhao, H., & Ni, J. (2008). Essential explanation of the strong mineralization performance of boron-doped diamond electrodes. *Environmental Science and Technology, 42*(13), 4914–4920. https://doi.org/10.1021/es800298p

CHAPTER

10 Digital twin and IoT-Enabled smart water treatment systems: Foundations, frameworks, and applications

Randeep Singh[1], Anwar Faizaan Reza[2], Arkadeep Mukherjee[3] and Saikat Sinha Ray[4]

[1]*Department of Basic Sciences, School of Sciences and Humanities, SR University, Warangal, Telangana, India;* [2]*Department of Computer Science and Engineering, School of Engineering and Sciences, SRM University-AP, Amaravati, Andhra Pradesh, India;* [3]*Department of Biosciences, Graphic Era (Deemed to Be) University, Dehradun, Uttarakhand, India;* [4]*Department of Environmental Science and Engineering, School of Engineering and Sciences, SRM University-AP, Amaravati, Andhra Pradesh, India*

1. Introduction

The trend of global urbanization indicates that close to 68% of the total human population will be living in urban areas by the year 2050 [(68% of the World Population Projected to Live in Urban Areas by 2050, Says UN, 2026)]. This poses serious stress on the essential infrastructure, namely, Water Distribution Networks (WDNs). The traditional water management practices, which were meant for the needs of the past decades, are currently unable to support the needs of the present era (Singh & Ahn, 2025). As a result, the water utilities across the globe are experiencing serious operational inefficiencies, with water losses above a quarter of the total water supply.

Simultaneously, water resources are also under increasingly complex pressures due to industrial demand, climate change-related hydrological variability, and pollution (Singh et al., 2025). Globally, it has been estimated that more than four billion people suffer from extreme water scarcity for at least 1 month in a year, and this problem is expected to worsen. Under these circumstances, the conventional approach to governance will be inadequate to ensure a sustainable water supply and equitable distribution. There is, therefore, an urgent need to adopt a paradigm shift in water management through Smart Water Management (SWM). SWM combines Industry 4.0 technologies to improve resilience and sustainability, which will directly contribute to the United Nations Sustainable Development Goals (SDGs), namely SDG 6 (Clean Water and Sanitation), SDG 9 (Industry, Innovation and Infrastructure), and SDG 11 (Sustainable Cities and Communities) (Tomaszewski & Kołakowski, 2023).

Digital transformation provides a way to bring about a paradigm shift in these systems by integrating high-performance computing and environmental engineering (Reza et al., 2024). At the vanguard of this paradigm shift are Digital Twins (DTs). DTs are digital replicas of physical entities that enable real-time monitoring, simulation, and stochastic modeling. When integrated with the

Sustainable Water Management. https://doi.org/10.1016/B978-0-443-44853-9.00015-0

Internet of Things (IoT) and Artificial Intelligence (AI), DTs become intelligent systems that can make autonomous decisions (Reza et al., 2026).

This technological convergence brings about a paradigm shift in urban management by enabling the seamless integration of data and predictive analytics. For urban planners and decision-makers, DTs are advanced tools that enable the visualization of infrastructure performance, simulation of intervention strategies, and resource allocation optimization. AI-driven analytics take this to the next level by predicting water quality variability, instant leak detection, and accurate demand forecasting (Mukherjee et al., 2025). This shift from reactive to predictive management is essential for making cities resilient to the challenges of population growth and resource scarcity.

Water safety is an important aspect of SWM. Although clean water is a basic human right, it remains a concern because of the degradation of water quality caused by urbanization, agricultural runoff, and industrial effluent. The World Health Organization recognizes that contaminated drinking water is a major cause of mortality worldwide, while nutrient pollution is a major cause of harmful algal blooms that decimate aquatic habitats.

Traditional monitoring systems, which involve grab sampling and laboratory analysis on a periodic basis, are labor-intensive and involve a high degree of latency (Singh & Ahn, 2025). These point-in-time sampling approaches are often inadequate to detect dynamic, transient contamination events such as pesticide runoff during storms or sudden network intrusions (Singh et al., 2025). To address this, the combination of IoT and Machine Learning (ML) is an effective approach. IoT networks involve the use of in-situ sensors to measure high-frequency data points for parameters such as pH, turbidity, and dissolved oxygen levels. ML algorithms are then used to analyze these data points to identify anomalies and predict microbial contamination or nutrient surges in real-time, with a much lower response time for health-related decisions.

However, the potential of these technologies remains untapped due to the fragmented nature of the academic and practical environment. Systematic literature reviews have shown that although research exists on the application of 3D models for urban decision-making and DTs Cities (DTC), adaptive frameworks are still very limited (A.-J. Wang et al., 2024). Moreover, the current state of research does not incorporate semantic interoperability or integrate virtual models with real-world governance processes. Additionally, most reviews on the topic tend to separate ML algorithms from the real-world limitations of IoT sensors, or vice versa, discuss hardware architectures without considering the analysis techniques necessary for the extraction of meaningful insights.

There is a lack of standardized datasets and evaluation protocols, which makes it difficult to reproduce results. Moreover, most studies do not properly quantify the social implications of these implementations. To fill this gap, a holistic approach is needed to overcome the current divide between sensor hardware, data analysis, and real-world urban governance to develop resilient water systems.

2. Conceptual foundations

2.1 Internet of things in water systems

For dealing with critical issues and problems like water scarcity, water loss, and effective resource management, IoTs can be applied on a large scale to overcome the problems associated with water management. These problems can be overcome by combining the use of IoT sensors that acquire real-time information about the water resource on different parameters of water quality, like pH, turbidity, pressure, and temperature. The information that was previously acquired is then sent through a low-power network layer protocol and analyzed at the application layer through cloud computing and

AI. This helps to acquire effective insights about water quality and management in a cost-effective manner. Figure 10.1 represents various techniques.

The implementation of this technology in water management with the use of IoT has many advantages, like real-time monitoring, leakages, and contamination, along with effective resource management. This provides us with the option of maximum resource recovery because real-time monitoring features enable us and allow us to detect problems. This creates a more reliable, secure, and economically efficient and practical water supply network for better water management and contributes to overall growth. It also opens the doors for interdisciplinary collaboration across various fields (Radhakrishnan & Wu, 2018).

The proposed system utilizes the IoTs to create a smart, real-time water quality monitoring and control system. Employing IoT sensors that are set to continuously measure factors such as pH,

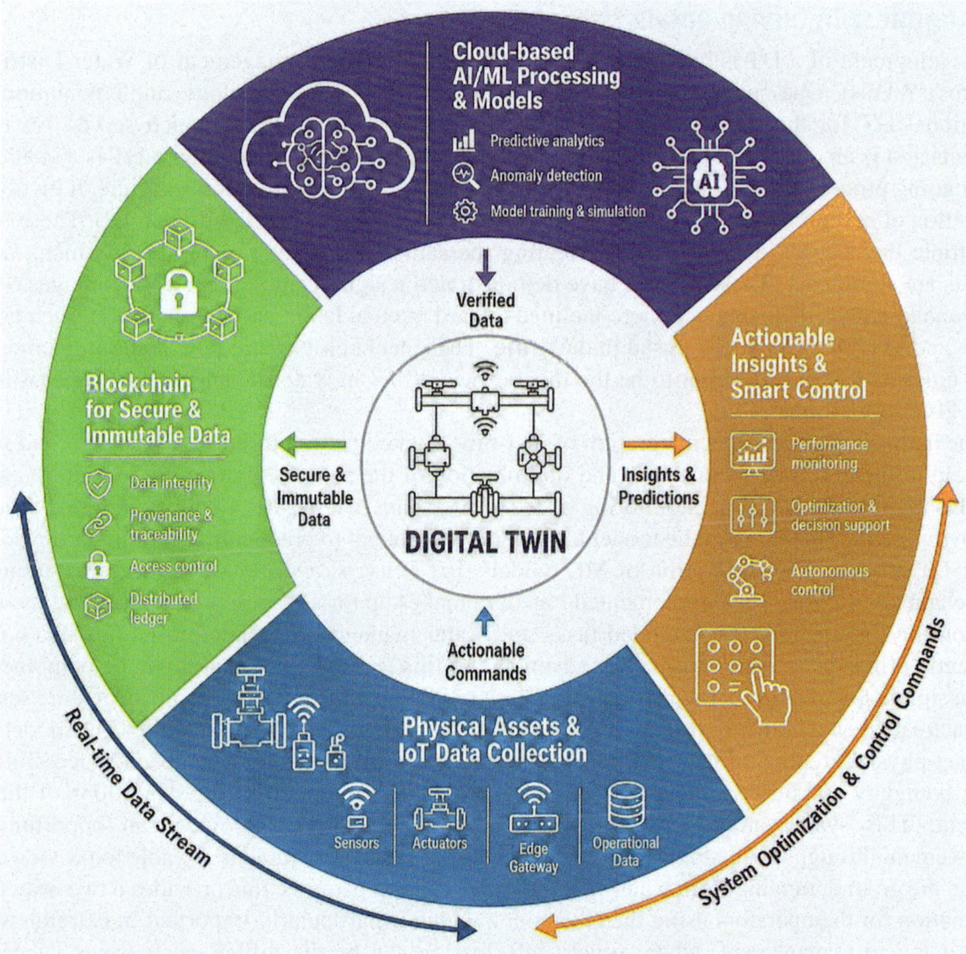

FIGURE 10.1

The representation of various DTs utilized for smart water treatment systems.

temperature, and turbidity, along with many other parameters that affect the water quality in real time, enables users to quickly assess the quality of their water supply. The IoT integrations main feature is its ability to make autonomous decisions without any support when combined with technologies like ML and deep learning (DL), activating filtration systems only when necessary to adjust water parameters to make efficient use of the system and to reduce the power consumption of the system while combined with different technologies are some of its valiant features that helps to automate and ease the process in day to day life. This efficient approach reduces costs and extends the lifespan of filters. The data that is collected from the system is then used for analysis and provides us with clear infographics. The main goal here is to improve public health and lower water treatment expenses by replacing the traditional method of manual monitoring, lab testing, and eliminating manual tasks, and this contributes to the overall growth toward sustainable water management (AlMetwally et al., 2020).

2.2 Digital twin fundamentals

The development of a DT is becoming essential for the effective management of Water Distribution Systems (WDSs). One such example of this groundbreaking DT technology implementation of an operational DT for the water network was observed in Valencia, Spain, which serves 1.6 million residents and is among the first of its kind for a major utility. Although creating a DT is a challenging and ongoing process of learning and adaptation that requires a lot of attention and time, it involves the integration of several technologies like blockchain, cloud computing, IoT, ML, and DL. These models can mimic the scenario in real time by creating the same virtual copy of the environment, and the benefits are significant. These models have demonstrated a significant number of advantages over the traditional methods that are much more inclined toward manual labor, and these models, such as DT's, have acted as a better tool when used in daily life. These technologies are now being implemented in smart cities and have turned out to be the most efficient tool in water management (Conejos Fuertes et al., 2020).

The main innovation is the integration of real-time data collection through IoT devices and sensors that help in the real time monitoring, and optimization of the process is done via hybrid ML models that also include the real-time monitoring of IoT sensor data that has been collected in real time from the environment. This elevates the model from being theoretical to practical. This unlocks the potential of the DT models. The integration of ML models, IoT sensors, and cloud architectures, along with blockchain algorithms, when implemented together, makes up for a successful and practical model that can not only be used for water-related tasks and water management-related tasks but also could be efficiently utilized for many others. The advantage of this technology is that it can be combined with any discipline that requires a lot of attention. These models are designed to operate on a large scale and are practical when it comes to mimicking the closest scenario in day-to-day real life. The model builds an exact physical replica of the environment digitally, and these models have been successfully outlining boundary conditions to ensure accuracy and reaction to the changing conditions of the environment. This system not only improves local models but also provides an opportunity for improvement through the feedback analysis of the models. As these models are able to provide critical data on the ground, ranging from water levels to precipitation patterns, this provides a two-way flow of information for the operators using the DT model, which is particularly important in extreme weather conditions and management, where timely information can be the difference between success and failure. The application of this multi-scale framework enables water authorities and organizations to optimize this technology for management-related tasks that are related to water distributed systems, water resource management, and water resource planning. In terms of the lower-scale application of this technology, farmers can also apply this technology to make informed decisions regarding irrigation, while emergency services can prepare for any potential crisis using data-driven strategies. Even citizens can apply this technology to take advantage of the localized information that enables them to understand their risk and take proactive steps to protect their homes and communities. This is because it directly addresses the need to reduce vulnerability and exposure to climate hazards and the proper utilization of resources. The integration of localized DTs within the systems shows us how technology can transform agricultural, aquaculture practices, and urban planning, and this stands out as an example that shows us how the technology could be used in an efficient manner. Ultimately, the system serves not only as a powerful example of how DT technology can be used to safeguard a nation's resources from the uncertainties of nature. It illustrates the potential for innovative technologies. By harnessing the capabilities of hybrid ML and real-time data collection, along with the help of IoT devices and sensors that are powered by the accessible cloud and secured by blockchain technologies, this not only safeguards but also helps us to protect the resources and resource-related data and opens the doors for advancements across all the disciplines (Henriksen et al., 2022).

2.3 Synergy of IoT and DT

The synergy between the IoTs and DTs represents a transformative approach toward aquaculture, agriculture, and industrial resource management. These advancements take a step forward in improving the accuracy and efficiency of the systems in real life when compared with the traditional process for effective water management. Furthermore, this integration allows for the evolution from reactive to predictive management for better resource management while being economical and sustainable at the same time. The system is also capable of predicting the problems in the future with respect to continuous changes in parameters in real time that are based on subtle changes in nature and in data trends. The model is self-efficient and can make changes accordingly. This makes the DT a fully functional one that works as an intelligent co-pilot for the user. The development of a DT is becoming an essential requirement for the effective management of WDSs. The pioneering work of implementing an operational DT for the water network in Valencia, Spain, which serves 1.6 million people, and is one of the first of its kind for a major utility, is one such example that proves the capability of this model in terms of how reliable and effective this model could be when it comes to practicality (Alves et al., 2019; Conejos Fuertes et al., 2020).

3. System architecture and frameworks

3.1 Multi-layered digital twin framework for water treatment

The proposed monitoring system for sustainable water management is a multi-layered DT framework. The main goal of this model is to create virtual replicas of the physical entities, and it processes them all by organizing them into layers. This technology offers a more practical approach to the challenges of manual assembly and resource management in the Industry 4.0 and 5.0 era, in scenarios where human and machine collaboration becomes important for achieving optimal results. By integrating the DT technology along with the features of computer vision and 3D point cloud processing we can achieve some beneficial results, with tools like blockchain for security and integrity of the data and ML for automation can also be combined with the existing model to achieve lower costs, higher accuracy and easy control with high security, this system can not only detect errors but can also track progress in real time while being cost-effective. This system could be particularly beneficial for small-batch monitoring systems, where the flexibility and precision are the key to judging the model's capabilities one of the major problems that could arise while doing so is to Maintaining the resources while being error prone can be challenging task, as it requires continuous monitoring and must be precisely done, with errors sometimes arising unpredictably due to human oversight and negligence but this model and the multi layered DT model along with the ML models stands as the exception is capable of detecting the errors on its own and can generate alerts when needed or when any error is found, This system addresses these challenges while minimizing disruption to workers and operators.

The concept of multi-layered DTs was created to tackle the problems at subsequent levels, which led to the multi-layer model, which comprises multiple layers to make the digital replica. At the assembly level, it checks on the multiple-part interaction of the models and how they actually fit in. At the part level, it focuses on individual components and monitoring individual parts for more data transparency. While these models work together, the layers create a living representation/replica digitally, where any kind of errors/discrepancies or abnormalities between the digital and physical environments can be seen as clear indicators for potential errors and errors that have already occurred, by allowing us for easier identification and correction during practical functioning of the model.

One of the main features of this architecture is the cloud reservation algorithm. This algorithm isolates the visible surfaces of a part and simulates how the parts should appear digitally, even when certain sections are tampered with. This allows for validation and allows us to correct the errors without requiring complete visibility of the environment. Another crucial element of this system is the two-way matching algorithm that deals with the potential risks of human error. Rather than predicting all the potential errors, this algorithm compares the point cloud of the real component to the digital model and vice versa. The process will only continue when the real and digital components match both comparisons and when all the required criteria are fulfilled. Any discrepancy in placement, orientation, or geometry can be easily made correctable, making it simpler to spot errors as they happen.

The primary and most important benefit of this system is that it only demands a very small amount of data about the scene to develop an actual replica of the existing systems. The system gives real-time notifications when there is something missing during the tracking process of the overall assembly to move ahead simultaneously. This piece of information can be utilized later to optimize the efficiency of the systems, which can significantly improve accuracy, and can be used for effective management. Many of the Experiments have also confirmed that the system effectively detects errors and monitors progress. By combining layered DTs, point cloud reservation, and bidirectional matching, this architecture offers a reliable and scalable solution for assembly and management-related tasks, thereby reducing the risks associated with human error. In the future, this system could be enhanced with augmented reality and virtual reality applications that could help guide operators during the monitoring phase, or can be expanded to manage more complex roles requiring extensive oversight, further strengthening the role of DTs in sustainable resource management, and can be used for better resource allocation (Pang et al., 2023).

3.2 Interoperability and standards

Previously, many investigations and experiments have been done with respect to the interoperability challenges of integrating DTs within Cyber-Physical Systems (CPS) under the frameworks of Industry 4.0 and 5.0, with a focus on edge-processed networks. DTs serve as virtual representations/replicas of physical assets and systems, enabling real-time monitoring, simulation, and optimization through data flows between DTs and CPS. These systems include technologies such as the Industrial Internet of Things (IIoT), edge computing, AI, ML, augmented reality (AR), virtual reality (VR), blockchain, and cloud computing, as can be observed in Fig. 10.2. These models can also utilize approaches and

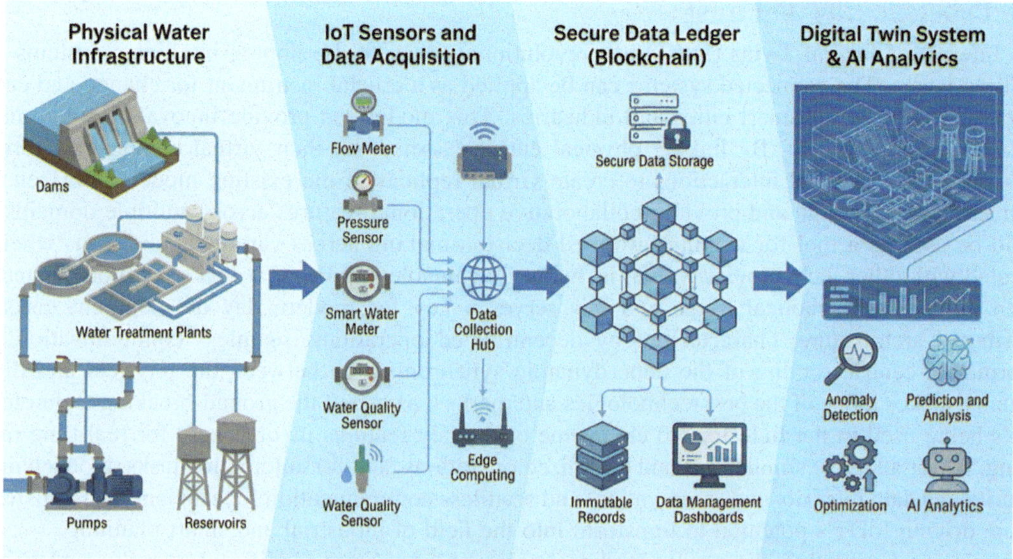

FIGURE 10.2

Demonstrates the interoperability features when integrated with multiple technologies.

integrations of new technologies, such as federated learning (a subset of ML) and cybersecurity concepts. Standards such as RAMI4.0, IIRA, and OPC-UA, along with emerging standards like STF-641 and DTDL, the aim here is to ensure seamless interoperability between the systems and help in transforming systems into interconnected frameworks among the functioning of the models. However, the increasing complexity of these technologies and their integration within DT-CPS setups with respect to interoperability issues hinders the scalability of the systems for the expansion of the systems on a large scale.

Some of the Key findings from the experimentation and testing phases of past research concluded that interoperability in DT-CPS is crucial for enabling seamless data exchange and control between physical and digital systems. The identified challenges include technical issues such as data redundancy, protocol mismatches, and latency in edge networks while transmitting data from the source to the destination machine. While doing so, some of the key barriers that were observed included inconsistent data and models lacking standardized interfaces for rapid data transfers. While edge-processed networks had redundancy and latency issues, they also introduced complexities in real-time data synchronization and system scalability. The proposed interoperability framework categorizes these challenges.

The implications and the significant impact are relevant with respect to industries that are adopting DT-CPS. The interoperability via standardized architecture and technology serves as the roadmap for the achievement of the best interoperability standards. Highlighting the need for interdisciplinary research. Future research should investigate new standards and technologies that could improve DT-CPS interoperability and integration of the systems, especially in dynamic large-scale systems and multi-scale systems, to ensure efficient and secure integration for applications such as smart manufacturing, healthcare, and heavy tasks that involve lots of monitoring, such as sustainable resource management and allocation (Acharya et al., 2024).

3.3 Cybersecurity and trust

The Internet of Digital Twins (IoDT) is a revolutionary way of developing intelligent systems with multiple layers. The connected systems can be applied as a crucial instrument for change and can be observed in action for smart cities and industries. This model can provide innovative solutions for sustainable management. By linking physical entities/assets with their virtual components through real-time cyber-physical interactions to create virtual replicas of the existing models, IoDT enables seamless data exchange and provides collaborative operational features across multiple domains and could be used as a tool for making informed decision-making across various applications when integrated with other technologies, which is one of its valuable features that make the model stand out when put through practical usage and this serves a key for modern day development tasks. Its distributed architecture, characterized by decentralized operations, seamless communication, and information-centric routing of the data, dynamic synchronization between the physical and digital systems, makes it one of the best technologies and marks it as one of the ground-breaking technologies that's being used in the Industry 5.0 era. Some of the key features involve IoTs for real-time monitoring, AI for adaptive simulations and optimized operations, and for automation tasks, blockchain for security and data integrity and preservation, and seamless communication for efficient data transfer via cloud, driving IoDT's potential to transform into the field of industrial and smart planning.

However, integrating these advanced technologies introduces significant security and privacy concerns that impact the widespread adoption of IoDT. It also highlights a range of threats and potential threats, including data tampering, data redundancy, impersonation, and privacy data leaks due to coordinated and targeted attacks. These risks, coupled with cyber-physical vulnerabilities that could translate digital breaches into physical harm, is one of the growing concerns with respect to digital well-being and digital security, Current solutions to this problem include advanced encryption, decentralized authentication, and privacy-preserving techniques, and these differential privacy methods/techniques are some of the best defense mechanisms to overcome the problem of data breaches, these methods are promising but struggle with scalability and performance trade-offs in large-scale IoDT networks.

Cloud-edge-end and space-air-ground integrated networks, which are innovative solutions with low latency, and 6G-enabled DTs, which can improve the real-time capabilities, are some of the promising solutions. Standardization of interoperability protocols and development of regulatory frameworks and policies that can harness AI and blockchain for governance are the keys to developing scalable and trustworthy systems. Further, integration of explainable AI capabilities can also improve the transparency and robustness of the models, which can also resolve the ethical and privacy issues. IoDT, which has addressed the security and privacy issues through adaptive and scalable solutions, is considered a revolutionary solution for sustainable management, implementation of smart city technology, and industrial management (Y. Wang et al., 2023).

4. Smart water treatment applications

4.1 Process optimization

The overall goal of contemporary water treatment is to convert raw information into intelligent control actions that guarantee a stable effluent output. In a conventional treatment plant, the control system usually consists of Proportional-Integral-Derivative (PID) controllers and fixed setpoints. Although effective, these legacy systems are ill-equipped to handle the strongly nonlinear and dynamic behavior of the influent water quality. The combination of DTs and AI turns this paradigm on its head by allowing for predictive process control instead of reactive control.

In a smart treatment system, the DT is essentially a virtual simulation environment that operates concurrently with the actual process. It continuously processes real-time information from IoT sensors measuring turbidity, pH, dissolved oxygen, and chemical oxygen demand. Machine learning algorithms process this information flow to forecast future process behavior. For instance, during the coagulation and flocculation phases, feed-forward control algorithms can process incoming raw water quality information and compute the optimal coagulant dose before the water even enters the mixing tank. This accurate dosing eliminates both under-dosing, which is unsafe, and over-dosing, which is wasteful and generates excessive sludge.

Moreover, the process optimization process is further extended to the complex biological treatment stages. The activated sludge process, which is known to be notoriously difficult to model because of the uncertainties involved in microbial activity, can be greatly aided using "soft sensors." These are virtual sensors that employ data-driven models to predict difficult-to-measure values such as biological oxygen demand (BOD) from easily measurable proxy variables. With a real-time estimate of biochemical demand, the system can automatically adjust aeration rates. This ensures that the microorganisms receive exactly what they need, just the right amount of oxygen for their metabolic activity, without the energy inefficiencies of constant high-rate aeration.

The integration of the physical and virtual worlds of the plant gives rise to a self-optimizing closed loop. As the system gathers historical data, the models learn to retrain themselves to identify seasonal changes or particular contamination incidents. This leads to a treatment system that not only maintains but actively pursues the most optimal operating mode. As a result, plant operators shift from fighting fires to managing daily operational disturbances to monitoring a highly autonomous and intelligent system that mathematically certifies water quality standards are met.

4.2 Asset lifecycle management

Water treatment plants are capital-intensive environments that consist of pumps, membranes, valves, and motors that are operated in harsh conditions. The conventional method of managing such assets was run-to-failure reactive maintenance or fixed preventative maintenance schedules according to generic manufacturer guidelines. Both methods are inefficient. Run-to-failed reactive maintenance results in expensive emergency repairs and down times, while fixed preventative maintenance schedules often involve replacing perfectly good assets well before their time. Smart Asset Lifecycle Management (ALM) utilizes the Industrial Internet of Things (IIoT) to change the paradigm from predictive and prescriptive maintenance.

The key to this strategy is the use of vibration sensors, acoustic detectors, and thermal cameras on strategic assets. The sensors provide continuous health information to the facility's DT. By using anomaly detection algorithms, the system can detect the subtle precursors of failure that cannot be detected by the naked eye. For example, a change in the vibration pattern of a high-service pump may indicate bearing deterioration weeks before a catastrophic failure. The system uses this information to calculate the Remaining Useful Life (RUL) of the asset, so that maintenance can be planned during a scheduled down time instead of during a crisis.

In addition, DT serves as a repository for the complete history of an asset, combining design parameters, maintenance records, and current operating data. This comprehensive perspective allows for "digital thread" traceability. If a part does fail, engineers can then trace the problem back to possible root causes, including design problems, installation issues, or operating conditions. This information drives long-term capital planning. Utilities can leverage simulation tools to compare the life-cycle costs of repairing existing infrastructure versus developing more efficient technologies. By optimizing the balance between Capital Expenditures (CAPEX) and Operating Expenditures (OPEX), smart Asset Lifecycle Management (ALM) ensures that water infrastructure is both reliable and cost-effective over its planned life cycle.

4.3 Energy and resource efficiency

The water-energy nexus is a challenge to sustainability because water treatment and distribution are among the biggest electricity users in a municipality. Traditional systems tend to operate pumps and aerators at fixed speeds, irrespective of actual demand, resulting in substantial energy losses. The smart water system breaks this inefficiency by integrating energy management and process control. Based on AI logic and Variable Frequency Drives (VFDs), the system dynamically manages energy use.

Intelligent algorithms process information on electricity tariffs, hydraulic demand, and storage tank levels to execute "peak shaving" algorithms. The system can choose to operate energy-hungry pumps during off-peak periods when electricity tariffs are low, storing water in storage reservoirs to satisfy hydraulic demand during peak hours through gravity feeding. In aeration tanks, which consume up to 60% of energy in a plant, ammonia-based aeration control systems modulate air blowers to precisely match biological demand. This eliminates the tendency to over-aerate and substantially reduces the carbon footprint of the plant.

Resource efficiency also includes the reduction of chemical use and waste production. In membrane filtration systems, biofouling is a problem that needs to be addressed periodically through chemical cleaning. The DT can simulate biofouling rates according to transmembrane pressure and permeate flux. This will optimize the cleaning process, ensuring that the cleaning process is started only when it is required, thus reducing the use of cleaning chemicals and increasing the life of the membrane modules. In addition, smart systems make it easier to achieve a circular economy. By accurately measuring sludge properties, plants can optimize anaerobic digesters to produce the maximum amount of biogas. This biogas can be used to power the plant, making the plant energy neutral and also allowing it to recover valuable nutrients such as phosphorus for agricultural use.

4.4 Resilience and emergency response

The urban water infrastructure is increasingly being threatened with extinction by climate change-related hydrological disasters and cyber-physical attacks. Resilience in this case is measured not only by the capacity to absorb a shock but by the ability to rapidly recover and continue essential operations. DT is the nervous system of resilience, providing a safe space to model disaster scenarios that cannot be replicated in the physical world.

By means of "what-if" scenario modeling, the operators can assess the effects of catastrophic events such as 100-year floods, or even the failure of the power grid and contamination intrusion events. For instance, in the case of a chemical spill that is reported upstream, the DT can model the distribution of the contaminant in the river and the intake structures. This enables the system to accurately determine the time of arrival and concentration of the contaminant at the treatment plant. The system can then automatically recommend or implement containment strategies, such as the shutdown of intake pumps or the diversion of flow to emergency holding basins.

In the world of cybersecurity, the integration of information technology (IT) and Operational Technology (OT) brings about new risks. A hacked sensor could provide erroneous information to the control system, potentially leading to physical harm or quality incidents. Smart systems employ physics-informed AI to check sensor information. When a sensor provides information that goes against the laws of physics or does not correlate with other parameters (for example, a sudden change in pressure without a corresponding change in flow), the system identifies it as a potential cyber incident or sensor malfunction. This feature enables the plant to shut down the affected area and go back to safe mode operations. Finally, the combination of these technologies enables the emergency response system to shift from a disorganized, manual process to a coordinated, data-driven defense plan that secures water security in an uncertain future.

5. Case studies and emerging implementations

5.1 Municipal wastewater treatment

The use of DT technology in municipal wastewater treatment has progressed from static modeling to dynamic synchronization, which has brought a paradigm shift in operational practices. This is evident in the latest use of hybrid digital twins (HDTs), which combine mechanistic biological models with data-driven ML algorithms. For example, Ge et al. developed an urban flooding digital twin platform that comprehensively manages flood events throughout the entire disaster management lifecycle (Ge & Qin, 2025). The proposed framework serves four essential functions. First, it accommodates various model generation methods and tools for constructing what-if scenarios. Second, it supports readiness simulations and pre-event evaluations. Third, it coordinates collaborative prevention and intervention services among multiple stakeholders during emergency responses. Fourth, it enables continuous monitoring and predictive assessment of flood risks during the recovery phase.

Furthermore, the integration of IoT sensor arrays for soft sensing solved the longstanding challenge of measuring process variables that are difficult to monitor in real time online, such as Biochemical Oxygen Demand (BOD) (Hassnain et al., 2025). Recently, it was demonstrated that ML models, particularly Extra Trees for prediction and Neural Networks for process simulation, can accurately estimate final effluent BOD in an industrial wastewater treatment facility using 19 years of historical data. The AI-driven approach offers a reliable alternative to conventional 5-day BOD testing, enabling real-time process control, optimized operational conditions, and enhanced treatment efficiency while reducing costs and environmental impact.

Singapore's implementation of DT technology at the Tuas Water Reclamation Plant serves as a model of a smart municipal facility functioning as a cyber-physical system (Innovation in Water Singapore, 2026). In this setup, predictive maintenance algorithms process vibration and acoustic data from pumps and blowers. This shift from volatile repairs to predictive maintenance decreased equipment downtime by approximately 30%, validating that digital maturity and its use are as critical as process capacity in modern municipal infrastructure.

5.2 Industrial water recycling

The error margin at an industrial level for water recycling is impossible, particularly when the recycled water is cycled back into sensitive production cycles. The IoT and DT are critical to the reliability of Zero Liquid Discharge (ZLD) processes. The semiconductor industry has led the way in the application of Digital Shadows (DSs), a simplified version of DTs, to monitor the passage of particular contaminants through the recycling process. Additionally, a study showcased a DT framework for WDNs, which has been developed using the EPANET-based Python library Water Network Tool for Resilience (WNTR) and optimized using the Differential Evolution algorithm on a benchmark hydraulic network (Brahmbhatt et al., 2023). The optimized DT is used as a real-time surrogate model for the generation of simulation data to solve two important operational problems: optimizing disinfectant injection using a neural network-based control strategy and locating pipe leaks with high accuracy using graph convolutional networks. The presented approach has been shown to be effective in water quality control and leakage detection. Thus, DT helps to recycle as well as save water from leakage.

Likewise, the textile industry has also started adopting the use of IoT-enabled smart meters to monitor variations in the discharge of dye-bath effluent, while other industries are adopting the technology to minimize the discharge of pollutants into the wastewater. Recently, a study presented APAH, a scalable and frugal IoT-based autonomous real-time monitoring system for industrial wastewater management, integrating multi-parameter sensors with edge and application layer architecture for continuous data acquisition and remote access (Chavhan et al., 2025). Validated across textile, dairy, and greywater treatment plants in Maharashtra, India, the system employs automated valve controls, real-time alerts, and machine learning capabilities to enable timely intervention and ensure regulatory compliance. APAH demonstrates significant potential for enhancing treatment efficiency and enabling sustainable water resource management in industrial environments facing acute pollution and water scarcity challenges.

Further, the chemical industry is using "soft sensors" in DT models to forecast the rates of membrane fouling (Shahouni et al., 2024). Based on the analysis of trans-membrane pressure (TMP) variations, coupled with feed water temperature and flux, it is possible to forecast the rate of irreversible fouling weeks in advance, planning Clean-in-Place (CIP) cycles only when absolutely necessary, thus prolonging the life of membranes.

5.3 Centralized versus decentralized systems

The contrast between centralized and decentralized systems is bridged by the "Internet of Water", where IoT provides the remote visibility vital to make decentralized solutions convenient at scale. In general, decentralized units failed due to a lack of timely maintenance; however, modern IoT-enabled plants function as autonomous nodes in a larger network. For example, in the rural outskirts of Bangalore, India, a pilot water treatment plant linked 50 decentralized membrane bioreactors (MBRs) to a single central command center via 5G telemetry (Tyagi et al., 2024). This "virtual centralization" allowed a small team of experts to monitor and troubleshoot dispersed units remotely, achieving uptime rates of 98%, comparable to large, centralized facilities, while avoiding the capital expenditure of extensive sewer networks.

Conversely, in dense urban centralized systems, DTs are being used to optimize the conveyance network itself, treating the sewer system as an active reactor rather than a passive pipe. Smart manholes equipped with level and flow sensors feed data into a hydraulic DT of the city's catchment area. Recently, in Paris, a new system utilized active control of automated sluice gates to optimize in-sewer storage capacity during storm events. This made it possible to control effectively the peak flow that reached the centralized treatment plant. The approach led to a reduction of 40% in combined sewer overflows (CSOs) (Lumley et al., 2024).

Today, the debate is no longer between centralization and decentralization but between integration and digitalization. Whether it is a large plant or a group of scale units, the common thread is the digital data layer that promotes compliance, maximizes resource recovery, and enables a responsive water infrastructure [(Digital Water By Technology Smart Water Meters, Internet Of Things, Artificial Intelligence & Machine Learning, Digital Twin, Geographic Information Systems, Others, By Solutions, By End-User Municipal Water Industries, Industrial, By Region - Global Forecast To 2029, 2026)].

6. Integration with emerging technologies

6.1 AI and ML for water quality

The latest technologies, like explainable AI using traditional ML algorithms, are being used extensively to improve and automate the water quality classification, and this can also be used to analyze the importance of different parameters to improve the overall efficiency and accuracy of the models. Some of the latest research and ongoing research in this field include the use of ML models to improve the accuracy of the traditional model, which has also included a comprehensive comparative study of different classification models like Logistic Regression, Support Vector Machine, Gaussian Naive Bayes, Decision Tree, and Random Forest. These models are being used to evaluate different important physicochemical parameters like pH, hardness, total dissolved solids, chloramines, sulfate, conductivity, organic carbon, trihalomethanes, and turbidity as predictive variables for classifying water potability into a binary class.

Not long ago, research was carried out involving physiochemical variables, and the findings indicated that the Random Forest classifier performed best, with outstanding performance metrics of 99.99% accuracy, 99.97% precision, 99.8% recall, and 99.99% F1-score (Nallakaruppan et al., 2024). It also had sensitivity and specificity of 99.8% and 99.9%, respectively; all these were measured during the research. This outstanding performance made it a suitable choice for the selection of XAI models to address the interpretability challenges associated with ML models. The combination of the traditional models with various disciplines and ML.

The SHAPLEY Additive exPlanations (SHAP) framework has been employed for the comprehensive visualization of the results, including force visualization, summary visualization, dependency visualization, and decision visualization for the models employed. The visualization of the results enables a clear interpretation of how various features affect the classification results. The feature importance analysis shows that the total dissolved solids are the most important parameter in determining the potability of water, followed by pH, sulfate, and hardness.

This framework can be extended to other hydro-climatic, physiochemical, and water management supply-related applications, as well as many other applications that are related to nature and industry, such as the prediction of groundwater potential, flood hazard risk, and could be used to determine environmental impacts. The use of technology through the implementation of AI and ML models can improve the efficiency and accuracy of the models beyond the industrial standards. However, the study does acknowledge some critical limitations, which include variations in geographic parameters, optimizations that are specific to the dataset, and the need to retrain the models in different environments. Future research directions include the use of DL models for more complex sedimentation analysis and validation of the models using geographically diverse datasets to ensure universal applicability in operational water quality monitoring systems.

6.2 5G/6G and edge AI

For effective communication between the models and from source to destination, especially in the case of 5G and future 6G communication systems, to fulfill the communication needs of the smart agriculture, forestry, biodiversity monitoring, and water management (AFBW) sectors. It fills an essential gap in the telecommunication industry, as there was no other comprehensive and domain-specific approach available to identify the sector-specific requirements to ensure seamless communication between the communication channels.

Numerous results have been formulated from various researches, which have concluded and shown that the AFBW sectors are under immense pressure from population explosion, climate change, environmental degradation, and geopolitical events, and it is difficult to maintain the proper and effective communication channels, which requires the immediate adoption of Industry 4.0 technology for improved development. All these sectors include agricultural, forest, and freshwater land, which account for about 88% of the Earth's habitable and usable surface. This creates a serious challenge for the Mobile Network Operators (MNOs), whose infrastructure development has been concentrated in the urban regions that account for only 1.4% of the habitable land.

By means of a systematic analysis of limited distinct use cases, starting from precise position sensing and field mapping up to drone operations and multimedia streaming research, some of the results have been accumulated that include Quality of Service requirements, including varying needs for bandwidth, latency, reliability, and device density. It unfolds and states the multiple gaps and open issues in current standardization efforts and states the issues that hinder its technological capabilities that must be addressed and are being addressed to enable effective delivery service for effective communication among all the communication channels.

Achieving success in this domain requires collective efforts in research and development across many disciplines, and it involves the standardization and implementation of standard phases. This must involve advancements, particularly toward 5G and 6G spectrum systems, that incorporate multiple mechanisms that facilitate the network capabilities exposure, integration with external entities to meet the communication requirements of smart agriculture, forestry, biodiversity monitoring, and water management (AFBW) sectors. This addresses a critical gap in the telecommunications industry, where no comprehensive domain-oriented approach previously existed to identify sector-specific requirements (Tomaszewski & Kołakowski, 2023).

6.3 AR/VR for training

Exploring a wide variety of options like VRAM (Virtual Reality Architecture in the Metaverse) could be regarded as one of the best future advancements made in this field. When deployed for resource-related and environmental-related tasks and management, this innovative framework is designed to enhance environmental conservation, particularly in the areas of water and soil resource management. Grounded in the Cognitive Theory of Multimedia Learning (CTML), VRAM transforms a traditional picture book on conservation concepts into sequential 3D VR video clips. This technology involves the integration of AR and VR.

This model engages the user in a quasi-experimental setting, where multiple gestures can be used for control and to reposition an interaction with the system, allowing the system to perform actions such as replaying videos, navigating through the clips, displaying text annotations, and answering questions according to the user's demands. Standard Interaction logs, which can be used for user-triggered events based on commands and active checkpoints, and these interactions with the system are sent in real-time to a cloud database via web services that enable seamless transfer from the device to the cloud and back. This process allows for subsequent behavioral breakdowns using big data techniques.

VRAM-assisted methods are now being used extensively for understanding traditional methods, revealing statistically significant improvements in participants and showcasing their flow experience. The results that had been observed from multiple experiments highlight VRAM's ability to enhance cognitive engagement and temporal perception.

While there are limitations to this technology, and there is indeed a lot of scope for improvement in certain areas that involve motion-induced interactions and low latency in real time, and the smoothness of gesture interactions, VRAM represents a promising approach to multimedia-driven education and can be used as a tool for monitoring the resources. It could also be used for building simulations (Lo & Tsai, 2022).

6.4 Blockchain-enabled trust

This formal method of analysis could be regarded as a decentralized method, a framework that blends edge computing, ML, and blockchain technology, which is combined with DTs. This approach increases efficiency, and as can be seen in Fig. 10.3, it is being used for water-related tasks and water resource management. This not only exhibits precision but also shows a promising and efficient intersection and integration of technology.

At the individual home and housing level, a Feed-Forward Neural Network (FNN), trained via Symbiotic Organism Search (SOS), depicts the day-to-day water usage based on varied segments. Meanwhile, a Mixture Density Network (MDN) estimates probabilistic historical demand forecasts. These projections are then compared against existing consumption within a blockchain intelligent contract that automatically allocates incentives for protection and takes care of excessive use, confirming transparent, secure, and timely usage of the resource.

Architecture considers each house a node in the blockchain, allowing device computation, which enhances speed, privacy, and scalability, as well as block-level abstraction for privacy. Architecture allows for rule-based rewards through smart contracts on Ethereum, ensuring behavioral alignment with conservation objectives using monetary incentives.

In terms of methodology, the FNN addresses multivariate feature-driven daily demand, while SOS adopts gradient-free training for low MSE with smooth convergence and high accuracy. MDN captures uncertainty and multimodality in past consumption data, outperforming the neural network approaches. The model primarily relies on two points: the daily point and the historical point (Thakur et al., 2021).

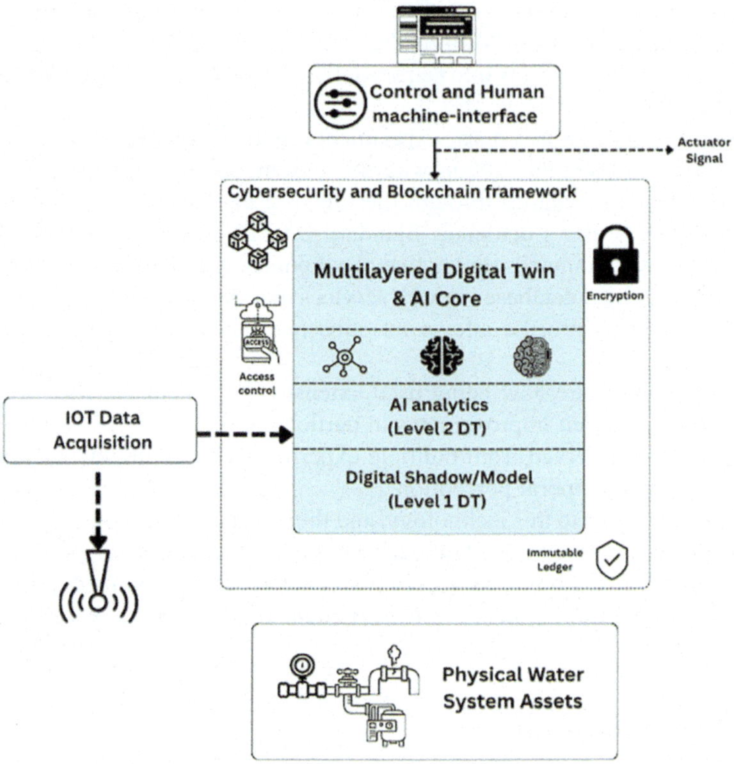

FIGURE 10.3

The graphical representation shows the level of abstraction and the implementation of blockchain technology along with multilayered DTs.

7. Challenges and research directions

7.1 Data governance and privacy

AI management and governing data has become a crucial task for data management across organizations. The main aim here is to ensure the responsible and ethical operation of Big Data Algorithmic Systems for efficient use of the resource and automated management.

AI management involves defining and implementing regulations from governing authorities that define control and responsibility over data and AI processes, as shown in Fig. 10.4. Key responsibilities of governance include establishing system-level controls and data, and building trusted frameworks for secure data sharing and efficient transfers. These steps help ensure legal and ethical measures, risks such as bias and privacy, and promote transparency and accountability in AI decision-making.

Effective governance measures enable collaboration among multiple organizations across multiple disciplines, and ethical governance helps to develop more trust among the users, organizations, and associations, enabling them to make informed decisions on a timely basis. And this opens a lot of

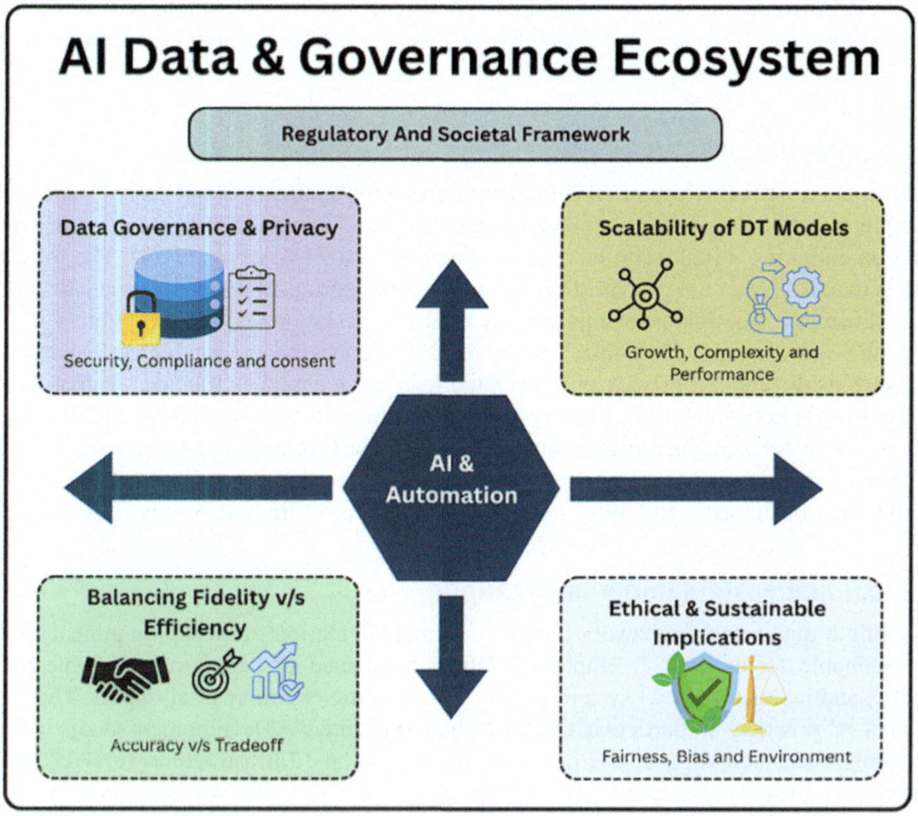

FIGURE 10.4

Schematics of a regulatory and societal framework for government accountability over data and AI processes.

interdisciplinary opportunities, and this can be used as an advantage to make better decisions. And this enables the development of trustworthy AI systems that serve interests securely and lawfully according to the user's demands (Janssen et al., 2020).

7.2 Scalability of DT models

Scaling of the DT models involves breaking down the complex, multi-scale systems into much simpler models using a 4C architecture that involves Composition, Context, Component, and Code. This process ensures each layer provides only the relevant data and functions for its specific scale and scenario while hiding the unnecessary details, while maintaining data integrity, and while being much more accurate when compared to other scaling models.

Components are standardized and are reusable, which allows for quick extension and replacement as systems evolve, contributing to the efficient scalability of the systems. This flexibility between layers promotes expansibility and interoperability across various tools for better scalability. Integration of attributes from different simple twins to maintain consistency. Multi-context interactions, such as superposition, transmission of the data, and coordinate behaviors to enable accurate and scalable system-wide simulations, which are some of the valuable features of the scalable DT technology (Jia et al., 2022).

7.3 Balancing fidelity versus efficiency

Responsible and trustworthy use of AI actually requires a lot of attention and careful balance between transparency, robustness, fairness, accountability, privacy, and human management for efficient model use, while maintaining a high state of efficiency and accuracy (Y. Wang, 2023). The most practical approach is to make better use of hybrid approaches, implementing targeted and efficient techniques to improve robustness. Continuous monitoring in real-time tasks increases the system's accuracy by enhancing the rate of adaptive learning. Privacy remains a concern, and it can be safeguarded through methods such as differential privacy and federated learning techniques. Besides, human designs can also be used for better scalability of the systems. The aim is to optimize across multiple constraints using adaptive governance, data lineage and linkages, and performance-centric practices. Through the optimization of the trade-offs in the systems, AI can be created that works well on a large scale while keeping the trust of the user, and all of this could be made to fit the user's expectations.

7.4 Ethical and sustainability implications

For designing a qualitative framework and for the implementation of designing ethical AI that promotes sustainable technology. It emphasizes the significance of transparency, righteousness, responsibility, and inclusivity in AI systems while also reducing environmental impact. The framework links ethical AI practices to particular United Nations Sustainable Development Goals (SDGs), specifically goals associated with Goal 9 Ind.ustry, Innovation, and Infrastructure, Goal 12 Responsible Consumption and Production, and Goal 13 Climate Action.

Most of the large tech and industrial corporations, including Google, Microsoft, IBM, and Tesla, are currently working on their initiatives for the development of energy-efficient data centers, which promote the overall growth [https://doi.org/10.34306/ajri.v6i2.1205]. The key developments emphasize the significance of data privacy and abstraction of the data, improvement in energy-efficient approaches to power up the data centers, and promotion of governance for all the key systems. It also recognizes the difficulties faced by individuals, small organizations, and businesses, and the need for the adoption of these ethical AI practices on a large scale.

8. Future outlook

8.1 Vision for autonomous plants

Industrial automation and management are a revolutionary trend that goes beyond application, even entering complex industries such as processing industries and power sources for better efficiency, as shown in Fig. 10.5. This approach emphasizes us with a focused six-level autonomy taxonomy, with a greater inclination toward the sustainable industry, which demonstrates its capabilities of understanding and problem-solving, along with the strategic use of AI technologies for automation and management.

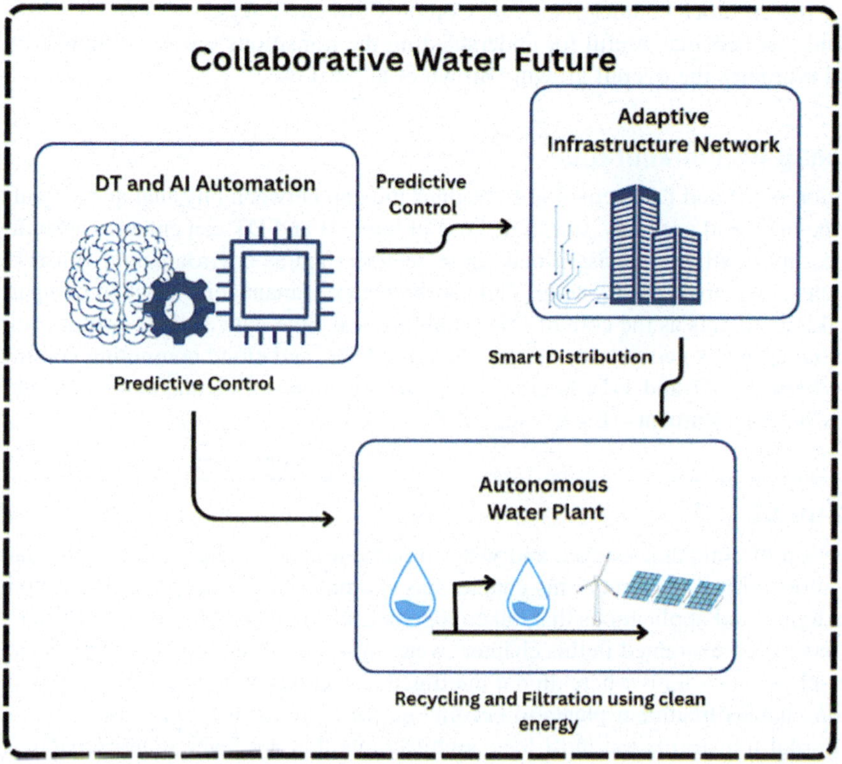

FIGURE 10.5
Representation of the current advancements in the field of sustainable infrastructure and smart management.

This emphasizes the major points that are associated with autonomy and can be adjusted according to the industry and standards, with a major focus on the operational points. This approach toward the automation of the systems is methodical and strategically data-driven to complete its leveled framework. AI stands as the main enabling technology. However, navigating to the desired levels of autonomy while balancing critical factors such as risk, benefits, and liability is where the potential use of AI could be better seen. This approach is far better when compared to traditional scalability approaches that are industry-specific and often fail to bring out the full potential of technology (Gamer et al., 2020).

8.2 Toward adaptive infrastructures

The interactions among system-level structures with respect to multiple disciplines like industrial, environmental, and auto industries can be seen, and the government can both shape the framework through rule changes or act as one agent within it to implement the standards to govern the organization, and this plays a vital role during the development. It provides us with the best value and collaborative approach for better and safer development. These approaches drive micro-interactions and produce macro-order characteristics of Complex Adaptive Systems, making CAS particularly emergent, and this becomes useful for understanding the transitions toward sustainability and development, and promotes the overall growth (Brown et al., 2012).

8.3 Roadmap for collaboration

The integration of AI and DT technologies bridges the gap between physical and virtual systems and increases efficiency and accuracy. Combined use of both AI and DT can enhance accuracy, predictive maintenance, and intelligent decision-making in sectors such as environmental conservation, autonomous vehicles, and smart city systems, and can be better at management when compared to human decision-making. AI acts as the core of DTs, enabling real-time data analysis, optimization, and self-learning within complex ecosystems. As big data, the IoTs, and cloud computing continue to evolve, the synergy between AI and DTs is crucial in creating more efficient, intelligent, and sustainable digital-industrial environments (Lv & Xie, 2024).

9. Conclusion

The intersection of computer science and water treatment is a paradigm shift in the management of humanity's most critical resource. This chapter has examined the conceptual underpinnings, technological paradigms, and applications that, taken together, form this paradigm shift. Primarily, three key takeaways have been examined in this chapter: water systems are no longer just physical systems but rather cyber-physical systems where digital intelligence coexists with water. The sensor, network, and model fusion enables treatment plants to become active systems that can sense, learn, and adapt. A successful digital transformation must have architectures that are layered and separate concerns but still integrate, and the multi-layered DT paradigm illustrated in this chapter provides a roadmap for building complex yet manageable systems with distinct boundaries between sensing, communication, modeling, and application layers.

The chapter also concludes that the connection between DTs and IoTs is an inseparable one. For example, the IoT devices generate continuous data streams that animate digital models, converting the static data into dynamic data. DTs provide interpretive intelligence that transforms raw data into actionable insights. Importantly, neither technology achieves its full potential alone. IoT without DTs produces overwhelming data without meaning, while DTs without IoTs are theoretical models that are not connected to reality.

The transformation investigated throughout this chapter finds that climate change exacerbates water issues, an aging infrastructure struggles to keep up with increasing demands, and new contaminants challenge traditional treatment methods. DTs provide the means to meet these challenges with the same complexity and sophistication. This transformation enables operators with unprecedented knowledge, communities with increased transparency, and society with the wisdom to manage water in a manner commensurate with its value. The systems that are built today will serve tomorrow's generations. The future course is clear. The frameworks are in place. The tools are available. What is needed is the shared commitment to build the smart, adaptive, and resilient water systems that the future requires. The water that flows from our taps will tell the story of this intelligence, a reflection of our growing ability to safeguard and preserve life's most precious resource, "water".

References

68% of the world population projected to live in urban areas by 2050, says UN. (2026). https://www.un.org/uk/desa/68-world-population-projected-live-urban-areas-2050-says-un.

Acharya, S., Khan, A. A., & Päivärinta, T. (2024). Interoperability levels and challenges of digital twins in cyber–physical systems. *Journal of Industrial Information Integration, 42*, 100714. https://doi.org/10.1016/j.jii.2024.100714

AlMetwally, S. A. H., Hassan, M. K., & Mourad, M. H. (2020). Real time internet of things (IoT) based water quality management system. *Procedia CIRP, 91*, 478–485. https://doi.org/10.1016/j.procir.2020.03.107

Alves, R. G., Souza, G., Maia, R. F., Tran, A. L. H., Kamienski, C., Soininen, J.-P., Aquino, P. T., & Lima, F. (2019). A digital twin for smart farming. In *2019 IEEE global humanitarian technology conference (GHTC)* (pp. 1–4). https://doi.org/10.1109/GHTC46095.2019.9033075

Brahmbhatt, P., Maheshwari, A., & Gudi, R. D. (2023). Digital twin assisted decision support system for quality regulation and leak localization task in large-scale water distribution networks. *Digital Chemical Engineering, 9*, 100127. https://doi.org/10.1016/j.dche.2023.100127

Brown, K., Furneaux, C., & Gudmundsson, A. (2012). Infrastructure transitions towards sustainability: A complex adaptive systems perspective. *International Journal of Sustainable Development, 15*(1/2), 54. https://doi.org/10.1504/IJSD.2012.044034

Chavhan, N., Bhattad, R., Khot, S., Patil, S., Pawar, A., Pawar, T., & Gawli, P. (2025). APAH: An autonomous IoT driven real-time monitoring system for industrial wastewater. *Digital Chemical Engineering, 14*, 100217. https://doi.org/10.1016/j.dche.2025.100217

Conejos Fuertes, P., Martínez Alzamora, F., Hervás Carot, M., & Alonso Campos, J. C. (2020). Building and exploiting a digital twin for the management of drinking water distribution networks. *Urban Water Journal, 17* (8), 704–713. https://doi.org/10.1080/1573062X.2020.1771382

Digital Water. (2026). *By technology (smart water meters, internet of things, artificial intelligence and machine learning, digital twin, geographic information systems, others), by solutions, by end-user (municipal water industries, industrial)*. By Region - Global Forecast To 2029. Retrieved February 22, 2026, from https://www.marketsandmarkets.com/Market-Reports/digital-water-market-248980683.html.

Gamer, T., Hoernicke, M., Kloepper, B., Bauer, R., & Isaksson, A. J. (2020). The autonomous industrial plant – Future of process engineering, operations and maintenance. *Journal of Process Control, 88*, 101–110. https://doi.org/10.1016/j.jprocont.2020.01.012

Ge, C., & Qin, S. (2025). Urban flooding digital twin system framework. *Systems Science & Control Engineering, 13*(1). https://doi.org/10.1080/21642583.2025.2460432

Hassnain, M., Lee, S. M. W., & Azhar, M. R. (2025). Soft sensing of biological oxygen demand in industrial wastewater using machine learning models. *Journal of Water Process Engineering, 78*, 108699. https://doi.org/10.1016/j.jwpe.2025.108699

Henriksen, H., Schneider, R., Koch, J., Ondracek, M., Troldborg, L., Seidenfaden, I., Kragh, S., Bøgh, E., & Stisen, S. (2022). A new digital twin for climate change adaptation, water management, and disaster risk reduction (HIP digital twin). *Water, 15*(1), 25. https://doi.org/10.3390/w15010025

Innovation in Water Singapore. (2026). Retrieved February 22, 2026, from https://www.pub.gov.sg/-/media/Images/Feature/Content-Pages/Resources/R-and-D/Issue-11Innovation-in-Water-SingaporeFull-PDF.pdf.

Janssen, M., Brous, P., Estevez, E., Barbosa, L. S., & Janowski, T. (2020). Data governance: Organizing data for trustworthy artificial intelligence. *Government Information Quarterly, 37*(3), 101493. https://doi.org/10.1016/j.giq.2020.101493

Jia, W., Wang, W., & Zhang, Z. (2022). From simple digital twin to complex digital twin part I: A novel modeling method for multi-scale and multi-scenario digital twin. *Advanced Engineering Informatics, 53*, 101706. https://doi.org/10.1016/j.aei.2022.101706

Lo, S.-C., & Tsai, H.-H. (2022). Design of 3D virtual reality in the metaverse for environmental conservation education based on cognitive theory. *Sensors, 22*(21), 8329. https://doi.org/10.3390/s22218329

Lumley, D. J., Polesel, F., Refstrup Sørensen, H., & Gustafsson, L.-G. (2024). Connecting digital twins to control collections systems and water resource recovery facilities: From siloed to integrated urban (waste)water management. *Water Practice and Technology, 19*(6), 2267–2278. https://doi.org/10.2166/wpt.2024.128

Lv, Z., & Xie, S. (2024). Artificial intelligence in the digital twins: State of the art, challenges, and future research topics. *Digital Twin, 1*(1). https://doi.org/10.12688/digitaltwin.17524.2

Mukherjee, A. S. S. K., Singh, R., Ahn, Y. H., & Ray, S. S. (2025). Tracing the footprint of microplastics: Transport mechanism, degradation, and remediation in marine environment. *Water, Air, and Soil Pollution, 236*(2), 134. https://doi.org/10.1007/s11270-025-07751-3

Nallakaruppan, M. K., Gangadevi, E., Shri, M. L., Balusamy, B., Bhattacharya, S., & Selvarajan, S. (2024). Reliable water quality prediction and parametric analysis using explainable AI models. *Scientific Reports, 14* (1), 7520. https://doi.org/10.1038/s41598-024-56775-y

Pang, J., Zheng, P., Li, S., & Liu, S. (2023). A verification-oriented and part-focused assembly monitoring system based on multi-layered digital twin. *Journal of Manufacturing Systems, 68*, 477–492. https://doi.org/10.1016/j.jmsy.2023.05.008

Radhakrishnan, V., & Wu, W. (2018). IoT technology for smart water system. In *2018 IEEE 20th international conference on high performance computing and communications; IEEE 16th international conference on smart city; IEEE 4th international conference on data science and systems (HPCC/SmartCity/DSS)* (pp. 1491–1496). https://doi.org/10.1109/HPCC/SmartCity/DSS.2018.00246

Reza, A. F., Singh, R., Das, K., Chen, S.-S., Ahn, Y. H., Kumar, R., Kumar, R., & Ray, S. S. (2026). IoT driven smart innovation for microplastic monitoring and detection: A multidimensional review. *Water Conservation Science and Engineering, 11*(1), 22. https://doi.org/10.1007/s41101-026-00496-y

Reza, A. F., Singh, R., Verma, R. K., Singh, A., Ahn, Y.-H., & Ray, S. S. (2024). An integral and multidimensional review on multi-layer perceptron as an emerging tool in the field of water treatment and desalination processes. *Desalination, 586*, 117849. https://doi.org/10.1016/j.desal.2024.117849

Shahouni, R., Abbasi, M., Dibaj, M., & Akrami, M. (2024). Utilising artificial intelligence to predict membrane behaviour in water purification and desalination. *Water, 16*(20), 2940. https://doi.org/10.3390/w16202940

Singh, R., & Ahn, Y.-H. (2025). Separation science in modern era. In *Advances in separation sciences* (pp. 19–36). Elsevier. https://doi.org/10.1016/B978-0-323-95292-7.00024-4

Singh, R., Ahn, Y.-H., & Ray, S. S. (2025). Water footprint: A comprehensive assessment of environmental impacts. In *Water footprints* (pp. 97–111). Elsevier. https://doi.org/10.1016/B978-0-443-30054-7.00010-0

Thakur, T., Mehra, A., Hassija, V., Chamola, V., Srinivas, R., Gupta, K. K., & Singh, A. P. (2021). Smart water conservation through a machine learning and blockchain-enabled decentralized edge computing network. *Applied Soft Computing, 106*, 107274. https://doi.org/10.1016/j.asoc.2021.107274

Tomaszewski, L., & Kołakowski, R. (2023). Mobile services for smart agriculture and forestry. *Biodiversity Monitoring, and Water Management: Challenges for 5G/6G Networks. Telecom, 4*(1), 67–99. https://doi.org/10.3390/telecom4010006

Tyagi, R. D., Pandey, A., Drogui, P., Yadav, B., Pilli, S., & Wong, J. W. C. (2024). *Decentralized sanitation and water treatment : Concept and technologies*. CRC Press, Taylor and Francis Group.

Wang, Y. (2023). Balancing trustworthiness and efficiency in artificial intelligence systems: An analysis of tradeoffs and strategies. *IEEE Internet Computing, 27*(6), 8–12. https://doi.org/10.1109/MIC.2023.3303031

Wang, A.-J., Li, H., He, Z., Tao, Y., Wang, H., Yang, M., Savic, D., Daigger, G. T., & Ren, N. (2024). Digital twins for wastewater treatment: A technical review. *Engineering, 36*, 21–35. https://doi.org/10.1016/j.eng.2024.04.012

Wang, Y., Su, Z., Guo, S., Dai, M., Luan, T. H., & Liu, Y. (2023). A survey on digital twins: Architecture, enabling technologies, security and privacy, and future prospects. *IEEE Internet of Things Journal, 10*(17), 14965–14987. https://doi.org/10.1109/JIOT.2023.3263909

PART V

Digital innovations and smart monitoring

CHAPTER

AI-driven predictive modeling and real-time optimization in water systems

11

N. Nandini Devi[1] and Kamal Das[2]

[1]*Bennett University, Greater Noida, Uttar Pradesh, India;* [2]*National Institute of Technology Agartala, Agartala, Tripura, India*

1. Introduction

Water is very important for the survival of human life, economic activity, ecological stability, and to support the overall well-being of society. Access to clean and safe water is getting constrained due to the factors population expansion, rapid urban development, and the growth of industrial sectors (Chunyang et al., 2021). It increased the demand on existing water resources and put stress on the water supply infrastructure, distributed networks, and treatment systems. This highlights the need for more efficient and resilient water management. Day by day, we are facing increasing freshwater scarcity due to poor water management, overconsumption, and pollution caused by the fast-growing urbanization and industrial sectors, including energy, manufacturing, textiles, and pharmaceuticals (Musie and Gonfa, 2025). Besides rising demand of fresh water the climate change has become a critical factor affecting global water availability (Biswas et al., 2025). Unpredictable rainfall patterns, extended droughts, melting glaciers, and extreme weather are reducing the quantity and quality of freshwater resources (Kazemi Garajeh et al., 2024). Water scarcity is a major issue for global risks, food security, public health, and economic stability (Dinar et al., 2019). Many countries are already struggling with serious water scarcity, and by 2050, billions of people could face high water stress (Schlosser et al., 2014).

Traditional water management systems mostly depends on manual measurements and periodic field inspections and react only after problems arise, making them inefficient in addressing the rapidly changing and uncertain nature of modern water systems. High distribution losses, contamination events, and inefficient operational strategies continue because of limited monitoring and delayed decision making (Al-washali et al., 2016). Therefore, there is an urgent need to shift from traditional, labor-intensive practices towards smarter, sustainable, and intelligent water management approaches to make better use of resources, improve reliability, lower operational costs, and ensure balanced water distribution (Palermo et al., 2022). Modern water systems are evolving from traditional water systems monitoring to highly interconnected smart automated systems where numerous smart Internet of Things (IoT) devices are deployed across water treatment plants, pipelines, distributed networks, groundwater systems, and wastewater facilities to continuously monitor the flow conditions, water

Sustainable Water Management. https://doi.org/10.1016/B978-0-443-44853-9.00008-3

quality, pressure variation, leakages, and energy consumption (Júnior et al., 2021; Ismail et al., 2019; Qiao et al., 2025). Real-time acquisition of data such as flow rate, pressure, water quality parameters, and equipment health from IoT devices along with artificial intelligence can provide a clear view of the entire water network. It can provide better system monitoring, fast decision-making and effective operational control. AI-based models can predict future water demand, detect faults early, and find hidden patterns in the data that may not be easily visible through traditional methods. This helps operators take preventive action before major failures occur. Real-time optimization can automatically adjust pumping, treatment, and pressure settings that can help to reduce water loss, lower operating costs, and to ensure a reliable water supply. Overall, these intelligent technologies improve system efficiency, enhance the ability to respond to disruptions, and increase the resilience of water systems under growing demand and changing environmental conditions (Alhussan et al., 2025; Lu et al., 2025; Mahmoud et al., 2025; Petri et al., 2017).

This chapter reviews existing research on AI-based predictive models and real-time optimization methods in different areas of the water sector. It introduces different types of water systems and their management challenges. The chapter also explains how the AI techniques are applied across various water-related applications. The chapter helps readers to understand recent developments and identify research gaps that can support the development of smarter and more resilient water infrastructure systems. The overall organization of the chapter is shown in Fig. 11.1, and the chapter concludes by highlighting key research gaps and possible future research directions.

2. Water systems and management challenges

This section presents a systematic classification of major water system types and examines the key challenges affecting their efficiency and sustainable management.

2.1 Classification of water systems

Water systems are essential parts of modern society, as they provide water for daily use, treat wastewater, and support water reuse. The way these systems are designed and operated depends on their size, purpose, and available technology. Water systems range from large urban networks that serve millions of people to small, community-based systems in rural areas, and each plays an important role in sustainable water management. In general, these systems can be grouped into closely connected categories such as urban water supply systems, rural and community water supplies, wastewater collection and treatment systems, stormwater drainage and flood control systems, irrigation and agricultural water distribution systems, groundwater extraction and recharge systems, surface water reservoirs and dams, water treatment and desalination facilities and recycled and reclaimed water systems. This classification helps explain how different parts of the water system are connected within the overall water cycle and provides a useful basis for developing smarter, AI supported water management approaches.

2.1.1 Urban water supply networks

Urban water supply networks are essential city infrastructures that provide clean drinking water to homes, industries, and businesses. These systems typically include water sources, treatment plants, storage facilities, pumps, and distribution pipelines. Their main objectives are to ensure a reliable

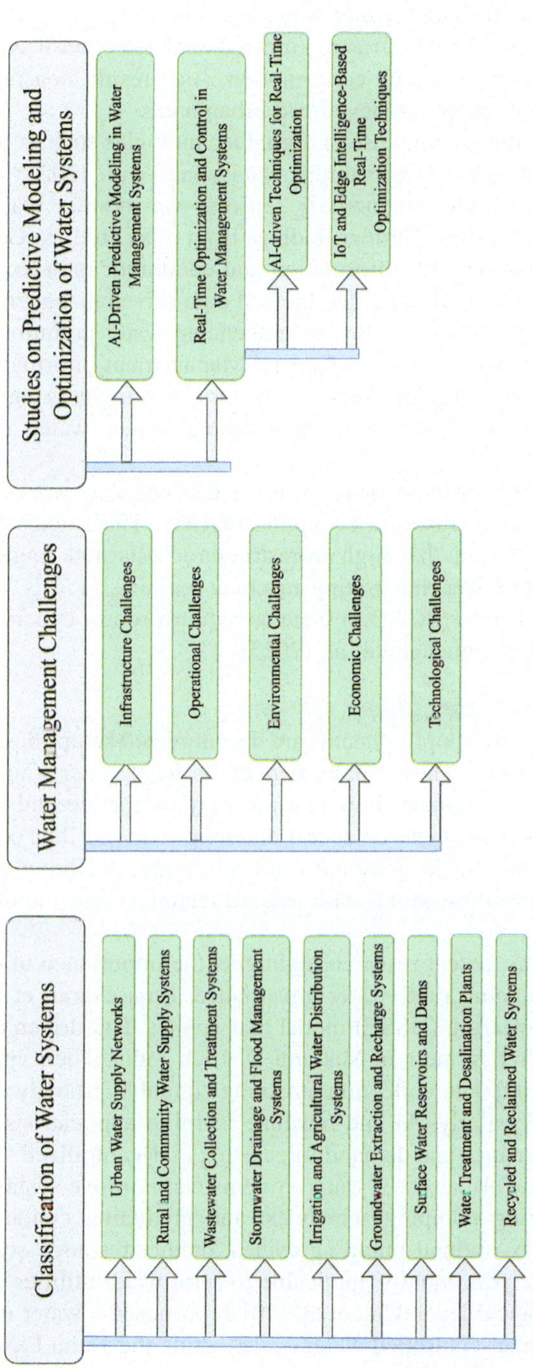

FIGURE 11.1
Overview of the chapter.

water supply, maintain efficient flow, and reduce water loss and energy use (Gaonkar & Kande, 2014). However, many cities still depend on old infrastructure and outdated control systems, which often lead to leaks, poor monitoring, and high energy consumption. As a result, non-revenue water remains a serious challenge for both developing and developed urban areas.

Recent research highlights the growing use of digital technologies such as IoT, AI, and automation in improving urban water networks. For example, Saravanan et al. (2023) showed that IoT-based sensors, including pH, turbidity, and conductivity sensors, can monitor water quality in real time and reduce the need for manual testing. Similar findings from other studies (Gaonkar & Kande, 2014) suggest that combining automation with information and communication technologies (ICTs) helps move traditional water systems toward more flexible and data-driven management. The integration of automation and computational intelligence is also influencing water pollution control and treatment database systems. Shi et al. (2010) developed a General Management Information System (GMIS) for storing, querying, and managing pollution control technologies, emphasizing the importance of data integration and decision support systems in maintaining urban water quality and regulatory compliance.

Despite these advances, the transition toward smart urban water networks (SWNs) remains challenged by data silos, privacy concerns, and economic barriers. The studies (Ojadi et al., 2025 and Gaonkar & Kande, 2014) highlight that high infrastructure costs and fragmented data ownership hinder full-scale adoption. Moreover, integrating machine learning models with operational Supervisory Control and Data Acquisition (SCADA) systems requires robust cybersecurity and governance frameworks, as identified in Egbemhenghe et al. (2023).

2.1.2 Rural and community water supply systems

Rural and community-based water supply systems are decentralized setups designed to meet the needs of small and scattered populations. They usually depend on local water sources such as bore wells, ponds, hand pumps, and small treatment units that are easy to operate and maintain. Unlike urban water networks, these systems focus more on access and long-term use than on advanced automation. However, they often face problems like seasonal water shortages, declining groundwater levels, and pollution from agricultural activities, along with limited technical and maintenance support (Lowe et al., 2022).

To address these limitations, recent research highlights the importance of digital engineering and AI-based innovations in enhancing rural water governance. Rajasekaran et al. (2018) developed a Digital Service Systems Engineering (DSSE) model that applies data-driven decision frameworks to promote water sustainability for farmers in Madurai District, India. Their approach integrates community participation with digital data collection, enabling equitable groundwater allocation and real-time tracking of water use. Similarly, AI and machine learning applications are increasingly being explored for predictive maintenance and demand forecasting in decentralized systems. Jin et al. (2025) emphasize that AI-driven transformation of water systems can improve sustainability by automating demand forecasting, optimizing pumping schedules, and predicting contamination events. These frameworks use AI to identify optimal pumping cycles, predict resource scarcity, and recommend conservation strategies—principles equally applicable to rural water utilities and cooperatives.

At the policy and hydrological level, Chu et al. (2012) proposed a water management framework based on both natural and human-controlled water cycles, using the Taihu Lake Basin as a case study. Mathematical and computational models also play a key role in optimizing small-scale water infrastructure. Kaviya et al. (2024) employed graph-theoretic optimization to improve the efficiency of water distribution and treatment networks in urban as well as rural water supplies. Complementing these computational and AI-based strategies, Lowe et al. (2022) stress the significance of community based governance frameworks for achieving water security in rural and semi-urban regions. Their study reveals that long-term success depends not only on technological upgrades but also on institutional capacity building, stakeholder inclusion, and adaptive management.

2.1.3 Wastewater collection and treatment systems

Wastewater systems encompass the critical infrastructure responsible for the collection, transport, treatment, and reuse of domestic, industrial, and commercial effluents. Their main components include sewer networks, pumping stations, primary and secondary treatment units, sludge digesters, and effluent discharge systems that ensure the safe management of liquid wastes. As highlighted by Gaonkar and Kande (20140, these systems form the backbone of urban sanitation but often struggle with aging assets, limited automation, and inefficient process control, which lead to operational inefficiencies and high energy consumption.

Modern wastewater management has evolved beyond conventional treatment and disposal; it now emphasizes resource recovery and circular economy integration, seeking to extract energy, nutrients, and reusable water from waste streams. Rahman et al. (2024) demonstrated this transformation within the industrial and mining sectors, proposing an integrated framework for sustainable water-waste management through the adoption of IoT sensors, process automation, and closed-loop reuse systems. Similarly, Tripathi et al. (2019) underscored that urban growth and changing land use patterns increase wastewater generation, which requires smarter systems that dynamically align water demand, wastewater production, and treatment capacity to achieve long-term sustainability.

2.1.4 Stormwater drainage and flood management systems

Stormwater drainage systems are essential infrastructures designed to collect, convey, and manage rainwater runoff from urban surfaces such as roads, rooftops, and pavements to mitigate flooding and waterlogging. The components typically include catch basins, culverts, retention ponds, and pumping stations, which together maintain urban hydrological balance and protect built environments from excessive stormwater accumulation. However, rapid urbanization and climate change have significantly altered hydrological responses, increasing runoff volumes and flood risk in densely built regions. The study (Tripathi et al., 2019) emphasized that uncontrolled land use and land cover changes directly increase impervious surfaces, resulting in higher runoff and decreased infiltration, which amplify the frequency and intensity of urban flooding. Traditional drainage designs based solely on historical rainfall patterns are proving inadequate to manage the intensity and unpredictability of current precipitation events, necessitating the adoption of dynamic, data-driven, and adaptive stormwater systems.

To address these challenges, the integration of automation and real-time monitoring technologies has gained prominence. Gaonkar and Kande (2014) highlighted the critical role of Information and Communication Technologies (ICTs) and automation in improving the responsiveness and reliability of water and wastewater systems, principles that extend naturally to stormwater management. The incorporation of IoT-based rainfall and flow sensors allows for continuous monitoring of runoff dynamics, while cloud-connected systems enhance situational awareness during heavy rainfall events (Saravanan et al., 2023). These IoT-enabled sensor networks, capable of measuring parameters such as water level, turbidity, and flow velocity, provide real-time insights that support proactive flood control and rapid response.

Furthermore, Rahman et al. (2024) emphasized that industrial and urban stormwater runoff contributes significantly to water pollution and must be integrated into a sustainable water-waste framework that incorporates advanced treatment and recycling mechanisms. The coupling of such frameworks with real-time hydrological sensors and decision-support models can reduce contamination risks and improve post-event recovery efficiency.

2.1.5 Irrigation and agricultural water distribution systems

Agricultural irrigation systems account for nearly 70% of global freshwater withdrawals, making their efficient management a vital priority for sustainable food production and water security. These systems—ranging from surface irrigation canals to pressurized sprinkler and drip networks—regulate water distribution based on crop type, soil properties, and climatic conditions. However, as (Gaonkar & Kande, 2014) note, traditional irrigation approaches suffer from inefficiencies linked to manual control, inadequate automation, and lack of real-time data, leading to over-irrigation and substantial water loss. To address these issues, the integration of Internet of Things (IoT) and artificial intelligence (AI) technologies is transforming irrigation management from reactive to predictive and adaptive. The study by Rajasekaran et al. (2018) introduced a Digital Service Systems Engineering (DSSE) framework for farmers in Madurai District, India, demonstrating how data-driven and IoT-enabled decision-making systems can enhance groundwater sustainability and crop productivity through holistic monitoring of soil moisture, weather conditions, and resource flows.

Building on this, Saravanan et al. (2023) developed a sensor-based water quality monitoring system using IoT technologies that can also be adapted for irrigation networks. Their system continuously monitors parameters such as pH, turbidity, and temperature, transferring data to cloud platforms for real-time assessment and feedback. Such sensing frameworks ensure that irrigation water meets quality thresholds essential for maintaining crop health and preventing soil degradation. Meanwhile, Ojadi et al. (2025) emphasized that AI-driven optimization frameworks—combining IoT sensors, predictive modeling, and real-time analytics—enable precision irrigation by assessing soil moisture and evapotranspiration rates to deliver water exactly when and where it is needed. Their research shows that machine learning models can predict crop water requirements based on historical data, climate variables, and plant health metrics, thereby improving yield while conserving water.

Complementary to these applications, Egbemhenghe et al. (2023) demonstrated how AI-based conversational systems such as ChatGPT can serve as decision-support tools for farmers, interpreting complex environmental datasets and offering actionable irrigation recommendations in accessible language. This integration of AI-enabled advisory systems represents an emerging frontier in democratizing precision agriculture technologies. Similarly, Rahman et al. (2024) explored the implementation of sustainable water-waste management practices in the mining sector, whose methodological framework—emphasizing resource efficiency, monitoring transparency, and circular reuse—offers transferable insights for agricultural irrigation systems, particularly in managing nutrient-rich wastewater for crop irrigation. Furthermore, Kaviya et al. (2024) introduced a graph-theoretic optimization model (TCCD-number) for improving water distribution network resilience, a concept that can be applied to irrigation canals and drip systems to optimize connectivity and minimize transmission losses. Tripathi et al. (2019) further emphasized the impact of land use and land cover (LULC) changes on regional water demand, noting that urban expansion and reduced agricultural land necessitate smarter irrigation planning and adaptive water allocation strategies to maintain agricultural output under increasing pressure.

2.1.6 Groundwater extraction and recharge systems

Groundwater systems form a critical component of global water security, serving as a dependable source during surface water shortages and playing a vital role in sustaining both rural and urban water supplies. These systems comprise borewells, tube wells, submersible pumps, and recharge structures such as percolation pits, recharge wells, and aquifer storage systems. However, the overexploitation of aquifers has resulted in declining groundwater tables, land subsidence, and deteriorating water quality across many regions. Tripathi et al. (2019) observed that rapid urbanization and uncontrolled land-use changes in cities like Dehradun have significantly increased water demand, intensifying pressure on groundwater reserves. Effective management therefore necessitates continuous monitoring of extraction and recharge rates to maintain aquifer sustainability and prevent long-term depletion.

Advances in automation and IoT technologies are revolutionizing groundwater monitoring and management. Gaonkar and Kande (2014) highlighted how automation and information communication technologies (ICTs) can enhance precision in data acquisition, improve pumping control, and reduce operational inefficiencies in water systems. These technologies underpin modern IoT-based groundwater monitoring systems, which employ sensors for real-time tracking of water table depth, pressure, and quality parameters such as pH, electrical conductivity, and total dissolved solids (TDS). Saravanan et al. (2023) demonstrated that IoT-enabled water quality monitoring platforms using sensors and cloud analytics provide low-cost and effective solutions for continuous environmental monitoring, a framework that can be adapted for groundwater quality assessment. Similarly, Rajasekaran et al. (2018) proposed a data-driven Digital Service Systems Engineering (DSSE) model that integrates distributed sensors and local data to promote sustainable groundwater extraction in India, demonstrating how systemic digital transformation can support rural water sustainability.

Artificial intelligence (AI) and machine learning (ML) models are also becoming integral to predictive groundwater management. Ojadi et al. (2025) emphasized that AI-driven optimization frameworks can forecast water demand patterns, analyze historical pumping data, and predict recharge needs under varying climatic conditions. Egbemhenghe et al. (2023) further explored the integration of AI-driven decision support systems such as ChatGPT in hydrological management, suggesting their capacity to process complex datasets and assist decision-makers in planning recharge interventions and preventing overextraction. Complementarily, Rahman et al. (2024) demonstrated that integrated sustainability frameworks developed for industrial water systems can be adapted to manage groundwater recharge processes—combining monitoring, reporting, and regulatory compliance to ensure environmental protection.

In addition to data and AI-driven management, innovative field-based remediation and recharge techniques are being developed to restore groundwater quality. Lisena et al. (2016) proposed a bioremediation-based restoration method employing electric boats equipped with GPS and sensor arrays to deliver microbial cultures into contaminated water bodies, illustrating how autonomous, sensor-driven water treatment technologies can be extended to groundwater recharge basins for improving aquifer health. The integration of renewable-powered monitoring systems within such approaches promotes energy-efficient and eco-friendly recharge operations.

2.1.7 Surface water reservoirs and dams

Surface water reservoirs and dams play a pivotal role in ensuring water storage, flood regulation, hydropower generation, and irrigation support, acting as strategic buffers against hydrological variability by storing excess water during wet seasons and releasing it during dry periods. Their efficient and safe operation requires accurate forecasting of inflows, evaporation losses, and downstream demands. However, conventional rule-curve-based management approaches and manual gate operations often result in suboptimal decisions, leading to over-spillage during floods or water shortages in dry conditions. The integration of automation, artificial intelligence (AI), and Internet of Things (IoT) technologies has transformed the paradigm of reservoir management from static to adaptive and predictive systems (Gaonkar & Kande, 2014).

Recent research has demonstrated that data-driven modeling and optimization significantly enhance reservoir efficiency and safety. Rajasekaran et al. (2018) proposed a Digital Service Systems Engineering (DSSE) approach using IoT and regional data integration for improving water resource sustainability; an approach applicable to reservoir networks where hydrological and operational data are continuously collected and analyzed to inform adaptive management. Similarly, Saravanan et al. (2023) developed a sensor-based IoT water monitoring system that measures key water quality parameters such as pH, turbidity, and temperature; integrating such real-time sensing technologies into reservoir systems enhances environmental monitoring and ensures compliance with safety and ecological standards.

Furthermore, AI-driven optimization frameworks and decision-support tools enable multi-objective reservoir management. Egbemhenghe et al. (2023) and Ojadi et al. (2025) highlighted how AI and ChatGPT-based models can analyze large hydrological datasets to forecast inflow patterns, optimize release schedules, and improve decision-making in water infrastructure operations. Reinforcement learning algorithms are increasingly being applied to balance competing objectives such as water supply, energy production, and flood control, allowing dynamic operational strategies under uncertainty. Complementary to these developments, Rahman et al. (2024) emphasized the necessity of integrated sustainability frameworks that couple water reuse, quality control, and risk management—principles that are critical to reservoir system governance and operational transparency.

From a broader resource management perspective, Tripathi et al. (2019) demonstrated through their Dehradun case study that changes in land use and land cover (LULC) directly affect surface water availability and urban water demand, reinforcing the need for smart water management systems that incorporate geospatial data and real-time hydrological modeling for reservoir inflow forecasting. At the same time, decision-analytic approaches, such as those proposed by Prajwal et al. (2023) using Fuzzy TOPSIS and Fuzzy AHP, provide structured methodologies for evaluating and selecting optimal water treatment or release strategies under uncertainty, offering a decision-making framework adaptable to dam operations.

Ensuring structural safety is another cornerstone of sustainable dam operation. IoT-based structural health monitoring systems are now being employed to track pressure, vibration, seepage, and deformation parameters in real time, enabling early warning and preventive maintenance. Such approaches align with the automation and data-integration strategies proposed by Gaonkar and Kande (2014) and reaffirm the potential of Industry 4.0 technologies in modern water infrastructure management. Collectively, the integration of AI, IoT, and decision-support analytics within surface water reservoir systems facilitates proactive, adaptive, and sustainable water management, ensuring optimal resource utilization while maintaining safety and environmental balance.

2.1.8 Water treatment and desalination plants

Water treatment plants (WTPs) and desalination facilities are central to ensuring the availability of safe and potable water, particularly in regions facing chronic water scarcity or saline groundwater intrusion. Conventional treatment methods—such as coagulation, sedimentation, filtration, and disinfection—are reliable but energy-intensive and often depend on static operational settings that fail to adapt to varying inflow and contamination levels. The increasing urbanization and industrialization have compounded these inefficiencies, calling for the integration of automation, artificial intelligence (AI), and Internet of Things (IoT) to improve efficiency and sustainability (Gaonkar & Kande, 2014).

AI-driven optimization frameworks are transforming the way modern WTPs operate. As shown by Egbemhenghe et al. (2023), AI-assisted decision-support systems, including ChatGPT-based analytical models, can identify process anomalies, optimize operational schedules, and dynamically adjust chemical dosing to maintain consistent effluent quality. Similarly, Ojadi et al. (2025) emphasized that AI-based predictive control improves process adaptability in smart city water networks by forecasting fluctuations in water quality parameters such as turbidity, pH, and residual chlorine. These approaches enable energy-efficient operations and minimize manual intervention, improving resilience across urban utilities.

The role of IoT-enabled sensing systems in water treatment has also expanded significantly. Saravanan et al. (2023) demonstrated a low-cost IoT water quality monitoring system that continuously measures parameters such as pH, temperature, turbidity, and conductivity, with data transmitted to cloud-based platforms for analysis and real-time response. When integrated into large-scale WTPs, these systems enhance real-time water quality assurance, predictive maintenance, and automation of filtration and disinfection cycles. Rahman et al. (2024) further reinforced the importance of such integration, showing how sustainability frameworks in industrial sectors—like mining—can be adapted to promote resource efficiency and circular water reuse in treatment plants.

Desalination facilities, particularly reverse osmosis (RO) plants, stand to benefit immensely from AI and machine learning (ML) techniques. Predictive models can anticipate membrane fouling, scaling potential, and energy consumption, enabling preemptive maintenance and improved membrane longevity. Kaviya et al. (2024) introduced a graph-theoretic optimization model (TCCD-number) that can inform the design of energy-efficient water treatment and purification networks, minimizing operational redundancy. Complementarily, Lisena et al. (2016) proposed a bioremediation-based water quality restoration system employing electric boats equipped with GPS and sensors, a technique that demonstrates the potential of renewable-powered, autonomous systems to maintain urban water quality while integrating with city-wide energy management systems.

2.1.9 Recycled and reclaimed water systems

Recycled and reclaimed water systems are pivotal in modern integrated water resource management, treating and redistributing wastewater or greywater for non-potable applications such as landscape irrigation, industrial cooling, and groundwater recharge. These systems embody the principles of a circular water economy, reducing the dependency on freshwater withdrawals while mitigating the pressure on urban wastewater infrastructure. According to Li et al. (2009), the separation and reuse of greywater and blackwater significantly enhance treatment efficiency and nutrient recovery, fostering a decentralized, sustainable approach to urban water management. Similarly, Chu et al. (2012) emphasized the dualistic hydrological cycle framework, integrating both natural and artificial water cycles to balance source reduction, pollution control, and wastewater reuse—a model particularly effective in highly urbanized and industrialized basins such as Taihu Lake in China.

Technological innovation plays a central role in optimizing reclaimed water systems. Chao et al. (2016) developed an optimization and distribution model for industrial water systems based on process system integration, enabling efficient reuse of treated wastewater streams through mathematical modeling and online correction technologies. This model minimizes both freshwater demand and wastewater discharge by employing real-time water quality modeling and dynamic allocation. Complementarily, Jian et al. (2011) proposed an information management system for wastewater treatment supervision, which enhances government oversight through real-time data analysis, supporting consistent water quality compliance and regulatory efficiency in urban treatment facilities.

From an analytical and computational standpoint, Kaviya et al. (2024) presented a graph-theoretic optimization approach—the TCCD-number model—to improve urban drainage purification and water reclamation networks. This mathematical framework enhances network connectivity and ensures redundancy, improving the reliability of recycled water distribution systems. The incorporation of AI and IoT-based technologies further strengthens reclaimed water management. As noted by Egbemhenghe et al. (2023), AI-driven models such as ChatGPT can analyze large operational datasets to detect anomalies in treatment performance and predict disinfection or filtration needs, while IoT sensors enable real-time visibility of flow rates, contamination risks, and infrastructure health across distributed reuse systems.

These integrated digital systems facilitate predictive quality assurance, dynamic process optimization, and risk informed allocation of reclaimed water resources. In parallel, Rahman et al. (2024) highlighted that sustainability frameworks and automation originally designed for the industrial and mining sectors can be effectively applied to reclaimed water systems, improving energy efficiency and promoting circular resource flows. Tripathi et al. (2019) further demonstrated that urbanization-induced changes in land use and land cover (LULC) directly affect wastewater generation and reuse potential, underscoring the need for adaptive modeling that links urban hydrology with reclaimed water distribution.

2.2 Water management challenges

Water management faces a wide range of challenges across different types of water systems. These challenges can be categorized into various dimensions such as infrastructure, operational, environmental, economic, and technical. An overview of these challenges is presented below. The combined impact of these factors shows the requirement of adopting data-driven and intelligent solutions to ensure reliable, efficient, and sustainable water management.

2.2.1 Infrastructure challenges

Due to aging assets, poor maintenance, insufficient funding, and a lack of integrated management approaches, the water management systems are facing significant infrastructure related challenges. Many water infrastructure systems, such as water supply, wastewater, and irrigation systems, are exceeding their design life, which can lead to frequent failure, leakage, and degraded service reliability. For instance, in South Korea, nearly 72.3% of water infrastructure is projected to deteriorate by 2035 due to poor maintenance, improper management, and lack of budget allocations, which can increase the risk of system failures (Kang, 2019). Similarly, in South Africa, inefficient investments are increasing infrastructure backlogs and capital funding gaps every year, which increase water losses and non-revenue water issues (Ruiters & Amadi-Echendu, 2022).

Rapid aging of water pipelines, treatment plants, and sewer networks causes them to struggle to fulfill the demands of growing populations and changing climates. Exceeding the design lifetime of many water systems are leading to frequent leakage, pipeline bursts, and contamination incidents, such as the Flint, Michigan crisis that exposed thousands of residents to lead-contaminated water due to corroded pipelines and lack of timely maintenance (Morris, 2017). In the United States, several water and wastewater systems operate with outdated facilities. They are required to reconstruct to maintain public health standards and service reliability (Pamidimukkala et al., 2021).

Moreover, due to inadequate maintenance and poor financial mechanisms, many developing regions' irrigation systems suffer from water wastage (Ward, 2010). The lack of integration between the interdependent systems, such as water, wastewater, and stormwater networks, makes maintenance coordination difficult (Araya & Vasquez, 2022). To address these infrastructure related challenges, we are required to adopt efficient investment and data-driven approaches for long-term sustainability of water systems.

2.2.2 Operational challenges

Operational challenges in water systems mainly arise from manual controls, decision-making delay, and weak coordination across different subsystems such as water supply, distribution, and treatment. Most of the water management systems still relies on human supervision, manual valve operation, and static scheduling without real-time feedback, which slows responses to dynamic conditions such as demand fluctuations, leaks, or equipment failures. For example, in intermittent supply networks due to manual monitoring and inadequate data integration, Arraiján, Panama, faces unpredictable supply interruptions, inefficient pumping schedules, and inequitable water distribution (Erickson et al., 2020). Due to the lack of automation, irregular maintenance, and communication gaps between treatment and distribution units, outages and unreliable services have been observed in the urban water utilities of Ghana (Amoah and Yahaya, 2025).

The coordination between different subsystems of the water system across different zones is another challenge. The absence of integrated decision-support systems makes proactive water management difficult. Predictions of river and reservoir flows often rely on scattered data and individual experience. As a result, responses to emergencies such as floods, droughts, or sudden changes in water availability are often slow and less accurate (Pagano et al., 2014).

2.2.3 Environmental challenges

Environmental challenges such as water scarcity, pollution, and climate change pose critical threats to the sustainability and resilience of modern water systems, including the availability and quality of freshwater resources. Rapid population growth, industrialization, and agricultural expansion are increasing water demand and putting pressure on the already limited freshwater supplies. Declining groundwater levels and drying surface water sources in many regions are creating severe water shortages that threaten ecosystems, food security, and human livelihoods (Cosgrove & Loucks, 2015; Gosling & Arnell, 2016; Mishra et al., 2021). The overexploitation of groundwater for irrigation has further intensified these crises (Dalin et al., 2017).

Water management is getting more challenging due to pollution. Many industries release untreated waste into rivers, lakes, and other water sources. Moreover, the agricultural firms add harmful fertilizers and pesticides into these water bodies (Kalmakhanova et al., 2025; Rad et al., 2022; Rajput et al., 2017). In recent years, pollutants such as medicines and microplastics have become a serious concern. These substances are difficult to remove because most traditional water treatment systems are not designed to handle them effectively (Aib et al., 2025; Arienzo & Donadio, 2023).

Moreover, climate change is causing unpredictable rainfall patterns. Some regions receive very heavy rainfall, while others face long periods of drought (Tahroudi, 2025). Rising temperatures also increase evaporation, which reduces the amount of available water (Ajjur & Al-Ghamdi, 2021; Al-Hasani et al., 2025). All these environmental problems create challenges for water management and require adaptive policies and climate-resilient technologies to address them.

2.2.4 Economic challenges

Economic challenges play a major role in limiting the development and sustainability of water systems. High energy and operating costs, inadequate infrastructure funding, and inefficient resource allocation create financial barriers to modernization and long-term efficiency. Water pumping, treatment, and distribution require large expenses for most water utilities, and inefficient systems make them even costlier, leaving little funding available for regular maintenance and the adoption of new technologies (Pedraza et al., 2025).

Limited financial resources and inconsistent government funding make it difficult to upgrade the old infrastructure and expand services in the underserved areas. As a result many water systems continue to operate with the old facilities that frequently break down and waste water due to leakage and energy losses (Erickson et al., 2020; Gurara et al., 2025).

2.2.5 Technological challenges

Many water system utilities still rely on manual control without automation or real-time feedback that slows responses and reduces operational flexibility. In addition, current water systems suffer from insufficient sensor coverage, meaning that key parameters such as flow, pressure, water quality, and leak detection are often under-monitored. As a result, problems in water distribution and treatment networks frequently go undetected until they escalate. Furthermore, the lack of end-to-end automation across different system components is a major challenge in water management. In many cases, sensor networks, cloud-based data analysis, and control systems do not work together smoothly. This weak coordination limits the effective use of technologies such as IoT and AI in water systems (Gaonkar & Kande, 2014; Kara et al., 2016).

These challenges clearly show the need for intelligent and more flexible solutions. Using AI-based predictive models and real-time optimization can help improve system efficiency, reliability, and long-term sustainability across different types of water systems.

3. AI-driven predictive modeling in water management systems

This section reviews existing studies on AI-driven predictive modeling across different water management systems. It highlights how machine learning and deep learning techniques are applied to address system-specific challenges, improve forecasting accuracy, support proactive operations, and enhance overall efficiency and reliability. The overall workflow of AI-driven predictive modeling in water management systems is illustrated in Fig. 11.2. It includes real-time data acquisition from sensors and IoT devices, data preprocessing and feature extraction, AI model training and validation,

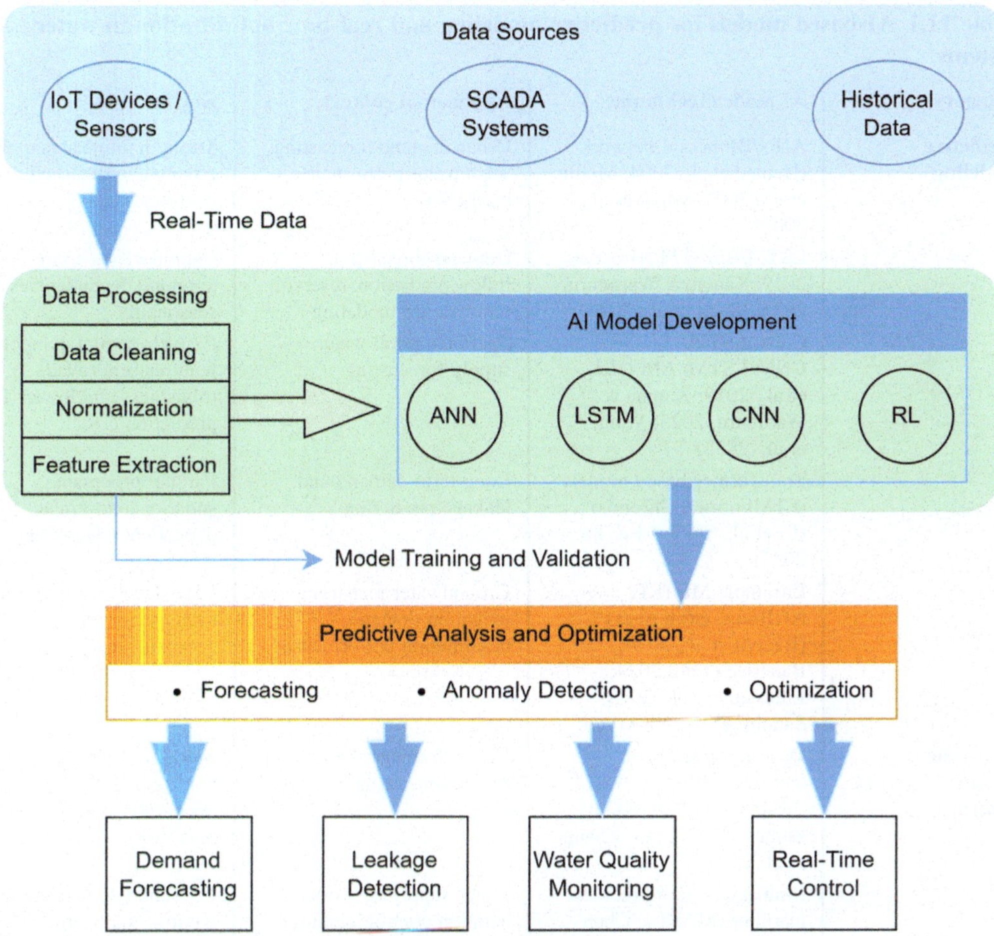

FIGURE 11.2

Workflow of AI-driven predictive modeling in water management systems.

predictive analytics, and decision-making for operational optimization. The framework highlights applications such as demand forecasting, fault detection, water quality monitoring, and real-time control of water distribution systems. Table 11.1 summarizing the existing techniques for AI-based models for predictive modeling and real-time optimization in water systems.

3.1 Works on urban water supply networks

AI-driven predictive modeling has become central to modern urban water management, enabling utilities to respond to growing uncertainty from climate variability, urbanization, and aging infrastructure. The study (Kamau & Waruguru, 2025) uses smart-meter data from Nairobi to model urban

Table 11.1 AI-based models for predictive modeling and real-time optimization in water systems.

Category	AI model/Technique	Application context	Key advantages
Predictive modeling	ANN/BP neural network (Hamed et al., 2004; Mjalli et al., 2007; Wang et al., 2009)	Urban demand forecasting, wastewater performance prediction	Strong nonlinear mapping; effective with structured historical data
	LSTM/BiLSTM (Hu et al., 2019; Kamau & Waruguru, 2025; Wang et al., 2022)	Demand forecasting, inflow prediction, reservoir temperature modeling	Captures long-term temporal dependencies and seasonality
	CNN–LSTM/ CNN–LSTM–AM (Hu et al., 2019; Kamau & Waruguru, 2025; Wang et al., 2022)	Spatiotemporal water supply forecasting	Extracts spatial–temporal features; improved robustness and lower prediction error
	Transformer-based models (LTMFormer, Chronos) (Li et al., 2025; Li & Fu, 2023)	Long-term demand and leakage prediction	Parallel processing; superior long-range dependency learning
	Ensemble ML (RF, XGBoost, extra trees) (Badem et al., 2024; Banerjee et al., 2024; Elhussiny et al., 2023; Saddiqi et al., 2024)	Groundwater recharge, irrigation demand, desalination performance	High accuracy; robustness to noise; strong generalization
Real-time optimization and control	Deep reinforcement learning (DRL) (Mullapudi et al., 2020; Zaman et al., 2023; Zhang et al., 2022)	Urban drainage, flood mitigation, pump control	Model-free adaptive control; handles uncertainty and dynamic conditions
	Multi-agent reinforcement learning (MARL) (Chen et al., 2021; Ma et al., 2024; Ni et al., 2013)	Water supply networks, aeration control, resource allocation	Coordinated distributed control; multi-objective optimization
	RL-based irrigation control (Ding & Du, 2024; Yang et al., 2020)	Smart irrigation scheduling	Minimizes water use while maintaining crop yield
	IoT-enabled monitoring + AI (Lambrou et al., 2014; Saravanan et al., 2018; Forhad et al., 2025)	Leak detection, contamination detection, smart desalination	Continuous sensing; rapid decision support
	IoT + edge intelligence frameworks (Olatinwo & Joubert, 2023; Premkumar & Sigappi, 2022; Das, 2024; Godase, 2025; Mehta & Choudhary, 2024)	Precision irrigation, water quality monitoring	Low latency; reduced bandwidth; scalable decentralized control

water consumption. By capturing non-linear daily and monthly usage patterns, the LSTM provides more accurate forecasts, demonstrating its suitability for managing water demand in rapidly growing Kenyan cities. Hu et al. (2019) develops a CNN–BiLSTM hybrid model to improve urban water demand forecasting by addressing nonlinear patterns, temperature effects, and holiday anomalies. Zhou et al. (2025) proposed an empirical mode decomposition-based model to enhance short-term prediction performance, significantly outperforming traditional methods and improving smart city water management. Wang et al. (2009) integrates PCA with a BP neural network to predict urban water demand in Hefei. PCA identifies economic development and population size as the primary influencing factors. Using these as model inputs, the BP network achieves high accuracy with prediction errors below 0.9. Fan et al. (2019) develops a two-level leakage diagnosis model for heating pipe networks using a Deep Belief Network (DBN). A hydraulic calculation model first identifies pressure changes caused by leakage. The first level DBN model accurately detects leaking pipe segments, outperforming Back Propagation (BP) neural networks and Support Vector Machines (SVMs). The second-level model then locates the exact leak position. Tests on branch and loop networks confirm high accuracy, demonstrating strong potential for intelligent fault detection. Li and Fu (2023) proposes a deep-learning approach for forecasting water demand in residential communities with secondary water supply systems. After cleaning outliers and integrating epidemic and meteorological factors, it introduces the Long and Short Time Transformer (LTMFormer), combining LSTM recursion with Transformer parallelism.

Li et al. (2025) addresses growing leakage challenges in urban water supply networks by proposing a flow prediction method based on the time series Large Language Model (LLM) Chronos. Chronos tokenizes scaled and quantized time-series data and uses Transformer architectures for long-term dependency modeling. Zhao et al. (2023) proposes a spatiotemporal deep learning model integrating Convolutional Neural Networks (CNN), Long Short-Term Memory (LSTM), and an Attention Mechanism (AM) for accurate urban daily water supply forecasting. CNN extracts spatial features, LSTM models temporal patterns with Bayesian-optimized parameters, and AM enhances feature weighting. The CNN-LSTM-AM model significantly lowers MAE, MSE, and RMSE compared to traditional models, demonstrating superior accuracy, robustness, and spatiotemporal feature extraction capabilities for urban water supply prediction. Liu et al. (2010) develops three shortterm water consumption forecasting models for Zhengzhou using the moving arithmetic mean, regression analysis, and Back Propagation (BP) neural networks. Using water supply and meteorological data, all models meet scheduling needs, but regression and BP neural networks provide higher accuracy. Wu, Wang, et al. (2019) addresses growing challenges in sustainable smart water supply systems and proposes a water quality risk analysis framework using industrial indicator data. An Adaptive Frequency Analysis (Adp-FA) method is introduced to capture frequency-domain relationships and enable explainable predictions. Tested on datasets from four Norwegian cities, Adp-FA outperforms Artificial Neural Networks and Random Forests in accuracy and efficiency, demonstrating strong potential for early water quality risk detection and decision support in modern urban water systems.

3.2 Works on rural and community water supply systems

AI-based predictive modeling is increasingly transforming rural and community water supply systems, where resource scarcity, limited infrastructure, and delayed detection of contamination or service failures pose persistent challenges. Ogore et al. (2021) demonstrates how lightweight, embedded AI models can provide early warning of cholera contamination directly at communal taps without relying on constant connectivity. This offline inference capability is particularly critical in remote environments where internet access is unreliable, enabling local operators to detect pathogens and take timely action to protect public health. Similarly, Paul et al. (2023) highlights the use of sensor networks and ML classifiers to detect changes in key water quality parameters, enabling timely intervention and reducing health risks in underserved communities. Beyond quality monitoring, Cao et al. (2024) uses automated machine learning (AutoML) to forecast domestic water consumption and sewage generation under varying socio-economic and climatic conditions. This enables rural planners to optimize treatment capacity, distribution schedules, and infrastructure investments. Leakage remains another major concern in decentralized rural supply networks. Addressing supply reliability, Li et al. (2022) introduces an AI-driven framework capable of identifying distribution pipe failures early, minimizing water loss and service interruptions. At the community scale, Wu, Lei, et al. (2019) applies advanced neural models to anticipate storage tank fluctuations, helping ensure stable and efficient water delivery. Together, these works highlight how AI enables safer, more resilient, and more efficient water management in rural settings.

3.3 Works on wastewater collection and treatment systems

AI-based predictive modeling has become an essential tool for improving reliability, efficiency, and decision-making in wastewater collection and treatment systems. A growing body of research demonstrates how machine learning, deep learning, and statistical models can forecast treatment performance, anticipate failures, and support proactive infrastructure management. Lin et al. (2022) showcases the ability of deep neural architectures to capture complex nonlinear biochemical processes within treatment reactors, enabling more accurate predictions of effluent quality and operational responses. Complementing this, Li et al. (2013) employs probabilistic modeling to interpret dependencies among operational variables, supporting uncertainty-aware forecasting and revealing causal relationships that help operators understand system behavior under varying conditions.

Several studies highlight the value of artificial neural networks (ANNs). The works (Hamed et al., 2004) and (Mjalli et al., 2007) demonstrate that ANNs can reliably predict key performance indicators such as COD, BOD, suspended solids, and nutrient removal efficiency. Their ability to learn from historical plant data makes them effective for optimizing chemical dosing, aeration control, and energy consumption. Inflow forecasting, crucial for preventing overloading and combined sewer overflow, is addressed in Zhou et al. (2019), which shows that ensemble learning models outperform traditional statistical methods in handling highly variable hydraulic loads. Expanding the scope to multi-plant environments, Ly et al. (2022) leverages large-scale datasets to generalize predictive models across diverse operating conditions, highlighting the scalability of ML approaches. On the infrastructure side, the studies (Younis & Knight, 2008 and Salman & Salem, 2012) apply analytical and regression-based techniques to predict pipe deterioration and failure risks. These tools support asset management, rehabilitation planning, and proactive maintenance in aging sewer networks.

3.4 Works on stormwater drainage and flood management systems

AI-based prediction modeling has become a vital component of modern stormwater drainage and flood management systems, offering enhanced accuracy, faster response, and greater resilience against extreme rainfall events. Several recent studies demonstrate how machine learning and deep learning models can forecast water levels, optimize drainage operations, and reduce flood risks in increasingly complex urban environments.

Huang and Lee (2023) highlights how data-driven models can outperform traditional hydraulic simulations by producing rapid, real-time water level predictions that support operational decision-making during storm events. Aderyani and Mousavi (2024) introduces a predictive real-time operation (PRTOP) model for urban drainage systems by integrating rainfall forecasting with optimal system control. Using LSTM and PSO-optimized SVR rainfall models linked to an SWMM–Harmony Search framework, the system continuously updates decisions during storms. Abou Rjeily et al. (2017) develop a flood forecasting system for urban drainage networks using a Nonlinear Auto Regressive with eXogenous inputs (NARX) neural network, which links rainfall intensity to water depth variations in critical manholes. Tested at the University of Lille campus, the model provides rapid, reliable flood-risk warnings, enabling proactive management and improved flood mitigation in urban areas.

Piadeh et al. (2023) presents a dynamic ensemble-based data mining model for urban flood forecasting, integrating event identification technique and rainfall feature extraction to build stacked weak learners. By incorporating rainfall history and seasonal timing, it accurately classifies flood events with minimal lag, reducing false alarms and improving real-time early warning in drainage systems. Chang et al. (2025) enhances urban flood resilience by integrating knowledge graphs and hybrid deep learning into Taipei's Real-Time Urban Drainage Early Warning System. The model reliably predicts sewer, forebay, and river levels under intense rainfall, enabling smarter pump operations, faster emergency response, and improved flood mitigation for the Zhongshan Pumping Station.

3.5 Works on irrigation and agricultural water distribution systems

AI-based prediction modeling is increasingly reshaping irrigation and agricultural water distribution systems, enabling more efficient resource allocation, improved crop productivity, and sustainable water management in data scarce rural environments. The study (Pulido-Calvo et al., 2003) evaluates daily irrigation water demand forecasting using multiple regression, time-series models, and Computational Neural Networks (CNNs). Using real data from southern Spain, CNNs incorporating past water demand, temperature, and crop factors outperform traditional methods. The best predictions occur when water demand and maximum temperature from the previous 2 days are used as inputs. Elhussiny et al. (2023) develop Random Forest, Extreme Gradient Boosting, and hybrid XGB–RF models to predict sprinkler irrigation water distribution uniformity using pressure, sprinkler height, discharge, nozzle size, and weather variables. Large-scale agricultural networks also benefit from predictive analytics. Perea et al. (2021) introduces Watergy Forecaster, a machine-learning tool that predicts weekly field-scale water and energy demand in large irrigation networks using satellite data and open climate datasets to support efficient irrigation management and resource savings.

Authors in Hedley and Yule (2009) develops a method to spatially predict daily soil water status using high-resolution electrical conductivity maps, regression models, and data from wireless soil-moisture sensors. The approach supports variable-rate irrigation by generating digital soil-water maps, improving irrigation efficiency when spatial variability is well captured. Water quality forecasting plays an equally important role in agricultural sustainability. The studies (El Bilali & Taleb, 2020 and Omeka, 2024) demonstrate that integrating GIS, environmental indicators, and ML algorithms can reliably predict salinity, nutrient levels, and contamination risks, ensuring safe and suitable water for crops. Damtie et al. (2021) develop supervised machine-learning models to predict irrigation water supply for ungated quaternary canals in Ethiopia's Koga Irrigation Scheme using geometric and hydraulic data from upstream regulated canals. Its interpretability and strong predictive ability make it well suited for estimating flows in lower-level irrigation channels, supporting improved water allocation and management across the scheme.

3.6 Works on groundwater extraction and recharge systems

AI-based predictive modeling has become an essential tool for understanding and managing groundwater extraction and recharge systems, especially under growing pressure from climate change, land-use transformation, and increasing water demand. The study (Mohan et al., 2018) analyzes groundwater recharge data from 715 global sites to identify key controlling factors and develop a large-scale empirical recharge model. Using multimodel inference and information theory, it finds precipitation, potential evapotranspiration, and land cover as the strongest predictors. The model supports large-scale groundwater planning under future water scarcity. Smerdon et al. (2009) develops a method for predicting groundwater recharge in mountainous watersheds by combining MIKE-SHE simulations in data-rich valley areas with seasonal water-budget modeling in data-limited uplands. Applied to the BX Creek watershed, recharge ranged from 0 to 50 mm/year. Integrated modeling with geochemical and isotopic analyses revealed dominant recharge contributions from upland snowmelt and highlighted key flow paths through alluvial fan aquifers, providing a reliable first-order understanding of watershed-scale groundwater dynamics.

Yaqub et al. (2022) develops predictive machine learning models to estimate the removal of pharmaceutical pollutants in a managed aquifer recharge (MAR) system. Using 1 year of experimental data, ensemble learners—decision tree, random forest, and XGBoost—were trained on water and soil quality parameters. Banerjee et al. (2024) forecasts groundwater recharge in Punjab, India, under future climate change scenarios using 34 years of data such as precipitation, land use land cover, soil type, land slope, temperature, potential evapotranspiration, and aridity index. Multiple AI and deep learning models were compared, with XGBoost delivering the highest accuracy. Pham et al. (2022) evaluates seven machine learning models, such as random tree, random forest, decision stump, M5P, support vector machine, locally weighted linear regression, and reduce error pruning tree to predict groundwater levels in drought-prone northwestern Bangladesh using long-term hydroclimatic data such as historical groundwater level, mean temperature, rainfall, and relative humidity.

Kayhomayoon et al. (2021) introduces a machine learning–based method to estimate groundwater storage loss across 178 major Iranian aquifers, particularly valuable for data-scarce regions. Using easily accessible water-budget variables, the Harris Hawks Optimization–Adaptive NeuroFuzzy Inference System (HHO-ANFIS) and LS-SVM emerged as the best predictors, with HHO-ANFIS performing most efficiently. The approach offers reliable, cost-effective GSL prediction and helps identify the most influential factors for sustainable groundwater management. This study (Igwebuike et al., 2025) evaluates machine learning and deep learning models—Random Forest, Support Vector Machine (SVM), SimpleRNN, and LSTM—for predicting groundwater levels in South Africa's West Coast Aquifer System. The findings confirm ML and DL as effective tools for groundwater-level forecasting and support future enhancements using detailed geologic information.

3.7 Works on surface water reservoirs and dams

AI-based predictive modeling has become a critical tool for managing surface water reservoirs and dams, enabling more reliable decision-making under increasing hydrological uncertainty. Recent studies demonstrate how machine learning (ML) and deep learning (DL) techniques enhance forecasting of reservoir dynamics, water quality, and ecosystem impacts. Badem et al. (2024) develop advanced models to accurately predict dam occupancy using historical reservoir data combined with physical parameters such as evapotranspiration, rainfall, solar radiation, and water use. Multiple methods—including Random Forest, Extra Trees, LSTM, OMP-CV, and Lasso-Lars—were evaluated, with statistical tests confirming the enhanced accuracy gained by integrating physical processes. The results identify the most reliable model for sustainable dam management. Wang et al. (2022) proposes an efficient LSTM model for predicting surface water temperature in large, deep reservoirs by replacing the computationally intensive Delft3D hydrodynamic model. Sensitivity analysis revealed strong influences from air temperature and water level, supporting improved reservoir operations under climate change.

Vishwakarma et al. (2022) predicts pre- and post-dam river water temperature changes in the Yangtze River at Cuntan using machine learning models, including M5P, Random Forest, Random Subspace, and REPTree. Results show significant dam-induced thermal alterations and demonstrate that ML models effectively support ecological planning, reservoir operation, and temperature forecasting using lagged water-level inputs. Allawi et al. (2021) develops a modified Coactive Neuro-Fuzzy Inference System (CANFIS) model to improve reservoir evaporation prediction, overcoming limitations of physical and traditional data-driven models. By enhancing feature extraction and updating membership rules automatically, the model outperformed ANFIS, SVR, and RBF-NN in two climatic regions (Aswan High Dam and Timah Tasoh Dam). Shah et al. (2021) compares Gene Expression Programming (GEP), Artificial Neural Networks (ANN), and Linear Regression Models for predicting total dissolved solids and electrical conductivity in the Upper Indus River using 30 years of data. Sensitivity analysis identified key ionic contributors, and external validation confirmed model robustness, demonstrating AI's effectiveness for reliable, cost-efficient water quality assessment and management. Babaei et al. (2019) predicts inflow to the Zayandehroud Dam using artificial neural networks and support vector machines with nine input patterns incorporating lagged inflow, time indices, and rainfall. The results highlight SVM's reliability for reservoir inflow forecasting.

3.8 Works on water treatment and desalination plants

AI-based prediction modeling has become a transformative tool in enhancing the efficiency, reliability, and cost-effectiveness of water treatment and desalination plants. Across recent studies, machine learning and optimization-driven predictive frameworks have been applied to improve plant performance, forecast operational behavior, and optimize treatment processes under varying environmental and operational conditions. The study (Al-Shayji & Liu, 2002) compares data based neural networks and model-based process simulations for predicting performance in large-scale multistage flash (MSF) and reverse osmosis (RO) desalination plants. Using real operational data, both approaches achieved high accuracy, with neural networks excelling at handling noisy data and predicting performance variables. While validated simulations better quantify variable impacts in MSF systems, neural networks are more effective for RO plants where process modeling is complex. Mahadeva et al.

(2022) introduces a Gray Wolf Optimizer–enhanced Artificial Neural Network (GWO-ANN) to improve prediction accuracy for RO desalination plant performance. Using key operational inputs, six hybrid models were developed. The method outperformed traditional ANN, RSM, PSO-ANN, and GA-ANN models, demonstrating superior optimization of weights and biases and offering enhanced stability, efficiency, and reduced operational costs for desalination plants.

Saddiqi et al. (2024) applies Artificial Neural Networks (ANN) and Extreme Gradient Boosting (XGBoost) to optimize membrane-based produced water treatment. Focusing on predicting trans-membrane pressure and total membrane resistance, both models demonstrated strong performance. The combined ML approach enhances operational control, improves process modeling, and supports more efficient and sustainable produced water treatment. This study introduces a Modified Whale Optimization Algorithm hybridized with Artificial Neural Networks to improve performance prediction in RO desalination plants. Using four key operational inputs, such as feed flow rate, evaporator inlet temperature, feed salt concentration and condenser inlet temperature, 10 hybrid models were developed. Giwa et al. (2025) develop a global machine learning model and web application to estimate the Optimum Cost-Effective Solution for desalination plant construction using data from over 10,000 projects, considering factors such as plant size, technology, location, and procurement models. The model supports improved planning and investment decisions for desalination under growing water scarcity. Kolluri et al. (2015) applies multivariate statistical methods to monitor and predict performance in a full-scale seawater reverse osmosis plant to supporting improved process control and optimization.

3.9 Works on recycled and reclaimed water systems

AI-driven prediction models are increasingly advancing the performance, reliability, and sustainability of recycled and reclaimed water systems. Recent studies apply deep learning, signal decomposition, probabilistic diagnosis, and sensor-based analytics to forecast reclaimed water volumes, assess system health, and predict microbial water quality in real time. Additionally, machine learning supports cost estimation for water reuse infrastructure, enabling informed planning and optimized system operation. Chen et al. (2024) analyzes seasonal and meteorological influences on reclaimed water potential and develops a hybrid forecasting model combining signal decomposition with CNN–LSTM deep learning. Results demonstrate its strong robustness and suitability for precise reclaimed water management, especially in regions with significant seasonal fluctuations. Zhang et al. (2024) develops a dynamic real-time forecasting method for reclaimed water volumes using a hybrid ML-CEEMDAN-LSTM model. By integrating decomposition, machine learning, and deep learning, the model overcomes boundary issues and outperforms traditional ML and CEEMDAN-LSTM approaches across seasons. Applied to Beijing's reclaimed water plants, it provides highly accurate predictions essential for urban river environmental management and wastewater reuse planning.

Roychoudhury et al. (2013) develops a physics-based model and an integrated diagnosis–prognosis framework for a water recycling system (WRS). The framework detects faults, identifies their root causes, assesses damage severity, and predicts system end-of-life under various scenarios. By enabling condition-based maintenance, the approach helps prevent unplanned outages and extends WRS operational lifespan, supporting future spacecraft water recycling technologies. Reynaert et al. (2023) demonstrates an effective method for predicting microbial safety in on-site water reuse systems using low-cost online sensors and machine learning. Logistic regression, support vector machines, and random forests were evaluated to predict the microbial water quality as "safe" or "unsafe" for reuse. Using only two key sensors (free chlorine and oxidation–reduction potential), the models achieved high accuracy, offering an interpretable, low-cost, and reliable solution for real-time microbial water quality monitoring.

4. Real-time optimization and control in water management systems

Real-time optimization and control play a crucial role in managing today's water systems, where operating conditions frequently change due to demand fluctuations, climate impacts, and infrastructure limitations. This section discusses AI-driven real-time optimization techniques, particularly reinforcement learning, and explores IoT and edge-intelligence-based approaches that support intelligent, efficient, and resilient water system operations. Table 11.2 presents a state-of-the-art summary of AI-based and IoT-edge-intelligence-based real-time optimization and control techniques in water management systems. The underlying methodologies and application contexts are discussed in detail in Sections 4.1 and 4.2.

4.1 AI-driven techniques for real-time optimization

Reinforcement Learning (RL) is becoming a popular AI method for real-time optimization in water systems. It allows systems to learn how to make better decisions by interacting with their environment and using past experience. RL can handle changing conditions and uncertainty and make it useful for managing urban drainage, irrigation, pumping operations, stormwater control, and wastewater treatment facilities.

The robustness of RL in challenging environments is evident in Zhang et al. (2022) where the authors investigate how sensor uncertainty affects DRL-driven drainage control. The study shows that DRL can still maintain efficient overflow reduction even when input signals are noisy, highlighting its resilience in real-world stormwater systems. In irrigation management, Yang et al. (2020) demonstrates how DRL autonomously learns optimal irrigation timings by modeling soil moisture fluctuations, while (Ding & Du, 2024) extends this work to real agricultural fields, confirming that RL can minimize water use without compromising crop yield.

In water supply network management, Ma et al. (2024) integrates physical hydraulic constraints into a Multi-Agent Reinforcement Learning (MARL) architecture, leading to significant reductions in pump energy cost and pressure oscillations. Likewise, Zaman et al. (2023) shows that DRL can handle large-scale urban water networks by learning pressure-based control strategies that improve service reliability while lowering system losses. Emergency response and multi-objective control are addressed in Hu et al. (2022), where RL agents jointly optimize response time, restoration stability, and supply continuity. For broader resource management, Ni et al. (2013) shows how multiple Q-learning agents can collaborate to ensure fair and efficient allocation across competing sectors.

Real-time pump and hydraulic optimization is explored in Hajgato et al. (2020), which reports substantial energy savings and dynamic adaptability. Similarly, Hu and Jiang (2025) highlights RL's ability to adjust valve settings and flow conditions without needing explicit hydrodynamic models. Large-scale storage and treatment processes also benefit from RL. Mahootchi et al. (2025) demonstrates that RL can balance release policies and flood control by learning seasonal inflow patterns.

Table 11.2 State-of-the-art AI, IoT, and edge intelligence–based real-time optimization and control techniques in water management systems.

References	Application domain	Methodology/ Tools	System scale	Optimization objective	Key outcomes
Zhang et al. (2022)	Urban drainage control	Deep reinforcement learning (DRL)	Urban stormwater network	Overflow reduction	Robust performance under sensor uncertainty
Yang et al. (2020)	Irrigation management	DRL with soil–water modeling	Agricultural fields	Optimal irrigation timing	Reduced water usage
Ding and Du (2024)	Precision agriculture	DRL (field deployment)	Agricultural fields	Water-use efficiency	Maintained crop yield with less water
Ma et al. (2024)	Water supply networks	Multi-agent RL	Urban water supply network	Energy and pressure control	Reduced pump energy cost
Zaman et al. (2023)	Urban water distribution	DRL-based pressure control	City-scale water distribution network	Loss reduction	Improved service reliability
Hu et al. (2022)	Emergency water management	Multi-objective RL	Water distribution network	Response and recovery	Improved resilience
Ni et al. (2013)	Multi-sector allocation	Multi-agent Q-learning	Regional scale	Fair allocation	Balanced resource distribution
Hajgato et al. (2020)	Pump optimization	RL-based adaptive control	Water distribution systems	Energy efficiency	Significant energy savings
Hu and Jiang (2025)	Hydraulic control	Reinforcement learning	Water distribution networks	Flow and valve control	No explicit hydraulic model required
Mahootchi et al. (2025)	Reservoir operation	RL-based policy learning	Large storage systems	Flood control	Improved seasonal handling
Chen et al. (2021)	Wastewater treatment	Multi-agent RL	Water treatment plants	Cost reduction	Maintained effluent standards
Mullapudi et al. (2020)	Flood management	RL with rainfall prediction	Stormwater systems	Flood risk reduction	Better than rule-based control
Barros (2024)	Desalination systems	IoT-based monitoring	Solar stills	Efficiency optimization	Improved real-time performance

Table 11.2 State-of-the-art AI, IoT, and edge intelligence–based real-time optimization and control techniques in water management systems.—cont'd

References	Application domain	Methodology/ Tools	System scale	Optimization objective	Key outcomes
Vianny et al. (2022)	Smart irrigation	IoT + ML (KNN, GBT, LSTM)	Agricultural systems	Water and energy optimization	Reduced consumption
Premkumar and Sigappi (2022)	Smart irrigation	IoT + edge intelligence	Agricultural systems	Latency reduction	Faster response time
Olatinwo and Joubert (2023)	Water quality monitoring	IoT + edge optimization	Sensor networks	Resource allocation optimization	Real-time contamination detection
Lambrou et al. (2014), Saravanan et al. (2018), Forhad et al. (2025)	Urban water monitoring	IoT, SCADA, PLC	Pipelines and plants	Leak detection, contamination detection	Early warning and monitoring
Das (2024), Godase (2025), Mehta and Choudhary (2024)	Integrated water management	AI + IoT frameworks	Multi-domain systems	Leakage detection	Improved water-use efficiency

Chen et al. (2021) presents MARL-based aeration and chemical control strategies that reduce operational costs while meeting effluent standards. Finally, Mullapudi et al. (2020) shows RL's capability to anticipate rainfall patterns and reduce flooding more effectively than static rule-based methods.

4.2 IoT and edge intelligence-based real-time optimization techniques

The integration of Internet of Things (IoT) technologies with real-time optimization and control mechanisms has emerged as a transformative approach in modern water management systems. IoT-enabled sensing infrastructures provide continuous, high-resolution monitoring of environmental and operational parameters, enabling timely and data-driven decision-making. For example, IoT-based monitoring of solar stills for water desalination has demonstrated how real-time measurements of temperature, humidity, and basin water levels can be used to optimize system efficiency and support sustainable water production in water-scarce regions (Barros, 2024). The availability of real-time dashboards and sensor feedback facilitates adaptive control and informed design decisions.

In agricultural water management, IoT-driven precision irrigation systems combined with machine learning techniques enable dynamic optimization of water and energy usage. Hybrid models incorporating KNN, Gradient Boosting Trees, LSTM, and correlation analysis have shown effective forecasting of irrigation requirements based on soil moisture, temperature, and weather conditions, resulting in significant reductions in water consumption and energy wastage (Vianny et al., 2022). Such approaches highlight the role of predictive analytics in real-time irrigation scheduling and resource efficiency.

Edge intelligence further strengthens real-time optimization by relocating computation closer to data sources. IoT-enabled edge computing models for smart irrigation demonstrate that deploying predictive algorithms at the edge significantly reduces latency, bandwidth usage, and cloud dependency while improving response time and computational efficiency (Premkumar & Sigappi, 2022). Similar benefits are observed in IoT-based water quality monitoring systems, where edge-enabled resource allocation optimizes power and bandwidth usage to ensure real-time contamination detection and energy efficient operation (Olatinwo & Joubert, 2023).In urban water systems, IoT sensor networks enable continuous monitoring of water quality, leakage, and contamination. Low-cost in-pipe sensor networks and SCADA-integrated IoT frameworks have demonstrated effective real-time detection of contaminants, leaks, and quality degradation, supporting early warning mechanisms and rapid response (Lambrou et al., 2014; Saravanan et al., 2018). IoT-based monitoring solutions deployed in water treatment plants further enhance operational efficiency through real-time alerts, remote monitoring, and PLC-based control integration (Forhad et al., 2025).

Comprehensive IoT-based water management systems across agricultural, urban, and industrial domains demonstrate notable improvements in water-use efficiency, leak detection speed, and sustainability outcomes (Das, 2024; Godase, 2025; Mehta & Choudhary, 2024). Collectively, these studies confirm that the convergence of IoT sensing, edge intelligence, and AI-driven optimization enables responsive, scalable, and sustainable real-time control in water management systems.

5. Research gaps and future directions

AI-driven predictive modeling and real-time optimization in water systems have shown considerable promise; however, several limitations continue to restrict their practical deployment. Limited adaptability of current AI-based water management systems remains a key research gap, as many existing models are trained offline and updated infrequently, reducing their effectiveness under rapidly changing environmental and operational conditions. Future work should focus on developing autonomous and self-learning water systems that continuously update predictive and control strategies using real-time data and system feedback to enhance resilience and efficiency (Ahmed et al., 2024; Hernández-del Olmo et al., 2025).

Constraints on data sharing due to privacy, security, and ownership concerns represent another important gap, leading to fragmented datasets and limited model generalization across utilities. Future research can address this challenge through federated learning frameworks that enable privacy preserving collaborative predictive modeling without requiring the exchange of raw or sensitive data (Jenifel & Linda, 2025; Wang et al., 2025).

The lack of real-time system representation and seamless integration between AI models and physical infrastructure limits the operational value of many existing solutions. Future work should emphasize the integration of AI with digital twins and next-generation IoT 2.0 technologies to support continuous system monitoring, simulation, and proactive decision-making (Ghorbani Bam et al., 2025).

Existing optimization approaches often focus on narrow objectives such as cost or energy efficiency, overlooking broader sustainability and reliability considerations. Future work should advance multi-objective optimization frameworks that jointly balance sustainability, operational cost, energy efficiency, and system reliability for long-term resilient water management (Jung et al., 2025; Boindala and Ostfeld, 2025).

A key challenge lies in the limited transferability of AI models across different hydrological and climatic regions (Slater et al., 2025). Most models are trained using localized datasets and calibrated for specific system configurations, resulting in performance degradation when applied elsewhere. Addressing this limitation requires the development of adaptive learning frameworks based on transfer learning, domain adaptation, and meta-learning, enabling models to generalize effectively across diverse water systems (Onyutha, 2025; Yao et al., 2023; Islam, 2025).

The lack of interpretability in complex AI models remains a significant barrier to trust, adoption, and regulatory acceptance. Deep learning and reinforcement learning frameworks often operate as black boxes, limiting their usability in safety-critical applications such as flood management and drinking water quality control. Integrating explainable artificial intelligence techniques can enhance transparency by revealing key drivers of system behavior and supporting human–AI collaboration in operational environments (Slater et al., 2025; Aderemi et al., 2025).

6. Conclusion

This chapter has reviewed recent work on AI-based predictive modeling and real-time optimization in water management systems. The studies show that machine learning and deep learning can improve prediction accuracy, system operation, and response to changes in water demand and environmental conditions. Combining AI with IoT sensors and real-time control systems can help improve monitoring, decision-making, and system flexibility. However, problems such as poor data quality, lack of model transparency, difficulty in scaling, and challenges in real-world use still exist. Future research should focus on simple and explainable AI models, better integration with current water infrastructure, and real-world testing to support the wider use of intelligent water management systems.

References

Abou Rjeily, Y., Abbas, O., Sadek, M., Shahrour, I., & Hage Chehade, F. (2017). Flood forecasting within urban drainage systems using narx neural network. *Water Science and Technology, 76*(9), 2401–2412.

Aderemi, I. A., Kehinde, T. O., Ugochukwu, D. O., Ahmad, K. H., Adjei, K. Y., & Chijioke, C. E. (2025). Beyond the black box: A systematic review of explainable ai for transparent and trustworthy water quality monitoring. *IEEE Sensors Reviews, 2*, 419–443. https://doi.org/10.1109/SR.2025.3595500

Aderyani, F. R., & Mousavi, S. J. (2024). Machine learning-based rainfall forecasting in real-time optimal operation of urban drainage systems. *Journal of Hydrology, 645*, Article 132118.

Ahmed, A. A., Sayed, S., Abdoulhalik, A., Moutari, S., & Oyedele, L. (2024). Applications of machine learning to water resources management: A review of present status and future opportunities. *Journal of Cleaner Production, 441*, Article 140715. https://doi.org/10.1016/j.jclepro.2024.140715

Aib, H., Parvez, Md S., Czédli, & Mária, H. (2025). Pharmaceuticals and microplastics in aquatic environments: A comprehensive review of pathways and distribution, toxicological and ecological effects. *International Journal of Environmental Research and Public Health, 22*(5), 799.

Ajjur, S. B., & Al-Ghamdi, S. G. (2021). Evapotranspiration and water availability response to climate change in the Middle East and North Africa. *Climatic Change, 166*(3), 28.

Al-Hasani, B., Abdellatif, M., Carnacina, I., Harris, C., Al-Quraishi, A. M. F., Shammari, A.-, & Muthanna, M. A. (2025). Forecasting evaporation trends amid climate change for sustainable water management in semi-arid regions. *Water, 17*(7), 1039.

Al-Shayji, K. A., & Liu, Y. (2002). Predictive modeling of large-scale commercial water desalination plants: Data-based neural network and model-based process simulation. *Industrial and Engineering Chemistry Research, 41*(25), 6460–6474.

Al-Washali, T., Sharma, S., & Kennedy, M. (2016). Methods of assessment of water losses in water supply systems: A review. *Water Resources Management, 30*(14), 4985–5001.

Alhussan, A. A., El-kenawy, El-Sayed, M., Khafaga, D. S., Alharbi, A. H., & Eid, M. M. (2025). Groundwater resource prediction and management using comment feedback optimization algorithm for deep learning. *IEEE Access, 13*, 169554–169593. https://doi.org/10.1109/ACCESS.2025.3614168

Allawi, M. F., Ahmed, M. L., Aidan, I. A., Deo, R. C., & El-Shafie, A. (2021). Developing reservoir evaporation predictive model for successful dam management. *Stochastic Environmental Research and Risk Assessment, 35*(2), 499–514.

Amoah, S. T., & Yahaya, A. (2025). Evolution of water systems and its challenges in the Wa municipal of Ghana. *Journal of Environment and Earth Science, 3*(7).

Araya, F., & Vasquez, S. (2022). Challenges, drivers, and benefits to integrated infrastructure management of water, wastewater, stormwater and transportation systems. *Sustainable Cities and Society, 82*, Article 103913.

Arienzo, M., & Donadio, C. (2023). Microplastic–pharmaceuticals interaction in water systems. *Journal of Marine Science and Engineering, 11*(7), 1437.

Babaei, M., Moeini, R., & Ehsanzadeh, E. (2019). Artificial neural network and support vector machine models for inflow prediction of dam reservoir (case study: Zayandehroud dam reservoir). *Water Resources Management, 33*(6), 2203–2218.

Badem, A. C., Yılmaz, R., Cesur, M. R., & Cesur, E. (2024). Advanced predictive modeling for dam occupancy using historical and meteorological data. *Sustainability, 16*(17), 7696.

Banerjee, D., Ganguly, S., & Kushwaha, S. (2024). Forecasting future groundwater recharge from rainfall under different climate change scenarios using comparative analysis of deep learning and ensemble learning techniques. *Water Resources Management, 38*(11), 4019–4037.

Barros, R. D. P. (2024). Design and implementation of an iot monitoring system for the optimization of solar stills for water desalination. *LatIA*, (2), 18.

Biswas, A., Sarkar, S., Das, S., Dutta, S., Choudhury, M. R., Giri, A., Bera, B., Bag, K., Mukherjee, B., & Banerjee. (2025). Koushik and others, water scarcity: A global hindrance to sustainable development and agricultural production–A critical review of the impacts and adaptation strategies. *Cambridge Prisms: Water, 3*, e4.

Boindala, S. P., & Ostfeld, A. (2025). Robust multi-objective design optimization of water distribution system under uncertainty. *Water, 14*(14). https://doi.org/10.3390/w14142199. URL https://www.mdpi.com/2073-4441/14/14/2199.

Cao, Y., Wang, Z., Li, P., Zhou, Z., Li, W., Zheng, T., Liu, J., Wu, W., Shi, Z., & Liu, J. (2024). Prediction of rural domestic water and sewage production based on automated machine learning in northern China. *Journal of Cleaner Production, 434*, Article 140016.

Chang, L.-C., Yang, M.-T., & Chang, F.-J. (2025). Flood resilience through hybrid deep learning: Advanced forecasting for taipei's urban drainage system. *Journal of Environmental Management, 379*, Article 124835.

Chao, D., Muyuan, Z., & Hongpeng, L. (2016). Research on the optimization and distribution technology of industrial water. In *2016 Chinese control and decision conference (CCDC)* (pp. 3992–3996). IEEE.

Chen, K., Wang, H., Valverde-Pérez, B., Zhai, S., Vezzaro, L., & Wang, A. (2021). Optimal control towards sustainable wastewater treatment plants based on multi-agent reinforcement learning. *Chemosphere, 279*, Article 130498.

Chen, Y., Zhang, H., Peng, J., Ma, S., Xu, T., & Tang, L. (2024). Forecasting the potential of reclaimed water using signal decomposition and deep learning. *Journal of Water Process Engineering, 65*, Article 105770.

Chu, J., Qin, D., & Shao, W. (2012). Dualistic hydrological cycle based water management policy screening: The case of Taihu lake basin. In *2012 international symposium on geomatics for integrated water resource management* (pp. 1–4). IEEE.

Chunyang, j. H., Liu, Z., Wu, J., Pan, X., Fang, Z., Li, J., & Bryan, B. A. (2021). *Future global urban water scarcity and potential solutions. 12* (p. 4667) (1).

Cosgrove, W. J., & Loucks, D. P. (2015). Water management: Current and future challenges and research directions. *Water Resources Research, 51*(6), 4823–4839.

Dalin, C., Wada, Y., Kastner, T., & Puma, M. J. (2017). Groundwater depletion embedded in international food trade. *Nature, 543*(7647), 700–704.

Damtie, M. T., Tilahun, S. A., Zimale, F. A., & Schmitter, P. (2021). Prediction of irrigation water supply using supervised machine learning models in Koga irrigation scheme, Ethiopia. In *International conference on advances of science and technology* (pp. 68–81). Springer.

Das, R. (2024). Smart urban water management: Integrating ai and iot for optimization and waste reduction. *Optimality, 1*(2), 309–317.

Dinar, A., Tieu, A., & Huynh, H. (2019). *Water scarcity impacts on global food production. 23* (pp. 212–226).

Ding, X., & Du, W. (2024). Optimizing irrigation efficiency using deep reinforcement learning in the field. *ACM Transactions on Sensor Networks, 20*(4), 1–34.

Egbemhenghe, A. U., Ojeyemi, T., Iwuozor, K. O., Emenike, Chizitere, E., Ogunsanya, T. I., Anidiobi, Ukamaka, S., Adeniyi, & George, A. (2023). Revolutionizing water treatment, conservation, and management: Harnessing the power of AI-driven ChatGPT solutions. *Environmental Challenges, 13*, Article 100782.

El Bilali, A., & Taleb, A. (2020). Prediction of irrigation water quality parameters using machine learning models in a semi-arid environment. *Journal of the Saudi Society of Agricultural Sciences, 19*(7), 439–451.

Elhussiny, K. T., Hassan, A. M., Habssa, A. A., & Mokhtar, A. (2023). Prediction of water distribution uniformity of sprinkler irrigation system based on machine learning algorithms. *Scientific Reports, 13*(1), Article 20885.

Erickson, J. J., Quintero, Y. C., & Nelson, K. L. (2020). Characterizing supply variability and operational challenges in an intermittent water distribution network. *Water, 12*(8), 2143.

Fan, Q., Guo, Y., Wu, S., & Liu, X. (2019). Two-level diagnosis of heating pipe network leakage based on deep belief network. *IEEE Access, 7*, 182983–182992.

Forhad, H., Uddin, M. R., Chakrovorty, R., Ruhul, A., Faruk, H., Kamruzzaman, S., Sharmin, N., Jamal, A. S. I. M., Haque, M. M.-U., & Morshed, A. M. (2025). Iot based real-time water quality monitoring system in water treatment plants (wtps). *Heliyon, 10*(23).

Gaonkar, P., & Kande, M. (2014). Challenges and opportunities of automation system for water and waste water applications. In *2014 IEEE international conference on industrial technology (ICIT)* (pp. 682–688). IEEE.

Ghorbani Bam, P., Rezaei, N., Roubanis, A., Austin, D., Austin, E., Tarroja, B., Takacs, I., Villez, K., & Rosso, D. (2025). Digital twin applications in the water sector: A review. *Water, 17*(20). https://doi.org/10.3390/w17202957. URL https://www.mdpi.com/2073-4441/17/20/2957.

Giwa, A., Ademola, H., & Yusuf, A. O. (2025). Machine learning and application for modeling and prediction of desalination cost globally. *Desalination, 608*, Article 118829.

Godase, V. (2025). Iot-mcu integrated framework for field pond surveillance and water resource optimization. *International Journal of Emerging IoT Technologies in Smart Electronics and Communication, 1*(1), 9–19.

Gosling, S. N., & Arnell, N. W. (2016). A global assessment of the impact of climate change on water scarcity. *Climatic Change, 134*(3), 371–385.

Gurara, D., Klyuev, M. V., Mwase, M. N., Presbitero, A., Xu, Cindy, X., Bannister, & Mr Geoffrey, J. (2025). *Trends and challenges in infrastructure investment in lowincome developing countries.*

Hajgato, G., Paál, G., & Gyires-Tóth, B. (2020). Deep reinforcement learning for real-time optimization of pumps in water distribution systems. *Journal of Water Resources Planning and Management, 146*(11), Article 04020079.

Hamed, M. M., Khalafallah, M. G., & Hassanien, E. A. (2004). Prediction of wastewater treatment plant performance using artificial neural networks. *Environmental Modelling and Software, 19*(10), 919–928.

Hedley, C., & Yule, I. (2009). A method for spatial prediction of daily soil water status for precise irrigation scheduling. *Agricultural Water Management, 96*(12), 1737–1745.

Hernández-del Olmo, F., Gaudioso, E., Duro, N., Dormido, R., & Gorrotxategi, M. (2025). Advanced control by reinforcement learning for wastewater treatment plants: A comparison with traditional approaches. *Applied Sciences, 13*(8). URL https://www.mdpi.com/2076-3417/13/8/4752.

Hu, W., & Jiang, Y. (2025). Intelligent control and optimization of hydraulic systems using reinforcement learning. *Neural Computing & Applications*, 1–16.

Hu, P., Tong, J., Wang, J., Yang, Y., & de Oliveira Turci, L. (2019). A hybrid model based on cnn and bi-lstm for urban water demand prediction. In *2019 IEEE congress on evolutionary computation (CEC)* (pp. 1088–1094). IEEE.

Hu, C., Wang, Q., Gong, W., & Yan, X. (2022). Multi-objective deep reinforcement learning for emergency scheduling in a water distribution network. *Memetic Computing, 14*(2), 211–223.

Huang, P.-C., & Lee, K. T. (2023). An alternative for predicting real-time water levels of urban drainage systems. *Journal of Environmental Management, 347*, Article 119099.

Igwebuike, N., Ajayi, M., Okolie, C., Kanyerere, T., & Halihan, T. (2025). Application of machine learning and deep learning for predicting groundwater levels in the west coast aquifer system, South Africa. *Earth Science Informatics, 18*(1), 6.

Islam, F. S. (2025). A multi-dimensional ai framework for sustainable drinking water management: Integrating federated learning, digital twins, and blockchain. *Journal of Engineering Research and Reports, 27*(6), 466–492.

Ismail, M. I. M., Dziyauddin, Rudzidatul, A., Salleh, N. A. A., Muhammad-Sukki, Firdaus, Bani, N. A., Izhar, M. A. M., Latiff, & Abdul, L. (2019). A review of vibration detection methods using accelerometer sensors for water pipeline leakage. *IEEE Access, 7*, 51965–51981.

Júnior, A. C. D. S., Munoz, R., Quezada, M. De L.Á., Neto, A. V. L., Hassan, M. M., Albuquerque, De, & Victor, H. C. (2021). Internet of water things: A remote raw water monitoring and control system. *IEEE Access, 9*, 35790–35800.

Jenifel, M. G., & Linda, M. M. (2025). Privacy-preserving water quality forecasting using federated learning across distributed water monitoring nodes and optimized rpart modelling. *Environmental Science: Water Research & Technology, 11*, 3143–3160. https://doi.org/10.1039/D5EW00758E

Jian, L., Qing, L., & Wu, Z. (2011). Study on the development of information management system of supervision of urban wastewater treatment plants. In *2011 international conference on electric technology and civil engineering (ICETCE)* (pp. 6903–6906). IEEE.

Jin, L., Huang, H., & Ren, H. (2025). AI-driven transformation of water treatment technology and industry: Toward a new era of comprehensive innovation. *Frontiers of Environmental Science and Engineering, 19*(8), 114.

Jung, B., Ko, D., & Kim, S. (2025). Transient-aware multi-objective optimization of water distribution systems for cost and fire flow reliability. *Sustainability, 17*(18). URL https://www.mdpi.com/2071-1050/17/18/8274.

Kalmakhanova, M. S., Kurtebayeva, A. A., Tleuova, Z. T., Satybaldiev, B., Orynbayev, S. A., Malakar, A., Gomes, H. T., & Snow, D. D. (2025). Environmental impact of wastewater on surface and groundwater in central Asia. *Sustainability, 17*(12), 5370.

Kamau, S., & Waruguru, M. L. (2025). Predicting domestic water consumption in urban areas using smart water meter data. In *2025 IEEE 4th international conference on computing and machine intelligence (ICMI)* (pp. 1–6). IEEE.

Kang, H. (2019). Challenges for water infrastructure asset management in South Korea. *Water Policy, 21*(5), 934–944.

Kara, S., Karadirek, I. E., Muhammetoglu, A., & Muhammetoglu, H. (2016). Real time monitoring and control in water distribution systems for improving operational efficiency. *Desalination and Water Treatment, 57*(25), 11506–11519.

Kaviya, S., Mahadevan, G., & Sivagnanam, C. (2024). Computing the TCCD-number for the strong product of graphs and its application in urban water treatment: Addressing water scarcity through drainage purification. In *2024 first international conference for women in computing (InCoWoCo)* (pp. 1–6). IEEE.

Kayhomayoon, Z., Azar, N. A., Milan, S. G., Moghaddam, H. K., & Berndtsson, R. (2021). Novel approach for predicting groundwater storage loss using machine learning. *Journal of Environmental Management, 296*, Article 113237.

Kazemi Garajeh, M., Haji, F., Tohidfar, M., Sadeqi, A., Ahmadi, R., & Kariminejad, N. (2024). Spatiotemporal monitoring of climate change impacts on water resources using an integrated approach of remote sensing and Google Earth Engine. *Scientific Reports, 14*(1), 5469.

Kolluri, S. S., Esfahani, I. J., Garikiparthy, P. S. N., & Yoo, C. (2015). Evaluation of multivariate statistical analyses for monitoring and prediction of processes in an seawater reverse osmosis desalination plant. *Korean Journal of Chemical Engineering, 32*(8), 1486–1497.

Lambrou, T. P., Anastasiou, C. C., Panayiotou, C. G., & Polycarpou, M. M. (2014). A low-cost sensor network for real-time monitoring and contamination detection in drinking water distribution systems. *IEEE Sensors Journal, 14*(8), 2765–2772.

Li, D., & Fu, Q. (2023). Deep learning model-based demand forecasting for secondary water supply in residential communities: A case study of shanghai city, China. *IEEE Access, 12*, 38745–38757.

Li, Z., Geiger, W., & Jin, X. (2009). New concept for water management in urban areas of China. In *International conference on management and service science* (pp. 1–4). IEEE.

Li, J., Huang, B., Gui, H., Liu, X., & Xu, J. (2025). Flow prediction of water supply network based on time series large language model. In *2025 10th international conference on computer and communication system (ICCCS)* (pp. 83–87). IEEE.

Li, X., Wu, X., Sun, M., Yang, S., & Song, W. (2022). A novel intelligent leakage monitoring-warning system for sustainable rural drinking water supply. *Sustainability, 14*(10), 6079.

Li, D., Yang, H. Z., & Liang, X. F. (2013). Prediction analysis of a wastewater treatment system using a Bayesian network. *Environmental Modelling and Software, 40*, 140–150.

Lin, W., Hanyue, Y., & Bin, L. (2022). Prediction of wastewater treatment system based on deep learning. *Frontiers in Ecology and Evolution, 10*, Article 1064555.

Lisena, V., Paschero, M., Gentile, V., Amicucci, P., Rizzi, A., & Mascioli, F. F. (2016). A new method to restore the water quality level through the use of electric boats. In *2016 IEEE international smart cities conference (ISC2)* (pp. 1–4). IEEE.

Liu, J., Zhang, R., & Wang, L. (2010). Prediction of urban short-term water consumption in zhengzhou city. In *2010 international conference on intelligent computation technology and automation* (Vol. 2, pp. 922–926). IEEE.

Lowe, M., Qin, R., & Mao, X. (2022). A review on machine learning, artificial intelligence, and smart technology in water treatment and monitoring. *Water, 14*(9), 1384.

Lu, T., Jiang, T., & Liu, Y. (2025). Research on an optimization model for water resource management based on deep learning. *IEEE Access, 13*, 147906–147924. https://doi.org/10.1109/ACCESS.2025.3600664

Ly, Q. V., Truong, V. H., Ji, B., Nguyen, X. C., Cho, K. H., Ngo, H. H., & Zhang, Z. (2022). Exploring potential machine learning application based on big data for prediction of wastewater quality from different full-scale wastewater treatment plants. *Science of the Total Environment, 832*, Article 154930.

Ma, H., Wang, X., & Wang, D. (2024). Pump scheduling optimization in urban water supply stations: A physics-informed multiagent deep reinforcement learning approach. *International Journal of Energy Research, 2024*(1), Article 9557596.

Mahadeva, R., Kumar, M., Patole, S. P., & Manik, G. (2022). Desalination plant performance prediction model using grey wolf optimizer based ann approach. *IEEE Access, 10*, 34550–34561.

Mahmoud, M. B. R., Hemdan, D. I. I., Alajmani, S. H., Alyami, R. Y., Elmarhomy, G., Hashim, H., & Atlam, E. (2025). Optimized AI and IoT-driven framework for intelligent water resource management. *IEEE Access 13* (2025): 97628–97646. https://doi.org/10.1109/ACCESS.2025.3572067

Mahootchi, M., Tizhoosh, H. R., & Ponnambalam, K. P. (2025). Reservoir operation optimization by reinforcement learning. *Journal of Water Management Modeling, 15*. https://doi.org/10.14796/JWMM.R227-08

Mehta, S., & Choudhary, S. (2024). Optimizing water resources: An iot smart management system analysis. In *2024 7th international conference on circuit power and computing technologies (ICCPCT)* (Vol. 1, pp. 472–476). IEEE.

Mishra, B. K., Kumar, P., Saraswat, C., Chakraborty, S., & Gautam, A. (2021). Water security in a changing environment: Concept, challenges and solutions. *Water, 13*(4), 490.

Mjalli, F. S., Al-Asheh, S., & Alfadala, H. (2007). Use of artificial neural network black-box modeling for the prediction of wastewater treatment plants performance. *Journal of Environmental Management, 83*(3), 329–338.

Mohan, C., Western, A. W., Wei, Y., & Saft, M. (2018). Predicting groundwater recharge for varying land cover and climate conditions—A global meta-study. *Hydrology and Earth System Sciences, 22*(5), 2689–2703.

Morris, J. C. (2017). Planning for water infrastructure: Challenges and opportunities. *Public Works Management and Policy, 22*(1), 24–30.

Mullapudi, A., Lewis, M. J., Gruden, C. L., & Kerkez, B. (2020). Deep reinforcement learning for the real time control of stormwater systems. *Advances in Water Resources, 140*, Article 103600.

Musie, W., & Gonfa, G. (2025). Fresh water resource, scarcity, water salinity challenges and possible remedies: A review. *Heliyon, 9*(8).

Ni, J., Liu, M., Ren, L., & Yang, S. X. (2013). A multiagent q-learning-based optimal allocation approach for urban water resource management system. *IEEE Transactions on Automation Science and Engineering, 11*(1), 204–214.

Ogore, M. M., Nkurikiyeyezu, K., & Nsenga, J. (2021). Offline prediction of cholera in rural communal tap waters using edge ai inference. In *2021 IEEE globecom workshops (GC Wkshps)* (pp. 1–6). IEEE.

Ojadi, J. O., Owulade, Akindele, O., Odionu, Somtochukwu, C., Onukwulu, & Cynthia, E. (2025). AI driven optimization of water usage and waste management in smart cities for environmental sustainability. *Engineering and Technology Journal, 10*(3), 4284–4306.

Olatinwo, S. O., & Joubert, T. H. (2023). Resource allocation optimization in iot-enabled water quality monitoring systems. *Sensors, 23*(21), 8963.

Omeka, M. E. (2024). Evaluation and prediction of irrigation water quality of an agricultural district, se Nigeria: An integrated heuristic GIS-based and machine learning approach. *Environmental Science and Pollution Research, 31*(41), 54178–54203.

Onyutha, C. (2025). Challenges of modelling climate change impacts on hydrology and water resources: Ai is the game changer—A review. *Environmental Research: Climate, 5*(1), 012001. https://doi.org/10.1088/2752-5295/ae2a60

Pagano, T. C., Wood, A. W., Ramos, Helena, M.-, Cloke, H. L., Pappenberger, F., Clark, M. P., & Cranston. (2014). Michael and Kavetski, Dmitri and Mathevet, Thibault and Sorooshian, Soroosh and others, challenges of operational river forecasting. *Journal of Hydrometeorology, 15*(4), 1692–1707.

Palermo, S. A., Maiolo, M., Brusco, A. C., Turco, M., Pirouz, B., Greco, E., Spezzano, G., & Piro, P. (2022). Smart technologies for water resource management: An overview. *Sensors, 22*(16), 6225.

Pamidimukkala, A., Kermanshachi, S., Adepu, N., & Safapour, E. (2021). Resilience in water infrastructures: A review of challenges and adoption strategies. *Sustainability, 13*(23), Article 12986.

Paul, T., Raj, R., Garg, P., & Jain, S. (2023). Real time monitoring of water quality for rural areas: A machine learning and internet of things approach. In *2023 4th international conference on intelligent engineering and management (ICIEM)* (pp. 1–6). IEEE.

Pedraza, A., Riquelme, R., & Méndez, P. (2025). *Energy efficiency in water utilities: The case of Guyana.*

Perea, R. G., Ballesteros, R., Ortega, J. F., & Moreno, M.Á. (2021). Water and energy demand forecasting in large-scale water distribution networks for irrigation using open data and machine learning algorithms. *Computers and Electronics in Agriculture, 188*, Article 106327.

Petri, I., Yuce, B., Kwan, A., & Rezgui, Y. (2017). An intelligent analytics system for real-time catchment regulation and water management. *IEEE Transactions on Industrial Informatics, 14*(9), 3970–3981.

Pham, Q. B., Kumar, M., Di Nunno, F., Elbeltagi, A., Granata, F., Islam, A. R. M. T., Talukdar, S., Nguyen, X. C., Ahmed, A. N., & Anh, D. T. (2022). Groundwater level prediction using machine learning algorithms in a drought-prone area. *Neural Computing & Applications, 34*(13), 10751–10773.

Piadeh, F., Behzadian, K., Chen, A. S., Kapelan, Z., Rizzuto, J. P., & Campos, L. C. (2023). Enhancing urban flood forecasting in drainage systems using dynamic ensemble-based data mining. *Water Research, 247*, Article 120791.

Prajwal, N., Rohini, B., Kushal, B., Gargi, N., Kodipalli, A., & Soma, B. (2023). Efficient selection of optimal sewage treatment method for industries using fuzzy topsis and fuzzy ahp. In *IEEE 3rd mysore sub section international conference (MysuruCon)* (pp. 1–7). IEEE.

Premkumar, S., & Sigappi, A. (2022). Iot-enabled edge computing model for smart irrigation system. *Journal of Intelligent Systems, 31*(1), 632–650.

Pulido-Calvo, I., Roldán, J., López-Luque, R., & Gutiérrez-Estrada, J. (2003). Demand forecasting for irrigation water distribution systems. *Journal of Irrigation and Drainage Engineering, 129*(6), 422–431.

Qiao, J., Xiao, C., & Fu, A. (2025). An edge cloud collaborative leakage detection approach for urban water distribution systems. *IEEE Transactions on Instrumentation and Measurement, 74*, 1–15. https://doi.org/10.1109/TIM.2025.3613911

Rad, S. M., Ray, Ajay, K., & Barghi, S. (2022). Water pollution and agriculture pesticide. *Clean Technologies, 4*(4), 1088–1102.

Rahman, S. M. M., Ojo, O. O., & Shah, S. (2024). Adopting sustainable waterwaste management operations in the mining sector: An integrated research. In *2024 IEEE international conference on technology management, operations and decisions (ICTMOD)* (pp. 1–6). IEEE.

Rajasekaran, R. H., Selvaraj, M. S., & Ramanathan, J. (2018). A novel data-driven DSSE method to achieve water sustainability for farmers in Madurai district, India. In *2018 9th international conference on computing, communication and networking technologies (ICCCNT)* (pp. 1–7). IEEE.

Rajput, R. S., Pandey, S., & Bhadauria, S. (2017). Status of water pollution in relation to industrialization in Rajasthan. *Reviews on Environmental Health, 32*(3), 245–252.

Reynaert, E., Steiner, P., Yu, Q., D'Olif, L., Joller, N., Schneider, M. Y., & Morgenroth, E. (2023). Predicting microbial water quality in on-site water reuse systems with online sensors. *Water Research, 240*, Article 120075.

Roychoudhury, I., Hafiychuk, V., & Goebel, K. (2013). Model-based diagnosis and prognosis of a water recycling system. In *2013 IEEE aerospace conference* (pp. 1–9). IEEE.

Ruiters, C., & Amadi-Echendu, J. (2022). Economic costs and investment challenges of water infrastructure in South Africa. *Infrastructure Asset Management, 9*(4), 194–206.

Saddiqi, H. A., Javed, Z., Ali, Q. M., & Ullah, A. (2024). Optimization and predictive modeling of membrane based produced water treatment using machine learning models. *Chemical Engineering Research and Design, 207*, 65–76.

Salman, B., & Salem, O. (2012). Modeling failure of wastewater collection lines using various section-level regression models. *Journal of Infrastructure Systems, 18*(2), 146–154.

Saravanan, K., Anusuya, E., Kumar, R., & Son, L. H. (2018). Real-time water quality monitoring using internet of things in scada. *Environmental Monitoring and Assessment, 190*(9), 556.

Saravanan, S., Guntuku, N., Akshaya, C., Prakash, B., Kumar, Gireesh, D., Goud, & Dastagiri, P. C. (2023). A novel sensor based water quality monitoring system using internet of things (IoT). In *2023 4th international conference on electronics and sustainable communication systems (ICESC)* (pp. 355–360). IEEE.

Schlosser, C. A., Strzepek, K., Gao, X., Fant, C., Blanc, É., Paltsev, S., Jacoby, H., Reilly, J., & Gueneau, A. (2014). The future of global water stress: An integrated assessment. *Earth's Future, 2*(8), 341–361.

Shah, M. I., Alaloul, W. S., Alqahtani, A., Aldrees, A., Musarat, M. A., & Javed, M. F. (2021). Predictive modeling approach for surface water quality: Development and comparison of machine learning models. *Sustainability, 13*(14), 7515.

Shi, C., Li, N., Zhou, B., & Song, W. (2010). Designing of water pollution control technology database platform based on general management information system. In *2010 international conference on mechanic automation and control engineering* (pp. 2172–2175). IEEE.

Slater, L., Blougouras, G., Deng, L., Deng, Q., Ford, E., Hoek van Dijke, A., Huang, F., Jiang, S., Liu, Y., Moulds, S., et al. (2025). Challenges and opportunities of ml and explainable ai in large-sample hydrology. *Philosophical Transactions of the Royal Society A: Mathematical, Physical and Engineering Sciences, 383* (2302).

Smerdon, B., Allen, D., Grasby, S., & Berg, M. (2009). An approach for predicting groundwater recharge in mountainous watersheds. *Journal of Hydrology, 365*(3–4), 156–172.

Tahroudi, N. (2025). Mohammad, Comprehensive global assessment of precipitation trend and pattern variability considering their distribution dynamics. *Scientific Reports, 15*(1), Article 22458.

Tripathi, R., Pingale, S. M., & Khare, D. (2019). Assessment of LULC changes and urban water demand for sustainable water management: A case study of Dehradun city. In *2019 IEEE international conference on smart cities model (ICSCM)* (pp. 1–6). IEEE.

Vianny, D. M. M., John, A., Mohan, S. K., Sarlan, A., Ahmadian, A., et al. (2022). Water optimization technique for precision irrigation system using iot and machine learning. *Sustainable Energy Technologies and Assessments, 52*, Article 102307.

Vishwakarma, D. K., Ali, R., Bhat, S. A., Elbeltagi, A., Kushwaha, N. L., Kumar, R., Rajput, J., Heddam, S., & Kuriqi, A. (2022). Pre-and postdam river water temperature alteration prediction using advanced machine learning models. *Environmental Science and Pollution Research, 29*(55), 83321–83346.

Wang, Y.-Q., Wang, H.-C., Wang, W.-Z., Yang, H.-L., Chen, J.-J., Fan, Y.-X., Yin, W.-X., Lv, J.-Q., Luo, X.-Q., Zhou, X., & Wang, A.-J. (2025). Federated machine learning enables risk management and privacy protection in water quality. *Environmental Science and Technology, 59*(21), 10310–10322.

Wang, Z., Xu, J., Lu, F., & Zhang, Y. (2009). Using the method combining pca with bp neural network to predict water demand for urban development. In *2009 fifth international conference on natural computation* (Vol. 2, pp. 621–625). IEEE.

Wang, L., Xu, B., Zhang, C., Fu, G., Chen, X., Zheng, Y., & Zhang, J. (2022). Surface water temperature prediction in large-deep reservoirs using a long short-term memory model. *Ecological Indicators, 134*, Article 108491.

Ward, F. A. (2010). Financing irrigation water management and infrastructure: A review. *International Journal of Water Resources Development, 26*(3), 321–349.

Wu, H., Lei, Z., Shi, Y., & Li, X. (2019). Water level prediction of community secondary water supply tank based on deep learning. In *IEEE 4th advanced information technology, electronic and automation control conference (IAEAC)* (Vol. 1, pp. 1336–1340). IEEE.

Wu, D., Wang, H., Mohammed, H., & Seidu, R. (2019). Quality risk analysis for sustainable smart water supply using data perception. *IEEE Transactions on Sustainable Computing, 5*(3), 377–388.

Yang, Y., Hu, J., Porter, D., Marek, T., Heflin, K., & Kong, H. (2020). Deep reinforcement learning-based irrigation scheduling. *Transactions of the ASABE, 63*(3), 549–556.

Yao, Y., Zhao, Y., Li, X., Feng, D., Shen, C., Liu, C., Kuang, X., & Zheng, C. (2023). Can transfer learning improve hydrological predictions in the alpine regions? *Journal of Hydrology, 625*, Article 130038.

Yaqub, M., Ngoc, N. M., Park, S., & Lee, W. (2022). Predictive modeling of pharmaceutical product removal by a managed aquifer recharge system: Comparison and optimization of models using ensemble learners. *Journal of Environmental Management, 324*, Article 116345.

Younis, R., & Knight, M. A. (2008). An empirical model for the prediction of structural behavior of wastewater collection systems. In *Pipelines 2008: Pipeline asset management: Maximizing performance of our pipeline infrastructure* (pp. 1–12).

Zaman, M., Tantawy, A., & Abdelwahed, S. (2023). Optimizing smart city water distribution systems using deep reinforcement learning. In *2023 IEEE 20th international conference on smart communities: Improving quality of life using AI, robotics and IoT (HONET)* (pp. 228–233). IEEE.

Zhang, L., Wang, C., Hu, W., Wang, X., Wang, H., Sun, X., Ren, W., & Feng, Y. (2024). Dynamic real-time forecasting technique for reclaimed water volumes in urban river environmental management. *Environmental Research, 248*, Article 118267.

Zhang, M., Xu, Z., Wang, Y., Zeng, S., & Dong, X. (2022). Evaluation of uncertain signals' impact on deep reinforcement learning-based real-time control strategy of urban drainage systems. *Journal of Environmental Management, 324*, Article 116448.

Zhao, Y., Xu, Y., Ye, J., Zhang, X., & Long, Z. (2023). Urban water supply forecasting based on CNN-LSTM-AM spatiotemporal deep learning model. *IEEE Access, 11*, 144204–144212.

Zhou, P., Li, Z., Snowling, S., Baetz, B. W., Na, D., & Boyd, G. (2019). A random forest model for inflow prediction at wastewater treatment plants. *Stochastic Environmental Research and Risk Assessment, 33*(10), 1781–1792.

Zhou, C., Zhang, Y., Kong, B., & Li, M. (2025). Short-term urban water demand forecasting model based on noise reduction processing. In *2025 7th international conference on internet of things, automation and artificial intelligence (IoTAAI)* (pp. 865–868). https://doi.org/10.1109/IoTAAI66837.2025.11213310

PART

VI

Contaminants of emerging concern and health risk management

CHAPTER

Treatment and risk mitigation of emerging water contaminants including PFAS, microplastics, and pharmaceuticals

12

Anindita Ganguly[a,1], Deblina Roy[a,2] and Siddhant Dash[3]

[1]*Department of Civil, Geodetic and Environmental Engineering, Division of Environmental Health Sciences, and Sustainability Institute, The Ohio State University, Columbus, OH, United States;* [2]*Department of Environmental Science and Engineering, School of Engineering and Sciences, SRM University–AP, Amaravati, Andhra Pradesh, India;* [3]*Department of Civil Engineering, School of Engineering and Sciences, SRM University-AP, Guntur, Andhra Pradesh, India*

1. Introduction

Emerging water contaminants (EWCs), also known as contaminants of emerging concern (CECs), are chemicals and pollutants that are not commonly regulated or routinely monitored in water systems but have been identified as presenting possible risks due to their environmental prevalence and toxicological impact. These substances may include synthetic chemicals such as pharmaceuticals, industrial byproducts, pesticides, personal care products, plasticizers, and nanomaterials (Kumar et al., 2022; Nishmitha et al., 2025). EWCs are distinct from traditional pollutants because many remain undetected or unaddressed in existing water treatment processes, and their environmental toxicology is often not fully understood or assessed. Many EWCs are anthropogenic and have only recently been recognized due to advances in detection methods and increased environmental monitoring (Essamuah-quansah, 2017, pp. 22–44). EWCs represent a growing challenge to the quality and safety of aquatic environments across the world. The scope of EWCs extends beyond municipal drinking water; these substances are present in surface waters, groundwater, wastewater, and sometimes even in agricultural runoff or urban stormwater. EWCs such as per- and polyfluoroalkyl substances (PFAS), microplastics, and pharmaceuticals have captured the attention of researchers, environmental agencies, and policymakers due to their persistence, complex behavior in water, and the potential consequences for both human health and ecosystems (Romano et al., 2025; Lee et al., 2025).

PFAS are synthetic chemicals used widely for their resistance to heat, water, and oil. They are found in products such as firefighting foams, non-stick cookware, water-resistant textiles, and food packaging. PFAS are known to be persistent ("forever chemicals") because they do not degrade easily in the environment (Crone et al., 2019; Perera et al., 2024). This stability allows them to travel and

[a]Both authors have contributed equally.

Sustainable Water Management. https://doi.org/10.1016/B978-0-443-44853-9.00007-1

accumulate in surface and groundwater, posing risks of bioaccumulation in aquatic organisms and eventual transfer through the food web. Microplastics are small plastic particles typically less than 5 mm in diameter, resulting from the degradation of larger plastic debris or manufactured as microbeads for use in products such as exfoliants, detergents, and toothpaste (Kumar, 2025; Yee et al., 2021). Microplastics are now found in nearly all aquatic settings, from urban rivers to remote oceanic locations, and are known not only as pollutants themselves but also as carriers for other contaminants, including PFAS and pharmaceutical residues. They can enter aquatic environments directly or indirectly, and their effects include physical harm to aquatic wildlife and potential vectors for chemical toxicity. Pharmaceuticals enter water systems from sources such as household waste, hospital effluent, agricultural runoff, and improper disposal of unused medications. Common examples detected in water include antibiotics, analgesics, anti-inflammatory drugs, and hormones. Pharmaceuticals are problematic because they are biologically active and can impact non-target organisms in aquatic ecosystems by influencing growth, reproduction, or microbial community structure. Many water treatment facilities are not designed to fully remove pharmaceutical residues, leading to their persistence in both surface and groundwater (Méndez et al., 2017; Ortúzar et al., 2022).

EWCs enter aquatic systems through multiple sources and pathways, often intersecting across human, industrial, and environmental activities (Fig. 12.1). Domestic wastewater can carry pharmaceuticals, personal care products, microplastics, and chemical residues, especially in areas with outdated or inadequate sewage treatment (Aib & Parvez, 2025). Industrial discharges add further loads of PFAS, solvents, and plasticizers that may be directly released into rivers and lakes. Agricultural practices contribute through runoff containing fertilizers, pesticides, animal pharmaceuticals, and veterinary medicines, which are particularly mobilized during rainfall or irrigation. Waste management sites and landfills also pose risks, as chemicals from solid waste, including legacy contaminants and poorly regulated materials, can leach into groundwater or surface water through percolation and runoff (Peter & Lee, 2025). In addition, atmospheric deposition plays a significant role, as certain EWCs volatilize into the air and return with precipitation, enabling long-range transport and contamination even in remote or pristine regions. Microplastics, for example, may sink and settle in aquatic sediments, while pharmaceuticals or PFAS often dissolve and travel downstream, potentially entering drinking water supplies (Wang & Good, 2024).

Addressing EWCs is critical for safeguarding human health and preserving the functioning of ecosystems. Many EWCs present known or suspected risks, such as endocrine disruption, reproductive toxicity, neurological effects, or carcinogenicity in humans. For example, PFAS exposure has been linked to immune system, hormones and reproductive disorders, and increased cancer risk (Factors et al., 2024, pp. 1–32; Li et al., 2025). Pharmaceuticals can disturb microbial and hormonal balance, while microplastics may physically harm the digestive tracts or gills of aquatic organisms and act as vectors for chemical pollutants (Air et al., 2023).

Ecosystems face structural and functional damage from EWCs. Toxic chemical emissions can cause species loss, degrade functional diversity, and disrupt ecosystem services such as water purification, food provision, and habitat maintenance. Aquatic organisms may suffer from bioaccumulation of toxic substances, altered reproductive success, or impaired growth, while contaminants may move through food webs to impact higher order species and humans (Adam et al., 2024; Amoatey & Baawain, 2019; Zhang et al., 2022). Some populations, such as children, pregnant women, and communities dependent on local water sources, are especially vulnerable to EWCs due to underlying health conditions, high exposure rates, or limited access to clean water (Bloomquist et al.,

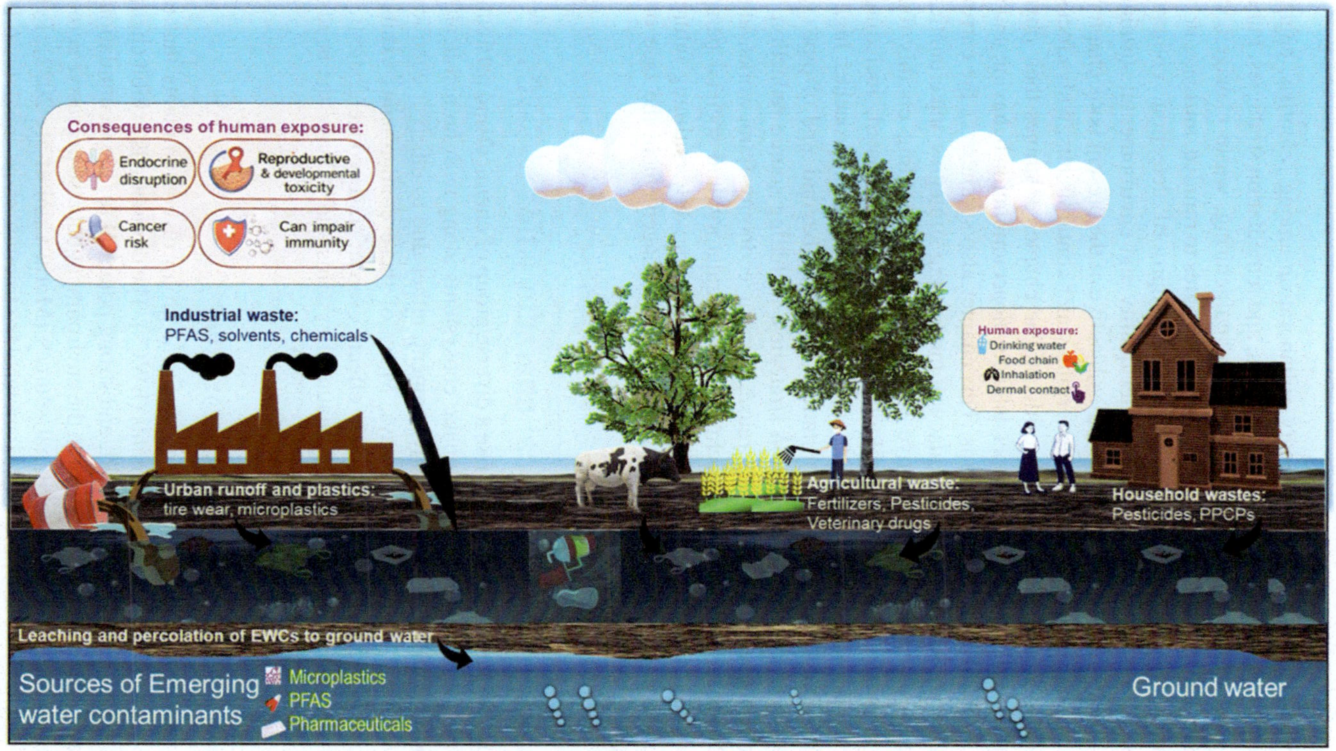

FIGURE 12.1

Schematic representation of the sources, environmental pathways, and health impacts of emerging water contaminants (EWCs).

2024). The risk is exacerbated in regions with limited water treatment infrastructure and greater industrial or agricultural activity.

Addressing EWCs requires cross-disciplinary collaboration among scientists, engineers, policymakers, and the public to advance detection, regulations, and remediation strategies. Public awareness and advocacy are vital for driving improvements to manufacturing, waste management, and water treatment, and for securing the development of new technologies to remove, mitigate, or monitor these pervasive contaminants. Future efforts must focus on improved detection, comprehensive risk assessment, and innovative solutions to ensure water quality and environmental sustainability in the face of these evolving challenges. In this context, the objective of the present book chapter is to provide a comprehensive and integrated assessment of emerging water contaminants, with a focused emphasis on PFAS, microplastics, and pharmaceutical residues. The chapter examines their physicochemical and biological characteristics, environmental persistence, bioaccumulation behavior, and toxicological implications across aquatic and human systems. This chapter also synthesizes the current challenges related to analytical detection limits, the adequacy of monitoring strategies, uncertainties in exposure pathways, and the complexities of cumulative risk assessment. Building on this foundation, the conventional and advanced treatment technologies are critically evaluated, including adsorption, membrane-based processes, advanced oxidation, and hybrid biological approaches, highlighting both their removal efficiencies and practical limitations. Moving beyond end-of-pipe solutions, the manuscript addresses integrated risk mitigation strategies, including source reduction, green chemistry, plastic waste management, wastewater reuse, circular economy frameworks, and public risk communication. By integrating scientific, technological, regulatory, and societal perspectives, this chapter aims to inform researchers, engineers, and policymakers while outlining future research directions needed to support sustainable and resilient water management in the face of increasingly complex contaminant challenges.

2. Physicochemical and biological characteristics of EWCs

Emerging water contaminants (EWCs) comprise a diverse group of chemical and particulate pollutants which are recognized as critical threats to aquatic ecosystems along with human health. A defining feature of EWCs is their persistence in the environment, either due to inherent chemical stability or resistance to conventional degradation pathways. Many of these contaminants exhibit strong tendencies for bioaccumulation and can exert toxic effects even at trace concentrations. Collectively, these properties challenge conventional water treatment systems and indicate the need for advanced approaches to risk mitigation (Morin-Crini et al., 2022).

2.1 Persistence, bioaccumulation, and toxicity

Persistent organic pollutants (POPs) represent a critical group of contaminants that exemplify the challenges posed by emerging water contaminants (EWCs). The compounds exhibit chemical stability, resistance to degradation, long-range transport, and strong potential for bioaccumulation and toxicity. Such properties make them not only persistent in the environment but also pervasive across aquatic and terrestrial ecosystems, far beyond their original sources (Morin-Crini et al., 2022). Persistence is a defining characteristic of POPs. The complex molecular structure resist breakdown by photolysis, hydrolysis, or microbial action, allowing them to remain in soils, sediments, and aquatic systems for decades. This persistence facilitates global transport, leading to detection of POPs in remote environments such as the Arctic, high-altitude lakes, and deep-sea sediments (Pan et al., 2020).

Bioaccumulation is closely tied to the lipophilic nature and low water solubility of POPs. Once released, these contaminants preferentially partition into fatty tissues of living organisms. Over time, they bio magnify through food chains, reaching higher trophic levels at concentrations many times greater than in the surrounding environment. Such bioaccumulation has been observed in fisheries of the North Pacific, in large freshwater systems such as Lake Superior and Lake Tanganyika, and in regional ecosystems like the Caspian Sea (Datta et al., 2018). Toxicity is another dimension of concern. Chronic exposure to POPs has been linked to cancers, congenital disorders, endocrine disruption, reproductive abnormalities, immune suppression, and neurological damage in humans and wildlife. Well-known examples include organochlorine pesticides such as DDT, industrial chemicals like polychlorinated biphenyls (PCBs), and polycyclic aromatic hydrocarbons (PAHs). More recently, brominated flame retardants and per- and polyfluoroalkyl substances (PFAS) have gained attention as "emerging POPs" due to their similar persistence and health effects. Studies in China and other regions have shown that POP burdens in aquatic ecosystems not only endanger biodiversity but also threaten human populations through dietary exposure (Ding et al., 2022). Their environmental half-lives, ability to travel across regions, and potential for long-term health risks underscore the urgency of developing advanced detection, monitoring, and treatment strategies (Morin-Crini et al., 2022).

2.2 PFAS: Chemical stability, hydrophobic/hydrophilic interactions

Per- and polyfluoroalkyl substances (PFASs) are a large class of synthetic organofluorine compounds characterized by strong carbon–carbon (C–C) and carbon–fluorine (C–F) bonds. The exceptional bond strength of C–F (>110 kcal/mol) confers extreme resistance to chemical, photolytic, and biological degradation, making these compounds highly persistent in the environment. Because of this stability, PFASs are widely referred to as "forever chemicals" (Saifur & Gardner, 2021). A defining feature of PFASs is their amphiphilic nature. They contain a hydrophobic fluorinated carbon chain that repels oil and water, coupled with a polar, hydrophilic functional head group. This duality creates complex hydrophobic/hydrophilic interactions, which make modeling their environmental transport and behavior challenging. For instance, long-chain PFASs with higher hydrophobicity (e.g., PFOS, PFOA) exhibit a strong tendency to adsorb onto sediments and particulate matter, whereas short-chain analogs such as PFHxS remain largely dissolved in water due to their greater polarity and solubility. At the same time, chain length and functional groups are key determinants of environmental partitioning and mobility (Kim & Kannan, 2007). Since their introduction in the 1940s, PFASs have been used in diverse applications, including nonstick cookware, firefighting foams, stain-resistant textiles, food packaging, and industrial coatings, owing to their chemical stability and surfactant properties. However, widespread use and poor degradation have resulted in extensive environmental release. PFASs and their precursors have been detected in urban runoff, groundwater, wastewater effluents, and biota across the globe (Nguyen et al., 2011). Studies from San Francisco, Minneapolis–St. Paul, and Tokyo have reported PFOS and PFOA concentrations in stormwater ranging from a few nanograms per liter to several tens of nanograms per liter, with even higher levels observed after precursor oxidation. These findings indicate that actual PFAS burdens may be substantially underestimated when precursors are not accounted for (Post, 2009).

Stormwater has been identified as a major transport pathway for PFASs into aquatic environments. In many urban areas, first-flush runoff events mobilize high concentrations of long-chain PFASs associated with traffic-related particulate matter, while atmospheric deposition contributes to contamination in residential regions. Point sources such as wastewater treatment plants, industrial discharges, landfill leachates, and firefighting foam residues at airports and military bases further amplify their presence. Transport studies have demonstrated that PFASs occur both in dissolved and particle-bound states, enabling long-range mobility across landscapes and into groundwater and surface waters, even at sites far removed from primary sources (Dong et al., 2020). The persistence and amphiphilic nature of PFASs also translate into significant biological implications. Unlike many hydrophobic pollutants that accumulate in fatty tissues, PFASs preferentially bind to proteins in blood, liver, and kidneys. Internal exposure has been linked to a range of adverse health outcomes, including reduced fertility, endocrine disruption, thyroid disease, immune suppression, developmental effects, and elevated risks of certain cancers. The combination of chemical stability, amphiphilic transport behavior, and bioactive interactions underscores the complexity of managing PFAS pollution in aquatic environments and highlights the urgent need for improved treatment and mitigation strategies (Lu et al., 2018).

2.3 Microplastics: Size, shape, polymer types, and surface adsorption properties

Microplastics represent a heterogeneous group of contaminants that vary widely in their physicochemical characteristics, which in turn govern their environmental fate, transport, and toxicity. Among the diverse types of microplastics, tire wear particles (TWPs) have emerged as a significant and often overlooked source. TWPs are generated from the mechanical abrasion between tires and road surfaces and are typically classified as microplastics due to their small size (<5 mm) (Peter et al., 2018). From a size perspective, TWPs exist in a broad distribution. Fine particles (<10 μm) are released into the atmosphere during tire abrasion, while larger particles (>10 μm) tend to accumulate on road surfaces. These deposited particles are subsequently mobilized by stormwater running off into soil and aquatic systems. The irregular shape of TWPs, often angular or fragmented, reflects their origin from frictional processes rather than controlled manufacturing. This morphology enhances their surface reactivity and facilitates interactions with co-contaminants (Wagner et al., 2018).

The polymeric composition of TWPs is chemically complex. In addition to rubber polymers, TWPs incorporate a wide array of additives such as resins, reinforcing agents, process oils, vulcanization agents, and antioxidants. Compounds such as hexamethoxymethyl melamine (HMMM), 1,3-diphenylguanidine (DPG), and 6PPD and its transformation product 6PPD-quinone exemplify the chemical heterogeneity present in these particles. This complexity distinguishes TWPs from other microplastics like polyethylene or polystyrene, as they represent composite materials combining both polymeric and non-polymeric fractions.

Importantly, TWPs exhibit high surface adsorption capacity (Paterson et al., 2024). The irregular morphology, roughened surfaces, and chemical functional groups facilitate the sorption of a range of environmental contaminants, including heavy metals, hydrocarbons, and road dust. This property allows TWPs to act as carriers of toxic co-contaminants, amplifying their environmental relevance beyond their intrinsic toxicity (Lee et al., 2020). Biologically, these physicochemical traits are small size, complex composition, and surface activity enhance the potential for TWPs to interact with organisms through ingestion, inhalation, or trophic transfer. The heterogeneity of TWPs also complicates predictions of bioavailability and toxicity, highlighting the need for targeted studies on their environmental and health impacts (Saifur & Gardner, 2021).

2.4 Pharmaceuticals: Solubility, polarity, and transformation by product

Pharmaceuticals and personal care products (PPCPs) constitute one of the most significant classes of emerging wastewater contaminants. Their environmental relevance is primarily linked to their wide range of physicochemical properties, particularly solubility and polarity, which strongly influence persistence, transport, and toxicity. Over the past several decades, the global per capita usage of PPCPs has increased substantially, leading to continuous discharge of active and inactive compounds into aquatic systems via stormwater, wastewater effluents, combined sewer overflows, and landfill leachates. Conventional wastewater treatment plants (WTPs) are not fully effective in eliminating these substances, resulting in their widespread occurrence in receiving water bodies, sediments, and even groundwater (Hopkins & Blaney, 2016).

Pharmaceuticals are generally more hydrophilic and less volatile than many other classes of contaminants, making long-range transport within aquatic environments highly feasible. Hydrophilic drugs such as metformin are frequently detected in surface waters, while more hydrophobic PPCPs, such as musk fragrances and steroidal hormones, exhibit strong sorption to organic matter, leading to bioaccumulation in aquatic and terrestrial organisms. The lipophilicity of certain PPCPs, reflected in their log Kow values (ranging from moderate to high), significantly contributes to their persistence and trophic transfer. For instance, nitro musks and polycyclic musks exhibit Kow values between 4.0 and 5.9, indicating considerable potential for bioaccumulation in organisms higher in the food chain (de Solla et al., 2016).

The transformation of pharmaceuticals in the environment adds further complexity to their risk profile. Many compounds undergo photolysis, hydrolysis, or microbial degradation, yielding transformation products that can retain or even exceed the toxicity of the parent molecules. Estrogenic compounds, including natural hormones (estrone, estradiol) and synthetic estrogens (such as 17α-ethinylestradiol), are frequently reported above threshold levels for endocrine disruption. These compounds interfere with reproductive physiology in aquatic organisms, leading to feminization, hermaphroditism, and decreased fertility, often at sub-lethal concentrations. Similarly, antidepressants such as selective serotonin reuptake inhibitors (SSRIs) have been shown to bioaccumulate significantly in aquatic biota, with sertraline exhibiting bioaccumulation factors as high as 32,000 and reproductive toxicity at very low concentrations (Niemuth et al., 2015). Another pressing concern is the widespread detection of antibiotics in stormwater, wastewater, and agricultural runoff. Their persistence in the environment not only exposes aquatic ecosystems to continuous chemical stress but also promotes the development of antimicrobial resistance. The increasing occurrence of antibiotic residues in stormwater highlights the interplay between non-point agricultural sources and urban WTP discharges (Fabio Kaczala; Shlomo E. Blum, 2016).

So, the environmental significance of pharmaceuticals lies in their dual nature: their solubility and polarity allow for long-range dispersal, while their transformation products and lipophilicity increase their persistence and toxicity. Their continuous release into aquatic systems, coupled with bioaccumulation and potential endocrine-disrupting effects, poses a growing challenge for both ecosystem and human health. Understanding these characteristics is essential for the development of stormwater and wastewater treatment strategies aimed at mitigating PPCP pollution (Saifur & Gardner, 2021) (Table 12.1).

3. Current challenges in monitoring and risk assessment

3.1 Detection limits and analytical techniques

Detection limits form the cornerstone of analytical monitoring in water quality assessment. They determine the minimum concentration at which a contaminant can be confidently detected and quantified, directly influencing regulatory thresholds such as reporting limits, monitoring trigger levels, and maximum contaminant levels (MCLs). In drinking water surveillance, the detection limit serves as the foundation upon which public health standards and compliance criteria are built (Draper et al., 1998). The accurate detection of water contaminants at trace concentrations requires highly sensitive analytical techniques supported by effective sample preparation strategies. Among

Table 12.1 Comparative physicochemical and biological characteristics of PFAS, microplastics, and pharmaceuticals (Morin-Crini et al., 2022; Saifur & Gardner, 2021).

Contaminants	Key properties	Persistence	Bioaccumulation	Toxicity
PFAS	Strong C–F bonds; amphiphilic with hydrophobic tail and hydrophilic head; highly stable	Extremely persistent ("forever chemicals"); resistant to hydrolysis, photolysis, and biodegradation	Accumulate in blood serum, liver, kidneys; bio magnify in food webs	Endocrine disruption, immunotoxicity, developmental toxicity, carcinogenicity, thyroid and metabolic effects
Microplastics	Small size (<5 mm); varied shapes (fibers, beads, fragments, films); polymer diversity (PE, PP, PS, PVC, PET); high adsorption capacity for pollutants	Resistance to natural degradation; fragment into nanoplastics, increasing mobility and environmental half-life	Ingested by aquatic and terrestrial organisms; accumulated in GI tracts and tissues; trophic transfer across food webs	Physical blockage, inflammation, oxidative stress, altered feeding, reproductive impairment; act as vectors for POPs, metals, pathogens
Pharmaceuticals	Solubility and polarity vary; often hydrophilic; undergo transformation into active/inactive metabolites and by-products	Moderate to high persistence depending on compound; incomplete removal in wastewater treatment; frequent detection in surface and groundwater	Some compounds bioaccumulate in sediments and biota; lipophilic drugs and transformation products may persist longer	Endocrine disruption, reproductive toxicity, antimicrobial resistance, behavioral/ neurological effects in aquatic organisms

conventional sample preparation approaches, liquid-liquid extraction (LLE) and solid-phase extraction (SPE) are the most commonly applied. LLE relies on analyte partitioning between immiscible liquid phases and is adaptable to a wide range of compounds; however, its high solvent consumption limits its applicability. In contrast, SPE provides improved selectivity, reduced solvent usage, and enhanced compatibility with chromatographic analysis (HPLC, GC, TLC) (Kadadou et al., 2024).

Recent advancements in SPE include the use of discs and novel sorbent materials, which offer improved mass transfer, reduced processing times, and higher extraction efficiencies for large-volume environmental samples. Carbon-based nanomaterials, particularly carbon nanotubes, have gained attention due to their high surface area and strong adsorption capacity, enabling enhanced recovery and lower detection limits. In parallel, miniaturized extraction techniques such as solid-phase microextraction (SPME), dispersive solid-phase extraction (dSPE), magnetic solid-phase extraction (MSPE), and dispersive liquid-liquid microextraction (DLLME) have been increasingly adopted to minimize solvent use while maintaining high analytical sensitivity (Nunes et al., 2012).

Following these extraction and pre-concentration steps, chromatographic techniques serve as the primary analytical tools for the separation and quantification of organic contaminants. Chromatographic separation coupled with advanced detection systems plays a central role in achieving low detection limits. Gas chromatography (GC) and liquid chromatography (LC) remain the primary separation techniques for organic contaminants. While GC offers excellent separation efficiency, its application is limited for polar and thermally unstable compounds. Consequently, LC-based techniques, particularly high-performance liquid chromatography (HPLC) and ultra-high-performance liquid chromatography (UHPLC/UPLC), have become dominant. The use of smaller particle sizes and higher operating pressures in UHPLC results in narrower peaks, enhanced resolution, and improved sensitivity (Mijangos et al., 2018, pp. 615–632).

The coupling of chromatography with mass spectrometry (MS) has significantly advanced contaminant detection. LC-MS and LC-MS/MS are now widely regarded as the most powerful tools for the analysis of persistent organic pollutants and emerging contaminants, offering high selectivity and low method detection limits. Tandem mass spectrometry provides superior discrimination against matrix interferences compared to single-stage MS, although at increased cost and instrumental complexity. Similarly, GC-MS and GC-MS/MS remain important for the analysis of volatile and semi-volatile compounds, with recent developments in ionization techniques further improving detection performance (Tomai et al., 2018).

For inorganic contaminants, analytical techniques such as atomic absorption spectrometry (AAS), atomic emission spectrometry (AES), UV-visible spectroscopy, fluorimetry, and ion chromatography (IC) are widely employed. These methods provide reliable quantification, low detection limits, and effective reduction of matrix interferences, particularly for ionic and metallic species. The integration of advanced extraction techniques with high-resolution chromatographic and mass spectrometric systems remains central to the effective monitoring of trace contaminants in water (Vega et al., 2005) (Table 12.2).

3.2 Uncertainties in exposure pathways

Understanding exposure pathways is essential for evaluating the risks associated with environmental contaminants. Contaminants rarely remain confined to a single compartment; instead, they are continuously transported, transformed, and redistributed through physical, chemical, and biological

Table 12.2 Analytical techniques for detection of emerging water contaminants (Zhang et al., 2020; Ishibashi et al., 2015).

Analytical method	Target contaminants	Typical detection range	Strengths	Limitations	Typical applications
LC–MS/MS	Pharmaceuticals, PFAS, pesticides, endocrine disruptors	ng/L–pg/L	High selectivity, sensitive	Matrix effects, complex prep, costly	Drinking water, wastewater, surface water
GC–MS	Volatile and semi-volatile organics (e.g., benzene, hydrocarbons, pesticides)	ng/L	Robust, precise	Derivatization needed, thermal degradation	Drinking water, sediments, industrial effluents
FTIR/Raman	Microplastics, dyes, polymer types	μg/g	Rapid, non-destructive	Low sensitivity	Water, sediment, microplastic characterization
UV-Vis spectroscopy	Organic dyes, colored contaminants	mg/L	Quick, simple	Low sensitivity, not trace-level	Screening, rapid assessment

processes. These cross-media interactions such as volatilization, runoff, deposition, sorption, and sedimentation create overlapping exposure routes that are difficult to trace with precision. Consequently, uncertainty persists in determining how, where, and to what extent humans and ecological receptors are exposed to pollutants (Linkov et al., 2001). Emerging contaminants (ECs) pose significant challenges to environmental quality and planetary health due to their persistence, mobility, bioaccumulative nature, and biological activity. Once released into the environment, ECs do not remain confined to a single compartment; instead, they circulate through interconnected systems involving soil, water, sediments, air, and biological organisms. This environmental connectivity complicates monitoring efforts and risk assessment, as contaminants can migrate across media and magnify impacts through ecological and food-web interactions (Wang et al., 2024).

In terrestrial environments, ECs such as pharmaceuticals, pesticides, PFAS, engineered nanomaterials, and microplastics are increasingly detected in soils. These substances can disrupt soil microbial communities, alter nutrient cycling, and impair soil fertility. Plants growing in contaminated soils may accumulate ECs in their tissues, enabling their transfer to herbivores and higher trophic levels. Soil fauna further contributes to the redistribution of contaminants through bioaccumulation and migration, strengthening links between soil contamination and food-chain exposure (Ferrão-filho & Kozlowsky-suzuki, 2011).

Aquatic systems represent major sinks and transmission pathways for ECs. Compounds including endocrine-disrupting chemicals, persistent organic pollutants, pharmaceuticals, cyanotoxins, nanomaterials, and microplastics have been widely reported in surface waters, groundwater, and sediments. Sediments act as long-term reservoirs for many ECs, sustaining chronic exposure to benthic organisms and enabling remobilization under changing environmental conditions. Bioaccumulation in aquatic organisms facilitates contaminant transfer through aquatic food webs, ultimately increasing risks to wildlife and human populations reliant on aquatic resources for food and drinking water (Carter et al., 2024). Microplastics and nanomaterials further exacerbate these risks by serving as carriers for metals, hydrophobic organic compounds, and pathogenic microorganisms. Their ingestion across trophic levels enhances contaminant mobility and exposure complexity. Similarly, biological contaminants, including antibiotic-resistant bacteria, viruses, and toxin-producing microorganisms, alter microbial communities in water and sediments, threatening ecosystem stability and public health (Mbachu et al., 2020). Atmospheric transport adds another layer of complexity, enabling long-range dispersion of ECs and bioaerosols. Airborne contaminants can deposit onto land and water surfaces, reinforcing cross-media contamination cycles. Human exposure occurs through multiple pathways, including ingestion of contaminated food and water, inhalation of polluted air, and dermal contact, complicating traditional exposure assessment frameworks (Abdallah & Harrad, 2022) (Table 12.3).

3.3 Toxicological gaps and cumulative exposure effects

Emerging water contaminants, including pharmaceuticals, PFAS, and microplastics, present complex challenges for both environmental monitoring and risk mitigation. Many of these substances persist in the aquatic environment due to incomplete removal in conventional wastewater treatment plants (WTPs), leading to continuous exposure of aquatic organisms to complex chemical mixtures (Richardson & Kimura, 2017). Unlike legacy pollutants, these contaminants often exhibit bioactivity by design, and their interactions in mixtures can result in additive or even synergistic toxic effects that

Table 12.3 Major exposure pathways, contaminants, and associated uncertainties in water systems (Hamburger et al., 2020; Levin et al., 2023).

Exposure pathway	Main contaminants	Sources	Key uncertainties	Human/ Ecological risk implications	Notes
Drinking water	Pharmaceuticals, PFAS, microplastics, metabolites	Wastewater, industrial discharges, runoff	Seasonal fluctuations, transformation products, treatment efficiency	Chronic low-dose exposure, endocrine disruption	Trace-level detection challenges
Food chain	POPs, PFAS, heavy metals, microplastics, endocrine disruptors	Water, soil, aquatic organisms	Bioaccumulation, biomagnification, dietary variation	Trophic transfer, human dietary intake risks	Species-specific differences
Sediments	Hydrophobic organics, metals, microplastics	Adsorption from water, deposition	Bioavailability, resuspension, site-specific factors	Secondary contamination source, benthic organism exposure	Long-term sink versus secondary source

are not predicted by single-compound studies. For instance, co-exposure to multiple pharmaceuticals has been shown to trigger stronger physiological responses in fish than individual compounds, emphasizing the need to consider cumulative exposure effects in risk assessment frameworks (Nilsen et al., 2019).

Despite increasing research, significant toxicological knowledge gaps remain. Data on the bioaccumulation, trophic transfer, and ecological consequences of mixtures are limited, particularly for environmentally relevant concentrations. While studies indicate that pharmaceuticals and PFAS may accumulate in higher trophic-level species such as piscivorous fish, evidence of biomagnification across food webs is inconsistent. Additionally, the sublethal effects of chronic, low-dose exposure including behavioral changes, endocrine disruption, and immunotoxicity are insufficiently understood. Such effects may alter reproductive output, growth, and population dynamics, highlighting the need to integrate sublethal endpoints into monitoring and risk assessment programs (Richardson & Kimura, 2017). Multigenerational and life cycle exposures further complicate assessment. Certain contaminants, such as PFAS and pharmaceuticals, may bioaccumulate in tissues or be maternally transferred to offspring, potentially inducing delayed or epigenetic effects across generations. Sensitive developmental windows such as embryonic or larval stages are particularly vulnerable, and effects may not be detectable until later life stages or subsequent generations. Understanding these dynamics is critical for designing treatment strategies that prevent long-term ecological impacts (Altenburger et al., 2018).

Addressing these gaps requires an approach linking toxicology with treatment and risk mitigation. First, improved analytical methods, including high-resolution mass spectrometry and non-targeted screening, are essential for detecting contaminants, their transformation products, and metabolites in complex matrices. Second, predictive models that consider mixture toxicity, bioaccumulation, and long-term exposure can guide monitoring programs and inform priorities for treatment interventions (Xie et al., 2017). Third, environmentally relevant experimental designs including mesocosms and chronic, multigenerational studies can provide insight into cumulative effects and support the development of mitigation strategies such as advanced oxidation, membrane filtration, or assimilative removal techniques (Gago-ferrero et al., 2017). Integrating toxicological knowledge with treatment technologies and risk assessment frameworks ensures that mitigation efforts are evidence-based and environmentally relevant. Therefore, both the biological impacts and the limitations of current removal methods, researchers and practitioners can better prioritize contaminants, optimize treatment processes, and reduce long-term ecological and human health risks associated with PFAS, microplastics, and pharmaceuticals (Brodin et al., 2017) (Table 12.4).

3.4 Regulatory frameworks and global guidelines

The global management of emerging water contaminants particularly per- and polyfluoroalkyl substances (PFAS), pharmaceuticals, and microplastics has evolved through a combination of scientific evidence, public awareness, and regulatory initiatives. Governments and international organizations are progressively establishing policies that address the detection, monitoring, and mitigation of these pollutants, emphasizing preventive risk management and transparency (Puri et al., 2023). The U.S. Environmental Protection Agency (EPA) has developed a comprehensive strategy through its PFAS Strategic Roadmap (2021–2024) to reduce human and ecological exposure to PFAS. The roadmap builds on three pillars Research, Restriction, and Remediation and integrates cross-agency collaboration with states and tribal governments. The EPA has established enforceable drinking-water

Table 12.4 Comparative summary of toxicological challenges and cumulative exposure risks of PFAS, microplastics, and pharmaceuticals (Nilsen et al., 2019; Richardson & Kimura, 2017).

Category	Key contaminants	Exposure pathways	Health effects	Toxicological gaps	Research needs
PFAS	PFOA, PFOS, GenX	Drinking water, food, dermal contact	Endocrine and liver toxicity	Limited chronic data; unknown mixture effects	Long-term studies; standardized detection
Microplastics	PE, PP, PS fragments	Ingestion, inhalation	Oxidative stress, inflammation	Size- and shape-dependent toxicity unclear	Size-based assays; bioaccumulation studies
Pharmaceuticals	Antibiotics, hormones, analgesics	Effluent discharge, runoff	Antibiotic resistance, hormonal imbalance	Lack of threshold data; unknown metabolites	Ecotoxicity profiling; real-time monitoring
Cumulative exposure	PFAS–plastic, pharma–metal complexes	Multiple exposure routes	Synergistic or antagonistic toxicity	No integrated mixture models	Holistic risk frameworks; combined exposure tests

standards for six PFAS compounds and continues to expand analytical methods for a broader chemical inventory (Manotas-duque et al., 2020). In addition, the agency is strengthening cleanup enforcement under the Comprehensive Environmental Response, Compensation, and Liability Act (CERCLA) and advancing national monitoring through the Unregulated Contaminant Monitoring Rule 5 (UCMR 5) program. These actions underscore a shift from voluntary management to legally binding control measures, ensuring accountability for contamination sources and promoting long-term remediation efforts. The EPA also supports innovation in PFAS destruction and removal technologies, such as granular activated carbon (GAC), ion exchange, and high-temperature oxidation, while coordinating research to understand cumulative exposure risks (Puri et al., 2023).

The European Union (EU) has taken a broader chemical-safety approach by integrating PFAS, pharmaceuticals, and microplastics within its Water Framework Directive (2000/60/EC), Drinking Water Directive (2020/2184), and REACH (Registration, Evaluation, Authorization, and Restriction of Chemicals) regulations. The EU's Zero Pollution Action Plan aims to achieve a toxic-free environment by 2050, prioritizing the phase-out of persistent and bioaccumulative chemicals. Under REACH, PFAS are being considered as a single chemical class for restriction, marking a significant step toward comprehensive risk management (Barbosa et al., 2016). The EU also enforces environmental-quality standards for priority substances, including pharmaceuticals such as diclofenac and estrogens, and is developing binding thresholds for microplastic discharge. Moreover, the Urban Wastewater Treatment Directive mandates advanced treatment upgrades, such as ozonation or activated-carbon filtration, to improve removal efficiencies for micropollutants before discharge into aquatic environments. These legislative measures highlight Europe's integrated, precautionary approach, balancing environmental protection with industrial accountability (Puri et al., 2023).

The World Health Organization (WHO) plays a key coordinating role by setting health-based guideline values for contaminants in drinking water and promoting risk-based water-safety plans. The 2022 WHO Guidelines for Drinking-Water Quality introduced provisional limits for PFAS, including PFOS and PFOA, emphasizing cumulative exposure assessment through multiple pathways. The WHO framework promotes a tiered approach source control, water treatment, and continuous monitoring adaptable to both developed and resource-limited regions. For pharmaceuticals and microplastics, WHO advocates improved data on occurrence, fate, and potential health risks, calling for harmonized analytical methods and intergovernmental data sharing to strengthen global response (Richards et al., 2021).

In conclusion, the EPA, EU, and WHO frameworks illustrate a convergence toward risk-based, life-cycle management of emerging contaminants. While regulatory maturity varies among regions, all emphasize research integration, public disclosure, and technology innovation as pillars of mitigation. Continued alignment of analytical standards, toxicity thresholds, and monitoring protocols is essential to support global comparability and guide investment in advanced treatment technologies. Such harmonization will facilitate not only compliance but also the broader goal of safeguarding water resources from persistent and emerging chemical threats (Richards et al., 2021) (Table 12.5).

Table 12.5 Overview of Global regulatory frameworks for managing emerging water contaminants (Barbosa et al., 2016; Manotas-duque et al., 2020; Puri et al., 2023; Richards et al., 2021).

Region/Body	Target contaminants	Key policies/ Programs	Main strategies	Highlights
U.S. EPA	PFAS, pharmaceuticals	PFAS strategic roadmap (2021–2024); UCMR 5	Research, restriction, remediation	Enforceable PFAS limits; cleanup under CERCLA; promotes PFAS destruction technologies
European Union (EU)	PFAS, pharmaceuticals, microplastics	Water and drinking water directives; REACH; Zero pollution action plan	Phase-out of persistent chemicals; advanced wastewater treatment	PFAS treated as one class; stricter discharge limits; industry accountability
World health organization (WHO)	PFAS, pharmaceuticals, microplastics	2022 drinking-water guidelines; water safety plans	Risk-based framework; global monitoring	Provisional PFAS limits; harmonized data and analytical standards
Global outlook	Emerging contaminants	Multi-agency collaboration	Life-cycle management; technology innovation	Converging policies promoting transparency and preventive risk control

4. Treatment technologies for PFAS, microplastics, and pharmaceuticals

Traditional water treatment systems, relying on coagulation, sedimentation, filtration, and chlorination, mainly target particulates and microbial pathogens but are poorly equipped to remove emerging contaminants such as PFAS, microplastics, and pharmaceuticals. PFAS are exceptionally persistent, escaping conventional units due to strong carbon–fluorine bonds and resistance to chemical degradation. Most pharmaceuticals and personal care products retain low biodegradability and diverse physicochemical profiles, resulting in their continued presence in treated effluents. Microplastics, particularly those smaller than 100 μm, frequently bypass filtration and persist in finished water, creating additional health and ecological concerns. Moreover, treatment sludge accumulates these contaminants, and subsequent disposal or land application risks secondary pollution through environmental release or thermal processes that may not fully eliminate PFAS. While advanced processes such as absorption, advanced oxidation, and membrane filtration can improve removal, their adoption remains limited by cost and infrastructure requirements. The lack of regulatory standards further complicates reliable removal, monitoring, and reporting of these substances in wastewater treatment operations.

4.1 Advanced treatment approaches

4.1.1 Adsorption technologies

Adsorption is a highly effective strategy for removing PFAS, pharmaceuticals, and microplastics from water, with granular and powdered activated carbon (GAC/PAC) widely employed due to high surface area and affinity for organic contaminants. Ion exchange resins, particularly those tailored for PFAS removal, utilize electrostatic interactions for targeted extraction, especially of longer-chain PFAS (Bharti, 2025). Innovative adsorbents such as surface-modified biochar (Lin et al., 2025), graphene-based composites (Zubair et al., 2024, pp. 3693–3720), and metal–organic frameworks (MOFs) (Jiang et al., 2024), are under active development, delivering enhanced selectivity, regeneration, and adsorption kinetics compared to conventional materials. Cost, reusability, and environmental compatibility remain critical research directions.

4.1.2 Membrane-based technologies

Nanofiltration (NF) and reverse osmosis (RO) are leading membrane processes for the rejection of PFAS and trace organic contaminants, offering high selectivity and robust removal efficiencies. Membrane distillation, hybrid membrane-adsorption systems, and alternative techniques like forward osmosis are being explored to enhance contaminant rejection and manage fouling. Despite their efficacy, membrane fouling, concentrate disposal, and operational costs present significant challenges, necessitating ongoing optimization and integration with pretreatment or polishing units (Alonso et al., 2024; Geng et al., 2024).

4.1.3 Advanced oxidation processes

Advanced Oxidation Processes (AOPs) deploy powerful oxidant species including hydroxyl radicals (·OH) for pollutant degradation, utilizing processes such as photolysis, UV/H_2O_2, ozonation, photocatalysis, and electrochemical oxidation to break down pharmaceuticals and partially degrade PFAS.

Complete mineralization of PFAS, especially short-chain variants, is limited; therefore, coupled treatment approaches combining AOPs with adsorption or biological degradation are increasingly employed to enhance overall removal and minimize transformation product risk (Sun et al., 2025; Yu et al., 2025).

4.1.4 Biological and hybrid approaches

Conventional biological treatment is generally inadequate for PFAS and microplastics, but recent advances in engineered microbial communities, immobilized enzymes (laccases, peroxidases), and biofilm reactors have demonstrated promising results for the selective degradation of pharmaceuticals and, to some degree, microplastics. Hybrid approaches, such as membrane bioreactors integrated with AOPs or adsorption units, provide synergistic mechanisms that improve contaminant removal and energy efficiency, although full-scale deployment remains limited by technical complexity and costs (Table 12.6 and Fig. 12.2).

5. Risk mitigation strategies beyond treatment

5.1 Source reduction and pollution prevention

Effective management of emerging water contaminants such as PFAS, pharmaceuticals, and microplastics requires a shift from reactive treatment to preventive control at the source. Source reduction and pollution prevention aim to minimize the generation and release of contaminants throughout their life cycle during design, manufacture, use, and disposal thereby reducing the burden on downstream treatment systems. This preventive approach not only supports environmental protection but also aligns with the principles of sustainable production and circular economy (Wezel et al., 2017).

At the upstream level, source reduction begins with green chemistry and product design. Replacing hazardous or persistent compounds with safer, biodegradable alternatives can significantly reduce environmental emissions. For pharmaceuticals, rational prescription, improved disposal practices, and "benign-by-design" formulations are emerging as viable strategies (Processing & Dynamics, 2025). Similarly, minimizing PFAS use in consumer and industrial applications through substitution with short-chain or non-fluorinated alternatives can reduce environmental persistence (Timoner et al., 2018). For microplastics, interventions such as banning microbeads, improving textile manufacturing, and promoting biodegradable polymers represent proactive measures at the design stage (Gomes, 2025). Industrial process optimization also plays a vital role in pollution prevention. Cleaner production technologies, closed-loop systems, and improved material efficiency can reduce chemical discharges before they reach wastewater streams. At the community level, public awareness, product labeling, and take-back programs for pharmaceuticals can prevent improper disposal and limit chemical entry into municipal systems (Rose et al., 2015). Beyond product and process innovations, policy-driven approaches reinforce source reduction. Regulatory instruments such as extended producer responsibility (EPR), green procurement policies, and eco-labeling encourage manufacturers to design products with lower environmental footprints. Economic incentives, voluntary agreements, and sector-specific guidelines further strengthen prevention strategies (Timoner et al., 2018).

Table 12.6 Comparison of advanced treatment technologies for the removal of PFAS, microplastics, and pharmaceuticals, highlighting mechanisms, efficiencies, advantages, and operational challenges.

Technology	Target contaminants	Typical removal efficiency	Mechanism	Advantages	Challenges	References
Activated carbon (PAC/GAC)	PFAS, pharmaceuticals	60%–95% (depends on chain length and compound type)	Adsorption via hydrophobic and electrostatic interactions	High surface area, simple operation	Requires frequent regeneration; poor for short-chain PFAS	Burdick et al. (2018), Rahman et al. (2013)
Ion-exchange resins	PFAS (long and short chain)	80%–99%	Ion-exchange with sulfonic or quaternary ammonium sites	High PFAS selectivity	Resin fouling; regeneration waste	Burdick et al. (2018), Wang et al. (2017)
Biochar/ Nanomaterials (e.g., graphene, MOFs)	PFAS, pharmaceuticals	70%–98%	Adsorption on high-surface nanostructures	Low-cost, regenerable	Leaching and scaling-up challenges	Lee et al. (2019), Burdick et al. (2018)
Nanofiltration (NF)	PFAS, pharmaceuticals, microplastics	70%–99%	Size and charge-based exclusion	High selectivity; lower energy use than RO	Concentrate disposal; fouling	Leckie (2014), Rahman et al. (2013)
Reverse osmosis (RO)	PFAS, pharmaceuticals	90%–99%	Dense membrane rejection	Broad contaminant removal	High energy demand; brine management issues	Leckie (2014), Wang et al. (2017)
Membrane distillation/ Hybrid membranes	PFAS, microplastics	85%–99%	Evaporative or combined filtration	High rejection efficiency	Scaling and high cost	Burdick et al. (2018), Lee et al. (2019)
UV/H_2O_2, ozonation, photocatalysis (AOPs)	Pharmaceuticals, some PFAS	60%–95% (depends on process conditions)	Hydroxyl radical oxidation	Effective for complex organics	Incomplete mineralization; costly	Anumol et al. (2016), Mohapatra et al. (2014)
Electrochemical oxidation (EO)	PFAS, pharmaceuticals	70%–100% (lab scale)	Direct and indirect oxidation	Rapid, chemical-free	Energy-intensive; electrode corrosion	Anumol et al. (2016), Burdick et al. (2018)
Enzyme-mediated degradation (e.g., laccase)	Pharmaceuticals	40%–80%	Catalytic breakdown via oxidative enzymes	Environmentally benign	Slow kinetics; enzyme instability	Mohapatra et al. (2014), Petrie et al. (2014)

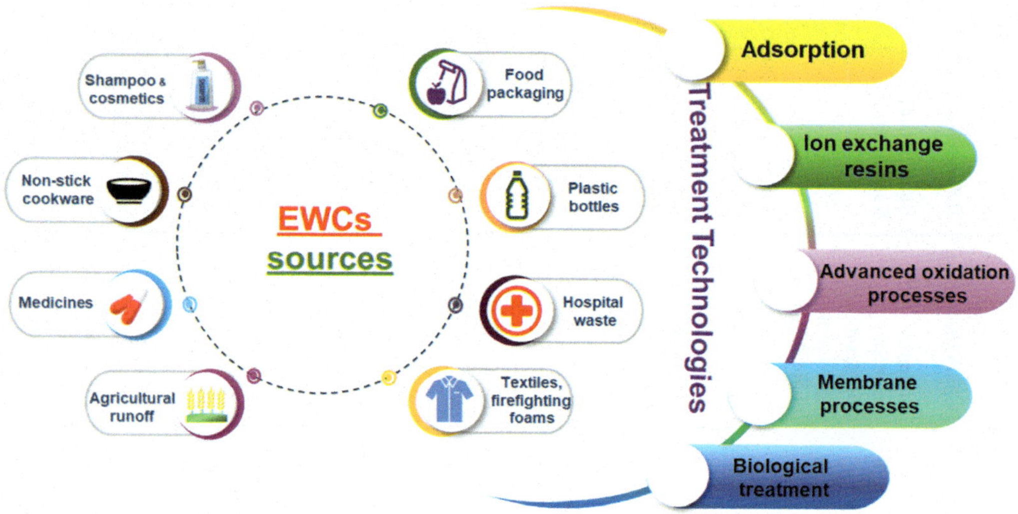

FIGURE 12.2

Schematic representation of major sources of emerging water contaminants (EWCs) and the treatment approaches.

5.2 Green chemistry and safer alternatives for PFAS and pharmaceuticals

The mitigation of emerging contaminants such as PFAS and pharmaceuticals requires a paradigm shift from traditional end-of-pipe treatment to preventive chemical design and substitution, rooted in the principles of green chemistry. Rather than focusing solely on removal after contamination occurs, green chemistry promotes the design of inherently safer and more sustainable chemicals that minimize persistence, toxicity, and bioaccumulation throughout their life cycle (Gomes, 2025).

Green chemistry approaches emphasize avoiding hazardous properties at the molecular level. For PFAS, this involves developing non-fluorinated or degradable fluorinated alternatives that maintain functional performance without long-term environmental persistence. Similarly, in the pharmaceutical sector, the "benign-by-design" approach focuses on synthesizing active compounds that retain therapeutic efficiency while degrading rapidly after use, thereby preventing accumulation in aquatic systems. These design strategies represent upstream interventions that reduce environmental loading before pollutants reach wastewater or surface waters (Kar et al., 2022). The adoption of safer alternatives also relies on life-cycle assessment and hazard-based screening tools to evaluate chemical substitutes. For instance, alternatives to PFAS in industrial coatings, textiles, and firefighting foams are being assessed not only for performance but also for environmental degradability and human health safety. In pharmaceuticals, the use of biodegradable excipients and green synthetic routes minimizes toxic by-products and solvent waste. These measures reflect a preventive framework that integrates chemical innovation with environmental stewardship (Nighojkar et al., 2022). Transitioning toward greener chemistry requires collaboration between regulators, industry, and researchers. In parallel, investment in green chemistry research and education enhances the discovery of functional yet environmentally benign molecules (Info, 2024). Ultimately, green chemistry provides a scientifically grounded and forward-looking pathway for mitigating the risks of PFAS and pharmaceuticals. By rethinking chemical design, production, and application, it is possible to balance innovation with ecological responsibility (Tickner et al., 2021).

5.3 Plastic waste reduction policies and biodegradable polymers

The persistence of plastics and microplastics in aquatic systems has driven global efforts toward sustainable waste management and regulatory reform. The United Nations Sustainable Development Goals (SDGs) advocate for cleaner production, responsible consumption, and protection of terrestrial and aquatic ecosystems (SDGs 12, 13, 14, and 15). In line with these goals, several nations have adopted stringent plastic waste reduction policies aimed at mitigating microplastic contamination in the environment (Filiciotto & Rothenberg, 2021).

A landmark event was China's 2017 waste import ban, which disrupted the global plastic trade and urged countries to develop local recycling and biodegradable alternatives. The Basel Convention (2019) reinforced accountability by restricting transboundary plastic waste movements, promoting domestic waste management solutions (Al-jabari, 2021). The European Union (EU) Green Deal (2020) further emphasized a circular economy through regulations targeting illegal exports, single-use plastics, and the establishment of clear standards for bio-based and biodegradable polymers (Filiciotto & Rothenberg, 2021). The United Kingdom's microbead ban (2018) curtailed intentional microplastic release into wastewater, while the United States Bioeconomy Blueprint (2012) (Filiciotto & Rothenberg, 2021) and Clean Water Act (1972) encourage biodegradable, renewable biomass-derived materials. Developing nations, such as Rwanda, have also demonstrated strong policy commitment by banning single-use plastics and advancing waste-free growth models (Tamkeen et al., 2020). International standardization bodies, including OECD, ISO, ASTM, and CEN, define test methods for assessing biodegradability under soil, freshwater, and marine conditions. Certification agencies such as TÜV Austria and DIN Certco provide eco-labels (*OK Compost*, *OK Biodegradable Marine*, *EU Ecolabel*), ensuring transparency in product claims and environmental compliance. These certifications rely on parameters such as CO_2 evolution, oxygen demand, and organic carbon degradation, while polymer characteristics like crystallinity, hydrophilicity, and functional group composition determine degradation efficiency (Shillito, 2022) (Fig. 12.3).

5.4 Wastewater reuse and risk considerations

Wastewater reuse has become an essential strategy to address global water scarcity and promote sustainable resource management. Treated or recycled wastewater finds application across multiple sectors such as agriculture, landscape irrigation, public parks, golf courses, and various industrial operations including cooling, processing, and construction. Among these, agricultural irrigation remains the most widespread form of reuse, offering not only an alternative water source but also an additional supply of nutrients like nitrogen and phosphorus that enhance soil fertility and crop yields (Trotta et al., 2024).

However, the reuse of wastewater introduces environmental, health, and socio-economic risks due to the potential accumulation of heavy metals, pathogens, and organic pollutants. Elements such as lead, copper, and chromium can enter the food chain, causing toxicity in plants, animals, and humans. To address these risks, advanced oxidation processes (AOPs) including ultrasound (US), ultraviolet

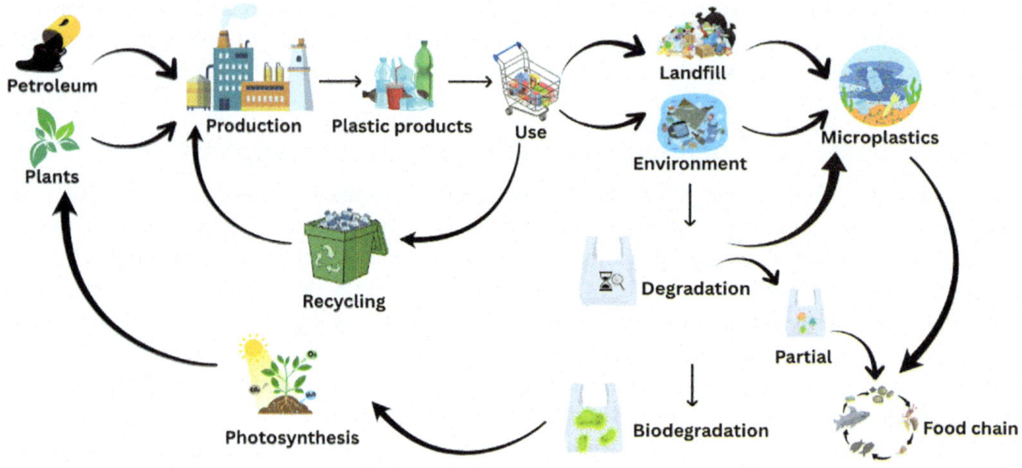

FIGURE 12.3

Overview of plastic waste reduction policies and the transition toward biodegradable polymer adoption for sustainable material management.

(UV) radiation, ozone (O_3), and nanoparticle-assisted treatments have been developed. These technologies generate reactive oxygen species that degrade complex organics, oxidize metals into less soluble forms, and effectively destroy pathogens. Combined US-O_3-UV systems have demonstrated high efficiency in producing water suitable for both irrigation and potential potable reuse (Kesari et al., 2021). The challenges linked with wastewater irrigation span economic, environmental, social, and health dimensions (Singh et al., 2018). Effective management requires reducing chemical and biological oxygen demand, suspended solids, and microbial loads through physical, chemical, and biological treatment processes (Iqbal et al., 2022). Techniques such as adsorption, photocatalysis, and electrochemical oxidation are being employed to eliminate diverse contaminants efficiently. Biological mitigation strategies, such as the use of endophytic plant growth-promoting fungi, can further enhance plant resilience and reduce stress caused by wastewater reuse (Ungureanu & Voicu, 2020). Advanced treatment systems, optimized irrigation methods, and collaborative research among scientists, engineers, and policymakers are critical to achieving safe and beneficial wastewater reuse (Trotta et al., 2024) (Fig. 12.4).

5.5 Risk communication and public awareness

Effective management of emerging water contaminants, including PFAS, microplastics, and pharmaceuticals, requires not only technological interventions but also active engagement with the public. Risk communication and public awareness are critical components in promoting responsible monitoring and mitigation of these contaminants. Citizen science initiatives, which leverage user-friendly sensing platforms and innovative monitoring technologies, have emerged as powerful tools for fostering public participation. Such platforms are specifically designed to be accessible to nonexperts, enabling communities to contribute meaningfully to environmental monitoring efforts (Ateia et al., 2024).

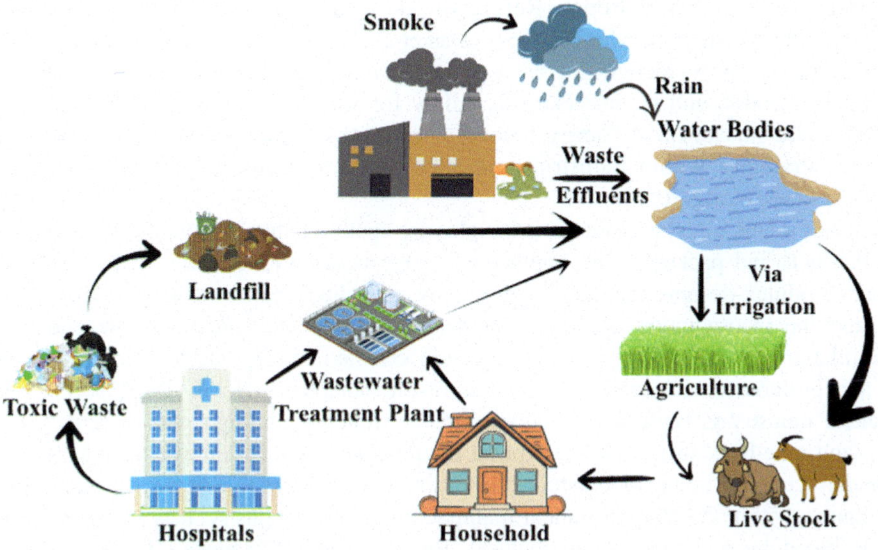

FIGURE 12.4

Schematic representation of wastewater generation, treatment, and reuse pathways across domestic, industrial, and agricultural systems.

Collaborations with schools, community organizations, and local stakeholders, along with structured workshops, training sessions, and webinars, can enhance understanding of the presence and risks associated with emerging contaminants. By integrating citizen-generated data into broader research projects and regulatory decision-making processes, the significance of public involvement in environmental monitoring is underscored. This participatory approach not only empowers citizens but also strengthens the overall knowledge base on contaminant occurrence, distribution, and impacts (Menger et al., 2021). Educational resources that clearly explain the importance of monitoring emerging contaminants and provide step-by-step guidance for citizen scientists are essential for broad outreach. Online platforms facilitating data sharing and collaboration among participants further enhance engagement, enabling knowledge exchange across diverse communities. Media coverage and public awareness campaigns complement these efforts by highlighting ongoing initiatives and emphasizing the relevance of emerging water contaminants to everyday life (Miller et al., 2023, pp. 1–12).

5.6 Circular economy approaches

Building on the discussion of wastewater reuse and its associated risks, the circular economy (CE) offers a broader framework for sustainable resource management. CE represents a systemic shift from the traditional linear "take, make, dispose" model toward an approach that emphasizes keeping materials and products in use, eliminating waste, and regenerating natural systems. Central to this approach is the concept of closing loops, which can be applied across water, plastics, and pharmaceuticals. Frameworks such as the 3R (Reduce, Reuse, Recycle) and the 9R hierarchy including

Refuse, Rethink, Reduce, Reuse, Repair, Refurbish, Remanufacture, Repurpose, and Recycle, provide structured strategies to guide circular practices (Das et al., 2025). In the context of water management, wastewater is increasingly recognized as a valuable resource. Advanced treatment technologies, including microfiltration and reverse osmosis, allow for safe reuse in both potable and non-potable applications, while nutrient and energy recovery from wastewater enhances its utility. Integrating circular water strategies within industrial and agricultural systems reduces reliance on freshwater resources (Kirchherr et al., 2023).

In the plastics sector, circular economy principles are operationalized through the design for circularity approach. Products and packaging are engineered to be recyclable, reusable, or biodegradable. Mono-material and modular designs facilitate high-quality recycling, lightweighting reduces material consumption, and the incorporation of post-consumer recycled content decreases dependence on virgin plastics. Refillable and reusable systems, along with certified biodegradable solutions, extend product lifecycles and promote environmentally safe end-of-life outcomes (Ho & Tang, 2025). Similarly, the pharmaceutical industry is beginning to adopt circular strategies through green chemistry, closed-loop production, and take-back programs for expired medications. In this context, the PM9R model adapts the 9R circular economy framework specifically designed to address to address sustainability challenges and opportunities within the pharmaceutical manufacturing (PM) industry. It is providing practical guidance by 9R hierarchy approach which includes Refuse, Rethink, Reduce, Reuse, Repair, Refurbish, Remanufacture, Repurpose, Recycle and Recover to manage resources by prioritizing actions that maximize product and resource utilization and minimize waste (Liang et al., 2021).

6. Case studies and global best practices

Globally, numerous full-scale and pilot initiatives illustrate both the technical feasibility and the practical challenges of mitigating per- and polyfluoroalkyl substances (PFAS) in contaminated water systems. In North America, long-term pump-and-treat systems coupled with granular activated carbon (GAC) adsorption, such as the Oakdale project in Minnesota and the Blades utility in Delaware, have proven highly effective in removing long-chain PFAS (>95%) but remain costly due to frequent media replacement and concentrate handling (Town et al., 2018). European utilities, notably in Uppsala, Sweden, and L'Ampolla, Spain, have demonstrated how GAC and ion-exchange resins can be integrated into conventional drinking-water treatment trains to achieve regulatory compliance while informing evolving EU standards (Ahrens et al., 2020). Meanwhile, innovative containment and destruction technologies are emerging: the Regenesis in-situ reactive barrier achieved >99% reduction within 1 year at a firefighting-foam-impacted site, and pilots such as PFASigator in the United States and Invicta Water at Jordan Lake show progress toward energy-intensive but potentially transformative "treat-to-destroy" pathways (PFAS Effectively Remediated at Former Naval Air Station Alameda |PFAS effectively remediated at former Naval air station Alameda | Regenesis, 2025). In the Asia–Pacific region, the Brisbane Airport program applied foam fractionation with GAC polishing to manage AFFF-derived runoff, setting a regional benchmark for industrial PFAS control (EHP Investigating Foam Spillage at Brisbane Airport—Ministerial Media Statements, 2025). Collectively, these cases highlight that while adsorption-based capture remains the most mature and scalable option, sustainable PFAS management ultimately requires coupling physical removal with safe destruction and residuals governance, supported by clear regulatory frameworks and long-term monitoring.

7. Future perspectives and research directions

Future research on emerging water contaminants must focus on developing integrated multi-contaminant treatment systems capable of addressing PFAS, microplastics, pharmaceuticals, and other pollutants simultaneously rather than through isolated approaches. Emphasis should be placed on low-cost, scalable, and sustainable technologies that are adaptable to diverse socio-economic and geographic contexts, ensuring applicability in both high-income and resource-limited settings. Rapid advances in nanotechnology, biotechnology, and machine learning present transformative opportunities for real-time contaminant monitoring, predictive modeling, and adaptive treatment optimization. Furthermore, effective mitigation will depend on strong policy frameworks, transparent governance mechanisms, and cross-sectoral collaborations among scientists, industries, regulators, and communities. Such integrative strategies will be essential to achieve safe, resilient, and equitable water systems in the face of evolving environmental challenges.

8. Conclusion

This chapter highlights the increasing complexity associated with emerging water contaminants (EWCs) and the challenges they pose to existing water treatment and management systems. The persistence, mobility, and potential toxicity of contaminants such as PFAS, microplastics, and pharmaceutical residues reveal clear limitations in conventional treatment infrastructures that were not designed to address these compounds. While advanced treatment technologies, including adsorption-based methods, membrane filtration, and advanced oxidation processes, offer improved removal efficiencies, their widespread application is often restricted by high operational costs, energy requirements, and long-term sustainability concerns. Addressing the risks posed by EWCs therefore requires more than technological advancement alone. Effective risk mitigation must be supported by coherent regulatory frameworks, improved monitoring strategies, and policies that account for cumulative exposure and long-term environmental impacts. The role of public awareness and community engagement is equally important, particularly in regions where water is reused, agricultural practices, and industrial activities intersect closely with drinking water sources. Looking ahead, sustainable water management strategies must adopt an integrated and interdisciplinary perspective that bridges scientific research, engineering solutions, policy development, and social considerations. The incorporation of data-driven decision tools, adaptive governance mechanisms, and equitable access to clean water resources will be critical in building resilient water systems. Finally, ensuring water safety in the face of emerging contaminants demands coordinated efforts that balance environmental protection, public health priorities, and socio-economic realities.

References

Abdallah, M. A., & Harrad, S. (2022). Dermal uptake: An important pathway of human exposure to perfluoroalkyl substances? *Environmental Pollution, 307*(April). https://doi.org/10.1016/j.envpol.2022.119478

Adam, A. B., Abubakar, M. Y., & Idris, A. M. (2024). *Soil remediation techniques for heavy metal contamination. A. August.*

Ahrens, L., Belkouteb, N., Franke, V., Mccleaf, P., & Stephan, K. (2020). Removal of per- and poly fluoroalkyl substances (PFASs) in a full-scale drinking water treatment plant: Long-term performance of granular activated carbon (GAC) and influence of flow-rate. *Water Research, 182*. https://doi.org/10.1016/j.watres.2020.115913

Aib, H., & Parvez, S. (2025). Pharmaceuticals and microplastics in aquatic environments: A comprehensive review of pathways and distribution. *Toxicological and Ecological Effects, 1–35*.

Air, W., Sarker, A., & Ram, K. (2023). *Microplastic toxicity in aquatic organisms and aquatic ecosystems: A review*. https://doi.org/10.1007/s11270-023-06062-9. January.

Al-Jabari, M. (2021). 1. International approaches to hazardous waste classification and listings. In *Handbook of advanced approaches towards pollution prevention and control*. INC. https://doi.org/10.1016/B978-0-12-822134-1.00001-4

Alonso, E., Sanchez-huerta, C., Ali, Z., Wang, Y., & Fortunato, L. (2024). Evaluation of nanofiltration and reverse osmosis membranes for efficient rejection of organic micropollutants. *Journal of Membrane Science, 693* (August 2023), Article 122357. https://doi.org/10.1016/j.memsci.2023.122357

Altenburger, R., Scholze, M., Busch, W., Escher, B. I., Jakobs, G., Krauss, M., Krüger, J., Neale, P. A., Aitaissa, S., Catarina, A., Seiler, T., Brion, F., Hilscherová, K., Hollert, H., Novák, J., Schlichting, R., Serra, H., Shao, Y., Tindall, A., … Kortenkamp, A. (2018). Mixture effects in samples of multiple contaminants—An inter-laboratory study with manifold bioassays. *Environment International, 114*, 95–106. https://doi.org/10.1016/j.envint.2018.02.013. December 2017.

Amoatey, P., & Baawain, M. S. (2019). Effects of pollution on freshwater aquatic organisms. *Water Environment Research*, 1272–1287. https://doi.org/10.1002/wer.1221

Anumol, T., Dagnino, S., Vandervort, D. R., & Snyder, S. A. (2016). Transformation of polyfluorinated compounds in natural waters by advanced oxidation processes. *Chemosphere, 144*, 1780–1787. https://doi.org/10.1016/j.chemosphere.2015.10.070

Ateia, M., Wei, H., & Andreescu, S. (2024). *Sensors for emerging water contaminants: Overcoming roadblocks to innovation*. https://doi.org/10.1021/acs.est.3c09889

Barbosa, M. O., Moreira, N. F. F., Ribeiro, A. R., & Pereira, M. F. R. (2016). *Occurrence and removal of organic micropollutants: An overview of the watch list of EU decision 2015/495* (p. 94). https://doi.org/10.1016/j.watres.2016.02.047

Bharti, S. (2025). PFAS in water systems: A critical review on contamination pathways, analytical methods, and treatment technologies. *Water Conservation Science and Engineering, 10*, Article 89. https://doi.org/10.1007/s41101-025-00415-7

Bloomquist, T. R., Spaur, M., Cerna-turoff, I., Kress, A. M., Burjak, M., Kupsco, A., Casey, J. A., Herbstman, J. B., & Nigra, A. E. (2024). Public drinking water contaminant estimates for birth cohorts in the environmental influences on child health outcomes (ECHO) cohort. *Journal of Exposure Science and Environmental Epidemiology*, 1–13. https://doi.org/10.1038/s41370-024-00699-2

Brodin, T., Nordling, J., Lagesson, A., Klaminder, J., Hellström, G., Christensen, B., & Fick, J. (2017). Environmental relevant levels of a benzodiazepine (oxazepam) alters important behavioral traits in a common planktivorous fish (Rutilus rutilus). *Journal of Toxicology and Environmental Health, Part A, 80*(16–18), 963–970. https://doi.org/10.1080/15287394.2017.1352214

Burdick, J., Mcdonough, J., Storch, P., Hurst, J., & Dasgupta, S. S. (2018). *A review of emerging technologies for remediation of PFASs* (pp. 101–126). https://doi.org/10.1002/rem.21553

Carter, L. J., Armitage, J. M., Brooks, B. W., Nichols, J. W., & Trapp, S. (2024). Predicting the accumulation of ionizable pharmaceuticals and personal care products in aquatic and terrestrial organisms. *Environmental Toxicology and Chemistry, 43*(3), 502–512. https://doi.org/10.1002/etc.5451

Crone, B. C., Speth, T. F., Wahman, D. G., Smith, S. J., Abulikemu, G., Kleiner, E. J., Pressman, J. G., Crone, B. C., Speth, T. F., Wahman, D. G., Smith, S. J., Abulikemu, G., Kleiner, E. J., & Pressman, J. G. (2019). Technology occurrence of per- and polyfluoroalkyl substances (PFAS) in source water and their treatment in drinking water. *Critical Reviews in Environmental Science and Technology, 49*(24), 2359–2396. https://doi.org/10.1080/10643389.2019.1614848

Das, S., Kant, C., Kushneet, S., & Sodhi, K. (2025). Circular economy approaches for water reuse and emerging contaminant mitigation: Innovations in water treatment. In *Environment, development and sustainability.* Netherlands: Springer. https://doi.org/10.1007/s10668-023-04183-z. Vol. 27, Issue 3.

Datta, A. R., Kang, Q., Chen, B., & Ye, X. (2018). Fate and transport modelling of emerging pollutants from watersheds to oceans: A review. In *Advances in marine biology.* Elsevier Ltd. https://doi.org/10.1016/bs.amb.2018.09.002, 1st ed., Vol. 81.

de Solla, S. R., Gilroy, A. M., Klinck, J. S., King, L. E., McInnis, R., Struger, J., Backus, S. M., & Gillis, P. L. (2016). Bioaccumulation of pharmaceuticals and personal care products in the unionid mussel Lasmigona costata in a river receiving wastewater effluent. *Chemosphere, 146*, 486–496. https://doi.org/10.1016/j.chemosphere.2015.12.022

Ding, Y., Zheng, X., Yu, L., Lu, R., Wu, X., Luo, X., & Mai, B. (2022). Occurrence and distribution of persistent organic pollutants (POPs) in amphibian species: Implications from biomagnification factors based on quantitative fatty acid signature analysis. *Environmental Science and Technology, 56*(5), 3117–3126. https://doi.org/10.1021/acs.est.1c07416

Dong, H., Lu, G., Yan, Z., Liu, J., Yang, H., Zhang, P., Jiang, R., Bao, X., & Nkoom, M. (2020). Distribution, sources and human risk of perfluoroalkyl acids (PFAAs) in a receiving riverine environment of the Nanjing urban area, East China. *Journal of Hazardous Materials, 381*, Article 120911. https://doi.org/10.1016/j.jhazmat.2019.120911

Draper, W. M., Dhoot, J. S., Dhaliwal, J. S., Remoy, J. W., Perera, S. K., & Baumann, F. J. (1998). Detection limits of organic contaminants in drinking water. *Journal of the American Water Works Association, 90*(6), 82–90. https://doi.org/10.1002/j.1551-8833.1998.tb08456.x

EHP investigating foam spillage at Brisbane airport, Ministerial Media Statements. (2025). Retrieved November 5, 2025, from https://statements.qld.gov.au/statements/80790.

Essamuah-Quansah, J. (2017). *Early warning systems: A review early warning systems: A review* (pp. 22–44).

Fabio, K., & Blum, S. F. (2016). *Chemistry Impact Factor: 1. 132* (pp. 169–182).

Factors, S., Ayodele, A., & Obeng-gyasi, E. (2024). Exploring the potential link between PFAS exposure and endometrial cancer: A review of environmental and sociodemographic factors. *Cancers*, 1–32.

Ferrão-filho, A. S., & Kozlowsky-suzuki, B. (2011). *Cyanotoxins: Bioaccumulation and effects on aquatic animals*, 2729–2772. https://doi.org/10.3390/md9122729

Filiciotto, L., & Rothenberg, G. (2021). *Biodegradable plastics: Standards, policies, and impacts* (pp. 56–72). https://doi.org/10.1002/cssc.202002044

Gago-ferrero, P., Gros, M., Ahrens, L., & Wiberg, K. (2017). Impact of on-site, small and large scale wastewater treatment facilities on levels and fate of pharmaceuticals, personal care products, artificial sweeteners, pesticides, and per fluoroalkyl substances in recipient waters. *Science of the Total Environment, 601–602*, 1289–1297. https://doi.org/10.1016/j.scitotenv.2017.05.258

Geng, H., Zhang, W., Zhao, X., Shao, W., & Wang, H. (2024). *Research on reverse osmosis (RO)/nanofiltration (NF) membranes based on thin film composite (TFC) structures: Mechanism, recent progress and application* (Vol. 14 (9)). MDPI. https://doi.org/10.3390/membranes14090190.

Gomes, I. B. (2025). Emerging contaminants affect the microbiome of water systems—Strategies for their mitigation. *npj Clean Water, 3*, Article 39. https://doi.org/10.1038/s41545-020-00086-y

Hamburger, T., Gering, F., Yevdin, Y., Schantz, S., Geertsema, G., & Vries, H. De (2020). *Uncertainty propagation from ensemble dispersion simulations through a terrestrial food chain and dose model* (Vol 55, pp. 69–74).

Ho, K., & Tang, D. (2025). An overview of circular economy approaches for plastics. *Recent Progress in Materials*. https://doi.org/10.21926/rpm.2503011

Hopkins, Z. R., & Blaney, L. (2016). An aggregate analysis of personal care products in the environment: Identifying the distribution of environmentally-relevant concentrations. *Environment International, 92–93*, 301–316. https://doi.org/10.1016/j.envint.2016.04.026

Info, A. (2024). Toxic-free environment: Forever chemicals removal from water and wastewater. *Environmental Toxicology and Ecology, 4*(2), 122–142. https://doi.org/10.59838/etoxec.1538643

Iqbal, M., Nauman, S., Ghafari, M., Parnianifard, A., Gomes, A., & Gomes, C. (2022). Treatment of wastewater for agricultural applications in regions of water scarcity. *Biointerface Research in Applied Chemistry, 12*(5), 6336–6360.

Ishibashi, M., Izumi, Y., Sakai, M., Ando, T., Fukusaki, E., & Bamba, T. (2015). High-throughput simultaneous analysis of pesticides by supercritical fluid chromatography coupled with high-resolution mass spectrometry. *Journal of Agricultural and Food Chemistry, 63*(18), 4457–4463. https://doi.org/10.1021/jf5056248

Jiang, L., Huang, Z., Xu, J., Zhang, L., & Du, Z. (2024). Selective adsorption of OBS (sodium p—Perfluorous nonenoxybenzenesulfonate) as an emerging PFAS contaminant from aquatic environments by fluorinated MOFs: Novel mechanisms of F–F exclusive attraction. *Chemical Engineering Journal, 484*, Article 149355. https://doi.org/10.1016/j.cej.2024.149355

Kadadou, D., Tizani, L., Alsafar, H., & Hasan, S. W. (2024). Analytical methods for determining environmental contaminants of concern in water and wastewater. *MethodsX, 12*(January), 1–20.

Kar, S., Sanderson, H., Roy, K., Benfenati, E., & Leszczynski, J. (2022). *Green chemistry in the synthesis of pharmaceuticals*. https://doi.org/10.1021/acs.chemrev.1c00631

Kesari, K. K., Soni, R., Mohammad, Q., & Jamal, S. (2021). *Wastewater treatment and reuse: A review of its applications and health implications*.

Kim, S. K., & Kannan, K. (2007). Perfluorinated acids in air, rain, snow, surface runoff, and lakes: Relative importance of pathways to contamination of urban lakes. *Environmental Science and Technology, 41*(24), 8328–8334. https://doi.org/10.1021/es072107t

Kirchherr, J., Yang, N. N., & Schulze-spüntrup, F. (2023). Conceptualizing the circular economy (revisited): An analysis of 221 definitions. *Resources, Conservation and Recycling, 194*, Article 107001. https://doi.org/10.1016/j.resconrec.2023.107001

Kumar, A. (2025). *Microplastics pollution and its remediation*.

Kumar, R., Qureshi, M., Kumar, D., & Al-ansari, N. (2022). A review on emerging water contaminants and the application of sustainable removal technologies. *Case Studies in Chemical and Environmental Engineering, 6*, Article 100219. https://doi.org/10.1016/j.cscee.2022.100219

Leckie, J. O. (2014). Effect of flux (transmembrane pressure) and membrane properties on fouling and rejection of reverse osmosis and nanofiltration membranes treating perfluorooctane sulfonate containing wastewater. *Environmental Science and Technology, 41*(6), 2008–2014.

Lee, J., Enfrin, M., & Dum, L. F. (2019). *Nano/microplastics in water and wastewater treatment processes e origin, impact and potential solutions* (p. 161). https://doi.org/10.1016/j.watres.2019.06.049

Lee, H., Ju, M., & Kim, Y. (2020). Estimation of emission of tire wear particles (TWPs) in Korea. *Waste Management, 108*, 154–159. https://doi.org/10.1016/j.wasman.2020.04.037

Lee, J. C., Smaoui, S., Duffill, J., Marandi, B., & Varzakas, T. (2025). *Forever chemicals PFAS global impact and activities, cascading consequences of colossal systems failure: Long-term health effects* (pp. 0–53). https://doi.org/10.20944/preprints202501.1698.v1

Levin, R., Villanueva, C. M., Beene, D., Cradock, A. L., Donat-vargas, C., Lewis, J., Martinez-morata, I., Minovi, D., Nigra, A. E., Olson, E. D., Schaider, L. A., Ward, M. H., & Deziel, N. C. (2023). *OPEN US drinking water quality: Exposure risk pro fi les for seven legacy and emerging contaminants*. https://doi.org/10.1038/s41370-023-00597-z. February.

Li, X., Hou, M., Zhang, F., Ji, Z., Cai, Y., & Shi, Y. (2025). *Per- and polyfluoroalkyl substances and female health concern: Gender-based accumulation differences, adverse outcomes, and mechanisms*. https://doi.org/10.1021/acs.est.4c08701

Liang, K., Toon, E., He, W., Dong, X., & Ramakrishna, S. (2021). *Sustainability framework for pharmaceutical manufacturing (PM): A review of research landscape and implementation barriers for circular economy transition* (Vol 280).

Lin, D., Lyu, H., Liu, Y., & Gong, Y. (2025). Rapid and efficient removal of aqueous perfluorooctane sulfonate at environmentally relevant concentrations using photoregenerable modified biochar. *Chemical Engineering Journal, 523*, Article 168638. https://doi.org/10.1016/j.cej.2025.168638. September.

Linkov, I., Stackelberg, K. E. Von, Burmistrov, D., & Bridges, T. S. (2001). *Uncertainty and variability in risk from trophic transfer of contaminants in dredged sediments.*

Lu, G. H., Jiao, X. C., Piao, H. T., Wang, X. C., Chen, S., Tan, K. Y., Gai, N., Yin, X. C., Yang, Y. L., & Pan, J. (2018). The extent of the impact of a fluorochemical industrial park in Eastern China on adjacent rural areas. *Archives of Environmental Contamination and Toxicology, 74*(3), 484–491. https://doi.org/10.1007/s00244-017-0458-x

Méndez, E., González-fuentes, M. A., Rebollar-perez, G., Torres, E., Méndez, E., González-fuentes, M. A., Rebollar-perez, G., Rebollar-perez, G., & Alia, M. (2017). Toxic/hazardous substances and environmental engineering emerging pollutant treatments in wastewater: Cases of antibiotics and hormones. *Journal of Environmental Science and Health, Part A, 52*(3), 235–253. https://doi.org/10.1080/10934529.2016.1253391

Manotas-duque, D., Ramírez-malule, H., & Qui, D. H. (2020). Emerging contaminants as global environmental hazards. A bibliometric analysis. *Emerging Contaminants, 6*, 179–193. https://doi.org/10.1016/j.emcon.2020.05.001

Mbachu, O., Jenkins, G., & Pratt, C. (2020). *A new contaminant superhighway ? A review of sources, measurement techniques and fate of atmospheric microplastics.*

Menger, R. F., Funk, E., Henry, C. S., & Borch, T. (2021). Sensors for detecting per- and polyfluoroalkyl substances (PFAS): A critical review of development challenges, current sensors, and commercialization obstacles. *Chemical Engineering Journal, 417*, Article 129133. https://doi.org/10.1016/j.cej.2021.129133

Mijangos, L., Ziarrusta, H., Olivares, M., & Zuloaga, O. (2018). *Simultaneous determination of 41 multiclass organic pollutants in environmental waters by means of polyethersulfone microextraction followed by liquid chromatography—Tandem mass spectrometry* (pp. 615–632).

Miller, M., Kisiel, A., Cembrowska-lech, D., Durlik, I., & Miller, T. (2023). *IoT in water quality monitoring—Are we really here ?* (pp. 1–12).

Mohapatra, D. P., Brar, S. K., Tyagi, R. D., Picard, P., & Surampalli, R. Y. (2014). Analysis and advanced oxidation treatment of a persistent pharmaceutical compound in wastewater and wastewater sludge-carbamazepine. *Science of the Total Environment, 470–471*, 58–75. https://doi.org/10.1016/j.scitotenv.2013.09.034. The.

Morin-Crini, N., Lichtfouse, E., Liu, G., Balaram, V., Ribeiro, A. R. L., Lu, Z., Stock, F., Carmona, E., Teixeira, M. R., Picos-Corrales, L. A., Moreno-Piraján, J. C., Giraldo, L., Li, C., Pandey, A., Hocquet, D., Torri, G., & Crini, G. (2022). Worldwide cases of water pollution by emerging contaminants: A review. *Environmental Chemistry Letters, 20*(4), 2311–2338. https://doi.org/10.1007/s10311-022-01447-4

Nguyen, V. T., Reinhard, M., & Karina, G. Y. H. (2011). Occurrence and source characterization of perfluorochemicals in an urban watershed. *Chemosphere, 82*(9), 1277–1285. https://doi.org/10.1016/j.chemosphere.2010.12.030

Niemuth, N. J., Jordan, R., Crago, J., Blanksma, C., Johnson, R., & Klaper, R. D. (2015). Metformin exposure at environmentally relevant concentrations causes potential endocrine disruption in adult male fish. *Environmental Toxicology and Chemistry, 34*(2), 291–296. https://doi.org/10.1002/etc.2793

Nighojkar, A., Sangal, V. K., Dixit, F., & Kandasubramanian, B. (2022). Value-added products: Future prospects and challenges with toxic per- and sustainable conversion of saturated adsorbents (SAs) from wastewater into value-added products: Future prospects and challenges with toxic per- and poly-fluoroalkyl substan. *Environmental Science and Pollution Research, 29*, 78207–78227. https://doi.org/10.1007/s11356-022-23166-7. October.

Nilsen, E., Smalling, K. L., Ahrens, L., Gros, M., Miglioranza, K. S. B., & Schoenfuss, H. L. (2019). Critical review: Grand challenges in assessing the adverse effects of contaminants of emerging concern on aquatic food webs. *Environmental Toxicology and Chemistry, 38*(1), 46–60. https://doi.org/10.1002/etc.4290

Nishmitha, P. S., Akhilghosh, K. A., Aiswriya, V. P., Ramesh, A., Muthuchamy, M., & Muthukumar, A. (2025). Advances Understanding emerging contaminants in water and wastewater: A comprehensive review on detection, impacts, and solutions. *Journal of Hazardous Materials Advances, 18*(April), Article 100755. https://doi.org/10.1016/j.hazadv.2025.100755

Nunes, J., Nardini, G., Merib, J., Neves, A., Martendal, E., & Carasek, E. (2012). Simultaneous determination of polycyclic aromatic hydrocarbons and benzene, toluene, ethylbenzene and xylene in water samples using a new sampling strategy combining different extraction modes and temperatures in a single extraction solid-phase microextraction-gas chromatography—Mass spectrometry procedure. *Journal of Chromatography A, 1233*, 22–29. https://doi.org/10.1016/j.chroma.2012.02.022

Ortúzar, M., Esterhuizen, M., & Olicón-hernández, D. R. (2022). Pharmaceutical pollution in aquatic environments: A concise review of environmental impacts and bioremediation systems. *Frontiers in Microbiology, 13*, 1–25. https://doi.org/10.3389/fmicb.2022.869332. April.

Pan, Y., Wang, J., Yeung, L. W. Y., Wei, S., & Dai, J. (2020). Analysis of emerging per- and polyfluoroalkyl substances: Progress and current issues. *TrAC, Trends in Analytical Chemistry, 124*, Article 115481. https://doi.org/10.1016/j.trac.2019.04.013

Paterson, K., Miller, E., Lin, D., Paterson, K., Miller, E., & Lin, D. (2024). *Microplastics monitoring and science strategy for San Francisco Bay 2024 revision* (Vol 1144). Prepared by.

Perera, D. C., Meegoda, J. N., Scherer, O., & Scherer, O. (2024). *Applied sciences PFAS: The journey from wonder chemicals to environmental nightmares and the search for solutions.*

Peter, L. G., & Lee, L. S. (2025). *Sources and pathways of PFAS occurrence in water sources: Relative contribution of land-applied biosolids in an agricultural dominated watershed.* https://doi.org/10.1021/acs.est.4c09490

Peter, K. T., Tian, Z., Wu, C., Lin, P., White, S., Du, B., McIntyre, J. K., Scholz, N. L., & Kolodziej, E. P. (2018). Using high-resolution mass spectrometry to identify organic contaminants linked to urban stormwater mortality syndrome in Coho Salmon. *Environmental Science and Technology, 52*(18), 10317–10327. https://doi.org/10.1021/acs.est.8b03287

Petrie, B., Barden, R., & Kasprzyk-hordern, B. (2014). A review on emerging contaminants in wastewaters and the environment: Current knowledge, understudied areas and recommendations for future monitoring. *Water Research, 72*(0), 3–27. https://doi.org/10.1016/j.watres.2014.08.053

PFAS effectively remediated at former Naval air station Alameda | Regenesis. (2025). Retrieved November 5, 2025, from https://regenesis.com/case-study/pfas-effectively-remediated-at-former-naval-air-station-alameda/.

Post, G. (2009). Development of health-based drinking water guidance for PFOA. *Reproductive Toxicology, 27* (3–4), 423. https://doi.org/10.1016/j.reprotox.2008.11.075

Processing, I., & Dynamics, S. (2025). *Circular economy in the pharmaceutical industry: Theoretical background and bibliometric analysis* (pp. 28–30). https://doi.org/10.1109/LOGISTIQUA66323.2025.11122786

Puri, M., Gandhi, K., & Kumar, M. S. (2023). Emerging environmental contaminants: A global perspective on policies and regulations. *Journal of Environmental Management, 332*, Article 117344. https://doi.org/10.1016/j.jenvman.2023.117344. November 2022.

Rahman, M. F., Peldszus, S., & Anderson, W. B. (2013). Behaviour and fate of perfluoroalkyl and polyfluoroalkyl substances (PFASs) in drinking water treatment: A review. *Water Research, 50*, 318–340. https://doi.org/10.1016/j.watres.2013.10.045

Richards, L. A., Kumari, R., White, D., Parashar, N., Kumar, A., Ghosh, A., Kumar, S., Chakravorty, B., Lu, C., Civil, W., Lapworth, D. J., Krause, S., Polya, D. A., & Gooddy, D. C. (2021). Emerging organic contaminants in groundwater under a rapidly developing city (Patna) in northern India dominated by high concentrations of lifestyle chemicals. *Environmental Pollution, 268*, Article 115765. https://doi.org/10.1016/j.envpol.2020.115765

Richardson, S. D., & Kimura, S. Y. (2017). Emerging environmental contaminants: Challenges facing our next generation and potential engineering solutions. *Environmental Technology and Innovation, 8*, 40–56. https://doi.org/10.1016/j.eti.2017.04.002

Romano, P., Zuffranieri, A., & Giacomo, G. Di (2025). *Energy valorization and resource recovery from municipal sewage sludge: Evolution, recent advances, and future prospects. 1–32.*

Rose, M., Fernandes, A., Mortimer, D., & Baskaran, C. (2015). Contamination of fish in UK fresh water systems: Risk assessment for human consumption. *Chemosphere, 122*, 183–189. https://doi.org/10.1016/j.chemosphere.2014.11.046

Saifur, S., & Gardner, C. M. (2021). Loading, transport, and treatment of emerging chemical and biological contaminants of concern in stormwater. *Water Science and Technology, 83*(12), 2863–2885. https://doi.org/10.2166/wst.2021.187

Shillito, R. (2022). International standards and guidelines. In *Application of sampling and detection methods in agricultural plant biotechnology.* Elsevier Inc. https://doi.org/10.1016/B978-0-323-99293-0.00004-0

Singh, S., Ghosh, N. C., & Gurjar, S. (2018). *Index-based assessment of suitability of water quality for irrigation purpose under Indian conditions* (Vol. 190, p. 29). Springer Nature. https://doi.org/10.1007/s10661-017-6407-3.

Sun, W., Liu, G., Zou, H., Wang, S., & Duan, X. (2025). Catalysts for co-catalysis in advanced oxidation processes. *Npj Materials Sustainability.* https://doi.org/10.1038/s44296-024-00046-4

Tamkeen, N., Rangoonwala, S., & Sanyal, K. (2020). Effect of pesticides on survival behavior, feeding pattern, growth and mortality of the marine edible fish Lepturacanthus savala. *International Journal of Creative Research Thoughts, 8*(6), 1522–1535.

Tickner, J. A., Simon, R. V., Jacobs, M., Pollard, L. D., & Saskia, K. (2021). The nexus between alternatives assessment and green chemistry: Supporting the development and adoption of safer chemicals. *Green Chemistry Letters and Reviews, 8253.* https://doi.org/10.1080/17518253.2020.1856427

Timoner, X., Alexandre, S., & Balc, L. (2018). Emerging contaminants and nutrients synergistically affect the spread of class 1 integron-integrase (intI1) and sul1 genes within stable streambed bacterial communities. *Water Research, 138*, 77–85. https://doi.org/10.1016/j.watres.2018.03.025

Tomai, P., Martinelli, A., Morosetti, S., Curini, R., Fanali, S., & Gentili, A. (2018). *Oxidized buckypaper for stir-disc solid phase extraction: Evaluation of several classes of environmental pollutants recovered from surface water samples.* https://doi.org/10.1021/acs.analchem.8b00927

Town, T., County, S., Testing, I., Substances, P., Pfoa, C. C., Levels, P., Epa, A., Advisory, H., February, O., Resources, N., & Health, P. (2018). *ASDWA PFAS water utility treatment case study town of blades. 1–4.* Delaware.

Trotta, V., Baaloudj, O., & Brienza, M. (2024). *Risks associated with wastewater reuse in agriculture: Investigating the effects of contaminants in soil, plants, and insects* (pp. 1–7). https://doi.org/10.3389/fenvs.2024.1358842

Ungureanu, N., & Voicu, G. (2020). *Water scarcity and wastewater reuse in crop irrigation.*

Vega, A. B., Frenich, A. G., & Mart, J. L. (2005). Monitoring of pesticides in agricultural water and soil samples from Andalusia by liquid chromatography coupled to mass spectrometry. *Analytica Chimica Acta, 538*, 117–127. https://doi.org/10.1016/j.aca.2005.02.003 (0003).

Wagner, S., Hüffer, T., Klöckner, P., Wehrhahn, M., Hofmann, T., & Reemtsma, T. (2018). Tire wear particles in the aquatic environment—A review on generation, analysis, occurrence, fate and effects. *Water Research, 139*, 83–100. https://doi.org/10.1016/j.watres.2018.03.051

Wang, Z., Dewitt, J. C., Higgins, C. P., & Cousins, I. T. (2017). *A never-ending story of per- and poly fl uoroalkyl substances (PFASs)?.* https://doi.org/10.1021/acs.est.6b04806

Wang, Y., & Good, K. D. (2024). Microplastics and PFAS air-water interaction and deposition. *Science of the Total Environment, 954*, Article 176247. https://doi.org/10.1016/j.scitotenv.2024.176247

Wang, F., Xiang, L., Leung, K. S., Elsner, M., Zhang, Y., Guo, Y., Pan, B., Sun, H., An, T., Wang, F., Xiang, L., Leung, K. S., Elsner, M., Zhang, Y., Guo, Y., Pan, B., Sun, H., & An, T. (2024). Emerging contaminants: A one health perspective. *The Innovation, 5*(4), Article 100612. https://doi.org/10.1016/j.xinn.2024.100612

Wezel, A. P. Van, Laak, L., Fischer, A., Bäuerlein, P. S., & Posthuma, L. (2017). *Environmental Sciences* (pp. 403–414). https://doi.org/10.1039/c7ew00077d

Xie, Z., Lu, G., Yan, Z., Liu, J., Wang, P., & Wang, Y. (2017). Bioaccumulation and trophic transfer of pharmaceuticals in food webs from a large freshwater lake. *Environmental Pollution, 222*, 356–366. https://doi.org/10.1016/j.envpol.2016.12.026

Yee, M. S., Hii, L., Looi, C. K., Lim, W., Wong, S., Kok, Y., Tan, B., Wong, C., & Leong, C. (2021). *Impact of microplastics and nanoplastics on human health.*

Yu, J., Gong, Z., Wang, S., Zhong, H., Tao, Y., Hou, Y., Fu, Q., Yang, H., Li, J., Wang, J., Hao, Z., & Ouyang, G. (2025). Two major deactivation mechanisms in carbon-based advanced oxidation processes (AOPs) dominated by electron-transfer pathway (ETP). *Applied Catalysis B: Environment and Energy, 364*(135). https://doi.org/10.1016/j.apcatb.2024.124850

Zhang, T., Jiang, B., Xing, Y., Ya, H., Lv, M., & Wang, X. (2022). Current status of microplastics pollution in the aquatic environment, interaction with other pollutants, and effects on aquatic organisms. In *Environmental science and pollution research.* Berlin Heidelberg: Springer. https://doi.org/10.1007/s11356-022-18504-8

Zhang, X., Robson, M., Jobst, K., Pena-Abaurrea, M., Muscalu, A., Chaudhuri, S., Marvin, C., Brindle, I. D., Reiner, E. J., & Helm, P. (2020). Halogenated organic contaminants of concern in urban-influenced waters of Lake Ontario, Canada: Passive sampling with targeted and non-targeted screening. *Environmental Pollution, 264*, Article 114733. https://doi.org/10.1016/j.envpol.2020.114733

Zubair, M., Roopesh, M. S., & Ullah, A. (2024). Nano ions and organic pollutants. *Environmental Science*, 3693–3720. https://doi.org/10.1039/d4en00143e

PART

VII

Water-use efficiency and circularity in industrial systems

CHAPTER

13 Circular economy in industrial water use: Approaches to water recovery, recycling, and sustainable effluent treatment

Aditya Tiwari[1], Ishita Sarkar[2], Chinmoy Kanti Deb[3], Anil Kumar Vinayak[4] and Mahesh Ganesapillai[3]

[1]*School of Engineering, Newcastle University, Newcastle Upon Tyne, United Kingdom;* [2]*Departament d'Enginyeria Química i Química Analítica, Universitat de Barcelona, Barcelona, Spain;* [3]*Mass Transfer Group, School of Chemical Engineering, Vellore Institute of Technology, Vellore, Tamil Nadu, India;* [4]*Department of Chemical and Materials Engineering, University of Alberta, Edmonton, AB, Canada*

1. Introduction

The global water crisis is at a critical juncture, intensified by ongoing population growth, rapid urbanization, climate change, and unsustainable practices. The escalating crisis presents significant humanitarian, economic, and geopolitical challenges that require urgent attention. Industrial activities, second only to agriculture in water consumption, significantly contribute to this crisis, especially in emerging economies. These industries release complex combinations of pollutants that exceed the capacity of natural systems, compromising the safety of water sources. Organic waste that comes from food processing and paper industries reduces oxygen levels, posing a threat to aquatic life. Toxic heavy metals, acids, and salts resulting from mining and manufacturing present significant toxicological risks. Excess nutrients from wastewater and agricultural runoff induce eutrophication and the proliferation of harmful algal blooms (Noor et al., 2020). Persistent Organic Pollutants accumulate within food webs, and emerging contaminants such as pharmaceuticals and microplastics present increasing challenges. Thermal discharges from energy production disrupt ecosystems by modifying temperature regimes and threatening species survival. The stressors result in contaminated water, ecosystem degradation, health risks, reduced agricultural productivity, and economic losses (Manasa & Mehta, 2020). The necessity for sustainable industrial water management is increasingly evident. Decisive action is necessary to protect water resources and ensure a sustainable future.

The standard linear "take-make-dispose" model of water utilisation is unsustainable in the face of global challenges. This approach encourages the loss of resources by regarding water as a disposable commodity. Industrial practices prioritize productivity and profitability, often neglecting environmental consequences, which leads to a focus on single-use disposal and dependence on end-of-pipe treatment instead of promoting reuse. Outcomes encompass freshwater scarcity, pollution of aquatic systems, and effects on ecosystem integrity and human health (Pandey, 2025). The linearity fosters a

Sustainable Water Management. https://doi.org/10.1016/B978-0-443-44853-9.00001-0

loss of embedded value in wastewater streams, since it does not use any of the resources that may be recovered. The model persists despite inefficiencies, primarily due to its legacy infrastructure and insufficient incentives for circular practices. The escalating global water crisis highlights the necessity for a transition to sustainable water management practices.

The Circular Economy Framework (CEF) offers a systematic and transformative strategy to address the limitations of the conventional linear water flow model, which has significantly contributed to the global water crisis. At its core, the circular economy seeks to decouple economic activity from the unsustainable patterns of resource utilization and waste generation (Giakoumis et al., 2020). In contrast to the traditional "take–make–dispose" paradigm, the circular economy operates on three foundational principles. First, it aims to design out waste and pollution by integrating sustainability into the design and production stages, ensuring that materials and processes minimize environmental harm from the outset. Second, it emphasizes maintaining the circulation of resources through enhanced durability, reuse, repair, and recycling, thereby extending the life cycle of materials and reducing the need for virgin resource extraction. Third, it focuses on regenerating natural systems by returning valuable materials and nutrients back to ecosystems, supporting their long-term health and resilience (Daga et al., 2025).

When applied to water management, the circular water economy prioritizes the recovery, reuse, and recycling of water within industrial systems, effectively closing the loop in the water cycle (Farfán Chilicaus et al., 2025; Nguyen et al., 2022). This approach encompasses several key applications that exemplify circular principles in practice. Water recovery and recycling initiatives (Karkou et al., 2024a) involve treating and reusing wastewater within industrial processes, thereby reducing dependence on freshwater sources. Resource recovery from wastewater (Zarei, 2020) focuses on extracting valuable by-products such as nutrients, metals, and energy, transforming waste streams into economic opportunities. Industrial symbiosis (Ramin et al., 2024) promotes collaboration among industries, enabling one facility's waste or by-product to serve as a resource for another, optimizing collective resource efficiency. Furthermore, closed-loop systems, including zero liquid discharge (ZLD) technologies (Abdelfattah & El-Shamy, 2024; Buyukada-Kesici et al., 2023), ensure that all wastewater is treated, recovered, and reused, achieving near-total elimination of liquid waste.

By embracing the principles of circularity, industries can transform water management from a traditional cost center into a strategic asset that enhances operational efficiency, strengthens resilience against water scarcity, and minimizes environmental pollution. The adoption of the Circular Economy Framework not only fosters sustainability and competitiveness but also aligns industrial operations with the broader goals of resource conservation, climate adaptation, and sustainable development. The shift toward a circular water economy is propelled by the dual forces of economic feasibility and ecological sustainability. This shift creates opportunities for investments in water recovery, recycling, and effluent treatment. The process of reclaiming nutrients, energy, and resources from industrial wastewater not only drives conservation objectives forward but also enhances the efficiency of industrial operations. The economic advantages encompass lower costs for water acquisition and discharge, in addition to income generated from reclaimed resources (Tomić & Schneider, 2020). Diverse reclamation sources enhance operational resilience, and adherence to environmental regulations mitigates financial risks. Taking proactive steps in environmental stewardship enhances a company's reputation and draws in investors who prioritize sustainability. The economic and environmental forces at play urge industries to rethink their connection with water, positioning circular strategies as essential for both business and ecological sustainability.

Thus, the current chapter explores how CE principles and emerging technologies can transform industrial water management into a more sustainable and resilient framework. It highlights the limitations of the conventional linear framework and advances CE as a robust alternative to address the intensifying global water crisis. The core objective is to articulate the foundational principles of CE within the context of industrial water systems, reconceptualizing water not as a disposable by-product but as a recoverable and valuable resource. The discussion examines the interplay of environmental pressures and economic drivers that are accelerating the adoption of CE-oriented water strategies. Subsequently, the chapter provides an overview of conventional and advanced approaches to water recovery, recycling, and reuse, with particular emphasis on sustainable effluent treatment and resource valorization. Attention is also given to the barriers and enabling conditions influencing CE implementation, alongside the role of governance, regulatory frameworks, and policy instruments.

2. Industrial water demand and environmental footprint

Industrial water supply and use patterns are influenced by both regional economic structures and water availability limits. In many water-scarce areas like the Middle East and parts of Australia, companies are rapidly supplementing or replacing freshwater withdrawals with non-traditional supplies such as desalinated saltwater, recycled wastewater, and brackish groundwater. This shift, prompted by both resource depletion and sustainability concerns, mirrors a larger worldwide trend of diversifying supply portfolios. The necessity for adaptable sourcing techniques is highlighted by the fact that, as shown in Fig. 13.1, agricultural and reservoir management uses continue to dominate worldwide sectoral water demand, with industrial and energy sectors accounting for a smaller but increasing proportion. Such adjustments provide crucial resilience against fluctuations in water supply while also decreasing burden on conventional freshwater supplies (Ferri & Bolelli, 2025).

2.1 Global and regional patterns of industrial water use

Industrial water use varies across regions based on water availability, industrial composition, and economic structures. Globally, 22% of freshwater withdrawals are industrial, though this varies between economies. In Western Europe and North America, industry accounts for 40%–50% of withdrawals due to manufacturing, power generation, and processing sectors that require water for cooling and operations (Sachidananda et al., 2016). In emerging economies across Asia, Latin America, and Africa, industrial water typically represents 10%–20% of withdrawals, though increasing with industrialization. In Malaysia's Kuantan River Basin, manufacturing comprised 32% of industrial activity in 2018, showing a 2% increase. China's sectoral variations show basic industries consuming 21.69 m^3 per 10,000 RMBWQ of production, versus 0.84 m^3 in high-technology industries and 736.66 m^3 in agriculture (Shah et al., 2023).

2.2 Functional categories of industrial water use

Industrial water consumption can be categorized into functional groups with distinct quality requirements (Shokri & Fard, 2023). Process water serves as an ingredient, reactant, or solvent in manufacturing, with purity needs ranging from ultrapure for microelectronics to potable for food processing. Cooling water forms the largest share of industrial use, exceeding half of total

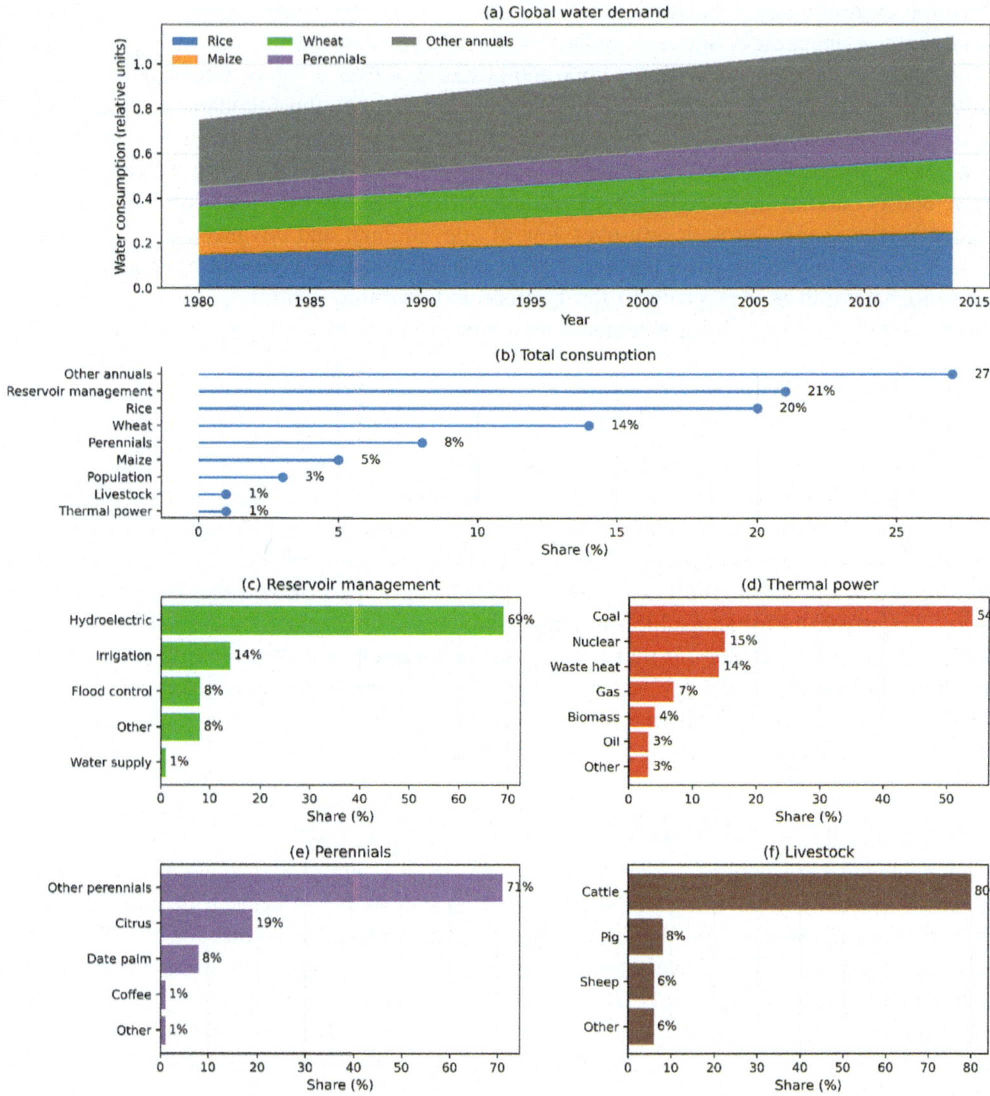

FIGURE 13.1

Sectoral breakdown of global water demand showing major contributors such as agriculture, livestock, and energy production.

consumption in petrochemicals and thermoelectric power generation for heat dissipation. Boiler feed water requires pretreatment including softening and demineralization to prevent scaling and corrosion. Clean-in-Place (CIP) water is essential in food, beverage, and pharmaceutical industries for meeting hygiene standards.

2.3 Sectoral water intensity

Several industrial sectors are characterized by high water intensity, with distinct consumption patterns. The iron and steel industry requires substantial volumes for cooling and quenching, with advances in cooling systems showing potential consumption reductions of up to 70%. The textile sector generates complex wastewater containing dyes, salts, and chemicals from bleaching, dyeing, and finishing processes. The chemical and petrochemical industries discharge organic contaminants requiring advanced treatment. The energy sector, especially thermoelectric plants, consumes vast water amounts for cooling, with thermal discharges affecting aquatic ecosystems. In food and beverage, high water quality and volume are essential, with water value generation exceeding 300 times the input in some cases. Mining and mineral processing produce metal-laden effluents requiring careful management. Data centers are gaining attention for cooling water needs in water-scarce regions, necessitating efficient thermal management solutions.

2.4 Environmental implications of industrial effluent discharge

Industrial wastewater discharges pose environmental risks due to their complex chemical compositions. High chemical oxygen demand (COD) and biological oxygen demand (BOD) accelerate dissolved oxygen depletion in aquatic systems, disrupting ecological balance. Nutrient-rich effluents promote eutrophication and hypoxic zones. High salinity in industrial effluents challenges agricultural irrigation, degrading soil structure and reducing crop productivity. Untreated effluents introduce hazardous pollutants into waterways, including pharmaceuticals, pesticides, and petroleum hydrocarbons, threatening water quality and bioaccumulating through food chains (Ahmad et al., 2024). Emerging contaminants like microplastics and per- and polyfluoroalkyl substances (PFAS) require advanced remediation approaches, such as high-performance adsorbents and membrane systems (Sanzana et al., 2025). Thermal pollution from heated cooling water alters temperature regimes and dissolved oxygen levels, impairing ecosystem function and degrading cold-water habitats. Heavy metals in industrial effluents, including chromium, nickel, copper, zinc, arsenic, and cadmium, cause physiological stress and bioaccumulation in aquatic organisms (Tabassum et al., 2024). Industrial water use is highly disproportionate, with facilities exceeding 20 million gallons annually representing 3% of operations but 95% of total use. This concentration presents opportunities for targeted interventions at large-scale facilities to reduce industrial water demand and advance environmental protection.

3. Water recycling and recovery strategies and practices

Industrial water recycling has evolved from being a mere sustainability choice to becoming an essential operational requirement, influenced by the challenges of water scarcity, stricter environmental regulations, and increasing cost pressures. This transformation has been facilitated by progress in treatment methodologies, the incorporation of various technologies, and tailored approaches to water quality management. The global market for industrial water recycling has experienced significant growth, especially with membrane-based systems, which have demonstrated water savings of 60%–90% across various sectors (de Melo Ribeiro & dos Santos Silva, 2020). This section delves into different treatment types, advanced processes, and operational strategies that characterize cutting-edge industrial water recycling and recovery methods.

3.1 Core treatment categories

In order to achieve the desired effluent quality, wastewater treatment systems are typically divided into three main technological categories: primary (physical), secondary (biological), and tertiary (advanced physicochemical) as illustrated in Fig. 13.2. Each of these category's functions as a sequential barrier that gradually eliminates suspended solids, biodegradable organics, and remaining micro-contaminants. Modern industrial water recycling systems use a multi-barrier treatment approach integrating physical, chemical, and biological processes to meet stringent reuse requirements. The physical pre-treatment stage removes suspended solids, oils, and large particulates, reducing loads on downstream processes. Fine mesh screening technologies, operating at apertures as small as 0.25 mm with automated cleaning systems, have improved reliability while minimizing maintenance. The use of stratified media in multimedia filters has enabled effective particle size fractionation, enhancing removal efficiency.

Clarification processes using lamella plate settlers achieve suspended solids removal efficiencies near 95%, reducing footprint compared to conventional clarifiers. Dissolved Air Flotation (DAF) systems remain essential for oil-rich effluents, achieving 90%–99% removal of free oils while producing effluents compatible with advanced treatment (Zabel, 1985). Chemical treatment forms the second layer of this hierarchical approach, focusing on dissolved contaminant removal and water chemistry optimization. Coagulation–flocculation processes, traditionally a major chemical source, are increasingly guided by real-time particle analysis and automated dosing control, reducing reagent use while maintaining performance (Alazaiza et al., 2025). Oxidation processes have gained prominence, with chlorine dioxide and ozone serving as potent disinfectants and oxidants for recalcitrant organic pollutants; ozone also provides color and odor abatement. pH control remains critical for preventing corrosion, minimizing scaling, and ensuring regulatory compliance, with modern automated systems maintaining setpoints within ±0.1 units (Kalandarov et al., 2024).

At the biological stage, efficient degradation of organic matter and nutrient removal underpin industrial wastewater recycling sustainability. Conventional activated sludge processes are being complemented by advanced variants like sequencing batch reactors and extended aeration systems,

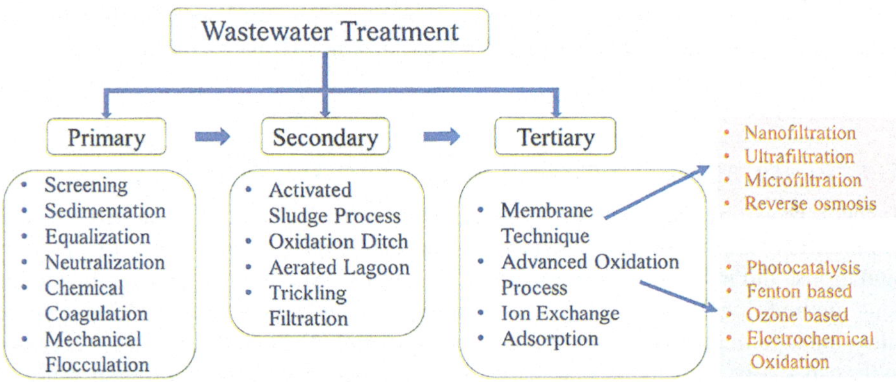

FIGURE 13.2

Classification of wastewater treatment based on fundamental technologies.

showing enhanced resilience under fluctuating conditions (Smetana & Grosser, 2024). The membrane bioreactor (MBR) couples' biological degradation with ultrafiltration, producing high-clarity effluents with BOD5 below 5 mg/L and turbidity under 0.1 NTU. Anaerobic technologies, particularly Upflow Anaerobic Sludge Blanket (UASB) reactors, enable 80%–90% COD removal while generating biogas, embedding resource recovery principles within treatment design (Khan et al., 2024). These developments represent a shift toward circular water management, where conventional processes are integrated into adaptive systems, reframing wastewater as a resource rather than liability.

3.2 Advanced and specialized processes

Adsorption remains as the most used treatment in removing trace contaminants, particularly persistent organic compounds and color bodies that resist conventional treatment. Activated carbon has been widely applied and demonstrates removal efficiencies up to 99% for chlorinated solvents and phenolic compounds. The development of advanced adsorbents, such as surface-modified zeolites, metal-organic frameworks (MOFs), and engineered biochars, has expanded adsorption capacity and selectivity for emerging contaminants (Hussain et al., 2025). Ion exchange processes are commonly used for removing hardness, nitrates, and heavy metals, with innovations in resin design enhancing selectivity and regeneration efficiency (Ahmed et al., 2022). Electrochemical technologies are gaining prominence in industrial wastewater treatment due to their versatility. Electrocoagulation generates coagulants in situ, effectively removing suspended solids and chemical oxygen demand (COD). Electrooxidation, using boron-doped diamond electrodes, achieves 90%–99% removal of recalcitrant contaminants like pharmaceuticals and synthetic dyes, making it powerful for treating challenging wastewaters (Rodrigo et al., 2010).

Membrane-based filtration technologies serve as the backbone of water recycling, offering scalable treatment solutions. Microfiltration (MF) and ultrafiltration (UF) remove suspended solids, bacteria, and viruses at low pressures, reducing fouling in downstream processes. Nanofiltration (NF) targets divalent ions and organic compounds while allowing partial salt passage, making it suitable for partial demineralization. Reverse osmosis (RO) remains the benchmark for high-purity water, achieving over 99% salt rejection and near-complete organic contaminant removal (Pérez-González et al., 2012). Forward osmosis (FO) and membrane distillation (MD) are advancing water treatment for high-salinity wastewaters. FO offers energy advantages, while MD handles extreme salinities beyond RO limits when integrated with waste-heat recovery. Ceramic membranes, known for chemical and thermal stability, are becoming cost-competitive for treating aggressive industrial effluents (Li et al., 2020).

The integration of membrane separation with biological treatment has transformed wastewater reclamation through balanced contaminant removal and operational sustainability. Membrane bioreactor–reverse osmosis (MBR–RO) systems combine high-purity water production with efficient organics removal, with biological pretreatment reducing fouling on RO membranes. MBR–nanofiltration combinations provide a balance between treatment quality and energy efficiency when ultrapure water isn't required (Nguyen et al., 2022). Advanced oxidation–membrane systems achieve 99% removal of recalcitrant contaminants while minimizing fouling. Photocatalytic membrane reactors integrate oxidative degradation with membrane separation, maintaining high flux while eliminating micropollutants (Ulinici et al., 2020). These developments show the shift toward integrated, efficient treatment frameworks for complex industrial wastewater.

3.3 Best practices in industrial water reuse

A fit-for-purpose approach in industrial water management matches treatment to end use, preventing excess costs. Water quality frameworks align reuse applications like boiler feed and cooling systems with TDS, turbidity, and microbial thresholds. Treatment design requires wastewater characterization and process simulations to determine optimal configurations. Reliability depends on fouling control, redundancy, and monitoring. Best practices include pretreatment optimization, permeability monitoring with automated cleaning, and modular architectures for maintenance.

Maximizing water recovery requires balancing production with concentrate management through technologies like crystallization and next-generation membrane systems, with facilities reporting recovery rates above 95% (Pérez-González et al., 2012). Modern treatment platforms integrate biological, chemical, physical, and membrane processes to deliver high-quality reuse water while reducing environmental impacts. Hybrid configurations and innovative materials are improving treatment performance and cost-effectiveness. Future developments will focus on automation, resource recovery, and resilient designs for varying feedwater conditions. Industries can meet operational and sustainability goals while reducing freshwater dependence through best practices.

4. Process optimization for water efficiency in industry

4.1 Water audits and flow analysis

Water auditing is a fundamental diagnostic tool for achieving water efficiency in industrial systems. It systematically evaluates water inflows, outflows, and consumption within defined boundaries using a mass balance framework. A comprehensive audit delineates water sources, categorizes usage, identifies reuse potential, and quantifies losses from leakage, evaporation, or inefficiencies (Neelofar et al., 2023). Water audits use flow meters for direct measurement and process mass balances for indirect estimations. In advanced applications, audits include water quality profiling to optimize reuse schemes based on contaminant loading and organic content (Ahmetović et al., 2017). A structured industrial water audit typically progresses through distinct stages, from source identification to continuous performance review. Each stage involves specific measurement tools, parameters, and outputs that collectively inform conservation strategies. Table 13.1 summarizes the technical framework for implementing an effective water audit in industrial settings, integrating both volumetric and quality-based assessments (Neelofar et al., 2023).

Flow analysis complements auditing by providing temporal and spatial resolution of water movement through process units. Methods range from manual flow-tracing using conservative tracers, to computational fluid dynamics (CFD)-based hydraulic modeling for complex distribution systems. In water-intensive sectors such as pulp and paper or petrochemicals, integrating dynamic flow analysis with statistical process control enables the identification of diurnal demand peaks, transient hydraulic losses, and non-revenue water (NRW) points (Ferrucci & Vocciante, 2021). Contemporary water auditing frameworks incorporate feedwater composition and process discharge variability. The water management model accounts for moisture content fluctuations in raw materials and products, reducing systematic errors in water use estimation. This aligns with ISO 46,001:2019 guidelines for water efficiency management (Ferrucci & Vocciante, 2021). Audit results inform decision-support systems for water conservation, prioritizing leak repair, retrofitting, and process redesign. Integration with IoT platforms enables continuous, real-time performance assessment rather than periodic manual audits.

Table 13.1 Technical framework for industrial water audits.

Audit step	Measurement/ Analysis tools	Parameters measured	Typical output	Applications	Ref.
Source identification	Ultrasonic/ Electromagnetic flow meters, intake records	Flow rate, totalized volume, water quality indices	Source water balance	Identification of primary and secondary water sources	Neelofar et al. (2023)
Process flow mapping	IoT-linked flow meters, tracer studies, hydraulic models	Instantaneous flow, pressure, temperature	Flow profiles, diurnal variability curves	Detection of high-demand processes and flow peaks	Ferrucci and Vocciante (2021)
Water quality characterization	Inline TOC, conductivity meters, turbidity sensors	Chemical oxygen demand, total suspended solids, salinity, temperature	Water quality matrix by process unit	Identification of reuse and recycle potential	Ahmetović et al. (2017)
Loss detection	Pressure logging, infrared thermography, leak correlators	Pressure drops, abnormal flows, surface temperature	Leak location and quantification	NRW reduction	Ferrucci and Vocciante (2021)
Mass balance computation	Spreadsheet-based balance, supervisory control and data acquisition (SCADA)-linked databases	Inflow-outflow volumes, evaporation losses, product water	Plant-wide water mass balance	Prioritization of conservation interventions	Sachidananda et al. (2016)
Continuous performance review	IoT dashboards, cloud analytics	Real-time consumption trends, efficiency KPIs	Continuous improvement reports	Integration into ISO 46001 water efficiency systems	Ferrucci and Vocciante (2021)

4.2 Process modifications for water minimization

Process modifications for water minimization can be implemented at three levels: unit operations, process integration, and plant-wide network synthesis. These modifications target reduced freshwater intake while maximizing reuse and regeneration potential, guided by water audits and flow analyses. At the unit operation level, changes include redesigning CIP cycles with volumetric flow control, optimizing backwash frequencies for filtration units, and adjusting evaporation parameters to minimize condensate loss. Such modifications are best quantified using a water mass balance, expressed as (Sachidananda et al., 2016):

$$Q_f = Q_p + Q_r + Q_l \tag{13.1}$$

where Qf is freshwater input ($m^3.day^{-1}$), Qp the water in product ($m^3.day^{-1}$), Qr the recycled water (m3.day-1), and Ql the water losses from evaporation, leakage, and blowdown ($m^3.day^{-1}$). Eq. (13.1) helps facilities identify freshwater demand and reductions after modifications (Sachidananda et al., 2016). At process integration level, cascading water reuse strategies allow effluent from high-grade processes to be reused in lower quality operations. Hybrid treatment trains combining physical, chemical, and biological methods produce water quality for reuse points. Minimum freshwater targets can be determined using water pinch analysis (Ferrucci & Vocciante, 2021):

$$W_{\min} = \frac{\sum_{i=1}^{n} (F_i \cdot C_i)}{C_{\text{target}}} \tag{13.2}$$

W_{min} is the minimum freshwater requirement ($m^3.day^{-1}$), F_i is the flow of stream I ($m^3.day^{-1}$), C_i the contaminant concentration in stream i (mgL^{-1}), and C_{target} the maximum permissible concentration for process use (mgL^{-1}). This formulation enables precise allocation of streams to minimize freshwater demand while maintaining product quality (Ahmetović et al., 2017; Ferrucci & Vocciante, 2021). At the plant-wide level, Water Network Synthesis (WNS) approaches are used to reallocate water streams based on contaminant constraints, cost functions, and flow continuity equations. To guide and evaluate these modifications, industries increasingly adopt Water Usage Efficiency (WUE) indicators (Esteves et al., 2025):

$$\text{WUE} = \frac{P_{\text{output}}}{Q_f} \tag{13.3}$$

where WUE is the water usage efficiency ($\text{tonnes}_{\text{product}} \cdot (m^3{}_{\text{freshwater}})^{-1}$), P_{output} is the total product output (tonnes.day^{-1}), and Q_f the freshwater input ($m^3.day^{-1}$). Higher WUE values indicate more efficient water utilization and are used for performance benchmarking and continuous improvement (Esteves et al., 2025). Successful process modifications require periodic reassessment to ensure they remain aligned with production demands and evolving raw material characteristics.

4.3 Real-time monitoring and control systems

Real-time monitoring and control are critical to sustaining water efficiency by ensuring audit outcomes and process modifications are continuously implemented. Modern systems integrate multi-point sensors with centralized data platforms for immediate detection of performance deviations (Esteves et al., 2025; Mutikanga et al., 2013). Instrumentation like ultrasonic flow meters, pressure transducers, and inline TOC analyzers monitor flow dynamics and water quality. These devices connect through IoT frameworks, enabling cloud-based dashboards and automated alerts for anomalies (Mutikanga et al., 2013). Control strategies range from PLCs to Advanced Model Predictive Control (MPC), allowing proactive operational adjustments. MPC optimizes membrane cleaning intervals in reverse osmosis processes, balancing flux recovery while minimizing water and chemical use (Ahmetović et al., 2017). Digital twin technologies create virtual replicas of water networks, allowing operators to simulate strategies under varying conditions.

Machine learning algorithms enable predictive leak detection and anomaly identification by recognizing system deviations before visible failures, improving reliability and reducing maintenance costs (Ferrucci & Vocciante, 2021; Mutikanga et al., 2013). The integration of energy-water monitoring frameworks with ISO 50,001 ensures efficiency improvements in one domain do not create inefficiencies in another (Esteves et al., 2025). The convergence of continuous monitoring, AI-driven analytics, and adaptive control mechanisms enables closed-loop water efficiency management. By incorporating real-time data into optimization strategies, industries can transition toward circular water use while maintaining operational resilience.

5. Sustainable industrial water management: A circular economy approach

The globally looming threat of water scarcity has become an existential crisis that requires a fundamental shift from the linear industrial water management system of "take-make-dispose and dilute" to a more circular approach of "recycle and reuse." This paradigm shift will redefine industrial wastewater from a mere waste stream to an asset with valuable resources. To enhance water and resource efficacy for industrial use, it is imperative to re-envision and remodel existing and new wastewater treatment facilities as "Resource surplus facilities" (Cairone et al., 2024). This would reduce environmental adversities, increase water efficiency, and create new value streams, all of which offer significant economic benefits alongside environmental stewardship and are pressing priorities (Rodríguez et al., 2022).

5.1 Zero liquid discharge systems (ZLD)

ZLDs aim to eliminate liquid effluent discharge from industrial water settings by maximizing water recovery and concentrating contaminants into solids. Water-intensive sectors such as power generation, pharmaceuticals, and textiles are adopting ZLD infrastructure (Yaqub & Lee, 2019). ZLD systems use multistage processes, starting with effluent pretreatment through physical, chemical, or biological methods, followed by advanced water treatment technologies, including membrane treatment (e.g., membrane distillation, ultrafiltration, and reverse osmosis) and thermochemical processes (e.g., crystallization and evaporation), to produce reusable water and solids (Abdelfattah & El-Shamy, 2024). ZLD offers key advantages over conventional discharge treatment: it reduces water pollution while enhancing reusable water availability, and concentrated solids can be treated or used as valuables (Tong & Elimelech, 2016). However, ZLD infrastructure requires high capital expenditure and energy demand. Membrane technologies face fouling and wetting challenges, requiring continuous maintenance (Panagopoulos & Haralambous, 2020). Despite these challenges, development of energy-efficient and alternative primary treatment technologies for ZLD with better waste stream management continues to grow.

5.2 Energy recovery from industrial wastewater

Industrial side-streams contain organic matter, reusable chemicals, and thermal, chemical, and hydraulic potential for developing energy-sufficient treatment facilities (Duarte et al., 2025). Energy recirculation in water treatment facilities reduces greenhouse gas emissions and supports circular economy goals (Zarei, 2020). Microbial fuel cells (MFCs), anaerobic digestion (AD), and biohydrogen production are preferred energy recovery methods (Guo et al., 2019). MFCs produce electricity from wastewater's organic components using electro-active microbes and have gained prominence for industrial use (AlSayed et al., 2020). Anaerobic digestion breaks down organic matter without oxygen to produce biogas and components like organic acids and hydrogen (Jain et al., 2020), which can generate heat and electricity (Negro et al., 2025). AD can integrate with membrane bioreactors and digest matter from multiple sources, though uncaptured methane can increase emissions (Stazi & Tomei, 2021). Biohydrogen production converts carbohydrate-rich effluents to hydrogen and VFAs, but has limited yield and produces residual VFAs (Tiwari & Nakamura, 2024). Converting WWTPs to water resource recovery facilities can transform operational models and may require energy partnerships (Liu et al., 2022).

5.3 Nutrient and material recovery from industrial effluents

In addition to water and energy, industrial effluents are rich sources of valuable nutrients and materials. If sustainably recovered, they can reduce environmental harm and enhance resource security. Key resources and their associated recovery technologies include:

5.3.1 Nutrients

Nitrogen and phosphorus are crucial raw materials for agriculture. The complete recovery and reuse of nutrients from wastewater can successfully close the fertility cycle. This could reduce freshwater contamination through eutrophication and increase resource security, offering a win-win scenario (Akter et al., 2024). Studies show that N and P in wastewater can be cost-effectively recovered using sorbents derived from agricultural and industrial waste (Simha & Ganesapillai, 2017). Chemical precipitation of mineral salts using additives can recover fertilizer-grade nutrients. Other methods, including electrochemical, recovery, and biological processes, are being developed and have gained attention for nutrient recovery from side streams industrially (Witek-Krowiak et al., 2022).

5.3.2 Metals

With the increase in industrialization, the demand for critical metals for industries such as semiconductors and electronics has increased. This has led to extensive mining of limited metal resources. However, industrial wastewater and mining trails, themselves are sources of these valuable critical metals. Membrane technologies, in particular, have proven to be effective in removing and concentrating metals such as chromium, cobalt, nickel, copper, zinc, cadmium, mercury, lead, arsenic, boron, and silica in high-purity forms (Staszak & Wieszczycka, 2023). Electrochemical methods such as reduction, electro-deionization, and electrodialysis have proven efficient in recovering precious metals like silver and gold (Devda et al., 2021).

5.3.3 Organic components

High-strength organic wastewaters can be converted into valuable commodities. The carboxylate platform uses microbiomes to synthesize mixed carboxylic acids, including short- and medium-chain fatty acids, from organic waste streams (Holtzapple et al., 2022). These acids are useful as food preservatives, solvents, plasticizers, and biofuels.18 The production of high-value biopolymers, such as those used in bioplastics manufacturing, can be achieved using these organic acids, hydrogen, and methane (Edo et al., 2025).

5.3.4 Nature-based solutions for industrial wastewater treatment

Nature-Based Solutions utilize natural processes and ecosystems for wastewater treatment, offering sustainable alternatives to traditional "gray" systems. These solutions use vegetation, soil, and microbial communities to eliminate contaminants (Afzal et al., 2024). Key NbS technologies include algae-based systems and constructed wetlands. Algal grass scrubbers grow algae to absorb nutrients and CO_2 while generating biomass for bioenergy and biofertilizer (Ugwuanyi et al., 2024). Constructed wetlands are engineered systems that treat effluents using microbial decomposition and plant absorption, effectively handling industrial wastewaters from petrochemical and textile facilities (Wu et al., 2015). These sustainable solutions require less maintenance than conventional technologies and serve as biodiversity centers while converting infrastructure from "gray" to "green" (Afzal et al., 2024).

6. Digitalization, artificial intelligence, and smart water management

The convergence of digital technologies, artificial intelligence (AI), and advanced analytics is reshaping industrial water management by enabling proactive, efficient, and sustainable practices. Through integration of sensors, Internet of Things (IoT) architectures, big data analytics, and cyber-physical systems, industries can transition toward smart water networks that are resilient, adaptive, and resource-efficient (Liu et al., 2021).

6.1 Sensors, role of IoT, data integration, big data, and smart infrastructure

Thc foundation of smart water management is deploying advanced sensing technologies across the water cycle. IoT-enabled sensor networks monitor parameters like flow rate, pressure, turbidity, temperature, pH, and dissolved oxygen in water streams. The IoT ecosystem comprises sensing hardware, data transmission networks, and data processing capabilities. This architecture functions like a neural system where sensors act as inputs, networks transmit signals, and analytics platforms act as the processing brain (Zekri et al., 2022). Data collection includes Automated Meter Reading (AMR) systems and SCADA platforms that monitor distribution networks and detect anomalies, as well as consumer-end sensors in fixtures and irrigation systems for consumption profiling.

Integration of datasets from multiple sensor modalities, known as data fusion, is critical for situational awareness. Multi-level fusion processes correlate heterogeneous data streams to produce insights, such as identifying pressure drops indicating leaks or correlating demand patterns with pump efficiency (Liu et al., 2021). Big Data frameworks provide scalable analytics for petabyte-scale water data. Cloud architectures with edge computing improve latency-sensitive operations like real-time leak detection and pumping optimization. Coupling IoT water data with cross-sector datasets enables integrated optimization, avoiding peak energy tariffs while maintaining performance (Farhaoui & El Allaoui, 2024). Smart water grids embed control capabilities, allowing automated valve actuation and flow rerouting based on predictive analytics. This transforms monitoring into adaptive, self-regulating infrastructure.

6.2 AI/ML for predictive maintenance and leak detection

Artificial intelligence (AI) and machine learning (ML) are transformative tools for enhancing operational reliability and reducing non-revenue water losses through predictive maintenance. By leveraging SCADA records, telemetry streams, and asset data, predictive models can forecast failures and enable targeted maintenance. These approaches extend asset lifespans and optimize resources (Masud et al., 2025). ML applications in water systems, particularly leak detection, outperform conventional methods. Techniques like support vector machines (SVM), artificial neural networks (ANN), and fuzzy logic models process multi-parameter datasets to identify anomalies. These models capture nonlinear patterns that characterize early-stage leaks while distinguishing them from normal fluctuations, improving detection accuracy (Chhipi-Shrestha et al., 2023).

The integration of AI with hydraulic model-based leak detection advances real-time fault localization by optimizing detection thresholds and accelerating responses. Reinforcement learning techniques add adaptability by adjusting valve settings to isolate leaks with minimal disruption, maintaining network integrity while reducing water losses. AI applications in cooling water systems demonstrate predictive analytics potential. By correlating variables like conductivity, temperature, and chemical dosing, AI models can forecast scaling events before impacting performance, enabling timely corrective measures (Masud et al., 2025). These integrated applications enhance predictive capabilities of water utilities while supporting resilient, data-driven frameworks for sustainable water system operations.

6.3 Digital twins for water networks

Digital twins, as virtual replicas of water networks, integrate sensor data with hydraulic models to enable system state replication and support decision-making. These platforms provide testing of operational strategies like emergency sequences and pump scheduling, ensuring system resilience. When integrated with artificial intelligence, digital twins compare predicted and observed behaviors to identify deviations from leaks or sensor drift in real time. Implementation is supported by model-based systems engineering (MBSE), which combines modeling and digital threads with Internet of Things (IoT) platforms for data acquisition. Derivative technologies including big data analytics, blockchain, and cloud computing enhance system intelligence and responsiveness.

As shown in Fig. 13.3, the interplay between modeling foundations and computational frameworks enables real-time monitoring, predictive analytics, and adaptive control of industrial water infrastructure (Wang et al., 2024a; 2024b). Digital twins contribute to lifecycle optimization by simulating impacts of asset replacement, chemical dosing, and climate stressors on water systems. By providing a platform that visualizes operational and environmental datasets, they support collaboration among engineers, managers, and regulators, ensuring technically robust and transparent decisions (Ramos et al., 2023). Digital twins represent a holistic framework for advancing sustainability, efficiency, and resilience in water infrastructure management.

6.4 Water use and sustainability concerns in data centers

The expansion of data centers poses challenges for water sustainability, as hyperscale facilities rely on evaporative cooling systems requiring high water use in water-stressed regions. In indirect evaporative

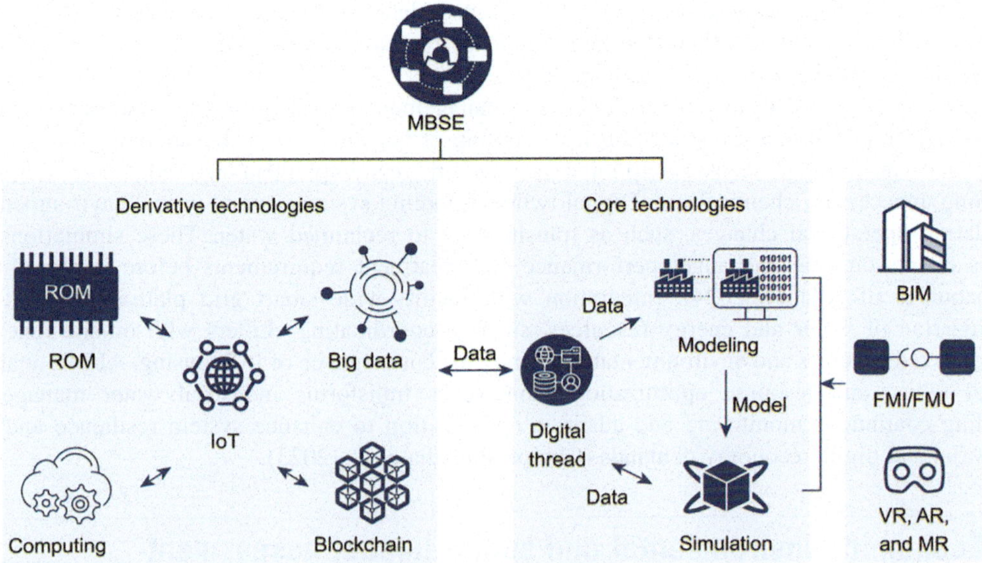

FIGURE 13.3

General key technology architecture of digital twins (Wang et al., 2024a; 2024b). Copyright 2024, Elsevier.

cooling, make-up water compensates for evaporative losses and blowdown, governed by cycles of concentration (CoC). Higher CoC reduces water consumption but increases scaling risk, requiring chemical treatment for reliability. These trade-offs demonstrate the water-energy nexus in data centers. While evaporative cooling minimizes electrical demand but increases water use, air-cooled systems reduce water needs but raise energy consumption and emissions. AI-driven thermal optimization models now adjust parameters like CoC, airflow, and chiller load to minimize environmental impact across water and energy dimensions (Al Kez et al., 2022).

IoT-enabled monitoring systems provide real-time visibility into water use through sub-system flow metering and cooling tower volume tracking. Combined with IT load metrics, these data enable adaptive load shifting to balance water and energy use. Reclaimed water sources and district cooling networks can reduce freshwater dependency, though requiring precise control for thermal reliability. Sustainable cooling requires holistic evaluation through lifecycle assessment to quantify benefits across operational parameters, water treatment, and infrastructure impacts. Through AI optimization, IoT monitoring, and lifecycle assessment, data centers can manage water-energy trade-offs while improving sustainability (Ahmetović et al., 2017; Al Kez et al., 2022).

6.5 Resource use optimization in industrial cooling

Industrial cooling systems in power generation, petrochemicals, and manufacturing are major water consumers, making optimization essential for sustainability. Effective strategies must balance reducing water withdrawal, chemical dosing, and thermal efficiency. Closed-loop systems with hybrid wet-dry cooling towers achieve this by switching modes based on ambient conditions and electricity prices. Predictive control algorithms use real-time sensor data to determine optimal setpoints, reducing blowdown volumes and maintaining system efficiency by maximizing cycles of concentration within scaling thresholds (Farhaoui & El Allaoui, 2024).

Big data analytics adds intelligence to system management, enabling condition-based maintenance by identifying performance degradation from fouling or biofilm growth. By analyzing heat transfer coefficients and pump power consumption trends, facilities can implement targeted cleaning regimens, avoiding unnecessary chemical use. Digital twins of cooling systems create virtual environments for simulating operational changes, such as transitioning to reclaimed water. These simulations help assess effects on heat exchanger performance and treatment requirements before implementation (Farhaoui & El Allaoui, 2024). Integration with facility-wide smart grid platforms enables co-optimization of water and energy resources, such as coordinating chillers with on-site renewable energy to reduce costs and environmental impacts. This convergence of IoT sensing, AI/ML analytics, digital twins, and resource optimization frameworks transforms industrial water management, enabling continuous monitoring and adaptive optimization to enhance system resilience and meet industrial and digital economy demands (Chhipi-Shrestha et al., 2023).

7. Economic, environmental and social impact assessment

Assessing the impact of industrial water circularity is pivotal for translating ambitious decisions into high-quality value propositions. This section delves into the multifaceted frameworks, methods, standards, and challenges in evaluating and optimizing water circularity in industrial settings. It highlights not only the direct benefits of water circularity, such as waste minimization and resource reutilization, but also its implications for social equity and environmental health (Delgado et al., 2024).

7.1 Economic viability, cost-benefit analysis of circular water practices

The economic assessment of industrial wastewater reuse requires a holistic methodology that accounts for costs, benefits, and limitations across financial, environmental, and social dimensions (Ghafourian et al., 2022). The Shadow Pricing Life Cycle Cost-Benefit (SPLCCB) approach has emerged as a comprehensive tool for evaluating circular water systems' economic viability (Ghafourian et al., 2021). The SPLCCB framework integrates externalities and conventional financial indicators, offering a realistic appraisal of system performance. Internal costs comprise Capital Expenditures (CAPEX) and Operational Expenditures (OPEX), while external costs include environmental and social impacts (Mbavarira & Grimm, 2021). Studies show circular water systems achieve economic viability when external dimensions are included, with ROI up to 13.04% (Orsini et al., 2023) and positive NPV returns with one to 3-year payback periods (Afzal et al., 2024). Benefits include reduced freshwater costs, carbon emissions, wastewater fees, and compliance costs, plus revenue from recovered resources (Das et al., 2025; Jones et al., 2021).

The inclusion of environmental considerations through nature-based solutions (NbS) enhances cost efficiency, with studies reporting savings of up to €1,984,335 compared to conventional alternatives (López-Serrano et al., 2023). The Renault-Nissan Tangier plant reported a 70% reduction in water consumption and operational cost savings through circular water practices (El Khatir, 2021). Such cases demonstrate the benefits of integrating circular industrial water systems (CIWS). However, to ensure positive economic, environmental, and social outcomes, sensitivity analyses must account for economic and societal variability (Hoareau & Tam, 2024). As shown in Table 13.2, multiple variables influence wastewater reuse viability, highlighting the need for multidimensional evaluation frameworks in sustainable industrial water management.

7.2 Life cycle assessment (LCA) of water systems

Life Cycle Assessment (LCA) serves as a tool for evaluating environmental impacts of circular industrial wastewater systems, providing a framework to assess sustainability across a system's lifespan, from raw material procurement through operation to end-of-life management (Curran, 2013). By quantifying environmental burdens, LCA identifies hotspots within the industrial water cycle and provides insights into system performance. Given the multi-sectoral nature of water use, LCAs frequently involve complex analyses that integrate approaches like material flow analysis to capture resource use and interdependencies (Farfán Chilicaus et al., 2025). The LCA framework minimizes burden shifting, where improvements in one process stage may increase impacts in another, ensuring a holistic environmental footprint. The reliability of these assessments depends on accurate datasets, making clean and comprehensive indicator databases essential (Nika et al., 2020). LCAs can be categorized into four types: cradle-to-grave, evaluating impacts from extraction through disposal; cradle-to-gate, assessing processes from extraction to final usage; gate-to-gate, focusing on individual processes; and cradle-to-cradle, emphasizing material reuse in circular economy principles (Nabavi-Pelesaraei et al., 2019).

Conventional applications of LCA in water systems have employed the cradle-to-grave approach to assess environmental performance (Samani, 2023). However, recent developments favor the cradle-to-cradle perspective, which reflects circular water management objectives by incorporating resource recovery and reuse. LCAs have been employed to evaluate tertiary treatment options for wastewater reuse and inform advanced circular wastewater systems, including city-wide initiatives such as the Hammarby Sjöstad eco-city in Sweden (Das et al., 2025).

7.3 Water-energy nexus and water-carbon footprint

The interdependencies between water and energy systems are captured by the Water–Energy Nexus, which emphasizes water's role as both an input for energy generation and a resource requiring substantial energy for its management (Castelli et al., 2024). Within circular industrial water utilization, this interaction is significant, as optimizing water use enhances efficiency and reduces water–carbon footprint. Industrial water circuits including pretreatment, transportation, and process water treatment are energy-intensive operations, with pumping accounting for 30%–40% of energy bills (Oliveira et al., 2019). The transition toward renewable technologies like hydrogen electrolysis further complicates this relationship due to their water-intensive nature (Lin et al., 2025).

Water reuse within industrial operations mitigates these interdependencies by reducing freshwater extraction and transportation demands. Recirculating process water in float glass plants reduces freshwater demand by 15%–65%, achieves energy savings up to 52%, and lowers operational costs by 13%–35%, reducing water–carbon footprint (Dakwala et al., 2014). Zero Liquid Discharge strategies maximize water recovery while minimizing energy requirements. Centralized recycling of industrial wastewater has shown potential carbon emission reductions of 1.06 kg per ton of product (Pahunang et al., 2021).

Table 13.2 Impact of economic and social variables on the viability of circular use of industrial water.

Economic variables	Investment flow	Unit/Proxy	Effect of variable ▲ on economic viability	References
CAPEX	Up-front/Capital/ One time investment on assets	Currency units	▼	Ilyas et al. (2021)
OPEX	Recurring/ Operating/ Maintenance cost	Currency units/ year	▼	Ilyas et al. (2021)
Inflation	Increase in price of materials and services	%/year	▼	Ilyas et al. (2021); Prateep Na Talamg et al. (2022)
Carbon price	Price of green house gas emission as credit or penalty	Currency Units/ tCO_2 eq	For penalty:▼ For credit:▲	Kuo et al. (2024)
NPV	Total present value of net benefits	Currency units	▲	Ćetković et al. (2022)
ROI	Return on relative investments	%	▲	Ćetković et al. (2022)
LCC	Cumulative cost over assets life span	Currency units	▼	Ilyas et al. (2021)

Social variables	Contribution	Unit/Proxy	Effect of variable ▲ on social viability	References
DALYs avoided	Lower disease burden	DALYs avoided/ year	▲	WHO (2006)
Social license	Public perception and support to reuse	Acceptance score	▲	Fielding et al. (2019)
Water security gains	Additional water supply	m^3/year added	▲	Garcia and Pargament (2015)
Equity impact	Fair charges and ability to pay	Bill to income ratio	▲	Zhang et al. (2022)
Jobs	Employment opportunities created	Full-time equivalent (FTE)	▲	Esteves et al. (2024)

#▲: increasing; ##▼: decreasing

7.4 Environmental performance indicators

Achieving the SDGs by 2030 requires monitoring progress, accountability mechanisms, and reinforcement of drivers supporting sustainable transitions. Environmental Performance Indicators (EPIs) serve as quantitative metrics for evaluating environmental outcomes and guiding evidence-based decisions (Nika et al., 2020). In transitioning to circular water systems, key performance indicators fall into four domains: Water Use Efficiency and Circularity, Pollutant Reduction and Water Quality, Resource Recovery and Material Circularity, and Energy and Carbon Footprint (Kakwani & Kalbar, 2020). Water Use Efficiency and Circularity indicators measure freshwater withdrawal per production unit, water recycling rates, water loss reductions, and dependency on external sources. These metrics help stakeholders evaluate circular principles in water management and identify optimization opportunities.

- Annual water savings ($m^3.year^{-1}$,%): This metric quantifies freshwater reduction through wastewater reuse and waterless solutions (Karkou et al., 2024a; 2024b).
- Water reuse rate (or % circular water consumption): This directly measures circularity by comparing reused water to freshwater intake (Pilipenets et al., 2025).
- Percentage of circular water discharge (%): Measures industry water discharged that can be recycled or reused (Kayal et al., 2019).
- Water footprint: An LCA-based approach assessing water consumption and its environmental impact on ecosystems, freshwater depletion, and human health (Weerasooriya et al., 2021).

Building on this framework, Pollutant Reduction and Water Quality indicators measure treatment systems' effectiveness in reducing contaminants and meeting discharge standards, reflecting progress in minimizing environmental impacts while protecting ecosystem and human health.

- Annual volume of wastewater treated ($m^3.year^{-1}$): This measures the volume of wastewater discharged and treated by an industry annually (Jones et al., 2021).
- Reduction in pollutant load (%): This metric is crucial for protecting environmental and human health, as it quantifies the amount of specific pollutants, such as BOD, heavy metals, nitrogen, and phosphorus, in discharged and reused water (Das et al., 2025).

Similarly, Resource Recovery and Material Circularity examine how industrial waste streams' valuable by-products like nutrients, metals, and energy are recovered and reintegrated, shifting from disposal to resource valorization for economic and environmental sustainability.

- Resource recovery rate (%): Measures the amount of resources, such as nutrients, critical raw metals, biogas, and biosolids, that can be recovered from industry-discharged water (Kakwani & Kalbar, 2020).
- Treatment and Reuse of Sewage Sludge (tons dry solids.$year^{-1}$, %): Quantifies the amount of sludge reused versus that generated by an industry (Rodríguez et al., 2022).

Finally, Energy and Carbon Footprint indicators measure energy consumption and greenhouse gas emissions from water processes. These indicators link water management with energy use and carbon output, highlighting opportunities for optimization and contributing to climate change mitigation.

- Energy density (or intensity) for water treatment ($kWh.m^{-3}$): Measures the amount of energy required to treat and distribute a cubic meter of wastewater (Oliveira et al., 2019).

- GHG Emission Reduction (tCO_2 eq.m^{-3}, %): This metric quantifies the reduction in greenhouse gas emissions resulting from enhancements in water management, specifically through improvements in energy efficiency and the utilization of biogas (Kuo et al., 2024).
- Percentage of Recyclable Energy Consumption (%): The proportion of energy derived from renewable sources used in water operations (Delgado et al., 2024).

Effective monitoring of these indicators requires real-time measurement and smart metering to accurately track water footprints and operational performance; therefore, it is imperative to integrate data systems within these indicators to increase the overall efficiency of water reuse.

7.5 Overcoming implementation challenges

The circular use of industrial water offers substantial benefits to environmental health, yet faces technological, economic, socio-cultural, and legislative challenges. From a technological perspective, advanced wastewater treatment systems require higher investments than conventional methods, limiting their adoption. The feasibility of such recycling solutions varies across industries, constraining widespread uptake. Studies indicate that 80% of wastewater globally is discharged untreated, due to unavailability or absence of suitable treatment technologies (Das et al., 2025; Tzanakakis et al., 2023). Economic barriers include substantial initial investments and limited short-term profitability. Non-revenue water accounts for 25%–50% of global water supply, reaching 80% in some urban areas (Delgado et al., 2024). To address these barriers, financial mechanisms like recovery-reuse business models, water credit systems, and performance-based contracting have been proposed, though their success depends on supportive government incentives and regulatory frameworks emphasizing water conservation (Rodríguez et al., 2022).

Legislative challenges pose critical obstacles, with complex regulatory systems, overlapping mandates, and insufficient stakeholder collaboration hampering progress. The process is time-consuming, costly, and influenced by political stability and ideology, affecting how circular water principles are integrated into governance (Farfán Chilicaus et al., 2025; Tzanakakis et al., 2023). Socio-cultural barriers, particularly public perception of recycled water systems, remain significant challenges. Acceptance of water reuse faces psychological and cultural reservations, requiring transparent communication about treatment processes and safety assurance (Tzanakakis et al., 2023). This is crucial as 4.8–5.7 billion people may face water scarcity by 2050, highlighting the need for public acceptance of recycled water (Weerasooriya et al., 2021).

8. Governance, regulations, and policy frameworks

Water governance comprises political, social, economic, and administrative structures controlling water resource allocation and management (Boran, 2018). This system determines water distribution rights and services, requiring regulatory frameworks that balance demands while ensuring sustainable management (Pandya & Sharma, 2022). Effective governance integrates policy tools and institutional frameworks that adapt to changing conditions. In industry, governance frameworks establish water quality standards, regulate usage, and enforce compliance.

8.1 National and multinational water reuse standards

International and national standards provide the technical framework for industrial water management, with varying approaches. International standards like ISO 14,046:2014 and WHO Guidelines establish recognized methodologies and health objectives, but remain voluntary and depend on national implementation. Regional systems show stronger enforcement but monitoring them is still a challenge. The European Union's Water Framework Directive (2000/60/EC) takes an ecosystem-based approach, emphasizing river basin management and ecological status rather than individual compliance parameters (Albiac et al., 2024). US EPA's National Primary Drinking Water Regulations mandating contaminant limits for potable water.

National frameworks adapt broader principles into local rules. In India, the Central Pollution Control Board (CPCB) sets discharge standards under Environment Protection Rules 1986, requiring industries to meet treatment requirements. Unlike US and EU's prescriptive systems, India's framework faces enforcement challenges but increasingly aligns with international standards in high-impact sectors. Sector-specific standards address operational contexts overlooked by general regulations. ASHRAE Standard 188–2018 establishes cooling water protocols for microbial risk, ASME Boiler Code specifies feedwater requirements (<0.3 mg/L dissolved oxygen), and ISO 16,075:2015 defines irrigation water standards. These guidelines bridge environmental regulations and industry needs. While global standards offer harmonization, national frameworks vary: US emphasizes contaminant limits, EU focuses on ecosystem sustainability, and India adopts a hybrid approach with implementation gaps. Sectoral standards ensure circular water reuse meets ecological and operational objectives (Jha & Tyagi, 2023).

8.2 Water quality and discharge regulations

Water quality laws establish numerical effluent limitations for parameters affecting ecological and human health. The Clean Water Act (CWA) mandates the National Pollutant Discharge Elimination System (NPDES) to enforce technology-based and water-quality-based criteria. To control eutrophication, regulatory thresholds include biochemical oxygen demand (BOD5) of 10–30 mg L^{-1}, chemical oxygen demand (COD) of 50–250 mg L^{-1}, and total suspended solids (TSS) of 10–30 mg L^{-1}. Nutrient discharges are managed with limits of 3–8 mg.L^{-1} for nitrogen and 0.1–1.0 mg L^{-1} for phosphorus (Graversgaard, 2024), preventing algal blooms and oxygen depletion in aquatic systems.

Sector-specific rules govern the discharge of toxic substances, particularly heavy metals. Under 40 CFR Part 433, the allowable daily maximum concentrations include 2.77 mg L^{-1} for chromium, 0.69 mg L^{-1} for lead, and 0.012 mg L^{-1} for mercury, reflecting concerns for bioaccumulate contaminants. Thermal discharges are regulated under 40 CFR Part 125, requiring that effluent releases do not increase receiving water temperatures by more than 2°C above ambient conditions to protect aquatic ecosystems. The regulatory scope now includes emerging pollutants like pharmaceuticals, endocrine-disrupting compounds, and per- and polyfluoroalkyl substances (PFAS). While these contaminants lack binding discharge limits, they are tracked through the US EPA's Contaminant Candidate List and the European Union's Watch List, generating data on occurrence and risk potential (Wilcox et al., 2016). These regulations demonstrate water quality governance's evolution from controlling established hazards toward addressing complex threats in industrial effluent management.

8.3 Regulatory frameworks and institutional roles

Governance frameworks for water management include regulations on allocation, basin planning, and environmental compliance, integrating legal and ecological dimensions to balance demands. The EU Water Framework Directive requires 6-year River Basin Management Plans with measures to achieve "good ecological status" of water bodies (Grigg, 2016). In the United States, water allocation follows the Federal Reclamation Act of 1902 and state-level Prior Appropriation Doctrine, prioritizing beneficial use in arid regions. Interstate compacts like the Colorado River Compact of 1922 regulate shared water resources across states. Regulatory responsibilities are distributed among multiple institutions. In the US, the Environmental Protection Agency (EPA) enforces the Clean Water Act (CWA), while the US Army Corps of Engineers manages Section 404 wetlands permits. National Pollutant Discharge Elimination System (NPDES) programs are delegated to state agencies under EPA approval. In the EU, environment ministries and river basin authorities ensure compliance with EU directives, implementing supranational legislation locally. Effective governance requires clear institutional responsibilities, coordinated agency interactions, and adaptability to evolving priorities. Water governance frameworks must enforce compliance while promoting resilience and innovation in resource management (Grigg, 2023).

8.4 Incentives and policy instruments

Beyond regulations, economic and voluntary processes promote industrial water reuse and efficiency. Tax credits (e.g., US Internal Revenue Code Section 48, which provides up to 30% for eligible recycling equipment), low-interest loans, and infrastructure upgrade subsidies are all examples of financial incentives. Market-based techniques, such as transferable water rights and water quality trade, enable higher-value allocation and cost-effective pollution reduction (Adeyeye, 2013; Keyes, 2005). Voluntary water stewardship initiatives encourage the business sector to go above and beyond the minimal compliance requirements, integrating corporate responsibility with environmental goals. Water price reforms can improve efficiency, but they must be carefully designed to guarantee fairness and affordability while paying operational and maintenance expenses.

8.5 Permitting, compliance, and enforcement

Permitting systems connect policy and practice. Abstraction licenses govern withdrawals, while discharge permits establish effluent quality and monitoring requirements. In the United States, NPDES licenses are valid for 5 years, with renewals due 180 days before expiration; EU permits must comply with Water Framework Directive goals (Wang et al., 2024a; 2024b). Compliance requires regular reporting, authorized analytical techniques (40 CFR Part 136), and quality assurance (Jha & Tyagi, 2023). Administrative fines (up to USD 37,500/day under CWA Section 309 (g)) and criminal penalties are available for enforcement, with citizen suit provisions enabling public oversight. Effective enforcement requires institutional capacity, technical competence, and political support. Strong governance, clear standards, and adaptable policy tools enable circular industrial water usage. Harmonizing international standards, integrating regulatory mechanisms, and enhancing enforcement can improve compliance and align with sustainability goals.

9. Challenges, opportunities, and key drivers in circular industrial water use

9.1 Overcoming implementation challenges

Transitioning from linear to circular water treatment systems presents multiple challenges. The primary hurdle is the initial investment needed for infrastructure upgrades and new facilities with advanced technologies(Lyu et al., 2020). Regional regulatory variations create uncertainty, inhibiting investments in circular water practices(Berbel et al., 2023). While technologies exist, implementation requires substantial process modifications. Organizational resistance and lack of skilled personnel further hinder deployment. Addressing these barriers requires economic incentives, regulations, and enhanced organizational capabilities.

9.2 Economic barriers

Economic constraints significantly impede the implementation of circular industrial water usage systems. The predominant reason that Industries, particularly those with moderate to limited financial resources, refrain from adopting advanced water treatment technologies for recovery and reuse primarily due to the substantial initial investment required (Chen et al., 2025). Although these expenditures may yield long-term operational benefits, the initial financial outlay can be prohibitive for many firms. The current low discounts on freshwater use and wastewater discharge often fail to reflect actual environmental costs, thereby diminishing the financial motivation for water conservation initiatives (Schellenberg et al., 2020). Furthermore, the absence of comprehensive cost-benefit analyses highlighting long-term economic advantages deters such investment decisions.

9.3 Regulatory and policy challenges, technological and operational hurdles

The shift toward circular industrial water systems is affected by inconsistencies and policies lacking sensitivity to reuse. Regulations focus on discharge limits rather than encouraging resource recovery and reuse, resulting in administrative uncertainties (Berbel et al., 2023). This ambiguity and hesitance to invest in advanced treatment technologies stems from unclear standards for reuse applications (Mukherjee & Jensen, 2020). Technological and operational challenges compound this situation. While advanced treatment technologies exist, their implementation varies by industrial wastewater needs, requiring specialized solutions. Operational challenges include managing system complexity, energy consumption, and addressing contaminants and micropollutants.

9.4 Economic and business benefits

The adoption of circular economic principles in industrial water management offers substantial economic advantages. Financial gains can be achieved by reducing freshwater consumption and wastewater discharge. Internal water treatment and reuse strategies reduce reliance on external sources, lowering operational expenses. The recovery of valuable resources from wastewater presents revenue streams and enhances resource efficiency (Rodríguez et al., 2022). Circular water systems enhance company reputation, attracting environmentally conscious consumers and investors (Zhang et al., 2021). Water self-sufficiency mitigates business risks and ensures operational resilience, aligning with sustainability goals and fostering innovation (Nika et al., 2020).

9.5 Technological and innovation opportunities

The need to achieve efficient water recovery and sustainable effluent treatment has driven innovative membrane technologies and advanced oxidation processes, which are crucial for contaminant elimination and high-quality reclaimed water (Shehata et al., 2023). Water quality management precision has improved through sensor technology and real-time monitoring systems, allowing industries to optimize water circuits and reduce waste. The circular economy model promotes industrial symbiosis through strategies where effluents from one industry serve as resources for another (Fussone et al., 2025). The implementation of digitalization and Industry 4.0 concepts, including artificial intelligence and big data analytics, can enhance water network efficiency, predict maintenance needs, and improve water efficiency in industrial environments (Howell et al., 2017).

9.6 Public-private partnerships, corporate social responsibility and stakeholder pressure

A significant correlation exists between circular industrial water usage implementation and collaborative public-private partnerships, alongside demand from corporate social responsibility (CSR) programs and stakeholder engagement. Public-private partnerships mobilize financial resources, expertise, and infrastructure needed for water circularity initiatives (Georgiou et al., 2023). By combining public sector regulatory authority with private sector technical capabilities, these partnerships facilitate information dissemination. The growing importance of CSR compels businesses to adopt sustainable practices. These enterprises recognize that responsible water stewardship is vital for maintaining their social license, brand reputation, and attracting ethical investors. Pressure from stakeholders, including environmental organizations and communities, urges industries to mitigate ecological impacts and adopt sustainable practices. This pressure promotes circular water solutions, demonstrating that business success is linked to environmental sustainability (Halidu et al., 2025).

10. Circular water: A dynamic future

10.1 Vision of the circular water economy

The circular water economy envisions an integrated industrial water cycle that minimizes freshwater use, treats wastewater as a resource, and maximizes reuse through closed-loop systems. This shift moves from "take-use-dispose" toward a regenerative framework optimizing water quality, quantity, and environmental impact. Industrial water use is embedded within water-energy-food (WEF) nexus considerations (Sauvé et al., 2021), treating water management as an interconnected system with energy recovery and nutrient reclamation. Circular water frameworks must adapt to regional constraints, cultural attitudes, and regulations across continents. The goal is achieving net water-positive operations through integrated technological and nature-based solutions (Rodriguez-Anton et al., 2019), while emphasizing resilience against climate variability and resource competition (Bosco et al., 2025; Sauvé et al., 2021).

10.2 Technological innovations for water reuse

Technological advancements drive circular water adoption through water quality restoration and resource recovery. Advanced treatment systems combine physical, chemical, and biological processes for site-specific requirements (Sauvé et al., 2021). Key technological innovations in industrial water reuse encompass a diverse array of advanced treatment and recovery systems designed to achieve both contaminant removal and resource valorization. Membrane-based processes, including reverse osmosis, nanofiltration, and ultrafiltration, are increasingly integrated with forward osmosis and membrane bioreactor configurations to achieve high purification efficiency across a broad spectrum of contaminants. The incorporation of hybrid membrane and adsorption systems further enhances the removal of emerging pollutants such as pharmaceuticals and trace organics. Complementing these are electrochemical technologies, including electrocoagulation, capacitive deionization, and electrodialysis, which effectively treat saline industrial effluents while enabling selective ion recovery for subsequent reuse. Biological systems, particularly anaerobic digestion, contribute to organic load reduction and biogas generation, while the use of engineered microbial consortia facilitates targeted nutrient recovery. Finally, resource recovery frameworks focus on extracting valuable constituents such as phosphorus, nitrogen, metals, and biopolymers from wastewater streams, thereby transforming waste management into a circular resource economy (De Armas-Calderón & Manjarrés-Henríquez, 2024; Sauvé et al., 2021). Innovation roadmaps emphasize localizing technologies for infrastructure and policy compatibility. IoT monitoring enables process optimization, reducing operational costs and energy use (De Armas-Calderón & Manjarrés-Henríquez, 2024).

10.3 Roadmap for industries to adopt circular water use

A structured adoption roadmap enables industries to implement circular water practices while maintaining operations. Based on global best practices, the roadmap comprises five strategic phases. The first phase involves baseline assessment through water audits and flow mapping to establish water balances and identify recovery potential, aligning with sustainability goals (Sauvé et al., 2021). The second phase selects and integrates treatment systems based on reuse specifications and system compatibility. The third phase implements digitalization through smart sensors and AI analytics for leak detection and process optimization. Digital twins can simulate plant performance before implementation (Bosco et al., 2025). The fourth phase involves stakeholder engagement and governance alignment, engaging regulators, communities, and supply chain partners early in the transition. Insights from the Indian Delphi study show governance alignment in wastewater reuse depends on trust, transparency, and clear communication (Breitenmoser et al., 2022). The final phase is scaling and continuous improvement, implementing pilot projects, collecting performance data, refining operational protocols, and scaling technologies plant-wide while ensuring alignment with evolving regulations and new contaminants (De Armas-Calderón & Manjarrés-Henríquez, 2024).

10.4 Water-positive manufacturing

Water-positive manufacturing extends beyond neutrality, where water withdrawn is replenished, to actively restoring ecosystems and supporting community water resilience. This concept integrates quantitative targets, such as achieving a net positive water balance, with qualitative measures that improve water quality in local watersheds. Achieving water positivity requires on-site high-efficiency recycling to reduce withdrawal intensity below industry benchmarks, complemented by off-site replenishment projects such as constructed wetlands, aquifer recharge, and riparian buffer restoration (Bosco et al., 2025; Rodriguez-Anton et al., 2019). Nutrient circularity is also central to this concept, with nitrogen and phosphorus recovered from industrial effluents being reused in agriculture, thereby reducing synthetic fertilizer production and mitigating eutrophication. Energy-water co-optimization is another critical aspect, where energy recovered from wastewater treatment offsets plant electricity consumption, reducing the overall water-energy-carbon nexus footprint. In addition to the environmental benefits, water-positive initiatives position companies as sustainability leaders, building consumer trust and providing a competitive advantage in sectors such as food and beverage, textiles, and electronics (Rodriguez-Anton et al., 2019).

10.5 Circular water for climate-resilient and SDG-aligned industry

The alignment of circular water strategies with UN SDGs provides policy guidance and performance benchmarks. Water reuse advances SDG 6 (Clean Water and Sanitation), while contributing to SDG 12 (Responsible Consumption and Production), SDG 13 (Climate Action), and indirectly to SDG 7 (Affordable and Clean Energy) and SDG 15 (Life on Land) (Rodriguez-Anton et al., 2019). A systemic approach integrating industrial water use into WEF nexus planning enables cross-sector benefits. Reclaimed water in agriculture reduces withdrawals, while nutrient recovery from industrial effluent reduces fertilizer emissions. Climate resilience within industrial water management is reinforced through a multifaceted approach that integrates technological innovation, infrastructural adaptability, and ecological stewardship. Diversification of water sources through reuse, stormwater harvesting, and renewable-powered desalination reduces dependence on conventional freshwater supplies and mitigates vulnerability to hydrological stress. System redundancy, achieved by deploying decentralized treatment units, ensures operational continuity during climatic or infrastructural disruptions and enhances overall system reliability and adaptive capacity. At the same time, watershed stewardship is promoted through active industry participation in catchment-level planning and conservation initiatives, fostering collaborative management of shared water resources. Collectively, these strategies strengthen industrial systems against climate-induced uncertainties while advancing sustainable water governance (Rodriguez-Anton et al., 2019). Industries adopting circular water frameworks are better positioned to meet discharge regulations, adapt to hydrological variability, and leverage environmental performance advantages. The future of circular water harmonizes technology, governance, and environmental stewardship. Industries will be evaluated on their ability to regenerate water resources, close material loops, and build resilient systems.

11. Conclusions

The findings highlight that industrial water use connects economic development, environmental sustainability, and social well-being. The linear "take-make-dispose" model worsens resource depletion and ecological degradation. Adopting a circular economy (CE) framework transforms industrial wastewater from a liability into assets of reusable water, energy, nutrients, and critical materials. This allows industries to mitigate water scarcity risks while establishing new revenue streams.

Technological advances enable this transition. Membrane systems, biological-chemical processes, and zero liquid discharge (ZLD) strategies enable near-complete recycling, while energy recovery technologies position facilities as net energy contributors. Nutrient and metal recovery technologies strengthen industrial supply security. Nature-based solutions like algae systems and wetlands operate alongside high-tech systems, providing biodiversity gains and carbon sequestration.

Digital technologies add analytical sophistication through IoT sensors, artificial intelligence, and digital twins, enabling real-time optimization and predictive maintenance. These tools help industries track water circularity within water-energy-carbon frameworks for sustainability reporting. However, barriers exist: high capital costs, technology complexity, and varying regulations create uncertainty. Public acceptance of recycled water remains crucial. Solutions require policy incentives, public-private partnerships, and stakeholder engagement. Case studies show circular water systems achieve cost savings in water procurement, compliance, and energy use, with returns within three to 5 years. Environmental benefits contribute directly to multiple SDGs. The path forward requires industries to implement an integrated roadmap: baseline assessments, strategic technology selection, digitalization, and alignment with governance frameworks. The vision aims for water positivity, where industries replenish more water than consumed, contributing to ecosystem health. This positions industrial water users as agents of regeneration in a sustainable future.

References

Abdelfattah, I., & El-Shamy, A. M. (2024). Review on the escalating imperative of zero liquid discharge (ZLD) technology for sustainable water management and environmental resilience. *Journal of Environmental Management, 351*, Article 119614.

Adeyeye, K. (2013). Water policy and regulations: A UK perspective. *Water efficiency in buildings: Theory and practice*, 5–23.

Afzal, M., Arslan, M., Younus, S., Müller, J. A., Usman, M., Yasin, M., Mehmood, M. A., Mehdi, T., Islam, E., Tauseef, M., & Iqbal, S. (2024). A nature-based closed-loop wastewater treatment system at vehicle-washing facilities: From linear to circular economy. *iScience, 27*(4).

Ahmad, H. W., Bibi, H. A., Chandrasekaran, M., Ahmad, S., & Kyriakopoulos, G. L. (2024). Sustainable wastewater treatment strategies in effective abatement of emerging pollutants. *Water, 16*(20), 2893.

Ahmetović, E., Grossmann, I. E., Kravanja, Z., & Ibrić, N. (2017). Water optimization in process industries. In *Sustainable utilization of natural resources* (pp. 487–512). CRC Press.

Akter, M. M., Hosen, M. A., Gilroyed, B. H., Akter Eti, S., Sarker, P., Kabir, M. M., & Mohinuzzaman, M. (2024). Nutrient recovery for fertilizer production and wastewater treatment for a circular economy. *ACS Sustainable Resource Management, 1*(4), 787–798.

Al Kez, D., Foley, A. M., Laverty, D., Del Rio, D. F., & Sovacool, B. (2022). Exploring the sustainability challenges facing digitalization and internet data centers. *Journal of Cleaner Production, 371*, Article 133633.

Alazaiza, M. Y., Alzghoul, T. M., Nassani, D. E., & Bashir, M. J. (2025). Natural coagulants for sustainable wastewater treatment: Current global research trends. *Processes, 13*(6), 1754.

Albiac, J., Calvo, E., & Esteban, E. (2024). A quarter of a century of the European water framework directive—The slow path towards sustainable water management. *Water Economics and Policy, 10*(02), Article 2430001.

AlSayed, A., Soliman, M., & Eldyasti, A. (2020). Microbial fuel cells for municipal wastewater treatment: From technology fundamentals to full-scale development. *Renewable and Sustainable Energy Reviews, 134*, Article 110367.

Berbel, J., Mesa-Pérez, E., & Simón, P. (2023). Challenges for circular economy under the EU 2020/741 wastewater reuse regulation. *Global Challenges, 7*(7), Article 2200232.

Boran, M. (2018). Water governance and its effectiveness. *Buletinul Ştiinţific al Universităţii Politehnica Timisoara–Seria Inginerie si Management, 4*(1), 28–35.

Bosco, C., Seglem, K. N., Sivertsen, E., Jovanović, O., & Helness, H. (2025). Developing a framework to assess water smartness and sustainability of circular economy solutions in the water sector. *Journal of Cleaner Production*, Article 145874.

Breitenmoser, L., Quesada, G. C., Bassi, N., Dkhar, N. B., Phukan, M., Kumar, S., Babu, A. N., Kierstein, A., Campling, P., & Hooijmans, C. M. (2022). Perceived drivers and barriers in the governance of wastewater treatment and reuse in India: Insights from a two-round Delphi study. *Resources, Conservation and Recycling, 182*, Article 106285.

Buyukada-Kesici, E., Topuz, E., Pala, B., Koseoglu-Imer, D. Y., & Aydiner, C. (2023). Implementation of zero liquid discharge policy in industrial water management. In *Resource recovery in industrial waste waters* (pp. 199–228). Elsevier.

Cairone, S., Hasan, S. W., Choo, K. H., Lekkas, D. F., Fortunato, L., Zorpas, A. A., Korshin, G., Zarra, T., Belgiorno, V., & Naddeo, V. (2024). Revolutionizing wastewater treatment toward circular economy and carbon neutrality goals: Pioneering sustainable and efficient solutions for automation and advanced process control with smart and cutting-edge technologies. *Journal of Water Process Engineering, 63*, Article 105486.

Castelli, C., Castellini, M., Gusperti, C., Romani, I. G., Ciola, E., & Vergalli, S. (2024). Exploring macroeconomic models in the water, energy, food, and ecosystem (WEFE) field: a comprehensive review. *Environmental Research Letters, 19*(5), Article 053003.

Ćetković, J., Knežević, M., Lakić, S., Žarković, M., Vujadinović, R., Živković, A., & Cvijović, J. (2022). Financial and economic investment evaluation of wastewater treatment plant. *Water, 14*(1), 122.

Chen, O., Mustafee, N., Evans, B., Khoury, M., Vamvakeridou-Lyroudia, L., Chen, A. S., Djordjević, S., & Savić, D. (2025). New business models for the circular economy of water: A hybrid simulation study of a mobile rental wastewater treatment service. *Journal of Cleaner Production, 495*, Article 145041.

Chhipi-Shrestha, G., Mian, H. R., Mohammadiun, S., Rodriguez, M., Hewage, K., & Sadiq, R. (2023). Digital water: Artificial intelligence and soft computing applications for drinking water quality assessment. *Clean Technologies and Environmental Policy, 25*(5), 1409–1438.

Curran, M. A. (2013). Life cycle assessment: A review of the methodology and its application to sustainability. *Current Opinion in Chemical Engineering, 2*(3), 273–277.

Daga, S., Yadav, K., Singh, P., & Mishra, V. (2025). Beyond the take-make-dispose model—unlocking the power of circular economy for an environmentally resilient future. *Circular Economy and Environmental Resilience: Solutions for a Sustainable Tomorrow, 1*, 1–11 (Cham: Springer Nature Switzerland).

Dakwala, M., Mohanty, B., & Bhargava, R. (2014). Simultaneous water and energy conservation through graphical and mathematical programming: A case study for float glass industry. *Journal of Cleaner Production, 78*, 15–34.

Das, S., Singh, C. K., Sodhi, K. K., & Singh, V. K. (2025). Circular economy approaches for water reuse and emerging contaminant mitigation: Innovations in water treatment. *Environment, Development and Sustainability, 27*(3), 5753–5794.

De Armas-Calderón, N., & Manjarrés-Henríquez, L. (2024). Technology roadmap: A strategic approach to innovation in water treatment in Colombia. *Procedia Computer Science, 241*, 514–519.

de Melo Ribeiro, F. H., & dos Santos Silva, Y. (2020). Auxiliary strategies for water management in industries: Minimization of water use and. *Innovation in Global Green Technologies, 2020*, 79.

Delgado, A., Rodriguez, D. J., Amadei, C. A., & Makino, M. (2024). Water in circular economy and resilience (WICER) framework. *Utilities Policy, 87*, Article 101604.

Devda, V., Chaudhary, K., Varjani, S., Pathak, B., Patel, A. K., Singhania, R. R., Taherzadeh, M. J., Ngo, H. H., Wong, J. W., Guo, W., & Chaturvedi, P. (2021). Recovery of resources from industrial wastewater employing electrochemical technologies: Status, advancements and perspectives. *Bioengineered, 12*(1), 4697–4718.

Duarte, I. C. D., Souza Chaves, D. H. D., Reis da Silva, S., Ferreira da Costa Junior, E., & Oliveira Souza da Costa, A. (2025). Sustainability of energy recovery from by-products in wastewater treatment: A thermo-economic and environmental review. *Biofuels*, 1–19.

Edo, G. I., Ndudi, W., Ali, A. M., Yousif, E., Jikah, A. N., Isoje, E. F., Igbuku, U. A., Mafe, A. N., Opiti, R. A., Madueke, C. J., & Essaghah, A. E. A. (2025). Biopolymers: An inclusive review. *Hybrid Advances, 9*, Article 100418.

El Khatir, N. (2021). The competitive positioning of Moroccan automotive industry: A diagnosis attempt. *International Journal of Innovation and Applied Studies, 33*(2), 385–394.

Esteves, F., Carlos Cardoso, J., Leitão, S., & Pires, E. J. S. (2025). Energy audit in wastewater treatment plant according to ISO 50001: Opportunities and challenges for improving sustainability. *Sustainability, 17*(5), 2145.

Esteves, F., Molina-Perez, E., Kalra, N., Syme, J., & Vogt-Schilb, A. (2024). Job creation and decarbonization synergies in Latin America: A simulation-based exploratory modeling analysis. *Frontiers in Climate, 6*, Article 1339877.

Farfán Chilicaus, G. C., Cruz Salinas, L. E., Silva León, P. M., Lizarzaburu Aguinaga, D. A., Vera Zelada, P., Vera Zelada, L. A., Luque Luque, E. O., Licapa Redolfo, R., & Ramos Farroñán, E. V. (2025). Circular economy and water sustainability: Systematic review of water management technologies and strategies (2018–2024). *Sustainability, 17*(14).

Farhaoui, Y., & El Allaoui, A. (April 2024). Sustainability in the internet of things: Insights, scope, and AI-driven optimized water management with big data integration. In *The international workshop on big data and business intelligence* (pp. 468–475). Cham: Springer Nature Switzerland.

Ferri, E. N., & Bolelli, L. (2025). Wastewater remediation treatments aimed at water reuse: Recent outcomes from pilot-and full-scale tests. *Applied Sciences, 15*(5), 2448.

Ferrucci, A., & Vocciante, M. (2021). Improved management of water resources in process industry by accounting for fluctuations of water content in feed streams and products. *Journal of Water Process Engineering, 39*, Article 101870.

Fielding, K. S., Dolnicar, S., & Schultz, T. (2019). Public acceptance of recycled water. *International Journal of Water Resources Development, 35*(4), 551–586.

Fussone, R., Sammatrice, C., Cannella, S., & Dominguez, R. (2025). Enhancing circular economy through industrial symbiosis: An agent-based simulation analysis of supply chain dynamics. *Sustainable Operations and Computers*.

Garcia, X., & Pargament, D. (2015). Reusing wastewater to cope with water scarcity: Economic, social and environmental considerations for decision-making. *Resources, Conservation and Recycling, 101*, 154–166.

Georgiou, I., Caucci, S., Morris, J. C., Guenther, E., & Krebs, P. (2023). Assessing the potential of water reuse uptake through a private–public partnership: A practitioner's perspective. *Circular Economy and Sustainability, 3*(1), 199–220.

Ghafourian, M., Nika, C. E., Mousavi, A., Mino, E., Al-Salehi, M., & Katsou, E. (2022). Economic impact assessment indicators of circular economy in a decentralised circular water system—Case of eco-touristic facility. *Science of the Total Environment, 822*, Article 153602.

Ghafourian, M., Stanchev, P., Mousavi, A., & Katsou, E. (2021). Economic assessment of nature-based solutions as enablers of circularity in water systems. *Science of the Total Environment, 792*, Article 148267.

Giakoumis, T., Vaghela, C., & Voulvoulis, N. (2020). The role of water reuse in the circular economy. *Advances in chemical pollution, environmental management and protection, 5*, 227–252 (Elsevier).

Graversgaard, M. (2024). Regulatory approaches in water policy and governance. In *Handbook on the governance and politics of water resources* (pp. 21–32). Edward Elgar Publishing.

Grigg, N. S. (2016). Water laws and regulations. In *Integrated water resource management: An interdisciplinary approach* (pp. 265–290). London: Palgrave Macmillan UK.

Grigg, N. S. (2023). State government roles in water governance: Time for an upgrade. *Public Works Management and Policy, 28*(2), 117–134.

Guo, Z., Sun, Y., Pan, S. Y., & Chiang, P. C. (2019). Integration of green energy and advanced energy-efficient technologies for municipal wastewater treatment plants. *International Journal of Environmental Research and Public Health, 16*(7), 1282.

Halidu, O. B., Awuah-Gyawu, M., Fianko, A. O., Gyamfi, B. A., & Asongu, S. A. (2025). Corporate governance and circular supply chains: Synergizing eco-adaptive organizational culture, leadership eco-innovation willingness, and perceived urgency for circularity. *Journal of Environmental Management, 392*, Article 126689.

Hoareau, C. E., & Tam, N. (2024). Evaluating the impact of government policies on circular economy adoption in the construction sector. *Civil and Sustainable Urban Engineering, 4*(2), 83–94.

Holtzapple, M. T., Wu, H., Weimer, P. J., Dalke, R., Granda, C. B., Mai, J., & Urgun-Demirtas, M. (2022). Microbial communities for valorizing biomass using the carboxylate platform to produce volatile fatty acids: A review. *Bioresource Technology, 344*, Article 126253.

Howell, S., Rezgui, Y., & Beach, T. (2017). Integrating building and urban semantics to empower smart water solutions. *Automation in Construction, 81*, 434–448.

Hussain, M., Riaz, A., Zeb, H., Ali, A., Mujahid, R., Ahmad, F., & Zafar, M. S. (2025). Paving the path to water security: The role of advanced adsorbents in wastewater treatment. *Journal of Water Process Engineering, 71*, Article 107333.

Ilyas, M., Kassa, F. M., & Darun, M. R. (2021). Life cycle cost analysis of wastewater treatment: A systematic review of literature. *Journal of Cleaner Production, 310*, Article 127549.

Jain, M., Majumder, A., Ghosal, P. S., & Gupta, A. K. (2020). A review on treatment of petroleum refinery and petrochemical plant wastewater: A special emphasis on constructed wetlands. *Journal of Environmental Management, 272*, Article 111057.

Jha, S., & Tyagi, G. (2023). Water reuse planning, policy, monitoring requirements and standards/criteria. In *Industrial wastewater reuse: Applications, prospects and challenges* (pp. 317–338). Singapore: Springer Nature Singapore.

Jones, E. R., Van Vliet, M. T., Qadir, M., & Bierkens, M. F. (2021). Country-level and gridded estimates of wastewater production, collection, treatment and reuse. *Earth System Science Data, 13*(2), 237–254.

Kakwani, N. S., & Kalbar, P. P. (2020). Review of circular economy in urban water sector: Challenges and opportunities in India. *Journal of Environmental Management, 271*, Article 111010.

Kalandarov, P., Zhunussov, K., Kassimov, A., Baibolov, B., Junussov, N., & Kaliyeva, K. (2024). Changes in pH during the ozonation process of surface water. *E3S Web of Conferences, 563*, Article 01020 (EDP Sciences).

Karkou, E., Angelis-Dimakis, A., Parlapiano, M., Savvakis, N., Siddique, O., Vyrkou, A., Sgroi, M., Fatone, F., & Arampatzis, G. (2024a). Process innovations and circular strategies for closing the water loop in a process industry. *Journal of Environmental Management, 370*, Article 122748.

Karkou, E., Teo, C. J., Savvakis, N., Poinapen, J., & Arampatzis, G. (2024b). Industrial circular water use practices through the application of a conceptual water efficiency framework in the process industry. *Journal of Environmental Management, 370*, Article 122596.

Kayal, B., Abu-Ghunmi, D., Abu-Ghunmi, L., Archenti, A., Nicolescu, M., Larkin, C., & Corbet, S. (2019). An economic index for measuring firm's circularity: The case of water industry. *Journal of Behavioral and Experimental Finance, 21*, 123–129.

Keyes, A. M. (2005). State water policy alternatives for water conservation/efficiency and reuse. In *Proceedings of the 2005 Georgia water resources conference* [np]. 2005.

Khan, M. J., Wibowo, A., Karim, Z., Posoknistakul, P., Matsagar, B. M., Wu, K. C. W., & Sakdaronnarong, C. (2024). Wastewater treatment using membrane bioreactor technologies: Removal of phenolic contaminants from oil and coal refineries and pharmaceutical industries. *Polymers, 16*(3), 443.

Kuo, T. C., Chen, H. Y., Chong, B., & Lin, M. (2024). Cost benefit analysis and carbon footprint of biogas energy through life cycle assessment. *Cleaner Environmental Systems, 15*, Article 100240.

Li, C., Sun, W., Lu, Z., Ao, X., & Li, S. (2020). Ceramic nanocomposite membranes and membrane fouling: A review. *Water Research, 175*, Article 115674.

Lin, N., Arzumanyan, M., Rodriguez Calzado, E., & Nicot, J. P. (2025). Water requirements for hydrogen production: Assessing future demand and impacts on Texas water resources. *Sustainability, 17*(2), 385.

Liu, Q., Li, R., Dereli, R. K., Flynn, D., & Casey, E. (2022). Water resource recovery facilities as potential energy generation units and their dynamic economic dispatch. *Applied Energy, 318*, Article 119199.

Liu, Q., Yang, L., & Yang, M. (2021). Digitalisation for water sustainability: Barriers to implementing circular economy in smart water management. *Sustainability, 13*(21), Article 11868.

López-Serrano, M. J., Lakho, F. H., Van Hulle, S. W., & Batlles-delaFuente, A. (2023). Life cycle cost assessment and economic analysis of a decentralized wastewater treatment to achieve water sustainability within the framework of circular economy. *Oeconomia Copernicana, 14*(1), 103–133.

Lyu, Y., Ye, H., Zhao, Z., Tian, J., & Chen, L. (2020). Exploring the cost of wastewater treatment in a chemical industrial park: Model development and application. *Resources, Conservation and Recycling, 155*, Article 104663.

Manasa, R. L., & Mehta, A. (2020). Wastewater: Sources of pollutants and its remediation. *Environmental Biotechnology, 2*, 197–219 (Cham: Springer International Publishing).

Masud, M. M., Shamem, A. S. M., Saif, A. N. M., Bari, M. F., & Mostafa, R. (2025). The role of artificial intelligence in sustainable water management in Asia: A systematic literature review with bibliographic network visualization. *International Journal of Energy and Water Resources, 9*(1), 247–265.

Mbavarira, T. M., & Grimm, C. (2021). A systemic view on circular economy in the water industry: Learnings from a Belgian and Dutch case. *Sustainability, 13*(6), 3313.

Mukherjee, M., & Jensen, O. (2020). Making water reuse safe: A comparative analysis of the development of regulation and technology uptake in the US and Australia. *Safety Science, 121*, 5–14.

Mutikanga, H. E., Sharma, S. K., & Vairavamoorthy, K. (2013). Methods and tools for managing losses in water distribution systems. *Journal of Water Resources Planning and Management, 139*(2), 166–174.

Nabavi-Pelesaraei, A., Kaab, A., Hosseini-Fashami, F., Mostashari-Rad, F., & Chau, K. W. (2019). Life cycle assessment (LCA) approach to evaluate different waste management opportunities. In *Advances in waste-to-energy technologies* (pp. 195–216). CRC Press.

Neelofar, M. R., Bhat, S. U., & Muslim, M. (2023). Water auditing and recycling as a tool for management of water resources: An Indian perspective. *Applied Water Science, 13*(9), 176.

Negro, V., Noussan, M., & Chiaramonti, D. (2025). Alternative options for biogas-to-energy: A comparison of electricity and biomethane generation based on the real operation of a production site. *Applied Energy, 377*, Article 124687.

Nguyen, M. L., Nakhjiri, A. T., Kamal, M., Mohamed, A., Algarni, M., Yu, S. T., Wang, F. M., & Su, C. H. (2022). State-of-the-art review on the application of membrane bioreactors for molecular micro-contaminant removal from aquatic environment. *Membranes, 12*(4), 429.

Nika, C. E., Vasilaki, V., Expósito, A., & Katsou, E. (2020). Water cycle and circular economy: Developing a circularity assessment framework for complex water systems. *Water Research, 187*, Article 116423.

Noor, T., Javid, A., Hussain, A., Bukhari, S. M., Ali, W., Akmal, M., & Hussain, S. M. (2020). Types, sources and management of urban wastes. In *Urban ecology* (pp. 239–263). Elsevier.

Oliveira, M. C., Iten, M., Matos, H. A., & Michels, J. (2019). Water–energy nexus in typical industrial water circuits. *Water, 11*(4), 699.

Orsini, L. P., Leardini, C., Danesi, L., Guerrini, A., & Frison, N. (2023). Circular economy in the water and wastewater sector: Tariff impact and financial performance of SMARTechs. *Utilities Policy, 83*, Article 101593.

Pahunang, R. R., Buonerba, A., Senatore, V., Oliva, G., Ouda, M., Zarra, T., Muñoz, R., Puig, S., Ballesteros Jr, F. C., Li, C. W., & Hasan, S. W. (2021). Advances in technological control of greenhouse gas emissions from wastewater in the context of circular economy. *Science of the Total Environment, 792*, Article 148479.

Panagopoulos, A., & Haralambous, K. J. (2020). Minimal liquid discharge (MLD) and zero liquid discharge (ZLD) strategies for wastewater management and resource recovery–Analysis, challenges and prospects. *Journal of Environmental Chemical Engineering, 8*(5), Article 104418.

Pandey, A. K. (2025). Sustainable water management through integrated technologies and circular resource recovery. *Environmental Science: Water Research and Technology, 11*(8), 1822–1846.

Pandya, A. B., & Sharma, P. (2022). Water governance—Role for assuring water security. *Irrigation and Drainage, 71*(5).

Pérez-González, A., Urtiaga, A. M., Ibáñez, R., & Ortiz, I. (2012). State of the art and review on the treatment technologies of water reverse osmosis concentrates. *Water Research, 46*(2), 267–283.

Pilipenets, O., Gunawardena, T., Hui, F. K. P., Mendis, P., & Aye, L. (2025). A novel circular economy framework: Assessing process circularity through resource flow and emissions analysis. *Resources, Conservation and Recycling, 215*, Article 108083.

Prateep Na Talang, R., Sirivithayapakorn, S., & Polruang, S. (2022). Life cycle impact assessment and life cycle cost assessment for centralized and decentralized wastewater treatment plants in Thailand. *Scientific Reports, 12*(1), Article 14540.

Ramin, E., Faria, L., Gargalo, C. L., Ramin, P., Flores-Alsina, X., Andersen, M. M., & Gernaey, K. V. (2024). Water innovation in industrial symbiosis-A global review. *Journal of Environmental Management, 349*, Article 119578.

Ramos, H. M., Kuriqi, A., Besharat, M., Creaco, E., Tasca, E., Coronado-Hernández, O. E., Pienika, R., & Iglesias-Rey, P. (2023). Smart water grids and digital twin for the management of system efficiency in water distribution networks. *Water, 15*(6), 1129.

Rodrigo, M. A., Cañizares, P., Buitrón, C., & Sáez, C. (2010). Electrochemical technologies for the regeneration of urban wastewaters. *Electrochimica Acta, 55*(27), 8160–8164.

Rodríguez, C., García, B., Pinto, C., Sánchez, R., Serrano, J., & Leiva, E. (2022). Water context in Latin America and the Caribbean: Distribution, regulations and prospects for water reuse and reclamation. *Water, 14*(21), 3589.

Rodriguez-Anton, J. M., Rubio-Andrada, L., Celemín-Pedroche, M. S., & Alonso-Almeida, M. D. M. (2019). Analysis of the relations between circular economy and sustainable development goals. *The International Journal of Sustainable Development and World Ecology, 26*(8), 708–720.

Sachidananda, M., Webb, D. P., & Rahimifard, S. (2016). A concept of water usage efficiency to support water reduction in manufacturing industry. *Sustainability, 8*(12), 1222.

Samani, P. (2023). Synergies and gaps between circularity assessment and life cycle assessment (LCA). *Science of the Total Environment, 903*, Article 166611.

Sanzana, S., Fenti, A., Iovino, P., & Panico, A. (2025). A review of PFAS remediation: Separation and degradation technologies for water and wastewater treatment. *Journal of Water Process Engineering, 74*, Article 107793.

Sauvé, S., Lamontagne, S., Dupras, J., & Stahel, W. (2021). Circular economy of water: Tackling quantity, quality and footprint of water. *Environmental Development, 39*, Article 100651.

Schellenberg, T., Subramanian, V., Ganeshan, G., Tompkins, D., & Pradeep, R. (2020). Wastewater discharge standards in the evolving context of urban sustainability–the case of India. *Frontiers in Environmental Science, 8*, 30.

Shah, W. U. H., Hao, G., Yasmeen, R., Yan, H., Shen, J., & Lu, Y. (2023). Role of China's agricultural water policy reforms and production technology heterogeneity on agriculture water usage efficiency and total factor productivity change. *Agricultural Water Management, 287*, Article 108429.

Shehata, N., Egirani, D., Olabi, A. G., Inayat, A., Abdelkareem, M. A., Chae, K. J., & Sayed, E. T. (2023). Membrane-based water and wastewater treatment technologies: Issues, current trends, challenges, and role in achieving sustainable development goals, and circular economy. *Chemosphere, 320*, Article 137993.

Shokri, A., & Fard, M. S. (2023). Principles, operational challenges, and perspectives in boiler feedwater treatment process. *Environmental Advances, 13*, Article 100389.

Simha, P., & Ganesapillai, M. (2017). Ecological sanitation and nutrient recovery from human urine: How far have we come? A review. *Sustainable Environment Research, 27*(3), 107–116.

Smetana, G., & Grosser, A. (2024). The application of an upflow anaerobic sludge blanket reactor in the treatment of brewery and dairy wastewater: A critical review. *Energies, 17*(6), 1504.

Staszak, K., & Wieszczycka, K. (2023). Recovery of metals from Wastewater—state-of-the-art solutions with the support of membrane technology. *Membranes, 13*(1), 114.

Stazi, V., & Tomei, M. C. (2021). Dissolved methane in anaerobic effluents: A review on sustainable strategies for optimization of energy recovery or internal process reuse. *Journal of Cleaner Production, 317*, Article 128359.

Tabassum, S., Kotnala, C. B., Salman, M., Tariq, M., Khan, A. H., & Khan, N. A. (2024). The impact of heavy metal concentrations on aquatic insect populations in the asan wetland of dehradun, Uttarakhand. *Scientific Reports, 14*(1), 4824.

Tiwari, A., & Nakamura, K. (2024). Closing the loop on biohydrogen production: A critical review on the post-fermentation broth management techniques. *International Journal of Hydrogen Energy, 81*, 595–614.

Tomić, T., & Schneider, D. R. (2020). Circular economy in waste management–socio-economic effect of changes in waste management system structure. *Journal of Environmental Management, 267*, Article 110564.

Tong, T., & Elimelech, M. (2016). The global rise of zero liquid discharge for wastewater management: Drivers, technologies, and future directions. *Environmental Science and Technology, 50*(13), 6846–6855.

Tzanakakis, V. A., Capodaglio, A. G., & Angelakis, A. N. (2023). Insights into global water reuse opportunities. *Sustainability, 15*(17), Article 13007.

Ugwuanyi, E. D., Nwokediegwu, Z. Q. S., Dada, M. A., Majemite, M. T., & Obaigbena, A. (2024). The role of algae-based wastewater treatment systems: A comprehensive review. *World Journal of Advanced Research and Reviews, 21*(02), 937–949.

Ulinici, S. C., Baisan, G., Vlad, G., Popa, A., & Toloman, D. (November 2020). Water decontamination hybrid processes using photocatalytic ultrafiltration membranes. *Proceedings, 57*(1), 82 (MDPI).

Wang, A. J., Li, H., He, Z., Tao, Y., Wang, H., Yang, M., Savic, D., Daigger, G. T., & Ren, N. (2024a). Digital twins for wastewater treatment: A technical review. *Engineering, 36*, 21–35.

Wang, Z., Zhan, A., Tao, Y., Jian, Y., & Yao, Y. (2024b). Sustainable governance of drinking water conservation areas based on adaptive thresholds. *Journal of Environmental Management, 351*, Article 119605.

Weerasooriya, R. R., Liyanage, L. P. K., Rathnappriya, R. H. K., Bandara, W. B. M. A. C., Perera, T. A. N. T., Gunarathna, M. H. J. P., & Jayasinghe, G. Y. (2021). Industrial water conservation by water footprint and sustainable development goals: A review. Environment. *Development and Sustainability, 23*(9), 12661–12709.

Wilcox, J., Bell, S., & Nasiri, F. (2016). Water reuse trajectories. *Urban Water Trajectories*, 69–80.

Witek-Krowiak, A., Gorazda, K., Szopa, D., Trzaska, K., Moustakas, K., & Chojnacka, K. (2022). Phosphorus recovery from wastewater and bio-based waste: An overview. *Bioengineered, 13*(5), 13474–13506.

World Health Organization. (2006). *Guidelines for the safe use of wastewater, excreta and greywater, 4*. World Health Organization.

Wu, S., Wallace, S., Brix, H., Kuschk, P., Kirui, W. K., Masi, F., & Dong, R. (2015). Treatment of industrial effluents in constructed wetlands: Challenges, operational strategies and overall performance. *Environmental Pollution, 201*, 107–120.

Yaqub, M., & Lee, W. (2019). Zero-liquid discharge (ZLD) technology for resource recovery from wastewater: A review. *Science of the Total Environment, 681*, 551–563.

Zabel, T. (1985). The advantages of dissolved-air flotation for water treatment. *Journal-American Water Works Association, 77*(5), 42–46.

Zarei, M. (2020). Wastewater resources management for energy recovery from circular economy perspective. *Water-Energy Nexus, 3*, 170–185.

Zekri, S., Jabeur, N., & Gharrad, H. (2022). Smart water management using intelligent digital twins. *Computing and Informatics, 41*(1), 135–153.

Zhang, X., González Rivas, M., Grant, M., & Warner, M. E. (2022). Water pricing and affordability in the US: Public vs. private ownership. *Water Policy, 24*(3), 500–516.

Zhang, L., Tang, Q., & Huang, R. H. (2021). Mind the gap: Is water disclosure a missing component of corporate social responsibility? *The British Accounting Review, 53*(1), Article 100940.

PART

VIII

Resilience building and public participation

CHAPTER

Hybrid intelligence–integrated water resources management: Addressing global challenges for climate-resilient strategies

14

Min Liu[1], Chenran Shao[1], Bakim Chandra[2], Jean-Luc Betrand-Krajewski[3], Gulomjon Umirzakov[4], Khan Shahzada[5], Gete Zeleke[6], Yazidhi Bamutaze[7], Anh Tien Do[8] and Mahboob Alam[9]

[1]*Nanjing University of Information Science and Technology, Nanjing, China;* [2]*National Institute of Technology Manipur, Imphal, Manipur, India;* [3]*INSA Lyon, Villeurbanne, France;* [4]*Hydrometeorological Research Institute, Central Asian University for Environmental and Climate Change Studies, Tashkent, Uzbekistan;* [5]*University of Engineering & Technology Peshawar, Peshawar, Pakistan;* [6]*Water and Land Research Center, Addis Ababa, Ethiopia;* [7]*Makerere University, Kampala, Uganda;* [8]*Vietnam Meteorological and Hydrological Administration, Hanoi, Vietnam;* [9]*Water & Sanitation Service Peshawar, Peshawar, Pakistan*

1. Introduction

Over the past four decades, rapid global urbanization and accelerating climate change have fundamentally reshaped the terrestrial water cycle, placing unprecedented pressure on water systems at urban, catchment, and regional scales. Urban population growth, land-use change, and infrastructure expansion have significantly altered hydrological processes, while rising temperatures, shifting precipitation patterns, and increasing climate extremes have intensified hydrological variability. Moreover, observational and remote sensing datasets from HadCRUT, GISTEMP, TRMM, GRACE, SMAP, etc., and projections by mainstream global and regional climate models, such as Coupled Model Intercomparison Project (CMIP5 and CMIP6), consistently indicate clear trends of climate change of clear warming trend, accelerating ice mass loss, increased frequency and intensity of heatwaves, heavier precipitation extremes, and higher risks of drought since the 1900s. The reciprocal effects of anthropogenic activities and climate change lead to intensified extreme hydrological events, deteriorated water pollution, accelerated groundwater depletion, posed enormous threat, and compound challenges to sustainable development.

These challenges are not uniformly distributed in Asia, Africa, and Latin America, which exhibit both common and distinct vulnerabilities driven by contrasting climate regimes, socioeconomic development pathways, governance structures, and hydrogeological conditions, and water management practices. Rapid urban expansion and industrial growth in parts of Asia have exacerbated water pollution and flood risks, while many regions in Africa struggle with water scarcity, infrastructure

Sustainable Water Management. https://doi.org/10.1016/B978-0-443-44853-9.00014-9

deficits, and high climate sensitivity. Understanding both shared and region-specific challenges is therefore critical for developing robust and transferable water management strategies. This manuscript synthesizes global urbanization and climate change impacts since the 1980s and assesses projected risks through 2050, providing a comparative perspective across continents.

Despite substantial investments in water infrastructure and management, existing approaches still struggle in understanding the water cycle processes. Conventional water management and modeling frameworks frequently fail to simulate floods, droughts, and groundwater dynamics with sufficient accuracy, particularly under nonstationary climate conditions, especially in data sparse region. Three fundamental limitations underpin these shortcomings. First, interactions among climate, meteorology, surface water, groundwater, and urban water systems are often inadequately represented, leading to fragmented understanding and decision-making. Second, hydrometeorological and hydrogeological observations remain sparse and unevenly distributed, especially in developing regions, constraining both process-based modeling and risk assessment. Third, current modeling practices are highly fragmented, with limited integration among land surface models, global and regional climate models, local hydrological and hydraulic models, and coupled urban water system models.

In this context, artificial intelligence (AI) has emerged as a transformative enabler for water management under climate change. Advances in machine learning (ML), deep learning (DL), and hybrid modeling provide new opportunities to extract value from heterogenous and incomplete datasets, improve prediction skills, and enhance system operation and planning. This review systematically evaluates existing AI applications and identifies potential applications in water management across five key domains: urban water supply, sewer system management, urban stormwater management, catchment-scale water resources management and flood control, and groundwater management. Both achievements and limitations of current AI-based approaches are critically assessed, with particular attention to their applicability in data-sparse and climate change conditions.

Building on this synthesis, the chapter proposes a forward-looking strategy for climate-resilient water management grounded in AI-enabled innovation. Key directions include the establishment of integrated hydrometeorological and groundwater observation networks, the advancement of interdisciplinary and unified modeling frameworks spanning climate to urban water systems, and the development of hybrid mechanical- and data-driven models, such as physics-informed neural networks (PINNs). Collectively, this chapter proposes an enhanced paradigm of integrated water resources management, referred to as hybrid intelligence integrated water resources management (HI-IWRM), which provides a roadmap for leveraging AI to support robust, adaptive, and climate-resilient water management strategies across diverse sociohydrological contexts worldwide.

2. Global change and challenges for water

2.1 Urbanization

Satellite-based analyses by Zhao et al. (2022) show that the global urban land footprint has expanded rapidly with the urbanized area being increased by more than three times from 1992 to 2020, totaling $\approx$980,000 km^2 of new urban area. The share of global urban dwellers increased from about 15% in 1900 to roughly 56% in 2025, with projections indicating a surge to 68% by 2050. In absolute numbers, this means that the urban population has grown from hundreds of millions in the early 20th century to over 4 billion people today, more than fourfold since 1960, according to the data of International Institute for Environment and Development. This transition is not merely a shift in population but a fundamental restructuring of terrestrial surfaces. Analysis of 1567 major cities across the globe reveals that urban expansion between 1985 and 2015 was characterized by decreasing morphological and population densities compared to pre-1985 structures (Taubenböck et al., 2024). This trend toward "urban sprawl" has systemic implications for water cycles, as it increases the extent of impervious surfaces and places a higher ecological burden on surrounding watersheds (Tables 14.1–14.3).

2.2 Climate change

Simultaneously, the global climate is witnessing a period of unprecedented variability, which has been robustly documented by multiple independent observational datasets, satellite products, and climate modeling frameworks. Global surface temperature records, including HadCRUT (UK Met Office/CRU), GISTEMP (NASA), and NOAA GlobalTemp, consistently show a clear and sustained warming trend since the mid-20th century (Ramil Abbasov et al., 2025). Reanalysis products such as ERA5 further confirm changes in atmospheric circulation, precipitation patterns, and land–atmosphere interactions. Satellite missions, including TRMM/GPM for precipitation, MODIS for land surface processes, and GRACE/GRACE-FO for terrestrial water storage, provide strong evidence of an intensifying hydrological cycle, accelerating ice mass loss, and widespread groundwater depletion. Future climate evolution is assessed using global climate models participating in the Coupled Model Intercomparison Project (CMIP5 and CMIP6), with regional refinement through CORDEX downscaling experiments. Climate change scenarios based on representative concentration pathways (RCPs) and shared socioeconomic pathways (SSPs) consistently project continued warming, increased frequency and intensity of heatwaves, heavier precipitation extremes, and higher risks of drought and flooding. Collectively, these sources lead to a clear scientific consensus that climate change is unequivocal, primarily anthropogenic, and will increasingly amplify water-related risks, with impacts varying strongly across regions and development contexts.

Table 14.1 Historical and projection of urbanization and water demand index.

Global demographic and urban indicators (1985–2050)	1985/1990 baseline	Current estimate (2020–2024)	2050 projection
Global urban population share	~41%	56%	68%
Total global population	~4.8 billion	~8.1 billion	~10 billion
Urban expansion rate	Rapid (morphological sprawl)	High density (Asia/Africa)	4500+ new African cities
Water demand index	Base 100	+20% increase	+163% (sub-Saharan Africa)

Table 14.2 Historical and projected climate change.

Parameter	Observed changes (global, recent decades)	Projected changes (2081–2100, relative to 1850–900 unless noted)
Global surface temperature	+1.1°C (2011–20); ~1.19°C (2014–2023 decade average)	SSP1-2.6 (low): ~1.8°C SSP2-4.5 (medium): ~2.7°C SSP5-8.5 (high): ~4.4°C
Precipitation	Slight increase over land (~0.1 mm/day per century); increased heavy precipitation events and extremes	Increases in global mean and monsoon precipitation; higher intensity/frequency of heavy events in most regions; regional drying in subtropics
Soil moisture	Drying trends in subtropics/Mediterranean; mixed elsewhere; anthropogenic signal emerging	Widespread drying in subtropics and mediterranean-like regions; increased drought risk, especially under higher scenarios
Terrestrial water storage (TWS) and groundwater	GRACE data: Declines in many regions; groundwater depletion accelerating (71% of aquifers declining rapidly >0.5 m/yr in dry cropland areas)	Continued/increased depletion in dry regions due to higher demand and variable recharge; higher risks under high emissions; some recovery possible in low-emission scenarios
Ice and glaciers	Glaciers (excl. Ice sheets): ~5%–6% mass loss since 2000; annual loss ~267–273 Gt/yr (2010–2020s); total ~6542 Gt (2000–23)	26% (±6%) mass loss at +1.5°C to 41% (±11%) at +4°C (by 2100 rel. To 2015); higher losses under SSP5-8.5; committed ongoing retreat

2.3 Interaction of urbanization and climate change

Urbanization and climate change are strongly coupled processes that interact through multiple biophysical and socioeconomic pathways, reinforcing risks to water, energy, and environmental systems. On one hand, urbanization influences climate change by altering land–atmosphere interactions and increasing greenhouse gas emissions. On the other hand, climate change reshapes urbanization pathways and urban system performance. While climate projections indicate surface air temperature increases of approximately 2.0°C from the 1990s to the 2070s under the IPCC A1B scenario, urbanization contributes an additional 0.5°C warming (Adachi et al., 2012). These findings underscore that areas experiencing major urban growth are largely coincident with climate change hotspots, particularly in the Global South, necessitating integrated management approaches that address both urbanization and climate adaptation simultaneously (Scheuer et al., 2017).

The interplay between urbanization-driven land-use change and climate-driven hydrometeorological shifts creates unfavorable conditions for water resources management, where traditional stationary assumptions in hydrological modeling are no longer valid. From a scientific perspective, the interaction between urbanization and climate change is nonlinear, multiscale, and feedback-driven, operating across local, regional, and global scales. Urban expansion amplifies climate signals, while climate change constrains and reshapes urban development trajectories. These reciprocal effects

Table 14.3 Summary of regional water conditions and challenges (2025).

Region	Water supply and sanitation	Water resource availability	Flood and dam management	Groundwater management
China	High basic access (93%–96%); primary challenge is rural–urban service gap.	Acute north–south imbalance; 20% of the country in acute drought in 2025.	Heavy reliance on megaprojects (three Gorges, SNWDP); dam safety risks.	Severe overexploitation in north China plain; land subsidence in megacities.
Middle Asia	Decaying Soviet-era infrastructure; seasonal imbalances in energy/water.	Transboundary river dependence (86% upstream); Aral sea depletion.	Conflicts over winter hydropower versus summer irrigation releases.	50% of irrigated land affected by salinization/ waterlogging.
South Asia	74% under high stress 1; infrastructure exists but service delivery is poor.	Critical dependency on melting Himalayan glaciers ("third pole").	Monsoon compression (40-day rains); catastrophic flash floods and landslides.	Massive overextraction for irrigation; arsenic contamination (e.g., Bangladesh).
Southeast Asia	40% supply–demand shortfall projected by 2030.	80% of river water polluted; high seasonal variability.	Urban sprawl on floodplains; rapid damming of the Mekong disrupts natural flow.	Coastal saltwater intrusion; land subsidence in Jakarta and Bangkok.
East Africa	Safely managed sanitation <24%; piped water access is regressing.	Highest demand growth globally (+163% by 2050) 1; "economic" water scarcity.	High casualties from dam collapses and riverine floods.	15% share of total water; critical for rural drought resilience.
Latin America	150m live in water-scarce areas; 400m lack safe sanitation.	High endowment but 25% of rivers/ aquifers polluted by mining/industry.	Recurrent floods and rapid retreat of Andean glaciers.	Overexploitation in megacities to compensate for surface drought; land subsidence.

underscore the need for integrated observation systems, coupled climate–hydrology–urban models, and hybrid physics-informed and data-driven approaches to support climate-resilient urban water management.

2.4 Water challenges in global south

The convergence of rapid global urbanization and the intensifying impacts of climate change has created dual pressure for the hydrological stability, which manifests as a multifaceted crisis. While total urban populations in these regions are projected to swell significantly by 2050, the "engineered water cycle"—comprising supply, sanitation, groundwater and flood management—is struggling to adapt to the nonstationarity of a changing climate and the rapid degradation of natural ecosystems.

2.4.1 Urban water supply

(1) Infrastructure design and optimization: Solving complex design problems for the expansion and rehabilitation of distribution networks with best hydraulic and cost conditions.
(2) Demand forecasting: Predicting spatiotemporal short-term and seasonal water needs under different weather and socio-economic factors.
(3) Water quality modeling: Monitoring and simulating pollutants and disinfection byproducts (DBPs) in both intermittent and continuous supply systems.
(4) Leakage detection and localization: Utilizing AI and IoT sensors for real-time detection of physical and apparent water losses, including nonrevenue water (NRW).
(5) Burst prediction and anomaly detection: Identifying pipe bursts and sensor faults through monitoring data and pattern recognition.
(6) Asset renovation planning: Developing predictive models to schedule maintenance and prioritize the rehabilitation of aging pipelines.
(7) Treatment process optimization: Enhancing energy efficiency and chemical dosing in water treatment plants.

2.4.2 Sewerage system management

(1) Structural deterioration modeling: Using image recognition and statistical models (Markov chain) to predict the condition of sewer pipes from CCTV inspection data.
(2) Blockage and defect detection: Automating the identification of structural cracks, roots, and deposits for proactive maintenance.
(3) Infiltration and inflow (I/I) analysis: Identifying extraneous water entering the system due to cracks or misconnections.
(4) Pump station operation: Improving real-time pump scheduling and energy optimization to handle dynamic inflows and prevent pipe overflows and treatment plant overloads.
(5) Combined sewer overflow (CSO) mitigation: Optimizing in-line storage and intercatchment wastewater transfer to reduce the volume and frequency of overflows.
(6) Effluent quality prediction: Continuous monitoring and forecasting of treated wastewater properties to ensure compliance with environmental standards.

2.4.3 Urban stormwater management

(1) Real-time control (RTC) optimization: Developing feedback controllers to dynamically maximize existing storage capacity.
(2) Low-impact development (LID) planning: Evaluating the cost-effectiveness of green infrastructures (e.g., green roofs, bioretention cells).
(3) Flood nowcasting and early warning: Integrating tides, groundwater, and rainfall data to provide real-time flood simulations for decision-making.
(4) Performance assessment of hybrid infrastructure: Integrating simulation-optimization modules to balance green and gray infrastructure performance.
(5) Impact of urbanization on runoff: Modeling how increased impervious surfaces diminish rainwater absorption and exacerbate flood risks.

2.4.4 Catchment water resources and flood control

(1) Transboundary flood forecasting: Developing regional early warning systems for international basins (e.g., Mekong, Nile) to coordinate disaster responses.
(2) Reservoir and dam operations: Optimizing water allocation and release strategies to balance competing demands for hydropower, irrigation, and flood control.
(3) Soil erosion and sediment forecasting: Predicting suspended sediment loads (SSL) and optimizing dredging or drawdown flushing to maintain reservoir efficiency.
(4) Modeling in data-sparse regions: Using satellite-based remote sensing and AI to bridge the "observational void" in watersheds with inadequate ground gauging.
(5) Water–energy–food nexus trade-offs: Analyzing the complex interdependencies between resource sectors for sustainable regional planning.
(6) Long-term climate impact assessment: Projecting the influence of climate change on streamflow, glacial melt, and future water availability through 2050.
(7) Water resources allocation: Predicting water resources availability under climate change uncertainty to optimize the water resources allocation for industry and agriculture to maximize the output.
(8) Integrated catchment flood prediction and management for EWC: Developing holistic modeling approach combining meteorological, hydrological, hydraulic processes and considering the human control modified elements in the catchment system.

2.4.5 Groundwater management

(1) Seawater intrusion (SWI) prediction: Simulating saline water dynamics in coastal aquifers due to overextraction and sea-level rise.
(2) Groundwater level and head forecasting: Capturing spatiotemporal variations in hydraulic heads under climate change and human extraction for both confined and unconfined aquifers using physics-informed neural networks (PINNs).
(3) Contaminant fate and transport: Identifying pollution sources and modeling the movement of nitrate, arsenic, or industrial plumes to optimize remediation.
(4) Overexploitation and land subsidence: Assessing risks associated with massive drawdown and the resulting compaction of aquifer systems in megacities.
(5) Managed aquifer recharge (MAR) optimization: Designing optimal groundwater recharge and replenishment strategies with simulated surface water availability through hydrometeorological modeling.
(6) Surface-groundwater integrated modeling: Addressing the dynamic exchange at the soil–water interface to improve the accuracy of regional water balance models.

As the western countries have established a complete water infrastructure with extensive monitoring and advanced management system, the transition into modern economy has placed the Global South Countries at the forefront of a hydrological crisis. These challenges are geographically heterogenous; for instance, while sub-Saharan Africa faces a projected 163% surge in water demand by midcentury, South Asia must contend with the thinning of "Third Pole" glaciers and compressed monsoon cycles that heighten the risk of catastrophic flash floods. While some countries, such as China and Brazil, have made significant strides in rural and urban water security, reaching high basic access rates, other countries such as Pakistan and east African countries have a strong shortage in infrastructure. The rapidly constructed infrastructure has also enormous problem of water leakage, pipe burst, infiltration/inflow, etc., which calls for renovation and refined management. Moreover, in many regions, the water sustainability is threatened by an acute spatial imbalance, for example, the water-rich south versus the water-scarce north in China leads to the massive South-to-North Water Diversion Project (SNWDP). Many global south countries rely heavily on groundwater supply due to the superior water quality requiring less treatment efforts and costs, which results in groundwater over-exploitation and triggered enormous problem, such as land subsidence and ecological disasters.

3. Current approaches and limitations

Managing catchment and urban water systems is an inherently complex task, as it involves the interplay of multiple interconnected systems including meteorology, hydrology and hydraulics, and groundwater system, which consequently are related to multifold problems in different sectors, such as flood control, irrigation, hydropower generation, pollution control, stormwater management, wastewater treatment, and drinking water supply. The interdependencies between these sectors mean that a decision made in one sector can have cascading effects on other sectors.

In this context, decision-making support models, particularly IWRM models, have become indispensable tools for decision-makers to evaluate and demonstrate the effectiveness of various management, operational, and design strategies. They provide a virtual platform to test "what-if" scenarios, quantify trade-offs between competing objectives (e.g., energy production vs. environmental flow requirements), and demonstrate the effectiveness of proposed solutions to stakeholders. However, the proliferation of available models with diverse capabilities, data requirements, and technical characteristics has posed significant challenges for users in selecting the most appropriate model for their specific objectives.

3.1 Hydrological modeling

Accurate streamflow forecasting is fundamental to flood risk mitigation, drought preparedness, reservoir operation, water allocation, and integrated water resource management. Recent trends have progressed hydrological forecasting as various approaches are used to enhance it continually. The type of hydrological model can be classified into statistical model, conceptual model, and physically based model.

3.1.1 Statistical model

For hydrological modeling, historical data and statistical approaches are used to predict and forecast future hydrological conditions (Kochanek & Markiewicz, 2022). These models rely on probabilistic and regression-based approaches to describe data relationships and extremes, in comparison with deterministic physical models. Well-known examples include autoregressive integrated moving average (ARIMA) models for time-series forecasting of river discharge, multiple linear regression (MLR) for rainfall-runoff correlations, and extreme value theory distributions such as Gumbel and Log-Pearson Type III for flood frequency analysis. Both linear models and nonlinear regression have been applied to streamflow forecasting (Masselot et al., 2016; Safari et al., 2020). Moreover, multiple regression analysis of hydrometeorological variables such as temperature, precipitation, humidity, and river flow was proposed by Baig et al. (2022), which has also shown promising results.

In general, statistical forecasting techniques are computationally efficient; however, their linear assumptions limit their ability to represent nonlinear hydrological responses. These constraints have prompted growing interest in more flexible ML and hybrid approaches capable of capturing complex dynamics, especially under extreme or rapidly changing climatic conditions. Therefore, physically based models explicitly simulate watershed processes using laws of physics rather than relying purely on historical data relationships gain importance.

3.1.2 Physically based model

Physically based hydrological models simulate catchment processes through fundamental physical equations, such as those governing mass, momentum, and energy conservation, often incorporating Darcy's law for subsurface flow and kinematic wave approximations for surface routing, enabling detailed spatial and temporal representations of water movement in different domains. Prominent examples include MIKE SHE (Abrahart & See, 2002), a fully integrated distributed model for coupled surface-groundwater dynamics; VIC (Liang et al., 1994), a macroscale land-atmosphere model emphasizing soil moisture and runoff at grid resolutions from kilometers to global scales; DHSVM (Wigmosta et al., 1994), tailored for mountainous terrains with explicit vegetation and topographic influences; GSSHA (Downer & Ogden, 2004), a high-resolution gridded model for event-based simulations including sediment transport; SWAT (Arnold et al., 1998), a semidistributed basin-scale tool for assessing agricultural impacts on water quality; ParFlow (Jones et al., 2006), a 3D variably saturated groundwater model integrated with surface hydrology; and HydroGeoSphere (Therrien et al., 2006), which extends to reactive transport in coupled surface–subsurface systems. However, models such as TOPMODEL (Beven, 2012) occupy an intermediate position, blending physical principles (e.g., topographic index for saturation zones) with conceptual simplifications, making it less strictly physics-based than fully distributed counterparts but more mechanistic than purely empirical approaches. Overall, these models offer advantages such as high predictive accuracy in well-characterized catchments, improved process understanding for scenario analysis (e.g., climate change impacts), and the ability to incorporate heterogenous data such as remote sensing. Disadvantages include high computational demands, extensive data requirements for parameterization (e.g., soil properties, DEMs), sensitivity to input errors, and challenges in upscaling from plot to global levels, often necessitating simplifications in line with conceptual models.

3.1.3 Conceptual model

Conceptual hydrological models represent a cornerstone in rainfall-runoff simulation by simplifying complex catchment processes into interconnected conceptual reservoirs and empirical transfer functions, striking a balance between physical interpretability and computational efficiency, particularly in regions with limited data availability. These lumped or semidistributed models have been widely applied for operational flood forecasting, water resources assessment, and climate impact studies. Several well-known conceptual models have been extensively validated and applied globally. The HBV model, originally developed by the Swedish Meteorological and Hydrological Institute (SMHI) in the early 1970s, is a classic bucket-type model that integrates snowmelt, soil moisture balance, and multireservoir runoff routing modules, requiring only daily precipitation and temperature as input data, and has been successfully applied in over 80 countries for hydropower operation, flood warning, and climate change impact assessment (Lindström et al., 1997). The XAJ model (Zhao & Liu, 1995), proposed by Zhao Renjun from Hohai University in 1984, is based on the saturation-excess runoff mechanism and features a three-source runoff separation structure (surface runoff, interflow, groundwater runoff), which is particularly suitable for humid and semihumid basins in China and has been continuously improved and extended for regional adaptability enhancement. The GR4J model (Perrin et al., 2001), a lumped rainfall-runoff model developed by Cemagref (now INRAE) with only four parameters, operates on a daily time step and uses conceptual reservoirs to simulate water storage and release processes, demonstrating excellent performance in ungauged basins and regionalization studies due to its simplicity and robustness. These conceptual models have laid the foundation for modern hydrological modeling, and their continuous refinement and integration with new technologies (e.g., remote sensing data assimilation, ML optimization) further expand their application scope and simulation accuracy.

Despite progress in methodologies, challenges still exist in the field of hydrological forecasting due to the changing environment and human activity dynamics. Long-term streamflow forecasting remains particularly challenging due to climatic variability and hydrological nonlinearity. Therefore, integrated hydrometeorological modeling exhibits its importance. The development of hydrological forecasting highlights a clear shift toward integrated, data-rich, and adaptive forecasting systems that combine physical principles with AI, reflecting the increasing need for resilience under climate uncertainty.

3.2 Urban water system modeling

3.2.1 Water supply models

Water supply network modeling involves hydraulic simulation, water quality analysis, demand forecasting, and optimization for planning, design, operation, and rehabilitation of distribution systems. These models integrate principles from fluid mechanics, graph theory, and systems engineering to represent complex infrastructures, including pipes, pumps, valves, reservoirs, and treatment facilities. Broadly categorized into academic/open-source and commercial variants, mainstream examples include EPANET (developed by the US EPA), a foundational open-source simulator for steady-state and extended-period hydraulic and water quality analysis, widely used in research for network calibration, contaminant tracking, and leakage detection. Commercial models such as WaterGEMS (Bentley Systems), InfoWater (Autodesk), and MIKE+ (DHI) offer advanced features such as real-time SCADA integration, energy optimization for pumping, and asset management, supporting large-scale utility operations with GIS compatibility and scenario-based planning.

These models play a pivotal role in the EHC system, which refers to human-modified components of the natural water cycle, such as abstraction, distribution, consumption, and wastewater return. By enabling precise simulation of flow pressures, flow rate, water age, and chemical residuals, they mitigate risks such as pipe bursts, contamination events, and inefficiencies, ensuring reliable supply amid growing urbanization and climate variability. In the broader hydrological context, they bridge natural processes (e.g., groundwater recharge) with engineered interventions, optimizing resource allocation and reducing nonrevenue water losses, which globally exceed 30% in many regions. Future developments in water supply modeling are poised to, facilitate demand management and resilience against droughts or floods, contributing to sustainable development goals such as SDG 6 (clean water and sanitation), and leverage AI and big data, enabling real-time predictive analytics for leakage detection, burst prediction and adaptive control.

3.2.2 Sewer network models

Sewer network models are specialized computational frameworks designed to simulate the hydraulic, hydrologic, and water quality dynamics within urban wastewater collection systems, including combined sewer overflows (CSOs) and sanitary sewers. These models enable analysis of flow routing, surcharge events, infiltration/inflow (I/I), pollutant transport, and system capacity under varying loads. Key models include the open-source SWMM, a versatile tool for 1D/2D hydrodynamic simulations of urban runoff, CSOs, and low-impact development (LID) practices, widely adopted in academic research for calibration, scenario testing, and climate impact assessments. Commercial models, such as InfoWorks ICM (Autodesk), which excels in detailed 2D overland flow coupling and asset management; and SewerGEMS (Bentley Systems), focused on sanitary sewer design, energy efficiency, and rehabilitation planning. These tools often incorporate Saint-Venant equations for unsteady flow and advection-dispersion models for water quality, supporting regulatory compliance with standards like those from the EU Water Framework Directive or US Clean Water Act. As integral components of the EHC, human-altered pathways for water abstraction, use, treatment, and return, sewer network models are crucial for managing urban water security.

Future developments in sewer network modeling are geared toward enhanced resilience amid climate change and digital transformation. Integration of AI, such as ML for predictive maintenance (e.g., pipe blockage forecasting via neural networks) and real-time optimization using reinforcement learning for RTC, will address data scarcity and computational demands.

3.2.3 Urban flood modeling

Urban flood modeling encompasses computational frameworks that simulate the complex interplay of hydrological, hydraulic, and infrastructural processes in densely built environments. As the downstream confluence of catchment-scale hydrological processes, urban flood events are inherently linked to upstream runoff generation and routing, making urban flood modeling a critical bridge between watershed-scale flood simulation and urban-specific risk assessment. For coastal cities, this complexity is further amplified by tidal dynamics, where the interaction between fluvial floods, storm surges, and astronomical tides often triggers compound flooding events with exacerbated inundation depth and extent, posing catastrophic threats to densely populated coastal areas.

These models integrate high-resolution topographic data (e.g., LiDAR-derived DEMs), land imperviousness, and drainage networks to predict flood extents, depths, velocities, and durations under extreme precipitation events. Core approaches include 1D hydrodynamic models for sewer/pipe flows (e.g., Saint-Venant equations), 2D overland flow simulations (e.g., shallow water equations), and coupled 1D-2D schemes for realistic urban representations. Prominent tools include the open-source SWMM, which excels in both pipe flow and river flow; HEC-RAS (US Army Corps of Engineers), with extensions for 2D urban inundation and dam breach scenarios; MIKE FLOOD offering integrated 1D-2D-3D coupling for real-time forecasting; and InfoWorks ICM (Autodesk), specialized in detailed network hydraulics and pollutant transport. Academic variants such as LISFLOOD-FP provide computationally efficient 2D approximations, while TUFLOW emphasizes GPU-accelerated high-fidelity simulations. These models often incorporate probabilistic elements, such as Monte Carlo methods for uncertainty quantification, and are calibrated using remote sensing (e.g., SAR imagery) or crowdsourced data.

The future development of urban flood modeling is oriented toward four couplings: first, the advancement of coupling hydrodynamic models (e.g., 1D-2D coupled models) to accurately simulate complex urban hydrological processes, including surface runoff, pipe network drainage, and fluvial-tidal interactions. Second, the seamless coupling of meteorological nowcasting model with hydrological/hydraulic models for accurate rainfall-runoff process. Third, the deep coupling of AI and ML techniques with physical models to enhance simulation efficiency and predictive accuracy, especially for handling the nonlinearity of urban flood processes. Fourth, the strengthening of climate-urban flood coupled modeling frameworks that consider future climate change scenarios (e.g., sea-level rise, increased frequency of extreme rainfall) to support long-term urban flood adaptation planning.

3.2.4 Groundwater modeling

Traditional groundwater modeling, fundamentally reliant on numerical methods such as MODFLOW to solve governing physical equations, faces persistent challenges in data-scarce regions. These include high uncertainty in hydrogeological parameters, difficulties in characterizing aquifer heterogeneity, and the computational cost of simulating coupled processes (e.g., land subsidence or surface water interactions). These limitations constrain reliable resource assessment and predictive management.

Recent advances are defined by the integration of physics-informed neural networks (PINNs, Raissi et al., 2019) and multisource data assimilation. PINNs address data scarcity by embedding physical laws directly into the loss function of neural networks, ensuring predictions remain physically consistent even with sparse observations (e.g., simulating hydraulic head with limited well data). Furthermore, the fusion of traditional models with diverse data streams significantly reduces uncertainty. For instance, inverse modeling frameworks (Carrera et al., 2005) that jointly assimilate hydraulic head data and interferometric synthetic aperture radar (InSAR)–derived land displacement measurements can more accurately calibrate parameters controlling groundwater flow and aquifer compaction (Chaussard et al., 2014), as demonstrated in studies of Mexico City and other subsiding basins. Similarly, deep generative models such as generative adversarial networks (GANs) are now used to create numerous stochastic realizations of complex aquifer structures from limited borehole data (Laloy et al., 2018), enhancing the representation of heterogeneity.

This evolution toward intelligent, physics-constrained, and data-integrated simulation is a cornerstone for the proposed HI-IWRM framework. By moving beyond stand-alone models to interconnected systems that learn from real-time, multimodal data while respecting physical principles, HI-IWRM enables more robust and adaptive groundwater management strategies within the broader engineered water cycle.

3.3 The emergence of AI in water

The evolution of data-driven models has significantly transformed hydrological, urban water, and groundwater modeling by AI techniques to address complex, nonlinear relationships fundamental in hydrological systems. Unlike conventional models that rely on explicit physical process formulations, AI-driven approaches leverage ML and DL techniques to learn patterns from historical and geospatial datasets, enabling more accurate and efficient predictions across diverse water system contexts. These approaches are particularly valuable in dynamic and heterogenous environments, such as urban catchments or complex aquifer systems where physical processes are difficult to parameterize explicitly.

3.3.1 Hydrological modeling and flood forecasting

AI has significantly advanced hydrological modeling by capturing the nonlinear relationships inherent in hydrological systems (Mosavi et al., 2018), while bypassing the need for full physical process characterization. This is particularly advantageous in data-scarce regions or areas with high climatic variability, where traditional models struggle due to parameter uncertainty. Core ML algorithms, including artificial neural networks (ANNs), random forests (RF), support vector machines (SVMs), and gradient boosting machines (GBMs), have been widely adopted in hydrological modeling tasks (Refadah, 2025). The study of Zamri et al. (2022) demonstrates respectively that unsupervised ML and supervised ML outperform the traditional hydrological models in terms of forecasting accuracy. ML methods excel at capturing nonlinear interactions, which are common in hydrological processes.

DL as an improved subset of ML, has further improved performance in time-series forecasting tasks. Algorithms, such as long short-term memory (LSTM), ConvLSTM, and encoder–decoder networks, outperform traditional models in capturing spatial–temporal hydrological patterns (Feng et al., 2020; Li et al., 2021; Monjardin et al., 2017; Sabzipour et al., 2023). Recent advances in remote sensing, GPU computing, and reanalysis datasets have further increased the application and accuracy of DL hydrological models.

3.3.2 Transboundary flood forecasting

AI is revolutionizing transboundary flood forecasting by addressing transboundary data availability challenges primarily through satellite-based data integration and alternative data sources that bypass the need for upstream country cooperation.

Multiple studies demonstrate AI's potential to overcome traditional hydrological modeling limitations and improve prediction accuracy for transboundary rivers, when data sharing between countries in the upper and lower reaches of a basin is insufficient (Asghar et al., 2019). Nearing et al. (2024) show AI can achieve flood prediction reliability in ungauged watersheds that matches or exceeds current same-day forecasting systems, with up to 5-day lead times. Bibi et al. (2025) specifically highlighted LSTM networks achieving 96% accuracy in river flow predictions. Bhattarai et al. (2024) further confirmed that ML techniques such as ANNs and SVMs can effectively map flood susceptibility in transboundary basins.

3.3.3 Reservoir and dam operations

ML has emerged as a powerful tool for transforming reservoir and dam operations, offering significant improvements in water resource management, flood prevention, and operational efficiency. Al-Nouti et al. (2024) highlight ML techniques such as neural networks, genetic algorithms, and LSTM networks as crucial for predicting reservoir inflows and optimizing water release schedules. Lekamge et al. (2024) showed practical implementations using LightGBM and convolutional neural networks (CNNs) for gate automation and crack detection, achieving high precision. Umar et al. (2025) reported impressive model performance, with K-nearest neighbors achieving an R^2 score of 0.9443 and classification models reaching 99.5% accuracy in water resource management.

3.3.4 Soil erosion and sediment forecasting

Multiple studies demonstrate exceptional predictive capabilities for soil erosion and sediment forecasting with accuracy rates ranging from 80% to 99%, consistently outperforming traditional methods with remarkable accuracy. The methods including XGBoost, RF, SVMs, ANNs, and decision trees have been used. The research spans multiple applications, which covers erosion susceptibility mapping, sediment transport modeling, watershed-scale erosion prediction and climate change impact assessment (Bezak et al., 2025; Nguyen et al., 2019). Bolsynbek et al. (2025) achieved 99% accuracy using XGBoost for erosion stage classification, while Bhattacharya et al. (2024) reported excellent capacity with XGBoost and bagging models. Dinh et al. (2021) found that relevance vector machine performed best with 92.22% training accuracy and 91.94% testing accuracy and 96% classification accuracy using social spider algorithm optimized MARS.

3.3.5 Drought prediction and management

ML models, such as SVM, RF, and hybrid approaches, have shown outstanding performance with promising results in predicting the drought (Tawfeek et al., 2025; Zhang, 2024;). Key strengths of ML models for drought prediction include handling complex environmental data, identifying influential variables, and capturing intricate relationships between meteorological variables. Emerging ensemble and hybrid models are particularly promising, delivering higher performance compared to single-method approaches (Liu et al., 2025; Mosavi & Ardabili, 2023). The research suggests these techniques can significantly improve drought mitigation strategies, offering crucial insights for policymakers, agricultural planners, and resource managers.

3.3.6 Irrigation and water resources allocation

In the field of optimization of water resources allocation, ML is revolutionizing irrigation and water distribution by enabling precise, data-driven water management strategies that significantly improve efficiency and sustainability. Gao et al. (2023) identify three main areas where ML improves irrigation: water scarcity diagnosis, water demand prediction, and irrigation decision-making. Automatic decision-making based on a smart sensor system has shown great potential for irrigation management (Dhyani et al., 2024). Multiple studies demonstrate ML's powerful capabilities in improving decision support systems across water resource management sectors, including groundwater management, water demand prediction, and irrigation scheduling (Ahmed et al., 2024; Blasi et al., 2021).

3.3.7 Urban water system design and optimization

By providing powerful data-driven approaches to complex infrastructure challenges, ML is rapidly transforming urban water system design and optimization. Evolutionary computation techniques such as genetic algorithms(GA) can assist engineers in solving complex design problems. Bahrami Chegeni et al. (2025) developed ML models that achieved 98% accuracy in predicting optimal water distribution network component diameters. Haochen & John (2022) used ANNs to optimize clarification basin geometrics, addressing costly design challenges. Garzón et al. (2022) noted that current ML approaches still face limitations like the "curse of dimensionality" and lack of generalizability. Key strengths include rapid computational analysis, ability to handle complex interactions, and potential for optimizing infrastructure design. However, researchers emphasize the need for more robust, transferable ML architectures to fully realize urban water system optimization potential.

3.3.8 Urban water supply

AI and ML are pivotal to optimize water supply system and enhance service delivery. There are eight ways appeared in the literature, which demonstrate the impact of AI on water supply, i.e., predictive maintenance of water distribution network (WDN), automatic asset management, water quality prediction, nonrevenue water management, real-time leak detection, demand forecasting, and treatment process optimization, energy saving of WDN (Alhendi et al., 2022; Jayakumar et al., 2024; Mehmood et al., 2020). Integrating ML models and sensors enhances real-time system diagnosis and prediction, which can detect flow and pressure anomalies and pinpoint the leak's location and scale, assisting to hasten repair services. AI-driven approach can augment this process to help companies accurately determine demand side supply. AI models process leakage sounds, flow, and pressure data from sensors. Algorithms such as deep neural network (DNN), convolution neural network (CNN), fuzzy inference systems, and Bayesian network can process data and pinpoint leakages and their scales, while other algorithms, such as LSTM, RF, ANN, and principal component analysis (PCA) can assist in demand forecasting.

3.3.9 Urban sewer system

Modern sewer management in 2026 has transitioned into a proactive, data-driven discipline by integrating high-accuracy AI models across the entire infrastructure life cycle. This starts with structural deterioration modeling, which utilizes Markov chain probabilities to predict pipe failure (Ana, 2009), alongside DL models such as YOLO for the automated detection of cracks and blockages in CCTV footage (Zhang et al., 2025; Zhao et al., 2024). To maintain system capacity, AI identifies infiltration and inflow (I/I) by spotting flow anomalies through LSTM and random forest frameworks (Wang et al., 2024), while reinforcement learning optimizes pump station schedules to reduce energy consumption and prevent overflows (Pei, 2025). During storm events, intelligent systems mitigate combined sewer overflows (CSO) by dynamically managing in-line storage and intercatchment transfers using multiobjective PSO algorithms (Liu & Zeng, 2025a; 2025b), all while attention–LSTM–XGBoost hybrid networks provide real-time forecasting of effluent quality to ensure strict environmental compliance (Xiong et al., 2025). Collectively, these technologies transform the "hidden" underground network into an optimized, self-sensing asset that minimizes both operational costs and environmental impact. AI-driven solutions for sewer system can enhance data analysis, allowing optimization of treatment parameters while minimizing operational costs and reducing energy consumption.

3.3.10 Groundwater modeling

AI has significantly enhanced groundwater modeling by integrating complex data-driven frameworks with process-based hydrogeological principles to address global water security challenges (Chen et al., 2025). Recent advancements in seawater intrusion (SWI) prediction utilize hybrid RF and ANN architectures to simulate saline encroachment along coastal aquifers, driven by both anthropogenic overextraction and climate-induced sea-level rise (Sanad et al., 2024). For contaminant fate and transport, attention-based graph neural networks (aGNNs) have emerged as robust tools for identifying pollution sources and modeling the movement of nitrate or arsenic plumes, even with sparse monitoring data (Pang et al., 2024). In megacities, AI-driven risk assessment frameworks monitor overexploitation and land subsidence by correlating massive drawdown with aquifer compaction through InSAR-integrated DL models. Furthermore, managed aquifer recharge (MAR) optimization now relies on ML to design replenishment strategies by simulating surface water availability alongside hydrometeorological drivers (Zakir-Hassan et al., 2025). Finally, surface–groundwater integrated modeling has been improved by externally coupled, physically based multimodel frameworks that address the dynamic exchange at the soil–water interface, providing more accurate regional water balances (Xie et al., 2025a; 2025b). Together, these innovations create a self-sensing management paradigm that enhances both the predictive accuracy and the operational sustainability of global groundwater systems.

4. Hybrid models addressing data scarcity

4.1 The development of hybrid model

The proliferation of AI and ML within the water sector has strongly relied on high-frequency, high-fidelity datasets that are frequently absent in real-world hydraulic environments. While purely data-driven models demonstrate exceptional predictive capabilities in controlled settings, their "black-box" nature and susceptibility to overfitting in data-sparse regions pose significant risks for high-stakes water management. Therefore, there is an emerging necessity for hybrid modeling frameworks, specifically physics-informed neural networks (PINNs), which bridge the gap between process-based mechanical modeling and flexible ML. PINNs embed governing physical laws, expressed as conservation equations and partial differential equations (PDEs), directly into neural network training. This integration enables models to respect fundamental conservation principles while learning from sparse or incomplete data, which allows the physics to act as a potent regularizer, ensuring that model outputs remain physically consistent even when training data is noisy or incomplete. Physics can be integrated through various strategies, including loss-term embedding to penalize physical residuals, architectural constraints that reflect hydraulic structures, and inverse parameter discovery to estimate latent variables like hydraulic conductivity or pipe roughness.

Recent research progress underscores the transformative impact of these hybrid models across the water cycle, moving from theoretical explorations to robust operational deployments. In urban pipe network modeling, hybrid frameworks have been successfully applied to enhance incorporate Hazen–Williams or Darcy-Weisbach equations to regulate the pressure gradient or flow to be physically feasible, effectively mitigating the challenges of sparse sensor networks (Xie et al., 2025a; 2025b; Bella et al., 2026). For hydrological modeling, PINNs have proven particularly adept at solving complex inverse problems and simulating spatiotemporal variations in hydraulic heads under climate-induced stress, often outperforming traditional numerical solvers in both speed and accuracy (Liu & Zeng, 2025a; 2025b; Kumar et al., 2024). In the Andean watersheds of Southern Ecuador, Farfán et al. (2020) combined outputs from physical models (WEAP and GR2M) with stand-alone ANNs as inputs to a secondary ANN, significantly improving streamflow forecast accuracy in data-scarce regions, which increase the Nash-Sutcliffe Efficiency (NSE) by capturing hidden nonlinear patterns while retaining physical insights, outperforming all individual models. Systematic reviews indicate that these models are increasingly utilized to simulate groundwater level and optimize MAR by coupling hydrometeorological drivers with real-time hydraulic constraints (Liu et al., 2025; Zakir-Hassan et al., 2025). By facilitating a more transparent and trustworthy decision-making process, hybrid models represent a critical milestone in the development of self-sensing, resilient water infrastructure capable of addressing global water security challenges in the mid-2020s.

4.2 Proposal of HI-IWRM concept

The escalating complexity of global water challenges, ranging from catastrophic urban flooding to the crisis of groundwater depletion, demands a paradigm shift from fragmented modeling to an integrated, multisystem approach. Water systems do not exist in isolation; the feedback loops between the atmosphere, catchment hydrology, urban infrastructure, and subsurface aquifers are deeply intertwined. For instance, urban flood mitigation is intrinsically linked to rainfall nowcasting and hydraulic network capacity, while long-term irrigation sustainability depends on the delicate balance between climate-driven recharge and groundwater extraction. Consequently, effective management requires a unified modeling framework that can integrate the regimes of atmosphere, surface, and subsurface and connect the disciplines of meteorology, hydrology, and hydraulics. Only through such integrated modeling can we capture the nonlinear interactions within the water cycle that dictate the resilience of both natural and anthropogenic environments.

Despite the conceptual strength of integrated modeling, its practical application is frequently hindered by the "data desert" problem. Traditional mechanistic models, while physically accurate, are computationally expensive and require dense parameterization that is often unavailable in real-world catchments. Conversely, purely data-driven approaches such as ML offer rapid predictive capabilities but often fail in extrapolation or produce physically inconsistent results due to a lack of high-quality training data. To overcome these limitations, the development of hybrid models is essential. By embedding physical laws, such as mass balance and fluid dynamics, into the architecture of AI, we can reduce the dependency on massive datasets. Hybrid models leverage the deterministic rigor of physics to guide the learning process, allowing for robust predictions even in data-scarce regions where sensor networks are sparse or subsurface characteristics remain unknown.

In light of these challenges, we propose the HI-IWRM framework. This concept represents a transformative evolution in water science, synthesizing the interpretability of process-based models with the adaptive power of AI. The HI-IWRM framework utilizes physics-informed neural network (PINN) to couple hydrometeorological forcing with hydraulic and groundwater systems. This allows for real-time urban flood forecasting and long-term resource assessment within a single, computationally efficient pipeline. By treating the water system as a holistic intelligence network, this approach uses synthetic data generation to fill observational gaps and employs intelligent agents to optimize irrigation and urban drainage dynamically. Ultimately, HI-IWRM provides a robust decision-support system that ensures physical consistency and predictive agility, offering a scalable solution for sustainable water governance in an era of increasing climatic uncertainty.

5. Conclusion

The chapter synthesizes the compounded impacts of accelerated urbanization and climate change on water systems, highlighting the progressive transformation of the natural hydrological regime to highly modified engineered water cycles. These transformations exacerbate spatially heterogenous vulnerabilities across regions, particularly in the Global South. Existing water management approaches face critical limitations due to inadequate system interactions accounting, sparse observations, and fragmented modeling. Recent advances in AI and ML offer transformative solutions, with proven efficacy in urban water supply, flood control, groundwater management, and other key domains. To address data scarcity and ensure physical consistency, hybrid models such as physics-informed neural networks (PINNs) have become pivotal. The proposed HI-IWRM framework integrates observation networks, coupled modeling, and hybrid mechanism-data-driven approaches, bridging process-based models' interpretability and AI's adaptability. This paradigm shift provides a scalable, resilient roadmap for navigating climate uncertainty, optimizing water resource allocation, and safeguarding global water security amid ongoing urbanization and environmental change.

References

Abbasov, R. (2025). The impact of climate change on urbanization processes: Comprehensive analysis. *International Journal of Science and Research Archive, 14*(3), 713–719.

Abrahart, R. J., & See, L. (2002). MIKE SHE: A review of its theoretical basis and applications. *Hydrological Processes, 16*(12), 2493–2513.

Adachi, S., Kimura, F., Kusaka, H., Inoue, T., & Ueda, H. (2012). *Comparison of the impact of global climate changes and urbanization on summertime future climate in the Tokyo metropolitan area.*

Ahmed, A. A., Sayed, S., Abdoulhalik, A., Moutari, S., & Oyedele, L. (2024). Applications of machine learning to water resources management: A review of present status and future opportunities. *Journal of Cleaner Production, 441.*

Al-Nouti, A. F., Fu, M., & Bokde, N. D. (2024). Reservoir operation based machine learning models: Comprehensive review for limitations, research gap, and possible future research direction. *Knowledge-Based Engineering and Sciences.*

Alhendi, A. A., Al-Sumaiti, A. S., Elmay, F. K., Wescaot, J., Kavousi-Fard, A., Heydarian-Forushani, E., & Alhelou, H. H. (2022). Artificial intelligence for water–energy nexus demand forecasting: A review. *International Journal of Low-Carbon Technologies, 17*, 730–744.

Ana, E. V. (2009). *Sewer asset management: Sewer structural deterioration modelling and multicriteria decision making in sewer rehabilitation projects prioritization [Doctoral dissertation.* Vrije Universiteit Brussel.

Arnold, J. G., Srinivasan, R., Muttiah, R. S., & Williams, J. R. (1998). Large area hydrologic modeling and assessment part I: Model development. *Journal of the American Water Resources Association, 34*(1), 73–89.

Asghar, M. R., Ushiyama, T., Riaz, M., & Miyamoto, M. (2019). Flood and inundation forecasting in the sparsely gauged transboundary chenab river basin using satellite rain and coupling meteorological and hydrological models. *Journal of Hydrometeorology, 20*, 2315–2330. https://doi.org/10.1175/JHM-D-18-0226.1

Bahrami Chegeni, I., Riyahi, M. M., Bakhshipour, A. E., Azizipour, M., & Haghighi, A. (2025). Developing machine learning models for optimal design of water distribution networks using graph theory-based features. *Water, 17*(11), 1654.

Baig, F., Sherif, M., & Faiz, M. A. (2022). Quantification of precipitation and evapotranspiration uncertainty in rainfall-runoff modeling. *Hydrology, 9*(3), 51.

Bella, A. di, Raissi, M., Santoro, D., & Roccaro, P. (2026). Physics-informed neural networks in water and wastewater systems: A critical review. *Water Research*, Article 125449.

Beven, K. J. (2012). Physically based models and model complexity. *Hydrological Processes, 26*(18), 2745–2752.

Bezak, N., Lebar, K., Bai, Y., et al. (2025). Using machine learning to predict suspended sediment transport under climate change. *Water Resources Management, 39*, 3311–3326.

Bhattacharya, R. K., Chatterjee, N. D., & Das, K. (2024). Modelling of soil erosion susceptibility incorporating sediment connectivity and export at landscape scale using integrated machine learning, InVEST-SDR and fragstats. *Journal of Environmental Management, 353.*

Bhattarai, Y., Duwal, S., Sharma, S., & Talchabhadel, R. (2024). Leveraging machine learning and open-source spatial datasets to enhance flood susceptibility mapping in transboundary river basin. *International Journal of Digital Earth, 17*(1). https://doi.org/10.1080/17538947.2024.2313857

Bibi, H., Saleem, S., Jalil, Z., Nasir, M., & Alsubait, T. (2025). *Riverine flood prediction and early warning in mountainous regions using artificial intelligence.*

Blasi, H., Abbadi, M. A., & Al-Huweimel, R. (2021). Machine learning approach for an automatic irrigation system in southern Jordan valley. *Engineering, Technology & Applied Science Research, 11*(1), 6609–6613.

Bolsynbek, M., Abdikerimova, G., Serikbayeva, S., Batyrkhanov, A., Shrymbay, D., Taszhurekova, Z., Zhidekulova, G., & Shraimanova, G. (2025). Soil erosion analysis based on machine learning method. *Bulletin of Electrical Engineering and Informatics, 14*, 4797–4811.

Carrera, J., Alcolea, A., Medina, A., Hidalgo, J., & Slooten, L. J. (2005). Inverse problem in hydrogeology. *Hydrogeology Journal, 13*(1), 206–222. https://doi.org/10.1007/s10040-004-0404-7

Chaussard, E., Wdowinski, S., Cabral-Cano, E., & Amelung, F. (2014). Land subsidence in central Mexico detected by ALOS InSAR time-series. *Remote Sensing of Environment, 140*, 94–106. https://doi.org/10.1016/j.rse.2013.08.038

Chen, Y., Li, H., Zhan, Z., Zhang, J., Chen, Y., Cao, X., Yeh, T.-C. J., Barceló, D., Weymer, B. A., Huang, Y., Wang, X., & Yaqiang, W. (2025). Artificial intelligence modeling for groundwater environments across spatial scales. *Environmental Science and Technology, 59*(42), 22351–22372.

Dhyani, Manne, N., Garg, J., Motwani, D., Shrivastava, A. K., & Sharma, M. (2024). A smart irrigation system powered by IoT and machine learning for optimal water management. *2024 4th International Conference on Advance Computing and Innovative Technologies in Engineering (ICACITE), Greater Noida* (pp. 1801–1805). India.

Dinh, T. V., Hoang, N.-D., & Tran, X.-L. (2021). Evaluation of different machine learning models for predicting soil erosion in tropical sloping lands of Northeast Vietnam. *Applied and Environmental Soil Science, 2021*, 1–14.

Downer, C. W., & Ogden, F. L. (2004). GSSHA: A computationally efficient model for simulating watershed hydrology and sediment transport. *Journal of Hydraulic Engineering, 130*(10), 998–1008.

Farfán, J. F., Palacios, K., Ulloa, J., & Avilés, A. (2020). A hybrid neural network-based technique to improve the flow forecasting of physical and data-driven models: Methodology and case studies in Andean watersheds. *Journal of Hydrology: Regional Studies, 27*.

Feng, D., Fang, K., & Shen, C. (2020). Enhancing streamflow forecast and extracting insights using long-short term memory networks with data integration at Continental scales. *Water Resources Research, 56*.

Gao, H., Zhangzhong, L., Zheng, W., & Chen, G. (2023). How can agricultural water production be promoted? A review on machine learning for irrigation. *Journal of Cleaner Production, 414*.

Garzón, Z. K., Langeveld, J., & Taormina, R. (2022). Machine learning-based surrogate modeling for urban water networks: Review and future research directions. *Water Resources Research, 58*(5).

Haochen, L., & John, S. (2022). Implementing machine learning to optimize the cost-benefit of urban water clarifier geometrics. *Water Research, 220*, Article 118685.

Jayakumar, D., Bouhoula, A., & Al-Zubari, W. K. (2024). Unlocking the potential of artificial intelligence for sustainable water management focusing operational applications. *Water, 16*(22), 3328.

Jones, J. E., Woodward, C. S., & Maxwell, R. M. (2006). ParFlow: A parallel watershed hydrology model with coupled overland flow. *Computers and Geosciences, 32*(9), 1314–1329.

Kochanek, K., & Markiewicz, I. (2022). Statistical approach to hydrological analysis. *Water, 14*, 1094.

Kumar, S., Choudhary, M. K., & Thomas, T. (2024). A hybrid technique to enhance the rainfall-runoff prediction of physical and data-driven model: A case study of upper narmada river Sub-basin, India. *Scientific Reports, 14*(1). https://doi.org/10.1038/s41598-024-77655-5

Laloy, E., Hérault, R., Lee, J., Jacques, D., & Linde, N. (2018). Inversion using a new low-dimensional representation of complex binary geological media based on a deep neural network. *Advances in Water Resources, 110*, 387–405. https://doi.org/10.1016/j.advwatres.2017.09.029

Lekamge, R. S. T., Samaranayake, N. P. T. D., Weerawarna, N. V., Dissanayake, D. M. J. C. B., & Rathnayake, H. (2024). Integrated smart dam control system using machine learning. In *International conference on automation and computing*.

Li, W., Kiaghadi, A., & Dawson, C. (2021). High temporal resolution rainfall–runoff modeling using long-short-term-memory (LSTM) networks. *Neural Computing & Applications, 33*, 1261–1278.

Liang, X., Wood, E. F., & Lettenmaier, D. P. (1994). A simple hydrologically based model of land surface water and energy fluxes for general circulation models. *Journal of Geophysical Research: Atmospheres, 99*(D7), 14415–14428.

Lindström, G., Johansson, B., Persson, M., et al. (1997). Development and test of the HBV-96 hydrological model. *Journal of Hydrology, 201*(1–4), 272–288.

Liu, J., Liu, T., Huang, L., Ren, Y., & He, P. (2025). Next-generation drought forecasting: Hybrid AI models for climate resilience. *Remote Sensing, 17*(20), 3402.

Liu, X., & Zeng, S. (2025a). Physics-informed neural networks for solving the two-dimensional shallow water equations with terrain topography and rainfall. *Journal of Hydrology, 642*, Article 131821.

Liu, X., & Zeng, S. (2025b). Multiobjective optimization control for multifacility coordination in integrated urban drainage systems. *Frontiers of Environmental Science and Engineering, 19*(9), 125.

Masselot, P., Dabo-Niang, S., Chebana, F., & Ouarda, T. B. M. J. (2016). Streamflow forecasting using functional regression. *Journal of Hydrology, 538*, 754–766.

Mehmood, Liao, D., & Mahadeo, K. (2020). A review of artificial intelligence applications to achieve water-related sustainable development goals. In *2020 IEEE/ITU international conference on artificial intelligence for good (AI4G)* (pp. 135–141). Geneva, Switzerland.

Monjardin, C. E. F., Uy, F. A. A., & Tan, F. J. (2017). Estimation of river discharge at ungauged catchment using GIS map correlation method as applied in sta. Lucia river in mauban, Quezon, Philippines. *IOP Conference Series: Materials Science and Engineering, 216*, Article 012045.

Mosavi, & Ardabili, S. (2023). Machine learning for drought prediction; review, bibliometric analysis, and models evaluation. In *2023 IEEE 27th international conference on intelligent engineering systems (INES)*. Nairobi, Kenya, pp. 000095–000104.

Mosavi, A., Ozturk, P., & Chau, K. W. (2018). Flood prediction using machine learning models: Literature review. *Water, 10*, 1536.

Nearing, G., Cohen, D., Dube, V., et al. (2024). Global prediction of extreme floods in ungauged watersheds. *Nature, 627*, 559–563.

Nguyen, K. A., Chen, W., Lin, B.-S., Seeboonruang, U., & Thomas, K. (2019). Predicting sheet and rill erosion of shihmen reservoir watershed in Taiwan using machine learning. *Sustainability, 11*(13), 3615.

Pang, M., Du, E., & Zheng, C. (2024). Contaminant transport modeling and source attribution with attention-based graph neural network. *Water Resources Research, 60*(6).

Pei, S. (2025). *Real-time control of urban water systems using machine learning*. Exeter University. PhD Thesis.

Perrin, C., Michel, C., & Andréassian, V. (2001). Improvement of a parsimonious model for streamflow simulation. *Journal of Hydrology, 242*(1–4), 275–289.

Raissi, M., Perdikaris, P., & Karniadakis, G. E. (2019). Physics-informed neural networks: A deep learning framework for solving forward and inverse problems involving nonlinear partial differential equations. *Journal of Computational Physics, 378*, 686–707. https://doi.org/10.1016/j.jcp.2018.10.045

Refadah, S. S. (2025). Development in flood forecasting: A comprehensive review of complex and machine learning models. *Physics and Chemistry of the Earth, 139*, Article 103975.

Sabzipour, B., Arsenault, R., Troin, M., Martel, J. L., Brissette, F., Brunet, F., & Mai, J. (2023). Comparing a long short-term memory (LSTM) neural network with a physically-based hydrological model for streamflow forecasting over a Canadian catchment. *Journal of Hydrology, 627*, Article 130380.

Safari, M. J. S., Rahimzadeh Arashloo, S., & Danandeh Mehr, A. (2020). Rainfall-runoff modeling through regression in the reproducing kernel hilbert space algorithm. *Journal of Hydrology, 587*, Article 125014.

Sanad, H., Mouhir, L., Zouahri, A., Moussadek, R., El Azhari, H., Yachou, H., Ghanimi, A., Oueld Lhaj, M., & Dakak, H. (2024). Assessment of groundwater quality using the pollution index of groundwater (PIG), nitrate pollution index (NPI), water quality index (WQI), multivariate statistical analysis (MSA), and GIS approaches: A case study of the mnasra region, gharb plain, Morocco. *Water, 16*(9), 1263.

Scheuer, S., Haase, D., & Volk, M. (2017). Integrative assessment of climate change for fast-growing urban areas: Measurement and recommendations for future research. *PLoS One*.

Taubenböck, H., Mast, J., Geiß, C., Wurm, M., Esch, T., & Seto, K. C. (2024). Global differences in urbanization dynamics from 1985 to 2015 and outlook considering IPCC climate scenarios. *Cities*.

Tawfeek, M. A., Alrashdi, I., Alruwaili, M., Elhady, G. F., Abohamama, A. S., & Shehab, A. (2025). Leveraging innovative hybrid machine learning models to predict drought and optimize water resources based on climate change. *Advances in Meteorology*.

Therrien, R., Sudicky, E. A., & McLaren, R. G. (2006). HydroGeoSphere: A three-dimensional numerical model describing fully integrated surface and subsurface flow and transport. *Computers and Geosciences, 32*(7), 983–999.

Umar, S., Harris, A. C., & Leghari, M. (2025). Efficient water resource management in dam basins using machine learning. In *Proceedings of 3rd WAPDA conference on applications of artificial intelligence for water resources planning and management*. Fusion Journal of Engineering and Sciences.

Wang, Y., Huang, B., & Zhu, D. Z. (2024). Assessment of rainfall-derived inflow and infiltration in sewer systems with machine learning approaches. *Water Science and Technology, 89*(8), 1928–1945.

Wigmosta, M. S., Vail, L. W., & Lettenmaier, D. P. (1994). A distributed hydrology-vegetation model for complex terrain. *Water Resources Research, 30*(6), 1665–1679.

Xie, Y., Liu, J., & Chen, H. (2025a). Integrated surface and groundwater modeling to enhance water resource sustainability in agricultural watersheds. *Journal of Hydrology, 630*, Article 130255.

Xie, Y., Liu, J., & Chen, H. (2025b). Using hybrid machine learning (Attention-LSTM-XGBoost) to predict wastewater effluent quality and ensure treatment plant stability. *Water, 17*(13), 1851.

Xiong, Z., Liu, X., Igou, T., Li, Z., & Chen, Y. (2025). Using hybrid machine learning to predict wastewater effluent quality and ensure treatment plant stability. *Water, 17*(13), 1851.

Zakir-Hassan, G., Punthakey, J. F., & Allan, C. (2025). Integrating groundwater modelling for optimized managed aquifer recharge strategies. *Water, 17*(14), 2159.

Zamri, N., et al. (2022). A comparison of unsupervised and supervised machine learning algorithms to predict water pollution. *Procedia Computer Science, 204*, 172–179.

Zhang, R. (2024). Construction of a Meteorological Data Support Vector Machine Drought Prediction Model Based on Machine Learning Algorithms. *2024 Second International Conference on Networks, Multimedia and Information Technology (NMITCON)*, 1–5. https://doi.org/10.1109/NMITCON62075.2024.10699072

Zhang, S. S., Wang, M., & Kumar, S. (2025). Deep learning–based automated detection of sewer defects in CCTV videos. *Journal of Computing in Civil Engineering, 39*(2), Article 04024001.

Zhao, M., Cheng, C., Zhou, Y., Li, X., Shen, S., & Song, C. (2022). A global dataset of annual urban extents (1992-2020) from harmonized nighttime lights. *Earth System Science Data, 14*(2), 517–534. https://doi.org/10.5194/essd-14-517-2022

Zhao, R. J., & Liu, X. R. (1995). The Xinanjiang model applied in China. *Journal of Hydrology, 174*(1-4), 371–381.

Zhao, X., Xiao, N., Cai, Z., & Xin, S. (2024). YOLOv5-Sewer: Lightweight sewer defect detection model. *Applied Sciences, 14*(5), 1869.

Further reading

An urbanising world | International Institute for Environment and Development. (2026). Retrieved January 8, 2026 https://www.iied.org/urbanising-world.

CHAPTER

15 Policy innovation, financing mechanisms, and community participation for scaling sustainable water solutions

Naveen Chand, Rajat Pant and Sanjeev Kumar Prajapati

Environment and Biofuel Research Laboratory, Department of Hydro and Renewable Energy (HRED), Indian Institute of Technology Roorkee, Roorkee, Uttarakhand, India

1. Introduction

Water crisis has exposed the entire planning, financing, managing, designing, and policy framework of water infrastructure. Declining trust and limitation of existing infrastructure and changes in the weather pattern (e.g., unpredictable precipitation, rise in temperature) have threatened water quality and its availability (Gosling & Arnell, 2016; Granata & Di Nunno, 2025). The decrease in water resources has resulted in severe water scarcity problems. It has been estimated that more than half of the world's population lives under water scarcity for minimum 1 month (Mekonnen & Hoekstra, 2016). These water issues rationalize the comprehensive universal demand for a collaborative approach for advanced sustainable infrastructure and investment support from all sectors (Jiang, 2023; Muller et al., 2015; Van Dijk et al., 2020). However, currently the challenges are not limited to infrastructure, finance, or policy. The challenges are beyond this, as the conventional approaches, for example, centralized wastewater treatment systems, have proven to be insufficient for addressing water problems and complexity.

The financial and water sector relationship is basically based on the infrastructure, demand, cost, management, and water reallocation. Water infrastructure has a wide range of operational and support systems (which incorporates digital, physical, distribution, treatment, operation, and management). Apart from these, infrastructure pipelines and dams also encompass smart management and monitoring frameworks (Granata & Di Nunno, 2025). Hence to achieve the target of Sustainable Development Goal 6 (SDG-6) combined with climate change requires shift toward new sustainable approaches for wastewater treatment with a strong policy and financing on urgent basis (Fig. 15.1). In fact, the urgency to shift toward sustainable solution cannot be overstated, since it has been indicated that wastewater treatment plant accounts for approximately 5% of total greenhouse gas (GHG) emissions via different processes (Maktabifard et al., 2023; Upadhyay et al., 2025). Also, water problems have a strong impact on food production, energy security, and social stability. To ensure broader sanitation coverage, wastewater treatment services should be expanded to remote areas beyond the reach of centralized infrastructure through decentralized wastewater treatment systems (Ventura et al., 2024).

Sustainable Water Management. https://doi.org/10.1016/B978-0-443-44853-9.00010-1

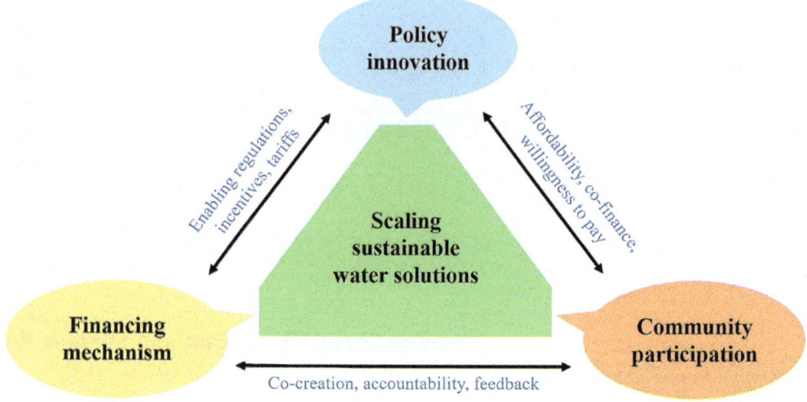

FIGURE 15.1

Tools to achieve a sustainable water solution.

Decentralized wastewater systems could become one of the most sustainable methods for purifying the wastewater across different regions. Subsequently these decentralized systems are now being regarded as resource which could be utilized to recover potable/nonpotable water, fertilizers, and energy (Guest et al., 2009; Sha et al., 2024). The emphasis over the years has now moved in favor of resource recovery approaches, for example, algal wastewater treatment, constructed wetlands. These nature-based decentralized systems can effectively treat wastewater with low energy requirement and nominal operational complication (Machado et al., 2017; Sha et al., 2024). In fact, decentralized systems are being adopted by both developed and developing countries to manage water and sanitation issues and promote sustainability practices. Hence by the virtue of which decentralized systems have emerged as low-cost, sustainable, and flexible alternative to centralized wastewater treatment systems (Sha et al., 2024).

Despite the potential benefits, several challenges have hindered the widespread adoption of decentralized water solutions (Sharma et al., 2010). One of the main barriers is financing—as the economic landscape in the water sector is characterized by challenges that make securing funding for new projects difficult (Hanak et al., 2014, p. 81), and these challenges can be exasperated for distributed solutions, due to their relatively smaller project size and/or inclusion of innovative technologies that can be perceived as risky to investors (Quesnel et al., 2017). However, there is a gap between the financial needs of service providers and their abilities to locally generate funds (Mehan and Kline 2012; Quesnel et al., 2017). Private investment that is difficult to secure for many conventional water projects, can be even more challenging to acquire for distributed infrastructure and technologies (Mukheibir et al., 2014).

Therefore, in this chapter, we aim to describe the sustainable wastewater treatment as an interreliant dimension: (1) policy innovation, (2) different and sustainable financing mechanisms, and (3) community participation. These three dimensions are not separate challenges to be addressed sequentially, but interconnected elements of a coherent strategy for water security.

2. Emerging sustainable technologies

Over the years, implementation of nature-based technologies for wastewater treatment has emerged as one of the potential methods for sustainable wastewater treatment. Since these nature-based technologies establish a socio-environmental framework, which directly influences the economic and societal variables (Kumar, Bhargawa, et al., 2024; Kumar, Mishra, et al., 2024), the nature-based system supports SDGs and are cost-effective and eco-friendly alternative. Nature-based systems, such as constructed wetlands (CWs) (Sezerino et al., 2015), polishing ponds and trickling filters (Chernicharo et al., 2018), and high rate algal pond systems (Craggs et al., 2014), have shown the potential to be one of the sustainable options for wastewater treatment. These approaches emphasize on efficacy of the system, circularity, environmental safety, and local adaptability (Silva, 2023). However, their widespread acceptance/adoption has been restricted by inflexible policy, inadequate financing, and insufficient community engagement. Hence, water management continues to be dependent on centralized, capital-intensive infrastructures.

It has been a well-established fact that decentralized technologies offer a multidimensional benefit that may lead to the broader implementation. Key drivers accelerate their adoption include factors such as sustainability and resilience, which clearly gives comparison to centralized counterparts (Table 15.1). Sustainability in water infrastructure includes designs and management practices that advance societal objectives across temporal scales while preserving ecological, environmental, and hydrological integrity (Kalbar & Lokhande, 2023; Marques et al., 2015).

Table 15.1 Comparative evaluation of scalability, cost, and sustainability of decentralized and centralized technologies.

Driver	Decentralized systems	Centralized systems
Economic benefits	• Energy savings from modular/plug and play builds • Lower excavation costs • Long-term cost-effective • Enables nature-based solutions	• High upfront capital for large plants • Energy-intensive pumping • Deep sewer networks expensive
Social benefits	• Reasonable supply • Local jobs (women/labor) • Community ownership sustains O&M	• Favors urban cores • Limited local employment • Top-down management
Market for recycled water	• Easy modular installs • Point-of-sale reuse cuts redistribution costs • Water exchange networks in mixed land-use	• Long pipe networks increase reuse costs • Centralized distribution challenges

Continued

Table 15.1 Comparative evaluation of scalability, cost, and sustainability of decentralized and centralized technologies.—cont'd

Driver	Decentralized systems	Centralized systems
Resource recovery	• P/N recovery from wetlands/algal ponds • By-products (fish feed, biomass, biodiesel) • Treats wastewater as resource	• Energy-intensive N-fertilizer production • Limited by-product recovery
Environmental impacts	• Lower GHG from reduced energy • Local circular economy/nutrient loops • Full capacity utilization	• Higher operational emissions • Nutrient leaching risks • Underutilized capacity during low flows
Ease of governance	• Fewer institutions/less complexity • No land acquisition delays • Faster tendering/implementation	• Multiagency coordination delays • Land/tendering bottlenecks • Complex stakeholder management

3. Policy innovation in sustainable water governance

Traditional water policies have historically followed centralized and supply-driven paradigms, prioritizing large-scale infrastructure development such as dams, centralized treatment plants, and extensive distribution networks (Onyena & Sam, 2025; Rudolph et al., 2020). While these approaches contributed to economic growth and urban expansion, they often resulted in ecological degradation and high operational costs (Daus et al., 2021). Therefore, policy innovation is required toward sustainability-oriented governance models that emphasize integration, decentralization, and adaptability (Fig. 15.2) (Onyena & Sam, 2025). In fact, integrated water resources management is one such example that advocates synchronized management of water, land, and related resources across sectors and scales. Even though the carrying out of IWRM has been uneven, it has contributed to a broader recognition of the interconnected nature of water challenges (UN-Water, 2007). Subsequently, decentralization technologies are emerging as one of the key dimensions of policy innovation (Ventura et al., 2024). Decentralized infrastructure enables the transfer of responsibility to local governments, utilities, and community-based institutions, context-specific water solutions which are more responsive to local needs. Decentralized policies may help small-scale wastewater treatment systems and nature-based solutions, which are often excluded under centralized regulatory frameworks.

Regulatory change is a very important aspect to facilitate sustainable water solutions (Table 15.2). Traditional regulatory frameworks have a predisposition to prescribe certain technologies or standards of infrastructure, which in turn may unintentionally limit innovation. On the other hand,

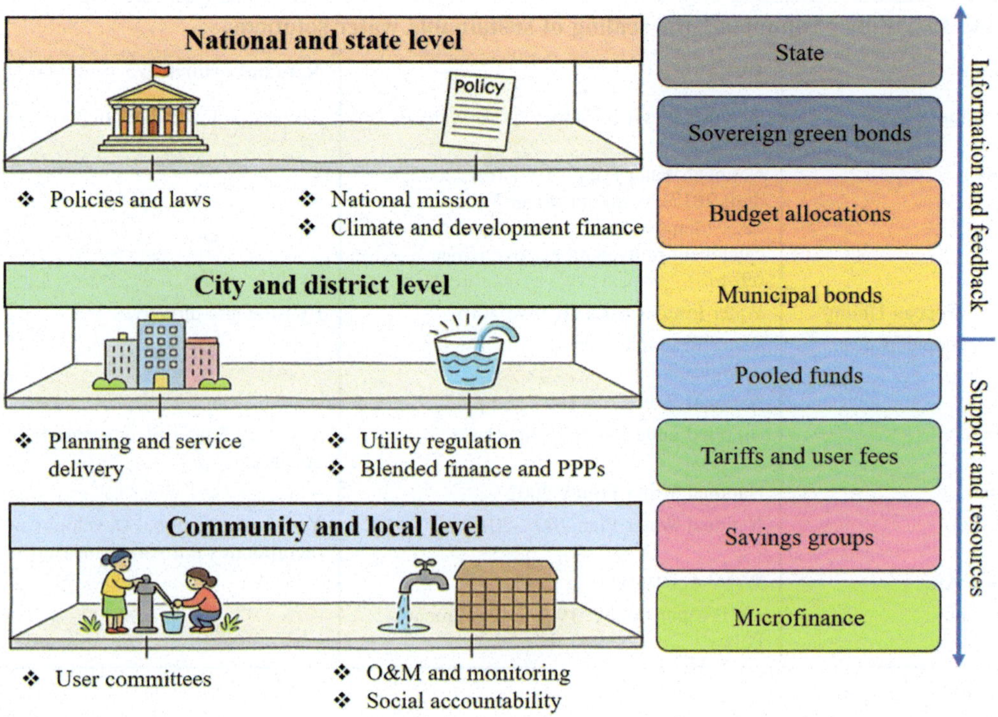

FIGURE 15.2

Policy innovation and sustainability-oriented governance models.

outcome-based or performance-based regulations emphasize the quality of water, health protection, and environmental outcomes, which implies the flexibility in the choice of the technologies. These regulatory measures are needed especially in the encouragement of wastewater reuse, decentralized treatment, and resource recovery schemes (Wadström et al., 2023). Adaptive governance is also capable of supporting policy innovation because it acknowledges the existence of uncertainty and change as part of the water systems. Climate change, socioeconomic transformation, and urban dynamics require policies, which can evolve due to constant monitoring, the learning process, and stakeholder feedback. Adaptive policies consequently have a focus on research, pilot projects, and iterative cycles of learning instead of hard and fast solutions.

3.1 Institutional integration and coordination

Policy innovation progressively puts emphasis on integration across various institutions. Water policy now openly links with climate action, agricultural development, energy security, and public

Table 15.2 Policy supporting the scaling of sustainable water solutions.

Country	Water	National priorities/5-year plan
Australia	National Water Initiative 2004; Water Act 2008	Climate and water key in 2007 election
India	National Water Policy 1987 (rev. 2002, draft 2012); National Water Policy 2021 (demand mgmt., water quality, catchment rejuvenation, governance reforms)	–
The European Union	Water Framework Directive 2001	EU strategic directives
England and Wales	Water Resources Strategy for England & Wales 2009	–
Brazil	National Water Act 1997 and plan	–
China	(Implied under energy/water law)	Yes, 11th 5-year plan; integration apparent
India	National Water Policy 2002	–
Mexico	National Water Plan 2007–2012	Yes, 5-year National Development Plan incorporates mitigation measures
South Africa	Water Act 1998	–
Tanzania	1991 Water Policy; 1997 Environment Policy	–

health programs. The recognition that "water action must be part of climate action, or else it will be part of the problem" has encouraged countries to align water policies with Nationally Determined Plans and climate action plans (Wambua et al., 2025). Institutional coordination extends to sectoral integration as well. Agricultural water use accounts for approximately 70% of global freshwater withdrawals, making agricultural water policy central to overall water security (Ingrao et al., 2023). Similarly, energy production and urban expansion have key water implications. Hence, innovative policies should include formal mechanisms for cross-sectoral coordination, establishing water security committees that consist of agriculture, energy, industry, and environment representatives (Fig. 15.3).

4. Financing mechanisms for scaling sustainable water solutions

Financing remains one of the utmost barriers to scaling-up of sustainable water solutions (Table 15.3) (Ahmed et al., 2023; Granata & Di Nunno, 2025). The water infrastructure projects are usually characterized by high capital investment, long payback period, and low revenue generation, which makes them unappealing to the private investors (Head, 2006). Public investors, though essential, often appear limited by budgetary constraints and other priorities to development. These

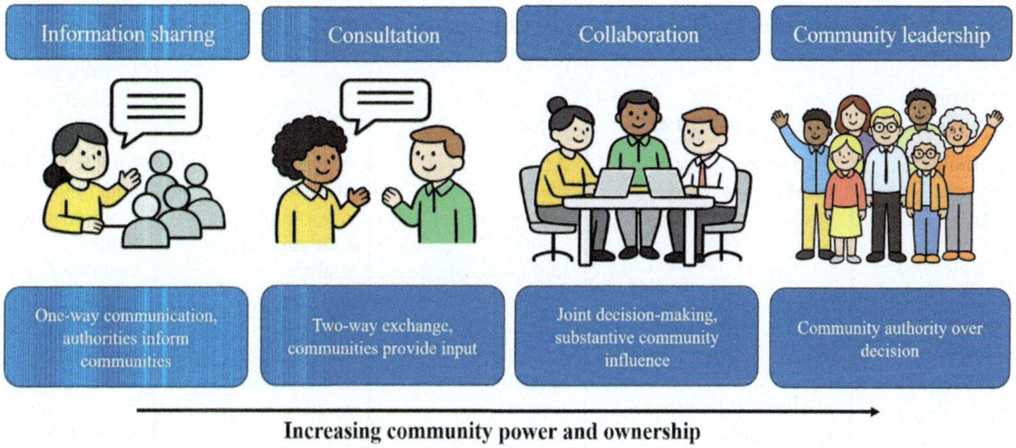

FIGURE 15.3

Cross-sectoral coordination, establishing water security.

Table 15.3 Financing mechanisms for sustainable water infrastructure development.

Financing source	Key characteristics	Strengths	Challenges	Community engagement level
Public investments	Government budgets, concessional loans from World Bank/ ADB, recovery programs (e. g., Next-Generation EU, WIFIA)	Foundational for large-scale projects; supports capacity-building; aligns with public policy	Insufficient alone for global needs; aging infrastructure vulnerabilities	Medium (policy-driven, some public programs involve local input)
Private investments	CSR, impact investing, institutional funds (pension/insurance) focusing on profitability + sustainability	Brings capital, expertise, innovation (e.g., smart systems, desalination)	Long horizons, high upfront costs, climate/ regulatory risks in unstable areas	Low (profit-focused, limited direct community ties unless via CSR)
Blended finance/ PPPs	Public derisking (guarantees) + private capital/efficiency; risk-sharing contracts	Scalable for capital-intensive projects; balances risks, fosters innovation	Regulatory complexity, tariff burdens if poorly managed; needs transparent governance	High (emphasizes stakeholder engagement, grievance mechanisms for acceptability)

Continued

Table 15.3 Financing mechanisms for sustainable water infrastructure development.—cont'd

Financing source	Key characteristics	Strengths	Challenges	Community engagement level
Sustainable funds and green bonds	ESG funds, green/blue bonds, water funds for environmental projects	Attracts ethical investors; funds nature-based solutions (e.g., watershed restoration)	Needs standardized reporting/ transparency for scaling	Medium (social benefits emphasized, some community-aligned via ESG criteria)
Emerging instruments	Water risk bonds, blue bonds, water funds pooling public/private resources	Risk transfer to investors; supports conservation over hard infrastructure	Early-stage adoption; tied to predefined climate thresholds	High (often nature-based, involves local ecosystem protection and participation)

limitations are particularly acute in the case of decentralized and nature-based solutions, the benefits of which, including improved health of the population, ecosystem services, and resistance to climate change, are hard to quantify (Biasin et al., 2024; Thompson et al., 2023). This is supported by the fact that the traditional financing models are more dominant, which usually favors centrally controlled schemes that are consistent with the traditional engineering paradigms. This, in turn, means that smaller-scale, community-based solutions have challenges with finding capital, even though the benefits are significant in both social and environmental terms. To correct this disparity, the use of innovative financing tools that can recognize the holistic worth of sustainable water solutions is required. Hence, over time, blended finance has emerged as a promising approach to mobilize resources for sustainable water projects (Leigland et al., 2016, p. 20). The risk is reduced with blended finance mechanisms, which are based on the synergistic relationship between the public capital, concessional loans, grants, and private investment (Mutambatsere & Schellekens, 2020). The early risks can be absorbed through strategic use of public or donor funds, draw in the private funds, and support pilot projects (Alaerts, 2019).

The water sector has been adopting public–private partnerships (PPP), more so in the area of operation and maintenance (O&M). PPPs when well-designed, can allow the efficiency and expertise of the private sector to be utilized and still maintain the oversight of the people (Lima et al., 2021). However, their success depends on transparent contracts, robust regulatory frameworks, and mechanisms to ensure affordability and equity. The performance-based funding and results-based financing models incorporate financial rewards with the measurable outcomes, e.g., better water quality and reliability of the services (Effah Ameyaw & Chan, 2013; Lima et al., 2021). These mechanisms promote responsibility and creativity besides reducing the chances of poor performance. Climate finance instruments, including green bonds and adaptation funds, have further expanded opportunities for financing water projects that contribute to climate resilience and environmental sustainability (Granata & Di Nunno, 2025).

Microfinance, revolving funds, and community savings schemes are critical at the household and community levels in facilitating the accessibility to water and sanitation services (Afrane & Adjei-Poku, 2014; Wachana, 2017). These systems enable the communities to participate in locally suitable solutions as well as promote money ownership and accountability. Subsequently, social equity needs to be in balance with financial sustainability. Affordability to the vulnerable populations can be ensured through tariff arrangements, subsidies, as well as cross-subsidization, which helps to recover costs (Effah Ameyaw & Chan, 2013; Lima et al., 2021). The life-cycle costing approach plays a critical role in the precise representation of the long term cost of operation and maintenance, as this aspect is often underestimated when planning a project.

5. Community participation and social dimensions of water management

The involvement of the community has been recognized as a core component of sustainable water management (Fig. 15.4) (Ali & Kamraju, 2024). Water systems are situated in sociocultural and economic conditions, and their effectiveness depends on the acceptance, trust, and behavior change of users (Contzen et al., 2023; Mankad & Tapsuwan, 2011). Without the significant involvement of the communities, projects tend to face resistance, misuse, or even abandonment. By incorporating the local knowledge, preferences, and priorities, participatory methodologies improve the legitimacy and sustainability (Ali & Kamraju, 2024; Smith, 2008). Participation can take multiple forms, ranging from information sharing and consultation to comanagement and community ownership. Significant participation includes empowering communities to influence decision-making, manage resources, and hold institutions accountable (Ali & Kamraju, 2024; Khaenamkhaew, 2025).

It is imperative to include in managing water inequities. Women, socially disadvantaged communities, and informal settlements are usually the ones who face an unequal amount of water insecurity, but they are underrepresented in the processes of governance. Gender-sensitive and socially inclusive approaches strengthen both equity and effectiveness. Capacity building is one of the key elements of community participation (Brewis et al., 2024). Technical operation, financial management, and institutional governance training to communities will equip them to maintain water systems in the long-term. The social learning processes also aid in change of behavior, especially in regard to waste water reuse, water conservation, and hygiene practices (Brewis et al., 2024; Wahid et al., 2024).

5.1 Integration of policy, finance, and community participation

Policy innovation, financing mechanisms, and community participation all work together to identify the scaling up of sustainable water solutions by their alignment and synergistic reinforcement. Policies are what create enabling conditions, financing provides economic sustainability, and involvement in the community provides social legitimacy and stability. The disintegration along these dimensions is often counterproductive. Integrated governance systems focus on multistakeholder cooperation, cross-sectoral integration, and adaptive learning and thus acknowledge water systems to be sociotechnical structures that require holistic responses. The successful studies of scaling practices have shown that

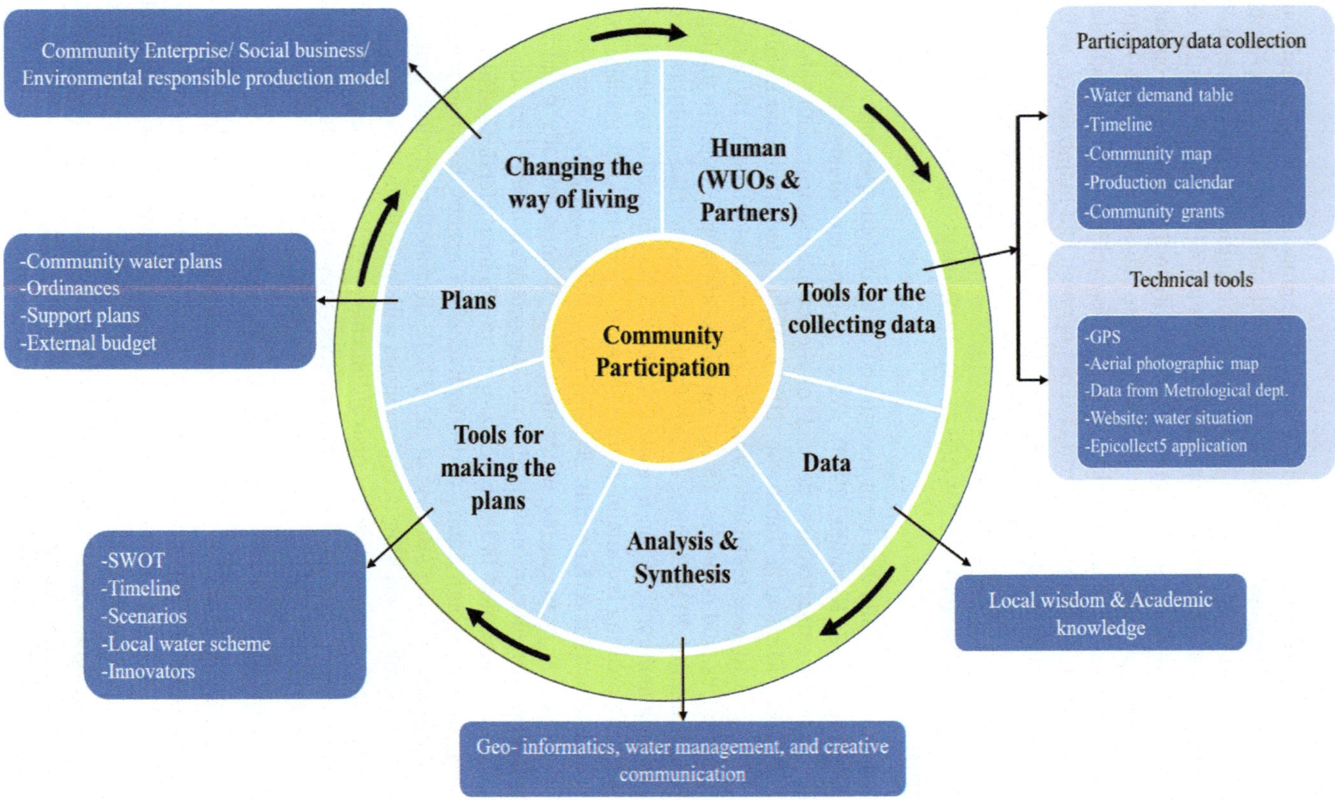

FIGURE 15.4

Community centered process for data collection, analysis planning, and implementation in sustainable water management.

policy support and innovation, combined with the financing mechanisms that reduce the risk and the active involvement of the community, will enable sustainable water solutions to transform the pilot projects to a regular mode.

6. Conclusion

Sustainable water solutions integration and scaling up require systematic transformation of the traditional methods to the newer contemporary sustainable methods, which involves the incorporation of technology-driven interventions in the integrated governance systems. Therefore, to achieve universal access to clean water and sanitation and to become resilient to climate change, there has to be a change in how society relates to water security. The basis of this transition is policy innovation, diversified financing mechanisms, and high involvement of the community. The alignment of these dimensions will allow societies to come up with environmental friendly, economically viable, and socially fair water systems, which will guarantee water security for the current and future generations.

References

Afrane, S. K., & Adjei-Poku, B. (2014). Expanding the frontiers of microfinance in the service of the poor: Experiment with water and sanitation. *International Journal of Academic Research in Business and Social Sciences, 4*(6), 133.

Ahmed, F., Johnson, D., Hashaikeh, R., & Hilal, N. (2023). Barriers to innovation in water treatment. *Water, 15*(4), 773.

Alaerts, G. J. (2019). Financing for water—water for financing: A global review of policy and practice. *Sustainability, 11*(3), 821.

Ali, M. A., & Kamraju, M. (2024). The role of community participation in sustainable integrated water resources management: Challenges, opportunities, and current perspectives. *Integrated Management of Water Resources in India: A Computational Approach: Optimizing for Sustainability and Planning*, 325–344.

Biasin, A., Toxopeus, H., Pettenella, D., Polzin, F., & Masiero, M. (2024). Financing urban nature-based solutions (NBS): A literature review from the perspective of funders. *Nature-Based Solutions, 6,* Article 100195.

Brewis, A., DuBois, L. Z., Wutich, A., Adams, E. A., Dickin, S., Elliott, S. J., … Korzenevica, M. (2024). Gender identities, water insecurity, and risk: Re-theorizing the connections for a gender-inclusive toolkit for water insecurity research. *Wiley Interdisciplinary Reviews. Water, 11*(2), Article e1685.

Chernicharo, C. A. L., Ribeiro, T. B., & Garcia, G. B. (2018). *Overview of sewage treatment in the South, Southeast and Midwest regions of Brazil: most employed technologies (Panorama do tratamento de esgoto sanitário nas regiões Sul, Sudeste e Centro-Oeste do Brasil: tecnologias mais empregadas)*. submitted: Revista DAE. In Portuguese.

Contzen, N., Kollmann, J., & Mosler, H. J. (2023). The importance of user acceptance, support, and behaviour change for the implementation of decentralized water technologies. *Nature Water, 1*(2), 138–150.

Craggs, R., Park, J., Heubeck, S., & Sutherland, D. (2014). High rate algal pond systems for low-energy wastewater treatment, nutrient recovery and energy production. *New Zealand Journal of Botany, 52*(1), 60–73.

Daus, M., Koberger, K., Koca, K., Beckers, F., Encinas Fernández, J., Weisbrod, B., … Wieprecht, S. (2021). Interdisciplinary reservoir management—A tool for sustainable water resources management. *Sustainability, 13*(8), 4498.

Effah Ameyaw, E., & Chan, A. P. (2013). Identifying public-private partnership (PPP) risks in managing water supply projects in Ghana. *Journal of Facilities Management, 11*(2), 152–182.

Gosling, S. N., & Arnell, N. W. (2016). A global assessment of the impact of climate change on water scarcity. *Climatic Change, 134*(3), 371–385.
Granata, F., & Di Nunno, F. (2025). Financing the future of water: Unlocking investment, innovation, and governance for resilient infrastructure in a changing climate. *Earth Systems and Environment*, 1–25.
Guest, J. S., Skerlos, S. J., Barnard, J. L., Beck, M. B., Daigger, G. T., Hilger, H., ... Love, N. G. (2009). *A new planning and design paradigm to achieve sustainable resource recovery from wastewater.*
Hanak, E., Gray, B., Lund, J. R., Mitchell, D., Chappelle, C., Fahlund, A., ... Suddeth, R. (2014). *Paying for water in California* (p. 81). San Francisco: Public Policy Institute of California.
Head, C. (2006). *The financing of water infrastructure: A review of case studies*. Washington DC: World Bank.
Ingrao, C., Strippoli, R., Lagioia, G., & Huisingh, D. (2023). Water scarcity in agriculture: An overview of causes, impacts and approaches for reducing the risks. *Heliyon, 9*(8).
Jiang, Y. (2023). Financing water investment for global sustainable development: Challenges, innovation, and governance strategies. *Sustainable Development, 31*(2), 600–611.
Kalbar, P. P., & Lokhande, S. (2023). Need to adopt scaled decentralized systems in the water infrastructure to achieve sustainability and build resilience. *Water Policy, 25*(4), 359–378.
Khaenamkhaew, D. (2025). Sustainable Knowledge, Understanding, and Public Participation in the Construction of Water-Retaining Structures: A Case Study of Kamphaeng Sao Subdistrict, Muang District, Nakhon Si Thammarat Province, Thailand. *Community and Social Development Journal, 26*(1), 140–165.
Kumar, A., Mishra, S., Singh, N. K., Yadav, M., Padhiyar, H., & Kumar, R. (2024). Ensuring carbon neutrality via algae-based wastewater treatment systems: Progress and future perspectives. *Journal of Environmental Management, 360*, Article 121182. https://doi.org/10.1016/j.jenvman.2024.121182
Kumar, S., Bhargawa, P. K., Chandra, S., & Dutta, V. (2024). Nature-based wastewater treatment technologies: Concept, understanding, opportunities and future prospectus. In *Nature-based wastewater treatment systems* (pp. 16–32). CRC Press.
Leigland, J., Trémolet, S., & Ikeda, J. (2016). *Achieving universal access to water and sanitation by 2030: The role of blended finance* (p. 20). Washington, DC: World Bank.
Lima, S., Brochado, A., & Marques, R. C. (2021). Public-private partnerships in the water sector: A review. *Utilities Policy, 69*, Article 101182.
Machado, A. I., Beretta, M., Fragoso, R., & Duarte, E. (2017). Overview of the state of the art of constructed wetlands for decentralized wastewater management in Brazil. *Journal of Environmental Management, 187*, 560–570.
Maktabifard, M., Al-Hazmi, H. E., Szulc, P., Mousavizadegan, M., Xu, X., Zaborowska, E., ... Mąkinia, J. (2023). Net-zero carbon condition in wastewater treatment plants: A systematic review of mitigation strategies and challenges. *Renewable and Sustainable Energy Reviews, 185*, Article 113638.
Mankad, A., & Tapsuwan, S. (2011). Review of socio-economic drivers of community acceptance and adoption of decentralised water systems. *Journal of Environmental Management, 92*(3), 380–391.
Marques, R. C., da Cruz, N. F., & Pires, J. (2015). Measuring the sustainability of urban water services. *Environmental Science & Policy, 54*, 142–151.
Mehan III, G. T., & Kline, I. (2012). Pricing as a demand-side management tool: Implications for water policy and governance. *Journal of the American Water Works Association, 104*(2), 61–66.
Mekonnen, M. M., & Hoekstra, A. Y. (2016). Four billion people facing severe water scarcity. *Science Advances, 2*, Article e1500323.
Mukheibir, P., Howe, C., & Gallet, D. (2014). What's getting in the way of a'one water' approach to water services planning and management? Water. *Journal of the Australian Water Association, 41*(3), 67–73.
Muller, M., Biswas, A., Martin-Hurtado, R., & Tortajada, C. (2015). Built infrastructure is essential. *Science, 349* (6248), 585–586.
Mutambatsere, E., & Schellekens, P. (2020). *The why and how of blended finance*. International Finance Corporation World Bank Group.

Onyena, A. P., & Sam, K. (2025). The blue revolution: Sustainable water management for a thirsty world. *Discover Sustainability, 6*(1), 63.

Quesnel, K. J., Ajami, N. K., & Wyss, N. (2017). Accelerating the integration of distributed water solutions: A conceptual financing model from the electricity sector. *Environmental Management, 60*(5), 867–881.

Rudolph, K. U., Boysen, B., Hilbig, J., Shalizi, F., Stroemer, K., & Walenzik, G. (2020). Drivers, challenges and solutions—Case studies for water reuse. *Advances in Chemical Pollution, Environmental Management and Protection, 5*, 189–225. Elsevier.

Sezerino, P. H., Bento, A. P., Decezaro, S. T., Magri, M. E., & Philippi, L. S. (2015). Brazilian experiences with constructed wetlands applied to wastewater treatment: Design parameters for horizontal systems. *Engenharia Sanitária e Ambiental, 20*, 151–158.

Sha, C., Shen, S., Zhang, J., Zhou, C., Lu, X., & Zhang, H. (2024). A review of strategies and technologies for sustainable decentralized wastewater treatment. *Water, 16*(20), 3003.

Sharma, A., Burn, S., Gardner, T., & Gregory, A. (2010). Role of decentralised systems in the transition of urban water systems. *Water Science and Technology: Water Supply, 10*(4), 577–583.

Silva, J. A. (2023). Wastewater treatment and reuse for sustainable water resources management: A systematic literature review. *Sustainability, 15*(14), Article 10940.

Smith, J. L. (2008). A critical appreciation of the "bottom-up" approach to sustainable water management: Embracing complexity rather than desirability. *Local Environment, 13*(4), 353–366.

Thompson, A., Bunds, K., Larson, L., Cutts, B., & Hipp, J. A. (2023). Paying for nature-based solutions: A review of funding and financing mechanisms for ecosystem services and their impacts on social equity. *Sustainable Development, 31*(4), 1991–2066.

UN-Water, G. W. P. (2007). *Roadmapping for advancing integrated water resources management (IWRM) processes*. Copenhagen, Denmark: UN-Water and Global Water Partnership.

Upadhyay, P., Prajapati, S. K., & Kumar, A. (2025). Water security and methane reduction in Ganga riverfront STPs: A comparative analysis of advance treatment technologies. *Journal of Water and Climate Change*. https://doi.org/10.2166/wcc.2025.134. jwc2025134.

Van Dijk, S., Lounsbury, A. W., Hoekstra, A. Y., & Wang, R. (2020). Strategic design and finance of rainwater harvesting to cost-effectively meet large-scale urban water infrastructure needs. *Water Research, 184*, Article 116063.

Ventura, J. R. S., Tulipan, J. U., Banawa, A., Umali, K. D., & Villanueva, J. A. L. (2024). Advancements and challenges in decentralized wastewater treatment: A comprehensive review. *Desalination and Water Treatment, 320*, Article 100830.

Wachana, J. K. (2017). *Impact of microfinance on increasing access to water and improved sanitation of households of slum dwellers in Kenya: A case study of Kibera Slum-Nairobi county*. Doctoral dissertation, University of Nairobi).

Wadström, C., Södergren, K., & Palm, J. (2023). Exploring total economic values in an emerging urban circular wastewater system. *Water Research, 233*, Article 119806.

Wahid, S., Cuddy, S. M., Bastola, A., Shrestha, A., & Almeida, A. (2024). *Gender equality, disability and social inclusion in water modelling: A practitioner's toolkit*. ICIMOD Toolkit.

Wambua, T. J., Jewitt, G., Sušnik, J., Masia, S., & Mabhaudhi, T. (2025). Exploring water-energy-food nexus connections between climate action and regional development in the East African community. *Climate Risk Management*, Article 100760.

PART

IX

Future horizons and transformational pathways in water treatment innovation

CHAPTER 16

Modeling groundwater extraction impacts on Pleistocene aquifer dynamics: A case study from the Eastern Desert, Egypt

Mohamed ElKashouty and Mohd Yawar Ali Khan

Department of Hydrogeology, Faculty of Earth Sciences, King Abdulaziz University, Jeddah, Saudi Arabia

1. Introduction

Groundwater modeling is one of the structural components for synthesizing hydrogeological information, assessing conceptual models, and providing quantitative predictions that guide the sustainable management of aquifer systems. To build a trustworthy numerical model, it is necessary to carefully combine site-specific empirical data with a set of known aquifer characteristics to produce a robust simulation that can satisfactorily answer important questions related to water resources (Condon et al., 2021; Khan et al., 2023; White et al., 2020). This undertaking begins with a clear definition of the system components, including the system boundary, groundwater budget, and prevailing processes in the system.

Geogenic emissions of fluoride and arsenic pose significant health risks, especially to children. The primary approach to minimize these dangers is establishing region-specific safety thresholds that account for local climate conditions and consumption habits, as generic standards are insufficient to fully mitigate these risks (Meena et al., 2025). Concurrently, increased pressure is being exerted on groundwater supplies; large-scale ecological restoration projects can inadvertently deplete aquifers by disturbing the hydrological balance, underscoring an important sustainability trade-off (Luan et al., 2025). Modern predictive tools, such as machine-learning models, have become vital for effectively identifying contamination hotspots, enabling targeted interventions (Zhao et al., 2024). Therefore, sustainable groundwater management must adopt a holistic approach, addressing water quality through rigorous monitoring and water quantity through careful resource planning. Shyam et al. (2024) assessed groundwater in Mewat, Haryana, revealing strong seasonal and spatial variations, with Na^+ and Cl^- as the dominant ions. The Water Quality Index indicated widespread unsuitability for drinking, particularly during the pre-monsoon period, and decreased irrigation viability during wetter seasons due to high salinity and sodium hazards. Hydrochemical diagrams identified rock weathering as the primary influence, with salinization intensifying during the wet season. The authors emphasize the need for improved irrigation practices and aquifer management to combat salinity and safeguard water resources in this semi-arid region.

Sustainable Water Management. https://doi.org/10.1016/B978-0-443-44853-9.00006-X

The PMWIN interface was used to simulate the groundwater flow in the current study, which is a popular portal to the MODFLOW code (Chiang & Kinzelbach, 1998). The modeling plan followed a two-step paradigm: a steady-state calibration to focus on hydraulic parameters under average conditions, and a transient exploration to forecast the system's evolution under potential stresses. A Pleistocene aquifer system was modeled using a discretized cell spacing that captured flow dynamics at the regional level. Parameters, including horizontal and vertical permeability, recharge, and effective porosity, were limited in the steady-state calibration. The boundary conditions were prescribed as constant hydraulic heads, placed at a reasonable distance from the main area of interest to eliminate artificial effects on the regional flow dynamics. The assignment of spatially distributed recharge was made in accordance with a synthesis of lithology, soil type, structural features, and exposure to the aquifer, with evapotranspiration estimates aligned with this. In the analysis of transience, storativity was used to depict the aquifer storage. The structural heterogeneity was realized by utilizing features such as fine-filled faults simulated using the horizontal flow barrier package (Hsieh & Freckleton, 1993).

Recent achievements highlight the importance of combining datasets and improving calibration methods to reduce model uncertainty and improve predictive accuracy. For example, integrating remote-sensing-based groundwater storage change data and land subsidence data has proven an effective approach for validating model outputs and calibrating storage parameters (Bockstiegel et al., 2024; Springer et al., 2025). In addition, recent research emphasizes the need to move beyond generic models and to use detailed geological structures to model interactions between groundwater and surface water more precisely, as well as the effects of climate variability (Abduljaleel et al., 2024; Qi & Zhan, 2022).

The goals of this modeling undertaking are three-fold: (1) to outline the current spatial distribution of the hydraulic head and map ground water flow direction; (2) to assess the implications of the envisaged pumping conditions on groundwater levels, drawdown, and aquifer sustainability; and (3) to measure changes in aquifer storage depletion alleviation under likely conditions of stress, thus, assessing the risks, such as land subsidence. It is hoped that the current research will provide a solid, evidence-based model of groundwater resource planning that incorporates the latest modeling methods and various data archives.

2. Study area and its hydrogeology

The research region (Fig. 16.1) is arid. Rainfall is low, averaging 20 mm/year over 1 year. The average diurnal air temperature is about 30°C during summer and 13°C during winter. Summer potential evapotranspiration is 12 mm/day. Figure 16.2 shows that the Pleistocene (Quaternary) aquifer is made of sand, gravel, and clay.

The aquifer thickness ranges from 100 to 400 m and increases toward the north (Haddad, 2002). The hydrogeological conditions range from unconfined to confined. The average hydraulic conductivity is 30 m/day, the average transmissivity is 1000 m^2/day, and the storage coefficient is 0.001 (Mawla, 1999). A dense clay layer underneath the aquifer is notable for its high rigidity and impermeability.

Groundwater flow is directed toward the north-eastern and northern regions (Fig. 16.3). The main sources of groundwater recharge in the study area include the River Nile, Ismailiya Canal, seepage from the sewage system, and agricultural drainage networks.

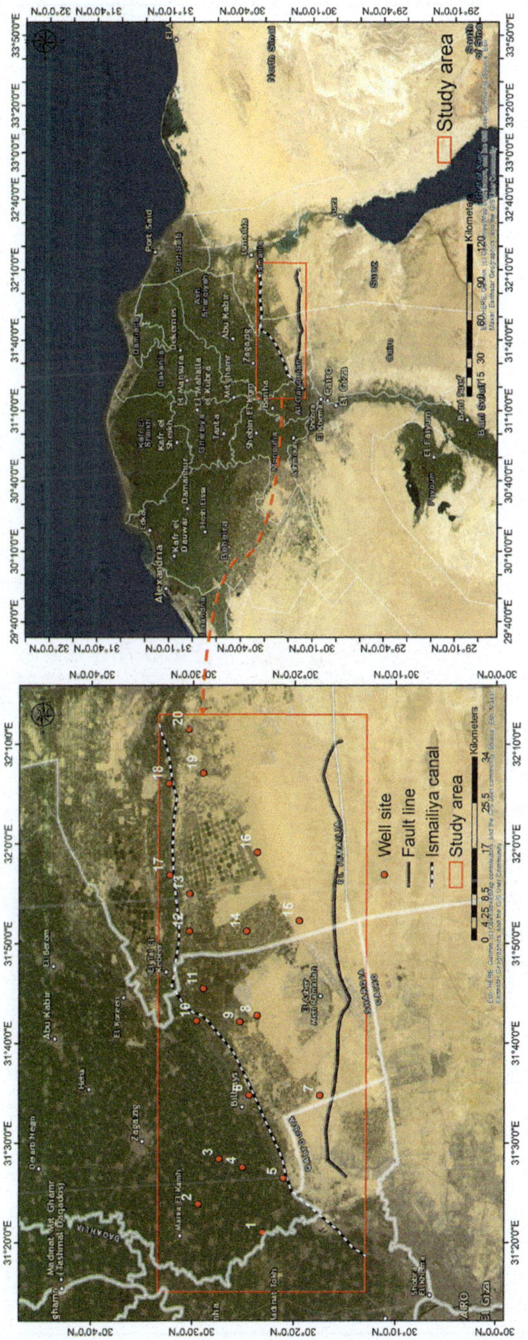

FIGURE 16.1
Map of the study area. ***Map lines delineate study areas and do not necessarily depict accepted national boundaries.***

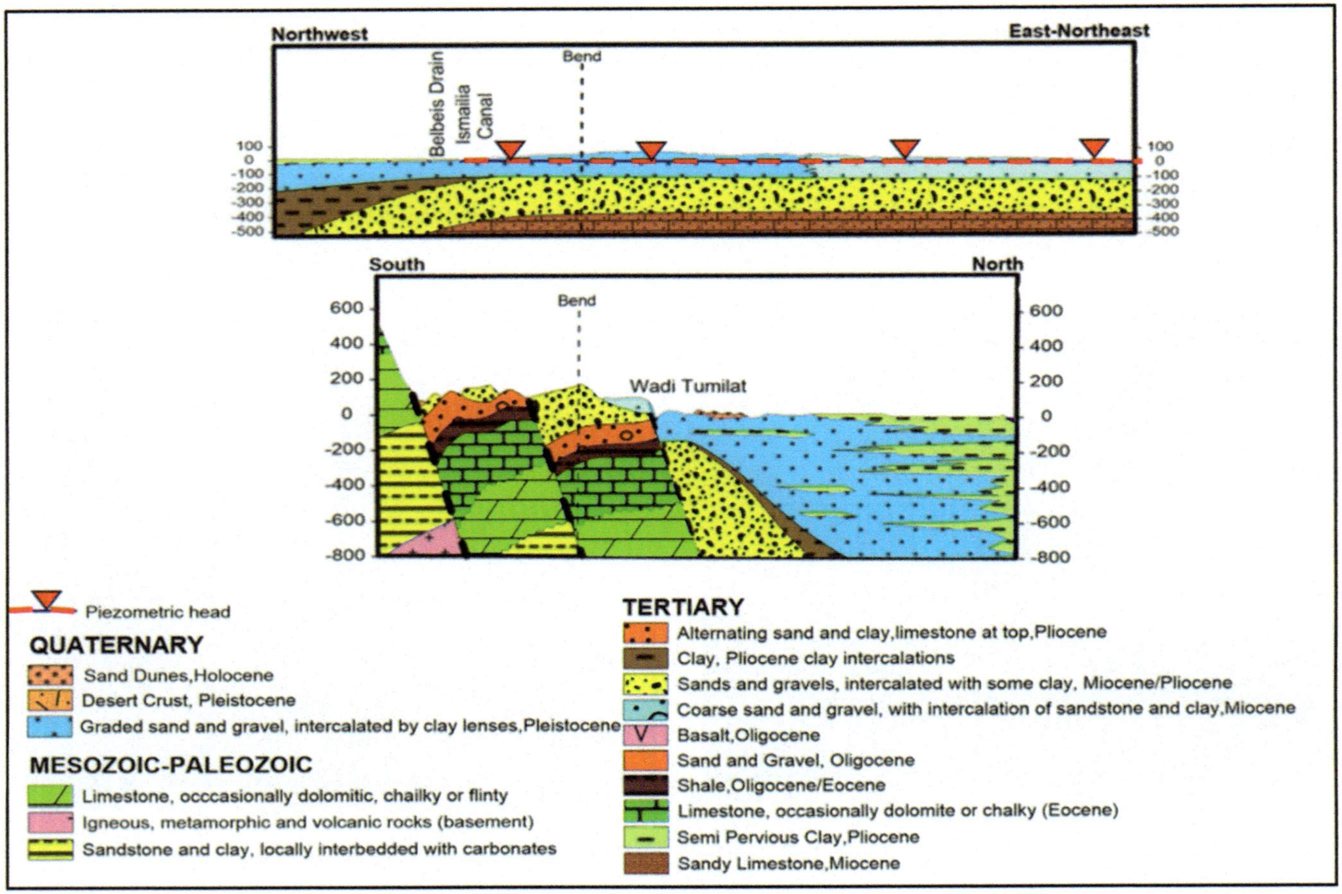

FIGURE 16.2

Hydrogeological profile of the study area.

Modified after Atta, H. S., Omar, M., Tawfik, A. (2022). Water quality index for assessment of drinking groundwater purpose case study: Area surrounding Ismailia Canal, Egypt. Journal of Engineering and Applied Science, 69, *83.*

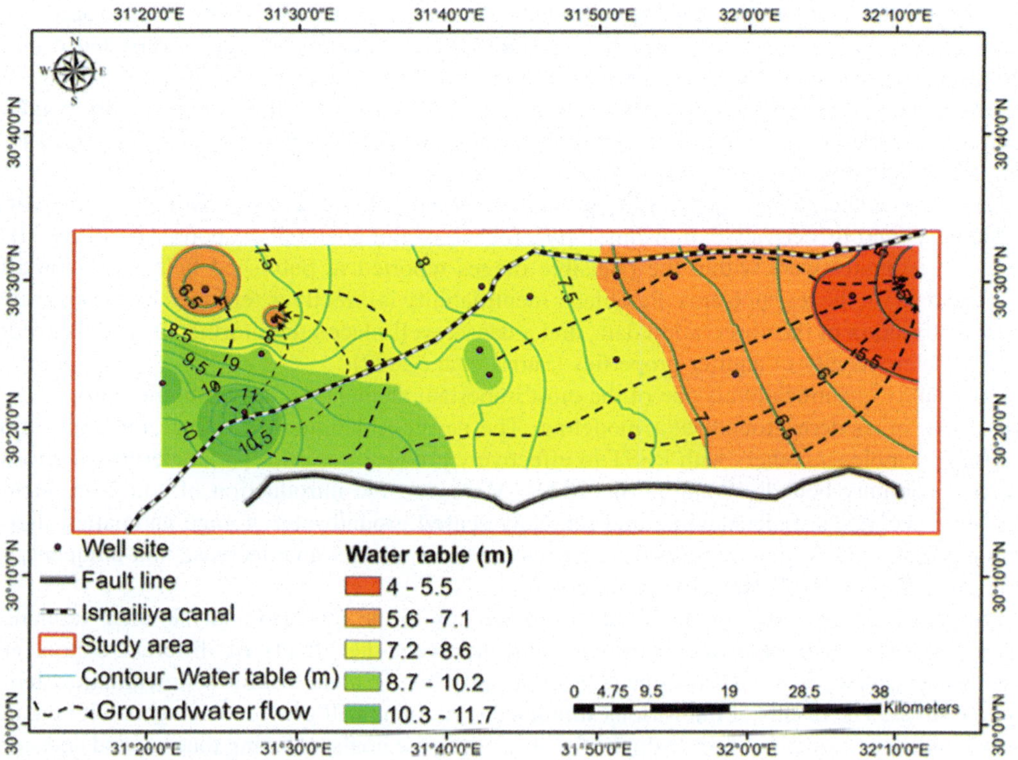

FIGURE 16.3

Present water table elevation of the Pleistocene aquifer.

3. Methodology

Calibration of a numerical groundwater flow model involves an iterative process of refining boundary conditions and aquifer parameters, the goal of which is to ensure that the simulation outputs, i.e., hydraulic heads and flows, match observed field values within a reasonable margin of error (Anderson et al., 2015, pp. 3–25). It is through these cycles that the model's credibility and predictive ability are determined. Before simulation, the observation-error standard deviations are estimated to assign suitable regression weights in an inverse model. In the case of groundwater level measurements, these errors include the errors in uncharacterized permeability heterogeneity, uncertainly due to multiple sources, and they include: (1) errors due to elevation error in topographic maps, (2) errors due to spot measurements of long-term average conditions, (3) errors due to localized anthropogenic effects (e.g., pumping), (4) errors due to the inadequate characterization of permeability heterogeneity and (5) errors due to vertical hydraulic gradients within open-hole monitoring wells.

Calibration was carried out using both manual and automated methods to maximize parameter estimation. The Parameter Estimation Suite (PEST) (Doherty, 2000a and b; Doherty, 2001) was an inversion framework widely used to automate calibration, minimizing a weighted least-squares objective function that assesses the difference between simulated and observed values to adjust model parameters. PEST is used to manage the MODFLOW executable and run it repeatedly until an appropriate parameter set is found. This hybrid calibration approach leverages the power of automated algorithms while allowing expert-directed optimization of model structure and boundary conditions, which is advisable to avoid the pitfalls of purely automated approaches to black-box calibration (Wali et al., 2024; White et al., 2020).

The estimated parameters in this process, in descending order of scaled sensitivity, were hydraulic conductivity (and its derivative, transmissivity), recharge, and potential evapotranspiration. All final parameter estimates were within the plausible ranges reported in published field studies and local empirical data for the research area. Parameter identifiability is directly determined by the quantity and spatial distribution of the observation data, and a dense, well-placed monitoring network is needed to limit the heterogeneity of aquifer properties (Sarris et al., 2022).

Uncertainty quantification is one of the most important elements of the calibration process, which is the focus of modern groundwater modeling. The newer techniques encourage the application of iterative ensemble smoothers with PEST to effectively produce posterior parameter distributions and predict uncertainty bounds (Wang et al., 2022). Moreover, the introduction of soft data, including qualitative geological understanding and remotely sensed groundwater storage anomalies, and conventional head measurements, helps increase parameter constraint and decrease non-uniqueness (Ali et al., 2022; Rodríguez-Castellanos et al., 2025).

The calibrated groundwater model serves as the key to the scenario analysis after validation. It allows water-resources planners to evaluate quantitatively the effects of different approaches to groundwater development and management, especially in arid areas with low precipitation and high resource stress. The results of the various simulated conditions will provide managers with the much-needed evidence to weigh and settle on the fighting water-use goals, ensuring that a sound decision can be made based on a sound understanding of how the system is likely to react to this choice.

4. Problem description

The expansion of urban and agricultural development demands many more boreholes. The installation and design of these wells are not scientifically verified; therefore, safe yields are often disregarded due to uncertainties surrounding the abstraction rates. Management and planning of groundwater resources lack adequate control, which may lead to the migration of the Miocene aquifer (TDS 500–6000 ppm) into the Pleistocene aquifer (TDS 400-$\geq$ 1500 ppm), as well as an increase in seawater intrusion from the Suez Canal (Fig. 16.1). These factors increase groundwater salinity, making it unsuitable for development, threatening ecosystems, and hindering urbanization in desert areas. No safeguards are in place to address these issues, since most pumping rates, irrigation areas, well numbers, and borehole design and installation are not regulated by a scientific body. The groundwater flow model is used to explain the aquifer system, predict future hydrogeological conditions, and develop management strategies. The Ismailiya Canal is 2.4 m deep, between 9 and 52 m wide, and 128 m long (Ghazaw, 2011; Mowafy, 2001). This system transfers water of the Nile to Ismailiya, Port Said and Suez at a rate of 50–135 m^3/s. It sprays 1,082,000 feddan with a flow of 5,000,000 m^3/s. About 90% of its length passes through sandy soils, which makes it easy to enter the Pleistocene aquifer (Mowafy, 2001). The Ismailiya canal seepage is about 3.3 m^3/s (Binnie & Partners, 1980). Water losses in the canal can reach up to 374 million m^3 per year (Afify, 2000; Geriesh, 1994). These seepage rates influence the simulation of increasing pumping rates from the Pleistocene aquifer to improve agricultural yields per capita.

5. Results and discussion

5.1 Simulation of groundwater level

Figure 16.4 shows the conceptual flow system and the location of observation wells. The flow system involves a confined or unconfined Pleistocene aquifer. The Ismailiya Canal, which crosses the aquifer, has a significant impact on it, and hydraulics partly connect it to the aquifer.

It is a permeable boundary condition that absorbs or discharges water into the aquifer across various canal regions. The stage, channel bed, and head of the Ismailiya Canal are derived from topography data and groundwater depth levels, which are incorporated into the model. The boundary conditions in the southern and upper western regions are a horizontal flow barrier and a no-flow boundary, respectively. These are obtained from groundwater flow networks and geological maps. The Horizontal-Flow Barrier Package (Hsieh & Freckleton, 1993) models slender, low-permeability geological structures, such as vertical faults or slurry walls that obstruct horizontal groundwater flow. These structures are represented as a sequence of horizontal-flow barriers at the boundaries between pairs of neighboring cells in the finite-difference grid. The General Head Barrier (GHB) is a head-dependent barrier where the volumetric flow is proportional to the head differential between the border and the adjacent model cell (Andersen, 1993; Melinda et al., 2004). GHB is prescribed in the north-eastern, north-western, and south-western regions (Fig. 16.4). The grid comprises 1827 cells of uniform area.

The transmissivity ranged from 10,000 to 20,000 m^2/d (Attwa et al., 2014; Eltarabily et al., 2020), which was integrated into the grid system on a per-cell basis using a transmissivity distribution map.

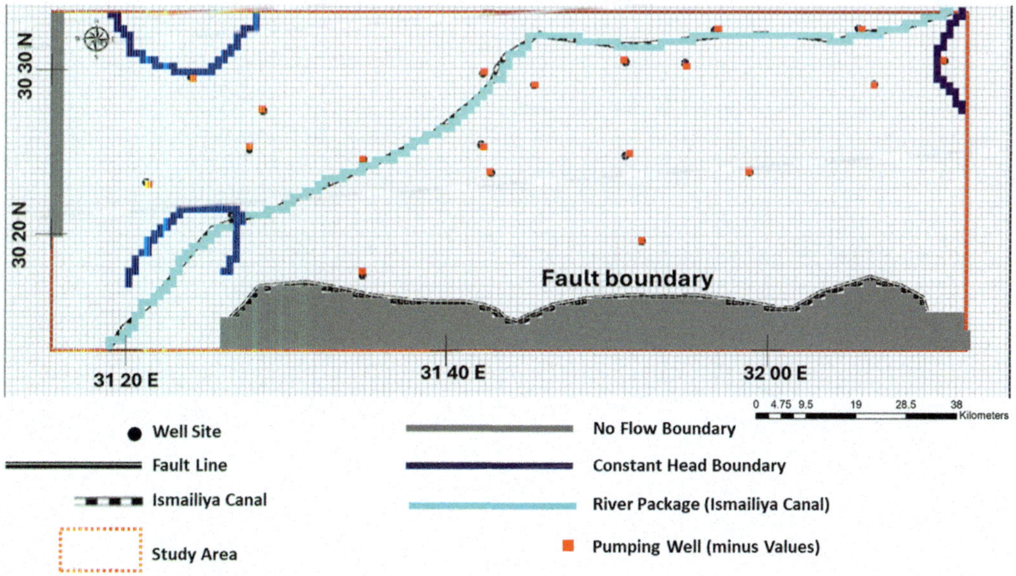

FIGURE 16.4

Configuration of the Pleistocene aquifer system in the study area.

Transmissivity increases toward the northern region. Recharge fluctuated between 0.000689 and 0.000576 m/d, while evapotranspiration varied from 0.007 to 0.0065 m/d. Elevation data for the Ismailiya Canal was calculated and assigned to each grid cell, with an effective porosity of 0.25. Groundwater flow models simulate a hydrological system or predict the impacts of future changes to it (Chiang & Kinzelbach, 1998). To ensure the model accurately reflects the behavior or characteristics of the flow system, it must be calibrated before being used for predictions. Manual calibration aims to find the optimal parameters that minimize the sum of squared errors between model outputs and observed data, known as the goal function (Fig. 16.5).

The PMWIN model is used to address hydrogeological issues expected from increased discharge rates to support development. It evaluates the reduction in groundwater levels caused by abstraction and performs a temporal simulation. Three discharge scenarios were selected for each well: −10,000, −30,000, and −50,000 m^3/d, with a simulation period of just 5 years to produce realistic results. Groundwater levels stay stable as the discharge rate rises to −10,000 m^3/d (Fig. 16.6).

The seepage from the Ismailiya canal into the aquifer is caused by this phenomenon. In the second and third scenarios (pumping rates increased to −30,000 and 50,000 m^3/d), the groundwater levels are minimally affected in the western region. Groundwater levels decline significantly in the eastern region, especially in the third scenario (Fig. 16.7A and B). The geographic location of the boreholes, their distance from the Ismailiya Canal, and the limited thickness of the Pleistocene aquifer contribute to their reduced storage capacity.

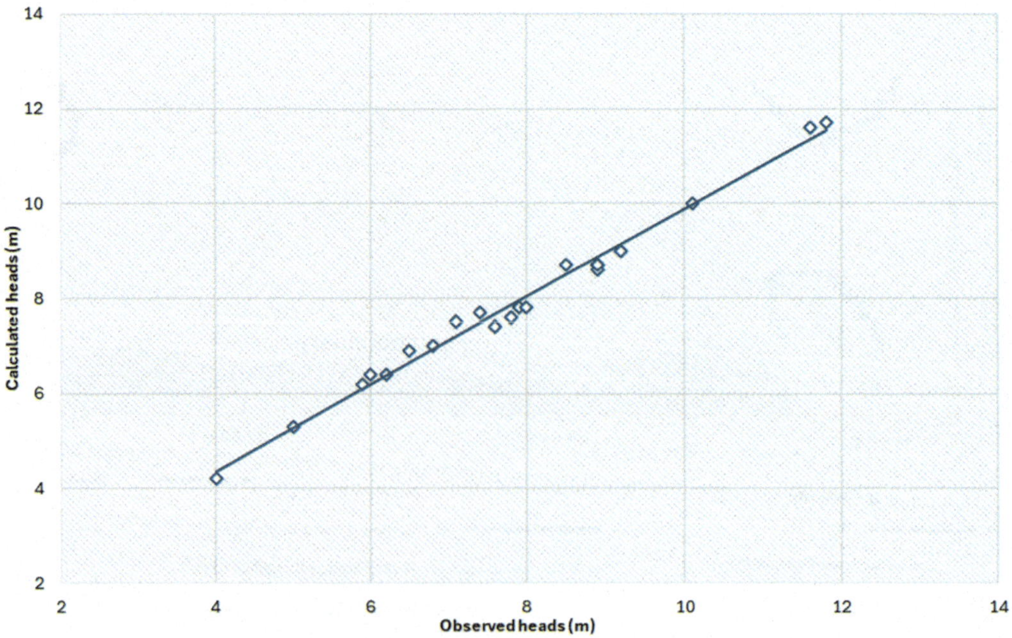

FIGURE 16.5

Calibration between measured and calculated hydraulic heads.

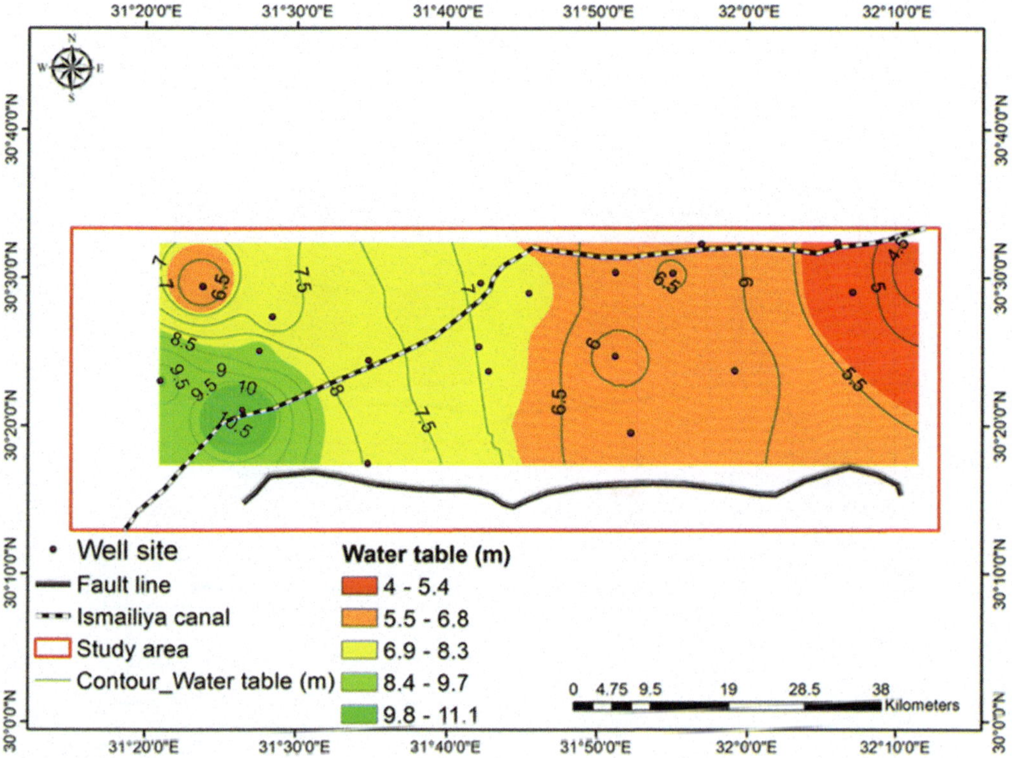

FIGURE 16.6

First simulation of the water table of the Pleistocene aquifer.

Groundwater levels in the western region are minimally influenced by increases in aquifer thickness (high aquifer storage) and by proximity to the Ismailiya seepage zones. The groundwater flow was directed toward the northeastern, eastern, and northwestern areas (Fig. 16.3). In contrast, the western region, which shows only a slight decline in the piezometric surface, is recognized as the recharge zone, consistently replenishing the other parts. The western region lies under unconfined conditions near the Ismailiya Canal, which greatly enhances infiltration into the aquifer and reduces the effects of pumping. In the eastern area, under restricted conditions, canal seepage into the aquifer was limited, leading to a notable decline in groundwater levels due to abstraction.

5.2 Simulation of the storage depletion

Groundwater abstraction can lead to the compaction of compressible fine sediments within porous aquifer materials. The pumped groundwater from the confined aquifer was linked to the compression of the clayey beds. Elastic compaction occurs when a piezometric drop does not cause the rearrangement of the aquifer materials, whereas inelastic compression is associated with the reorganization of the skeletal aquifer. The latter results from additional water released by aquitard beds, causing permanent compaction that cannot be reversed once pumping decreases or stops. Compaction of aquifers due to declining potentiometric levels can impact infrastructure and well installation and design; early land subsidence may go unnoticed

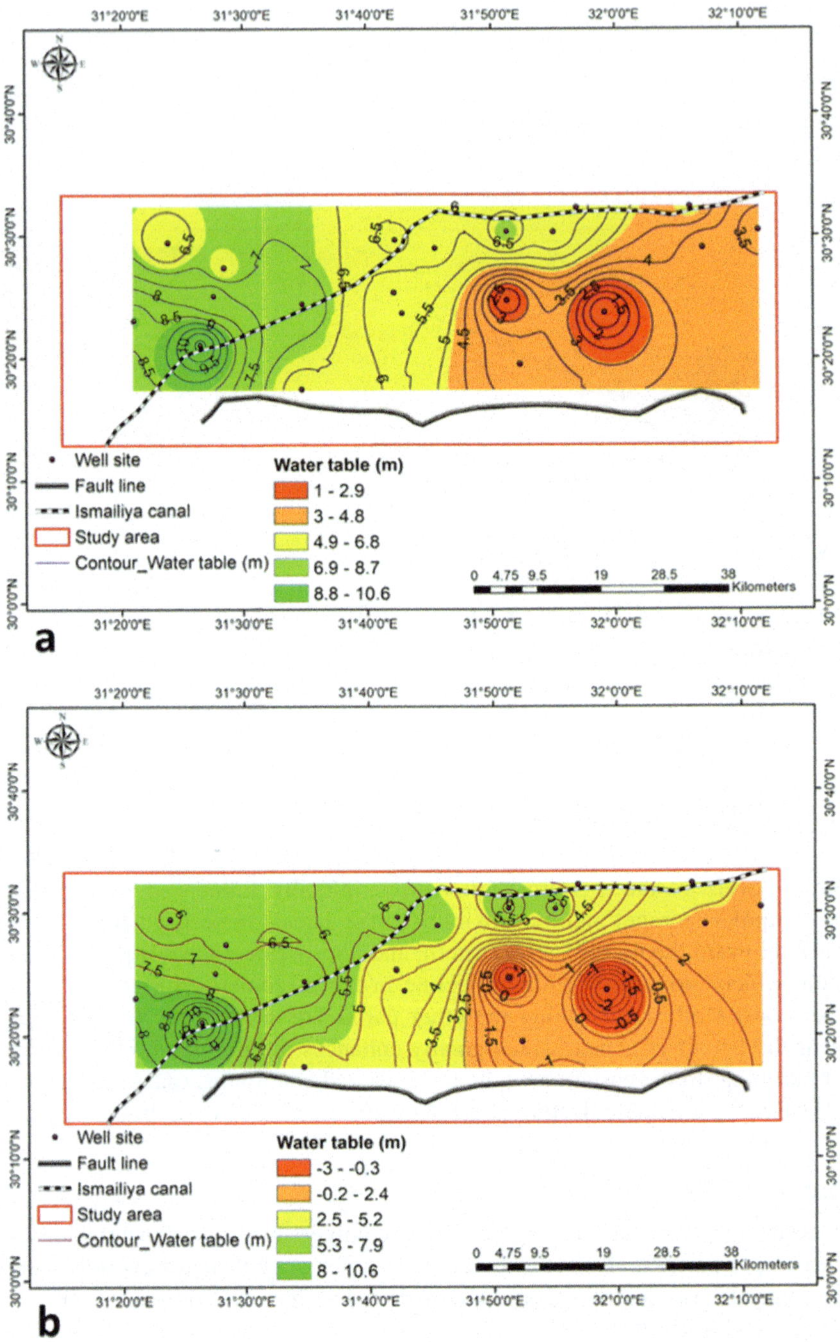

FIGURE 16.7

(A) Second and (B) third water table simulation of the Pleistocene aquifer.

but can increase later and be detected. Managing groundwater resources sustainably involves balancing the hazards of subsidence against the benefits of water extraction (Galloway & Sneed, 2013). The subsidence was attributed to a transient state involving hydromechanical effects. Consequently, the approach considers both steady-state conditions and transient states (post-development). Extraction from the unconsolidated aquifer (Pleistocene) results in permanent compaction of fine-grained sediments and subsequent land surface lowering, known as land subsidence (Leake & Prudic, 1991).

The average measured rate of land subsidence in Greater Cairo and the northeastern Nile Delta is 7 and 8 mm/y, respectively (Ali et al., 2009; Becker & Sultam, 2008). The Interbed Storage (IBS1) Package (Leake & Prudic, 1991) is applied to the Pleistocene aquifer. The compressible interbeds consist of aquitard clay beds (Fig. 16.2). The amount of water released from interbed storage (aquitards) is 8.3×10^5 and 1.3×10^5 m^3/d within 100 and 1000 days, respectively (Table 16.1).

The compaction ranged from 1×10^{-4} to 6.5×10^{-3} m in the 100- and 1000-day simulations for the observation wells. The compaction rate and the resulting land subsidence (especially if undetected) should be considered when designing and installing wells. The screen pipelines should be placed in front of porous sediments, while the casing pipelines face low-permeability deposits. If compaction affects the pipelines, the design should shift the screen tube before the aquitard clay, which can increase the risk of well failure and damage. Well failure may occur when the screen pipelines partially project into the clay sediments and, as a result, trap suspended materials. This may lead to decreased pump efficiency and farm productivity.

The results of this study have wide global relevance, as groundwater depletion caused by intensive extraction is a growing phenomenon affecting aquifer systems worldwide. Recent global assessments show that groundwater levels are declining in many monitored aquifers, mainly due to unsustainable pumping and climate-driven reductions in recharge. Similar to the trends observed in the Pleistocene aquifer of the Eastern Desert, excessive extraction elsewhere has been linked to land subsidence and irreversible loss of aquifer storage capacity, with global estimates indicating annual permanent storage losses of tens of cubic kilometers. In coastal regions, groundwater decline greatly increases seawater intrusion, a process expected to worsen globally under continued overextraction and sea-level rise, threatening freshwater supplies for agriculture and domestic use. These findings highlight the vital importance of groundwater flow and compaction modeling as tools for assessing sustainable extraction limits and reducing long-term hydrogeological, environmental, and socio-economic risks in arid and semi-arid regions worldwide.

Table 16.1 Volumetric budget for storage depletion (m^3/d).

1000 days of simulation time			100 days of simulation time			
IN-OUT	OUT	IN	IN-OUT	OUT	IN	FLOW TERM
24,600	1040	25,600	196,000	10,400	207,000	STORAGE
347,000	350,000	698,000	24,800	545,000	570,000	CONSTANT HEAD
424	0	424	424	0	424	RECHARGE
−4300	4300	0	−4300	4300	0	Evapotranspiration
−497,000	612,000	114,000	−1,040,000	1,060,000	18,500	RIVER LEAKAGE
129,000	0	129,000	825,000	0	825,000	INTERBED STORAGE
−1	967,000	967,000	−1	1,620,000	1,620,000	SUM

6. Conclusion and recommendation

Leakage of water from the Ismailiya Canal into the Pleistocene aquifer increases pumping rates, thereby boosting per-capita income. The Quaternary aquifer, which is mainly sandy, gravelly and clay-filled, is very thick in the north. Its mean permeability and transmissivity are 30 m/day and 1000 m^2/day, respectively. The decrease in groundwater levels sharply increases seawater intrusion through the Suez Canal, raising salinity in the aquifer, making it unsuitable for irrigation. Land subsidence in Greater Cairo and the northeastern Nile Delta averages 7–8 mm/year, and it poses a potential danger to infrastructure and people. The groundwater model forecasts future hydrogeological conditions under varying pumping and investment strategies, and evaluates aquifer depletion and its effects on land subsidence. As demonstrated by the PMWIN model, a sustainable abstraction rate is $-10{,}000$ m^3/d, but a sharp decrease in groundwater levels is observed as pumping rates exceed $-30{,}000$ and $-50{,}000$ m^3/d. Simulated aquifer compaction ranges from 1×10^{-4} to 6.5×10^{-3} m over a period of 100–1000 days, indicating that continued overexploitation could significantly worsen land subsidence and related hazards.

CRediT authorship contribution statement

Mohamed ElKashouty: Writing—original draft, Writing—review & editing, Supervision, Methodology, Investigation, Data curation, Formal analysis, Conceptualization. **Mohd Yawar Ali Khan:** Writing—original draft, Writing—review & editing, Methodology.

Declaration of generative AI in scientific writing

During the manuscript preparation, the Jenni AI (Subscribed version) tool was exclusively used for conducting the literature review. Nonetheless, content generated by Jenni AI has been paraphrased before inclusion in this article. Grammarly AI (Subscribed version) software was employed to improve the writing, focusing on grammar and paraphrasing, to minimize content similarity within the paper. No AI tool of any kind was used in developing the concept, methods, results, or discussion sections of this manuscript.

Declaration of competing interest

The authors declare that they have no known competing financial interests or personal relationships that could have influenced the work reported in this paper.

References

Abduljaleel, Y., Amiri, M., Amen, E. M., Salem, A., Ali, Z. F., Awd, A., ... Ghzal, M. (2024). Enhancing groundwater vulnerability assessment for improved environmental management: Addressing a critical environmental concern. *Environmental Science and Pollution Research, 31*(13), 19185–19205.

Afify, H. F. (2000). *Studies on water resources and pollution in some areas of the Suez Canal Province*. M. Sc. Thesis. Ismailia, Egypt: Suez Canal University.

Ali, M. Z., Chu, H. J., Tatas, & Burbey, T. J. (2022). Estimation of annual groundwater changes from InSAR-derived land subsidence. *Water and Environment Journal, 36*(4), 622–632.

Ali, M., Zabker, H., Giardino, R., & Klein, A. (2009). Permanent scattered investigation of land subsidence in greater cairo, Egypt. *Geophysical Journal International, 178*(3), 1238–1245. https://doi.org/10.1111/j.1365-246X.2009.04250.x

Andersen, P. (1993). *A manual of instructional problems for the USGS. MODFLOW model*. Washington, DC: Scientific Software Group.

Anderson, M. P., Woessner, W. W., & Hunt, R. J. (2015). *Applied groundwater modeling: Simulation of flow and advective transport* (2nd ed., pp. 3–25).

Attwa, M., Basokur, A. T., & Akca, I. (2014). Hydraulic conductivity estimation using direct current (DC) sounding data: A case study in East Nile Delta Egypt. *Hydrogeology Journal, 22*(5), 1163–1178.

Becker, R., & Sultam, M. (2008). Land subsidence in the Northeastern Nile Delta as measured by radar inter-feometry. In *Conference: GSA annual meeting, at Houston, Texas* (Vol 40).

Binnie and Partners. (1980). *Ismailia Canal Study, Final Report, Cairo, May, 1980*.

Bockstiegel, M., Richard-Cerda, J. C., Muñoz-Vega, E., Haghighi, M. H., Motagh, M., Lalehzari, R., & Schulz, S. (2024). Simulation of present and future land subsidence in the Rafsanjan plain, Iran, due to groundwater overexploitation using numerical modeling and InSAR data analysis. *Hydrogeology Journal, 32*(1), 289–305.

Chiang, W. H., & Kinzelbach, W. (1998). *Processing Modflow: A Simulation System for Modeling Groundwater Flow and Pollution. Software Instruction Book*. Hamburg-Zurich.

Condon, L. E., Kollet, S., Bierkens, M. F., et al. (2021). Global groundwater modeling and monitoring: Opportunities and challenges. *Water Resources Research, 57*(12). e2020WR029500.

Doherty, J. (2000a). *PEST model independent parameter estimation*. Bethesda; MD: Watermark Numerical Computing; S.S. Papadopoulos & Associates Inc.

Doherty, J. (2000b). *PEST utilities for MODFLOW and MT3D parameter estimation*. Bethesda: Watermark Numerical Computing; S.S. Papadopoulos & Associates Inc.

Doherty, J. (2001). *PEST groundwater data utilities*. Brisbane; Australia: Watermark Numerical Computing.

Eltarabily, M., Moghazy, H., AbdelFattah, S., & Negm, A. (2020). The use of numerical modeling to optimize the construction of lined sections for a regionally significant irrigation canal in Egypt. *Environmental Earth Sciences, 79*, 80.

Galloway, D. L., & Sneed, M. (2013). Analysis and simulation of regional subsidence accompanying groundwater abstraction and compaction of susceptible aquifer systems in the USA. *Boletin de la Sociedad Geologica Mexicana, 65*(1), 123–136.

Geriesh, M. H. (1994). *Hydrogeological and hydrogeochemical evaluation of groundwater resources in the Suez Canal region*. Fac. Sci. Suez Canal University. Ph. D. thesis.

Ghazaw, Y. M. (2011). Design and analysis of a canal section for minimum water loss. *Alexandria Engineering Journal, 50*(4), 337–344. https://doi.org/10.1016/j.aej.2011.12.002

Haddad, I. M. (2002). *Hydrogeological studies and their environmental impact on future management and sustainable development of the new communities and their surroundings, east of the Nile Delta, Egypt*. PhD Thesis. Egypt: Faculty of Science, Mansoura University.

Hsieh, A., & Freckleton, F. (1993). Documentation of a computer program to simulate horizontal-flow barriers using the U. S. Geological Survey's modular three-dimensional finite-difference groundwater flow model. *U.S. Geological Survey, Open-File Report*, 92–477.

Khan, M. Y. A., ElKashouty, M., Abdellattif, A., Egbueri, J. C., Taha, A. I., Al Deep, M., & Shaaban, F. (2023). Influence of natural and anthropogenic factors on the hydrogeology and hydrogeochemistry of Wadi Itwad Aquifer, Saudi Arabia: Assessment using multivariate statistics and PMWIN simulation. *Ecological Indicators, 151*, 110287.

Leake, S. A., & Prudic, D. E. (1991). Documentation of a computer program to simulate aquifer system compaction using the modular finite-difference ground-water flow model. *U.S. Geological Survey*.

Luan, J., Li, X., Naeem, S., Kumar, A., Miao, P., Wang, R., … Ma, N. (2025). Substantial decline in the groundwater in the ten Kongduis Basin in the Loess Plateau during 2001–2020. *Ecohydrology, 18*(1), e2764.

Mawla, F. (1999). Assessment of industrial activities impact on drinking wells at Shoubra El Khima. *International Conference on Intergrated Management of Water Resources*, 379–392.

Meena, V. P., Ahmad, N., Singh, R., Vyas, A., Kumar, A., Kumar, S., & Kumar, M. (2025). Fluoride variability in groundwater: Hydro-geochemical influences and risk assessment in an arid region of Northwestern India. *Physics and Chemistry of the Earth, Parts A/B/C*, 104029.

Melinda, A., Christian, D., & Swain, E. (2004). *Assinfing boundary conditions to the southern inland and coastal systems (SICS) model using results from the south Florida water management model (SFWMM)* (p. 30p). Reston, Virginia: USGS.

Mowafy, M. H. (2001). Seepage losses in Ismailia canal. In *6th international water technology conference (IWTC), Alexandria, Egypt*. https://www.iwtc.info/2001_pdf/03-3.pdf.

Qi, C., & Zhan, H. (2022). Soil property and subsurface heterogeneity control on groundwater recharge of vadose zone injection wells. *Journal of Hydrologic Engineering, 27*(3), 04021052.

Rodríguez-Castellanos, J. M., Sánchez-Gómez, A., Bieger, K., Martínez-Pérez, S., Senent-Aparicio, J., & Molina-Navarro, E. (2025). Calibrating streamflow and hydrological processes in geological regions using a combined soft and hard calibration approach. *JAWRA Journal of the American Water Resources Association, 61*(2), e70008.

Sarris, T. S., Kenny, A., Scott, D. M., & Close, M. E. (2022). Aquifer heterogeneity controls to quality monitoring network performance for the protection of groundwater production wells. *Water Research, 218*, 118485.

Shyam, R., Krishan, G., Kheraj, & Kumar, A. (2024). Evaluation of groundwater quality for life-supporting activities: A case study of Haryana, India. *International Journal of River Basin Management, 22*(2), 297–308.

Springer, A., De Lannoy, G., Rodell, M., Ewerdwalbesloh, Y., Gerdener, H., Khaki, M., … Kusche, J. (2025). *A review of current best practices and future directions in assimilating GRACE/-FO terrestrial water storage data into numerical models* (Vol 2025, pp. 1–51). EGUsphere.

Wali, S. U., Usman, A. A., & Usman, A. B. (2024). Resolving challenges of groundwater flow modeling for improved water resources management: A narrative review. *International Journal of Hydrology, 8*, 175–193.

Wang, Y., Zechner, M., Mern, J. M., Kochenderfer, M. J., & Caers, J. K. (2022). A sequential decision-making framework with uncertainty quantification for groundwater management. *Advances in Water Resources, 166*, 104266.

White, J. T., Hunt, R. J., Fienen, M. N., & Doherty, J. E. (2020). *Approaches to highly parameterized inversion: PEST++ version 5, a software suite for parameter estimation, uncertainty analysis, management optimization and sensitivity analysis (U.S. Geological Survey techniques and methods, book 7, chapter C26)*. U.S. Geological Survey.

Zhao, Z., Kumar, A., & Wang, H. (2024). Predicting arsenic contamination in groundwater: A comparative analysis of machine learning models in coastal floodplains and inland basins. *Water, 16*(16), 2291.

CHAPTER

Next-generation smart water solutions for water security and achieving sustainable development goals

17

Van-Truc Nguyen[1], Quoc-Hoang Do[1], Vu-Anh Le[2], Van-Anh Thai[3] and Thi-Dieu-Hien Vo[4]

[1]*Faculty of Engineering and Technology, Saigon University, Ho Chi Minh City, Vietnam;* [2]*School of Chemical and Biomolecular Engineering, University of Sydney, Darlington, NSW, Australia;* [3]*HUTECH Institute of Applied Sciences, HUTECH University, Ho Chi Minh City, Vietnam;* [4]*Faculty of Applied Science and Technology, Nguyen Tat Thanh University, Ho Chi Minh City, Vietnam*

1. Introduction

Water is an irreplaceable element in daily life, essential for the growth, production, maintenance, and functioning of all living beings (Babuna et al., 2023). However, with increasing demand driven by population growth, climate change, and economic expansion across multiple sectors, the availability of water resources has become increasingly uncertain. The worsening deterioration of water quality poses a serious threat to water security, particularly as rising water consumption increases pressure on resources, exacerbating water scarcity in rural and economically disadvantaged areas. Furthermore, in many cases, these risks exceed the capacity of existing water systems and government responses, leading to overexploitation of resources, ecosystem degradation, and increased flood risk. Significant efforts have been made to better understand the scale of global water security challenges. Assessing water security requires a multidimensional approach that accounts for water quantity, quality, and accessibility for human use. An estimate in the study by Makarigakis and Jimenez-Cisneros (2019) indicated that 450 million people in 29 countries suffer from severe water shortages. By 2050, it is predicted that nearly 40% of the world's population will live in river basins facing severe water stress, especially in Africa and Asia. Up to five billion people worldwide could face serious water quality problems (Chaplin-Kramer et al., 2019).

SDG 6 is one of the most important Sustainable Development Goals (SDGs), as it specifically targets water and sanitation. However, it is also integral to the achievement of many other SDGs, as water underpins both human development and ecosystem health (Vörösmarty et al., 2018). Besides, SDG 6 is a key component of the United Nations 2030 Agenda, which encompasses 17 interrelated SDGs. The 2030 Agenda was built on decades of global efforts to address water resources, supply, and sanitation. The United Nations first placed water on its agenda in 1977, with growing international concern emerging in the early 1990s. This momentum continued with the Millennium Declaration in

Sustainable Water Management. https://doi.org/10.1016/B978-0-443-44853-9.00005-8

2000, which included Millennium Development Goal (MDG) 7C, aimed at achieving sustainable access to safe drinking water and sanitation by 2015. Initially focused solely on drinking water, the goal was expanded in 2002 to include sanitation (Mujtaba et al., 2024).

Advances in sensor technologies and high-performance computing have reshaped how water systems are studied, designed, and managed, shifting the focus toward Big Data-driven solutions. The advent of affordable remote control devices is encouraging urban water plants to modernize. Within the broader framework of Industry 4.0, the integration of Internet of Things (IoT) technologies into smart grids presents new opportunities for advancing the management of water networks and related systems. This trend is driven by the need to track both the types and accessibility of existing water consumption data in the rapidly evolving field of smart metering. This may indicate that smart water solutions have great potential for practical implementation to contribute to reducing pressure on water security and achieving relevant SDGs. Therefore, this chapter aims to provide a more comprehensive overview of smart water solutions, including their concepts, characteristics, applied technologies, benefits, and challenges for sustainable water management strategies.

2. Overview of next-generation smart water technologies

2.1 Characteristics of smart water systems

Industrial and densely populated cities are facing the pressure of water pollution and the need for clean water. The development of information technology has opened up new potential for water resource management and solving urban water challenges. Smart cities are known as cities that integrate information technology into industry and urban infrastructure (Yang et al., 2021). Smart water management seeks to apply tools such as smart sensors, IoT, and cloud computing to enable intelligent, systematic, and equitable allocation of water resources. This vision is embodied in smart water systems (SWSs), which integrate multiple digital components and processes to achieve real-time, intelligent, and science-based comanagement of water resources. Although current SWSs share similar architectures, they differ in interfaces and implementation details. As illustrated in Fig. 17.1, the backbone of an SWS consists of a data layer and an integrated application layer, supported by robust network security and maintenance mechanisms to safeguard against cyber threats (Zhang et al., 2024). However, despite the expectation that SWSs would enable a complete cycle of urban water resource management, from monitoring to decision-making and ultimately to user-oriented presentation, in practice, most SWSs remain limited to monitoring basic hydrological indicators.

2.2 Evolution from traditional to smart water management

To date, automated control technology (ACT) and information and communication technology (ICT) have been applied to address challenges in water distribution networks, where both play vital roles in large-scale applications. Adapting existing systems with smart components, such as sensors, controllers, and data centers, enables real-time monitoring, transmission, and control, providing decision-makers with timely information. This integration offers a more cost-effective and sustainable approach to overcoming contemporary water management challenges (Li et al., 2020). Moreover, real-time measurements enhance the accuracy of hydraulic model calibration and forecasting, while real-time control is widely used in pump operation, valve regulation, and scheduling. Although ACT

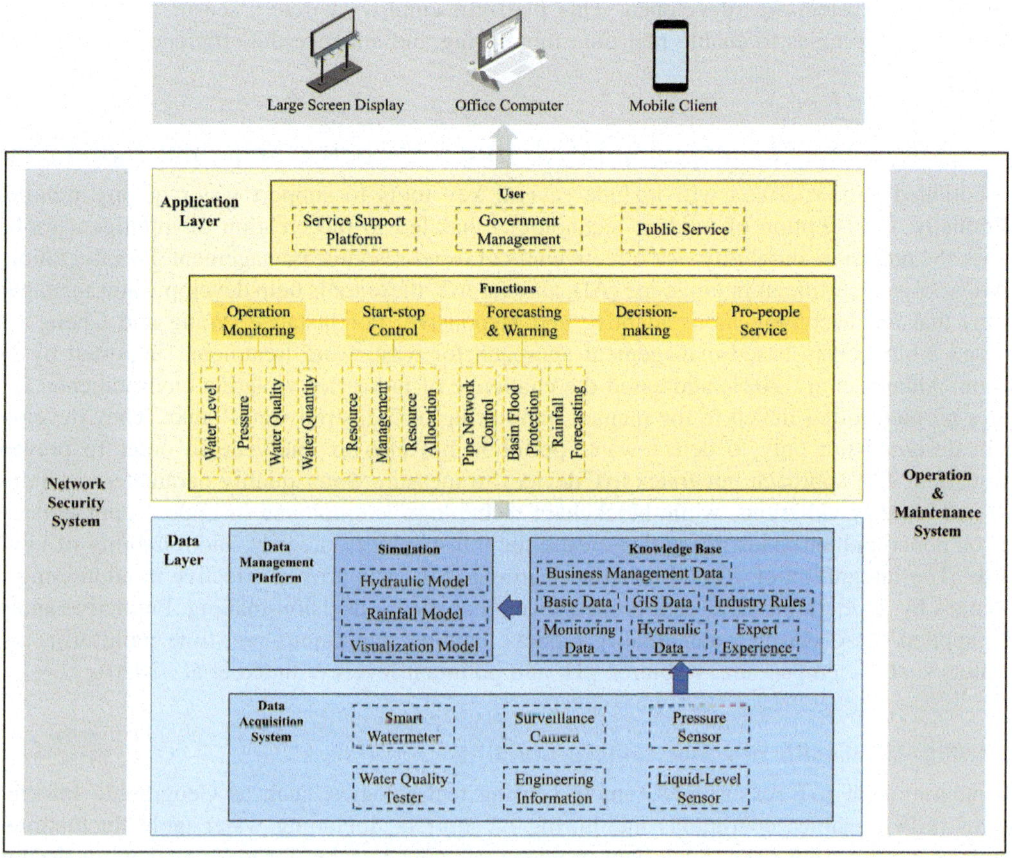

FIGURE 17.1

The general architecture of a smart water system.

Adapted with permission from Zhang, N., Khu, S.-T., Wang, J. (2024). A conceptual method to evaluate the smart water system based on fit-for-purpose concept. Environmental Impact Assessment Review, 105, *107447, https://doi.org/10.1016/j.eiar.2024.107447.*

significantly improves water supply efficiency, its energy efficiency, particularly in electricity use, still requires optimization for practical application. When coupled with effective ICT and ACT solutions within an SWS, urban water challenges can be more effectively managed and addressed at the city scale (Liu et al., 2019). Within SWS, advancements are achieved through technologies such as smart metering, which enables continuous monitoring and data transmission to service providers, and intelligent control, which allows for immediate feedback and operational response. An example of this can be seen in California's Western Municipal Water District (WMWD), where the SCADA platform is employed to handle real-time alerts and to automate the operation of treatment facilities and distribution networks (Li et al., 2020). Singapore provides another example of an SWS application, where the WaterWiSe system was developed. This platform employs wireless sensor networks and data acquisition technologies to enable real-time monitoring and enhance the efficiency of water supply operations.

2.3 Core technologies including IoT, AI, Big Data, blockchain, and digital twins

As mentioned earlier, SWSs will include several key tools to support water quality monitoring. Additionally, the adoption of modern technologies has led to the creation of intelligent tools that improve the precision, durability, and effectiveness of water resource management. By integrating data analytics, sensors, artificial intelligence (AI), and the IoT, these tools help develop water management systems that are becoming more adaptive and responsive. For instance, Wang and Cheng (2020) proposed a blockchain-based management approach for wastewater treatment, supported by AI. In addition, Alharbi et al. (2021) addressed the challenge of industries violating environmental regulations, a problem often linked to the manual monitoring of industrial water tanks. They developed a system designed not only to detect water pollution but also to intervene in order to prevent its continuation. The approach integrates IoT devices to measure water quality parameters in industrial tanks and identify violations, while blockchain technology is employed to impose appropriate penalties on noncompliant industries and to ensure the transparency, integrity, and reliability of violation records. The integration of AI and machine learning (ML) has proven effective in addressing these challenges by accelerating treatment processes and enhancing decision-making. Predictive analytics, when applied to AI-driven water quality models, enables continuous real-time monitoring of key indicators such as temperature, turbidity, pH, and pollutant levels (Ahmed et al., 2020).

2.4 Integration with GIS and remote sensing

The integration of IoT sensors and remote sensing technologies, such as Geographic Information Systems (GIS), enables continuous monitoring of aquifers, including water table fluctuations and potential contamination risks. Such data provides valuable insights for policymakers to design and implement strategies and groundwater conservation initiatives. By integrating GIS methodologies with aerial imagery from drones, Ramírez-Cuesta et al. (2018) successfully identified spatial differences in agricultural water demand. Beyond mapping, drones are increasingly applied for the early recognition of crop water stress. Serrano et al. (2018) linked vegetation indices captured through unmanned aerial vehicle (UAV) sensors with soil moisture levels and biomass yield, showcasing their potential in precision irrigation. Collectively, these technological advances have made drones a vital component of sustainable water management practices in agriculture.

In addition, Mandal et al. (2025) emphasize that the effective application of advanced technologies such as AI, GIS, and remote sensing can enable data-driven irrigation management, leading to higher agricultural productivity and reduced water wastage. Sensor-based irrigation systems, a core element of precision agriculture, optimize water use by integrating soil moisture measurements and weather data to determine optimal irrigation schedules, thereby preventing overwatering and minimizing runoff. Complementary tools such as drones and satellite imagery further support crop health monitoring and the early detection of water stress, allowing for targeted interventions. As climate change intensifies water scarcity, precision agriculture offers a promising pathway to safeguarding food security while ensuring the sustainable management of water resources.

3. Smart water monitoring and management systems

3.1 Real-time water quality monitoring

The limitation of existing water quality monitoring technologies lies in their dependence on manual sample collection at various locations, followed by laboratory analyses to determine water quality based on water quality index (WQI) parameters. This approach is time-consuming, inefficient, and fails to provide real-time information, as samples are collected only periodically. Therefore, the development of a real-time water quality monitoring system is essential, offering a more efficient and convenient means of analyzing water quality. For instance, Demetillo et al. (2019) introduced a low-cost, real-time water quality monitoring system designed for use in remote aquatic environments such as lakes, rivers, and other water bodies. The setup integrates a microcontroller, standard electrochemical sensors, a customized buoy, and a wireless communication unit. It can measure key parameters, such as water temperature, dissolved oxygen, and pH, at scheduled intervals. The collected data are then transmitted in both tabular and graphical formats through a dedicated web portal and directly to registered mobile devices, ensuring timely access for end-users.

In addition, Hamid et al. (2020) designed a real-time monitoring system called the Smart Water Quality Monitoring System for swimming pools, focusing on key parameters such as temperature and pH. Statistical tools, including Design of Experiments and analysis of variance (ANOVA), were employed to evaluate the influence of different factors on water quality. The results indicated that variables such as pool volume, time, and their interactions had no significant impact on pH, whereas time of day was found to influence water temperature. The developed system demonstrated the capability to automatically update water quality status in real time via IoT integration and to adjust pH levels as necessary. Furthermore, the system required minimal maintenance, underscoring its potential as an efficient and reliable solution for automated water quality management in recreational facilities.

3.2 Smart meters and consumption analytics

Integrating ML in SWSs to automatically make decisions during operation brings many benefits, such as being less time-consuming, cost-efficient, and effective. For instance, a Smart Water Quality Monitoring System based on IoT using cloud and deep learning (DL) methods was implemented in the study by Ajith et al. (2020). The system can monitor water quality across diverse water resources by employing IoT devices integrated with a Node-MCU microcontroller. The built-in Wi-Fi module of the Node-MCU enables seamless internet connectivity, allowing sensor data to be transmitted to a cloud platform for storage and analysis. Multiple sensors are incorporated to measure a range of parameters relevant to water quality assessment, and the collected data are subsequently processed using DL techniques to predict potability. In addition to potable water evaluation, the system also has potential applications in wastewater treatment, where it can automatically assess water quality and optimize chemical dosing in real time.

Additionally, the model proposed by Gupta et al. (2021) is particularly interesting because it automates the evaluation of key water quality parameters, including turbidity, pH, and temperature. Moreover, its low power consumption and built-in Wi-Fi make it well-suited for underwater communication. The integration of communication modules, turbidity meters, and pH sensors highlights how IoT can be leveraged in practical applications. The low-cost design, coupled with high-speed Wi-Fi for underwater communication, makes this system not only self-sufficient but also efficient in addressing real-world water quality challenges. Once fully developed, the system has the potential to be applied in water supply plants, where it could automatically regulate coagulant dosing for effective turbidity control.

3.3 Leak detection and pressure management

Leak detection and localization are critical priorities in the management of drinking water systems, as water loss not only leads to environmental and socio-economic costs but also poses significant risks to urban infrastructure. With advancements in hydraulic sensors and data acquisition technologies, SWSs offer an innovative approach to real-time leak detection, enabling the rapid identification of water losses and reducing leak duration. As one of the pioneers, Farah and Shahrour (2017) enhanced the application of the minimum night flow method, which operates by establishing flow thresholds; a leak alarm is triggered when the night flow surpasses these thresholds. This analytical approach enables the rapid detection of pipe bursts, thereby minimizing leakage duration. Implementation of the improved method successfully identified 25 previously unreported leaks and resulted in a 36% reduction in nonrevenue water. Farah and Shahrour (2017) explored leak detection and prediction in smart water management systems by integrating computational fluid dynamics (CFD) with DL techniques. In their approach, various DL models were trained and evaluated using datasets generated through CFD simulations, where SolidWorks was employed for pipe design and ANSYS Workbench for simulating water flow and producing training data. Model performance was compared to identify the most effective algorithms for both leak prediction and localization. The findings demonstrate the potential of this hybrid CFD–DL framework to improve the efficiency and reliability of leak detection in smart water infrastructures, thereby supporting water conservation efforts and mitigating risks of infrastructure damage.

3.4 Predictive maintenance of water infrastructure

The rapid growth of urban populations and the increasing demand for infrastructure systems present significant challenges to cities worldwide. In response, the transition toward smart cities has introduced advanced technologies aimed at enhancing infrastructure efficiency, sustainability, and resilience. With the integration of AI, predictive maintenance has emerged as a transformative approach, enabling both retrospective analysis and real-time optimization. By leveraging data from embedded sensors, AI-driven systems can detect faults as they occur, thereby improving infrastructure monitoring, reducing maintenance costs, and enhancing the overall reliability of construction and urban services. For instance, Bejarano et al. (2018) employed ensemble learning techniques, specifically XGBoost and Random Forest models, to predict the operational status of water pumps. Their findings demonstrated that these models achieved superior predictive performance across different water extraction methods. A comprehensive feature analysis was conducted to evaluate the influence of various feature groups, such as geographic and management attributes, on model performance in predicting pump functionality, as well as water quality and quantity. Furthermore, a fine-grained analysis revealed that, among individual features, spatial location (x, y, z coordinates) exerted the greatest impact on predictive accuracy. Additionally, Boulos (2017) presented a smart water network (SWN) framework integrated with GIS to facilitate the development of conveyance networks, two-dimensional surface mesh generation, and spatial placement (Fig. 17.2). The spatial and temporal

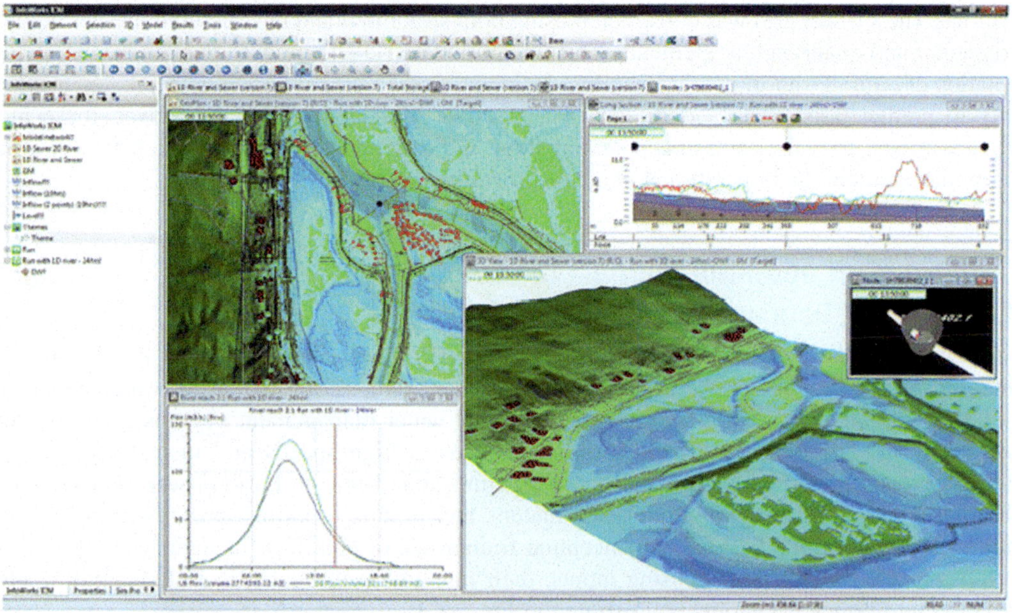

FIGURE 17.2

The integration of 1D–2D and GIS models.

Adapted with permission from Boulos, P. F. (2017). Smart water network modeling for sustainable and resilient infrastructure. Water Resources Management, 31, *3177–3188. https://doi.org/10.1007/s11269-017-1699-1.*

variations in water depths and flows were determined by numerically solving the Shallow Water Equations. The surface grid was constructed using both structured and unstructured meshes composed of triangular and quadrangular elements. This integration of 1D–2D models enabled the simultaneous simulation of rainfall–runoff processes, urban drainage, river flow, and flooding, thereby enhancing the ability to predict flood risks and design cost-effective drainage systems. Moreover, these models provide a robust tool for assessing potential flood hazards and formulating reliable, cost-efficient design, management, and operational strategies for urban water infrastructure.

4. Data-driven decision support for water governance

4.1 Role of Big Data analytics in water planning

Modern water governance has been revolutionized by Big Data analytics, which delivers useful information that helps policymakers, engineers, and managers tackle increasing water management issues more effectively. Big Data surpasses traditional water management systems because it unites diverse data streams that come from multiple sources at high speeds and in real time. The data collection process incorporates IoT sensors in pipelines together with Earth observation satellite and hydrological monitoring stations, industrial meters, and mobile reporting systems from community members (Solanki et al., 2025). The conceptual framework of data flow in smart water governance, encompassing IoT and remote sensing inputs, their integration through Big Data analytics, and the translation into policy decisions, is illustrated in Fig. 17.3.

Big Data serves multiple functions when it comes to water planning. The system uses Big Data to create accurate demand forecasts, which help utilities predict short-term and long-term water usage patterns across various conditions. The system proves essential for urban areas that experience fast population growth because it helps manage water consumption changes caused by population shifts and climate patterns. The system detects water leaks through sensor reading anomalies, which helps identify loss points, as demonstrated in Sadeghioon's study. A detection method combining a relative pressure sensor mounted on the external surface of the pipe, temperature differential measurements between the pipe wall and the surrounding soil, and an anomaly detection algorithm was evaluated through a 6-month field test. The validation results showed that all known leaks were successfully identified with an accuracy of 98.45%, with the anomaly detection algorithm achieving the highest performance (Sadeghioon et al., 2018). The combination of flow data with pressure readings and acoustic signals enables pipeline burst detection at an early stage, which minimizes water loss through nonrevenue channels. The system enhances water quality surveillance through integrated analysis of chemical and biological and hydrological data, which enables precrisis detection of contamination incidents. Tinelli and Juran (2019) demonstrated an AI-based smart monitoring system that enables near-real-time detection of chemical and biological contamination in water, using support vector machines (SVMs) and artificial neural network (ANN)-based sensors to identify and assess anomalies.

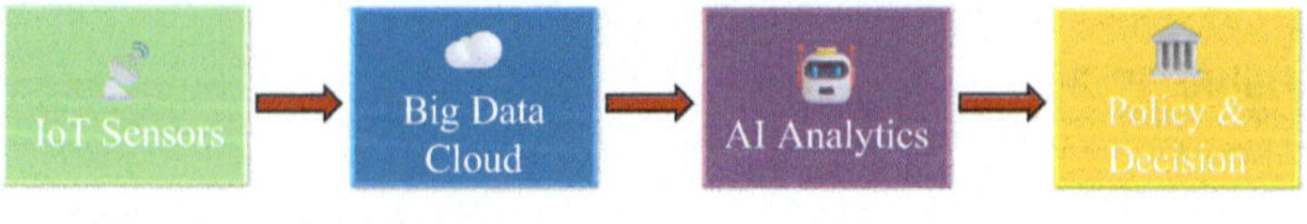

FIGURE 17.3

Conceptual framework of data flow in smart water governance: From IoT and remote sensing inputs to Big Data analytics and policy decisions.

The Public Utilities Board (PUB) of Singapore shows how Big Data operates in real-world applications. The distribution network of PUB operates with thousands of smart sensors that track flow rates, pressure levels, and water quality metrics in real time. The system uses ML to process data for leak detection and pumping schedule optimization, which helps maintain efficient operations in Singapore's limited water supply environment. The European Space Agency uses Earth observation satellite information together with hydrological data from specific regions to enhance irrigation systems and manage reservoirs effectively in various climate zones. Big Data analytics in water planning brings various advantages to the field, but it encounters multiple implementation obstacles. The management of diverse data types needs sophisticated interoperability systems, while real-time data processing requires expensive computing systems. The use of citizen-provided data for water management requires immediate solutions to protect privacy and maintain ethical standards. The implementation of Big Data in water governance has become vital because it helps achieve the United Nations SDGs, especially Goal 6, which focuses on Clean Water and Sanitation (Abouchabaka & Bentaleb, 2024).

4.2 AI-based forecasting models

The modeling of complex nonlinear water system interactions through AI technology leads to better forecasting abilities. AI algorithms process extensive unorganized data collections to generate enhanced predictive models, which improve their accuracy through additional data input. Representative AI models, including ANNs, SVMs, and long short-term memory (LSTM) networks, have been applied in smart water governance for flood, drought, and demand management, as illustrated in Fig. 17.4. In addition, Table 17.1 provides a summary of AI applications in water management by country, highlighting regional advancements and practical implementations.

AI models, including ANNs, SVMs, and DL networks like LSTM networks, perform flood forecasting tasks by modeling rainfall–runoff systems and predicting river discharge and urban flood events in real time. The Yangtze River Basin in China serves as a major testing ground for AI models that combine rainfall radar data with satellite imagery and hydrological sensor readings to generate flood predictions that result in thousands of saved lives and substantial economic benefits (Wu et al., 2023). AI tools for drought prediction use climate data, including temperature, precipitation, evapotranspiration, and soil moisture information, together with agricultural statistics to create water scarcity forecasts. The application of AI models by India enables district-level drought risk evaluation, which helps farmers make strategic decisions about their planting choices and water management techniques (Manikandan et al., 2023). AI systems use weather forecasts together with social data and consumption patterns to create precise short-term and long-term water demand predictions. European and North American utility companies use AI models to optimize their daily water pumping operations and detect upcoming peak usage times, which helps them minimize their energy and water consumption (Calatayud Asensi et al., 2025).

The implementation of AI models faces ongoing challenges to achieve transparent and explainable systems that maintain accountability. DL methods function as uninterpretable systems, which create doubts about their reliability among decision-makers (Scorzato, 2024). The development of interpretable AI models and hybrid systems that unite ML with physical hydrological models works to resolve these problems (Chin & Lai, 2025). AI technology transforms water governance through enhanced predictive capabilities and enhanced climate extreme resistance (Islam, 2025).

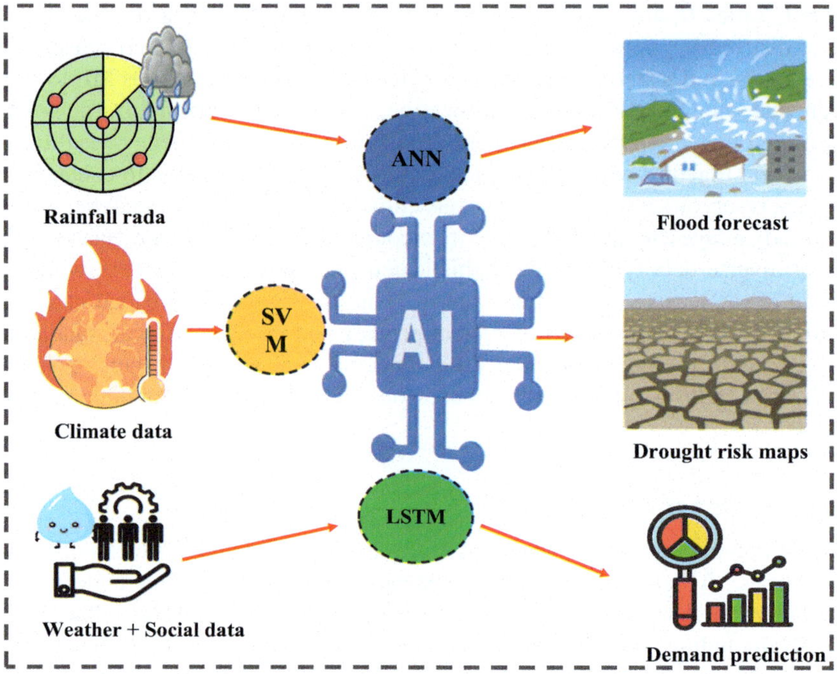

FIGURE 17.4

Application of AI-based models in smart water governance: Artificial neural network (ANN), support vector machine (SVM), and long short-term memory (LSTM) for flood, drought, and demand management.

4.3 Decision support systems for policymakers

Decision support systems (DSS) create organized platforms that unite data with models and simulations to help policymakers choose the best water management decisions. The DSS platform combines hydrological models with socio-economic indicators and policy scenarios through interactive dashboards and digital water system replicas (Yu et al., 2018). The virtual water infrastructure model, known as a digital twin, enables policymakers to evaluate how different scenarios affect water distribution, energy usage, and economic expenses (Zhao et al., 2025). The simulated models help users understand how different water distribution methods affect both energy usage and financial cost. The European Union operates the Water Framework Directive system as a DSS that unites ecological data with hydrological information and socio-economic statistics to support water management between countries (Boeuf & Fritsch, 2016). The Bureau of Reclamation in the United States employs DSS to determine Colorado River Basin water distribution between agricultural needs, industrial requirements, urban water consumption, and environmental preservation.

The scenario-based analysis capabilities of DSS systems improve both the transparency and accountability of decision-making processes. Through DSS, policymakers gain the ability to

Table 17.1 Summary of AI applications in water management by country.

Country/ Region	AI Model (s)	Application	Benefits	Limitations	References
China	ANN, LSTM	Flood forecasting (river discharge, urban floods)	- Real-time flood predictions - Reduces human and property losses - Optimizes emergency response	- Requires high-resolution rainfall and sensor data - Complex model training - Black-box nature limits interpretability	Zhao et al. (2022)
India	ANN, SVM	Drought risk assessment (district-level)	- Supports strategic agricultural decisions - Assists in water resource allocation - Enables early warning systems	- Limited climate and soil data in rural regions - Requires technical expertise - May not capture local microclimate variations	Oyounalsoud et al. (2024), Singh and Ahmed (2021)
EU/North America	ANN, LSTM	Water demand forecasting (short- and long-term)	- Optimizes pumping and energy use - Reduces water losses - Supports urban planning	- Privacy concerns with consumption and social data - Integration with legacy systems can be challenging - Heterogeneous data requires preprocessing	Kühnert et al. (2021), Zhang et al. (2025)

present their choices to stakeholders and obtain their agreement while securing funding for major infrastructure projects (Al Nahyan et al., 2018). The implementation of DSS faces obstacles because it needs specialized technical knowledge, faces compatibility problems, and sometimes encounters resistance from political leaders. The complete implementation of DSS for sustainable water governance depends on resolving existing technical and political challenges.

4.4 Community-based monitoring and citizen science

Community-based monitoring (CBM) and citizen science have become essential tools for water governance because they help areas with restricted institutional monitoring capabilities. The methods allow local people to gather water data, which they can share and understand better to improve both their knowledge and their ability to hold others accountable (Carroll et al., 2019). Local communities can monitor water quality parameters through low-cost IoT sensors and mobile applications that measure pH levels, turbidity, and dissolved oxygen concentrations (Lal et al., 2024). The combination of crowdsourced reporting systems enables citizens to document pollution events, unauthorized water extraction, and flooding occurrences. The combination of citizen science initiatives with NGO support in Kenya enables community members to monitor river water quality, which gets directly incorporated into national environmental database systems.

The implementation of CBM provides three main advantages, which include enhanced transparency, better stakeholder participation, and equal water management opportunities for all. Water agencies achieve better trust relationships with citizens through their involvement, which leads to behavioral changes in water consumption. Official datasets receive additional value from citizen-collected data, which helps fill monitoring gaps that governments cannot reach (Mvulirwenande & Wehn, 2019, pp. 167–207). The process of maintaining accurate and secure citizen-generated data faces ongoing difficulties. Standardized protocols together with data validation tools and training programs must be implemented to improve the quality of collected data (Vadapalli et al., 2024). Official decision-making processes need institutional changes to successfully incorporate citizen science data. The implementation of CBM and citizen science methods faces multiple obstacles, but they create essential pathways for water governance that support SDG participation (Jacquin et al., 2023).

5. Smart water solutions for agriculture and industry

5.1 Precision irrigation and smart farming

The combination of IoT sensors with satellite imagery and advanced analytics in precision irrigation systems enables farmers to optimize their water resources for agriculture. Real-time monitoring of soil moisture, evapotranspiration, and weather conditions enables precision irrigation systems to provide water only to specific areas that require it (Vadapalli et al., 2024). Smart controllers linked to drip irrigation systems achieve water savings of 40% better than traditional irrigation practices (Ghilan et al., 2024). Precision irrigation systems in Israel combine satellite monitoring with automated systems and soil sensors to achieve maximum crop yields in their arid climate. Mobile irrigation advisory systems in India now help smallholder farmers access real-time recommendations through local weather and soil data. Precision farming achieves threefold benefits through its water-saving capabilities, its ability to prevent fertilizer loss and decrease pumping energy needs, and enhance agricultural yields. The implementation of precision farming faces barriers in developing nations because of expensive setup costs, insufficient digital infrastructure, and insufficient technical knowledge among farmers (Mizik, 2023). The adoption of precision farming depends on government financial support and public–private collaborations and specialized training for farmers (Khaspuria et al., 2024).

5.2 Water reuse and recycling technologies in industry

The practice of industrial water reuse and recycling serves two essential purposes: decreasing water extraction from freshwater sources and protecting the environment. The combination of membrane bioreactors (MBRs), reverse osmosis (RO), and advanced oxidation processes (AOPs) allows industries to extract clean water from their wastewater streams. The water-intensive sectors of textiles, power generation, and chemicals have adopted zero liquid discharge (ZLD) as their main water management approach (Fig. 17.5). ZLD systems use various technologies to extract almost all water from wastewater streams, which results in solid waste as the only remaining output. The textile industry in India uses ZLD facilities to meet environmental standards while minimizing water contamination through these facilities (Mohan et al., 2021).

Industrial symbiosis functions as an innovative model that enables wastewater from one industry to serve as a resource for another operation (Chen et al., 2025). The wastewater treated by breweries and food processing facilities can be utilized for agricultural purposes and power plant cooling systems (Hamam et al., 2023). The implementation of advanced treatment systems faces three main obstacles, which include expensive equipment costs, high energy needs, and difficulties in managing concentrated brine waste. The practice of water recycling has become essential for businesses because freshwater resources are dwindling, and environmental protection standards are becoming more stringent.

5.3 Integration of renewable energy with water systems

The connection between water systems and energy networks allows renewable energy integration into water infrastructure, which leads to improved sustainability. Solar-powered desalination units serve remote areas that lack freshwater resources but receive plenty of sunlight (Gevorkov et al., 2025). The combination of hydro and solar power systems through hybrid technology enables resource management between seasonal variations (Balan et al., 2024). Rural communities across sub-Saharan Africa now access drinking water and irrigation through decentralized solar-powered pump systems. The Middle East construction of Massi server-powered desalination facilities serves two purposes by decreasing fossil fuel usage and fulfilling increasing water requirements. The implementation of renewable energy systems decreases both environmental pollution and utility operational expenses

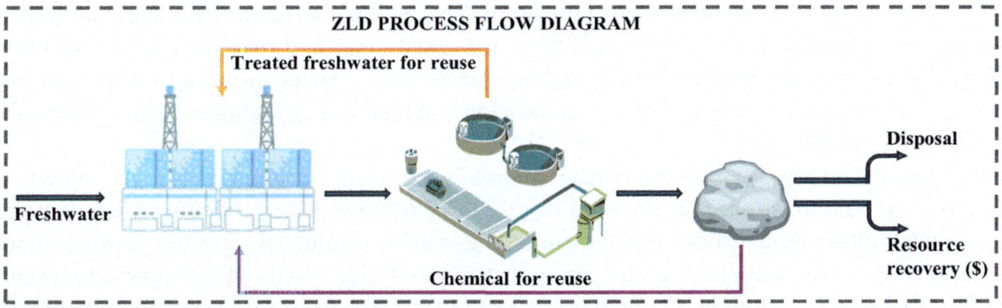

FIGURE 17.5

Schematic representation of the zero liquid discharge (ZLD) system.

(Darlington E. E. and Peter S., 2024). However, variability in renewable generation requires hybrid solutions and energy storage systems to ensure reliability.

5.4 Circular water economy models

The circular water economy works to transform linear "take—use—dispose" operations into self-sustaining systems that promote continuous resource utilization and waste reduction (Vinayagam et al., 2024). The framework combines water reuse with resource recovery and energy production to create a complete system. Modern wastewater treatment facilities now operate as resource recovery centers, which extract water resources and nutrients (nitrogen and phosphorus) and produce energy through biogas production. The Netherlands leads the way in circular water management through its practice of converting sewage sludge into fertilizers while using biogas to operate treatment facilities. The implementation of circular water economy models needs fundamental changes in regulatory frameworks, financial systems, and technological infrastructure. The transition to circular water management faces three main obstacles, which stem from separate governance structures, insufficient financial rewards, and negative public attitudes toward water recycling (Frijns et al., 2024). The shift toward circularity represents an essential step for establishing sustainable water security in the long term.

6. Case studies and best practices

The Big Data analysis in Singapore enables the production of high-quality drinking water from wastewater, which now supplies 40% of the country's water needs (Hannah Lee and Thai Pin Tan, 2017). The distribution network of smart sensors detects leaks in real time. The NEWater initiative in Singapore demonstrates how urban smart water management should be implemented. The combination of membrane technology, which helps Singapore maintain less than nonrevenue water loss. Nagpur, India, uses smart pressure management systems to reduce water leaks from its deteriorating water infrastructure. The city uses sensor data from thousands of points to control water pressure dynamically, which helps stop pipe bursts while using less energy (Gupta & Kulat, 2018). The examples demonstrate how technological advancements unite with proper governance to create sustainable urban water systems.

The arid regions of Israel benefit from the innovative drip irrigation and satellite-guided farming systems, which achieve maximum water efficiency. The national irrigation advisory services of the country provide farmers with immediate guidance for scheduling their irrigation needs. Mobile-based platforms that use AI and IoT technology enable smallholder farmers in India to track soil moisture levels for better irrigation management that protects agricultural economics from water shortages (Ramanan et al., 2025).

Blockchain technology enables the management of water rights through its features, which include transparency, decentralization, and immutability. The distributed ledger system of blockchain technology enables transparent water allocation tracking, which minimizes disputes and maintains full accountability. Smart contracts operate automated systems that verify that water extraction stays within legal boundaries (Thuy et al., 2025). The Murray–Darling Basin in Australia runs blockchain-based pilot programs to monitor water rights trading, which enhances both trust levels and operational efficiency. Blockchain technology shows great potential, yet its implementation faces multiple barriers that include system scalability issues, regulatory hurdles, and difficulties with system integration (Zeng et al., 2023).

The Mekong, Nile, and Danube rivers need collaborative governance systems because they exist across international borders. Smart water technologies through shared data platforms and digital twins help countries operate reservoirs jointly while managing floods and distributing water resources fairly. The International Commission for the Protection of the Danube River enables European countries to share data across borders through its Danube River Basin program. The Mekong River Commission implements satellite data and modeling tools to enhance cooperation between its member states. The implemented programs show how technology can build trust between nations while fostering international cooperation (Gao et al., 2022).

7. Challenges and barriers to implementation

Many regions still utilize old water infrastructure that is not upgradable to newer smart technology. Retrofitting such a system is capital and technically expensive. IoT-based solutions also depend on consistent electricity and internet access, which is limited in poor countries (Rasheed & Tambe, 2024). The virtualization of water utilities exposes them to cybersecurity breaches, including SCADA hacking and manipulation of sensor data. Data security is another challenge, particularly if the data is provided by citizens. Another problem is interoperability among devices of different suppliers, requiring standardization of data protocols.

The investment in smart water solutions involves enormous initial costs in sensors, treatment plants, and data network facilities. LDCs (least developed countries) may experience financial hardships to fund such investments (Adhikari, 2022). PPPs (public–private partnerships), foreign aid, and new forms of financing such as green bonds help to alleviate these bottlenecks. Water management is frequently decentralized across various agencies with overlapping mandates. The legal frameworks may be unable to keep pace with new technologies that keep changing rapidly, such as blockchain and AI, and thus provide legal uncertainties (Akpobome, 2024). Institutional reforms, capacity building, and innovative policies can foster an enabling environment to trigger smart water solutions. Table 17.2 summarizes the key challenges and potential solutions in smart water implementation, offering insights into how barriers can be addressed effectively.

8. Roadmap for scaling smart water innovations

8.1 Policy recommendations and regulatory frameworks

Policies underpinning SWSs primarily focus on optimizing the use of limited water resources, mitigating environmental degradation, and fostering data-driven governance (Tosin Michael Olatunde et al., 2024). Beyond these goals, SWSs enhance environmental carrying capacity and contribute to sustainable economic development. Bouramdane (2023) highlighted that meaningful progress can be achieved through regulations promoting sustainable water practices, such as water-use quotas, pricing mechanisms, penalties for excessive consumption, and incentives for efficiency-enhancing technologies. Smart water management initiatives often collaborate with academic institutions, technology entrepreneurs, NGOs, and water utilities to pool expertise, resources, and funding for effective

Table 17.2 Challenges and potential solutions in smart water implementation.

Challenge category	Specific challenge	Potential solution	Notes/Examples	References
Technical	Aging infrastructure incompatible with smart technologies	Investment in IoT upgrades, sensor deployment	Retrofitting old pipelines with sensors; Singapore PUB smart network	Gupta et al. (2020)
Economic	High initial cost of smart water solutions	Public–private partnerships (PPP), green bonds, foreign aid	Funding for smart irrigation or ZLD facilities	Kühnert et al. (2021)
Regulatory	Decentralized water governance, outdated legal frameworks	Institutional reforms, policy updates, capacity building	Enable blockchain or AI implementation for water rights management	Akhmouch et al. (2020)
Data and privacy	Cybersecurity threats, citizen data privacy concerns	Data encryption, anonymization protocols, cybersecurity standards	Protect SCADA systems; secure crowdsourced data	US EPA (2025)
Social/ Capacity	Limited technical knowledge among staff or farmers	Training programs, workshops, public–private collaboration	Precision irrigation adoption in developing countries	Bhushan Mani et al. (2023)

solutions. Within the context of water scarcity, Lee et al. (2015) proposed recycling domestic wastewater effluent for industrial applications, thereby supporting zero-discharge objectives while reducing overall consumption. Moreover, ensuring access to both conventional and alternative water sources is essential to establishing a resilient water cycle. In urban areas characterized by high demand and intensive socioeconomic activities, the use of water sources with higher production costs may also represent a viable option. Finally, water supply assets and resource risk management form a core structural component of SWSs. Given that climate variability and land-use changes exert substantial impacts on water demand, these parameters must be integrated into risk management frameworks to ensure adaptive and effective strategies.

8.2 Multistakeholder engagement and capacity building

Stakeholder engagement is a defining feature of the shift from purely governmental control toward more inclusive models of water resource management, ownership, and administration that involve nongovernmental actors. It is widely recognized as a fundamental component of effective management. Facilitating and sustaining constructive dialogue encourages greater participation from both governmental and nongovernmental stakeholders, enabling them to engage in the challenging yet valuable process of mutual listening and learning (Hove et al., 2021). For example, the Muzza Canal in

Northern Italy illustrates how stakeholder engagement can support multifunctional water management amid shifting socio-economic demands. Following severe droughts, the formation of a water alliance (patto per l'acqua) in Lombardy brought together 64 diverse stakeholders, including authorities, unions, NGOs, user associations, and universities, to foster dialogue, balance competing water uses, and promote integrated governance. Innovative participatory methods, such as the "*dialogue between masks*," facilitated open expression, mutual understanding, and the transformation of conflicting demands into practical proposals, thereby strengthening trust and cooperation in water governance. Similarly, in the Odra River basin, stakeholder engagement has expanded through alliances such as the Environment and Agriculture network in Mecklenburg-Vorpommern, which unites agencies, farmers, consultants, and conservation groups to address nutrient pollution. Although the involvement of associations like Weltfriedensdienst has increased collaboration and allowed NGOs to participate in decision-making, participation remains uneven, with established water actors retaining greater influence. This process reflects a predominantly top-down, conservative governance approach that values continuity and established expertise over more ambitious or inclusive restructuring (Kochskämper et al., 2016).

8.3 Role of international organizations and donors

International organizations and global donors play a central role in the development and diffusion of SWSs, particularly in low- and middle-income countries. In Eastern Europe, the Caucasus, and Central Asia, international donors have acted as key agents of policy transfer, especially when recipient governments adopt donor-promoted reforms with the expectation of securing grant funding (Karini, 2025). Beyond direct financial support, international policy networks have significantly shaped water governance agendas. Since its establishment in 1996, the Global Water Partnership (GWP) has been particularly influential in promoting the conceptual framework of Integrated Water Resources Management (IWRM) and facilitating its global dissemination (Mardiste, 2024). The use of ICTs represents a more recent approach, closely aligned with ongoing technological advancements. Mvulirwenande and Wehn (2019, pp. 167–207) analyzed 24 ICT-focused water innovation partnerships (ICT-WIPs) in Africa to examine how SWS innovations can be advanced through collaboration. Their study provides two key insights. First, these partnerships enable the effective exchange of complementary resources and the cocreation of ICT-based water solutions in a cost-efficient and timely manner, although collaboration is often complicated by the diversity of partners. Second, the findings underscore the importance of both relational (trust-based) and structural (legally binding) governance mechanisms to minimize opportunism and clarify partners' roles and responsibilities. Overall, the research indicates that carefully designed and managed ICT-WIPs can enhance the capacity of developing countries to advance the SWS agenda.

8.4 Education, awareness, and community involvement

Several key factors are critical to the success of SWSs, with community awareness driven by strong educational initiatives being the most fundamental. The first step involves raising awareness of local water challenges and their associated health impacts. The second step focuses on educating communities about available options for ensuring safe water access. Together, awareness and education enable individuals to understand their challenges and develop context-specific solutions. The third step is to provide adequate resources to support water-related action. By combining awareness, education, and resources, water stakeholders, including households, can be empowered to actively participate in the safe and sustainable management of drinking water systems (Zimmermann, 2024). Successfully strengthening water user organizations (WUOs) in Thailand requires a comprehensive strategy integrating community engagement, institutional development, financial sustainability, technical expertise, and policy support (Chuenchum et al., 2024). This strategy involves raising public awareness through outreach and education programs, fostering inclusive decision-making platforms, and establishing conflict resolution mechanisms to ensure transparency and collaboration among stakeholders. Financial stability can be achieved by diversifying funding sources, implementing transparent budgeting, and optimizing resource allocation. Institutional capacity can be enhanced through leadership training, clear governance frameworks, and accountability measures, while technical expertise can be developed via specialized training and partnerships with academic institutions, NGOs, and experts. In parallel, engaging policymakers to advocate for supportive legal frameworks and providing legal education for WUO members will help navigate the policy landscape effectively. Finally, adopting adaptive water management plans and investing in resilient infrastructure will enable WUOs to address environmental and seasonal variability (Pot, 2023). Collectively, these measures can empower WUOs in Thailand to manage water resources more sustainably and equitably, contributing to long-term water security and improved governance.

9. Contribution of smart water solutions to SDGs

9.1 Direct impacts on SDG 6—Clean water and sanitation

IoT—IoT sensors, real-time monitoring, AI—AI analytics, digital twins, smart meters, and automated control systems were integrated into an innovative water solution, directly contributing to multiple targets under SDG 6. IoT-based sensors quantify, without interruption, main values such as pH, turbidity, chlorine dosage, and pollutants to assess water quality in real time. Alerts are instantly triggered if water quality falls below requirements, enabling fast replies and declining health risks. Communities, especially in remote or underserved areas, gain reliable access to potable water that meets World Health Organization (WHO) requirements (Addow & Jimale, 2023). And IoT technologies also supplied innovative solutions for sewer blockage management by facilitating observation of blockages and prohibiting SSOs (sanitary sewer overflows) with limited human participation. This system integrates 4G telemetry and low-energy-level sensors to evaluate sewer blockages with low computational expenses precisely (Faris et al., 2024).

An innovative water system concentrates on control and automated distribution, ensuring that water is distributed to all customers via Remote Supervision of Water Distribution. The managers can prioritize supply to scarce areas or reroute water dynamically during shortages, improving approachability. In particular, the combination of Android and IoT support for the water scarcity issue allows the manager to specify the specific location and schedule for water circulation. After these values are done, water is distributed as planned, and residents in the designated areas are informed of the distribution timetable through Short Message Service (Ramamoorthy et al., 2024). Moreover, the smart water solution abundantly contributed to SDG 6 by upgrading water quality and decreasing pollution through an advanced contamination surveillance and prevention system (Durgun, 2024); boosting water-use efficiency (smart leak detection and water loss reduction (Ali et al., 2022), as well as predictive demand planning and smart supply adjustment (Kavya et al., 2023); integrating water resource management (Slany et al., 2025); and conserving the water ecosystem (Jayanthi & Arun Kumar, 2024).

9.2 Indirect contributions to other SDGs

The combination of IoT, AI, and innovative management technologies has revolutionized smart water solutions and created a range of indirect impacts that advance many other SDGs, including (i) Zero Hunger (SDG 2) forward to the end hunger, food security, and sustainable agriculture: smart irrigation for yield enhancement and resource efficiency (Bwambale et al., 2022), treated wastewater reuse for sustainable agriculture (Köksal et al., 2025), and smart water observation for crop cycle optimization (Jamal et al., 2023); (ii) Industry, Innovation, and Infrastructure (SDG 9) build strong infrastructure, promoting sustainable industrialization processes, and enhancing innovation: IoT-enabled water networks and digital twin modernization (Mohammed et al., 2024) and intelligent water management for industrial processes (Kalyanam, 2023); (iii) Sustainable Cities and Communities (SDG 11) construct cities that are safe, resilient, and sustainable: data-driven water supply planning for droughts and floods (Liya et al., 2025), smart coordination of urban water infrastructure (Kuok et al., 2024), and intelligent rainwater harvesting and reuse management (Maldonado Benitez et al., 2025).

Besides, integrating IoT, AI, and innovative management technologies also plays a vital role in SDG 12 (responsible consumption and production) via demand management—wastewater decreasing—wastewater recycling; SDG 13 (climate action) such as sustainability, energy efficiency, and sustainable water cycle; and SDG 14 (life below water and life on land) through pollution prohibition, ecosystem conservation, and wastewater reuse. IoT, AI, and automated control systems upgrade water quality, optimize distribution, stop blockages, and conserve resources. They accelerate the achievement of SDG 6 by enabling water management that is safe, efficient, and sustainable. Besides SDG 6, smart water solutions also support others, including zero hunger (SDG 2), industry–innovation–infrastructure (SDG 9), sustainable cities and communities (SDG 11), responsible consumption and production (SDG 12), climate action (SDG 13), and ecosystem protection (SDG 14). These innovations are a cornerstone for advancing sustainable, resilient, and equitable water frameworks.

9.3 Indicators for measuring success and progress

To evaluate how smart water solutions contribute to the SDGs, crucial indicators should measure technical performance, governance effectiveness, sustainability impacts, and social equity (Kavya et al., 2023; Maldonado Benitez et al., 2025). The proportion of the population with access to safely managed water services, the percentage reduction in nonrevenue water, the rate of water reuse, and the level of compliance with established water quality standards are crucial factors for water access and efficiency (SDG 6). Innovation and governance (SDG 9) is evaluated through factors such as the adoption rate of smart water technologies, the number of open-access water data platforms, the degree of policy alignment with SDG targets, and the level of investment in research and development for smart water solutions. The lead time of drought and flood early warnings, the annual volume of groundwater recharge, the per capita water storage capacity, and the climate resilience index of water infrastructure are considered as signals for the assessment of climate resilience and sustainability (SDG 13). Social and equity outcomes (SDG 10–11) are assessed via key signals like the decrease in urban–rural water access gaps, the proportion of vulnerable populations, and the citizen satisfaction index with water services. In conclusion, these indices present a clear and all-encompassing framework for tracking how smart water solutions drive progress toward global water security, climate resilience, and equitable, sustainable development.

10. Conclusion and future outlook

Overall, to contribute to the long-term vision of next-generation smart water solutions for water security and achieving the SDGs, it is necessary to develop and implement policies that support investment in smart water infrastructure and boost water reuse. Propose IWRM frameworks that are aligned with SDGs and guarantee logic between national water strategies and global sustainability objectives. Advance the knowledge of policymakers, engineers, and communities through training in smart water technologies. Enlarge public awareness of water conservation issues and encourage the application of smart technologies to instigate a culture of sustainable water use. Collaboration among policymakers, research facilities, and businesses should be raised to accelerate changes in water resource management. Support the development of new financing mechanisms. For example, green bonds and climate funds will mobilize resources and build up smart water infrastructure projects. Set up open-access water data platforms to lift transparency, allow policy-making from evidence, and support research and revolution. Use AI and ML to facilitate predictive planning, optimize resource allocation, and improve system resilience. Declining inequalities in water access through the smart water solutions that are designed and deployed to benefit marginalized populations and rural communities. In summary, promoting sustainable water futures requires integrating policy innovation, administration capacity building, collaborative governance, intelligent data analytics, and inclusive implementation—enabling smart water solutions to be powerful catalysts for achieving global water security and the SDGs.

References

Abouchabaka, J., & Bentaleb, A. (2024). The impact of big data on the sustainable development goals water and energy. *E3S Web of Conferences, 477*, Article 00062. https://doi.org/10.1051/e3sconf/202447700062

Addow, M. A., & Jimale, A. D. (2023). IoT-based real-time water quality monitoring for sustainable water management: A case study in Somalia. *International Journal of Electrical and Electronics Engineering, 10*, 170–175. https://doi.org/10.14445/23488379/IJEEE-V10I8P116

Adhikari, R. (2022). Leveraging aid for trade to mobilize climate finance in the least developed countries. *Glob Policy, 13*, 547–553. https://doi.org/10.1111/1758-5899.13136

Ahmed, Z., Mohamed, K., Zeeshan, S., & Dong, X. (2020). Artificial intelligence with multi-functional machine learning platform development for better healthcare and precision medicine. *Database*. https://doi.org/10.1093/database/baaa010

Ajith, J. B., Manimegalai, R., & Ilayaraja, V. (2020). An IoT based smart water quality monitoring system using cloud. In *2020 international conference on emerging trends in information technology and engineering (Ic-ETITE)* (pp. 1–7). IEEE. https://doi.org/10.1109/ic-ETITE47903.2020.450

Akhmouch, A., Clavreul, D., Hendry, S., Megdal, S., Nickum, J., Nunes-Correia, F., & Ross, A. (2020). *OECD principles on water governance: From policy standards to practice*. Routledge.

Akpobome, O. (2024). The impact of emerging technologies on legal frameworks: A model for adaptive regulation. *International Journal of Research Publication and Reviews, 5*, 5046–5060. https://doi.org/10.55248/gengpi.5.1024.3012

Al Nahyan, M. T., Hawas, Y. E., Mohammad, M. S., & Basheerudeen, B. (2018). A decision-support system for identifying the best contractual delivery methods of mega infrastructure developments. In *ICEIS 2018—Proceedings of the 20th international conference on enterprise information systems* (pp. 407–414). SciTePress. https://doi.org/10.5220/0006694704070414

Alharbi, N., Althagafi, A., Alshomrani, O., Almotiry, A., & Alhazmi, S. (2021). A blockchain based secure IoT solution for water quality management. In *2021 international congress of advanced technology and engineering (ICOTEN)* (pp. 1–8). IEEE. https://doi.org/10.1109/ICOTEN52080.2021.9493474

Ali, A. S., Abdelmoez, M. N., Heshmat, M., & Ibrahim, K. (2022). A solution for water management and leakage detection problems using IoTs based approach. *Internet of Things, 18*, Article 100504. https://doi.org/10.1016/j.iot.2022.100504

Babuna, P., Yang, X., Tulcan, R. X. S., Dehui, B., Takase, M., Guba, B. Y., Han, C., Awudi, D. A., & Li, M. (2023). Modeling water inequality and water security: The role of water governance. *Journal of Environmental Management, 326*, Article 116815. https://doi.org/10.1016/j.jenvman.2022.116815

Balan, M. H., Camargo, L. A. S., Ramos, D. S., Castro, R., Leonel, L. D., Pulcherio, E. S., & Melendez, J. (2024). Hydro–solar hybrid plant operation in a hydropower plant cascade: Optimizing local and bulk system benefits. *Water (Basel), 16*, 2053. https://doi.org/10.3390/w16142053

Bejarano, G., Jain, M., Ramesh, A., Seetharam, A., & Mishra, A. (2018). Predictive analytics for smart water management in developing regions. In *2018 IEEE international conference on smart computing (SMARTCOMP)* (pp. 464–469). IEEE. https://doi.org/10.1109/SMARTCOMP.2018.00047

Bhushan, M., Kumar, P., & Bhusan Sahoo, B. (2023). Advancements in precision irrigation techniques for sustainable water management. *Farming for the Future Smart Agriculture Innovations.*

Boeuf, B., & Fritsch, O. (2016). Studying the implementation of the water framework directive in Europe: A meta-analysis of 89 journal articles. *Ecology and Society, 21*, art19. https://doi.org/10.5751/ES-08411-210219

Boulos, P. F. (2017). Smart water network modeling for sustainable and resilient infrastructure. *Water Resources Management, 31*, 3177–3188. https://doi.org/10.1007/s11269-017-1699-1

Bouramdane, A.-A. (2023). Optimal water management strategies: Paving the way for sustainability in smart cities. *Smart Cities, 6*, 2849–2882. https://doi.org/10.3390/smartcities6050128

Bwambale, E., Abagale, F. K., & Anornu, G. K. (2022). Smart irrigation monitoring and control strategies for improving water use efficiency in precision agriculture: A review. *Agricultural Water Management, 260*, Article 107324. https://doi.org/10.1016/j.agwat.2021.107324

Calatayud Asensi, C., Martínez, J. V. B., Arnau Muñoz, L., & Maciá Pérez, F. (2025). Predictive optimization of energy consumption in drinking water systems using parallel algorithms. *IEEE Access, 13*, 156824–156838. https://doi.org/10.1109/ACCESS.2025.3606573

Carroll, J. M., Beck, J., Boyer, E. W., Dhanorkar, S., & Gupta, S. (2019). Empowering community water data stakeholders. *Interactive Computing, 31*, 492–506. https://doi.org/10.1093/iwcomp/iwz032

Chaplin-Kramer, R., Sharp, R. P., Weil, C., Bennett, E. M., Pascual, U., Arkema, K. K., Brauman, K. A., Bryant, B. P., Guerry, A. D., Haddad, N. M., Hamann, M., Hamel, P., Johnson, J. A., Mandle, L., Pereira, H. M., Polasky, S., Ruckelshaus, M., Shaw, M. R., Silver, J. M., Vogl, A. L., & Daily, G. C. (2019). Global modeling of nature's contributions to people. *Science, 366*, 255–258. https://doi.org/10.1126/science.aaw3372

Chen, O., Mustafee, N., Evans, B., Khoury, M., Vamvakeridou-Lyroudia, L., Chen, A. S., Djordjević, S., & Savić, D. (2025). Supporting decision-making for industrial symbioses using a hybrid modelling approach and its application to wastewater treatment. *Water Science and Technology, 91*, 501–523. https://doi.org/10.2166/wst.2025.022

Chin, R. J., & Lai, S. H. (2025). Editorial: Integration of hydrological models and machine learning techniques for water resources management. *Journal of Civil Engineering, Science and Technology, 16*, 1–5. https://doi.org/10.33736/jcest.9191.2025

Chuenchum, P., Meneesrikum, C., Teerapanuchaikul, C., & Sriariyawat, A. (2024). Community participation and effective water management: A study on water user organizations (WUOs) in Thailand. *World Development Perspectives, 34*, Article 100589. https://doi.org/10.1016/j.wdp.2024.100589

Darlington, E. E., & Peter, S. (2024). The intersection of renewable energy and environmental health: Advancements in sustainable solutions. *International Journal of Applied Research in Social Sciences, 6*, 1103–1132. https://doi.org/10.51594/ijarss.v6i6.1175

Demetillo, A. T., Japitana, M. V., & Taboada, E. B. (2019). A system for monitoring water quality in a large aquatic area using wireless sensor network technology. *Sustainable Environment Research, 29*, 12. https://doi.org/10.1186/s42834-019-0009-4

Durgun, Y. (2024). Real-time water quality monitoring using AI-enabled sensors: Detection of contaminants and UV disinfection analysis in smart urban water systems. *Journal of King Saud University Science, 36*, Article 103409. https://doi.org/10.1016/j.jksus.2024.103409

Farah, E., & Shahrour, I. (2017). Leakage detection using smart water system: Combination of water balance and automated minimum night flow. *Water Resources Management, 31*, 4821–4833. https://doi.org/10.1007/s11269-017-1780-9

Faris, N., Zayed, T., Aghdam, E., Fares, A., & Alshami, A. (2024). Real-time sanitary sewer blockage detection system using IoT. *Measurement, 226*, Article 114146. https://doi.org/10.1016/j.measurement.2024.114146

Frijns, J., Smith, H. M., & Makropoulos, C. (2024). Enabling the uptake of circular water solutions. *Water Policy, 26*, 94–110. https://doi.org/10.2166/wp.2024.167

Gao, J., Castelletti, A., Burlado, P., Wang, H., & Zhao, J. (2022). Soft-cooperation via data sharing eases transboundary conflicts in the Lancang-Mekong river basin. *Journal of Hydrology (Amsterdam), 606*, Article 127464. https://doi.org/10.1016/j.jhydrol.2022.127464

Gevorkov, L., Domínguez-García, J. L., & Trilla, L. (2025). The synergy of renewable energy and desalination: An overview of current practices and future directions. *Applied Sciences, 15*, 1794. https://doi.org/10.3390/app15041794

Ghilan, A., Afou, Y. E. L., Merras, M., & El Akkad, N. (2024). Data-driven precision agriculture advanced irrigation system for sustainable smart farming. In *2024 4th international conference on innovative research in applied science, engineering and technology (IRASET)* (pp. 01–06). IEEE. https://doi.org/10.1109/IRASET60544.2024.10548267

Gupta, S., Kohli, M., Kumar, R., & Bandral, S. (2021). IoT based underwater robot for water quality monitoring. *IOP Conference Series: Materials Science and Engineering, 1033*, Article 012013. https://doi.org/10.1088/1757-899X/1033/1/012013

Gupta, A. D., & Kulat, K. (2018). Leakage reduction in water distribution system using efficient pressure management techniques. Case study: Nagpur, India. *Water Supply, 18*, 2015–2027. https://doi.org/10.2166/ws.2018.023

Gupta, A. D., Pandey, P., Feijóo, A., Yaseen, Z. M., & Bokde, N. D. (2020). Smart water technology for efficient water resource management: A review. *Energies, 13*, 6268. https://doi.org/10.3390/en13236268

Hamam, M., Spina, D., Raimondo, M., Di Vita, G., Zanchini, R., Chinnici, G., Tóth, J., & D'Amico, M. (2023). Industrial symbiosis and agri-food system: Themes, links, and relationships. *Frontiers in Sustainable Food Systems, 6*. https://doi.org/10.3389/fsufs.2022.1012436

Hamid, S. A., Rahim, A. M. A., Fadhlullah, S. Y., Abdullah, S., Muhammad, Z., & Leh, N. A. M. (2020). IoT based water quality monitoring system and evaluation. In *2020 10th IEEE international conference on control system, computing and engineering (ICCSCE)* (pp. 102–106). IEEE. https://doi.org/10.1109/ICCSCE50387.2020.9204931

Hannah, L., & Tan, T. P. (2017). Singapore's experience with reclaimed water: NEWater. In *Water reuse policies for potable use*.

Hove, J., D'Ambruoso, L., Twine, R., Mabetha, D., van der Merwe, M., Mtungwa, I., Khoza, S., Kahn, K., & Witter, S. (2021). Developing stakeholder participation to address lack of safe water as a community health concern in a rural province in South Africa. *Global Health Action, 14*. https://doi.org/10.1080/16549716.2021.1973715

Islam, F. A. S. (2025). A multi-dimensional AI framework for sustainable drinking water management: Integrating federated learning, digital twins, and blockchain. *Journal of Engineering Research and Reports, 27*, 466–492. https://doi.org/10.9734/jerr/2025/v27i61558

Jacquin, C., Merodio Gómez, P., Arriaga, V., & Santiago, A. (2023). Citizen science interactions with official geospatial information; case studies from Mexico. *Frontiers in Environmental Science, 10*. https://doi.org/10.3389/fenvs.2022.800247

Jamal, A., Cai, X., Qiao, X., Garcia, L., Wang, J., Amori, A., & Yang, H. (2023). Real-time irrigation scheduling based on weather forecasts, field observations, and human-machine interactions. *Water Resources Research, 59*. https://doi.org/10.1029/2023WR035810

Jayanthi, J., & Arun Kumar, K. (2024). AI-driven restoration: Enhancing biodiversity conservation and ecosystem resilience. In *Explainable AI (XAI) for sustainable development* (pp. 180–193). Chapman and Hall/CRC.

Köksal, D. D., Ahi, Y., & Todorovic, M. (2025). Assessing agricultural reuse potential of treated wastewater: A hybrid machine learning approach. *Agronomy, 15*, 703. https://doi.org/10.3390/agronomy15030703

Kühnert, C., Gonuguntla, N. M., Krieg, H., Nowak, D., & Thomas, J. A. (2021). Application of LSTM networks for water demand prediction in optimal pump control. *Water (Basel), 13*, 644. https://doi.org/10.3390/w13050644

Kalyanam, A. K. (2023). Water management and its industrial impact (A comprehensive overview of water management and the role of IoT). *Journal of Artificial Intelligence, Machine Learning and Data Science, 2023*, 1984–1989. https://doi.org/10.51219/JAIMLD/ashok-kumar

Karini, A. (2025). Policy learning for national advantage and sovereign governance in non-western contexts: Comparing public management reforms in Azerbaijan, Malaysia, and Kazakhstan. *Journal of Public Affairs, 25*. https://doi.org/10.1002/pa.2959

Kavya, M., Mathew, A., Shekar, P. R., & Sarwesh, P. (2023). Short term water demand forecast modelling using artificial intelligence for smart water management. *Sustainable Cities and Society, 95*, Article 104610. https://doi.org/10.1016/j.scs.2023.104610

Khaspuria, G., Khandelwal, A., Agarwal, M., Bafna, M., Yadav, R., & Yadav, A. (2024). Adoption of precision agriculture technologies among farmers: A comprehensive review. *Journal of Scientific Research and Reports, 30*, 671–686. https://doi.org/10.9734/jsrr/2024/v30i72180

Kochskämper, E., Challies, E., Newig, J., & Jager, N. W. (2016). Participation for effective environmental governance? Evidence from water framework directive implementation in Germany, Spain and the United Kingdom. *Journal of Environmental Management, 181*, 737–748. https://doi.org/10.1016/j.jenvman.2016.08.007

Kuok, K. K., Chiu, P. C., Bakri, M. K. B., Rahman, MdR., & Yun, C. M. (2024). Industrial revolution 4.0 in water supply, wastewater and stormwater management: Opportunities, challenges, and impacts. *Environmental Technology Reviews, 13*, 143–167. https://doi.org/10.1080/21622515.2023.2291422

Lal, K., Menon, S., Noble, F., & Arif, K. M. (2024). Low-cost IoT based system for lake water quality monitoring. *PLoS One, 19*, Article e0299089. https://doi.org/10.1371/journal.pone.0299089

Lee, S. W., Sarp, S., Jeon, D. J., & Kim, J. H. (2015). Smart water grid: The future water management platform. *Desalination and Water Treatment, 55*, 339–346. https://doi.org/10.1080/19443994.2014.917887

Li, J., Yang, X., & Sitzenfrei, R. (2020). Rethinking the framework of smart water system: A review. *Water (Basel), 12*, 412. https://doi.org/10.3390/w12020412

Liu, X., Tian, Y., Lei, X., Wang, H., Liu, Z., & Wang, J. (2019). An improved self-adaptive grey wolf optimizer for the daily optimal operation of cascade pumping stations. *Applied Soft Computing, 75*, 473–493. https://doi.org/10.1016/j.asoc.2018.11.039

Liya, A., Stantec, R., & Carolina, N. (2025). *The growing role of artificial intelligence in tomorrow's urban hydrological infrastructure*.

Makarigakis, A. K., & Jimenez-Cisneros, B. E. (2019). UNESCO's contribution to face global water challenges. *Water (Basel), 11*, 388. https://doi.org/10.3390/w11020388

Maldonado Benitez, V. M., Morales Matamoros, O., & Moreno Escobar, J. J. (2025). Towards resilient cities: Systematic review of the literature on the sue of AI to optimize water harvesting and mitigate scarcity. *Water (Basel), 17*, 1978. https://doi.org/10.3390/w17131978

Mandal, S., Yadav, A., Panwar, R., Kumar, S. S., Karthick, A., Priya, A., Vijayakumar, R., & Ganesh, S. S. (2025). Smart water management for SDG 6: A review of AI and IoT-Enabled solutions. *Water Conservation Science and Engineering, 10*, 83. https://doi.org/10.1007/s41101-025-00407-7

Manikandan, G., Bhuvaneswari, G., & Joel, M. R. (2023). Artificial intelligence to the assessment, monitoring, and forecasting of drought in developing countries. In *2023 international conference on circuit power and computing technologies (ICCPCT)* (pp. 886–892). IEEE. https://doi.org/10.1109/ICCPCT58313.2023.10245072

Mardiste, P. (2024). International donors as agents of policy transfer in influencing water legislation: The adoption of the river basin principle by transition economies in Eastern Europe, Caucasus and Central Asia. *Environmental Science & Policy, 157*, Article 103766. https://doi.org/10.1016/j.envsci.2024.103766

Mizik, T. (2023). How can precision farming work on a small scale? A systematic literature review. *Precision Agriculture, 24*, 384–406. https://doi.org/10.1007/s11119-022-09934-y

Mohammed, M. A., Lakhan, A., Abdulkareem, K. H., Abd Ghani, M. K., Marhoon, H. A., Kadry, S., Nedoma, J., Martinek, R., & Zapirain, B. G. (2024). Industrial internet of water things architecture for data standarization based on blockchain and digital twin technology. *Journal of Advanced Research, 66*, 1–14. https://doi.org/10.1016/j.jare.2023.10.005

Mohan, S., Oke, N., & Gokul, D. (2021). Conventional and zero liquid discharge treatment plants for textile wastewater through the lens of carbon footprint analysis. *Journal of Water and Climate Change, 12*, 1392–1403. https://doi.org/10.2166/wcc.2020.100

Mujtaba, G., Shah, M. U. H., Hai, A., Daud, M., & Hayat, M. (2024). A holistic approach to embracing the United Nation's sustainable development goal (SDG-6) towards water security in Pakistan. *Journal of Water Process Engineering, 57*, Article 104691. https://doi.org/10.1016/j.jwpe.2023.104691

Mvulirwenande, S., & Wehn, U. (2019). *Promoting smart water systems in developing countries through innovation partnerships: Evidence from VIA water-supported projects in Africa* (pp. 167–207). https://doi.org/10.1007/698_2019_422

Oyounalsoud, M. S., Yilmaz, A. G., Abdallah, M., & Abdeljaber, A. (2024). Drought prediction using artificial intelligence models based on climate data and soil moisture. *Scientific Reports, 14*, Article 19700. https://doi.org/10.1038/s41598-024-70406-6

Pot, W. (2023). Deciding for resilience: Utilizing water infrastructure investments to prepare for the future. *WIREs Water, 10*. https://doi.org/10.1002/wat2.1661

Ramírez-Cuesta, J. M., Mirás-Avalos, J. M., Rubio-Asensio, J. S., & Intrigliolo, D. S. (2018). A novel ArcGIS toolbox for estimating crop water demands by integrating the dual crop coefficient approach with multi-satellite imagery. *Water (Basel), 11*, 38. https://doi.org/10.3390/w11010038

Ramamoorthy, R., Manasa, S. M., & Smitha, J. A. (2024). Android app-oriented smart supervision of water distribution using internet of things. *Information Systems Engineering and Management*, 223–237. https://doi.org/10.1007/978-3-031-61475-0_18

Ramanan, S. J., Ajai Krishna, T. S., Kavishvaran, S., & Velvizhy, P. (2025). Smart agriculture: IoT-driven soil moisture monitoring system for enhanced crop management. In *2025 international conference on inventive computation technologies (ICICT)* (pp. 1851–1856). IEEE. https://doi.org/10.1109/ICICT64420.2025.11005293

Rasheed, D. H., & Tambe, S. B. (2024). Advancing energy efficiency with smart grids and IoT-based solutions for a sustainable future. *ESTIDAMAA*, 36–42. https://doi.org/10.70470/ESTIDAMAA/2024/006

Sadeghioon, A. M., Metje, N., Chapman, D., & Anthony, C. (2018). Water pipeline failure detection using distributed relative pressure and temperature measurements and anomaly detection algorithms. *Urban Water Journal, 15*, 287–295. https://doi.org/10.1080/1573062X.2018.1424213

Scorzato, L. (2024). Reliability and interpretability in science and deep learning. *Minds and Machines, 34*, 27. https://doi.org/10.1007/s11023-024-09682-0

Serrano, J., Shahidian, S., & Marques da Silva, J. (2018). Monitoring seasonal pasture quality degradation in the mediterranean Montado ecosystem: Proximal versus remote sensing. *Water (Basel), 10*, 1422. https://doi.org/10.3390/w10101422

Singh, M., & Ahmed, S. (2021). IoT based smart water management systems: A systematic review. *Materials Today Proceedings, 46*, 5211–5218. https://doi.org/10.1016/j.matpr.2020.08.588

Slany, V., Krcalova, E., Balej, J., Zach, M., Kucova, T., Prauzek, M., & Martinek, R. (2025). Smart water-IoT: Harnessing IoT and AI for efficient water management. *ACM Computing Surveys, 57*, 1–36. https://doi.org/10.1145/3744338

Solanki, R. K., Rajawat, A. S., Goyal, S. B., & Meesala, S. K. (2025). Advanced water resource management using IoT and big data analytics. *Journal of Computers, Mechanical and Management, 4*, 19–26. https://doi.org/10.57159/jcmm.4.3.25198

Thuy, T. L., Nguyen, M. K., Bui, T. D., Yen, H. P. H., Hoai, N. T., Ngan, N. V. C., Khedulkhar, A. P., Van, D. P., Halog, A., & Hoang, T. D. (2025). Blockchain-enabled water quality monitoring: A comprehensive review of digital innovations and challenges. *Water (Basel), 17*.

Tinelli, S., & Juran, I. (2019). Artificial intelligence-based monitoring system of water quality parameters for early detection of non-specific bio-contamination in water distribution systems. *Water Supply, 19*, 1785–1792. https://doi.org/10.2166/ws.2019.057

Tosin Michael Olatunde, Adeshina Adelani, F., & Sikhakhane, Z. Q. (2024). A review of smart water management systems from Africa and the United States. *Engineering Science & Technology Journal, 5*, 1231–1242. https://doi.org/10.51594/estj.v5i4.1014

US EPA. (2025). *EPA cybersecurity for the water sector.*

Vörösmarty, C. J., Rodríguez Osuna, V., Cak, A. D., Bhaduri, A., Bunn, S. E., Corsi, F., Gastelumendi, J., Green, P., Harrison, I., Lawford, R., Marcotullio, P. J., McClain, M., McDonald, R., McIntyre, P., Palmer, M., Robarts, R. D., Szöllösi-Nagy, A., Tessler, Z., & Uhlenbrook, S. (2018). Ecosystem-based water security and the sustainable development goals (SDGs). *Ecohydrology and Hydrobiology, 18*, 317–333. https://doi.org/10.1016/j.ecohyd.2018.07.004

Vadapalli, J., Gupta, S., Karki, B., & Tsai, C.-H. (2024). Incorporating citizen-generated data into large language models. In *Proceedings of the 25th annual international conference on digital government research* (pp. 1023–1025). New York, NY, USA: ACM. https://doi.org/10.1145/3657054.3659119

Vinayagam, V., Sikarwar, D., Das, S., & Pugazhendhi, A. (2024). Envisioning the innovative approaches to achieve circular economy in the water and wastewater sector. *Environmental Research, 241*, Article 117663. https://doi.org/10.1016/j.envres.2023.117663

Wang, M., & Cheng, J. C. P. (2020). A unified convolutional neural network integrated with conditional random field for pipe defect segmentation. *Computer-Aided Civil and Infrastructure Engineering, 35*, 162–177. https://doi.org/10.1111/mice.12481

Wu, X., Zhang, Z., Xiong, S., Zhang, W., Tang, J., Li, Z., An, B., & Li, R. (2023). A near-real-time flood detection method based on deep learning and SAR images. *Remote Sensing, 15*, 2046. https://doi.org/10.3390/rs15082046

Yang, F., Wen, X., Aziz, A., & Luhach, A. K. (2021). The need for local adaptation of smart infrastructure for sustainable economic management. *Environmental Impact Assessment Review, 88*, Article 106565. https://doi.org/10.1016/j.eiar.2021.106565

Yu, Y., Chen, X., Huttner, P., Hinnenthal, M., Brieden, A., Sun, L., & Disse, M. (2018). Model based decision support system for land use changes and socio-economic assessments. *Journal of Arid Land, 10*, 169–182. https://doi.org/10.1007/s40333-018-0091-1

Zeng, H., Dhiman, G., Sharma, A., Sharma, A., & Tselykh, A. (2023). An IoT and Blockchain-based approach for the smart water management system in agriculture. *Expert Systems, 40*. https://doi.org/10.1111/exsy.12892

Zhang, N., Khu, S.-T., & Wang, J. (2024). A conceptual method to evaluate the smart water system based on fit-for-purpose concept. *Environmental Impact Assessment Review, 105*, Article 107447. https://doi.org/10.1016/j.eiar.2024.107447

Zhang, Z., Wang, D., Mei, Y., Zhu, J., & Xiao, X. (2025). Developing an explainable deep learning module based on the LSTM framework for flood prediction. *Frontiers in Water, 7*. https://doi.org/10.3389/frwa.2025.1562842

Zhao, G., Liu, R., Yang, M., Tu, T., Ma, M., Hong, Y., & Wang, X. (2022). Large-scale flash flood warning in China using deep learning. *Journal of Hydrology (Amsterdam), 604*, Article 127222. https://doi.org/10.1016/j.jhydrol.2021.127222

Zhao, T., Song, C., Yu, J., Xing, L., Xu, F., Li, W., & Wang, Z. (2025). Leveraging immersive digital twins and AI-driven decision support systems for sustainable water reserves management: A conceptual framework. *Sustainability, 17*, 3754. https://doi.org/10.3390/su17083754

Zimmermann, K. (2024). End-user participation in drinking water management through awareness, education, and resources: Tools for water partnerships. *Water Policy, 26*, 1222–1233. https://doi.org/10.2166/wp.2024.045

CHAPTER

18

Long-term variability of rainfall and temperature-driven drought in Chhattisgarh's agro-climatic zones for sustainable water management

Surendra Kumar Chandniha[1], Ramchandra Taram[2], Diwakar Naidu[2], Arun Mondal[3], Sananda Kundu[4], Anupam Panday[3] and Rupanshu Gupta[5]

[1]*Department of Soil and Water Conservation Engineering, BRSM College of Agricultural Engineering and Technology & Research Station, IGKV, Mungeli, Chhattisgarh, India;* [2]*Faculty of Agricultural Engineering, SV College of Agricultural Engineering and Technology & Research Station, IGKV, Raipur, Chhattisgarh, India;* [3]*Department of Geography, University of Allahabad, Prayagraj, Uttar Pradesh, India;* [4]*Department of Geography, Manipur University, Imphal, Manipur, India;* [5]*Department of Civil Engineering, National Institute of Technology, Raipur, Chhattisgarh, India*

1. Introduction

Water is one of the most essential inputs for crop production. It plays a vital role in regulating photosynthesis, nutrient absorption, nutrient utilisation, respiration, translocation, and cell division, along with several other physiological processes. India is an agrarian economy where more than 50% of people depend on agriculture and the allied sector for their livelihood. An English speaker is saying that Indian agriculture is a gamble of the monsoon, which means the agriculture in India totally depends on the monsoon rainfall. Agriculture accounts for approximately 58% of total employment in our country. In addition, our country's agricultural sector contributes to about 18% of GDP. As per the first climate change assessment report for the Indian region, 6% decrease was noticed in the summer monsoon over the Indo Gangetic plains, whereas the frequency of localized high rainfall events increased from 1951 to 2015. The occurrence of daily precipitation extremes exceeding 150 mm per day increased by nearly 75% over central India between 1950 and 2015. India's temperature has risen by 0.7°C from 1901 to 2018 (Krishnan et al., 2019). Various studies have been done all around the globe to understand the spatiotemporal variability of rainfall and drought occurrence. The 135 years of monthly rainfall time series for 30 sub-regions over India were analyzed by Kumar et al. (2010) and summarized the long-term rainfall variability using trend analysis. Prabhakar et al. (2019) found a declining trend of rainfall over the state of Odisha after 1945. The rainfall variability may have a hazardous impact on crops. Ahmed et al. (2019) assessed the variability of rainfall on groundnut yield in the Bundelkhand region of India. Similar studies on rainfall variability assessment have been done

Sustainable Water Management. https://doi.org/10.1016/B978-0-443-44853-9.00013-7

over Chhattisgarh state. Nema et al. (2018) have noticed strong interannual rainfall variability and a declining trend of rainfall over Chhattisgarh. District-wise changes in rainfall variability and trend were noticed in Chhattisgarh (Praveen et al., 2020).

Characterization of drought using SPI and SPEI was done in various parts of the world, i.e., Turkey (Mehr & Vaheddoost, 2020), China (Li et al., 2021), Iran (Mesbahzadeh et al., 2020) and Germany (Bachmair et al., 2016). In the Indian context, drought characterization has been carried out using several indices, among which the SPI, SPEI and evaporation index are most commonly used to assess drought trends and variability (Mallya et al., 2016). The drought frequency of India was found to be increasing in recent decades, and a significant variability is expected in the future precipitation and temperature distribution, which may further influence future drought conditions (Aadhar & Mishra, 2018; Mukhopadhyay et al., 2014). According to Krishnan et al. (2019) SPEI is better suited for the assessment of the influence of increasing temperature on drought occurrence. Singh and Shukla (2020), found a correlation between ENSO episodes and drought events using SPI and SPEI indices. A multimodal approach has been adopted by Saharwardi and Kumar to ascertain the future changes in drought and associated uncertainties. They analyzed drought characteristics of future climate using the SPI and SPEI indices and found that drought frequency and drought intensity have increased in recent decades. SPI and some other indices were used by Swain et al. (2017) to assess drought over Marathwada division of Maharashtra using 3-hourly gridded rainfall data. Drought conditions were projected for 21st century using precipitation and temperature data obtained from Regional Climate models by Gupta and Jain (2018). They calculated SPI and SPEI at 12-month time scale for drought characterization and found that in far future, most parts of India, except few southern states are expected to face an increase in drought severity and frequency, and the extent of area under drought conditions was found to increase. Earlier attempts have been made in the context of drought characterization over Chhattisgarh. Forty-three years of rainfall data were used by Bhelawe et al. (2015) to characterize meteorological drought in Raipur district of Chhattisgarh. Drought frequency, intensity and severity have been analyzed in Chhattisgarh along with other states in central India like Madhya Pradesh, Maharashtra and the southern parts of Uttar Pradesh by calculating percentage departure of rainfall from normal value (Amrit et al., 2018).

SPI and SPEI are two common drought assessment indices used worldwide. SPEI takes into account temperature along with precipitation, while SPI considers precipitation only. In this context, the difference between these indices needs to be addressed for applying in various conditions. SPI and SPEI indices calculated for 1, 3, 6 and 12-month time scale was compared in terms of their consistency and applicability over Mongolia regions of China (Pei et al., 2020). They found out that, at the shortest time scale, the difference between SPI and SPEI was greater, while at longer time scales, the drought reflected by SPI and SPEI was consistent. The magnitude of drought intensity predicted using both indices showed some difference, i.e., as per the drought conditions predicted for future, around 60% mild and 6% severe drought will occur as per SPI, but as per SPEI, 64% mild and 5% severe drought was predicted (Bera et al., 2021). By considering the temperature, SPEI can assess the influence of global warming on drought indices to some extent (Mathbout et al., 2018; Tefera et al., 2019).

Drought characterization is important in the context of agricultural and water resource planning (Khare et al., 2017; Mondal et al., 2025; Pandey et al., 2023). Various studies have been conducted all around the world in this context using various methods and indices (Kundu et al., 2017; Mondal et al., 2016, 2017, 2022). Some of the common indices used widely are Standardized Precipitation Index (SPI) and Standardized Precipitation and Evapotranspiration Index (SPEI). Standardized precipitation has been developed by McKee et al. (1993) based on precipitation data. It is suitable for drought characterization over various time scales. The standardized precipitation and evapotranspiration index has been proposed by Vicente-Serrano et al. (2010) based on precipitation and temperature. Both these indices can be used to characterize drought intensity, frequency and drought severity. The changes in climatic and hydrological processes, such as rising temperature, dropping the ratio of snowfall to total precipitation, and falling runoff in the basins, have increased the adverse effects of drought. Hence, regular monitoring and assessment of drought conditions is to be done for identifying areas susceptible to drought, drought preparedness, drought warning and agricultural planning (Mondal et al., 2012, 2015). Nowadays, due to climate variability, the rainfall distribution is getting unpredictable; in this condition the drought characterization is helpful in adopting contingency plans toward sustainable water management for future prospects.

2. Study area

Chhattisgarh state spreads over about 13.5 million hectares in central India. As a state of plateau region, it experiences a very high variety and diversity of climatic conditions. Geographically, it is located between 17° 46′ and 24° 5′ N, of northern latitudes and 80° 15′ and 84° 20′ E, of the eastern longitudes. The entire state can be divided into three different agro-climatic zones, such as Chhattisgarh plains (50.52%), Bastar plateau (28.62%) and Northern hills (20.86%). A large variation in topography can be seen in the state. The average rainfall for the state is 1200–1600 mm, but it shows a large spatial variability. High variation of total precipitation is observed in the Chhattisgarh plains (1000–1200 mm), Bastar plateau (1200–1400 mm) and northern hills (1400–1600 mm).

2.1 Salient feature

The geographical area of Chhattisgarh is approximately 135,192 km^2, with a 44% of forest area, and it has contributed approximately 12% of the forest to the Indian subcontinent. Chhattisgarh has been classified as one of India's biodiversity-rich states with densest forests, numerous wildlife and most significantly over 200 non-timber forest products with enormous value-adding potential (Kansal et al., 2015). Chhattisgarh State has been blessed with natural resources. Chhattisgarh is also known as mono-cropped state. The most popular foods consumed by tribes are rice and rice byproducts (Ramrao et al., 2005). The climate in Chhattisgarh is tropical, with significant regional variation. In the summer, temperatures reach up to 46°C; in general, temperatures range from 28°C to 46°C (Kansal et al., 2015).

In this study, Chhattisgarh state is considered as a study area, which is situated in the central zone of India. Chhattisgarh State lies between 17° 46′ and 24° 5′ N, of northern latitudes and 80° 15′ and 84° 20′ E, of the eastern longitudes (Fig. 18.1). Chhattisgarh State is also known as "rice bowl of India" due to high production of rice and favorable climatic conditions. The state is surrounded by six other states, i.e., Madhya Pradesh, Jharkhand, Uttar Pradesh, Orissa, Maharastra and Andhra Pradesh. There are four seasons defined by IMD, i.e., Winter (December–February), Pre-Monsoon (March–May), Monsoon (June–September) and Post-Monsoon (October–November). Chhattisgarh state is divided into 27 Districts and 149 Blocks. Annual mean precipitation of the state is about 1400 mm.

A Digital Elevation Model (DEM) is a digital presentation of earth surface. Sometimes, it includes (Digital Surface Model) the elevation of vegetation, buildings and other surface objects. Sometimes, it

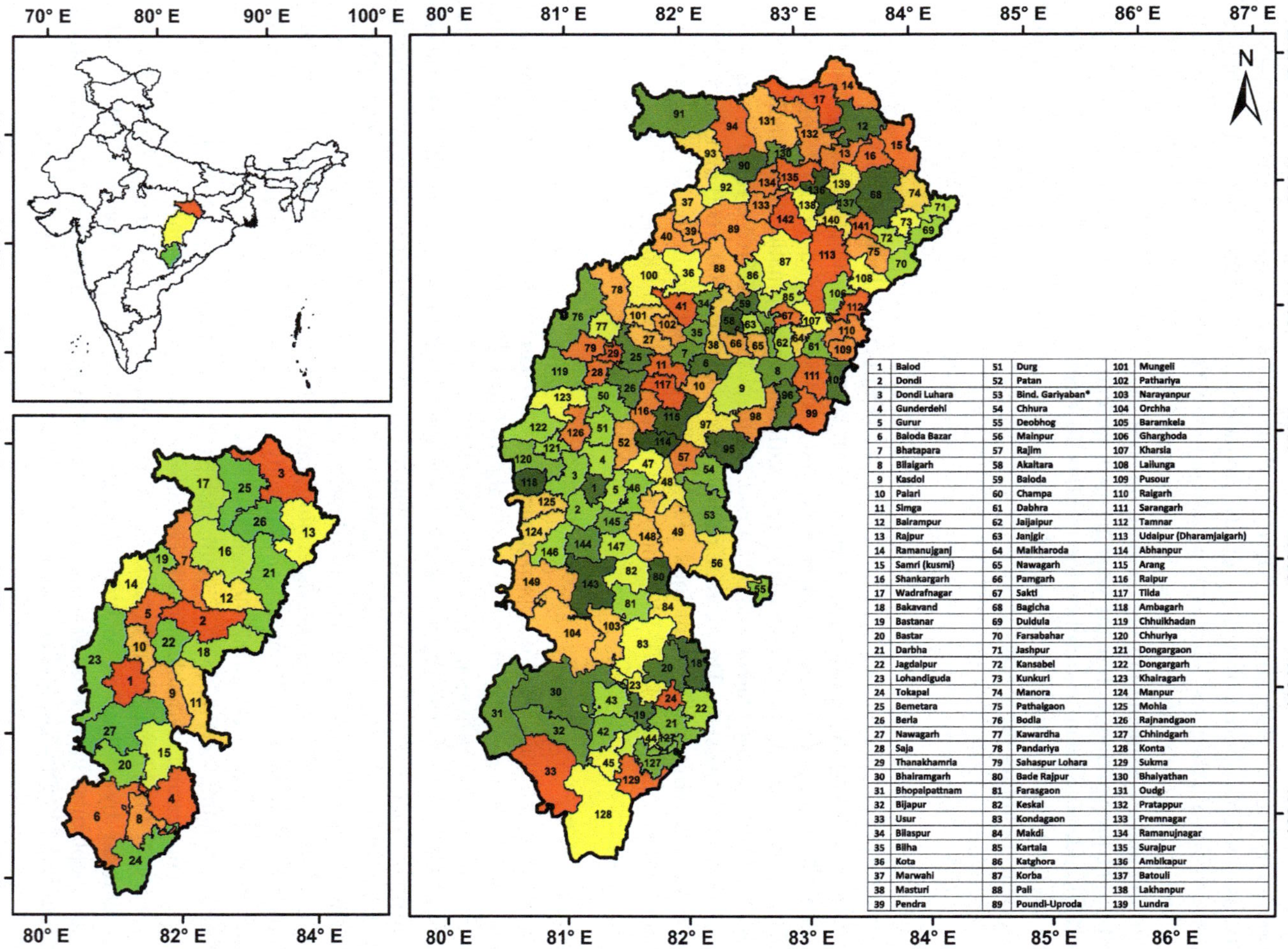

1	Balod	51	Durg	101	Mungeli
2	Dondi	52	Patan	102	Pathariya
3	Dondi Luhara	53	Bind. Gariyaban*	103	Narayanpur
4	Gunderdehi	54	Chhura	104	Orchha
5	Gurur	55	Deobhog	105	Baramkela
6	Baloda Bazar	56	Mainpur	106	Gharghoda
7	Bhatapara	57	Rajim	107	Kharsia
8	Bilaigarh	58	Akaltara	108	Lailunga
9	Kasdol	59	Baloda	109	Pusour
10	Palari	60	Champa	110	Raigarh
11	Simga	61	Dabhra	111	Sarangarh
12	Balrampur	62	Jaijaipur	112	Tamnar
13	Rajpur	63	Janjgir	113	Udaipur (Dharamjaigarh)
14	Ramanujganj	64	Malkharoda	114	Abhanpur
15	Samri (kusmi)	65	Nawagarh	115	Arang
16	Shankargarh	66	Pamgarh	116	Raipur
17	Wadrafnagar	67	Sakti	117	Tilda
18	Bakavand	68	Bagicha	118	Ambagarh
19	Bastanar	69	Duldula	119	Chhuikhadan
20	Bastar	70	Farsabahar	120	Chhuriya
21	Darbha	71	Jashpur	121	Dongargaon
22	Jagdalpur	72	Kansabel	122	Dongargarh
23	Lohandiguda	73	Kunkuri	123	Khairagarh
24	Tokapal	74	Manora	124	Manpur
25	Bemetara	75	Pathalgaon	125	Mohla
26	Berla	76	Bodla	126	Rajnandgaon
27	Nawagarh	77	Kawardha	127	Chhindgarh
28	Saja	78	Pandariya	128	Konta
29	Thanakhamria	79	Sahaspur Lohara	129	Sukma
30	Bhairamgarh	80	Bade Rajpur	130	Bhaiyathan
31	Bhopalpattnam	81	Farasgaon	131	Oudgi
32	Bijapur	82	Keskal	132	Pratappur
33	Usur	83	Kondagaon	133	Premnagar
34	Bilaspur	84	Makdi	134	Ramanujnagar
35	Bilha	85	Kartala	135	Surajpur
36	Kota	86	Katghora	136	Ambikapur
37	Marwahi	87	Korba	137	Batouli
38	Masturi	88	Pali	138	Lakhanpur
39	Pendra	89	Poundi-Uproda	139	Lundra

FIGURE 18.1

Location map of the study area (administrative boundary of Chhattisgarh state). ***Map lines delineate study areas and do not necessarily depict accepted national boundaries.***

only considers the bare earth surface (Digital Terrain Model). For this study, DEM (SRTM) is collected from the USGS website. Maximum and minimum elevations are in Northern Hills (about 1250 m from msl) and Bastar Plateau (about 198 m from msl), respectively. The Chhattisgarh plain has almost similar topography as compared to the Northern Hills and Bastar Plateau. Northen-East part of the Northern Hills is showing highest topography, whereas Southern part of Bastar plateau showing minimum elevation, as shown in Fig. 18.2.

2.2 Data source and extraction

In present study, gridded rainfall and temperature data have been downloaded from the India Meteorological Department (IMD). Both datasets have been downloaded in netcdf (.nc) format. For data extraction, python code was utilized. Further, daily data has been converted into monthly data sets as per requirement. Block-wise conversion of data to be finalized using the weighted average approach as per the grid size and location under blocks. Hence, entire block-wise weighted average data has been made using the MATLAB code and application of ArcGIS attributes of area. Data sources are summarized in Table 18.1.

3. Materials and methods

3.1 Standardized precipitation index (SPI)

McKee et al. (1993) proposed standardized precipitation index (SPI) to quantify the precipitation deficit (droughtiness) or excess (wetness) situation on multiple time scales (monthly, tri-monthly, six-monthly, yearly and two-yearly). To calculate SPI (Eq. 18.1), the available precipitation data need to be transformed into datasets at different time scales as required. A value of zero represents normal conditions, whereas positive and negative values indicate non-drought and drought conditions, respectively (Guttman, 1999; McKee et al., 1993).

$$SPI = \frac{x_i - \overline{x}}{\sigma} \quad (18.1)$$

where, x_i is the individual fitted precipitation values; $\overline{x}$ is the mean value and σ is the standard deviation.

3.1.1 Probability density function (PDF) used in SPI

The main objective of fitting any distribution across different time scales is to determine the time series behavioral pattern or trend. Generally, three types of distributions—Gamma, Weibull, and Log-normal—are used to calculate SPI, with the Gamma distribution considered the most suitable for SPI estimation (Guttman, 1999; Hayes et al., 1999; McKee, 1995; McKee et al., 1993; Shiau, 2006). The Weibull distribution is identified as the heavy-tailed distribution in precipitation fitting (Vicente-Serrano et al., 2012), whereas the Log-normal distribution is found to be the best fit for subsequent SPI computation (Lloyd-Hughes & Saunders, 2002; Loukas & Vasiliades, 2004; Mishra & Desai, 2005; Yusof et al., 2014). The Gamma distribution has been recommended by the majority of researchers as the best for SPI. To calculate the PDF and CDF (cumulative distribution function), Eq. (18.2) is applied.

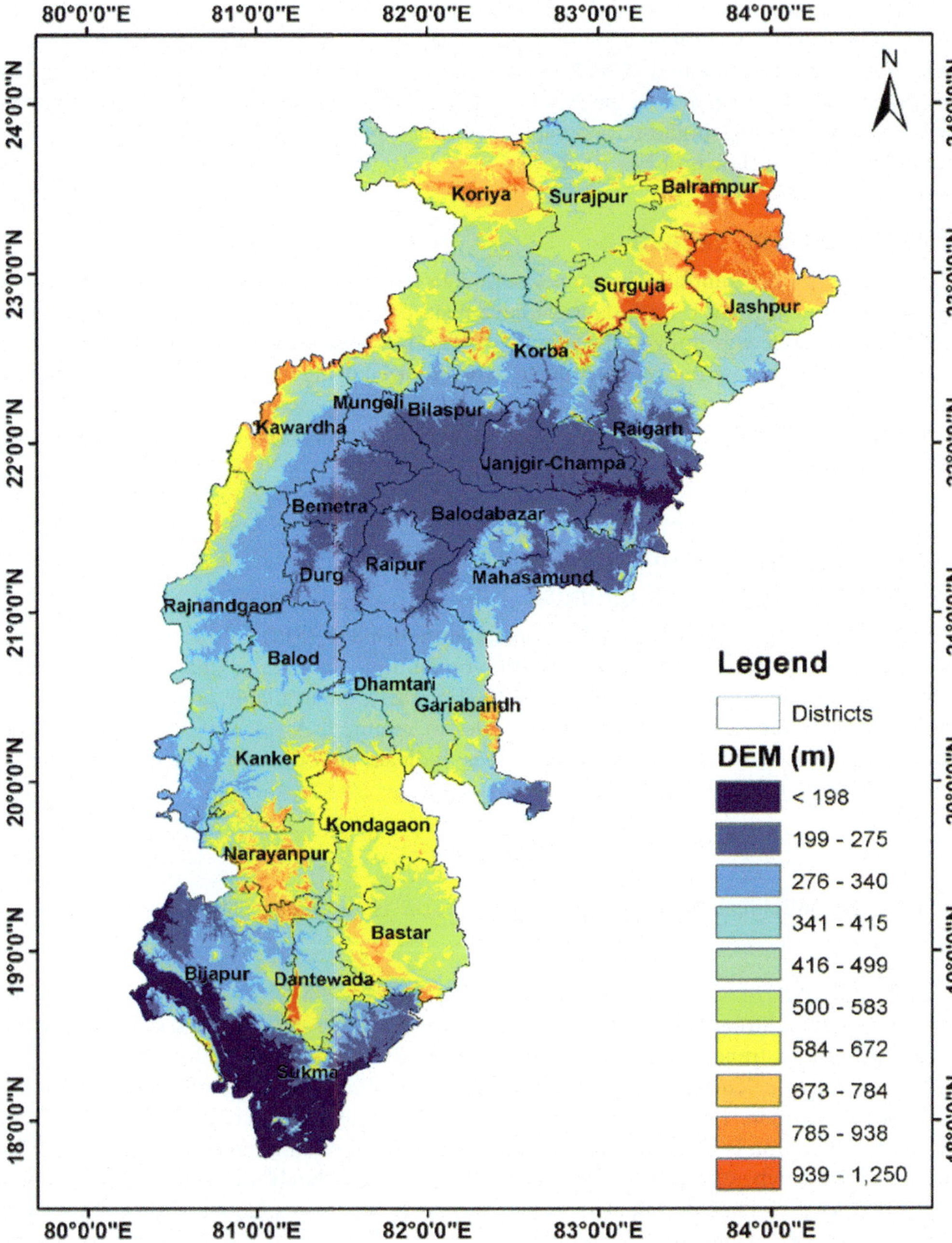

FIGURE 18.2

Digital elevation model (DEM) of the Chhattisgarh state. ***Map lines delineate study areas and do not necessarily depict accepted national boundaries.***

Table 18.1 Description of data sources.

Data	Source
Gridded rainfall & temperature	India Meteorological Department (IMD), Pune (India) Grid Resolution of (0.25 × 0.25 deg) & (1.0 × 1.0 deg) Data Length: 1901–2021 http://www.imdpune.gov.in/Clim_Pred_LRF_New/Grided_Data_Download.html
Observed rainfall	Department of Agrometeorology, IGKV, Raipur
Satellite Imageries	Landsat-7, Landsat-8 https://earthexplorer.usgs.gov/
Digital elevation Model (DEM)	Aster DEM https://lpdaac.usgs.gov/tools/opendap/
Population Census Data	Population data and projected population data https://censusindia.gov.in/census.website/
Soil Map	FAO soil map https://www.fao.org/soils-portal/data-hub/soil-maps-and-databases/faounesco-soil-map-of-the-world/en/ Indian soil map (Purchased) NBSS-LUP, Nagpur https://nbsslup.icar.gov.in/

$$f(x;\ \alpha,\beta) = x^{(\alpha-1)}\ \frac{e\frac{x}{\beta}}{\beta^{\beta}\Gamma\ (\alpha)}\ for\ x > 0$$

$$f(x;\ x,\beta) = \int_0^x f(u;\ \alpha,\beta)\ du = \frac{\gamma\left(\alpha,\frac{x}{\beta}\right)}{\Gamma(\alpha)}$$

$$for\ x > 0\ and\ \alpha,\beta > 0\ where,\Gamma(\alpha) = \int_0^\infty x^{\alpha-1}\ e^{-x}\ dx$$

$$and\ \gamma(s,) = \int_0^x t^{s-1}e^{-t}\ dt \tag{18.2}$$

is the lower incomplete gamma function, α is the shape parameter and β is the scale parameter.

3.1.2 Shape and scale parameter estimation

The SPI is calculated by applying a gamma probability density function to a time series frequency distribution. Annually, the shape (α) and scale (β) parameters of the gamma probability density function are estimated for each station (Eqs. 18.3 and 18.4). The highest likelihood solutions (Stacy & Mihram, 1965) are used to estimate α and β optimize.

$$\hat{\alpha} = \frac{1}{4A}\left(1 + \sqrt{1 + \frac{4A}{3}}\right) \tag{18.3}$$

$$\hat{\beta} = \frac{\overline{x}}{\hat{\alpha}} \tag{18.4}$$

where,

$$A = 1n\,(\overline{x}) - \frac{\sum \ln(x)}{n} \tag{18.5}$$

n = number of precipitation observations

3.1.3 Goodness-of-fit tests

A statistical goodness-of-fit (GoF) test is usually recommended for determining the compatibility of a model and data. GoF typically summarizes the difference between observed and predicted levels in the model in question. The best-fitted distribution is selected based on the minimum error generated. The following techniques can be used to measure these errors.

The Kolmogorov–Smirnov (KS) test has been used to calculate the maximum difference between the hypothesized and empirical distributions. The statistics for the KS test are as follows:

$$D = \max\left(\left|F(x_i) - \frac{i-1}{N}\right|, \left|\frac{i-1}{N} - F(x_i)\right|\right)$$

where, x_i denotes the increasing order of the data, F denotes the theoretical cumulative distribution, and N denotes the sample size. The model fitting between an observed CDF and an expected CDF was compared using Anderson-Darling (AD). The following Eq. (18.6) is the definition of the AD test:

$$A^2 = -n - \frac{1}{n}\sum_{i-1}^{n}(2i-1)[1n\,F(x_i) + 1n\,(1 - F(X_{n+1-i}))] \tag{18.6}$$

where, F is the CDF of the specified distribution and X_i are the ordered data.

3.1.4 Categorization of SPI

SPI uses only one climatic parameter that is precipitation. It does not count other climatic parameters, such as temperature, soil and air moisture and evapotranspiration. The SPI index assists in assessing drought severity by offering early warning signals (Byun & Wilhite, 1999). Its value lies between +3 and −3. Positive sign indicates water excess (wetness) situation (>1), whereas negative sign is used for water scarcity (droughtiness) (<−1). The value between +1 and −1 denotes near-normal conditions (NNC). Based on the SPI score, droughtiness and wetness have been classified into seven categories. These are described in Table 18.2, and a graphical representation is shown in Fig. 18.3.

3.2 Standardized precipitation evapotranspiration index (SPEI)

The computation process of SPEI is similar to that of SPI. But the main difference is about the datasets used for the study. For the calculation of SPEI, precipitation data along with potential evapotranspiration (PET) are required. Vicente- Serrano et al. (2010) designed the calculation process of the SPEI.

Table 18.2 Different categorization of drought and wet conditions as per SPI & SPEI values.

S. No.	Categories	SPI/SPEI ranges
1	Extreme wet	>2.00
2	Severe wet	1.50 to 1.99
3	Moderate wet	1.00 to 1.49
4	Near normal	−0.99 to 0.99
5	Moderate drought	−1.00 to −1.49
6	Severe drought	−1.50 to −1.99
7	Extreme drought	< −2.00

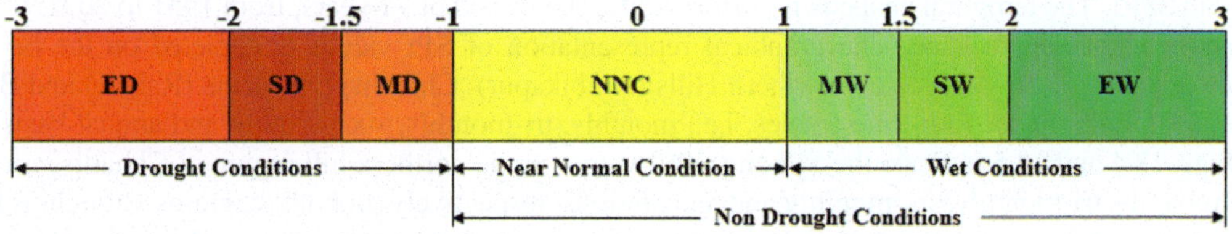

FIGURE 18.3

Criteria for evaluation of different drought and wet conditions as per SPI/SPEI values.

SPEI is used to detect the water balance and drought situation on different time scales (Wable et al., 2019). Potential evapotranspiration was estimated using the Penman–Monteith method, which incorporates various climatic variables such as temperature, rainfall, wind speed, and insolation (Vicente-Serrano et al., 2012). The drought severity classification values in SPEI are similar to the SPI categorized values. The SPEI is calculated using the following equations.

$$D_i = P_i - PET_i \tag{18.7}$$

here, P signifies the rainfall in a particular time frame, and PET means the rate of evapotranspiration over the same period. The possibility distribution function follows the given formula.

$$F(x) = 1 + \left[\left(\frac{\alpha}{X - Y}\right)\beta\right] - 1 \tag{18.8}$$

The computed $F(x)$ value has been transformed into Z standardized value, and after that SPEI can be intended by this formula.

$$SPEI = W - \frac{C_0 - C_1 W + C_2 W^2}{1 + d_1 W + d_2 W^2 + d_3 W^3} \tag{18.9}$$

where

$$C_0 = 2.5155, C_1 = 0.8082, C_2 = 0.0103, d_1 = 1.4327.\ d_2 = 0.1892 \textit{ and } d_3 = 0.0013$$

3.2.1 Categorization of SPEI

The SPEI index can help assess drought severity by providing early warnings (Byun & Wilhite, 1999). Index value is similar to the SPI. Drought and wetness conditions have been classified into seven categories based on SPI/SPEI values, as presented in Table 18.2 and illustrated in Fig. 18.3.

4. Results

The assessment of drought in Chhattisgarh using the SPI and SPEI yields valuable findings. These findings are discussed in the sub-sections below.

4.1 Temporal variability of SPI and SPEI for different ACZs during 1951–2021

In this study, the monthly rainfall and temperature time series are used as input parameters for estimating SPI and SPEI. The temporal analysis is performed for the period of 71 years, from 1951 to 2021, for 149 blocks of Chhattisgarh state. The graphical representation of SPI values is depicted on the basis of selected blocks under ACZs, i.e., Northern Hills (Ambikapur), Chhattisgarh Plains (Raipur) and Bastar Plateau (Bastar), under four-time frames, i.e., monthly, tri-monthly, six-monthly and annual basis.

Figs. 18.4 and 18.5 indicate the SPI and SPEI values in the northern hill region of Chhattisgarh state on a monthly, tri-monthly, six-monthly and annual basis, respectively. In both diagrams, drought is more prominent at the annual and 6-month scales than at the monthly and tri-monthly scales. However, calculated drought indices using both methods show a similar pattern of drought. From the years 1960–70 and 2000–10, the frequency of moderate drought is high in both indices, but the SPI gives more underrated values than the SPEI values. The occurrence of extreme drought is more prominent 6 months and yearly assessments (SPI-6, SPI-12, SPEI-6 and SPEI-12). Both methods indicate extreme drought conditions in the years 1962–63 and 2008–12 at six-monthly and twelve-monthly time frames.

Similar to Fig. 18.4 and Figs. 18.5–18.7 illustrate the SPI and SPEI indices at the Chhattisgarh Plains (Raipur) of Chhattisgarh state on a monthly, tri-monthly, six-monthly and annual basis, respectively. In both diagrams, moderate to severe (1–2) drought is observed. However, some extreme drought indices are experienced in short periods (monthly and 3-month). The sixth-month and twelfth-month SPI and SPEI indices also show that after the year 1965, several moderate and severe droughts have been experienced in the region. Among them, the droughts in 1965–70, 1987–90 and 1996–2009 were the most prominent.

Figs. 18.8 and 18.9 show the SPI and SPEI indices at the Chhattisgarh Plains (Raipur) of Chhattisgarh state on a monthly, tri-monthly, six-monthly and annual basis, respectively. In the Bastar region, several droughts (SPI and SPEI) have occurred on a short-term basis (monthly and tri-monthly), but the frequency of moderate drought increased after 1990. However, similar to the other region, the SPI provides an underestimated value compared to SPEI. In both indices, moderate to severe drought was experienced in the years 1963–67, 1972–75, 1982–85, 1997–99 and some scattered years. However, very few years (1962–63, 1965–66 and 2002–03) show extreme droughts in SPI and SPEI indexes.

4.2 Spatiotemporal variability of SPI and SPEI for different ACZs during 2017–21

For the analysis of the spatial pattern of drought, the entire region is categorized into seven water surplus and deficit categories, i.e., extreme wet, severe wet, moderate wet, near normal, moderate drought,

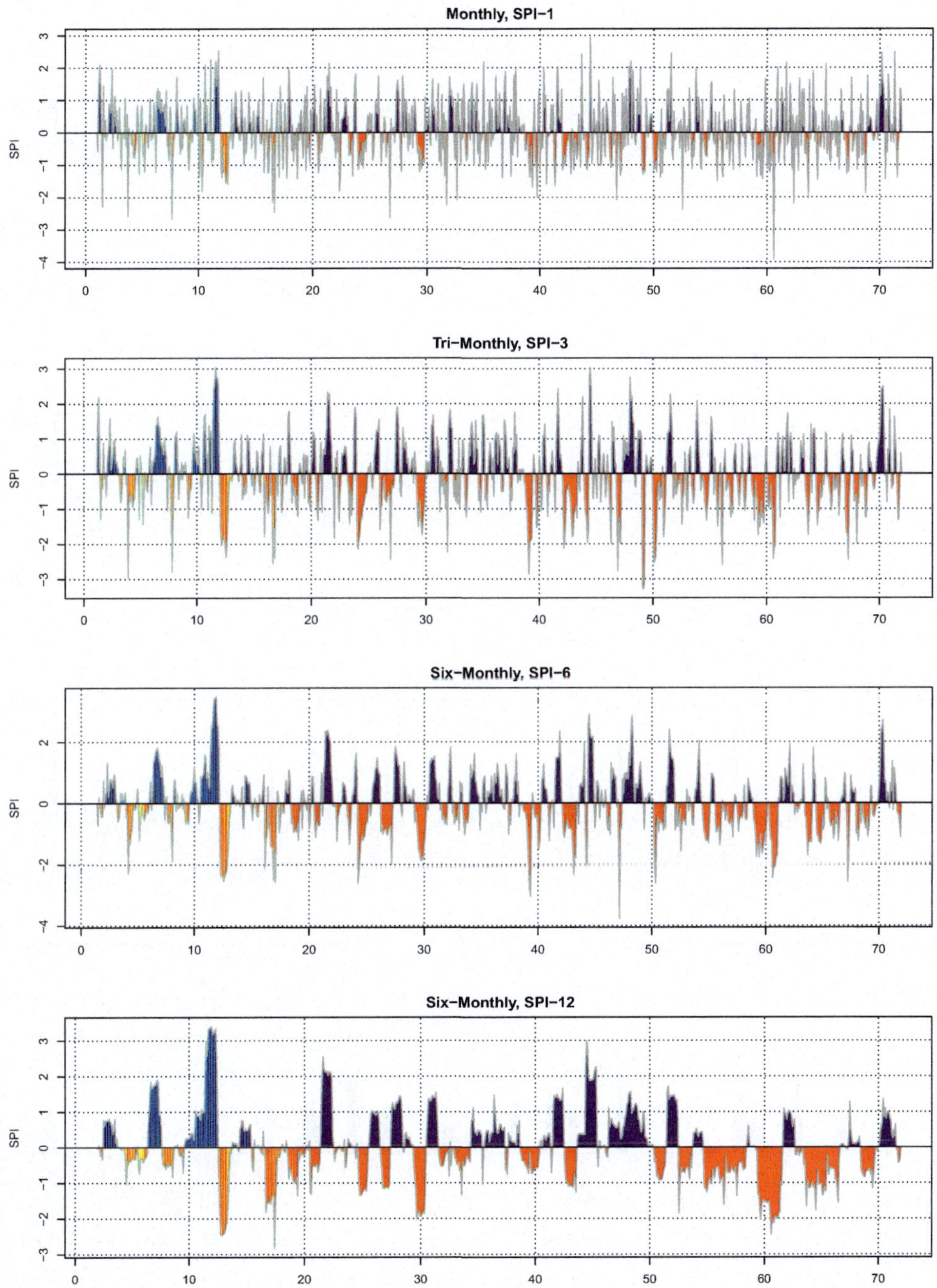

FIGURE 18.4

Graphical representation of SPI values at monthly, tri-monthly, six-monthly and annual basis for Northern Hills (Ambikapur) of Chhattisgarh state.

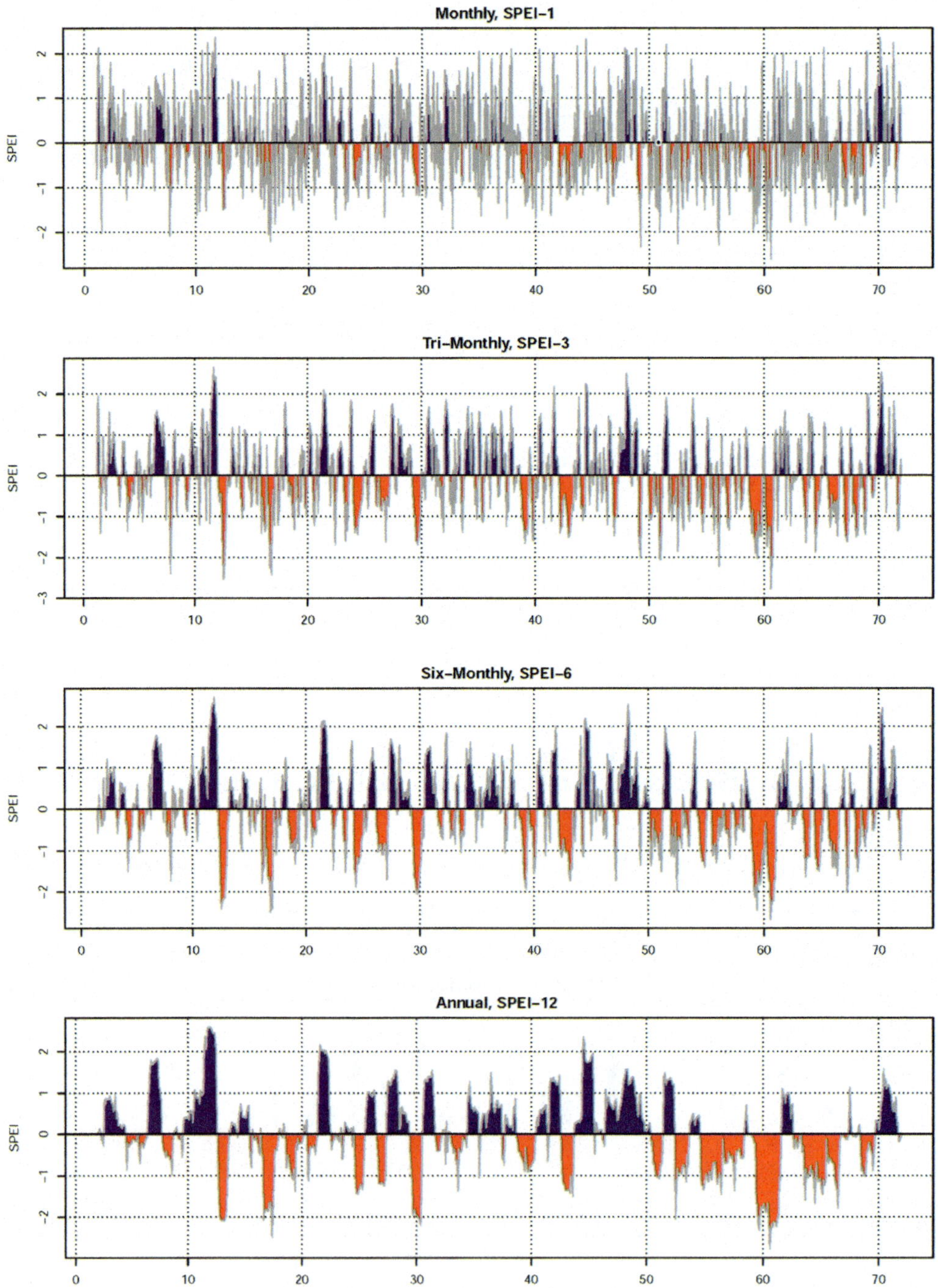

FIGURE 18.5

Graphical representation of SPEI values at monthly, tri-monthly, six-monthly and annual basis for Northern Hills (Ambikapur) of Chhattisgarh state.

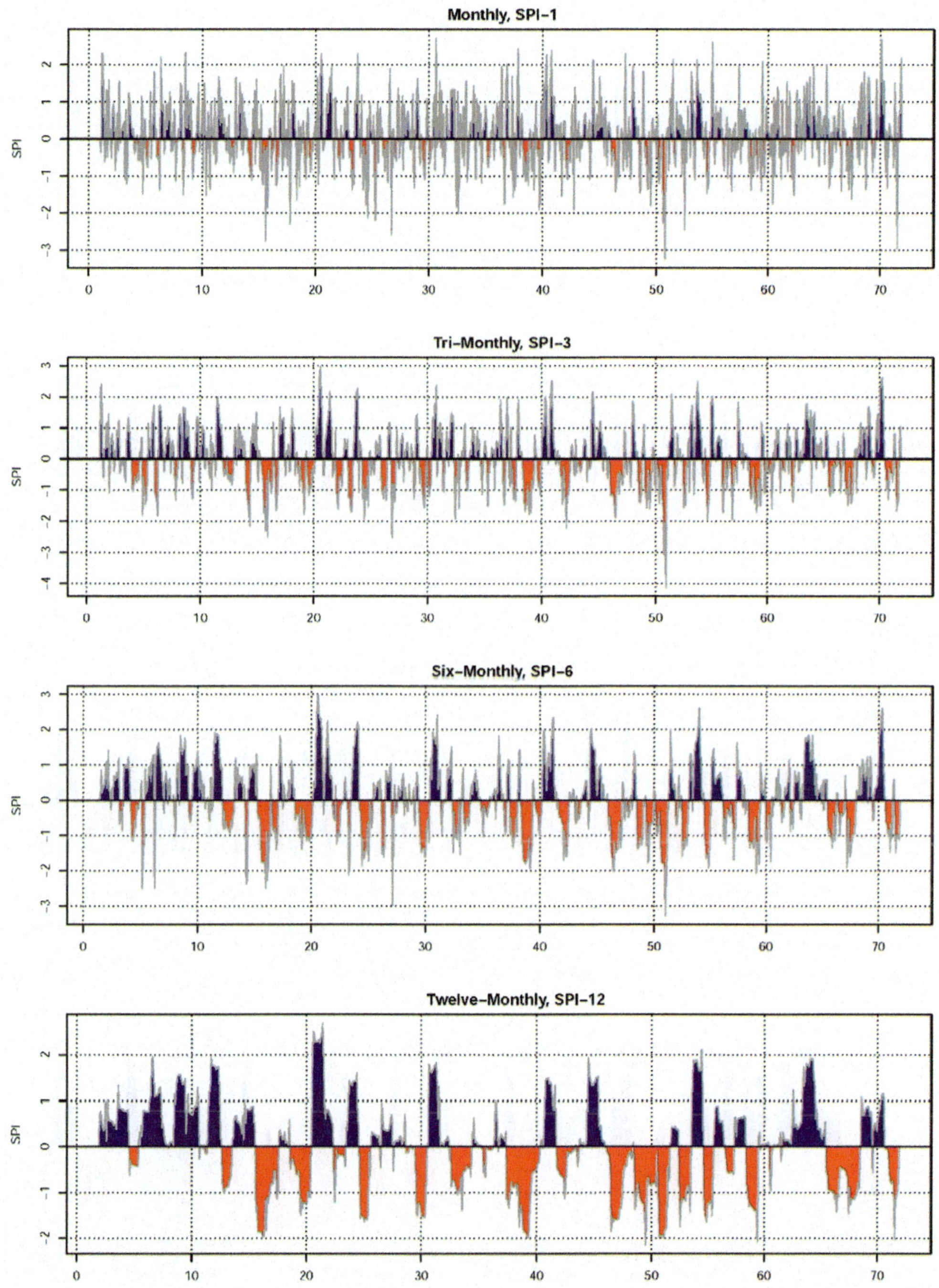

FIGURE 18.6

Graphical representation of SPI values at monthly, tri-monthly, six-monthly and annual basis for Chhattisgarh Plains (Raipur) of Chhattisgarh state.

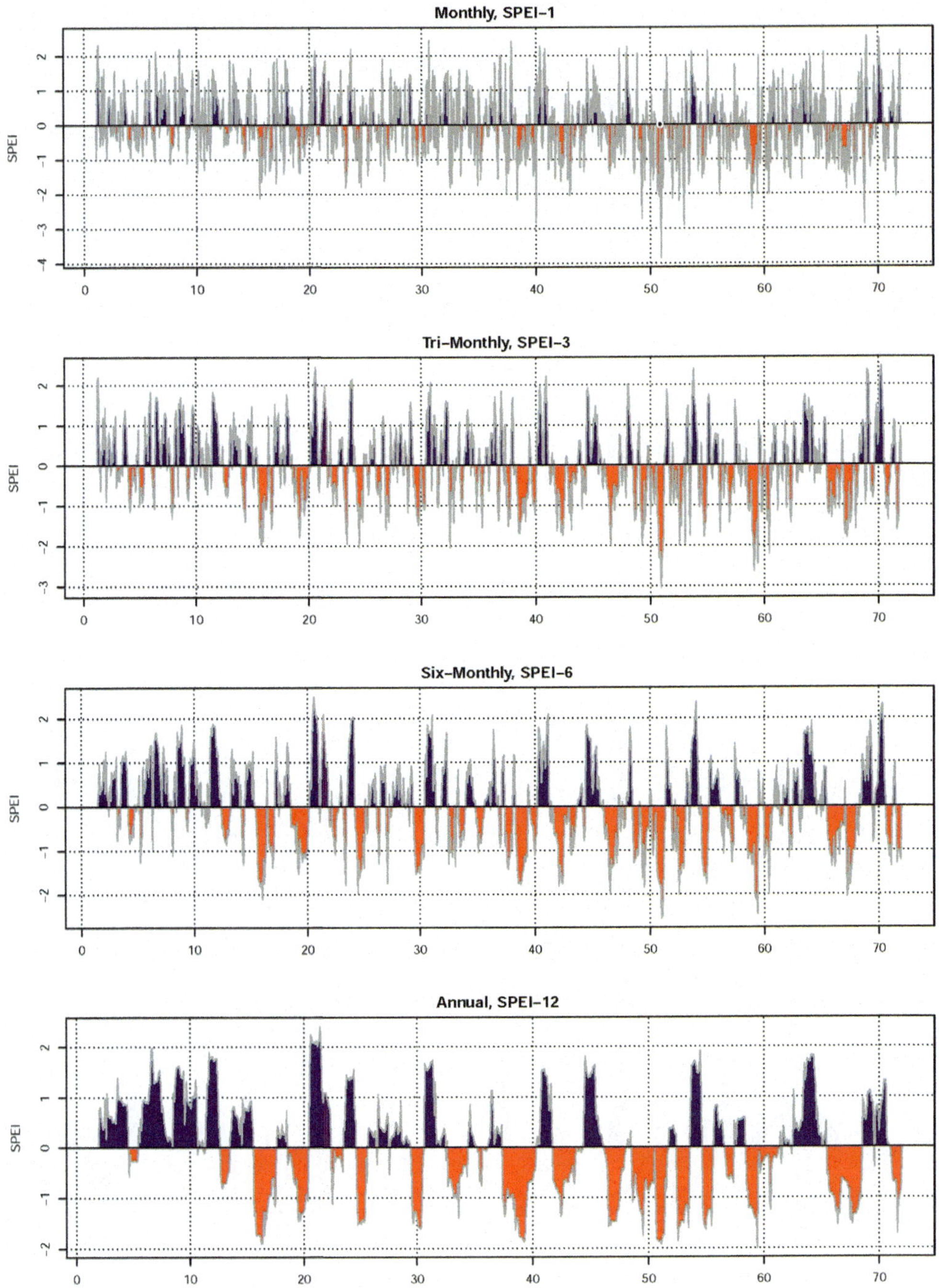

FIGURE 18.7

Graphical representation of SPI values at monthly, tri-monthly, six-monthly and annual basis for Chhattisgarh Plains (Raipur) of Chhattisgarh state.

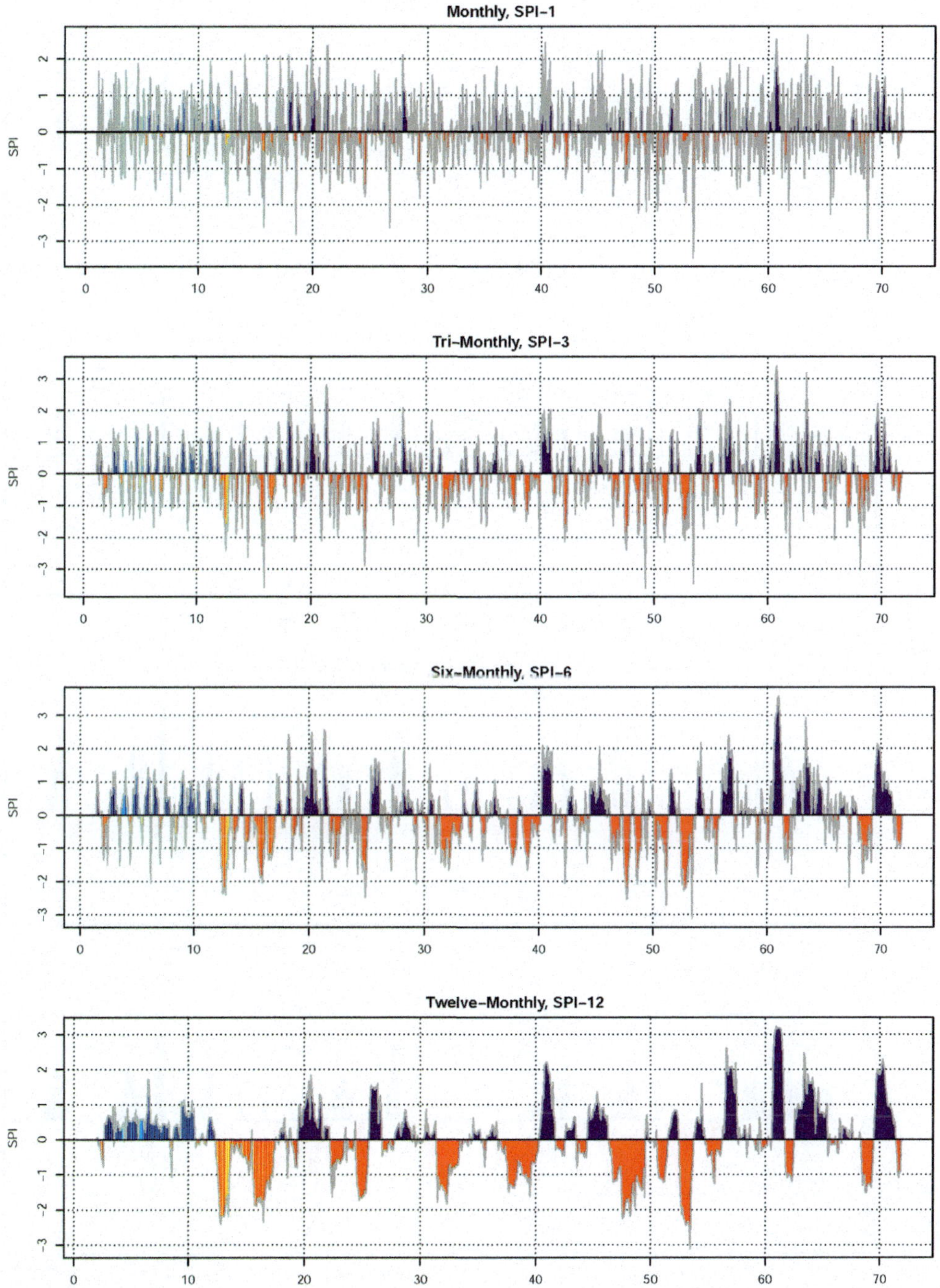

FIGURE 18.8

Graphical representation of SPI values at monthly, tri-monthly, six-monthly and annual basis for Bastar Plateau (Bastar) of Chhattisgarh state.

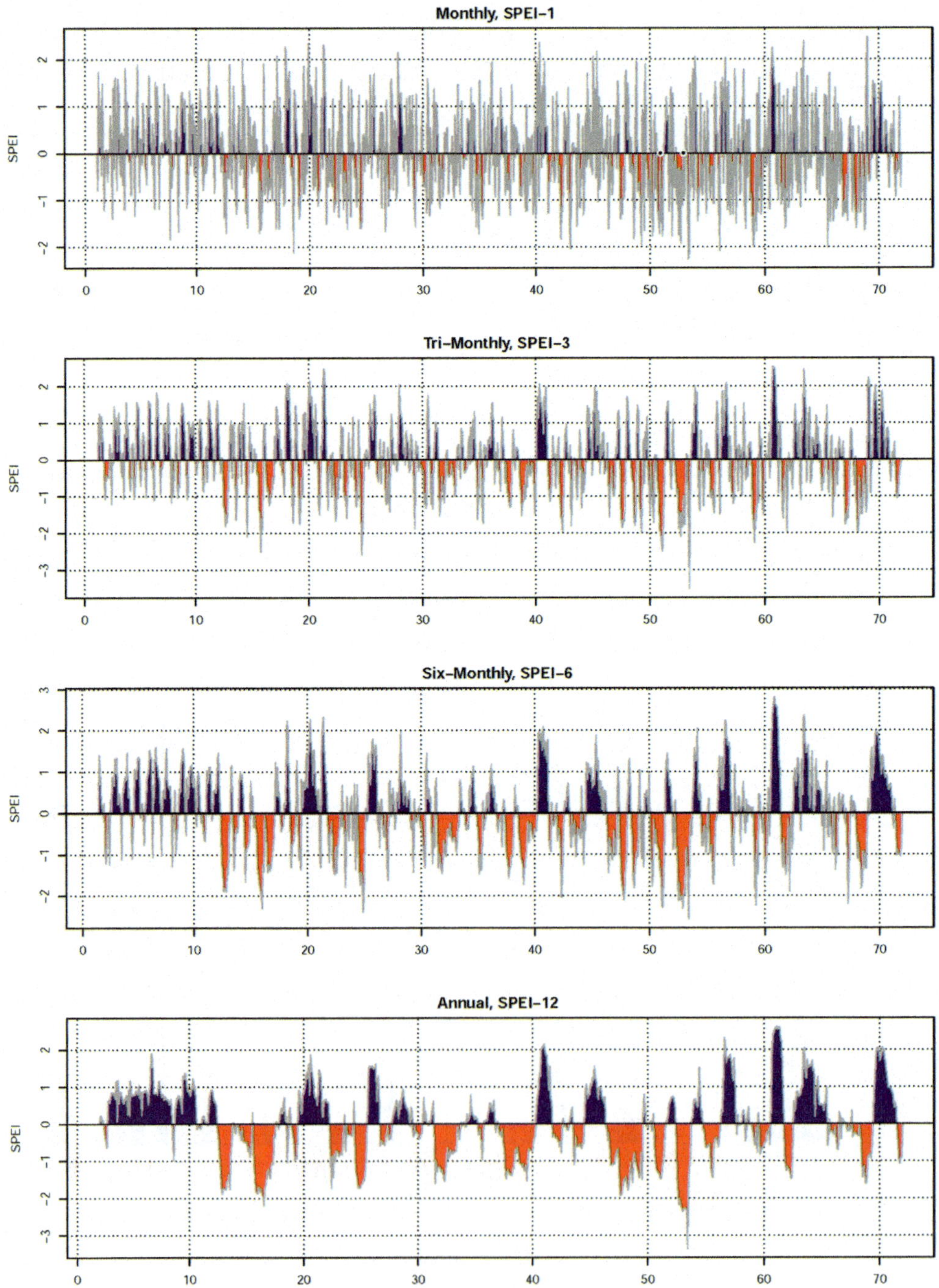

FIGURE 18.9

Graphical representation of SPEI values at monthly, tri-monthly, six-monthly and annual bases for Bastar Plateau (Bastar) of Chhattisgarh state.

severe drought, and extreme drought conditions. However, all the results of SPI and SPEI are evaluated using block-level data. Along with the monsoon season (June–September), the post-monsoon (October–November) season was also considered in this study to show the late-monsoon water stress. Spatiotemporal variability of SPI and SPEI during 2017–2021 (June–Nov) is described below.

Figs. 18.10 and 18.11 show the spatio-temporal variation of SPI and SPEI indices for a monthly time-step of three ACZs of Chhattisgarh district, respectively. The spatial pattern of drought is almost

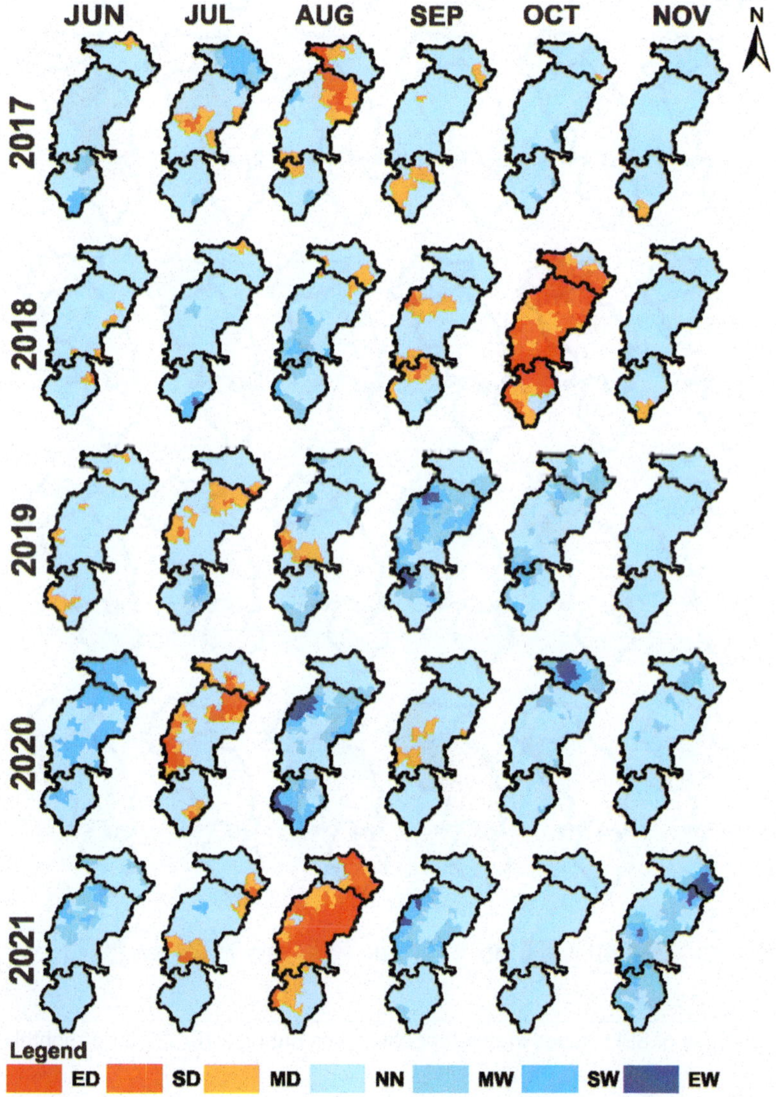

FIGURE 18.10

Spatiotemporal variability of SPI values in different categories during 2017–21 for monthly time-step.

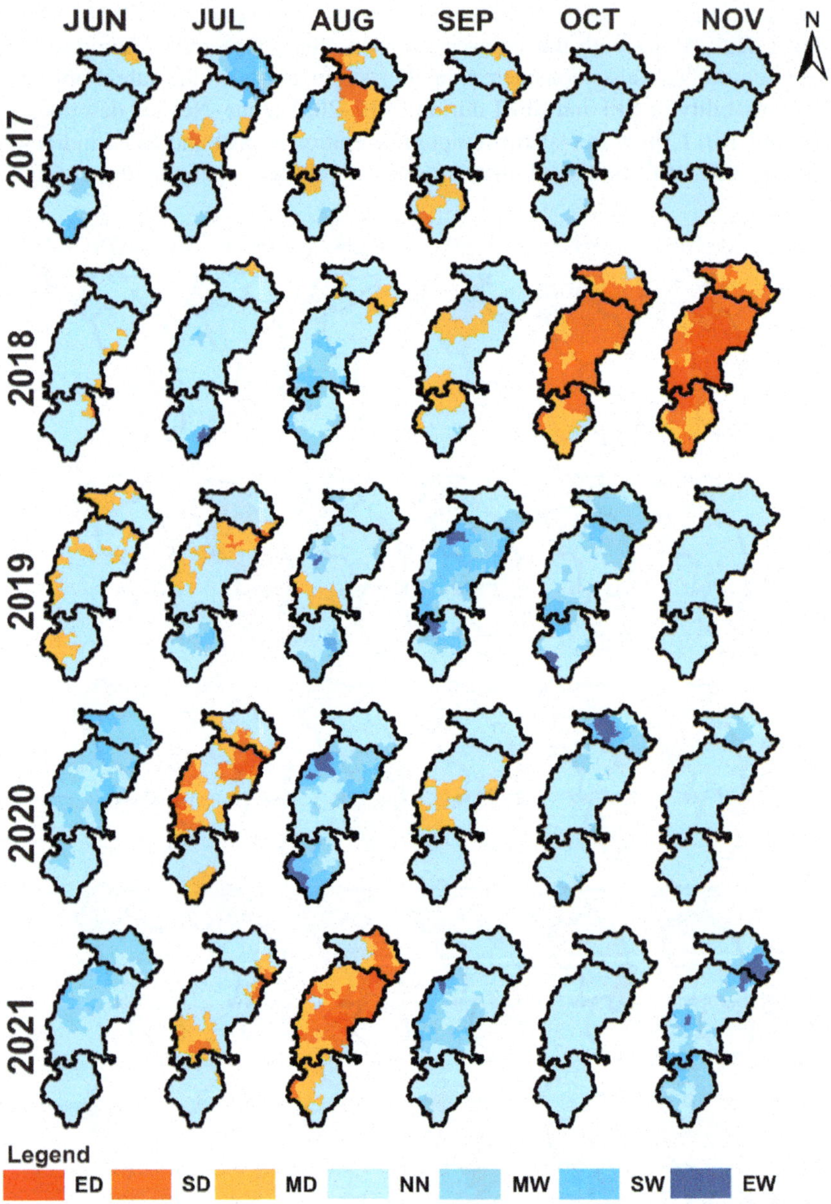

FIGURE 18.11

Spatiotemporal variability of SPEI values in different categories during 2017–21 for a monthly time-step.

similar to both indices. In August 2017, October 2018, July 2020 and August 2021, very intense (severe drought to extreme drought) drought occurred. Some scattered moderate drought is also observed in both images (September 2017, September 2018, July–August 2015, September 2018 and July 2021). However, the SPEI provides higher values of drought than the SPI. A very prominent difference was observed in November 2018, when the SPEI showed moderate to extreme drought, but a very small portion (southern) of the entire state showed moderate drought in the SPI indices.

Figs. 18.12 and 18.13 show the spatio-temporal variation of SPI and SPEI indices for a tri-monthly time-step of three ACZs of the Chhattisgarh district, respectively. Both diagrams show similar results of drought. In the years 2017, 2018 and 2021, both figures show intense drought (moderate to extreme drought). Whereas the intensity of drought was very low in the year 2020. Most of the intense drought occurred from the months August to November. In June and July, a very small scattered area shows a moderate drought region (except 2019) in both diagrams.

Figs. 18.14 and 18.15 show the spatio-temporal variation of SPI and SPEI indices for a six-monthly time-step of three ACZs of the Chhattisgarh district, respectively. In both figures, a very intense drought was observed in the months of 2017, 2018 and 2019. But, in the northern hills, only SPEI indices show moderate to severe drought conditions in June, July and August 2017. Except for this, both diagrams show a very similar pattern of drought. Very scattered and moderate drought conditions were observed in 2019 and 2021. Very small patches of drought were observed in the Chhattisgarh plain in 2020.

Figs. 18.16 and 18.17 show the spatio-temporal variation of SPI and SPEI indices for the annual time-step of three ACZs of the Chhattisgarh district, respectively. On the annual scale, both diagram shows the similar pattern of drought. In the years 2017 and 2018, the northern hilly region showed drought conditions. Whereas, the plain Chhattisgarh region shows drought conditions in the years 2017, 2018, 2018 and 2021. In the Bastar region, a very small region, experienced moderate drought conditions in a scattered manner (2017, 2018 and 2021). Almost drought-free year was 2020, but two very small patches of drought were found in October and November in the plain Chhattisgarh region.

The annual average rainfall of Chhattisgarh State is estimated as 1257 mm during 1991–2020. The state gets maximum rainfall during the monsoon season (83.32% of total rainfall), followed by the pre-monsoon season. The lowest rainfall received was during winter (3.11% of total annual rainfall) during 1991–2020. The maximum value of average monthly rainfall was noticed during July (314 mm), followed by September (298 mm), and the minimum value was noticed in January (2 mm). Drought characterization has been done by calculating SPI and SPEI values for a time period from 2017 to 2021 for a monthly, tri-monthly, semi-annual and annual time scales. The computation of the SPI index for the monthly timescale revealed that the critical conditions in terms of drought occurrence for Chhattisgarh state were noticed during October 2018 and August 2021. Bastar plateau was found to be least affected, while Chhattisgarh Plains was found to be more affected. As per the SPI computed for a tri-monthly timescale, the months of November 2018 and August 2021 were critical for Chhattisgarh state in terms of drought occurrence. Drought conditions in Chhattisgarh under the six-monthly timescale were less prominent than other time scales. The extent of extreme drought was confined to a small area of the Chhattisgarh Plains during August 2019. June 2020 was marked by wet events in major parts of the state. Chhattisgarh state was in wet conditions during 2020, and only some parts of the state were affected by drought as per the computed value of SPI under the annual timescale. As per the SPEI calculated for a monthly timescale during the period from 2017 to 2021, it revealed that drought was more pronounced in the year 2018. During 2018, the entire Chhattisgarh state experienced

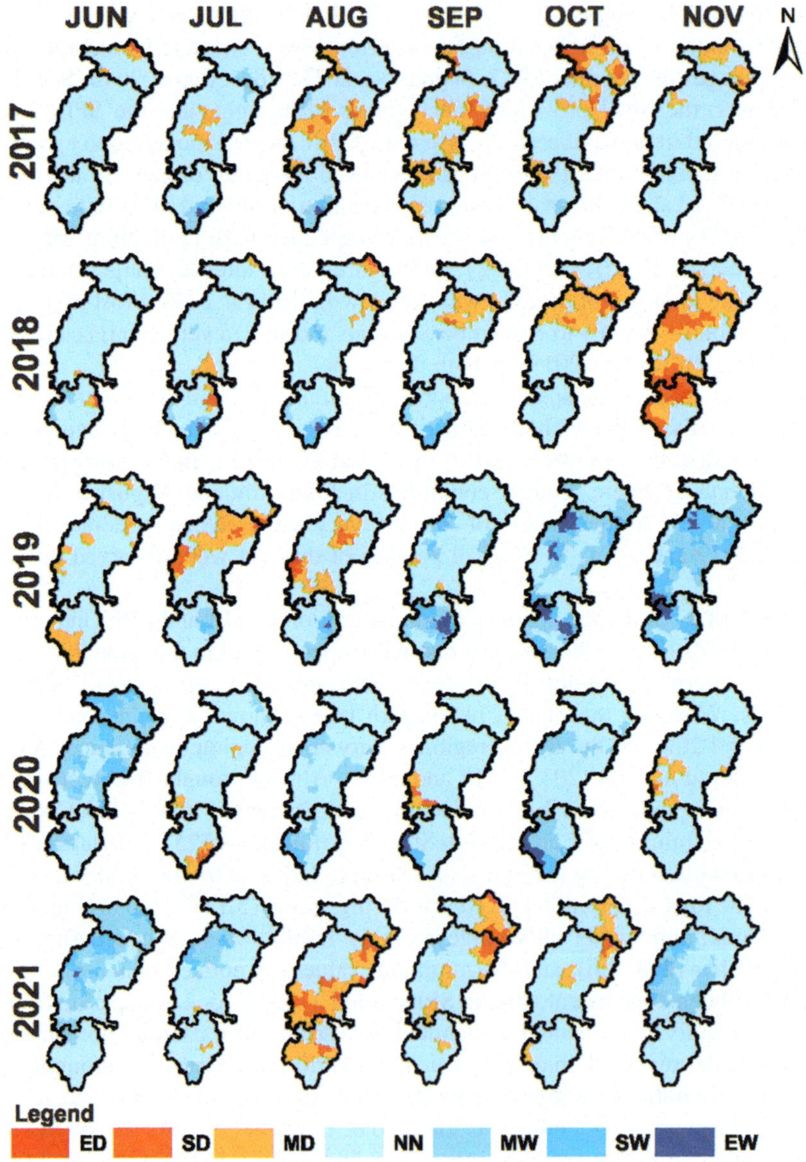

FIGURE 18.12

Spatiotemporal variability of SPI values in different categories during 2017–21 for tri-monthly time-step.

drought conditions during October and November. The computation of the SPEI for a tri-monthly timescale for the period of 2017–21 revealed that major parts of Chhattisgarh state experienced moderate and severe drought during November 2018. As per the analysis of drought using the SPEI for

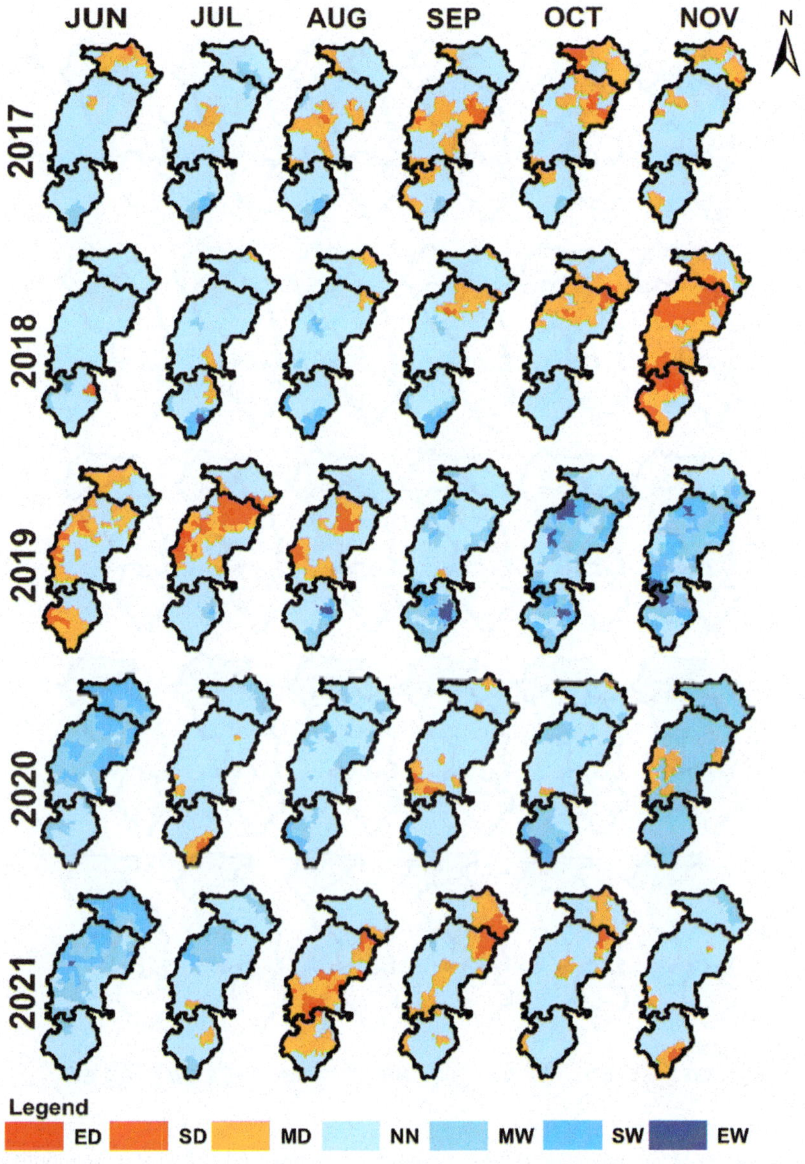

FIGURE 18.13

Spatiotemporal variability of SPEI values in different categories during 2017–21 for tri-monthly time-step.

6 monthly timescale, the extent of area under drought was more pronounced during 2017. During 2017, most of the regions of northern hilly zone were found to be experiencing severe and moderate drought during June, July and August. A few regions in the eastern part of the northern hilly zone

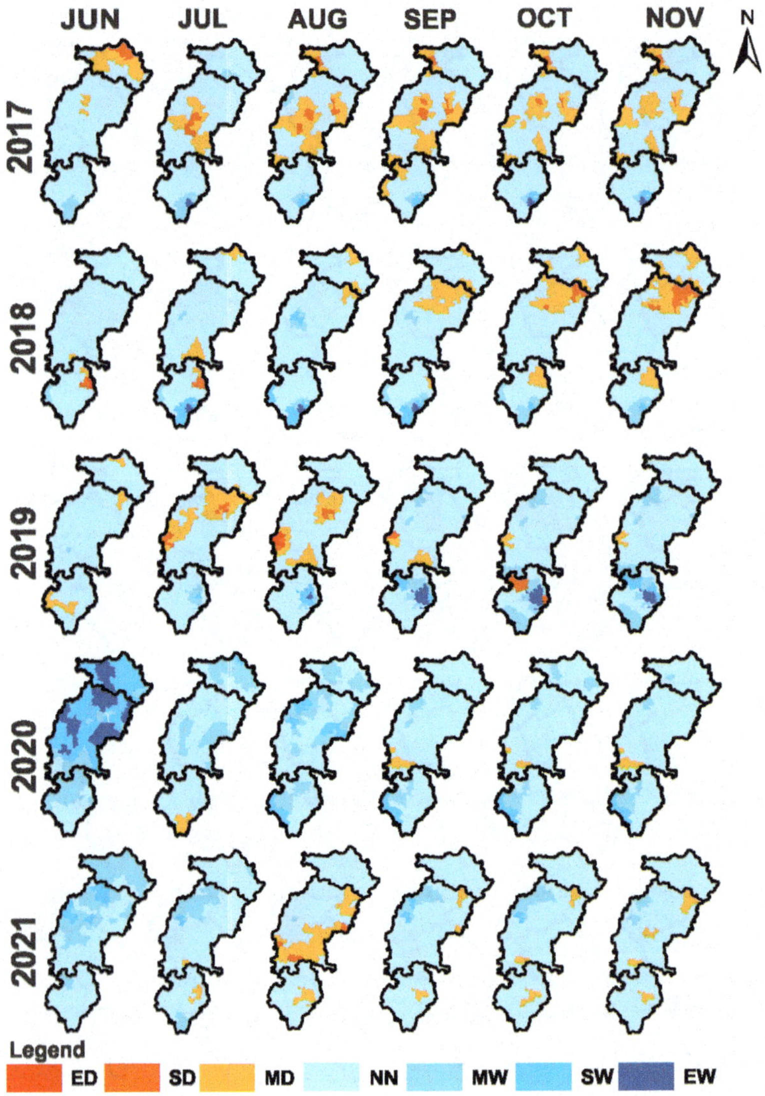

FIGURE 18.14

Spatiotemporal variability of SPI values in different categories during 2017–21 for a six-monthly time-step.

experienced moderate and severe drought during the same time period and the months of June to November 2018. The computation of the SPEI index for the annual timescale revealed that the state experienced severe and moderate drought during 2017 and 2018. A large area in the state was in severe and extreme wet conditions during June, July and August month of 2020.

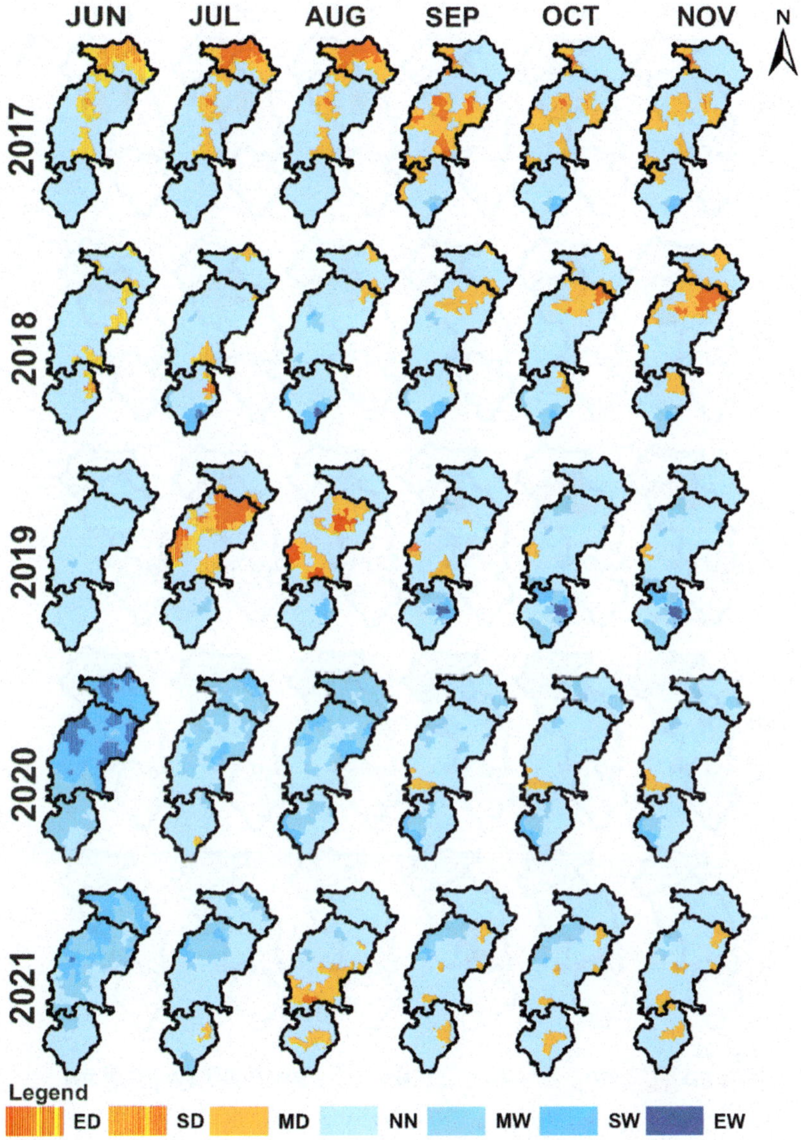

FIGURE 18.15

Spatiotemporal variability of SPEI values in different categories during 2017–21 for a six-monthly time-step.

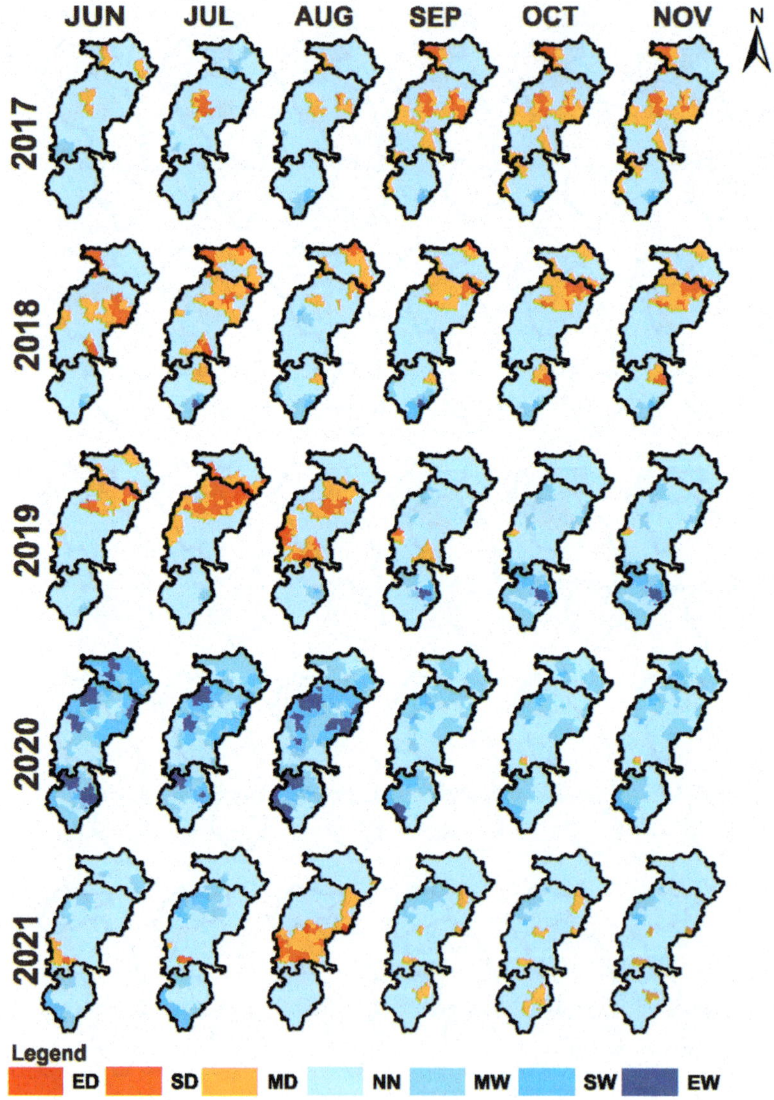

FIGURE 18.16

Spatiotemporal variability of SPI values in different categories during 2017–21 for annual time-step.

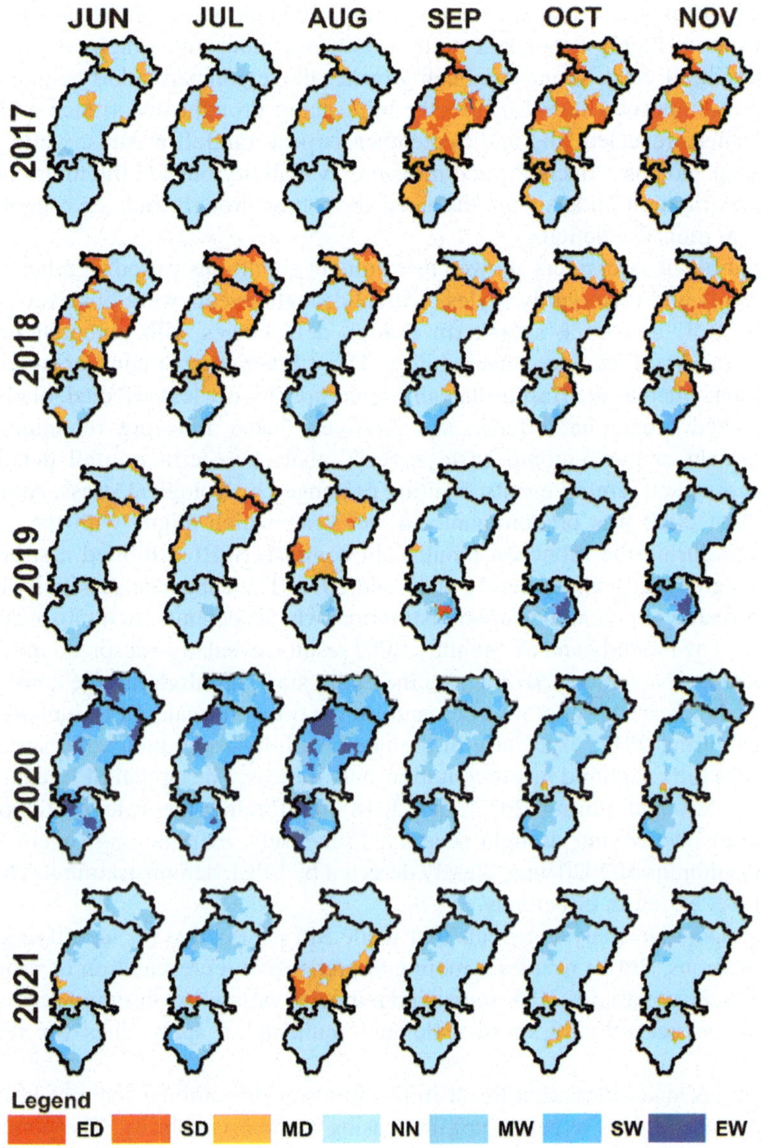

FIGURE 18.17

Spatiotemporal variability of SPEI values in different categories during 2017–21 for annual time-step.

5. Discussion

The present study provides a comparative assessment of meteorological drought over Chhattisgarh State using SPI and SPEI at multiple temporal scales, highlighting their effectiveness for drought monitoring and sustainable water management planning. The long-term rainfall analysis (1991–2020) indicates that the state is highly dependent on the southwest monsoon, which contributes more than 83% of the annual rainfall. Such strong seasonality makes the region particularly vulnerable to rainfall anomalies during the monsoon months, directly influencing drought occurrence and severity. The monthly rainfall pattern revealed July and September as peak rainfall months, while extremely low rainfall during winter months reflects limited moisture availability outside the monsoon season. This uneven temporal distribution underscores the need for robust drought indices capable of capturing short- and long-term moisture deficits.

The SPI-based drought assessment showed that drought conditions varied significantly across time scales. At the monthly and tri-monthly scales, critical drought events were observed during October 2018 and August 2021, indicating short-term rainfall deficiencies with potential implications for rainfed agriculture and soil moisture availability. The Chhattisgarh Plains emerged as the most drought-prone region, whereas the Bastar Plateau was comparatively less affected, likely due to higher forest cover, localized orographic effects, and relatively better moisture retention. The reduced prominence of drought at the 6-month scale suggests that short-term rainfall deficits were often compensated by subsequent rainfall events, limiting prolonged hydrological stress. Annual SPI results further indicated that 2020 was predominantly a wet year, which aligns with the widespread wet conditions observed during the monsoon months. In contrast, SPEI exhibited a stronger and more persistent drought signal, particularly during 2017 and 2018. The enhanced sensitivity of SPEI can be attributed to the inclusion of potential evapotranspiration, which accounts for atmospheric demand and temperature effects. The monthly and tri-monthly SPEI results revealed widespread moderate to severe drought during October–November 2018 across the entire state, highlighting the compounding effect of rainfall deficit and increased evaporative demand. At the 6-month scale, SPEI identified pronounced drought conditions during 2017, especially in the northern hilly zone, indicating sustained moisture stress that may not be fully captured by precipitation-only indices. The annual SPEI analysis confirmed severe to moderate drought during 2017 and 2018, reinforcing the role of temperature-driven evapotranspiration in intensifying drought severity. Conversely, extreme and severe wet conditions during the monsoon months of 2020 were clearly detected by SPEI, demonstrating its ability to capture both drought and wet extremes effectively.

Overall, the comparative analysis reveals that while SPI is effective for identifying precipitation-based drought conditions, SPEI provides a more comprehensive representation of drought dynamics under changing climatic conditions. The spatial and temporal variability observed across agro-climatic zones emphasizes the necessity of multi-scale and multi-index approaches for reliable drought monitoring.

The findings of this study have significant implications at the national scale of India, particularly for monsoon-dependent regions experiencing increasing climate variability. The comparative evaluation of SPI and SPEI demonstrates the necessity of integrating precipitation- and temperature-based drought indices for comprehensive drought monitoring under warming conditions. Given rising evapotranspiration trends, reliance solely on rainfall-based metrics may underestimate climatic water stress (Gupta et al., 2025; Upadhyay et al., 2025). The approach supports improved drought early warning systems, climate-resilient agricultural planning, and adaptive water resource management (Patel et al., 2024; Pradhan et al., 2025). Furthermore, long-term gridded datasets from the India Meteorological Department can facilitate standardized, region-specific drought assessments nationwide.

6. Conclusion

This study successfully evaluated the applicability of SPI and SPEI for meteorological drought monitoring over Chhattisgarh State at multiple temporal scales. The rainfall analysis confirmed the monsoon-dominated nature of the regional climate, making the state highly susceptible to rainfall variability and short-term drought events. Both indices effectively captured drought occurrences; however, notable differences were observed in drought intensity, spatial extent, and temporal persistence. SPI primarily reflected precipitation anomalies and was more responsive to short-term drought conditions, particularly at monthly and tri-monthly scales. In contrast, SPEI identified more severe and widespread drought conditions during 2017 and 2018, highlighting the critical role of evapotranspiration and temperature in drought intensification. The Chhattisgarh Plains were identified as the most vulnerable region, while the Bastar Plateau exhibited comparatively lower drought severity. The occurrence of widespread wet conditions during 2020 further demonstrates the high interannual variability of hydroclimatic conditions in the state.

The findings underscore the importance of integrating both SPI and SPEI for comprehensive drought assessment and early warning systems. SPEI, in particular, is more suitable for long-term drought monitoring and climate change impact assessment, whereas SPI remains valuable for rapid detection of rainfall deficits. The combined use of these indices can support sustainable water resource management, agricultural planning, and drought mitigation strategies in monsoon-dependent regions like Chhattisgarh.

Acknowledgment

The author sincerely acknowledges the India Meteorological Department (IMD), Pune, for providing free access to the gridded meteorological data through its data portal, which formed an essential input for this research. The author also gratefully acknowledges Indira Gandhi Krishi Vishwavidyalaya (IGKV), Raipur, for providing the necessary scientific workspace and laboratory facilities required for carrying out the M. Tech. (Soil and Water Engineering) dissertation work.

Declaration of competing interest

The authors declare that they have no known competing financial interests or personal relationships that could have appeared to influence the work reported in this chapter.

References

Aadhar, S., & Mishra, V. (2018). Impact of climate change on drought frequency over India. In *Book climate change and water resources in India publisher: Ministry of environment, forest and climate change (MoE-FandCC)*. Government of India.

Ahmed, A., Deb, D., & Mondal, S. (2019). Assessment of rainfall variability and its impact on groundnut yield in Bundelkhand region of India. *Current Science, 117*(5), 794–803.

Amrit, K., Pandey, R. P., & Mishra, S. K. (2018). Assessment of meteorological drought characteristics over Central India. *Sustainable Water Resources Management, 4*(4), 999–1010.

Bachmair, S., Svensson, C., Hannaford, J., Barker, L. J., & Stahl, K. (2016). A quantitative analysis to objectively appraise drought indicators and model drought impacts. *Hydrology and Earth System Sciences, 20*(7), 2589–2609.

Bera, B., Shit, P. K., Sengupta, N., Saha, S., & Bhattacharjee, S. (2021). Trends and variability of drought in the extended part of Chhota Nagpur plateau (Singbhum Protocontinent), India applying SPI and SPEI indices. *Environmental Challenges, 5*, 100310.

Bhelawe, S., Chaudhary, J. L., Manikandan, N., & Deshmukh, R. (2015). Meteorological drought assessment in Raipur district of Chhattisgarh state, India. *Plant Archives, 15*(1), 465–469.

Byun, H. R., & Wilhite, D. A. (1999). Objective quantification of drought severity and duration. *Journal of Climate, 12*(9), 2747–2756.

Gupta, V., & Jain, M. K. (2018). Investigation of multi-model spatiotemporal mesoscale drought projections over India under climate change scenario. *Journal of Hydrology, 567*, 489–509.

Gupta, P., Shekhar, M. S., Singh, G. P., Gupta, D. S., Singh, A., Kumar, A., Kumar, R., & Tomar, D. S. (2025). High-resolution analysis and prediction of heavy precipitation-induced GLOF events in North Sikkim Himalayas using the WRF model. *Physics and Chemistry of the Earth*. https://doi.org/10.1016/j.pce.2025.103968

Guttman, N. B. (1999). Accepting the standardized precipitation index: A calculation algorithm 1. *JAWRA Journal of the American Water Resources Association, 35*(2), 311–322.

Hayes, M. J., Svoboda, M. D., Wiihite, D. A., & Vanyarkho, O. V. (1999). Monitoring the 1996 drought using the standardized precipitation index. *Bulletin of the American Meteorological Society, 80*(3), 429–438.

Kansal, M. L., Chandniha, S. K., & Tyagi, A. (2015). Distance based water sustainability assessment using SPI for the state of Chhattisgarh in India. In *World environmental and water resources congress 2015* (pp. 2300–2319).

Khare, D., Mondal, A., Kundu, S., & Mishra, P. K. (2017). Climate change impact on soil erosion in the Mandakini River Basin, North India. *Applied Water Science, 7*(5), 2373–2383.

Krishnan, V. U., Mirajkar, A. B., & Nair, S. C. (2019). Comparative study on drought in Marathwada using SPI and SPEI. In *In proceedings of 24th hydro 2019, international conference* (p. 7p). Hyderabad, India: Osmania University.

Kumar, V., Jain, S. K., & Singh, Y. (2010). Analysis of long-term rainfall trends in India. *Hydrological Sciences Journal–Journal des Sciences Hydrologiques, 55*(4), 484–496.

Kundu, S., Khare, D., & Mondal, A. (2017). Past, present and future land use changes and their impact on water balance. *Journal of Environmental Management, 197*, 582–596.

Li, T., Jiang, Z., Treut, H. L., Li, L., Zhao, L., & Ge, L. (2021). Machine learning to optimize climate projection over China with multi-model ensemble simulations. *Environmental Research Letters, 16*(9), 094028.

Lloyd-Hughes, B., & Saunders, M. A. (2002). A drought climatology for Europe. *International Journal of Climatology, 22*(13), 1571–1592.

Loukas, A., & Vasiliades, L. (2004). Probabilistic analysis of drought spatiotemporal characteristics inThessaly region, Greece. *Natural Hazards and Earth System Sciences, 4*(5/6), 719–731.

Mallya, G., Mishra, V., Niyogi, D., Tripathi, S., & Govindaraju, R. S. (2016). Trends and variability of droughts over the Indian monsoon region. *Weather and Climate Extremes, 12*, 43–68.

Mathbout, S., Lopez-Bustins, J. A., Martin-Vide, J., Bech, J., & Rodrigo, F. S. (2018). Spatial and temporal analysis of drought variability at several time scales in Syria during 1961–2012. *Atmospheric Research, 200*, 153–168.

McKee, T. B. (1995). Drought monitoring with multiple time scales. *Proceedings of 9th Conference on Applied Climatology*. Boston.

McKee, T. B., Doesken, N. J., & Kleist, J. (1993). The relationship of drought frequency and duration to time scales. *Proceedings of the 8th Conference on Applied Climatology, 17*(22), 179–183.

Mehr, D. A., & Vaheddoost, B. (2020). Identification of the trends associated with the SPI and SPEI indices across Ankara, Turkey. *Theoretical and Applied Climatology, 139*(3), 1531–1542.

Mesbahzadeh, T., Mirakbari, M., Mohseni Saravi, M., Soleimani Sardoo, F., & Miglietta, M. M. (2020). Meteorological drought analysis using copula theory and drought indicators under climate change scenarios (RCP). *Meteorological Applications, 27*(1), e1856.

Mishra, A., & Desai, V. (2005). Spatial and temporal drought analysis in the Kansabati river basin, India. *International Journal of River Basin Management, 3*(1), 31–41.

Mondal, A., Khare, D., & Kundu, S. (2015). Spatial and temporal analysis of rainfall and temperature trend of India. *Theoretical and Applied Climatology, 122*(1), 143–158.

Mondal, A., Khare, D., & Kundu, S. (2016). Change in rainfall erosivity in the past and future due to climate change in the central part of India. *International Soil and Water Conservation Research, 4*(3), 186–194.

Mondal, A., Khare, D., Kundu, S., Mukherjee, S., Mukhopadhyay, A., & Mondal, S. (2017). Uncertainty of soil erosion modelling using open source high resolution and aggregated DEMs. *Geoscience Frontiers, 8*(3), 425–436.

Mondal, A., Kundu, S., & Mukhopadhyay, A. (2012). Rainfall trend analysis by Mann-Kendall test: A case study of north-eastern part of Cuttack district, Orissa. *International Journal of Geology, Earth and Environmental Sciences, 2*(1), 70–78.

Mondal, A., Le, M. H., & Lakshmi, V. (2022). Land use, climate, and water change in the Vietnamese Mekong Delta (VMD) using earth observation and hydrological modelling. *Journal of Hydrology: Regional Studies, 42*, 101132.

Mondal, A., Pandey, A., Mondal, S., & Kundu, S. (2025). Impact assessment of glacial Lake outburst floods (GLOFs) in the himalaya using satellite imageries over the last 25 years (2000-2024): A comprehensive review. *Remote Sensing Applications: Society and Environment*, 101845.

Mukhopadhyay, A., Mondal, A., Mukherjee, S., Khatua, D., Ghosh, S., Mitra, D., & Ghosh, T. (2014). Forest cover change prediction using hybrid methodology of geoinformatics and Markov chain model: A case study on sub-Himalayan town Gangtok, India. *Journal of Earth System Science, 123*(6), 1349–1360.

Nema, M. K., Khare, D., Adamowski, J., & Chandniha, S. K. (2018). Spatio-temporal analysis of rainfall trends in Chhattisgarh State, Central India over the last 115 years. *Journal of Water and Land Development, 36*.

Pandey, A., Mondal, A., Guha, S., Singh, D., & Kundu, S. (2023). Analysis of the variability in land surface temperature due to land use/land cover change for a sustainable urban planning. *Journal of Landscape Ecology, 16*(3), 20–35.

Patel, D. K., Thakur, T. K., Saini, S., Patel, A., Bhatt, S., Kumar, A., Kumar, R., & Husain, F. (2024). Morphometric indices assessment: Implications for active tectonics in the upper Narmada Basin, central India. *Physics and Chemistry of the Earth*. https://doi.org/10.1016/j.pce.2024.103746

Pei, Z., Fang, S., Wang, L., & Yang, W. (2020). Comparative analysis of drought indicated by the SPI and SPEI at various timescales in Inner Mongolia, China. *Water, 12*(7), 1925.

Prabhakar, A. K., Singh, K. K., Lohani, A. K., & Chandniha, S. K. (2019). Assessment of regional-level long-term gridded rainfall variability over the Odisha State of India. *Applied Water Science, 9*, 1–15.

Pradhan, A., Sao, A., Thakur, T. K., Anderson, J. T., Chandel, G., Kumar, A., et al. (2025). Wetlands as climate-sensitive hotspots: Evaluating greenhouse gas emissions in Southern Chhattisgarh. *Water, 17*(10), 1553. https://doi.org/10.3390/w17101553

Praveen, B., Talukdar, S., Mahato, S., Mondal, J., Sharma, P., Islam, A. R. M., & Rahman, A. (2020). Analyzing trend and forecasting of rainfall changes in India using non-parametrical and machine learning approaches. *Scientific Reports, 10*(1), 1–21.

Ramrao, W. Y., Tiwari, S. P., & Singh, P. (2005). Crop-livestock integrated farming system for augmenting socio-economic status of smallholder tribal farmers of Chhattisgarh in Central India. *Livestock Research for Rural Development, 17*(8), 252–259.

Shiau, J. T. (2006). Fitting drought duration and severity with two-dimensional copulas. *Water Resources Management, 20*, 795–815.

Singh, R. M., & Shukla, P. (2020). Drought characterization using drought indices and El Niño effects. *National Academy Science Letters, 43*, 339–342.

Stacy, E. W., & Mihram, G. A. (1965). Parameter estimation for a generalized gamma distribution. *Technimetrics, 7*(3), 349–358.

Swain, S., Patel, P., & Nandi, S. (2017). Application of SPI, EDI and PNPI using MSWEP precipitation data over Marathwada, India. In *2017 IEEE international geoscience and remote sensing symposium (IGARSS)* (pp. 5505–5507). IEEE.

Tefera, A. S., Ayoade, J. O., & Bello, N. J. (2019). Comparative analyses of SPI and SPEI as drought assessment tools in Tigray Region, Northern Ethiopia. *SN Applied Sciences, 1*(10), 1–14.

Upadhyay, P., Prajapati, S. K., & Kumar, A. (2025). Spatiotemporal variability and underlying drivers of greenhouse gas emissions from the subtropical Ganga River network, India. *ACS ES&T Water*. https://doi.org/10.1021/acsestwater.5c00967

Vicente-Serrano, S. M., Beguería, S., & López-Moreno, J. I. (2010). A multiscalar drought index sensitive to global warming: The standardized precipitation evapotranspiration index. *Journal of Climate, 23*(7), 1696–1718.

Vicente-Serrano, S. M., López-Moreno, J. I., Beguería, S., Lorenzo-Lacruz, J., Azorin-Molina, C., & Morán-Tejeda, E. (2012). Accurate computation of a stream flow drought index. *Journal of Hydrologic Engineering, 17*(2), 318–332.

Wable, P. S., Jha, M. K., & Shekhar, A. (2019). Comparison of drought indices in a semi-arid river basin of India. *Water Resources Management, 33*(1), 75–102.

Yusof, F., Hui-Mean, F., Suhaila, J., Yusop, Z., & Ching-Yee, K. (2014). Rainfall characterization by application of standardised precipitation index (SPI) in Peninsular Malaysia. *Theoretical and Applied Climatology, 115* (3–4), 503–516.

Index

Note: 'Page numbers followed by "*f*" indicate figures and "*t*" indicate tables.'

C

H

I

L

M

R

U

W

Z

Printed and bound by CPI Group (UK) Ltd, Croydon, CR0 4YY
14/09/2026
02216410-0016